# LONDON MATHEMATICAL SOCIETY LECTURE NOTE SERIES

Managing Editor: Professor M. Reid, Mathematics Institute,
University of Warwick, Coventry CV4 7AL, United Kingdom

The titles below are available from booksellers, or from Cambridge University Press at www.cambridge.org/mathematics

London Mathematical Society Lecture Note Series: 375

# Triangulated Categories

Edited by

THORSTEN HOLM
*Leibniz Universität Hannover, Germany*

PETER JØRGENSEN
*University of Newcastle upon Tyne*

RAPHAËL ROUQUIER
*University of Oxford*

CAMBRIDGE UNIVERSITY PRESS

CAMBRIDGE UNIVERSITY PRESS
Cambridge, New York, Melbourne, Madrid, Cape Town, Singapore,
São Paulo, Delhi, Dubai, Tokyo

Cambridge University Press
The Edinburgh Building, Cambridge CB2 8RU, UK

Published in the United States of America by Cambridge University Press, New York

www.cambridge.org
Information on this title: www.cambridge.org/9780521744317

© Cambridge University Press 2010



First published 2010

*A catalogue record for this publication is available from the British Library*

*Library of Congress Cataloguing in Publication data*
Triangulated categories / edited by Thorsten Holm, Peter Jørgensen, Raphaël Rouquier.
p.   cm. – (London Mathematical Society lecture note series ; 375)
Includes index.
ISBN 978-0-521-74431-7 (pbk.)
1. Triangulated categories.   I. Holm, Thorsten, 1965–   II. Jørgensen, Peter, 1970–
III. Rouquier, Raphaël.   IV. Title.   V. Series.
QA169.T685   2010
512′.62 – dc22      2010012362

ISBN 978-0-521-74431-7 Paperback

# Contents

# Preface

This volume grew out of a Workshop on Triangulated Categories held at the University of Leeds in August 2006. The meeting, a Satellite of the International Congress of Mathematicians 2006, has been generously supported by the Leverhulme Foundation (via the network Algebras, Representations and Applications), the London Mathematical Society (Conference Grant Ref. 1438) and the University of Leeds.

Over the past decades, triangulated categories have made their way into many different parts of mathematics, to the extent that today, they can be viewed as a unifying theory underlying major parts of modern mathematics. The Leeds workshop has brought together researchers from many parts of mathematics who all use triangulated methods but would not usually meet at more specialized conferences, with the aim to promote cross fertilization leading to new applications of triangulated categories.

The present book collects surveys by leading experts reflecting a broad range of important topics covered at the workshop. However, it is not a proceedings volume recording precisely the talks given at the conference and it does not claim to be a comprehensive coverage of all the numerous applications of triangulated categories throughout mathematics.

There are contributions dealing with fundamental general aspects of triangulated categories as well as articles covering important applications, e.g. in algebraic geometry, algebraic topology, commutative algebra, algebraic analysis, K-theory or representation theory.

We wish to express our sincere thanks to the authors of the contributions, as well as to the referees.

We think that the interdisciplinary spirit of the successful Leeds workshop and the many fruitful discussions having taken place there are well reflected by the articles and we hope that specialists and non-specialists alike will benefit from the broad perspective on triangulated categories and their applications provided by the surveys.

We are very grateful to the staff at Cambridge University Press for their help, their patience and constant support in bringing this book together.

Hannover, Newcastle, and Oxford, January 2010

Thorsten Holm, Peter Jørgensen, and Raphaël Rouquier

# Triangulated categories: definitions, properties, and examples

THORSTEN HOLM AND PETER JØRGENSEN

Triangulated categories were introduced in the mid 1960's by J.L. Verdier in his thesis, reprinted in [16]. Axioms similar to Verdier's were independently also suggested in [2]. Having their origins in algebraic geometry and algebraic topology, triangulated categories have by now become indispensable in many different areas of mathematics. Although the axioms might seem a bit opaque at first sight it turned out that very many different objects actually do carry a triangulated structure. Nowadays there are important applications of triangulated categories in areas like algebraic geometry (derived categories of coherent sheaves, theory of motives) algebraic topology (stable homotopy theory), commutative algebra, differential geometry (Fukaya categories), microlocal analysis or representation theory (derived and stable module categories).

It seems that the importance of triangulated categories in modern mathematics is growing even further in recent years, with many new applications only recently found; see B. Keller's article in this volume for one striking example, namely the cluster categories occurring in the context of S. Fomin and A. Zelevinsky's cluster algebras which have been introduced only around 2000.

In this chapter we aim at setting the scene for the survey articles in this volume by providing the relevant basic definitions, deducing some elementary general properties of triangulated categories, and providing a few examples.

Certainly, this cannot be a comprehensive introduction to the subject. For more details we refer to one of the well-written textbooks on triangulated categories, e.g. [4], [5], [7], [8], [12], [17], and for further topics also to the surveys in this volume.

This introductory chapter should be accessible for a reader with a good background in algebra and some basic knowledge of category theory and homological algebra.

# 1. Additive categories

In this first section we shall discuss the fundamental notion of an additive category and provide some examples. In particular, the category of complexes over an additive category is introduced which will play a fundamental role in the sequel.

**Definition 1.1.** *A category $\mathcal{A}$ is called an* additive category *if the following conditions hold:*

(A1) *For every pair of objects $X, Y$ the set of morphisms $\mathrm{Hom}_{\mathcal{A}}(X, Y)$ is an abelian group and the composition of morphisms*

$$\mathrm{Hom}_{\mathcal{A}}(Y, Z) \times \mathrm{Hom}_{\mathcal{A}}(X, Y) \to \mathrm{Hom}_{\mathcal{A}}(X, Z)$$

    *is bilinear over the integers.*

(A2) *$\mathcal{A}$ contains a zero object $0$ (i.e. for every object $X$ in $\mathcal{A}$ each morphism set $\mathrm{Hom}_{\mathcal{A}}(X, 0)$ and $\mathrm{Hom}_{\mathcal{A}}(0, X)$ has precisely one element).*

(A3) *For every pair of objects $X, Y$ in $\mathcal{A}$ there exists a coproduct $X \oplus Y$ in $\mathcal{A}$.*

**Remark 1.2.**

(i) A category satisfying $(A1)$ and $(A2)$ is called a *preadditive* category.

(ii) We recall the notion of coproduct from category theory. Let $\mathcal{C}$ be a category and $X, Y$ objects in $\mathcal{C}$. A coproduct of $X$ and $Y$ in $\mathcal{C}$ is an object $X \oplus Y$ together with morphisms $\iota_X : X \to X \oplus Y$ and $\iota_Y : Y \to X \oplus Y$ satisfying the following universal property: for every object $Z$ in $\mathcal{C}$ and morphisms $f_X : X \to Z$ and $f_Y : Y \to Z$ there is a unique morphism $f : X \oplus Y \to Z$ making the following diagram commutative.

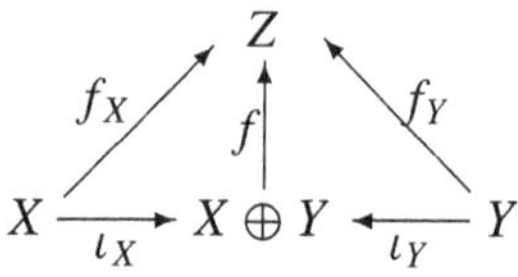

**Example 1.3.**

(i) Let $R$ be a ring and consider $R$ as a category $\mathcal{C}_R$ with only one object. The unique morphism set is the underlying abelian group and composition of morphisms is given by ring multiplication. Then $\mathcal{C}_R$ satisfies (A1) and (A2), thus preadditive categories can be seen as generalizations of rings. But $\mathcal{C}_R$ is not additive in general; in fact the coproduct of the unique object with itself would have to be again this object together with fixed ring elements $\iota_1, \iota_2$, and the universal property would mean that for arbitrary ring elements $f_1$, $f_2$ there existed a unique element $f$ factoring them as $f_1 = f \iota_1$ and $f_2 = f \iota_2$.

(ii) Let $R$ be a ring (associative, with unit element). Then the category **R-Mod** of all $R$-modules is additive. Similarly, the category **R-mod** of finitely generated $R$-modules is additive. In particular, the categories **Ab** of abelian groups and $\mathbf{Vec_K}$ of vector spaces over a field $K$ are additive.

(iii) The full subcategory of **Ab** of free abelian groups is additive.

(iv) For a ring $R$ the full subcategory **R-Proj** of projective $R$-modules is additive; similarly for **R-proj**, the category of finitely generated projective $R$-modules.

## 1.1. The category of complexes

Let $\mathcal{A}$ be an additive category. A *complex over* $\mathcal{A}$ is a family $X = (X_n, d_n^X)_{n \in \mathbb{Z}}$ where $X_n$ are objects in $\mathcal{A}$ and $d_n^X : X_n \to X_{n-1}$ are morphisms such that $d_n \circ d_{n+1} = 0$ for all $n \in \mathbb{Z}$. Usually, a complex is written as a sequence of objects and morphisms as follows.

$$\cdots \to X_{n+1} \xrightarrow{d_{n+1}} X_n \xrightarrow{d_n} X_{n-1} \longrightarrow \cdots$$

Let $X = (X_n, d_n^X)$ and $Y = (Y_n, d_n^Y)$ be complexes over $\mathcal{A}$. A *morphism of complexes* $f : X \to Y$ is a family of morphisms $f = (f_n : X_n \to Y_n)_{n \in \mathbb{Z}}$ satisfying $d_n^Y \circ f_n = f_{n-1} \circ d_n^X$ for all $n \in \mathbb{Z}$, i.e. we have the following commutative diagram.

$$
\begin{array}{ccccccc}
\cdots \longrightarrow & X_{n+1} & \longrightarrow & X_n & \longrightarrow & X_{n-1} & \longrightarrow \cdots \\
& \downarrow{\scriptstyle f_{n+1}} & & \downarrow{\scriptstyle f_n} & & \downarrow{\scriptstyle f_{n-1}} & \\
\cdots \longrightarrow & Y_{n+1} & \longrightarrow & Y_n & \longrightarrow & Y_{n-1} & \longrightarrow \cdots
\end{array}
$$

The complexes over an additive category $\mathcal{A}$ together with the morphisms of complexes form a category $\mathbf{C}(\mathcal{A})$, the *category of complexes over* $\mathcal{A}$.

**Proposition 1.4.** *Let $\mathcal{A}$ be an additive category. Then the category of complexes* $\mathbf{C}(\mathcal{A})$ *is again additive.*

*Proof.* (A1) Addition of morphisms is defined degreewise, i.e. for two morphisms $f = (f_n)_{n \in \mathbb{Z}}$ and $g = (g_n)_{n \in \mathbb{Z}}$ from $X$ to $Y$ their sum is $f + g := (f_n + g_n)_{n \in \mathbb{Z}}$. Using the additive structure of $\mathcal{A}$ it is then easy to check that (A1) holds.

(A2) The zero object in $\mathbf{C}(\mathcal{A})$ is the complex $(0_\mathcal{A}, d)$ where $0_\mathcal{A}$ is the zero object of the additive category $\mathcal{A}$ and all differentials are the unique (zero) morphism on the zero object.

(A3) The coproduct of two complexes $X = (X_n, d_n^X)$ and $Y = (Y_n, d_n^Y)$ is defined degreewise by using the coproduct in the additive category $\mathcal{A}$. More precisely $X \oplus Y = (X_n \oplus Y_n, d_n)_{n \in \mathbb{Z}}$ where the differential is obtained by the universal property as in the following diagram.

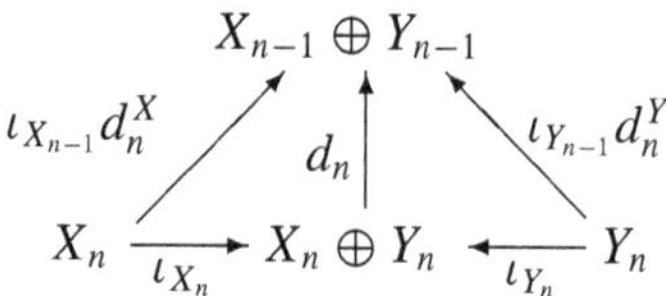

From uniqueness in the universal property applied to

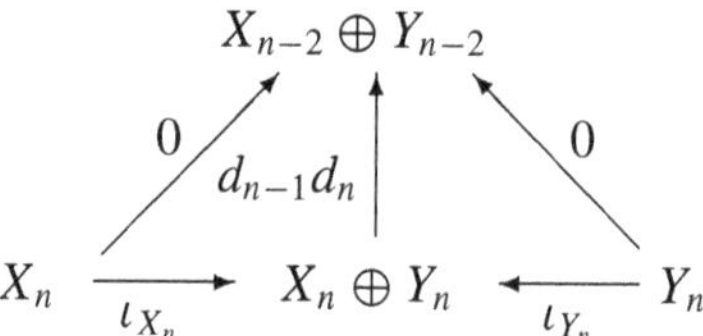

it follows that $d_{n-1} \circ d_n = 0$. This complex indeed satisfies the properties of a coproduct in the category of complexes $\mathbf{C}(\mathcal{A})$, with morphisms of complexes $\iota_X = (\iota_{X_n})_{n \in \mathbb{Z}} : X \to X \oplus Y$ and $\iota_Y = (\iota_{Y_n})_{n \in \mathbb{Z}} : Y \to X \oplus Y$. For checking the universal property let $Z$ be an arbitrary complex and let $f_X : X \to Z$ and $f_Y : Y \to Z$ be arbitrary morphisms. The unique morphism of complexes satisfying $f_X = f \circ \iota_X$ and $f_Y = f \circ \iota_Y$ is $f = (f_n)_{n \in \mathbb{Z}} : X \oplus Y \to Z$, where $f_n$ is obtained from the universal property in degree $n$ as in the following diagram.

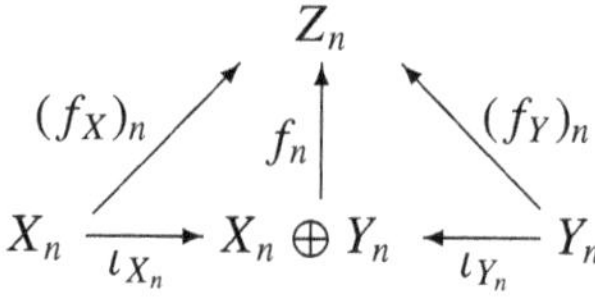

$\square$

**Remark 1.5.** For complexes over $\mathcal{A} = \mathbf{R\text{-}Mod}$ where $R$ is a ring with unit (and other similar examples) the coproduct of two complexes is more easily be described on elements as $X \oplus Y = (X_n \oplus Y_n, d_n)_{n \in \mathbb{Z}}$ where the differential is given by $d_n(x_n, y_n) = (d_n^X(x_n), d_n^Y(y_n))$ for $x_n \in X_n$ and $y_n \in Y_n$, and with morphisms $\iota_X : X \to X \oplus Y$ and $\iota_Y : Y \to X \oplus Y$ being the inclusion maps. The unique morphism of complexes satisfying $f_X = f \circ \iota_X$ and $f_Y = f \circ \iota_Y$ is then given by $f_n(x_n, y_n) = f_X(x_n) + f_Y(y_n)$.

## 1.2. The homotopy category of complexes

Let $\mathcal{A}$ be an additive category. Morphisms $f, g : X \to Y$ in the category $\mathbf{C}(\mathcal{A})$ of complexes are called *homotopic*, denoted $f \sim g$, if there exists a family $(s_n)_{n \in \mathbb{Z}}$ of morphisms $s_n : X_n \to Y_{n+1}$ in $\mathcal{A}$, satisfying $f_n - g_n = d^Y_{n+1} s_n + s_{n-1} d^X_n$ for all $n \in \mathbb{Z}$.

In particular, setting $g$ to be the zero morphism, we can speak of morphisms being *homotopic to zero*.

It is easy to check that $\sim$ is an equivalence relation. Moreover, if $f \sim g : X \to Y$ are homotopic and $\alpha : W \to X$ is an arbitrary morphism of complexes, then also the compositions $f\alpha \sim g\alpha$ are homotopic. In fact, $(s_n \alpha_n)_{n \in \mathbb{Z}}$ are homotopy maps since

$$(f_n - g_n)\alpha_n = (d^Y_{n+1} s_n + s_{n-1} d^X_n)\alpha_n = d^Y_{n+1}(s_n \alpha_n) + (s_{n-1}\alpha_{n-1})d^W_n.$$

Similarly, if $f, g : X \to Y$ are homotopic and $\beta : Y \to Z$ is a morphism of complexes then $\beta f \sim \beta g$ are homotopic.

This implies that we have a well-defined composition of equivalence classes of morphisms modulo homotopy by defining the composition on representatives.

**Definition 1.6.** *Let $\mathcal{A}$ be an additive category. The* homotopy category $\mathbf{K}(\mathcal{A})$ *has the same objects as the category $\mathbf{C}(\mathcal{A})$ of complexes over $\mathcal{A}$. The morphisms in the homotopy category are the equivalence classes of morphisms in $\mathbf{C}(\mathcal{A})$ modulo homotopy, i.e.*

$$\mathrm{Hom}_{\mathbf{K}(\mathcal{A})}(X, Y) := \mathrm{Hom}_{\mathbf{C}(\mathcal{A})}(X, Y)/\sim.$$

**Proposition 1.7.** *Let $\mathcal{A}$ be an additive category. Then the homotopy category $\mathbf{K}(\mathcal{A})$ is again an additive category.*

*Proof.* Addition of morphisms in $\mathbf{K}(\mathcal{A})$ is defined via addition on representatives (it is an easy observation that this is well-defined) and then the sets of morphisms $\mathrm{Hom}_{\mathbf{K}(\mathcal{A})}(X, Y)$ inherit the structure of an abelian group from the category $\mathbf{C}(\mathcal{A})$ of complexes, and also bilinearity of composition. Moreover, the zero object is the same as in $\mathbf{C}(\mathcal{A})$.

It remains to be checked that the universal property of the coproduct $X \oplus Y$ in $\mathbf{C}(\mathcal{A})$ (cf. Proposition 1.4) also carries over to the homotopy category. In fact, the equivalence classes of the morphisms $\iota_X$, $\iota_Y$ and $f$ still make the relevant diagram (cf. Remark 1.2) commutative; for uniqueness we observe that if there is another morphism $g$ making the diagram for the universal property commutative in $\mathbf{K}(\mathcal{A})$, i.e. up to homotopy, then this gives a homotopy between $f$ and $g$. $\qquad\square$

## 2. Abelian categories

In this section we shall review the fundamental definition of an abelian category, including the necessary background on the categorical notions of kernels and cokernels. The prototype example of an abelian category will be the category **R-Mod** of modules over a ring $R$; but we will also see other examples in due course.

We first recall some notions from category theory. Let $\mathcal{A}$ be an additive category; in particular for every pair of objects $X, Y$ there is a zero morphism, namely the composition of the unique morphisms $X \to 0 \to Y$ involving the zero object of $\mathcal{A}$.

The *kernel* of a morphism $f : X \to Y$ is an object $K$ together with a morphism $k : K \to X$ such that

(i) $f \circ k = 0$,
(ii) (universal property) for every morphism $k' : K' \to X$ such that $f \circ k' = 0$, there is a unique morphism $g : K' \to K$ making the following diagram commutative.

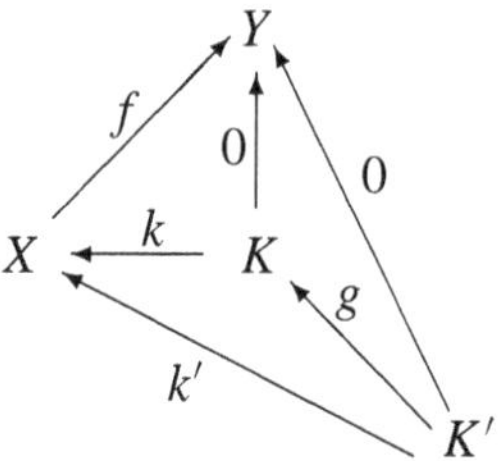

By the usual universal property argument, the kernel, if it exists, is unique up to isomorphism; notation: $\ker f$.

Dually, the *cokernel* of a morphism $f : X \to Y$ is an object $C$ together with a morphism $c : Y \to C$ such that

(i) $c \circ f = 0$,
(ii) (universal property) for every morphism $c' : Y \to C'$ such that $c' \circ f = 0$, there is a unique morphism $g : C \to C'$ making the following diagram commutative.

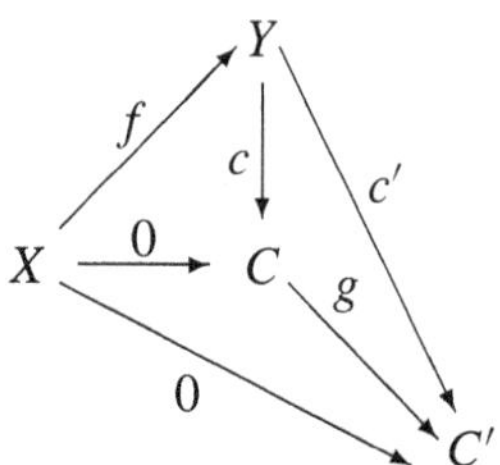

Again, the cokernel, if it exists, is unique up to isomorphism; notation: coker $f$.

If the above morphism $k : \ker f \to X$ has a cokernel in $\mathcal{A}$, this is called the *coimage* of $f$, and it is denoted by coim $f$.

If the above morphism $c : Y \to$ coker $f$ has a kernel in $\mathcal{A}$, this is called the *image* of $f$ and it is denoted by im $f$.

**Example 2.1.** Let $R$ be a ring. In the category **R-Mod** of all $R$-modules the categorical kernels and cokernels are the usual ones, i.e., for a morphism $f : X \to Y$ we have $\ker f = \{x \in X \mid f(x) = 0\}$ and coker $f = Y/\operatorname{im} f$ where $\operatorname{im} f = \{f(x) \mid x \in X\}$ is the usual image of $f$.

**Remark 2.2.** Suppose that for a morphism $f$ both the coimage and the image exist. Then we claim that it follows from the universal properties that there is a natural morphism coim $f \to$ im $f$.

In fact, the image of $f$ is the kernel of $c : Y \to$ coker $f$, hence there is a morphism $\tilde{k} : \operatorname{im} f \to Y$ such that $c \circ \tilde{k} = 0$ and by the universal property there exists a unique morphism $\tilde{g} : X \to \operatorname{im} f$ making the following diagram commutative.

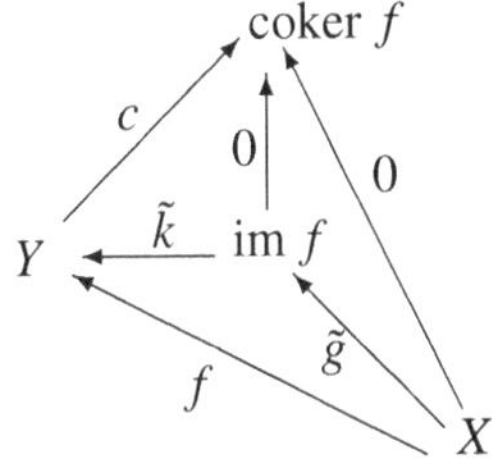

Note that $\tilde{k} \circ \tilde{g} \circ k = f \circ k = 0$, which implies that $\tilde{g} \circ k : \ker f \to \operatorname{im} f$ must be zero, by using the uniqueness in the following diagram.

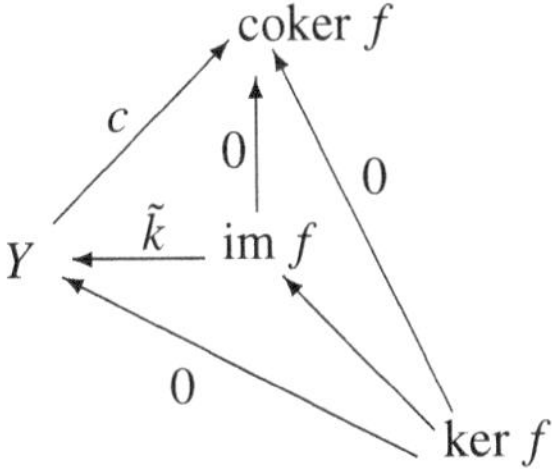

Then we can consider the following diagram for the universal property of the coimage

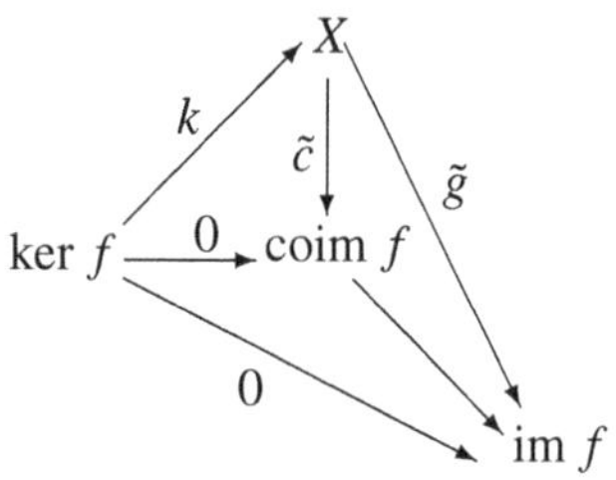

and deduce that there is a unique morphism coim $f \to$ im $f$, as desired.

**Definition 2.3.** *An additive category $\mathcal{A}$ is called an* abelian category *if the following axioms are satisfied:*

(A4) *Every morphism in $\mathcal{A}$ has a kernel and a cokernel.*

(A5) *For every morphism $f : X \to Y$ in $\mathcal{A}$, the natural morphism* coim $f \to$ im $f$ *is an isomorphism.*

**Example 2.4.**

(i) Let $R$ be a ring. The category **R-Mod** of all $R$-modules is an abelian category. In fact, (A5) follows directly from the isomorphism theorem for $R$-modules.

However, the subcategory **R-mod** of finitely generated modules is not abelian in general since kernels of homomorphisms between finitely generated modules need not be finitely generated. Indeed we have that **R-mod** is an abelian category if and only if $R$ is Noetherian.

In particular, the category of finite-dimensional vector spaces over a field is abelian, and the category of finitely generated abelian groups is abelian.

(ii) The subcategory of **Ab** consisting of free abelian groups is not abelian.

On the other hand, for a prime number $p$, the abelian $p$-groups form an abelian subcategory of **Ab** (an abelian group is called a $p$-group if for every element $a$ we have $p^k a = 0$ for some $k$).

(iii) For finding examples of additive categories satisfying (A4) but failing to be abelian, the following observation can be useful. Suppose $f : X \to Y$ is a morphism with ker $f = 0$ and coker $f = 0$, i.e. a monomorphism and an epimorphism. Then the coimage of $f$ is the identity on $X$, the image of $f$ is the identity on $Y$ and hence the natural morphism coim $f \to$ im $f$ is just $f$ itself. So in this special case the axiom (A5) states that a morphism which is a monomorphism and an epimorphism must be invertible.

(iv) Explicit examples of additive categories where axiom (A5) fails for the above reason are the category of topological abelian groups (with continuous group homomorphisms) or the category of Banach complex vector spaces (with continuous linear maps). In such categories the cokernel of a morphism $f : X \to Y$ is of the form $Y/\overline{\mathrm{im}}_f$ where $\overline{\mathrm{im}}_f$ is the *closure* of the usual set-theoretic image of $f$. In particular, the natural morphism $\mathrm{coim}\, f \to \mathrm{im}\, f$ is the inclusion of the usual image of $f$ into its closure, and this is in general not an isomorphism.

**Proposition 2.5.** *Let $\mathcal{A}$ be an abelian category. Then the category of complexes* $\mathbf{C}(\mathcal{A})$ *is also abelian.*

*Proof.* We have seen in Proposition 1.4 that $\mathbf{C}(\mathcal{A})$ is an additive category, so it remains to verify the axioms (A4) and (A5).

(A4) Let $f : X \to Y$ be a morphism in $\mathbf{C}(\mathcal{A})$, i.e. $f = (f_n)_{n\in\mathbb{Z}}$ with $f_n : X_n \to Y_n$ morphisms in $\mathcal{A}$. We show the existence of a kernel and leave the details of the dual argument for the cokernel as an exercise.

Since $\mathcal{A}$ is abelian, each morphism $f_n : X_n \to Y_n$ has a kernel $K_n := \ker f_n$ in $\mathcal{A}$, coming with a morphism $k_n : K_n \to X_n$ satisfying the above universal property. Note that for every $n \in \mathbb{Z}$ we have $f_{n-1} \circ d_n^X \circ k_n = d_n^Y \circ f_n \circ k_n = 0$. Then it follows by the universal property of kernels that there is a unique morphism $d_n^K : K_n \to K_{n-1}$ such that $k_{n-1} \circ d_n^K = d_n^X \circ k_n$. Note that

$$k_{n-1} \circ d_n^K \circ d_{n+1}^K = d_n^X \circ k_n \circ d_{n+1}^K = d_n^X \circ d_{n+1}^X \circ k_{n+1} = 0$$

since $X$ is a complex. By uniqueness of the map in the universal property of $K_{n-1}$ it follows that $d_n^K \circ d_{n+1}^K = 0$, i.e. $(K_n, d_n^K)$ is a complex.

Combining the universal properties of the kernels $K_n$ it easily follows that the complex $(K_n, d_n^K)$ indeed satisfies the universal property for the kernel of $f$ in $\mathbf{C}(\mathcal{A})$.

(A5) The crucial observation is that a morphism of complexes $f = (f_n) : X \to Y$ is an isomorphism in $\mathbf{C}(\mathcal{A})$ if and only if each $f_n$ is an isomorphism in $\mathcal{A}$. In fact, if each $f_n$ is an isomorphism, with inverse $g_n$, then the family $g = (g_n)$ is automatically a morphism of complexes (and hence clearly an inverse to $f$ in $\mathbf{C}(\mathcal{A})$): for all $n \in \mathbb{Z}$ we have

$$d_{n+1}^X \circ g_{n+1} = g_n \circ f_n \circ d_{n+1}^X \circ g_{n+1} = g_n \circ d_{n+1}^Y \circ f_{n+1} \circ g_{n+1} = g_n \circ d_{n+1}^Y.$$

The reverse implication is obvious.

For axiom (A5) now consider the natural morphism $\mathrm{coim}\, f \to \mathrm{im}\, f$. In the proof of (A4) above we have seen that kernels and cokernels in $\mathbf{C}(\mathcal{A})$, and hence also the morphism coim $f \to$ im $f$, are obtained degreewise. But since $\mathcal{A}$ is abelian by assumption, we know that for every $n$ the natural morphism coim $f_n \to$ im $f_n$ in $\mathcal{A}$ is indeed an isomorphism. Then, by the introductory remark, the morphism of complexes (coim $f_n \to$ im $f_n)_{n \in \mathbb{Z}}$ is an isomorphism in $\mathbf{C}(\mathcal{A})$.                                                    $\square$

An important observation is that the homotopy category $\mathbf{K}(\mathcal{A})$ is not abelian in general, even if $\mathcal{A}$ is abelian.

**Example 2.6.** We provide an explicit example for the failure of axiom (A4) in a homotopy category. Consider the abelian category $\mathcal{A} = \mathbf{Ab}$ of abelian groups.

Let $f : X \to Y$ be the following morphism of complexes of abelian groups, with non-zero entries in degrees 1 and 0,

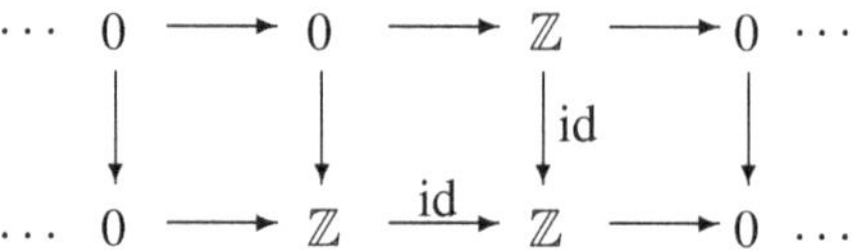

In the category $\mathbf{C}(\mathbf{Ab})$ of complexes $f$ is non-zero and has the zero complex as kernel (cf. the proof of Proposition 2.5). However, $f$ is homotopic to zero (with the identity as homotopy map), i.e. $f = 0$ in the homotopy category $\mathbf{K}(\mathbf{Ab})$.

We claim that in the homotopy category $f$ has no kernel. Recall the categorical definition of the kernel of a morphism $f : X \to Y$ from Section 2.

Suppose for a contradiction that our morphism $f$ had a kernel in $\mathbf{K}(\mathbf{Ab})$. So there is a complex $\ldots \to K_1 \to K_0 \to K_{-1} \to \ldots$ and a morphism $k = k_0 : K_0 \to \mathbb{Z}$ of abelian groups (in all other degrees the map $k$ has to be zero since $X$ is concentrated in degree 0). The image of $k$, being a subgroup of $\mathbb{Z}$, has the form $r\mathbb{Z}$ for some fixed $r \in \mathbb{Z}$. Now choose $K' = X$ and consider the morphisms $l : K' \to X$ given by multiplication with $l$ for any $l \in \mathbb{Z}$. Clearly, $f \circ l = 0$ in $\mathbf{K}(\mathbf{Ab})$ since $f = 0$ in $\mathbf{K}(\mathbf{Ab})$. According to the universal property of a kernel, there must exist (unique) morphisms $u_l : \mathbb{Z} \to K_0$ such that $k \circ u_l = l$ up to homotopy. However, these maps are from $K' = X$ to $X$ and this complex is concentrated in degree 0. Thus there are no non-zero homotopy maps and so $k \circ u_l = l$ as morphism of abelian groups. But the image of $k \circ u_l$ is contained in the image of $k$ which is $r\mathbb{Z}$ for a fixed $r$, so $k \circ u_l = l$ can not hold for arbitrary $l \in \mathbb{Z}$, a contradiction.

Hence axiom (A4) fails and therefore the homotopy category $\mathbf{K}(\mathbf{Ab})$ is not an abelian category.

# 3. Definition of triangulated categories

We have seen in the previous section that the homotopy category of complexes is not abelian in general. We shall see in Section 6 below that $\mathbf{K}(\mathcal{A})$ carries the structure of a triangulated category, a concept which we are going to define in this section. Roughly, one should think of the distinguished triangles occurring in this context as a replacement for short exact sequences (which do not exist in general since $\mathbf{K}(\mathcal{A})$ is not abelian). However, for an additive category to be abelian is purely an inherent property of the category. On the other hand a triangulated structure is an extra piece of data, consisting of a suspension functor and a set of distinguished triangles chosen suitably to satisfy certain axioms. In particular, an additive category can have many different triangulated structures; see [1] for more details and examples.

A functor $\Sigma$ between additive categories is called an *additive functor* if for every pair of objects $X, Y$ the map $\mathrm{Hom}(X, Y) \to \mathrm{Hom}(\Sigma(X), \Sigma(Y))$ is a homomorphism of abelian groups.

Let $\mathcal{T}$ be an additive category and let $\Sigma : \mathcal{T} \to \mathcal{T}$ be an additive functor which is an automorphism (i.e. it is invertible, thus there exists a functor $\Sigma^{-1}$ on $\mathcal{T}$ such that $\Sigma \circ \Sigma^{-1}$ and $\Sigma^{-1} \circ \Sigma$ are the identity functors).

A *triangle* in $\mathcal{T}$ is a sequence of objects and morphisms in $\mathcal{T}$ of the form

$$X \xrightarrow{u} Y \xrightarrow{v} Z \xrightarrow{w} \Sigma X.$$

A *morphism of triangles* is a triple $(f, g, h)$ of morphisms such that the following diagram is commutative in $\mathcal{T}$

$$
\begin{array}{ccccccc}
X & \xrightarrow{u} & Y & \xrightarrow{v} & Z & \xrightarrow{w} & \Sigma X \\
\downarrow{f} & & \downarrow{g} & & \downarrow{h} & & \downarrow{\Sigma f} \\
X' & \xrightarrow{u'} & Y' & \xrightarrow{v'} & Z' & \xrightarrow{w'} & \Sigma X'
\end{array}
$$

If in this situation, the morphisms $f, g$ and $h$ are isomorphisms in $\mathcal{T}$, then the morphism of triangles is called an *isomorphism of triangles*.

**Definition 3.1.** *A triangulated category is an additive category $\mathcal{T}$ together with an additive automorphism $\Sigma$, the translation or shift functor, and a collection of distinguished triangles satisfying the following axioms*

(TR0) *Any triangle isomorphic to a distinguished triangle is again a distinguished triangle.*

(TR1) *For every object $X$ in $\mathcal{T}$, the triangle $X \xrightarrow{\mathrm{id}} X \to 0 \to \Sigma X$ is a distinguished triangle.*

(TR2) *For every morphism $f : X \to Y$ in $T$ there is a distinguished triangle of the form $X \xrightarrow{f} Y \to Z \to \Sigma X$.*

(TR3) *If $X \xrightarrow{u} Y \xrightarrow{v} Z \xrightarrow{w} \Sigma X$ is a distinguished triangle, then also $Y \xrightarrow{v} Z \xrightarrow{w} \Sigma X \xrightarrow{-\Sigma u} \Sigma Y$ is a distinguished triangle, and vice versa.*

(TR4) *Given distinguished triangles $X \xrightarrow{u} Y \xrightarrow{v} Z \xrightarrow{w} \Sigma X$ and $X' \xrightarrow{u'} Y' \xrightarrow{v'} Z' \xrightarrow{w'} \Sigma X'$, then each commutative diagram*

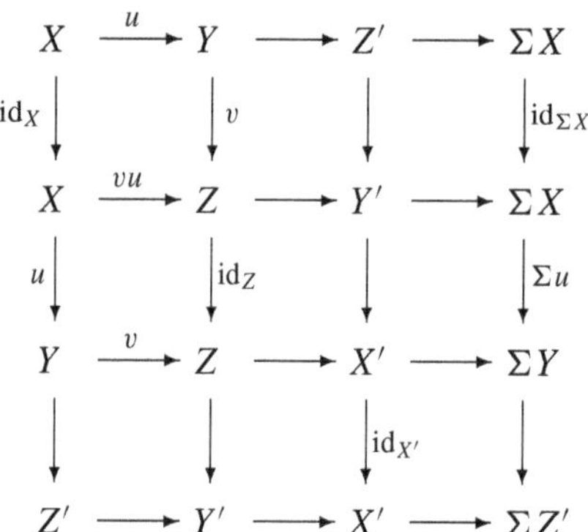

*can be completed to a morphism of triangles (but not necessarily uniquely).*

(TR5) *(Octahedral axiom) Given distinguished triangles $X \xrightarrow{u} Y \to Z' \to \Sigma X$, $Y \xrightarrow{v} Z \to X' \to \Sigma Y$ and $X \xrightarrow{vu} Z \to Y' \to \Sigma X$, there exists a distinguished triangle $Z' \to Y' \to X' \to \Sigma Z'$ making the following diagram commutative*

$$
\begin{array}{ccccccc}
X & \xrightarrow{u} & Y & \longrightarrow & Z' & \longrightarrow & \Sigma X \\
\mathrm{id}_X \downarrow & & \downarrow v & & \downarrow & & \downarrow \mathrm{id}_{\Sigma X} \\
X & \xrightarrow{vu} & Z & \longrightarrow & Y' & \longrightarrow & \Sigma X \\
u \downarrow & & \downarrow \mathrm{id}_Z & & \downarrow & & \downarrow \Sigma u \\
Y & \xrightarrow{v} & Z & \longrightarrow & X' & \longrightarrow & \Sigma Y \\
\downarrow & & \downarrow & & \downarrow \mathrm{id}_{X'} & & \downarrow \\
Z' & \longrightarrow & Y' & \longrightarrow & X' & \longrightarrow & \Sigma Z'
\end{array}
$$

**Remark 3.2.** The above version (TR5) of the octahedral axiom is taken from the book by Kashiwara and Schapira [7, Sec. 1.4]. There are various other versions appearing in the literature which are equivalent to (TR5), see for instance A. Neeman's article [13] or his book [12]; a short treatment can also be found in A. Hubery's notes [6] (which are based on the former references).

We shall only mention two variations here. Mainly a reformulation of the axiom (TR5) is the following. Note that in (TR5) the given three distinguished triangles are placed in the first three rows, whereas in (TR5') below they are placed in the first two rows and the second column.

(TR5') Given distinguished triangles $X \xrightarrow{u} Y \to Z' \to \Sigma X$, $Y \xrightarrow{v} Z \to X' \xrightarrow{l} \Sigma Y$ and $X \xrightarrow{vu} Z \to Y' \xrightarrow{s} \Sigma X$, then there exists a distinguished triangle $Z' \to Y' \xrightarrow{v'} X' \to \Sigma Z'$ making the following diagram commutative and satisfying $(\Sigma u)s = lv'$.

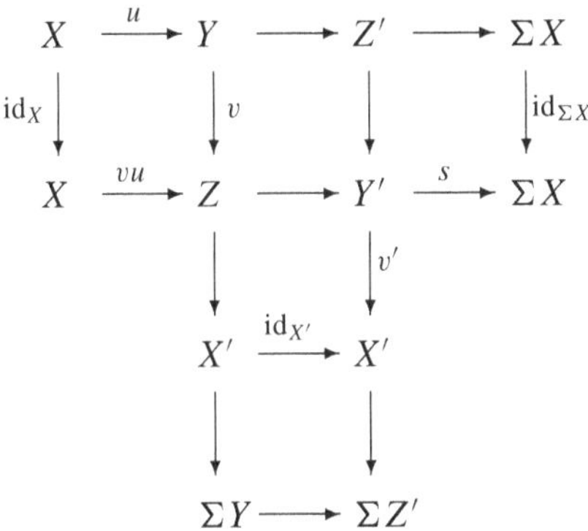

It is not difficult to check that (TR5) and (TR5') are indeed equivalent; we leave this verification as an exercise to the reader.

The following version (TR5") of the octahedral axiom can be found in Neeman's book [12, Prop. 1.4.6]. It is less obvious that it is equivalent to (TR5); for details on this we refer the reader to [12], [13] and [6].

(TR5") Given distinguished triangles $X \xrightarrow{u} Y \to Z' \to \Sigma X$, $Y \xrightarrow{v} Z \to X' \to \Sigma Y$ and $X \xrightarrow{vu} Z \to Y' \to \Sigma X$, then there exists a distinguished triangle $Z' \to Y' \to X' \to \Sigma Z'$ making the following diagram commutative in which every row and every column is a distinguished triangle.

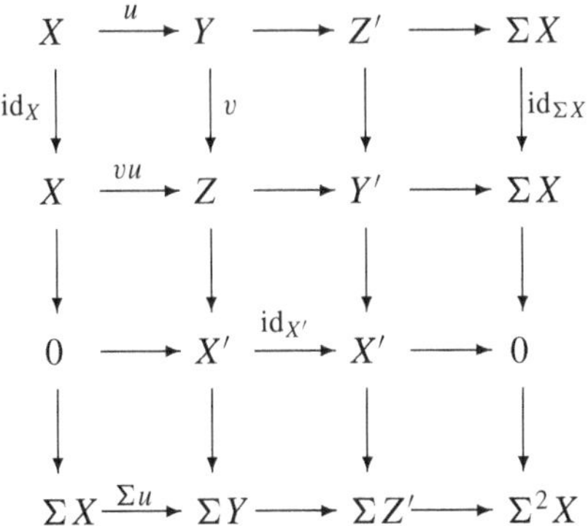

# 4. Some formal properties of triangulated categories

We shall draw some first consequences from the definition. Let $\mathcal{T}$ be a triangulated category with translation functor $\Sigma$.

**Proposition 4.1** (Composition of morphisms). *Let $X \xrightarrow{u} Y \xrightarrow{v} Z \xrightarrow{w} \Sigma X$ be a distinguished triangle. Then $v \circ u = 0$ and $w \circ v = 0$, i.e. any composition of two consecutive morphisms in a distinguished triangle vanishes.*

*Proof.* By the rotation property (TR3) it suffices to show that $v \circ u = 0$. Also by (TR3) we have a distinguished triangle $Y \xrightarrow{v} Z \xrightarrow{w} \Sigma X \xrightarrow{-\Sigma u} \Sigma Y$. By (TR1) and (TR4) the following diagram can be completed to a morphism of triangles.

$$
\begin{array}{ccccccc}
Y & \xrightarrow{\ v\ } & Z & \xrightarrow{\ w\ } & \Sigma X & \xrightarrow{-\Sigma u} & \Sigma Y \\
\downarrow{\scriptstyle v} & & \downarrow{\scriptstyle \mathrm{id}} & & & & \downarrow{\scriptstyle \Sigma v} \\
Z & \xrightarrow{\ \mathrm{id}\ } & Z & \xrightarrow{\ 0\ } & 0 & \xrightarrow{\ 0\ } & \Sigma Z
\end{array}
$$

In particular, $-\Sigma(v \circ u) = -\Sigma v \circ \Sigma u = 0$ which implies $v \circ u = 0$ since $\Sigma$ is an automorphism. $\qquad\square$

**Proposition 4.2** (Long exact sequences). *Let $X \xrightarrow{u} Y \xrightarrow{v} Z \xrightarrow{w} \Sigma X$ be a distinguished triangle. For any object $T \in \mathcal{T}$ there is a long exact sequence of abelian groups*

$$
\ldots \to \mathrm{Hom}_{\mathcal{T}}(T, \Sigma^i X) \xrightarrow{\Sigma^i u_*} \mathrm{Hom}_{\mathcal{T}}(T, \Sigma^i Y) \xrightarrow{\Sigma^i v_*} \mathrm{Hom}_{\mathcal{T}}(T, \Sigma^i Z)
$$

$$
\xrightarrow{\Sigma^i w_*} \mathrm{Hom}_{\mathcal{T}}(T, \Sigma^{i+1} X) \to \ldots
$$

*Proof.* For abbreviation we denote by $f_* := \mathrm{Hom}_{\mathcal{T}}(T, f)$ the morphism induced by $f$ under the functor $\mathrm{Hom}_{\mathcal{T}}(T, -)$ on the additive category $\mathcal{T}$.

By the rotation property, it suffices to show that

$$
\mathrm{Hom}_{\mathcal{T}}(T, \Sigma^i X) \xrightarrow{\Sigma^i u_*} \mathrm{Hom}_{\mathcal{T}}(T, \Sigma^i Y) \xrightarrow{\Sigma^i v_*} \mathrm{Hom}_{\mathcal{T}}(T, \Sigma^i Z)
$$

is an exact sequence of abelian groups.

By Proposition 4.1 we have $\Sigma^i v \circ \Sigma^i u = 0$ and hence also $\Sigma^i v_* \circ \Sigma^i u_* = 0$, i.e. the image of $\Sigma^i u_*$ is contained in the kernel of $\Sigma^i v_*$.

Conversely, take $f$ in the kernel of $\Sigma^i v_*$. Consider the following diagram whose rows are distinguished triangles by (TR1) and (TR3).

$$
\begin{array}{ccccccc}
\Sigma^{-i} T & \xrightarrow{\ 0\ } & 0 & \xrightarrow{\ 0\ } & \Sigma^{-i+1} T & \xrightarrow{-\mathrm{id}} & \Sigma^{-i+1} T \\
\downarrow{\scriptstyle \Sigma^{-i} f} & & \downarrow{\scriptstyle 0} & & & & \downarrow{\scriptstyle \Sigma^{-i+1} f} \\
Y & \xrightarrow{\ v\ } & Z & \xrightarrow{\ w\ } & \Sigma X & \xrightarrow{-\Sigma u} & \Sigma Y
\end{array}
$$

The left hand square is commutative by assumption on $f$. By (TR4) there exists a morphism $h : \Sigma^{-i+1}T \to \Sigma X$ completing the above diagram to a morphism of triangles. In particular, $\Sigma^{-i+1}f = \Sigma u \circ h$ and hence $f = \Sigma^i u \circ \Sigma^{i-1}h$ is in the image of $\Sigma^i u_*$ as desired. $\qquad\square$

**Proposition 4.3** (Triangulated 5-lemma). *Suppose we are given a morphism of distinguished triangles as in the following diagram.*

$$
\begin{array}{ccccccc}
X & \xrightarrow{u} & Y & \xrightarrow{v} & Z & \xrightarrow{w} & \Sigma X \\
\downarrow{f} & & \downarrow{g} & & \downarrow{h} & & \downarrow{\Sigma f} \\
X' & \xrightarrow{u'} & Y' & \xrightarrow{v'} & Z' & \xrightarrow{w'} & \Sigma X'
\end{array}
$$

*If $f$ and $g$ are isomorphisms then also $h$ is an isomorphism.*

*Proof.* We apply the functor $\mathrm{Hom}(Z', -) := \mathrm{Hom}_T(Z', -)$ to the distinguished triangles. By Proposition 4.2 this leads to the following commutative diagram whose rows are exact sequences of abelian groups.

$$
\begin{array}{ccccccccc}
\mathrm{Hom}(Z', X) & \to & \mathrm{Hom}(Z', Y) & \to & \mathrm{Hom}(Z', Z) & \to & \mathrm{Hom}(Z', \Sigma X) & \to & \mathrm{Hom}(Z', \Sigma Y) \\
\downarrow{f_*} & & \downarrow{g_*} & & \downarrow{h_*} & & \downarrow{\Sigma f_*} & & \downarrow{\Sigma g_*} \\
\mathrm{Hom}(Z', X') & \to & \mathrm{Hom}(Z', Y') & \to & \mathrm{Hom}(Z', Z') & \to & \mathrm{Hom}(Z', \Sigma X') & \to & \mathrm{Hom}(Z', \Sigma Y')
\end{array}
$$

By assumption, $f$ and $g$ are isomorphisms and hence also $f_*$, $g_*$, $\Sigma f_*$ and $\Sigma g_*$ are isomorphisms. So we can appeal to the usual 5-lemma in the category of abelian groups to deduce that $h_*$ is an isomorphism. In particular the identity $\mathrm{id}_{Z'}$ has a preimage, i.e. there exists a morphism $q \in \mathrm{Hom}_T(Z', Z)$ such that $h \circ q = \mathrm{id}_{Z'}$.

A similar argument using the functor $\mathrm{Hom}_T(-, Z')$ produces a left inverse to $h$, thus $h$ is an isomorphism. $\qquad\square$

**Proposition 4.4** (Split triangles). *Let $X \xrightarrow{u} Y \xrightarrow{v} Z \xrightarrow{w} \Sigma X$ be a distinguished triangle where $w = 0$ is the zero morphism. Then the triangle splits, i.e. $u$ is a split monomorphism and $v$ is a split epimorphism.*

**Remark 4.5.** The notion of split monomorphism is synonymous with that of a section, and a split epimorphism is also known as a retraction.

*Proof.* We first show that $u$ is a split monomorphism, i.e. there exists a morphism $u'$ such that $u' \circ u = \mathrm{id}_X$. We have the following commutative diagram of distinguished triangles.

$$X \xrightarrow{u} Y \xrightarrow{v} Z \xrightarrow{0} \Sigma X$$
$$\downarrow{\scriptstyle \mathrm{id}} \qquad\qquad \downarrow{\scriptstyle 0} \qquad \downarrow{\scriptstyle \mathrm{id}}$$
$$X \xrightarrow{\mathrm{id}} X \xrightarrow{0} 0 \xrightarrow{0} \Sigma X$$

By (TR3) and (TR4) it can be completed to a morphism of triangles, i.e. there exists $u' : Y \to X$ such that $u' \circ u = \mathrm{id}$.

Similarly, one can show that $v$ is a split epimorphism, i.e. there is a morphism $v' : Z \to Y$ such that $v \circ v' = \mathrm{id}$. $\qquad\qquad\qquad\qquad\square$

## 5. Abelian categories vs. triangulated categories

As an application of the formal properties in the previous section we shall compare the notions of abelian categories and triangulated categories.

**Definition 5.1.** *An abelian category $\mathcal{A}$ is called* semisimple *if every short exact sequence in $\mathcal{A}$ splits.*

**Example 5.2.**
 (i) *Let $R$ be a semisimple ring. Then the module categories* **R-Mod** *and* **R-mod** *are semisimple. In particular, the category of vector spaces* **Vec$_K$** *over a field $K$ is semisimple.*
(ii) *The category* **Ab** *of abelian groups is not semisimple. For instance, the short exact sequence* $0 \to \mathbb{Z}/2\mathbb{Z} \xrightarrow{\cdot 2} \mathbb{Z}/4\mathbb{Z} \xrightarrow{\cdot 1} \mathbb{Z}/2\mathbb{Z} \to 0$ *does not split.*

The following result illustrates that the concepts of abelian and triangulated categories overlap only slightly.

**Theorem 5.3.** *Let $\mathcal{T}$ be a category which is triangulated and abelian. Then $\mathcal{T}$ is semisimple.*

*Proof.* Let $0 \to X \xrightarrow{f} Y \xrightarrow{g} Z \to 0$ be a short exact sequence in $\mathcal{T}$. We have to show that it splits; to this end it suffices to show that $f$ is a section, i.e. there exists a morphism $f' : Y \to X$ such that $f' \circ f = \mathrm{id}_X$.

By (TR2) and (TR3), $f$ can be embedded into a distinguished triangle

$$\Sigma^{-1}V \xrightarrow{u} X \xrightarrow{f} Y \xrightarrow{v} V.$$

The composition of consecutive morphisms in a distinguished triangle is always zero by Proposition 4.1, in particular $f \circ u = 0$. But $f$ is a monomorphism in $\mathcal{T}$ since it is the first map in a short exact sequence, hence $u = 0$. Thus we have a distinguished triangle

$$X \xrightarrow{f} Y \xrightarrow{v} V \xrightarrow{\Sigma u} \Sigma X$$

where $\Sigma u = 0$. Now the triangle splits by Proposition 4.4. $\qquad\square$

We shall see in the next section that the homotopy category $\mathbf{K}(\mathcal{A})$ of complexes over an additive category $\mathcal{A}$ is a triangulated category. This, together with the preceding theorem, will then give a more structural explanation of the earlier observation that $\mathbf{K}(\mathbf{Ab})$ is not abelian in Example 2.6, where we have used an ad-hoc argument to show that morphisms do not necessarily have a kernel.

## 6. The homotopy category of complexes is triangulated

Let $\mathcal{A}$ be an additive category, with corresponding category of complexes $\mathbf{C}(\mathcal{A})$ and homotopy category $\mathbf{K}(\mathcal{A})$.

As discussed above, the homotopy category $\mathbf{K}(\mathcal{A})$ is in general not abelian, even if $\mathcal{A}$ is abelian. We shall explain in this section how the homotopy category $\mathbf{K}(\mathcal{A})$ becomes a triangulated category.

We first need an additive automorphism on $\mathbf{K}(\mathcal{A})$ which serves as translation functor. This functor can already be defined on the level of the category $\mathbf{C}(\mathcal{A})$.

**Definition 6.1.** *In* $\mathbf{C}(\mathcal{A})$ *we construct a translation functor* $\Sigma = [1]$ *by shifting any complex one degree to the left. More precisely, for an object* $X = (X_n, d_n^X)_{n \in \mathbb{Z}}$ *in* $\mathbf{C}(\mathcal{A})$ *we set*

$$X[1] := (X[1]_n, d_n^{X[1]})_{n \in \mathbb{Z}} \text{ with } X[1]_n = X_{n-1} \text{ and } d_n^{X[1]} = -d_{n-1}^X.$$

*For a morphism of complexes* $f = (f_n)_{n \in \mathbb{Z}}$ *in* $\mathbf{C}(\mathcal{A})$ *we set*

$$f[1] := (f[1]_n)_{n \in \mathbb{Z}} \text{ where } f[1]_n = f_{n-1}.$$

**Remark 6.2.**
  (i) The sign appearing in the differential of $X[1]$ might look arbitrary; it will become clear later when discussing the triangulated structure of the homotopy category why this sign is needed.
  (ii) The functor $\Sigma = [1]$ defined above is an additive functor and moreover an automorphism of the category $\mathbf{C}(\mathcal{A})$.
  (iii) Note that the above definitions are compatible with homotopies so we have a well-defined induced functor $\Sigma = [1]$ on the homotopy category $\mathbf{K}(\mathcal{A})$.

The next step for getting a triangulated structure on the homotopy category is to find a suitable set of distinguished triangles. To this end, the following construction of mapping cones is crucial.

**Definition 6.3.** *Let $f$ be a morphism between complexes $X = (X_n, d_n^X)$ and $Y = (Y_n, d_n^Y)$. The* mapping cone $M(f)$ *is the complex in* $\mathbf{C}(\mathcal{A})$ *defined by*

$$M(f)_n = X_{n-1} \oplus Y_n \quad and \quad d_n^{M(f)} := \begin{pmatrix} -d_{n-1}^X & 0 \\ f_{n-1} & d_n^Y \end{pmatrix}.$$

**Remark 6.4.**

(i) There are canonical morphisms in $\mathbf{C}(\mathcal{A})$ as follows

$$\alpha(f) : Y \to M(f), \quad \alpha(f)_n := (0, \mathrm{id}_{Y_n})$$

and

$$\beta(f) : M(f) \to X[1], \quad \beta(f)_n := (\mathrm{id}_{X_{n-1}}, 0).$$

Note that $\beta(f)$ is a morphism of complexes because the differential in $X[1]$ carries a sign. From the above definitions we get a short exact sequence of chain complexes

$$0 \to Y \xrightarrow{\alpha(f)} M(f) \xrightarrow{\beta(f)} X[1] \to 0.$$

(ii) Let $f : X \to Y$ be a morphism of complexes. The short exact sequence $0 \to Y \xrightarrow{\alpha(f)} M(f) \xrightarrow{\beta(f)} X[1] \to 0$ splits (i.e. there is a morphism of complexes $\sigma : X[1] \to M(f)$ such that $\beta(f) \circ \sigma = \mathrm{id}_{X[1]}$) if and only if $f$ is homotopic to zero. In fact, a splitting map is given by $\sigma(x) := (x, -s(x))$ where $s$ is a homotopy map.

**Example 6.5.**

(i) For any complex $X$ consider the zero map $f : X \to 0$ to the zero complex. Then the mapping cone is $M(f) = X[1]$. On the other hand, the mapping cone of $g : 0 \to Y$ is just $M(g) = Y$ itself.

(ii) Let $A$ and $B$ be objects in $\mathcal{A}$ and view them as complexes $X_A$ and $X_B$ concentrated in degree 0. Any morphism $f : A \to B$ in $\mathcal{A}$ induces a morphism of complexes $f : X_A \to X_B$. Its mapping cone is the complex

$$\ldots \to 0 \to A \xrightarrow{f} B \to 0 \to \ldots$$

where $A$ is in degree 1 and $B$ in degree 0.

(iii) Let $X = (X_n, d_n^X)$ be any complex in $\mathbf{C}(\mathcal{A})$. The mapping cone of the identity morphism $\mathrm{id}_X$ has degree $n$ term equal to $X_{n-1} \oplus X_n$ and differential

$$\begin{pmatrix} -d_{n-1}^X & 0 \\ \mathrm{id}_{X_{n-1}} & d_n^X \end{pmatrix} : X_{n-1} \oplus X_n \to X_{n-2} \oplus X_{n-1}.$$

The identity morphism on the mapping cone $M(\mathrm{id}_X)$ is homotopic to zero, via the map $s = (s_n)_{n \in \mathbb{Z}}$ where $s_n = \begin{pmatrix} 0 & \mathrm{id}_{X_n} \\ 0 & 0 \end{pmatrix}$. Thus, in the homotopy category $\mathbf{K}(\mathcal{A})$ the identity $\mathrm{id}_{M(\mathrm{id}_X)}$ is equal to the zero map. As a consequence, in the homotopy category, the mapping cone $M(\mathrm{id}_X)$ is isomorphic to the zero complex.

It is easy to check that the morphisms $\alpha(f)$ and $\beta(f)$ are also well-defined in the homotopy category $\mathbf{K}(\mathcal{A})$ (i.e. independent on the choice of representatives of the equivalence class of morphisms). This leads to the following definition.

**Definition 6.6.** *A sequence of objects and morphisms in the homotopy category* $\mathbf{K}(\mathcal{A})$ *of the form*

$$X \xrightarrow{f} Y \xrightarrow{\alpha(f)} M(f) \xrightarrow{\beta(f)} X[1]$$

*is called a* standard triangle.

*A distinguished triangle in* $\mathbf{K}(\mathcal{A})$ *is a triangle which is isomorphic (in* $\mathbf{K}(\mathcal{A})$*!) to a standard triangle.*

With this class of distinguished triangles the homotopy category obtains a triangulated structure as we shall show next. Due to the technical nature of the axioms of a triangulated category, the proof that a certain additive category is indeed triangulated is usually rather long, can be partly tedious and can still be quite involved. In this introductory chapter we want to present such a proof at least once in detail.

**Theorem 6.7.** *Let $\mathcal{A}$ be an additive category. Then the homotopy category of complexes* $\mathbf{K}(\mathcal{A})$ *is a triangulated category.*

*Proof.* We have to show that with the above translation functor [1] and the set of distinguished triangles just defined, the axioms (TR0)-(TR5) are satisfied.

The axioms (TR0) and (TR2) hold by Definition 6.6.

(TR1) From the mapping cone construction there is a standard triangle

$$X \xrightarrow{\mathrm{id}_X} X \longrightarrow M(\mathrm{id}_X) \longrightarrow X[1].$$

By Example 6.5 above, $M(\mathrm{id}_X)$ is isomorphic to the zero complex in the homotopy category. Hence we indeed have a distinguished triangle

$$X \xrightarrow{\mathrm{id}_X} X \longrightarrow 0 \longrightarrow X[1].$$

(TR3) Because the rotation property is compatible with isomorphisms of triangles, it suffices to prove (TR3) for a standard triangle

$$X \xrightarrow{f} Y \xrightarrow{\alpha(f)} M(f) \xrightarrow{\beta(f)} X[1].$$

We shall show that the rotated triangle

$$Y \xrightarrow{\alpha(f)} M(f) \xrightarrow{\beta(f)} X[1] \xrightarrow{-f[1]} Y[1]$$

is isomorphic in $\mathbf{K}(\mathcal{A})$ to the following standard triangle for $\alpha(f)$,

$$Y \xrightarrow{\alpha(f)} M(f) \xrightarrow{\alpha(\alpha(f))} M(\alpha(f)) \xrightarrow{\beta(\alpha(f))} Y[1].$$

For constructing an isomorphism between the latter two triangles we take the identity maps for the first, second and fourth entries. Moreover, we define morphisms

$$\phi = (\phi_n) : X[1] \to M(\alpha(f)) \ \text{ by setting } \ \phi_n = (-f_{n-1}, \mathrm{id}_{X_{n-1}}, 0)$$

and conversely

$$\psi = (\psi_n) : M(\alpha(f)) \to X[1] \ \text{ by setting } \ \psi_n = (0, \mathrm{id}_{X_{n-1}}, 0).$$

These yield morphisms of triangles since by definition $\beta(\alpha(f)) \circ \phi = -f[1]$, and $\phi \circ \beta(f) \sim \alpha(\alpha(f))$ via the homotopy given by

$$\begin{pmatrix} 0 & -\mathrm{id} \\ 0 & 0 \\ 0 & 0 \end{pmatrix} : M(f)_n = X_{n-1} \oplus Y_n \to M(\alpha(f))_{n+1} = Y_n \oplus X_n \oplus Y_{n+1}.$$

Similarly, $\psi$ is a morphism of triangles since $\beta(f) = \psi \circ \alpha(\alpha(f))$ by definition and $-f[1] \circ \psi \sim \beta(\alpha(f))$ via the homotopy $(0, 0, -\mathrm{id}) : M(\alpha(f))_n \to Y[1]_n$.

Finally, and most importantly for proving (TR3), the above morphisms are isomorphisms in $\mathbf{K}(\mathcal{A})$ because we have $\psi \circ \phi = \mathrm{id}_{X[1]}$ (by definition) and $\phi \circ \psi \sim \mathrm{id}_{M(\alpha(f))}$ via the homotopy map

$$\begin{pmatrix} 0 & 0 & -\mathrm{id} \\ 0 & 0 & 0 \\ 0 & 0 & 0 \end{pmatrix} : M(\alpha(f))_n \to M(\alpha(f))_{n+1}$$

(recall that $M(\alpha(f))_n = Y_{n-1} \oplus X_{n-1} \oplus Y_n$).

(TR4) Again it suffices to prove the axiom for standard triangles. By assumption we have a diagram

$$\begin{array}{ccccccc}
X & \xrightarrow{u} & Y & \xrightarrow{\alpha(u)} & M(u) & \xrightarrow{\beta(u)} & X[1] \\
\downarrow{f} & & \downarrow{g} & & & & \downarrow{f[1]} \\
X' & \xrightarrow{u'} & Y' & \xrightarrow{\alpha(u')} & M(u') & \xrightarrow{\beta(u')} & X'[1]
\end{array}$$

where the left square commutes in $\mathbf{K}(\mathcal{A})$, i.e. there exist homotopy maps $s_n : X_n \to Y'_{n+1}$ such that $g_n u_n - u'_n f_n = d^{Y'}_{n+1} s_n + s_{n-1} d^X_n$ for all $n \in \mathbb{Z}$. For completing the diagram to a morphism of triangles we define $h = (h_n)_{n\in\mathbb{Z}} : M(u) \to M(u')$ by setting

$$h_n = \begin{pmatrix} f_{n-1} & 0 \\ s_{n-1} & g_n \end{pmatrix} : M(u)_n = X_{n-1} \oplus Y_n \to M(u')_n = X'_{n-1} \oplus Y'_n.$$

This is indeed a morphism of complexes because of the homotopy property of $s$ given above. Moreover, the completed diagram commutes since by definition we have that $h \circ \alpha(u) = \alpha(u') \circ g$ and $\beta(u') \circ h = f[1] \circ \beta(u)$; note that these are proper equalities, not only up to homotopy.

(TR5) Again it suffices to prove the octahedral axiom for standard triangles. From the assumptions we already have the following part of the relevant diagram

$$\begin{array}{ccccccc}
X & \xrightarrow{u} & Y & \xrightarrow{\alpha(u)} & M(u) & \xrightarrow{\beta(u)} & X[1] \\
\| & & \downarrow{v} & & & & \| \\
X & \xrightarrow{vu} & Z & \xrightarrow{\alpha(vu)} & M(vu) & \xrightarrow{\beta(vu)} & X[1] \\
\downarrow{u} & & \| & & & & \downarrow{u[1]} \\
Y & \xrightarrow{v} & Z & \xrightarrow{\alpha(v)} & M(v) & \xrightarrow{\beta(v)} & Y[1] \\
\downarrow{\alpha(u)} & & \downarrow{\alpha(vu)} & & \| & & \downarrow{\alpha(u)[1]} \\
M(u) & & M(vu) & & M(v) & & M(u)[1]
\end{array}$$

We now define the missing morphisms as follows. Let $f = (f_n) : M(u) \to M(vu)$ be given in degree $n$ by $f_n = \begin{pmatrix} \mathrm{id}_{X_{n-1}} & 0 \\ 0 & v_n \end{pmatrix}$ and set $g = (g_n) : M(vu) \to M(v)$ to be given by $g_n = \begin{pmatrix} u_{n-1} & 0 \\ 0 & \mathrm{id}_{Z_n} \end{pmatrix}$. Finally define $h :$

$M(v) \to M(u)[1]$ as the composition $\alpha(u)[1] \circ \beta(v)$, i.e. it is given by the matrix $\begin{pmatrix} 0 & 0 \\ \mathrm{id}_{Y_{n-1}} & 0 \end{pmatrix}$. Then it is easy to check from the definitions that all squares in the completed diagram commute (not only up to homotopy).

For proving (TR5) it now remains to show that the bottom line

$$M(u) \xrightarrow{\ f\ } M(vu) \xrightarrow{\ g\ } M(v) \xrightarrow{\ h\ } M(u)[1]$$

is a distinguished triangle in $\mathbf{K}(\mathcal{A})$. To this end we construct an isomorphism to the standard triangle

$$M(u) \xrightarrow{\ f\ } M(vu) \xrightarrow{\alpha(f)} M(f) \xrightarrow{\beta(f)} M(u)[1].$$

Note that only the third entries in the triangles are different. So it suffices to find morphisms $\sigma = (\sigma_n) : M(v) \to M(f)$ and $\tau = (\tau_n) : M(f) \to M(v)$ leading to commutative diagrams (in $\mathbf{K}(\mathcal{A})$!), i.e. we need that $\beta(f) \circ \sigma = h$, $h \circ \tau = \beta(f)$, $\sigma \circ g = \alpha(f)$ and $\tau \circ \alpha(f) = g$, up to homotopy. Moreover, we have to show that they are isomorphisms in the homotopy category. We set

$$\sigma_n := \begin{pmatrix} 0 & 0 \\ \mathrm{id}_{Y_{n-1}} & 0 \\ 0 & 0 \\ 0 & \mathrm{id}_{Z_n} \end{pmatrix} \quad \text{and} \quad \tau_n := \begin{pmatrix} 0 & \mathrm{id}_{Y_{n-1}} & u_{n-1} & 0 \\ 0 & 0 & 0 & \mathrm{id}_{Z_n} \end{pmatrix}.$$

First, let us check that $\sigma$ and $\tau$ give commutative diagrams. Directly from the definitions we get that $\tau \circ \alpha(f) = g$; in fact both are given in degree $n$ by the map $\begin{pmatrix} u_{n-1} & 0 \\ 0 & \mathrm{id}_{Z_n} \end{pmatrix} : X_{n-1} \oplus Z_n \to Y_{n-1} \oplus Z_n$. Also by definition we see that $\beta(f) \circ \sigma = h$, both given by $\begin{pmatrix} 0 & 0 \\ \mathrm{id}_{Y_{n-1}} & 0 \end{pmatrix} : Y_{n-1} \oplus Z_n \to X_{n-2} \oplus Y_{n-1}$. The remaining commutativities will now only hold up to homotopy. Note that $\alpha(f) - \sigma \circ g : M(vu) \to M(f)$ is given in degree $n$ by

$$\begin{pmatrix} 0 & 0 \\ -u_{n-1} & 0 \\ \mathrm{id}_{X_{n-1}} & 0 \\ 0 & 0 \end{pmatrix} : X_{n-1} \oplus Z_n \to X_{n-2} \oplus Y_{n-1} \oplus X_{n-1} \oplus Z_n.$$

We claim that $\alpha(f) - \sigma \circ g$ is homotopic to zero, i.e. $\alpha(f) = \sigma \circ g$ in $\mathbf{K}(\mathcal{A})$. In fact, a homotopy map $s = (s_n)$ where $s_n : M(vu)_n \to M(f)_{n+1}$ is given by

$$\begin{pmatrix} \mathrm{id}_{X_{n-1}} & 0 \\ 0 & 0 \\ 0 & 0 \\ 0 & 0 \end{pmatrix} : X_{n-1} \oplus Z_n \to X_{n-1} \oplus Y_n \oplus X_n \oplus Z_{n+1}.$$

For verifying the details recall that the differential of the mapping cone $M(f)$ is given by

$$d_n^{M(f)} = \begin{pmatrix} d_{n-2}^X & 0 & 0 & 0 \\ -u_{n-2} & -d_{n-1}^Y & 0 & 0 \\ \mathrm{id}_{X_{n-2}} & 0 & -d_{n-1}^X & 0 \\ 0 & v_{n-1} & (vu)_{n-1} & d_n^Z \end{pmatrix}.$$

Finally, consider $\beta(f) - h \circ \tau : M(f) \to M(u)[1]$ which in degree $n$ is given by

$$\begin{pmatrix} \mathrm{id}_{X_{n-2}} & 0 & 0 & 0 \\ 0 & 0 & -u_{n-1} & 0 \end{pmatrix} : X_{n-2} \oplus Y_{n-1} \oplus X_{n-1} \oplus Z_n \to X_{n-2} \oplus Y_{n-1}.$$

This can be seen to be homotopic to zero by using the homotopy map $s = (s_n)$ where

$$s_n = \begin{pmatrix} 0 & 0 & \mathrm{id}_{X_{n-1}} & 0 \\ 0 & 0 & 0 & 0 \end{pmatrix} : X_{n-2} \oplus Y_{n-1} \oplus X_{n-1} \oplus Z_n \to X_{n-1} \oplus Y_n.$$

For the straightforward verification again use the differential of $M(f)$ as given above.

For completing the proof it now remains to show that $\sigma$ and $\tau$ are isomorphisms in the homotopy category. We have $\tau \circ \sigma = \mathrm{id}_{M(v)}$ by definition. Conversely, the composition $\sigma \circ \tau$ is in degree $n$ given by

$$\begin{pmatrix} 0 & 0 & 0 & 0 \\ 0 & \mathrm{id}_{Y_{n-1}} & u_{n-1} & 0 \\ 0 & 0 & 0 & 0 \\ 0 & 0 & 0 & \mathrm{id}_{Z_n} \end{pmatrix}$$

If we then define homotopy maps $s_n : M(f)_n \to M(f)_{n+1}$ by setting

$$s_n := \begin{pmatrix} 0 & 0 & -\mathrm{id}_{X_{n-1}} & 0 \\ 0 & 0 & 0 & 0 \\ 0 & 0 & 0 & 0 \\ 0 & 0 & 0 & 0 \end{pmatrix}$$

then we have $\sigma \circ \tau - \mathrm{id}_{M(f)} = d_{n+1}^{M(f)} \circ s_n + s_{n-1} \circ d_n^{M(f)}$ which is easily checked using the differential of $M(f)$ as given above.

Thus $\sigma \circ \tau = \mathrm{id}_{M(f)}$ in the homotopy category $\mathbf{K}(\mathcal{A})$ and we have proved the octahedral axiom for $\mathbf{K}(\mathcal{A})$. $\qquad\square$

**Remark 6.8.** We have seen that for every standard triangle

$$X \xrightarrow{f} Y \xrightarrow{\alpha(f)} M(f) \xrightarrow{\beta(f)} X[1]$$

in $\mathbf{K}(\mathcal{A})$ there is a corresponding short exact sequence

$$0 \to Y \xrightarrow{\alpha(f)} M(f) \xrightarrow{\beta(f)} X[1] \to 0$$

in $\mathbf{C}(\mathcal{A})$. On the other hand, it is not true that any short exact sequence in $\mathbf{C}(\mathcal{A})$ would lead to a distinguished triangle in the homotopy category $\mathbf{K}(\mathcal{A})$.

As an example, consider the short exact sequence of abelian groups

$$0 \to \mathbb{Z}/2\mathbb{Z} \xrightarrow{\cdot 2} \mathbb{Z}/4\mathbb{Z} \xrightarrow{\cdot 1} \mathbb{Z}/2\mathbb{Z} \to 0,$$

and consider the abelian groups as complexes concentrated in degree 0. There is no corresponding distinguished triangle

$$\mathbb{Z}/2\mathbb{Z} \xrightarrow{\cdot 2} \mathbb{Z}/4\mathbb{Z} \xrightarrow{\cdot 1} \mathbb{Z}/2\mathbb{Z} \xrightarrow{w} \mathbb{Z}/2\mathbb{Z}[1].$$

in $\mathbf{K}(\mathbf{Ab})$. In fact, suppose for a contradiction that such a distinguished triangle existed. The morphisms in $\mathbf{K}(\mathbf{Ab})$ are just equivalence classes of morphisms of complexes modulo homotopy. But since $\mathbb{Z}/2\mathbb{Z}$ is a complex concentrated in a single degree, there are no nonzero morphisms $\mathbb{Z}/2\mathbb{Z} \to \mathbb{Z}/2\mathbb{Z}[1]$ to its shifted version. Thus we must have $w = 0$.

By Proposition 4.4 a triangle with a zero map is a split triangle. Hence $\mathbb{Z}/4\mathbb{Z} \cong \mathbb{Z}/2\mathbb{Z} \oplus \mathbb{Z}/2\mathbb{Z}$ in $\mathbf{K}(\mathbf{Ab})$, i.e. there must exist a homotopy equivalence between these complexes. However, all these complexes are complexes concentrated in a single degree, hence there are no nonzero homotopy maps. So the above isomorphism would have to be an isomorphism already in $\mathbf{Ab}$ which is impossible, a contradiction.

We shall later see that this phenomenon disappears when passing from $\mathbf{K}(\mathcal{A})$ to the derived category. There every short exact sequence does lead to a distinguished triangle; see Section 7.6 below for details.

# 7. Derived categories

A very important class of triangulated categories is formed by derived categories. They occur frequently in many different areas of mathematics and have found numerous applications. In this section we shall provide the relevant constructions leading from the homotopy category to the derived category.

## 7.1. Homology and quasi-isomorphisms

In this short section we shall introduce the notion of quasi-isomorphism which is fundamental for derived categories.

Although one could set up a homology theory in a categorical manner in every abelian category we shall restrict from now on to categories of modules and to complexes over them. This considerably simplifies the presentation in certain parts of this section since we can then use element-wise arguments and hence avoid technical overload which might obscure the fundamental ideas underlying the definition of a derived category.

For the remainder of this section we let $\mathcal{A}$ be a category of modules over a ring.

**Definition 7.1.**
  (i) *Let $X = (X_n, d_n^X)$ be a complex in $\mathbf{C}(\mathcal{A})$. The $n$-th homology of the complex $X$ is defined as the following object from $\mathcal{A}$,*

$$H_n(X) := \ker d_n^X / \operatorname{im} d_{n+1}^X$$

  *(where the kernel and the image are the usual set-theoretic kernel and image, respectively).*
  (ii) *The complex $X$ is called* exact *if $H_n(X) = 0$ for all $n \in \mathbb{Z}$.*
  (iii) *Let $f : X \to Y$ be a morphism of complexes. We define an induced map on the level of homology by setting*

$$H_n(f) : H_n(X) \to H_n(Y), \quad x + \operatorname{im} d_{n+1}^X \mapsto f_n(x) + \operatorname{im} d_{n+1}^Y.$$

  *In this way we get homology functors $H_n : \mathbf{C}(\mathcal{A}) \to \mathcal{A}$ where $n \in \mathbb{Z}$.*

**Remark 7.2.** Note that the induced map on homology is well-defined; in fact let $x' = d_{n+1}^X(x'') \in \operatorname{im} d_{n+1}^X$; then

$$f_n(x') = f_n(d_{n+1}^X(x'')) = d_{n+1}^Y(f_{n+1}(x'')) \in \operatorname{im} d_{n+1}^Y.$$

**Proposition 7.3.** *Let $f, g : X \to Y$ be morphisms in $\mathbf{C}(\mathcal{A})$ which are homotopic. Then they induce the same map in homology, i.e. $H_n(f) = H_n(g)$ for all $n \in \mathbb{Z}$.*

*As a consequence, the homology functors on $\mathbf{C}(\mathcal{A})$ induce well-defined homology functors on the homotopy category $\mathbf{K}(\mathcal{A})$.*

*Proof.* By assumption there is a homotopy map $s = (s_n)_{n \in \mathbb{Z}}$ such that $f_n - g_n = d_{n+1}^Y s_n + s_{n-1} d_n^X$ for all $n \in \mathbb{Z}$. Let $x + \operatorname{im} d_{n+1}^X \in H_n(X)$, in particular $x \in \ker d_n^X$. Then it follows that

$$
\begin{aligned}
H_n(f)(x + \operatorname{im} d_{n+1}^X) &= f_n(x) + \operatorname{im} d_{n+1}^Y \\
&= (d_{n+1}^Y s_n + s_{n-1} d_n^X + g_n)(x) + \operatorname{im} d_{n+1}^Y \\
&= g_n(x) + \operatorname{im} d_{n+1}^Y = H_n(g)(x + \operatorname{im} d_{n+1}^X).
\end{aligned}
$$

Hence $H(f) = H(g)$. $\qquad\square$

**Definition 7.4.** *A morphism* $f : X \to Y$ *of complexes in* $\mathbf{C}(\mathcal{A})$ *is called a* quasi-isomorphism *if it induces isomorphisms in homology, i.e.* $H_n(f) : H_n(X) \to H_n(Y)$ *are isomorphisms for all* $n \in \mathbb{Z}$.

**Example 7.5.**

 (i) (Projective resolutions) As we are restricting in this section to categories **R-Mod** of modules over a ring any object $X$ in $\mathcal{A}$ has a projective resolution, i.e. a sequence

$$\ldots \to P_2 \to P_1 \to P_0 \to 0$$

where all $P_i$ are projective objects, together with a morphism $\epsilon : P_0 \to X$ such that the following augmented sequence is exact

$$\ldots \to P_2 \to P_1 \to P_0 \xrightarrow{\epsilon} X \to 0.$$

This gives rise to a morphism of complexes, also denoted $\epsilon : P \to X$,

$$
\begin{array}{ccccccccc}
\ldots & \to & P_2 & \to & P_1 & \to & P_0 & \to & 0 \\
 & & \downarrow & & \downarrow & & \downarrow{\scriptstyle \epsilon} & & \\
\ldots & \to & 0 & \to & 0 & \to & X & \to & 0
\end{array}
$$

where $X$ is supposed to be in degree 0. Then $\epsilon$ is a quasi-isomorphism. In fact, in non-zero degrees both complexes have zero homology, and in degree 0 we have isomorphisms $H_0(P) \cong X \cong H_0(X)$ induced by $\epsilon$.

(ii) (Injective resolutions) Dually, every object has an injective resolution, i.e. there is a sequence

$$0 \to I_0 \to I_{-1} \to I_{-2} \to \ldots$$

where all $I_j$ are injective objects, and a morphism $\iota : X \to I_0$ such that the following augmented sequence is exact

$$0 \to X \xrightarrow{\iota} I_0 \to I_{-1} \to I_{-2} \to \ldots$$

This also induces a morphism of complexes $\iota$ which is a quasi-isomorphism.

Our next goal is to characterize quasi-isomorphisms in terms of mapping cones. To this end we shall use the following standard result on long exact sequences; for a proof we refer for instance to Weibel's book [17, section 1.3].

**Proposition 7.6.** *Let* $0 \to X \xrightarrow{f} Y \xrightarrow{g} Z \to 0$ *be a short exact sequence of complexes in* $\mathbf{C}(\mathcal{A})$. *Then there are connecting morphisms* $\delta_n : H_n(Z) \to H_{n-1}(X)$ *giving rise to the following long exact homology sequence*

$$\ldots \xrightarrow{H_n(g)} H_{n+1}(Z) \xrightarrow{\delta_{n+1}} H_n(X) \xrightarrow{H_n(f)} H_n(Y) \xrightarrow{H_n(g)} H_n(Z) \xrightarrow{\delta_n} H_{n-1}(X) \xrightarrow{H_n(g)} \ldots$$

For the definition of mapping cones recall Definition 6.3.

**Proposition 7.7.** *In* $\mathbf{C}(\mathcal{A})$ *a morphism* $f : X \to Y$ *is a quasi-isomorphism if and only if the mapping cone complex* $M(f)$ *is exact.*

*Proof.* By Remark 6.4 we have an exact sequence of complexes

$$0 \to Y \xrightarrow{\alpha(f)} M(f) \xrightarrow{\beta(f)} X[1] \to 0.$$

The corresponding long exact homology sequence has the form

$$\ldots H_{n+1}(X[1]) \xrightarrow{\delta_{n+1}} H_n(Y) \xrightarrow{H_n(\alpha(f))} H_n(M(f)) \xrightarrow{H_n(\beta(f))} H_n(X[1]) \xrightarrow{\delta_n} H_{n-1}(Y) \ldots$$

But $H_n(X[1])$ can be identified with $H_{n-1}(X)$ for all $n \in \mathbb{Z}$ and it can be checked that then in our situation $\delta_n = H_{n-1}(f)$, so the above long exact sequence takes the form

$$\ldots H_n(X) \xrightarrow{H_n(f)} H_n(Y) \xrightarrow{H_n(\alpha(f))} H_n(M(f)) \xrightarrow{H_n(\beta(f))} H_{n-1}(X) \xrightarrow{H_{n-1}(f)} H_{n-1}(Y) \ldots$$

For necessity, suppose that $f$ is a quasi-isomorphism. Then $H_n(f)$ are isomorphisms by assumption, hence by exactness we have $H_n(\alpha(f)) = 0$ and $H_n(\beta(f)) = 0$ for all $n \in \mathbb{Z}$. But then again by exactness we deduce that $H_n(M(f)) = 0$ for all $n \in \mathbb{Z}$, i.e. the mapping cone $M(f)$ is exact.

Conversely, suppose that $M(f)$ is exact. Then the long exact sequence takes the form

$$\ldots 0 \to H_n(X) \xrightarrow{H_n(f)} H_n(Y) \longrightarrow 0 \longrightarrow H_{n-1}(X) \xrightarrow{H_{n-1}(f)} H_{n-1}(Y) \to 0 \ldots$$

from which it immediately follows by exactness that $H_n(f)$ is an isomorphism for all $n \in \mathbb{Z}$, i.e. $f$ is a quasi-isomorphism. $\square$

**Remark 7.8.** Let $X \xrightarrow{f} Y \xrightarrow{g} Z \xrightarrow{h} X[1]$ be a distinguished triangle in $\mathbf{K}(\mathcal{A})$. Then it follows from the previous proposition that $f$ is a quasi-isomorphism if and only if $Z$ is exact (i.e. $H_n(Z) = 0$ for all $n \in \mathbb{Z}$).

## 7.2. Localisation of categories

Derived categories of abelian categories are obtained from the homotopy categories of complexes by localising with respect to quasi-isomorphisms, i.e. by a formal process inverting all quasi-isomorphisms.

We shall not aim in this introductory chapter to provide a general account of localisation of categories. For a thorough treatment of this topic see H. Krause's article in this volume [10].

Instead we shall concentrate here on the special case leading from the homotopy category $\mathbf{K}(\mathcal{A})$ of complexes to the derived category $\mathbf{D}(\mathcal{A})$. We first want to give an elementary construction of a derived category, following the approach in the book by Gelfand and Manin [4, III.2]. In the following sections we shall also give alternative equivalent descriptions of the derived category which are perhaps more common and more suitable for explicit computations.

**Remark 7.9.** From now on we are following a time-honoured tradition by ignoring some set-theoretical issues.

**Theorem 7.10.** *Let $\mathcal{A}$ be an abelian category. Then there exists a category $\mathbf{D}(\mathcal{A})$, called the* derived category *of $\mathcal{A}$, and a functor $L : \mathbf{K}(\mathcal{A}) \to \mathbf{D}(\mathcal{A})$ satisfying the following properties:*

(L1) *For every quasi-isomorphism $q$ in $\mathbf{K}(\mathcal{A})$, $L(q)$ is an isomorphism in $\mathbf{D}(\mathcal{A})$.*

(L2) *Every functor $F : \mathbf{K}(\mathcal{A}) \to \mathcal{D}$ (where $\mathcal{D}$ is any category), having the property that quasi-isomorphisms are mapped to isomorphisms, factors uniquely through $L$.*

Property (L2) implies in particular that the category $\mathbf{D}(\mathcal{A})$, if it exists, is unique up to equivalence of categories.

*Proof.* The objects in $\mathbf{D}(\mathcal{A})$ are defined to be the same as in $\mathbf{K}(\mathcal{A})$, i.e. complexes over $\mathcal{A}$. But the morphisms have to be changed in order for quasi-isomorphisms to become isomorphisms. For each quasi-isomorphism $q$ in $\mathbf{K}(\mathcal{A})$ we introduce a formal variable $q^{-1}$. We then consider 'words' in $f$'s and $q^{-1}$'s, i.e. formal compositions of the form

$$(*) = f_1 \circ q_1^{-1} \circ f_2 \circ q_2^{-1} \circ \ldots \circ f_r \circ q_r^{-1}$$

where $r \in \mathbb{N}_0$, the $f_i$ are morphisms and the $q_j$ are quasi-isomorphisms in $\mathbf{K}(\mathcal{A})$. This has to be read so that some $f_i$ or some $q_j^{-1}$ can be the identity and then can be deleted, i.e. consecutive subexpressions $f_i \circ f_{i+1}$ or $q_j^{-1} \circ q_{j+1}^{-1}$ are also allowed in $(*)$.

As usual we read compositions from right to left; so if $f_1 : X_1 \to Y_1$ then $Y_1$ is called the end point of $(*)$ and if $q_r : X_r \to Y_r$ then $Y_r$ is called the starting point of $(*)$. The length of $(*)$ is the total number of $f_i$'s and $q_j^{-1}$'s occurring. For each object $X$ there is an empty expression of length 0 representing the identity on $X$.

We call two such expressions equivalent if they have the same starting and end point and if one can be obtained from the other by a sequence of the following operations

(i) for any composable morphisms $f, g$ in $\mathbf{K}(\mathcal{A})$ replace $f \circ g$ by their composition $(f \circ g)$;

(i') for any composable quasi-isomorphisms $q, r$ in $\mathbf{K}(\mathcal{A})$ replace $q^{-1} \circ r^{-1}$ by $(r \circ q)^{-1}$;

(ii) for any quasi-isomorphism $q$ in $\mathbf{K}(\mathcal{A})$ replace $q \circ q^{-1}$ or $q^{-1} \circ q$ by id.

The morphisms in $\mathbf{D}(\mathcal{A})$ are defined as equivalence classes of expressions of the form $(*)$. The composition of morphisms is induced by concatenating expressions of the form $(*)$ if the starting point of the second matches the end point of the first, and zero otherwise.

The crucial localisation functor $L$ is now defined as follows: on objects, $L$ is just the identity; a morphism $f$ in $\mathbf{K}(\mathcal{A})$ is sent by $L$ to its equivalence class in $\mathbf{D}(\mathcal{A})$.

In particular, if $q$ is a quasi-isomorphism in $\mathbf{K}(\mathcal{A})$ then $L(q)$ becomes invertible in $\mathbf{D}(\mathcal{A})$ with inverse $q^{-1}$ (because $q \circ q^{-1}$ is equivalent to the empty expression of length 0, representing the identity). Thus, axiom (L1) is satisfied.

For proving axiom (L2) let a functor $F : \mathbf{K}(\mathcal{A}) \to \mathcal{D}$ be given (where $\mathcal{D}$ is any category) which sends quasi-isomorphisms to isomorphisms. We need to define a functor $G : \mathbf{D}(\mathcal{A}) \to \mathcal{D}$ such that $G \circ L = F$. First we note that there is at most one possibility to define such a functor, namely setting $G(X) = F(X)$ on objects, defining $G(f) = F(f)$ for morphisms $f$ in $\mathbf{K}(\mathcal{A})$ and $G(q^{-1}) = F(q)^{-1}$ for quasi-isomorphisms $q$ in $\mathbf{K}(\mathcal{A})$ (and then extending $G$ to arbitrary compositions, in particular $G(\mathrm{id}) = \mathrm{id}$ for the empty composition). Note that the latter makes sense since $F(q)$ is an isomorphism by assumption.

It only remains to check that this functor is well-defined, i.e. compatible with the equivalence relation defining morphisms in $\mathbf{D}(\mathcal{A})$. For instance, for part (ii) of the above equivalence relation we have in $\mathbf{D}(\mathcal{A})$ that

$$G(q \circ q^{-1}) = G(q) \circ G(q^{-1}) = F(q) \circ F(q)^{-1} = \mathrm{id} = G(\mathrm{id})$$

showing well-definedness. The other parts also follow easily from the definition. $\qquad\square$

## 7.3. Morphisms in the derived category

The above description of morphisms in the derived category as equivalence classes of expressions of the form

$$(*) = f_1 \circ q_1^{-1} \circ f_2 \circ q_2^{-1} \circ \ldots \circ f_r \circ q_r^{-1}$$

is pretty inconvenient. We shall describe in this section a 'calculus of fractions' which will lead to a simpler description of the morphisms in the derived category. To this end we shall make use of certain useful properties of the class of quasi-isomorphisms in the homotopy category.

**Lemma 7.11.** *Let $\mathcal{A}$ be an abelian category. The class $Q$ of quasi-isomorphisms in the homotopy category $\mathbf{K}(\mathcal{A})$ satisfies the following properties:*

(Q1) *For every object $X$ in $\mathbf{K}(\mathcal{A})$ the identity $\mathrm{id}_X$ is in $Q$.*

(Q2) *$Q$ is closed under composition.*

(Q3) *(Ore condition) Given a quasi-isomorphism $q \in Q$ and a morphism $f$ in $\mathbf{K}(\mathcal{A})$ (with the same target) then there exist an object $W$, a morphism $g$ and a quasi-isomorphism $t \in Q$ such that the following diagram is commutative.*

$$
\begin{array}{ccc}
W & \xrightarrow{\ g\ } & Z \\
{\scriptstyle t}\downarrow & & \downarrow{\scriptstyle q} \\
X & \xrightarrow{\ f\ } & Y
\end{array}
$$

*Similarly, given a quasi-isomorphism $q \in Q$ and a morphism $f$ in $\mathbf{K}(\mathcal{A})$ (with the same range) then there exist an object $V$, a morphism $h$ and a quasi-isomorphism $r \in Q$ such that the following diagram is commutative.*

$$
\begin{array}{ccc}
Y & \xrightarrow{\ f\ } & X \\
{\scriptstyle q}\downarrow & & \downarrow{\scriptstyle r} \\
X & \xrightarrow{\ h\ } & V
\end{array}
$$

(Q4) *For any morphisms $f, g : X \to Y$ in $\mathbf{K}(\mathcal{A})$ the following are equivalent:*
  (i) *There exists a quasi-isomorphism $q : Y \to Y'$ in $Q$ (for some $Y'$) such that $q \circ f = q \circ g$.*
  (ii) *There exists a quasi-isomorphism $t : X' \to X$ in $Q$ (for some $X'$) such that $f \circ t = g \circ t$.*

**Remark 7.12.** The class of quasi-isomorphisms does not in general satisfy the conditions of the preceding lemma already in $\mathbf{C}(\mathcal{A})$; it is crucial first to pass to the homotopy category. In fact, in the proof below we shall make heavy use of the triangulated structure of the homotopy category (which has been proven in Theorem 6.7).

*Proof.* (Q1) and (Q2) are clear.

For (Q3) we have given a morphism $f$ and a quasi-isomorphism $q$. By axiom (TR2) for the triangulated category $\mathbf{K}(\mathcal{A})$ there exists a distinguished triangle $Z \xrightarrow{q} Y \xrightarrow{u} U \xrightarrow{v} Z[1]$. Similarly, considering $uf : X \to U$ there is a distinguished triangle $W \xrightarrow{t} X \xrightarrow{uf} U \xrightarrow{w} W[1]$. Applying axioms (TR4) and (TR3) we can deduce the existence of the morphism $g$ (and $g[1]$) in the following commutative diagram

$$W \xrightarrow{\;t\;} X \xrightarrow{\;uf\;} U \xrightarrow{\;w\;} W[1]$$
$$\downarrow g \qquad \downarrow f \qquad \downarrow \mathrm{id} \qquad \downarrow g[1]$$
$$Z \xrightarrow{\;q\;} Y \xrightarrow{\;u\;} U \xrightarrow{\;v\;} Z[1]$$

Since $q$ is a quasi-isomorphism by assumption, the long exact homology sequence applied to the bottom row yields that $H_n(U) = 0$ for all $n \in \mathbb{Z}$. And then the long exact homology sequence for the top row implies that $t$ must be a quasi-isomorphism, as desired (cf. Remark 7.8).

The symmetrical second claim in (Q3) is shown similarly.

Finally, let us prove (Q4). We will prove the direction (i)$\Rightarrow$(ii), the converse is proved similarly. For simplicity, set $h := f - g$, thus $q \circ h = 0$ by assumption. By (TR2) and (TR3) there exists a distinguished triangle $Z \xrightarrow{\;u\;} Y \xrightarrow{\;q\;} Y' \xrightarrow{\;w\;} Z[1]$. Since $q$ is a quasi-isomorphism, $Z[1]$ and hence $Z$ is exact (cf. Remark 7.8). By (TR1), (TR3) and (TR4) there exists a morphism $v$ making the following diagram commutative

$$X \xrightarrow{\;\mathrm{id}\;} X \longrightarrow 0 \longrightarrow X[1]$$
$$\downarrow v \qquad \downarrow h \qquad \downarrow \qquad \downarrow v[1]$$
$$Z \xrightarrow{\;u\;} Y \xrightarrow{\;q\;} Y' \xrightarrow{\;w\;} Z[1]$$

Now again by (TR2) and (TR3) $v$ can be embedded in a distinguished triangle $X' \xrightarrow{\;t\;} X \xrightarrow{\;v\;} Z \longrightarrow X'[1]$. Here $t$ is a quasi-isomorphism since $Z$ is exact (cf. Remark 7.8). Moreover, $v \circ t = 0$ since the composition of any consecutive maps in a distinguished triangle vanishes. It follows that $h \circ t = u \circ v \circ t = 0$, i.e. $f \circ t = g \circ t$, as desired. $\qquad\square$

The properties satisfied by the family $Q$ of quasi-isomorphisms in $\mathbf{K}(\mathcal{A})$ has useful consequences for the description of morphisms in the derived category $\mathbf{D}(\mathcal{A})$. As described above, a morphism in $\mathbf{D}(\mathcal{A})$ is an equivalence class of an expression of the form

$$(*) = f_1 \circ q_1^{-1} \circ f_2 \circ q_2^{-1} \circ \ldots \circ \circ f_r \circ q_r^{-1}$$

where $f_i$ are morphisms in $\mathbf{K}(\mathcal{A})$ and $q_i$ are quasi-isomorphisms in $Q$.

Property (Q3) above states that $q^{-1} \circ f = g \circ t^{-1}$ for some morphism $g$ and quasi-isomorphism $t \in Q$, and $f \circ q^{-1} = r^{-1} \circ h$ with $r \in Q$, respectively. This means that in the above expression $(*)$ we can move all 'denominators' $q_i$ to the right (or to the left). This means that any morphism in the derived category can be represented by an expression of the form $f \circ q^{-1}$ with a quasi-isomorphism $q$ and a morphism $f$. This can be conveniently visualised as a 'roof'.

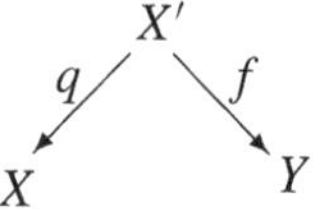

We shall use this description frequently in the sequel and hence want to make this more precise. Again, we follow the approach in the book by Gelfand and Manin [4, section III.2]. We shall define a category $\widetilde{\mathbf{D}}(\mathcal{A})$ where morphisms are represented by such roofs, and then show that this category is indeed equivalent to the derived category $\mathbf{D}(\mathcal{A})$ introduced in Theorem 7.10.

For computing with these roofs we need to introduce a suitable notion of equivalence for roofs. Two roofs $(q, f)$ and $(t, g)$ are called equivalent if there exists another roof $(r, h)$ making the following diagram commutative.

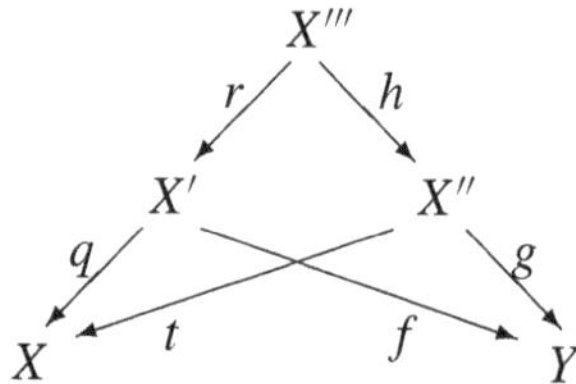

We leave it to the reader to verify that this indeed defines an equivalence relation. The non-obvious property is transitivity, see [4, Lemma III.2.8] for a detailed proof; actually, in the proof of transitivity the property (Q4) of Lemma 7.11 is used.

For the composition of roofs one makes use of the Ore condition (Q3) above. Namely, given roofs $(q, f) : X \to Y$ and $(t, g) : Y \to Z$ we can find by (Q3) an object $W$ and a roof $(t', g')$ making the following diagram commutative

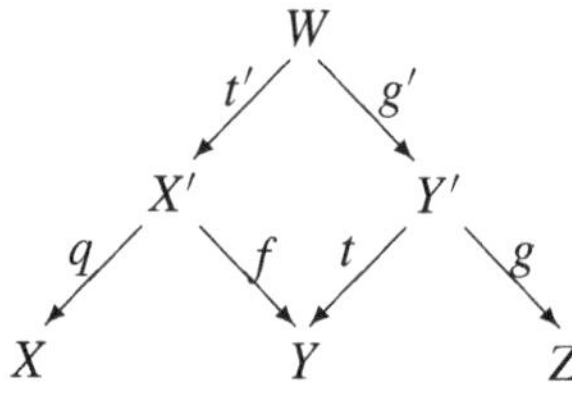

The composition of roofs $(t, g) \circ (q, f)$ is then defined to be the equivalence class represented by the roof $(q \circ t', g \circ g') : X \to Z$. It is not difficult to verify that this is well-defined, i.e. independent of the representatives of the roofs involved.

Note that the identity morphism for an object $X$ is represented by the roof $(\mathrm{id}_X, \mathrm{id}_X)$.

The category $\widetilde{\mathbf{D}}(\mathcal{A})$ is defined as having the same objects as $\mathbf{D}(\mathcal{A})$ (and hence as the homotopy category $\mathbf{K}(\mathcal{A})$), namely complexes over $\mathcal{A}$.

The morphisms in $\widetilde{\mathbf{D}}(\mathcal{A})$ are defined to be the equivalence classes of roofs, with the above composition.

**Proposition 7.13.** *Let $\mathcal{A}$ be an abelian category. Then the category $\widetilde{\mathbf{D}}(\mathcal{A})$ satisfies the universal property of the derived category $\mathbf{D}(\mathcal{A})$ given in Theorem 7.10. In particular, the categories $\mathbf{D}(\mathcal{A})$ and $\widetilde{\mathbf{D}}(\mathcal{A})$ are equivalent.*

*Proof.* We first define a functor $\widetilde{L} : \mathbf{K}(\mathcal{A}) \to \widetilde{\mathbf{D}}(\mathcal{A})$ as the identity on objects and on morphisms by sending $f$ to the roof $(\mathrm{id}, f)$. Clearly, $\widetilde{L}$ maps the identity to the identity. As for composition of morphisms, a composition $g \circ f$ is on the one hand sent by $\widetilde{L}$ to the roof $(\mathrm{id}, g \circ f)$; on the other hand the composition $\widetilde{L}(g) \circ \widetilde{L}(f)$ is given by the roof obtained from the commutative diagram

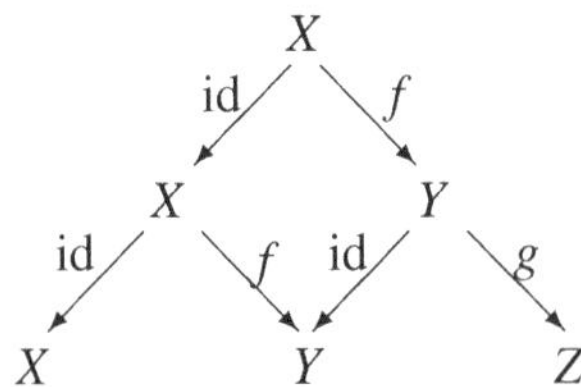

Thus, $\widetilde{L}(g \circ f) = \widetilde{L}(g) \circ \widetilde{L}(f)$ and $\widetilde{L}$ is indeed a functor.

Now it remains to prove that the category $\widetilde{\mathbf{D}}(\mathcal{A})$, together with the functor $\widetilde{L}$, satisfies the properties (L1) and (L2) from Theorem 7.10.

For (L1), any quasi-isomorphism $q$ in $\mathbf{K}(\mathcal{A})$ is mapped to the roof $(\mathrm{id}, q)$. It is immediate from the above composition of morphisms that $(q, \mathrm{id}) \circ (\mathrm{id}, q) = (\mathrm{id}, \mathrm{id})$, and that $(\mathrm{id}, q) \circ (q, \mathrm{id}) = (q, q)$; but the latter roof is equivalent to $(\mathrm{id}, \mathrm{id})$. Thus, $\widetilde{L}$ maps quasi-isomorphisms to isomorphisms and (L1) is satisfied.

For proving (L2), let $F : \mathbf{K}(\mathcal{A}) \to \mathcal{D}$ ($\mathcal{D}$ any category) be a functor which maps quasi-isomorphisms to isomorphisms. We have to show that there is a unique functor $\widetilde{F} : \widetilde{\mathbf{D}}(\mathcal{A}) \to \mathcal{D}$ such that $\widetilde{F} \circ \widetilde{L} = F$.

We first deal with uniqueness. On objects $X$, the only choice is $\widetilde{F}(X) = F(X)$ since $\widetilde{L}$ is the identity on objects. Now consider a morphism in $\widetilde{\mathbf{D}}(\mathcal{A})$, represented by a roof $(q, f)$. In $\widetilde{\mathbf{D}}(\mathcal{A})$ we have that

$$(q, f) \circ \widetilde{L}(q) = (q, f) \circ (\mathrm{id}, q) = (\mathrm{id}, f) = \widetilde{L}(f).$$

Since $\widetilde{F}$ has to be a functor with $\widetilde{F} \circ \widetilde{L} = F$ we can deduce that

$$F(f) = \widetilde{F}(\widetilde{L}(f)) = \widetilde{F}((q, f) \circ \widetilde{L}(q)) = \widetilde{F}((q, f)) \circ \widetilde{F}(\widetilde{L}(q))$$
$$= \widetilde{F}((q, f)) \circ F(q).$$

By assumption, $F(q)$ is an isomorphism, so the only possibility to define $\widetilde{F}$ on morphisms is to set

$$\widetilde{F}((q, f)) = F(f) \circ F(q)^{-1}.$$

Hence, the functor $\widetilde{F}$, if it exists, is unique.

For existence, we actually define $\widetilde{F}$ by the properties just exhibited, i.e. $\widetilde{F}(X) = F(X)$ on objects and $\widetilde{F}((q, f)) = F(f) \circ F(q)^{-1}$ on morphisms. Of course, we now have to prove that this indeed defines a functor.

We claim that the definition is well-defined, i.e. independent of the choice of the representative. In fact, let $(q, f)$ and $(t, g)$ be equivalent roofs, i.e. we have a commutative diagram of the form

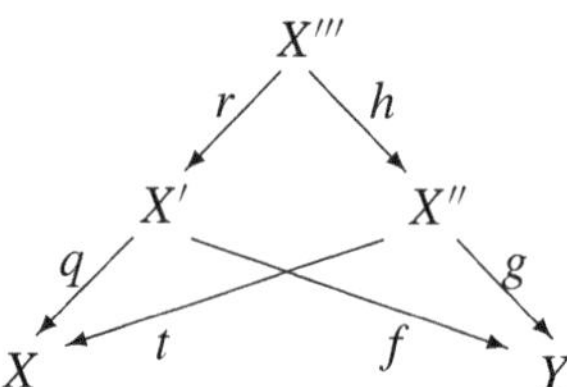

where $r$ is a quasi-isomorphism. Note that since $r, q, t$ are quasi-isomorphisms and $q \circ r = t \circ h$, also $h$ must be a quasi-isomorphism. Then we get from the functoriality of $F$ and the assumption that $F$ sends quasi-isomorphisms to isomorphisms that

$$\begin{aligned}
\widetilde{F}((q, f)) &= F(f) \circ F(q)^{-1} = F(f) \circ F(r) \circ F(r)^{-1} \circ F(q)^{-1} \\
&= F(f \circ r) \circ (F(q \circ r))^{-1} = F(g \circ h) \circ (F(t \circ h))^{-1} \\
&= F(g) \circ F(h) \circ F(h)^{-1} \circ F(t)^{-1} = F(g) \circ F(t)^{-1} = \widetilde{F}((t, g)).
\end{aligned}$$

By definition, $\widetilde{F}$ maps identity morphisms $(\mathrm{id}, \mathrm{id})$ in $\widetilde{\mathbf{D}}(\mathcal{A})$ to identity morphisms in $\mathcal{D}$. Finally, consider a composition $(t, g) \circ (q, f)$ in $\widetilde{\mathbf{D}}(\mathcal{A})$; this is represented by a roof $(q \circ t', g \circ g')$ coming from a commutative diagram

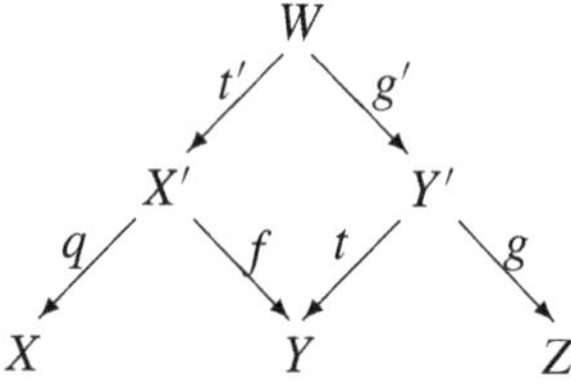

Since $f \circ t' = t \circ g'$ we get $F(f) \circ F(t') = F(t) \circ F(g')$ and since $F$ sends quasi-isomorphisms to isomorphisms $F(t)^{-1} \circ F(f) = F(g') \circ F(t')^{-1}$. This

implies that

$$\widetilde{F}((t, g) \circ (q, f)) = \widetilde{F}((q \circ t', g \circ g')) = F(g \circ g') \circ F(q \circ t')^{-1}$$
$$= F(g) \circ F(g') \circ F(t')^{-1} \circ F(q)^{-1}$$
$$= F(g) \circ F(t)^{-1} \circ F(f) \circ F(q)^{-1}$$
$$= \widetilde{F}((t, g)) \circ \widetilde{F}((q, f)).$$

Thus $\widetilde{F}$ is indeed a functor and this completes the proof of the universal property (L2). $\qquad\square$

**Remark 7.14.** In the sequel we shall denote the derived category exclusively by $\mathbf{D}(\mathcal{A})$ even if we usually use the more convenient equivalent version $\widetilde{\mathbf{D}}(\mathcal{A})$ just described.

**Proposition 7.15.** *Let $\mathcal{A}$ be an abelian category. Then the derived category* $\mathbf{D}(\mathcal{A})$ *is an additive category.*

*Proof.* Following Definition 1.1 we have to show the properties (A1), (A2) and (A3).

(A1) We first describe addition of morphisms. Let two morphisms $F, G$ from $X$ to $Y$ be represented by roofs $(q, f)$ and $(q', f')$. By the Ore condition (Q3) there exists an object $W$, a morphism $g$ and a quasi-isomorphism $t$ making the following diagram commutative

$$
\begin{array}{ccc}
W & \overset{g}{\longrightarrow} & X' \\
{\scriptstyle t}\downarrow & & \downarrow{\scriptstyle q} \\
X'' & \overset{q'}{\longrightarrow} & X
\end{array}
$$

Since $q, q'$ and $t$ are quasi-isomorphisms, also $g$ must be a quasi-isomorphism. From the definition of equivalence it is easy to check that the roof $(q, f)$ is equivalent to the roof $(q \circ g, f \circ g)$, and that $(q', f')$ is equivalent to the roof $(q' \circ t, f' \circ t) = (q \circ g, f' \circ t)$. Thus we have found a 'common denominator' and can set $F + G$ to be the roof represented by $(q \circ g, f \circ g + f' \circ t)$.

We leave it to the reader to verify that this addition is well-defined (i.e. independent of the representatives) and that the addition of morphisms is bilinear.

Note that in the derived category there is for any objects $X, Y$ a zero morphism in $\mathbf{D}(\mathcal{A})$ which is represented by the roof $(\mathrm{id}_X, 0_{X,Y})$ where $0_{X,Y}$ is the zero morphism of complexes from $X$ to $Y$.

(A2) The zero object in $\mathbf{D}(\mathcal{A})$ is the zero complex (i.e. it is the same zero object as in the homotopy category). We have to show that for every object $X$ the morphism sets $\mathrm{Hom}_{\mathbf{D}(\mathcal{A})}(X, 0)$ and $\mathrm{Hom}_{\mathbf{D}(\mathcal{A})}(0, X)$ contain only the morphism represented by the roof $(\mathrm{id}_X, 0)$ and $(0, \mathrm{id}_X)$, respectively. In fact, any morphism from $X$ to the zero complex is represented by a roof $(q, 0)$ where $q : Z \to X$ is a quasi-isomorphism. But it easily follows from the definition that the roof $(q, 0)$ is equivalent to $(\mathrm{id}_X, 0)$, thus $\mathrm{Hom}_{\mathbf{D}(\mathcal{A})}(X, 0)$ contains precisely one element. The assertion for $\mathrm{Hom}_{\mathbf{D}(\mathcal{A})}(0, X)$ is shown similarly.

(A3) For the coproduct of two objects $X$ and $Y$ in $\mathbf{D}(\mathcal{A})$ one uses the image of the coproduct $X \oplus Y$ in $\mathbf{K}(\mathcal{A})$ under the localisation functor $L$ (which is the identity on objects and maps a morphism $f$ in $\mathbf{K}(\mathcal{A})$ to the roof $(\mathrm{id}, f)$ in $\mathbf{D}(\mathcal{A})$). The corresponding maps $L(\iota_X) : X \to X \oplus Y$ and $L(\iota_Y) : X \to X \oplus Y$ are given by the roofs $(\mathrm{id}, \iota_X)$ and $(\mathrm{id}, \iota_Y)$, respectively, where $\iota_X$ and $\iota_Y$ are the embeddings (or more precisely, their equivalence classes in $\mathbf{K}(\mathcal{A})$).

We have to show that the universal property (A3) is satisfied. So let $\tilde{f}_X$ and $\tilde{f}_Y$ be arbitrary morphisms in $\mathbf{D}(\mathcal{A})$ from $X$ and $Y$ to some object $Z$. They are represented by roofs of the form

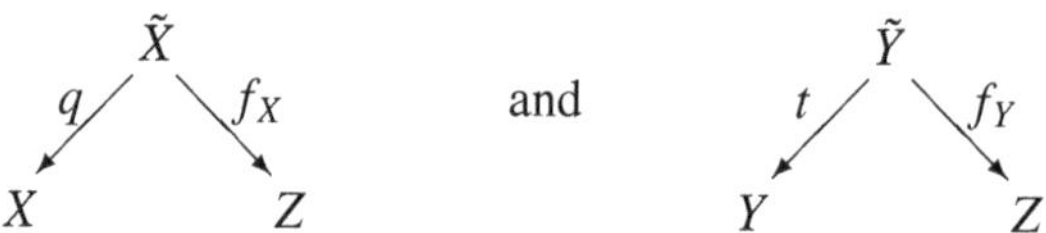

The required morphism $\tilde{f}$ from $X \oplus Y$ to $Z$ can then be defined as being represented by the roof $(q \oplus t, (f_X, f_Y))$ (where the notation $\oplus$ on maps between direct sums of complexes means componentwise application). Then indeed we have that the composition $\tilde{f} \circ L(\iota_X)$ in $\mathbf{D}(\mathcal{A})$ is the roof $(q, f_X)$, as can be seen from the diagram

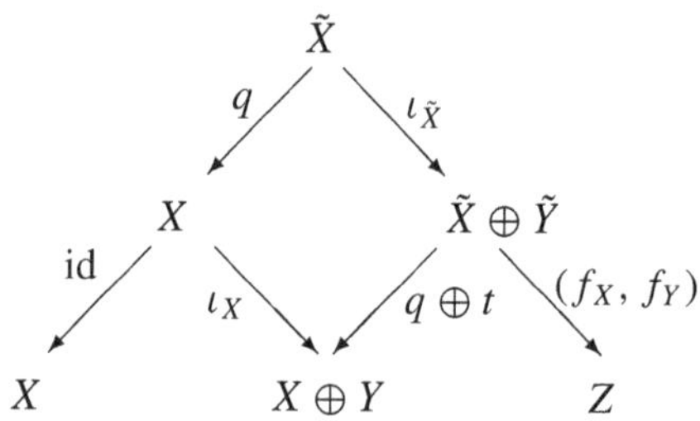

A similar diagram shows that $\tilde{f} \circ L(\iota_Y) = (t, f_Y)$ in $\mathbf{D}(\mathcal{A})$.

We leave it as an exercise to show that the morphism $\tilde{f}$ with these properties is actually unique, as required in (A3). $\qquad\square$

## 7.4. Derived categories are triangulated

Recall that the derived category has been obtained by localising the homotopy category with respect to the class of quasi-isomorphisms. In particular, there is a functor $L : \mathbf{K}(\mathcal{A}) \to \mathbf{D}(\mathcal{A})$ sending quasi-isomorphisms in $\mathbf{K}(\mathcal{A})$ to isomorphisms in $\mathbf{D}(\mathcal{A})$ (and satisfying a universal property). We have seen earlier that the homotopy category is triangulated, with distinguished triangles being the triangles isomorphic in $\mathbf{K}(\mathcal{A})$ to the standard triangles coming from mapping cones $X \xrightarrow{f} Y \xrightarrow{\alpha(f)} M(f) \xrightarrow{\beta(f)} X[1]$.

For obtaining a triangulated structure on the derived category the idea is to transport the triangulated structure on the homotopy category via the localisation functor $L$.

**Definition 7.16.** *The translation functor on $\mathbf{D}(\mathcal{A})$ is defined as the shift $[1]$ on objects, and for a morphism $F$ in $\mathbf{D}(\mathcal{A})$ represented by a roof $(q, f)$ we set $F[1]$ to be the equivalence class of the roof $(q[1], f[1])$.*

*A triangle in $\mathbf{D}(\mathcal{A})$ is a distinguished triangle if it is isomorphic (in $\mathbf{D}(\mathcal{A})$!) to the image of a distinguished triangle from $\mathbf{K}(\mathcal{A})$ under the localisation functor $L$.*

**Remark 7.17.** When passing from $\mathbf{K}(\mathcal{A})$ to $\mathbf{D}(\mathcal{A})$ all quasi-isomorphisms become isomorphisms, i.e. there are 'more' isomorphisms in $\mathbf{D}(\mathcal{A})$ than in $\mathbf{K}(\mathcal{A})$. This in turn means that in $\mathbf{D}(\mathcal{A})$ it is easier for a triangle to become isomorphic to a standard triangle than in $\mathbf{K}(\mathcal{A})$, i.e. the derived category contains 'more' distinguished triangles than the homotopy category.

As a crucial observation we shall see in the next section that the derived category has the property that every short exact sequence of complexes in $\mathbf{C}(\mathcal{A})$ gives rise to a corresponding distinguished triangle in the derived category. This is not yet the case in the homotopy category, see Remark 6.8 above for an example.

With the above definitions one can then show the main structural property of derived categories. Unfortunately, the proof of the axioms for a triangulated category will become very technical so that we shall refrain from providing a proof in this introductory chapter.

**Theorem 7.18.** *Let $\mathcal{A}$ be an abelian category. Then the derived category $\mathbf{D}(\mathcal{A})$ is triangulated.*

## 7.5. Comparing morphisms

Let $\mathcal{A}$ be an abelian category. We have now constructed various categories from $\mathcal{A}$:

$$\mathcal{A} \quad \mapsto \quad \mathbf{C}(\mathcal{A}) \quad \mapsto \quad \mathbf{K}(\mathcal{A}) \quad \mapsto \quad \mathbf{D}(\mathcal{A})$$

abelian        abelian        triangulated        triangulated

**Proposition 7.19.** *We have the following implications for a morphism $f$ in* $\mathbf{C}(\mathcal{A})$.

$$f = 0 \text{ in } \mathbf{C}(\mathcal{A}) \Rightarrow f = 0 \text{ in } \mathbf{K}(\mathcal{A}) \Rightarrow f = 0 \text{ in } \mathbf{D}(\mathcal{A}) \Rightarrow H_n(f)$$
$$= 0 \text{ for all } n \in \mathbb{Z}$$

*Proof.* The first two implications are obvious. For the third, $f : X \to Y$ as a morphism in $\mathbf{D}(\mathcal{A})$ is represented by the roof $(\mathrm{id}_X, f)$. For $f$ being 0 in $\mathbf{D}(\mathcal{A})$ means being equivalent to the roof $(\mathrm{id}_X, 0_{X,Y})$, i.e. in $\mathbf{K}(\mathcal{A})$ there is a commutative diagram

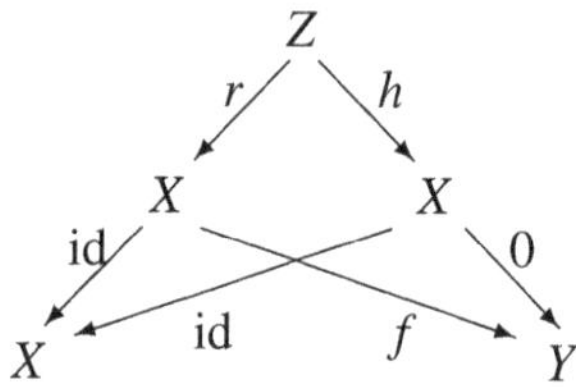

with a quasi-isomorphism $r$. By commutativity, $f \circ r = 0$ (in $\mathbf{K}(\mathcal{A})$) and passing to homology we get $H_n(f) \circ H_n(r) = H_n(f \circ r) = H_n(0) = 0$ for all $n \in \mathbb{Z}$. But $r$ is a quasi-isomorphism, i.e. all $H_n(r)$ are isomorphisms, thus we conclude that $H_n(f) = 0$ for all $n \in \mathbb{Z}$. $\qquad\qquad\square$

**Remark 7.20.**

(1) Note that a morphism $f$ in $\mathbf{K}(\mathcal{A})$ becomes zero in $\mathbf{D}(\mathcal{A})$ if and only if there exists a quasi-isomorphism $r$ such that $f \circ r$ is homotopic to zero.

(2) All implications given in Proposition 7.19 are strict. Let us give examples for each case. We consider the category $\mathcal{A} = \mathbf{Ab}$.

For the first implication, consider the following morphism of complexes

$$\cdots \to 0 \to 0 \longrightarrow \mathbb{Z} \to 0 \to \cdots$$
$$\downarrow \mathrm{id}$$
$$\cdots \to 0 \to \mathbb{Z} \xrightarrow{\ \mathrm{id}\ } \mathbb{Z} \to 0 \to \cdots$$

Clearly, this morphism is zero in $\mathbf{K}(\mathcal{A})$, but nonzero in $\mathbf{C}(\mathcal{A})$.

For the second implication, consider the identity map on the (exact) complex

$$0 \to \mathbb{Z} \xrightarrow{\cdot 2} \mathbb{Z} \xrightarrow{\pi} \mathbb{Z}/2\mathbb{Z} \to 0.$$

This morphism is not homotopic to zero (i.e. nonzero in $\mathbf{K}(\mathcal{A})$) because $\mathrm{Hom}_{\mathbb{Z}}(\mathbb{Z}/2\mathbb{Z}, \mathbb{Z}) = 0$ and hence the identity on $\mathbb{Z}/2\mathbb{Z}$ can not factor through a homotopy. However, it is zero in $\mathbf{D}(\mathcal{A})$ because we can find a quasi-isomorphism $r$ such that $f \circ r$ is homotopic to zero. In fact, let $r$ be the morphism of complexes

$$
\begin{array}{ccccccccc}
0 & \to & 0 & \to & \mathbb{Z} & \xrightarrow{\mathrm{id}} & \mathbb{Z} & \to & 0 \\
& & & & \downarrow{\mathrm{id}} & & \downarrow{\pi} & & \\
0 & \to & \mathbb{Z} & \xrightarrow{\cdot 2} & \mathbb{Z} & \xrightarrow{\pi} & \mathbb{Z}/2\mathbb{Z} & \to & 0
\end{array}
$$

Since both complexes are exact, $r$ is a quasi-isomorphism. Moreover, $f \circ r$ is homotopic to zero (a homotopy map is given by 0 and id in the two relevant degrees). This implies that $f$ is zero when considered as a morphism in $\mathbf{D}(\mathcal{A})$.

For the third implication we consider the morphism $f : X \to Y$ of complexes given as follows

$$
\begin{array}{ccccccc}
0 & \to & \mathbb{Z} & \xrightarrow{\cdot 2} & \mathbb{Z} & \to & 0 \\
& & \downarrow{\mathrm{id}} & & \downarrow{\pi} & & \\
0 & \to & \mathbb{Z} & \xrightarrow{\cdot 2} & \mathbb{Z}/3\mathbb{Z} & \to & 0
\end{array}
$$

The homology of $X$ is given by $H_0(X) = \mathbb{Z}/2\mathbb{Z}$ and $H_1(X) = 0$, whereas $H_0(Y) = 0$ and $H_1(Y) = 3\mathbb{Z}$ (and all other being zero). In particular, $H_n(f) = 0$ for all $n \in \mathbb{Z}$.

However, we claim that $f$ is nonzero in the derived category. Suppose for a contradiction that $f = 0$ in $\mathbf{D}(\mathcal{A})$, i.e. there exist a complex $R = (R_n, d_n^R)$ and a quasi-isomorphism $r : R \to X$ such that $f \circ r : R \to Y$ is homotopic to zero. Since $r$ is a quasi-isomorphism, we have that $H_n(R) \cong H_n(X) = 0$ for $n \neq 0$ and $H_0(R) = H_0(X) \cong \mathbb{Z}/2\mathbb{Z}$. Choose a generator of $H_0(R)$, i.e. $z_0 \in \ker d_0^R \setminus \mathrm{im}\, d_1^R$. Since $r$ is a quasi-isomorphism, $r_0(z_0)$ must not be in the image of $d_1^X$, i.e. $r_0(z_0) \notin 2\mathbb{Z}$. On the other hand, $f \circ r$ is homotopic to zero, thus there exist homotopy maps $s_0 : R_0 \to \mathbb{Z}$ and $s_{-1} : R_{-1} \to \mathbb{Z}/3\mathbb{Z}$ such that $(f \circ r)_0 = \pi \circ r_0 = 2s_0 + s_{-1} \circ d_0^R$. Applied to the generator $z_0$ (which is in the kernel of $d_0^R$) this yields

$$(\pi \circ r_0)(z_0) = (2s_0 + s_{-1} \circ d_0^R)(z_0) = 2s_0(z_0) \in 2(\mathbb{Z}/3\mathbb{Z}).$$

But then also $r_0(z_0) \in 2\mathbb{Z}$, a contradiction to the earlier conclusion.

Hence there is no such quasi-isomorphism $r$, i.e. $f \neq 0$ in $\mathbf{D}(\mathcal{A})$.

## 7.6. Short exact sequences vs. triangles

In this section we shall explain the crucial observation that a short exact sequence in $\mathbf{C}(\mathcal{A})$ induces a distinguished triangle in $\mathbf{D}(\mathcal{A})$. Recall that we have seen earlier that this does not yet happen in the homotopy category $\mathbf{K}(\mathcal{A})$ (cf. Remark 6.8).

In this subsection we will again use our assumption that the abelian category $\mathcal{A}$ is a category of modules over a ring, which allows us to define maps on elements.

### 7.6.1. *Mapping cylinders*

Let $f : X \to Y$ be a morphism of complexes in $\mathbf{C}(\mathcal{A})$. The *mapping cylinder of $f$* is the complex $Cyl(f)$ having degree $n$ part equal to $X_n \oplus X_{n-1} \oplus Y_n$ and the differential is given by

$$d_n^{Cyl(f)}(x, x', y) := (d_n^X x - x', -d_{n-1}^X x', f_{n-1}(x') + d_n^Y y).$$

In perhaps more convenient matrix notation,

$$d_n^{Cyl(f)} = \begin{pmatrix} d_n^X & -\operatorname{id}_{X_{n-1}} & 0 \\ 0 & -d_{n-1}^X & 0 \\ 0 & f_{n-1} & d_n^Y \end{pmatrix}.$$

It is now easily checked that $Cyl(f)$ is indeed a complex, i.e. $d_{n-1}^{Cyl(f)} \circ d_n^{Cyl(f)} = 0$.

We next aim at comparing the mapping cylinder with the mapping cone, as defined in Definition 6.3. We consider the following morphisms of complexes

$$\iota : X \to Cyl(f) \quad \text{given in degree } n \text{ by} \quad \iota_n = (\operatorname{id}_{X_n}, 0, 0),$$

$$\pi : Cyl(f) \to M(f) \quad \text{given in degree } n \text{ by} \quad \pi_n = \begin{pmatrix} 0 & \operatorname{id}_{X_{n-1}} & 0 \\ 0 & 0 & \operatorname{id}_{Y_n} \end{pmatrix}.$$

We leave the straightforward verification to the reader that these maps indeed commute with the differentials. Clearly, the resulting sequence

$$0 \to X \xrightarrow{\iota} Cyl(f) \xrightarrow{\pi} M(f) \to 0$$

is a short exact sequence in $\mathbf{C}(\mathcal{A})$.

**Lemma 7.21** (*Mapping cylinder vs. mapping cone*). *Let $f : X \to Y$ be a morphism of complexes. Then there are morphisms of complexes $\sigma : Y \to Cyl(f)$ with $\sigma_n = (0, 0, \operatorname{id}_{Y_n})$ and $\tau : Cyl(f) \to Y$ with $\tau_n = (f_n, 0, \operatorname{id}_{Y_n})$ such that the following holds*

(i) *The following diagram with exact rows is commutative in the category* $\mathbf{C}(\mathcal{A})$ *of complexes.*

$$
\begin{array}{ccccccccc}
0 & \longrightarrow & Y & \xrightarrow{\alpha(f)} & M(f) & \xrightarrow{\beta(f)} & X[1] & \longrightarrow & 0 \\
 & & \downarrow{\scriptstyle \sigma} & & \downarrow{\scriptstyle \mathrm{id}} & & & & \\
0 & \longrightarrow & X & \xrightarrow{\iota} & Cyl(f) & \xrightarrow{\pi} & M(f) & \longrightarrow & 0 \\
 & & \downarrow{\scriptstyle \mathrm{id}} & & \downarrow{\scriptstyle \tau} & & & & \\
 & & X & \xrightarrow{f} & Y & & & &
\end{array}
$$

(ii) $\tau \circ \sigma = \mathrm{id}_Y$ *and* $\sigma \circ \tau$ *is homotopic to the identity* $\mathrm{id}_{Cyl(f)}$, *i.e.* $Y$ *and* $Cyl(f)$ *are isomorphic in the homotopy category, and hence also in the derived category* $\mathbf{D}(\mathcal{A})$.

(iii) $\sigma$ *and* $\tau$ *are quasi-isomorphisms.*

*Proof.* (i) It is immediately checked from the definitions (for $\alpha(f)$ see Remark 6.4) that $\sigma$ and $\tau$ are indeed morphisms of complexes and that all squares in the diagram commute (in $\mathbf{C}(\mathcal{A})$, not only up to homotopy).

(ii) By definition we have $\tau \circ \sigma = \mathrm{id}_Y$. On the other hand, $\sigma \circ \tau$ is homotopic to the identity via the homotopy map $s = (s_n)$ with

$$
s_n = \begin{pmatrix} 0 & 0 & 0 \\ \mathrm{id}_{X_n} & 0 & 0 \\ 0 & 0 & 0 \end{pmatrix}.
$$

(iii) By (ii) the compositions $\tau \circ \sigma$ and $\sigma \circ \tau$ are homotopy equivalences, in particular they induce the identity in homology. Thus,

$$
H_n(\sigma) \circ H_n(\tau) = H_n(\sigma \circ \tau) = H_n(\mathrm{id}_{Cyl(f)}) = \mathrm{id}_{H_n(Cyl(f))}
$$

and similarly $H_n(\tau) \circ H_n(\sigma) = \mathrm{id}_{H_n(Y)}$ which implies that $H_n(\sigma)$ and $H_n(\tau)$ are isomorphisms, i.e. $\sigma$ and $\tau$ are quasi-isomorphisms. $\qquad\square$

As a consequence we can now state the main result of this section.

**Corollary 7.22.** *To any short exact sequence* $0 \to X \xrightarrow{f} Y \xrightarrow{g} Z \to 0$ *in* $\mathbf{C}(\mathcal{A})$ *there exists a corresponding distinguished triangle in* $\mathbf{D}(\mathcal{A})$ *of the form*

$$
X \xrightarrow{f} Y \xrightarrow{g} Z \longrightarrow X[1].
$$

*Proof.* Using the notations of the previous lemma we consider the following diagram with exact rows

$$
\begin{array}{ccccccccc}
0 & \to & X & \xrightarrow{\iota} & Cyl(f) & \xrightarrow{\pi} & M(f) & \to & 0 \\
& & \downarrow{\scriptstyle \mathrm{id}} & & \downarrow{\scriptstyle \tau} & & \downarrow{\scriptstyle \gamma} & & \\
0 & \to & X & \xrightarrow{f} & Y & \xrightarrow{g} & Z & \to & 0
\end{array}
$$

where $\gamma = (\gamma_n)$ is defined by setting $\gamma_n(x, y) := g_n(y)$; this is easily checked to be a morphism of complexes (using that $g$ is a morphism of complexes and that $g \circ f = 0$ since they are consecutive maps in a short exact sequence). Also one immediately deduces from the definitions that the above diagram is commutative (where the left hand square already appeared in the previous lemma).

Since id and $\tau$ (by the previous lemma) are both quasi-isomorphisms it follows from the long exact homology sequences and the usual 5-lemma that also $\gamma$ must be a quasi-isomorphism, hence an isomorphism in the derived category (but not necessarily in $\mathbf{K}(\mathcal{A})$, see the following remark). So we have a morphism of triangles in $\mathbf{D}(\mathcal{A})$

$$
\begin{array}{ccccccc}
X & \xrightarrow{f} & Y & \xrightarrow{\alpha(f)} & M(f) & \xrightarrow{\beta(f)} & X[1] \\
\downarrow{\scriptstyle \mathrm{id}} & & \downarrow{\scriptstyle \mathrm{id}} & & \downarrow{\scriptstyle \gamma} & & \downarrow{\scriptstyle \mathrm{id}} \\
X & \xrightarrow{f} & Y & \xrightarrow{g} & Z & \xrightarrow{\beta(f)\circ\gamma^{-1}} & X[1]
\end{array}
$$

where the inverse $\gamma^{-1}$ exists in the derived category. For the commutativity of the second square note that by the previous lemma and the above definition of $\gamma$ we have

$$
\gamma \circ \alpha(f) = \gamma \circ \pi \circ \sigma = g \circ \tau \circ \sigma = g \circ \mathrm{id} = g.
$$

Moreover, since $\gamma$ is an isomorphism in $\mathbf{D}(\mathcal{A})$ this morphism of triangles is indeed an isomorphism of triangles, i.e. the bottom line

$$
X \xrightarrow{f} Y \xrightarrow{g} Z \xrightarrow{\beta(f)\circ\gamma^{-1}} X[1]
$$

is isomorphic to the image of a standard triangle from $\mathbf{K}(\mathcal{A})$ under the localisation functor, hence a distinguished triangle in $\mathbf{D}(\mathcal{A})$. $\qquad\square$

**Remark 7.23.** The crucial quasi-isomorphism $\gamma$ occurring in the previous proof is in general not a homotopy equivalence (i.e. not an isomorphism in $\mathbf{K}(\mathcal{A})$) and hence the isomorphism of triangles exists only in the derived category but not yet in the homotopy category.

As an example, consider $\mathcal{A} = \mathbf{R\text{-}mod}$ for a ring $R$ and consider $R$-modules $X, Y$ as complexes concentrated in degree 0. The mapping cone of a (module) morphism $f : X \to Y$ is just $0 \to X \xrightarrow{f} Y \to 0$, and the morphism $\gamma : M(f) \to Y/X$ is just the natural projection. If this $\gamma$ is a homotopy equivalence then $f$ must be a *split* monomorphism. In fact, up to homotopy there exists an inverse $\rho : Y/X \to Y$; for $\rho \circ \gamma$ to be homotopic to the identity on $M(f)$ there must be a homotopy map $s : Y \to X$ which (when looking in degree 1) in particular satisfies $s \circ f = \mathrm{id}_X$, i.e. $f$ splits.

This explains again the earlier example (cf. Remark 6.8) of the short exact sequence of abelian groups

$$0 \to \mathbb{Z}/2\mathbb{Z} \xrightarrow{\cdot 2} \mathbb{Z}/4\mathbb{Z} \xrightarrow{\cdot 1} \mathbb{Z}/2\mathbb{Z} \to 0$$

which does not have a corresponding distinguished triangle in $\mathbf{K}(\mathcal{A})$, because clearly the map $\mathbb{Z}/2\mathbb{Z} \xrightarrow{\cdot 2} \mathbb{Z}/4\mathbb{Z}$ does not split.

## 8. Frobenius categories and stable categories

In the earlier sections we have considered categories of complexes leading to derived categories which form an important source of examples of triangulated categories.

In this section we shall briefly describe another source for triangulated categories, namely stable categories of Frobenius algebras. The aim is to give the relevant definitions and constructions and to provide some examples, in order to prepare the ground for the later articles in this book.

For more details, in particular for a complete proof of the triangulated structure of the stable category of a Frobenius algebra we refer the reader for instance to the well-written chapter on Frobenius categories in D. Happel's book [5].

We start by defining an exact category, a concept introduced by D. Quillen, which generalizes abelian categories, in the sense that an exact category has a certain class of 'exact triples' as a replacement for short exact sequences without having to be abelian itself.

**Definition 8.1** *(Exact category). Let $\mathcal{A}$ be an abelian category, and let $\mathcal{B}$ be an additive subcategory of $\mathcal{A}$ which is full and closed under extensions (i.e. if $0 \to X \to Y \to Z \to 0$ is an exact sequence in $\mathcal{A}$ where $X$ and $Z$ are objects in $\mathcal{B}$ then $Y$ is isomorphic to an object of $\mathcal{B}$). Take $\mathcal{E}$ to be the class of all triples $X \to Y \to Z$ in $\mathcal{B}$ whose corresponding sequences $0 \to X \to Y \to Z \to 0$ in $\mathcal{A}$ are exact.*

*Then the pair $(\mathcal{B}, \mathcal{E})$ is called an exact category.*

**Example 8.2.**

(i) Every abelian category is an exact category; in fact, take for $\mathcal{E}$ the class of all short exact sequences in $\mathcal{A}$.

(ii) Let $\mathcal{A} = \mathbf{Ab}$ be the category of abelian groups and $\mathcal{B} := \mathbf{tf\text{-}Ab}$ the full subcategory of torsionfree abelian groups. Then $\mathcal{B}$ is closed under extensions; in fact, let $0 \to X \xrightarrow{\alpha} Y \xrightarrow{\beta} Z \to 0$ be a short exact sequence with $X$ and $Z$ torsionfree. Suppose $ny = 0$ for some $y \in Y$ and $n \in \mathbb{Z}$. Then $n\beta(y) = \beta(ny) = 0$ and hence $\beta(y) = 0$ since $Z$ is torsionfree. By exactness of the sequence it follows that $y$ is in the image of $\alpha$, say $y = \alpha(x)$. But then $\alpha(nx) = n\alpha(x) = ny = 0$ which implies $nx = 0$ since $\alpha$ is a monomorphism. From the torsionfreeness of $X$ we deduce that $x = 0$ and thus also $y = 0$, i.e. $Y$ is also torsionfree. So the pair $(\mathbf{tf\text{-}Ab}, \mathcal{E})$ is an exact category.

   However, note that $\mathbf{tf\text{-}Ab}$ is not an abelian category (e.g. the morphism $\mathbb{Z} \xrightarrow{\cdot 2} \mathbb{Z}$ does not have a cokernel in $\mathbf{tf\text{-}Ab}$).

(iii) Similarly to the preceding example, consider the category $\mathbf{t\text{-}Ab}$ with objects the abelian groups containing torsion elements and the trivial group. This is a full subcategory of $\mathbf{Ab}$ which is closed under extensions; in fact, if $0 \to X \to Y \to Z \to 0$ is a short exact sequence of abelian groups and $X$ and $Z$ contain torsion elements, then $Y$ contains the torsion elements of $X$ if $X$ is nonzero, and otherwise $Y$ has the torsion elements of $Z$. So, $(\mathbf{t\text{-}Ab}, \mathcal{E})$ is also an exact category.

**Definition 8.3.** *An exact category $(\mathcal{B}, \mathcal{E})$ is called a* Frobenius category *if the following holds:*

(i) *Projective and injective objects coincide (where an object $I$ of $\mathcal{B}$ is called injective if every exact triple in $\mathcal{E}$ of the form $I \to Y \to Z$ splits; and an object $P$ of $\mathcal{B}$ is called projective if every exact triple in $\mathcal{E}$ of the form $X \to Y \to P$ splits).*

(ii) *The category has enough projective objects and enough injective objects, i.e. for every object $X$ in $\mathcal{B}$ there exist triples from $\mathcal{E}$ of the form $X' \to I \to X$ and $X \to I' \to X''$ where $I$ and $I'$ are injective.*

**Example 8.4.**

(i) A (finite-dimensional) algebra $\Lambda$ over a field $K$ is called a *Frobenius algebra* if there exists a non-degenerate associative bilinear form on $\Lambda$. Equivalently, if there exists a linear form $\pi : \Lambda \to K$ such that the kernel of $\pi$ does not contain any nonzero left ideal of $\Lambda$.

The notion of Frobenius algebra is closely related to that of *selfinjective* algebras (for which by definition projective and injective modules coincide). In fact, every Frobenius algebra is selfinjective, and every basic selfinjective algebra is Frobenius. For more details we refer to the notes by R. Farnsteiner [3].

Then, for any Frobenius algebra $\Lambda$ the category of finite-dimensional $\Lambda$-modules is a Frobenius category. For instance, for a finite group $G$, the category of finitely generated modules over the group algebra $KG$ is a Frobenius category.

(ii) Consider the abelian category $\mathcal{B} = \mathbf{Ab}$ of abelian groups.

An object $A$ in $\mathbf{Ab}$ is injective if and only if it is divisible (i.e. for every $n \in \mathbb{Z} \setminus \{0\}$ the multiplication map $A \xrightarrow{\cdot n} A$ is surjective); e.g. $\mathbb{Q}$, $\mathbb{R}$ or $\mathbb{Q}/\mathbb{Z}$ are injective abelian groups. On the other hand, abelian groups are nothing but modules over $\mathbb{Z}$ and since $\mathbb{Z}$ is a principal ideal domain, the projective objects are precisely the free objects, i.e. direct sums of copies of $\mathbb{Z}$. In particular, in $\mathbf{Ab}$, projective and injective objects do not coincide and hence $\mathbf{Ab}$ can not be a Frobenius category.

**Definition 8.5.** *Let $(\mathcal{B}, \mathcal{E})$ be a Frobenius category. For objects $X$ and $Y$ in $\mathcal{B}$ let* $\mathrm{Inj}(X, Y)$ *denote those morphisms from $X$ to $Y$ which factor through some injective object.*

*The stable category $\underline{\mathcal{B}}$ of the Frobenius category $(\mathcal{B}, \mathcal{E})$ has the same objects as $\mathcal{B}$; the morphisms are equivalence classes of morphisms modulo those factoring through injective objects, i.e.*

$$\mathrm{Hom}_{\underline{\mathcal{B}}}(X, Y) := \underline{\mathrm{Hom}}(X, Y) = \mathrm{Hom}_{\mathcal{B}}(X, Y)/\mathrm{Inj}(X, Y).$$

**Example 8.6** (Dual numbers). Let $K$ be a field and consider the algebra $\Lambda = K[X]/(X^2)$. This 2-dimensional algebra is a Frobenius algebra; in fact, $\pi(a + bX) := b$ defines a linear form on $\Lambda$ such that its kernel does not contain any nonzero left ideal.

The category of finitely generated $\Lambda$-modules is then a Frobenius category. It has only two indecomposable objects, $\Lambda$ itself and the one-dimensional simple module $K$, where $\Lambda$ is the only injective (and projective) module.

In the corresponding stable module category several module homomorphisms vanish, e.g. we have that $\underline{\mathrm{Hom}}(\Lambda, \Lambda) = 0$, $\underline{\mathrm{Hom}}(\Lambda, K) = 0$ and $\underline{\mathrm{Hom}}(K, \Lambda) = 0$; on the other hand $\underline{\mathrm{Hom}}(K, K)$ remains 1-dimensional since the isomorphism can not factor through the injective module $\Lambda$.

Our main aim is to describe how the stable category of a Frobenius category $(\mathcal{B}, \mathcal{E})$ carries the structure of a triangulated category. To this end we shall briefly describe the construction of a suspension functor and then of the distinguished triangles.

For every object $X$ in $\mathcal{B}$ we choose an exact triple $X \xrightarrow{\iota_X} I(X) \xrightarrow{\pi_X} \Sigma X$ with entries from $\mathcal{B}$ where $I(X)$ is an injective object, i.e. in the ambient abelian category $\mathcal{A}$ there is a short exact sequence $0 \to X \to I(X) \to \Sigma X \to 0$. Thus, on objects of $\mathcal{B}$ (and hence also on objects of $\underline{\mathcal{B}}$) we have a map $X \mapsto \Sigma X$ and this will be the candidate for the suspension functor on objects.

For defining $\Sigma$ on morphisms, let $u : X \to Y$ be a morphism and consider the chosen exact triples $X \xrightarrow{\iota_X} I(X) \xrightarrow{\pi_X} \Sigma X$ and $Y \xrightarrow{\iota_Y} I(Y) \xrightarrow{\pi_Y} \Sigma Y$. Since $I(Y)$ is injective there exists a morphism $I(u) : I(X) \to I(Y)$ such that $I(u)\iota_X = \iota_Y u$. By considering the corresponding short exact sequences in the ambient abelian category the morphism $I(u)$ induces a morphism $\Sigma(u) : \Sigma X \to \Sigma Y$ satisfying $\Sigma(u)\pi_X = \pi_Y I(u)$.

**Lemma 8.7.** *The morphisms $\Sigma(u)$ are well-defined in the stable category $\underline{\mathcal{B}}$, i.e. they are independent of the choice of the lifting morphisms $I(u)$ and of the representatives of the morphisms $u$.*

*Proof.* For a morphism $u$ the morphism $\Sigma(u)$ has been defined above as indicated in the following diagram

$$
\begin{array}{ccccc}
X & \xrightarrow{\iota_X} & I(X) & \xrightarrow{\pi_X} & \Sigma X \\
\downarrow{\scriptstyle u} & & \downarrow{\scriptstyle I(u)} & & \downarrow{\scriptstyle \Sigma(u)} \\
Y & \xrightarrow{\iota_Y} & I(Y) & \xrightarrow{\pi_Y} & \Sigma Y
\end{array}
$$

Now let $I(u)$ and $\tilde{I}(u)$ be two liftings for $u$, with corresponding induced morphisms $\Sigma(u)$ and $\tilde{\Sigma}(u)$ from $\Sigma X$ to $\Sigma Y$. Then

$$(I(u) - \tilde{I}(u))\iota_X = I(u)\iota_X - \tilde{I}(u)\iota_X = \iota_Y u - \iota_Y u = 0.$$

By exactness of the top row there exists a morphism $\sigma : \Sigma X \to I(Y)$ such that $\sigma\pi_X = I(u) - \tilde{I}(u)$. This implies that

$$(\Sigma(u) - \tilde{\Sigma}(u))\pi_X = \pi_Y I(u) - \pi_Y \tilde{I}(u) = \pi_Y \sigma \pi_X.$$

Since $\pi_X$ is an epimorphism we can deduce that $\Sigma(u) - \tilde{\Sigma}(u) = \pi_Y \sigma$. Hence $\Sigma(u) - \tilde{\Sigma}(u)$ factors through the injective object $I(Y)$, i.e. $\Sigma(u) = \tilde{\Sigma}(u)$ in the stable category $\underline{\mathcal{B}}$.

A similar argument shows that if $u$ factors through an injective object then also $\Sigma(u)$ factors through an injective object, i.e. $\Sigma(u)$ is independent of the representative of the morphism $u$ in $\underline{\mathcal{B}}$. We leave the details to the reader. $\qquad\square$

Our above construction of the objects $\Sigma X$ and of the morphisms $\Sigma(u)$ used a fixed choice of exact triples $X \to I(X) \to \Sigma X$. The following lemma shows that this construction does not depend on the choice of the exact triples, more precisely, a different choice leads to naturally isomorphic functors. In particular, the object $\Sigma X$ is uniquely defined up to isomorphism in the stable category $\underline{B}$.

**Lemma 8.8.** *For any object $X$ let $X \to I(X) \to \Sigma X$ and $X \to I'(X) \to \Sigma'X$ be exact triples in $B$ where $I(X)$ and $I'(X)$ are injective. Then $\Sigma X$ and $\Sigma'X$ are isomorphic in the stable category $\underline{B}$. Moreover, there is a natural transformation $\beta : \Sigma \to \Sigma'$ such that each $\beta_X : \Sigma X \to \Sigma'X$ is an isomorphism, i.e. the functors $\Sigma$ and $\Sigma'$ are isomorphic.*

*Proof.* In the ambient abelian category $\mathcal{A}$ we have short exact sequences

$$0 \to X \xrightarrow{\iota_X} I(X) \xrightarrow{\pi_X} \Sigma X \to 0 \quad \text{and} \quad 0 \to X \xrightarrow{\iota'_X} I'(X) \xrightarrow{\pi'_X} \Sigma'X \to 0.$$

Since $I(X)$ and $I'(X)$ are injective in $B$ there are morphisms $\alpha_X : I(X) \to I'(X)$ and $\alpha'_X : I'(X) \to I(X)$ making the left hand squares in the following diagram commutative; moreover since the rows are short exact sequences these morphisms induce morphisms $\beta_X$ and $\beta'_X$ also making the right hand squares commutative.

$$
\begin{array}{ccccccccc}
0 & \longrightarrow & X & \xrightarrow{\iota_X} & I(X) & \xrightarrow{\pi_X} & \Sigma X & \longrightarrow & 0 \\
& & \Big\| & & \Big\downarrow{\alpha_X} & & \Big\downarrow{\beta_X} & & \\
0 & \longrightarrow & X & \xrightarrow{\iota'_X} & I'(X) & \xrightarrow{\pi'_X} & \Sigma'X & \longrightarrow & 0 \\
& & \Big\| & & \Big\downarrow{\alpha'_X} & & \Big\downarrow{\beta'_X} & & \\
0 & \longrightarrow & X & \xrightarrow{\iota_X} & I(X) & \xrightarrow{\pi_X} & \Sigma X & \longrightarrow & 0
\end{array}
$$

By commutativity it follows that

$$(\alpha'_X\alpha_X - \mathrm{id}_{I(X)})\iota_X = \alpha'_X\iota'_X - \iota_X = \iota_X - \iota_X = 0.$$

Using exactness of the top row implies the existence of a morphism $\sigma_X : \Sigma X \to I(X)$ such that $\sigma_X\pi_X = \alpha'_X\alpha_X - \mathrm{id}_{I(X)}$. Again using commutativity of the above diagram we then have that

$$\pi_X\sigma_X\pi_X = \pi_X\alpha'_X\alpha_X - \pi_X = (\beta'_X\beta_X - \mathrm{id}_{\Sigma X})\pi_X.$$

Since $\pi_X$ is an epimorphism we deduce $\pi_X \sigma_X = \beta'_X \beta_X - \mathrm{id}_{\Sigma X}$, i.e. $\beta'_X \beta_X - \mathrm{id}_{\Sigma X}$ factors through the injective object $I(X)$ and thus $\beta'_X \beta_X = \mathrm{id}_{\Sigma X}$ in the stable category $\underline{\mathcal{B}}$.

An analogous argument shows that also $\beta_X \beta'_X = \mathrm{id}_{\Sigma' X}$ in $\underline{\mathcal{B}}$. Thus, $\Sigma X$ and $\Sigma' X$ are isomorphic in $\underline{\mathcal{B}}$, as claimed.

For naturality, consider different choices of exact triples for objects $X$ and $Y$ and a morphism $u : X \to Y$, as in the following commutative diagrams

$$
\begin{array}{ccccc}
X & \xrightarrow{\iota_X} & I(X) & \xrightarrow{\pi_X} & \Sigma X \\
\downarrow{\scriptstyle u} & & \downarrow{\scriptstyle I(u)} & & \downarrow{\scriptstyle \Sigma(u)} \\
Y & \xrightarrow{\iota_Y} & I(Y) & \xrightarrow{\pi_Y} & \Sigma Y
\end{array}
\qquad
\begin{array}{ccccc}
X & \xrightarrow{\iota'_X} & I'(X) & \xrightarrow{\pi'_X} & \Sigma' X \\
\downarrow{\scriptstyle u} & & \downarrow{\scriptstyle I'(u)} & & \downarrow{\scriptstyle \Sigma'(u)} \\
Y & \xrightarrow{\iota'_Y} & I'(Y) & \xrightarrow{\pi'_Y} & \Sigma' Y
\end{array}
$$

We claim that $\beta$ induces a natural isomorphism between the functors $\Sigma$ and $\Sigma'$ resulting form different choices of exact triples. From the first part of the proof we already know that each $\beta_X$ is an isomorphism in $\underline{\mathcal{B}}$. It remains to show that we indeed have a natural transformation, i.e. we have to show that $\beta_Y \Sigma(u) = \Sigma'(u)\beta_X$ in the stable category.

Note that by commutativity of the above diagrams we have

$$(\alpha_Y I(u) - I'(u)\alpha_X)\iota_X = \alpha_Y \iota_Y u - I'(u)\iota'_X = \iota'_Y u - I'(u)\iota'_X = 0.$$

Hence by exactness there exists a morphism $\tau : \Sigma X \to I'(Y)$ such that $\tau \pi_X = \alpha_Y I(u) - I'(u)\alpha_X$. It then follows that

$$
\begin{aligned}
\pi'_Y \tau \pi_X &= \pi'_Y(\alpha_Y I(u) - I'(u)\alpha_X) = \beta_Y \pi_Y I(u) - \Sigma'(u)\pi'_X \alpha_X \\
&= \beta_Y \Sigma(u)\pi_X - \Sigma'(u)\beta_X \pi_X = (\beta_Y \Sigma(u) - \Sigma'(u)\beta_X)\pi_X.
\end{aligned}
$$

Since $\pi_X$ is an epimorphism this implies that $\pi'_Y \tau = \beta_Y \Sigma(u) - \Sigma'(u)\beta_X$, i.e. $\beta_Y \Sigma(u) - \Sigma'(u)\beta_X$ factors through the injective object $I'(Y)$ and hence we have $\beta_Y \Sigma(u) = \Sigma'(u)\beta_X$ in the stable category $\underline{\mathcal{B}}$. $\qquad\square$

Hence we have a well-defined functor $\Sigma$ on the stable category $\underline{\mathcal{B}}$. This can be shown to be an autoequivalence. Under certain assumptions it is even an automorphism; for details on this subtle issue see Happel's book [5, Section I.2].

We now describe the construction of distinguished triangles in the stable category $\underline{\mathcal{B}}$. Let $X$, $Y$ be objects in $\mathcal{B}$ and let $u : X \to Y$ be a morphism. For $X$ we have from the above construction an exact triple $X \xrightarrow{\iota} I(X) \xrightarrow{\pi} \Sigma X$ where $I(X)$ is injective.

In the additive category $\mathcal{B}$ there exists a coproduct $I \oplus Y$, with morphisms $\iota_I : I \to I \oplus Y$ and $\iota_Y : Y \to I \oplus Y$ satisfying the universal property given in Remark 1.2. Now we form the pushout of the morphisms $\iota : X \to I$ and $u : X \to Y$. More precisely, the pushout $M(u)$ is defined as the cokernel of the morphism $\iota_I \iota - \iota_Y u : X \to I \oplus Y$. By definition, this cokernel is an object $M(u)$ together with a morphism $c : I \oplus Y \to M(u)$ such that $c(\iota_I \iota - \iota_Y u) = 0$ and satisfying the universal property for cokernels given in Section 2. In particular, the left hand square in the following diagram is commutative

$$
\begin{array}{ccccc}
X & \xrightarrow{\iota} & I(X) & \xrightarrow{\pi} & \Sigma X \\
\downarrow{\scriptstyle u} & & \downarrow{\scriptstyle c\iota_I} & & \downarrow{\scriptstyle \mathrm{id}} \\
Y & \xrightarrow{c\iota_Y} & M(u) & & \Sigma X
\end{array}
$$

We wish to complete this diagram with a morphism $w : M(u) \to \Sigma X$. To this end recall that from the properties of a coproduct there is a unique morphism $f : I \oplus Y \to \Sigma X$ such that $f\iota_I = \pi$ and $f\iota_Y = 0$. Using this morphism $f$ in the universal property for the cokernel $M(u)$ we deduce the existence of a (unique) morphism $w : M(u) \to \Sigma X$ such that $wc = f$. It follows that $wc\iota_I = f\iota_I = \pi$, i.e. the following diagram of objects and morphisms in $\mathcal{B}$ is commutative

$$
\begin{array}{ccccc}
X & \xrightarrow{\iota} & I(X) & \xrightarrow{\pi} & \Sigma X \\
\downarrow{\scriptstyle u} & & \downarrow{\scriptstyle c\iota_I} & & \downarrow{\scriptstyle \mathrm{id}} \\
Y & \xrightarrow{c\iota_Y} & M(u) & \xrightarrow{w} & \Sigma X
\end{array}
$$

Note that since $\mathcal{B}$ is closed under extensions the cokernel will again be an object in $\mathcal{B}$ (and not only in the ambient abelian category $\mathcal{A}$).

The images in the stable category $\underline{\mathcal{B}}$ of any triangles of the form

$$
X \xrightarrow{u} Y \xrightarrow{c\iota_Y} M(u) \xrightarrow{w} \Sigma X
$$

are called standard triangles in $\underline{\mathcal{B}}$. As usual, the set of distinguished triangles in $\underline{\mathcal{B}}$ is formed by the set of all triangles in $\underline{\mathcal{B}}$ which are isomorphic (in $\underline{\mathcal{B}}$ !) to a standard triangle.

The main structural result on stable categories of Frobenius categories is then the following; for a detailed proof we refer to Happel's book [5, Chapter I.2].

**Theorem 8.9.** *Let $(\mathcal{B}, \mathcal{E})$ be a Frobenius category. With the above suspension functor $\Sigma$ and the collection of distinguished triangles just defined, the stable category $\underline{\mathcal{B}}$ is a triangulated category.*

**Remark 8.10.** A triangulated category is called *algebraic* (in the sense of B. Keller, see [9]) if it is equivalent as a triangulated category to the stable category of a Frobenius category. For more details on algebraic and non-algebraic triangulated categories we refer to S. Schwede's article in this volume [14], see also [15]. Strikingly, there are triangulated categories which are neither algebraic nor topological [11].

# References


[1] P. Balmer, *Triangulated categories with several triangulations.* Available from the author's web page at http://www.math.ucla.edu/~balmer/ research/Pubfile/TriangulationS.pdf

[2] A. Dold, D. Puppe, *Homologie nicht-additiver Funktoren.* Ann. Inst. Fourier Grenoble 11 (1961) 201–312.

[3] R. Farnsteiner, *Self-injective algebras: I. The Nakayama permutation, II. Comparison with Frobenius algebras, III. Examples and Morita equivalence, IV. Frobenius algebras and coalgebras.* Available at http://www.mathematik.uni-bielefeld.de/~sek/selected.html

[4] S.I. Gelfand, Y.I.Manin, *Methods of Homological Algebra.* Second edition. Springer Monographs in Mathematics. Springer-Verlag, Berlin, 2003.

[5] D. Happel, *Triangulated categories in the representation theory of finite-dimensional algebras.* London Mathematical Society Lecture Note Series, 119. Cambridge University Press, Cambridge, 1988.

[6] A. Hubery, *Notes on the octahedral axiom.* Available from the author's web page at http://www.maths.leeds.ac.uk/~ahubery/Octahedral.pdf

[7] M. Kashiwara, P. Schapira, *Sheaves on manifolds.* Grundlehren der Mathematischen Wissenschaften, 292. Springer-Verlag, Berlin, 1990.

[8] M. Kashiwara, P. Schapira, *Categories and sheaves.* Grundlehren der Mathematischen Wissenschaften 332. Springer-Verlag, 2006.

[9] B. Keller, *On differential graded categories.* International Congress of Mathematicians. Vol. II, 151–190, Eur. Math. Soc., Zürich, 2006.

[10] H. Krause, *Localization theory for triangulated categories.* This volume.

[11] F. Muro, S. Schwede, N. Strickland, *Triangulated categories without models.* Invent. Math. 170 (2007), 231-241.

[12] A. Neeman, *Triangulated categories.* Annals of Mathematics Studies, 148. Princeton University Press, Princeton, NJ, 2001.

[13] A. Neeman, *New axioms for triangulated categories.* J. Algebra 139 (1991), 221–255.

[14] S. Schwede, *Algebraic versus topological triangulated categories.* This volume.

[15] S. Schwede, *Torsion invariants for triangulated categories.* Preprint (2009), available from the author's web page at http://www.math.uni-bonn.de/~schwede.

[16] J.-L. Verdier, *Des catégories dérivées des catégories abéliennes.* Astérisque No. 239 (1996).

[17] C.A. Weibel, *An introduction to homological algebra.* Cambridge Studies in Advanced Mathematics, 38. Cambridge University Press, Cambridge, 1994.

THORSTEN HOLM
LEIBNIZ UNIVERSITÄT HANNOVER, INSTITUT FÜR ALGEBRA, ZAHLENTHEORIE UND DISKRETE MATHEMATIK, WELFENGARTEN I, D-30167 HANNOVER, GERMANY
*E-mail address*: holm@math.uni-hannover.de
*URL*: http://www.iazd.uni-hannover.de/~tholm

PETER JØRGENSEN
SCHOOL OF MATHEMATICS AND STATISTICS, NEWCASTLE UNIVERSITY, NEWCASTLE UPON TYNE NE1 7RU, UNITED KINGDOM
*E-mail address*: peter.jorgensen@ncl.ac.uk
*URL*: http://www.staff.ncl.ac.uk/peter.jorgensen/

# Cohomology over complete intersections via exterior algebras


LUCHEZAR L. AVRAMOV AND
SRIKANTH B. IYENGAR


*To Karin Erdmann on her 60th birthday.*


ABSTRACT. A general method for establishing results over a commutative complete intersection local ring by passing to differential graded modules over a graded exterior algebra is described. It is used to deduce, in a uniform way, results on the growth of resolutions of complexes over such local rings.


## Introduction

This paper concerns homological invariants of modules and complexes over complete intersection local rings. The goal is to explain a method by which one can establish in a uniform way results over such rings by deducing them from results on DG (that is, differential graded) modules over a graded exterior algebra, which are often easier to prove. A secondary purpose is to demonstrate the use of numerical invariants of objects in derived categories, called 'levels', introduced in earlier joint work with Buchweitz and Miller [5]; see Section 1. Levels allow one to track homological *and* structural information under changes of rings or DG algebras, such as those involved when passing from complete intersections to exterior algebras.

We focus on the complexity and the injective complexity of a complex $M$ over a complete intersection ring $R$. These numbers measure, on a polynomial scale, the rate of growth of the minimal free resolution and the minimal injective resolution of $M$, respectively. The relevant basic properties are established in Section 2.

Complexities can be expressed as dimensions of certain algebraic varieties, attached to $M$ in [2], and earlier proofs of key results relied on that theory. In


*Date*: April 28, 2010.
2000 *Mathematics Subject Classification*. 13D02, 16E45; 13D07, 13D25, 13H10, 20J06.
*Key words and phrases*. Complexity, exterior algebra, complete intersection local ring, differential graded Hopf algebra, Bernstein-Gelfand-Gelfand correspondence.
Research partly supported by NSF grants DMS 0803082 (L.L.A) and DMS 0602498 (S.B.I). Part of the work of the second author was done at the University of Paderborn on a visit supported by a Forschungspreis from the Humboldtstiftung.

Section 6 we deduce them from results on differential graded Hopf algebras, presented with complete proofs in Section 4. These pleasingly simple proofs build on nothing more than basic homological algebra, summarized in Section 3.

In the last three sections of this article the goal is to link the complexity of $M$ and the Loewy length of its homology modules. In [5] such a result is deduced from a more general statement, which applies to arbitrary local rings. From that paper we import DG versions of the New Intersection Theorem, recalled in Section 7, and of the Bernstein-Gelfand-Gelfand correspondence, which we refine in Section 8. This allows us to establish a result on complexities over exterior algebras, which we then translate in Section 9 into the desired link between Loewy length and complexity over complete intersection local rings.

# 1. Levels

In this section $A$ denotes a DG algebra.

We write $\mathsf{D}(A)$ for the derived category of DG $A$-modules; it is a triangulated category with shift functor denoted $\Sigma$. The underlying graded object of a DG object $X$ is denoted $X^\natural$. Rings are treated as DG algebras concentrated in degree zero; over a ring DG modules are simply complexes, and modules are DG modules concentrated in degree zero; see [5, §3] for more details and references. Unless specified otherwise, all DG modules have left actions.

A non-empty subcategory $\mathsf{C}$ of $\mathsf{D}(A)$ is said to be *thick* if it is an additive subcategory closed under retracts, and every exact triangle in $\mathsf{D}(A)$ with two vertices in $\mathsf{C}$ has its third vertex in $\mathsf{C}$; thick subcategories are triangulated.

Given a DG $A$-module $X$, we write $\mathsf{thick}_A(X)$ for the smallest thick subcategory containing $X$. The existence of such a subcategory can be seen by realizing it as the intersection of all thick subcategories of $\mathsf{D}(A)$ that contain $X$. Alternatively, the objects of $\mathsf{thick}_A(X)$ can be built from $X$ in a series of steps, described below.

**1.1.** For every $X$ in $\mathsf{D}(A)$ and each $n \geq 0$ we define a full subcategory $\mathsf{thick}_A^n(X)$ of $\mathsf{thick}_A(X)$, called the *$n$th thickening* of $X$ in $\mathsf{A}$, as follows. Set $\mathsf{thick}_A^0(X) = \{0\}$; the objects of $\mathsf{thick}_A^1(X)$ are the retracts of finite direct sums of shifts of $X$. For each $n \geq 2$, the objects of $\mathsf{thick}_A^n(X)$ are retracts of those $U \in \mathsf{D}(A)$ that appear in some exact triangle $U' \to U \to U'' \to \Sigma U'$ with $U' \in \mathsf{thick}_A^{n-1}(X)$ and $U'' \in \mathsf{thick}_A^1(X)$.

Every thickening of $X$ is clearly embedded in the next one; it is also clear that their union is a thick subcategory of $\mathsf{A}$ containing $X$, which is therefore equal to $\operatorname{thick}_A(X)$. Thus, $\operatorname{thick}_A(X)$ is equipped with a filtration

$$\{0\} = \operatorname{thick}_A^0(X) \subseteq \operatorname{thick}_A^1(X) \subseteq \cdots \subseteq \bigcup_{n \in \mathbb{N}} \operatorname{thick}_A^n(X) = \operatorname{thick}_A(X)\,.$$

To each object $U$ in $\mathsf{D}(A)$ we associate the number

$$\operatorname{level}_A^X(U) = \inf\{n \in \mathbb{N} \mid U \in \operatorname{thick}_A^n(X)\}$$

and call it the *X-level of U in* $\mathsf{D}(A)$. It measures the number of extensions needed to build $U$ out of $X$. Evidently, $\operatorname{level}_A^X(U) < \infty$ is equivalent to $U \in \operatorname{thick}_A(X)$.

We refer the reader to [5, §§2–6] for a systematic study of levels. The properties used explicitly in this paper are recorded below.

**1.2.** Let $U$, $X$, and $Y$ be objects in $\mathsf{D}(A)$.

(1) If $\operatorname{thick}_A^1(X) = \operatorname{thick}_A^1(Y)$, then one has $\operatorname{thick}_A^n(X) = \operatorname{thick}_A^n(Y)$ for all $n$, and hence an equality $\operatorname{level}_A^X(U) = \operatorname{level}_A^Y(U)$.

(2) If $B$ is a DG algebra and $\mathsf{j}\colon \mathsf{D}(A) \to \mathsf{D}(B)$ is an exact functor, then

$$\operatorname{level}_A^X(U) \geq \operatorname{level}_B^{\mathsf{j}(X)}(\mathsf{j}(U))\,.$$

Equality holds when $\mathsf{j}$ is an equivalence.

A level of interest in this paper is related to the notion of Loewy length.

**1.3.** Let $B$ be a ring and $k$ a simple $B$-module. For each $B$-module $H$, $\ell\ell_B^k H$ denotes the smallest integer $l \geq 0$ such that $H$ has a filtration by $B$-submodules

$$0 = H^0 \subseteq H^1 \subseteq \cdots \subseteq H^{l-1} \subseteq H^l = H$$

with every $H^i/H^{i-1}$ isomorphic to a sum of copies of $k$; if no such filtration exists, we set $\ell\ell_B^k H = \infty$. If $B$ has a unique maximal left ideal $\mathfrak{m}$, then $k$ is isomorphic to $B/\mathfrak{m}$ and $\ell\ell_B^k H$ is equal to the *Loewy length* of $H$, defined by the formula

$$\ell\ell_B H = \inf\{l \in \mathbb{N} \mid \mathfrak{m}^l H = 0\}\,;$$

see [5, 6.1.3]. If $\operatorname{length}_B H$ is finite, then so is $\ell\ell_B H$, and the converse holds when $H$ is noetherian; see [5, 6.2(4)].

We say that the DG algebra $A$ is *non-negative* if $A_i = 0$ for $i < 0$. When this is the case, there is a canonical morphism of DG algebras $A \to \mathsf{H}_0(A)$, called the *augmentation* of $A$; it turns every $\mathsf{H}_0(A)$-module into a DG $A$-module.

**1.4.** When $A$ is non-negative and $k$ is a simple $H_0(A)$-module one has:

(1) The number $\operatorname{level}_A^k(U)$ is finite if and only if the $H_0(A)$-module $\bigoplus_{n\in\mathbb{Z}} H_n(U)$ admits a finite filtration with subfactors isomorphic to $k$.

(2) When $\operatorname{level}_A^k(U)$ is finite the following inequality holds:

$$\operatorname{level}_A^k(U) \leq \sum_{n\in\mathbb{Z}} \ell\ell_{H_0(A)}^k H_n(U).$$

(3) When $H_0(A) = k$ and the $k$-module $H(U)$ is finite, one has an inequality

$$\operatorname{level}_A^k(U) \leq \operatorname{card}\{n \in \mathbb{Z} \mid H_n(U) \neq 0\}.$$

Indeed, (1) and (2) are contained in [5, 6.2(3)], while (3) is extracted from [5, 6.4].

## 2. Complexities

In this section we introduce a notion of complexity for DG modules over a suitable class of DG algebras, and establish some of its elementary properties. We begin with a reminder of the construction of derived functors on the derived category of DG modules over a DG algebra; see [6, §1] for details.

**2.1.** Let $A$ be a DG algebra and $A^\circ$ its opposite DG algebra.

A *semifree filtration* of a DG $A$-module $F$ is a filtration

$$0 = F^0 \subseteq F^1 \subseteq \cdots \subseteq F^{n-1} \subseteq F^n \subseteq \cdots$$

by DG submodules with $\bigcup_{n\geq 0} F^n = F$ and each DG module $F^n/F^{n-1}$ isomorphic to a direct sum of suspensions of $A$; when one exists $F$ is said to be semifree. When $F$ is semifree its underlying graded $A^\natural$-module $F^\natural$ is free, and the functors $\operatorname{Hom}_A(F, -)$ and $(-\otimes_A F)$ preserve quasi-isomorphisms of DG $A$-modules.

A *semifree resolution* of a DG module $U$ is a quasi-isomorphism $F \to U$ of DG modules with $F$ semifree; such a resolution always exists. If $U \to V$ is a quasi-isomorphism of DG $A$-modules and $G \to V$ is a semifree resolution, then there is a unique up to homotopy morphism $F \to G$ of DG $A$-modules such that the composed maps $F \to G \to V$ and $F \to U \to V$ are homotopic.

These properties imply that after choosing a resolution $F_U \to U$ for each $U$, the assignments $(U, V) \mapsto \operatorname{Hom}_A(F_U, V)$ and $(W, U) \mapsto W \otimes_A F_U$ define exact functors

$$\operatorname{RHom}_A(-, -)\colon \mathsf{D}(A)^\circ \times \mathsf{D}(A) \to \mathsf{D}(\mathbb{Z}) \quad \text{and}$$
$$- \otimes_A^{\mathsf{L}} -\colon \mathsf{D}(A^\circ) \times \mathsf{D}(A) \to \mathsf{D}(\mathbb{Z}),$$

respectively. In homology, they define graded abelian groups

$$\mathrm{Ext}_A(U, V) = \mathrm{H}(\mathrm{RHom}_A(U, V)) \quad \text{and} \quad \mathrm{Tor}^A(W, U) = \mathrm{H}(W \otimes_A^{\mathbf{L}} U).$$

In case of modules over rings these are the classical objects.

We want to measure, on a polynomial scale, how the 'size' of $\mathrm{Ext}_A^n(U, V)$ grows when $n$ goes to infinity. In order to do this we use the notion of *complexity* of a sequence $(b_n)_{n \in \mathbb{Z}}$ of non-negative numbers, defined by the following equality

$$\mathrm{cx}\,(b_n) := \inf \left\{ d \in \mathbb{N} \,\middle|\, \begin{array}{l} \text{there is a number } a \in \mathbb{R} \text{ such that} \\ b_n \leq an^{d-1} \text{ holds for all } n \gg 0 \end{array} \right\}.$$

Throughout the rest of this section, $(A, \mathfrak{m}, k)$ denotes a *local DG algebra*, by which we mean that $A$ is non-negative, $A_0$ is a noetherian ring contained in the center of $A$, $\mathfrak{m}$ is the unique maximal ideal of $A_0$, and $k$ is the field $A_0/\mathfrak{m}$.

The *complexity* of a pair $(U, V)$ of DG $A$-modules is the number

$$\mathrm{cx}_A\,(U, V) = \mathrm{cx}\left(\mathrm{rank}_k(\mathrm{Ext}_A^n(U, V) \otimes_{A_0} k)\right).$$

The *complexity* and the *injective complexity* of $U$ are defined, respectively, by

$$\mathrm{cx}_A\, U = \mathrm{cx}_A\,(U, k) \quad \text{and} \quad \mathrm{inj}\,\mathrm{cx}_A\, U = \mathrm{cx}_A\,(k, U).$$

When $A$ is a local ring $\mathrm{cx}_A\, U$ is the polynomial rate of growth of $\mathrm{rank}_A F_n$, where $F$ is a minimal free resolution of $U$, while $\mathrm{inj}\,\mathrm{cx}_A\, U$ is that of the multiplicity of an injective envelope of $k$ in $I^n$, where $I$ is a minimal injective resolution of $U$.

The following properties of these invariants have been observed before for modules, and for complexes, over local rings. We extend them to handle DG modules over DG algebras, sometimes with alternative proofs.

**Lemma 2.2.** *If $U$ is in* $\mathrm{thick}_A(X)$ *for some DG $A$-module $X$, then one has*

$$\mathrm{cx}_A\, U \leq \mathrm{cx}_A\, X \quad \text{and} \quad \mathrm{inj}\,\mathrm{cx}_A\, U \leq \mathrm{inj}\,\mathrm{cx}_A\, X.$$

*Equalities hold in case* $\mathrm{thick}_A(U) = \mathrm{thick}_A(X)$.

*Proof.* We verify the inequality for complexities; a symmetric argument yields the one for injective complexities.

The number $h = \mathrm{level}_A^X(U)$ is finite; we induce on it. When $h$ equals one $U$ is a retract of $\bigoplus_{j=1}^s \Sigma^{i_j} X$ for some integers $i_1, \ldots, i_s$, so $\mathrm{Ext}_A^n(U, k)$ is a direct summand of $\bigoplus_{j=1}^s \mathrm{Ext}_A^{n+i_j}(X, k)$, and the desired inequality is clear. For $h \geq 1$ one has an exact triangle $U' \to U \to U'' \to \Sigma U'$ with $\mathrm{level}_A^X(U') \leq h - 1$

and $\operatorname{level}_A^X(U'') = 1$; the associated cohomology exact sequence shows that the induction hypothesis implies the desired inequality for complexities. $\qquad\square$

**Lemma 2.3.** *If $F$ is a finite free complex of $A_0$-modules with $\mathrm{H}(F) \neq 0$, then*

$$\operatorname{cx}_A (U \otimes_{A_0} F) = \operatorname{cx}_A U \quad and \quad \operatorname{inj\,cx}_A (U \otimes_{A_0} F) = \operatorname{inj\,cx}_A U .$$

*Proof.* Since $F$ is a finite free complex, in $\mathsf{D}(A)$ one has isomorphisms

$$\mathsf{RHom}_A(k, U \otimes_{A_0} F) \simeq \mathsf{RHom}_A(k, U) \otimes_{A_0}^{\mathsf{L}} F \simeq \mathsf{RHom}_A(k, U) \otimes_k^{\mathsf{L}} (k \otimes_{A_0} F) .$$

This yields, for each integer $n$, an isomorphism of $k$-vectorspaces

$$\operatorname{Ext}_A^n(k, U \otimes_{A_0} F) \cong \bigoplus_{i \in \mathbb{Z}} \operatorname{Ext}_A^{n+i}(k, U) \otimes_k \mathrm{H}_i(k \otimes_{A_0} F) .$$

Since $\mathrm{H}_i(k \otimes_{A_0} F)$ is finite for each $i$, and zero for $|i| \gg 0$ but not for all $i$, the equality of injective complexities follows. The argument for complexities is similar. $\qquad\square$

**2.4.** For a morphism of DG algebras $\varphi \colon A \to B$, let $\varphi_*$ be the functor forgetting the action of $B$. One then has an adjoint pair of exact functors

$$\mathsf{D}(A) \xrightleftharpoons[\varphi_*]{(B \otimes_A^{\mathsf{L}} -)} \mathsf{D}(B)$$

that are inverse equivalences if $\varphi$ is a quasi-isomorphism; see [5, 3.3.1, 3.3.2]. If, furthermore, $U$ is a DG $A$-module, $V$ is a DG $B$-module, and $\mu \colon U \to V$ is a $\varphi$-equivariant quasi-isomorphism, then by [5, 3.3.3] one has canonical isomorphisms

$$U \simeq \varphi_*(V) \qquad and \qquad B \otimes_A^{\mathsf{L}} U \simeq V .$$

A DG algebra $B$ is said to be *quasi-isomorphic* to $A$ if there exists a chain

$$A \xrightarrow{\ \simeq\ } A^0 \xleftarrow{\ \simeq\ } A^1 \xrightarrow{\ \simeq\ } \cdots \xleftarrow{\ \simeq\ } A^{i-1} \xrightarrow{\ \simeq\ } A^i \xleftarrow{\ \simeq\ } B$$

of quasi-isomorphisms of DG algebras. Such a chain induces a unique equivalence

$$\mathsf{j} \colon \mathsf{D}(A) \to \mathsf{D}(B) ,$$

of triangulated categories; if $B_i = 0$ for $i < 0$, then [5, 3.6] yields

$$\mathsf{j}(\mathrm{H}_0(A)) \simeq \mathrm{H}_0(B) .$$

**Lemma 2.5.** *If $(B, \mathfrak{n}, l)$ is a local DG algebra quasi-isomorphic to $A$, then each equivalence of derived categories* $\mathsf{j} \colon \mathsf{D}(A) \to \mathsf{D}(B)$ *as in 2.4 induces an isomorphism*

$$\mathsf{j}(k) \simeq l$$

*in $\mathsf{D}(B)$, and for all $U$ and $V$ in $\mathsf{D}(A)$ there is an equality*

$$\mathrm{cx}_A(U, V) = \mathrm{cx}_B(\mathsf{j}(U), \mathsf{j}(V)).$$

*Proof.* As $\mathrm{H}_0(A)$ is equal to $A_0/\partial(A_1)$, it is a local ring with maximal ideal $\mathfrak{m}/\partial(A_1)$. Similarly, $\mathrm{H}_0(B) = B_0/\partial(B_1)$ is a local ring with maximal ideal $\mathfrak{n}/\partial(B_1)$. The isomorphism of 2.4 maps $\mathsf{j}(\mathfrak{m}/\partial(A_1))$ to $\mathfrak{n}/\partial(B_1)$, and so induces $\mathsf{j}(k) \simeq l$.

Since $\mathsf{j}$ is an equivalence, in $\mathsf{D}(B)$ there is an isomorphism

$$\mathsf{j}(\mathrm{RHom}_A(U, V)) \simeq \mathrm{RHom}_B(\mathsf{j}U, \mathsf{j}V).$$

Passing to homology one obtains an isomorphism of graded modules

$$\mathrm{Ext}_A(U, V) \cong \mathrm{Ext}_B(\mathsf{j}U, \mathsf{j}V)$$

that is equivariant over the isomorphism $\mathrm{H}(A) \cong \mathrm{H}(B)$ of graded algebras induced by the chain of quasi-isomorphisms inducing $\mathsf{j}$. This yields an isomorphism

$$\mathrm{Ext}_A(U, V) \otimes_{\mathrm{H}_0(A)} k \cong \mathrm{Ext}_B(\mathsf{j}U, \mathsf{j}V) \otimes_{\mathrm{H}_0(B)} l$$

of graded vector spaces that is equivariant over the isomorphism $k \cong l$.   $\square$

Sometimes there exist an alternative way for computing complexity:

**Lemma 2.6.** *If the ring $A_0$ is artinian, then there is an equality*

$$\mathrm{cx}_A(U, V) = \mathrm{cx}\,(\mathrm{length}_{A_0} \mathrm{Ext}_A^n(U, V)).$$

*Proof.* Set $r_n = \mathrm{rank}_k(\mathrm{Ext}_A^n(U, V) \otimes_{A_0} k)$. By hypothesis $\mathfrak{m}^s = 0$ holds for some integer $s$, so by Nakayama's Lemma the $A_0$-module $\mathrm{Ext}_A^n(U, V)$ is minimally generated by $r_n$ elements. Thus, one has surjective homomorphisms

$$k^{r_n} \leftarrow \mathrm{Ext}_A^n(U, V) \leftarrow A_0^{r_n}$$

of $A_0$-modules. They yield inequalities

$$r_n \leq \mathrm{length}_{A_0} \mathrm{Ext}_A^n(U, V) \leq \mathrm{length}_{A_0}(A_0) r_n$$

that evidently imply $\mathrm{cx}\,(r_n) = \mathrm{cx}\,(\mathrm{length}_{A_0} \mathrm{Ext}_A^n(U, V))$.   $\square$

## 3. Composition products

Let $A$ denote a DG algebra, and let $U$, $V$, and $W$ be DG $A$-modules. In this section we recall the construction of products in cohomology of DG $A$-modules.

**Construction 3.1.** For all integers $i$, $j$ there exist *composition products*

$$\mathrm{Ext}_A^j(V, W) \times \mathrm{Ext}_A^i(U, V) \to \mathrm{Ext}_A^{i+j}(U, W).$$

Indeed, choose a semifree resolution $F \to U$; thus $\mathrm{Ext}_A^i(U, V) = \mathrm{H}^i(\mathrm{Hom}_A(F, V))$. An element $[\alpha] \in \mathrm{Ext}_A^i(U, V)$ is then the homotopy class of a degree $-i$ chain map $\alpha \colon F \to V$ of DG modules. Similarly, let $[\beta] \in \mathrm{Ext}_A^j(V, W)$ be the class of a chain map $\beta \colon G \to W$ of degree $-j$, where $G \to V$ is a semifree resolution. As $F$ is semifree and $F \to U$ is a quasi-isomorphism there is a unique up to homotopy chain map $\widetilde{\alpha} \colon F \to G$ whose composition with $G \to V$ is equal to $\alpha$. One defines the product $[\beta][\alpha]$ in $\mathrm{Ext}_A^{i+j}(U, W)$ to be the class of $\beta \circ \widetilde{\alpha} \colon F \to W$.

It follows from their construction that composition products are associative in the obvious sense. In particular, $\mathrm{Ext}_A(U, U)$ and $\mathrm{Ext}_A(V, V)$ are graded DG algebras, and that $\mathrm{Ext}_A(U, V)$ is a graded left-right $\mathrm{Ext}_A(V, V)$-$\mathrm{Ext}_A(U, U)$-bimodule.

Let $A$ and $B$ be DG algebras over a commutative ring $k$. Their tensor product is a DG algebra, with underlying complex $A \otimes_k B$ and multiplication defined by

$$(a \otimes b) \cdot (a' \otimes b') = (-1)^{|b||a'|}(aa' \otimes bb'),$$

where $|a|$ denotes the degree of $a$. For each DG $B$-module $X$ the complexes $U \otimes_k X$ and $\mathrm{Hom}_k(X, U)$ have canonical structures of DG module over $A \otimes_k B$, given by

$$(a \otimes b) \cdot (u \otimes x) = (-1)^{|b||u|}(au \otimes bx);$$
$$\big((a \otimes b)(\gamma)\big)(x) = (-1)^{|b||\gamma|}a\gamma(bx).$$

**Construction 3.2.** Let $k$ be a field, and let $A$ and $B$ be DG algebras with

$$A_i = 0 = B_i \quad \text{for} \quad i < 0 \quad \text{and} \quad A_0 = k = B_0.$$

Let $F \to k$ and $G \to k$ be semifree resolutions over $A$ and $B$, respectively, and

$$\omega \colon \mathrm{Hom}_A(F, k) \otimes_k \mathrm{Hom}_B(G, k) \to \mathrm{Hom}_{A \otimes_k B}(F \otimes_k G, k)$$

be the morphism given by $\big(\omega(\alpha \otimes \beta)\big)(a \otimes b) = (-1)^{|\beta||a|}\alpha(a)\beta(b)$. Composing $\mathrm{H}(\omega)$ with the *Künneth isomorphism* $[\alpha] \otimes_k [\beta] \mapsto [\alpha \otimes \beta]$ one obtains a homomorphism

$$\mathrm{H}(\mathrm{Hom}_A(F, k)) \otimes_k \mathrm{H}(\mathrm{Hom}_B(G, k)) \to \mathrm{H}(\mathrm{Hom}_{A\otimes_k B}(F \otimes_k G, k))$$

The map $F \otimes_k G \to k \otimes_k k = k$ evidently is a semifree resolution over $A \otimes_k B$, so the preceding homomorphisms defines a *Künneth homomorphism*

$$\kappa : \mathrm{Ext}_A(k, k) \otimes_k \mathrm{Ext}_B(k, k) \to \mathrm{Ext}_{A\otimes_k B}(k, k).$$

It is follows from the definition that this is a homomorphism of $k$-algebras.

**Lemma 3.3.** *If $\mathrm{rank}_k A_i$ is finite for each $i$, then $\kappa$ is an isomorphism.*

*Proof.* Under the hypothesis on $A$, one can choose an $F$ that has a semifree filtration where each free graded $A^\natural$-module $(F^n/F^{n-1})^\natural$ of finite rank; for instance, take $F$ to be the classical bar-construction, see [12]. In this case the map $\omega$ from Construction 3.2 is bijective, so $\mathrm{H}(\omega)$ is an isomorphism. $\square$

## 4. Differential Graded Hopf algebras

In this section $k$ is a field and $A$ is a *DG Hopf algebra* over $k$; that is, $A$ is a non-negative DG algebra with $A_0 = k$, and with a morphism $\Delta : A \to A \otimes_k A$ of DG $k$-algebras, called the *comultiplication*, satisfying $(\varepsilon^A \otimes A)\Delta = \mathrm{id}^A = (A \otimes_k \varepsilon^A)\Delta$, where $\varepsilon^A : A \to k$ is the canonical surjection.

The principal results in this section, Theorems 4.7 and 4.9, establish symmetry properties of complexities. The arguments in the proof exploit operations on DG modules provided by the Hopf algebra structure on $A$.

**4.1.** When $A$ is a DG Hopf algebra the DG $(A \otimes_k A)$-modules $U \otimes_k V$ and $\mathrm{Hom}_k(U, V)$ acquire structures of DG $A$-modules through $\Delta$. The isomorphisms

$$U \otimes_k k \cong U \cong k \otimes_k U \tag{4.1.1}$$

$$\mathrm{Hom}_k(k, V) \cong V \tag{4.1.2}$$

$$\mathrm{Hom}_A(U \otimes_k V, W) \cong \mathrm{Hom}_A(U, \mathrm{Hom}_k(V, W)) \tag{4.1.3}$$

are compatible with the respective DG $A$-structures.

**Lemma 4.2.** *Let $F \to k$ be a semifree resolution and $G$ a semifree DG module. The induced morphism $\pi : F \otimes_k G \to k \otimes_k G \cong G$ is a homotopy equivalence.*

*Proof.* The morphism $\pi$ is a quasi-isomorphism because $k$ is a field. As $G$ is semifree, the identity map on $G$ lifts through $\pi$ to give a morphism

$\kappa : G \to F \otimes_k G$ such that $\pi \kappa$ is homotopic to $\mathrm{id}^G$; see 2.1. Since $F$ is semifree it follows from (4.1.3) that the functor $\mathrm{Hom}_A(F \otimes_k G, -)$ preserves quasi-isomorphisms. Noting that $\pi \kappa \pi$ is homotopic to $\pi \, \mathrm{id}^{F \otimes_k G}$, one thus gets that $\kappa \pi$ is homotopic to $\mathrm{id}^{F \otimes_k G}$. $\square$

**Construction 4.3.** Let $A$ be a DG Hopf algebra over $k$ and set $S = \mathrm{Ext}_A(k, k)$. Let $V$ be a DG $A$-module. Choose semifree resolutions $F \to k$ and $G \to V$ over $A$. The assignment $\psi \mapsto \psi \otimes_k G$ defines a morphism

$$\mathrm{Hom}_A(F, F) \to \mathrm{Hom}_A(F \otimes_k G, F \otimes_k G)$$

of DG algebras. Since $F \otimes_k G$ is homotopy equivalent to $G$, by Lemma 4.2, in homology it induces a homomorphism of graded $k$-algebras

$$\zeta_V : S \to \mathrm{Ext}_A(V, V).$$

The graded center of a graded algebra $B$ consists of the elements $c$ in $B$ that satisfy $bc = (-1)^{|b||c|}cb$ for all $b \in B$. When every $c \in B$ has this property $B$ is said to be *graded-commutative*; every graded right $B$-module $X$ then has a canonical structure of left $B$-module, defined by setting $bx = (-1)^{|b||x|}xb$.

**Proposition 4.4.** *Let $A$ be a DG Hopf $k$-algebra, and let $U, V$ be DG $A$-modules.*

*The algebra $S = \mathrm{Ext}_A(k, k)$ is then graded-commutative and $\zeta_V(S)$ is contained in the graded center of $\mathrm{Ext}_A(V, V)$. Moreover, the $S$-module structures on $\mathrm{Ext}_A(U, V)$ induced via $\zeta_V$ and $\zeta_U$ coincide.*

*Proof.* We verify the second assertion; the first one follows, as $\zeta_k = \mathrm{id}^S$.

We may assume $U$ and $V$ are semifree DG $A$-modules. Let $F \to k$ be a semifree resolution. By Lemma 4.2, the homology of the complex $\mathrm{Hom}_A(F \otimes_k U, F \otimes_k V)$ is $\mathrm{Ext}_A(U, V)$. It implies also that any chain map $F \otimes_k U \to F \otimes_k V$ is homotopy equivalent to one of the form $F \otimes_k \mu$. For any chain map $\sigma : F \to F$ the compositions

$$F \otimes_k U \xrightarrow{\sigma \otimes U} F \otimes_k U \xrightarrow{F \otimes \mu} F \otimes_k V$$

$$F \otimes_k U \xrightarrow{F \otimes \mu} F \otimes_k V \xrightarrow{\sigma \otimes V} F \otimes_k V$$

are $\sigma \otimes \mu$ and $(-1)^{|\mu||\sigma|}\sigma \otimes \mu$, respectively. Hence the $S$-actions on $\mathrm{Ext}_A(U, V)$ induced via $\zeta_U$ and $\zeta_V$ coincide up to the usual sign. This is the desired result. $\square$

**Proposition 4.5.** *Let $A$ be a DG Hopf $k$-algebra, and $U$ and $V$ be DG $A$-modules.*

*If the $k$-algebra $S = \mathrm{Ext}_A(k, k)$ is finitely generated, and the $k$-vector spaces $\mathrm{H}(U)$ and $\mathrm{H}(V)$ are finite, then $\mathrm{Ext}_A(U, V)$ is finite over $S$, $\mathrm{Ext}_A(U, U)$, and $\mathrm{Ext}_A(V, V)$.*

*Proof.* Since $S$ acts on $\mathrm{Ext}_A(U, V)$ through $\mathrm{Ext}_A(U, U)$, finiteness over the former implies finiteness over the latter. The same reasoning applies with $\mathrm{Ext}_A(U, U)$ replaced by $\mathrm{Ext}_A(V, V)$, so it suffices to prove finiteness over $S$.

We claim that the following full subcategory of $\mathsf{D}(A)$ is thick:

$$\mathsf{C} = \{W \in \mathsf{D}(A) \mid \text{the } S\text{-module } \mathrm{Ext}_A(W, k) \text{ is noetherian}\}.$$

Indeed, it is clear that $\mathsf{C}$ is closed under retracts and shifts; furthermore, every exact triangle $W' \to W \to W'' \to \Sigma W'$ in $\mathsf{D}(A)$ induces an exact sequence

$$\mathrm{Ext}_A(W'', k) \to \mathrm{Ext}_A(W, k) \to \mathrm{Ext}_A(W', k)$$

of graded $S$-modules, so if $W'$ and $W''$ are in $\mathsf{C}$, then so is $W$.

The condition $\mathrm{rank}_k \mathrm{H}(U) < \infty$ implies $U \in \mathrm{thick}_A(k)$; see 1.4(1). The finitely generated $k$-algebra $S$ is graded-commutative by Proposition 4.4, and thus noetherian; this implies $k \in \mathsf{C}$. Now from the thickness of $\mathsf{C}$ we conclude $U \in \mathsf{C}$.

From a similar argument, now considering the subcategory

$$\{W \in \mathsf{D}(A) \mid \text{the } S\text{-module } \mathrm{Ext}_A(U, W) \text{ is noetherian}\},$$

one deduces that the $S$-module $\mathrm{Ext}_A(U, V)$ is finitely generated. $\qquad\square$

The preceding result allows one to draw conclusions about complexities from classical results in commutative algebra. This may not be immediately clear to casual users of the subject, as standard textbooks leave out important parts of the story by focusing early on on 'standard graded' algebras. In the following discussion we refer to Smoke's very readable and self-contained exposition in [15].

**4.6.** Let $S$ be a graded-commutative $k$-algebra, generated over $k$ by finitely many elements of positive degrees, and $C$ a finitely graded $S$-module. Replacing generators by their squares, one sees that $C$ is also finite over a finitely generated *commutative* $k$-algebra $S'$, so there exists a Laurent polynomial $p_C(t) \in \mathbb{Z}[t^{\pm 1}]$, such that

$$\sum_{n \in \mathbb{Z}} \mathrm{rank}_k(C_n) t^n = p_C(t) \Big/ \prod_{j=1}^{c}(1 - t^{d_j});$$

this is the Hilbert-Serre Theorem, see [15, 4.2] or [13, 13.2]. By [15, 5.5], the order of the pole at $t = 1$ of the rational function above is equal to the *Krull dimension* of $C$; that is, to the supremum of the lengths of chains of homogeneous prime ideals in $S'$ containing the annihilator of $C$. The order of the pole is independent of the choice of $S'$, so one gets a well defined notion of Krull dimension of $C$ over $S$; let $\dim_S C$ denote this number, and set $\dim S = \dim_S S$. By decomposing the rational function above into prime fractions, see [3, §2] for details, one easily obtains

$$\dim_S C = \operatorname{cx}\left(\operatorname{rank}_k(C_n)\right).$$

**Theorem 4.7.** *If $A$ is a DG Hopf $k$-algebra with $S = \operatorname{Ext}_A(k, k)$ finitely generated as $k$-algebra, and $U$ a DG $A$-module with $\mathrm{H}(U)$ finite over $k$, then one has*

$$\operatorname{inj}\operatorname{cx}_A U = \operatorname{cx}_A(U, U) = \operatorname{cx}_A U \leq \operatorname{cx}_A k = \dim S.$$

*If, furthermore, $V$ is a DG $A$-module with $\mathrm{H}(V)$ finite over $k$, then one has*

$$\dim_S(\operatorname{Ext}_A(U, V)) = \operatorname{cx}_A(U, V) \leq \min\{\operatorname{cx}_A(U, U), \operatorname{cx}_A(V, V)\}.$$

*Proof.* The expression for $\operatorname{cx}_A(U, V)$ comes from Propositions 4.4, 4.5, and 4.6. By Proposition 4.5, $\operatorname{Ext}_A(U, V)$ is a finite module over $\operatorname{Ext}_A(U, U)$ and $\operatorname{Ext}_A(V, V)$, whence the upper bounds on $\operatorname{cx}_A(U, V)$. For $(k, U)$ and $(U, k)$ they yield

$$\operatorname{inj}\operatorname{cx}_A U \leq \operatorname{cx}_A(U, U) \geq \operatorname{cx}_A U \leq \operatorname{cx}_A k.$$

To prove $\operatorname{cx}_A(U, U) \leq \operatorname{cx}_A U$ we show that the full subcategory

$$\mathsf{D} = \{W \in \mathsf{D}(A) \mid \operatorname{cx}_A(U, W) \leq \operatorname{cx}_A U\}$$

of $\mathsf{D}(A)$ contains $U$. One evidently has $k \in \mathsf{D}$, and 1.4(1) gives $U \in \operatorname{thick}_A(k)$, so it suffices to prove that $\mathsf{D}$ is thick. Closure under direct summands and shifts is clear. An exact triangle $W' \to W \to W'' \to \Sigma W'$ in $\mathsf{D}(A)$ yields an exact sequence

$$\operatorname{Ext}_A^n(U, W') \to \operatorname{Ext}_A^n(U, W) \to \operatorname{Ext}_A^n(U, W'')$$

for every $n \in \mathbb{Z}$. They imply inequalities

$$\operatorname{rank}_k\left(\operatorname{Ext}_A^n(U, W)\right) \leq \operatorname{rank}_k\left(\operatorname{Ext}_A^n(U, W')\right) + \operatorname{rank}_k\left(\operatorname{Ext}_A^n(U, W'')\right).$$

Thus, if $W'$ and $W''$ are in $\mathsf{D}$, then, Lemma 2.6 gives

$$\operatorname{cx}_A(U, W) \leq \max\{\operatorname{cx}_A(U, W'), \operatorname{cx}_A(U, W'')\} \leq \operatorname{cx}_A U.$$

A similar argument, now using $\operatorname{Ext}_A(-, U)$, yields $\operatorname{inj}\operatorname{cx}_A U = \operatorname{cx}_A(U, U)$.

The equality $\mathrm{cx}_A\, k = \dim S$ comes from the Hilbert-Serre theorem, see 4.6. $\qquad\square$

To compare $\mathrm{cx}_A(U, V)$ and $\mathrm{cx}_A(V, U)$ we need one more lemma.

**Lemma 4.8.** *Let $A$ be a DG Hopf $k$-algebra. For all $U, V \in \mathsf{D}(A)$ the assignment $(u \otimes v) \mapsto [\alpha \mapsto (-1)^{|v||\alpha|} u\alpha(v)]$ defines a morphism*

$$U \otimes_k V \to \mathrm{Hom}_k(V^*, U)$$

*in $\mathsf{D}(A)$; it is an isomorphism when $\mathrm{rank}_k\, \mathrm{H}(V)$ is finite.*

*Proof.* It is a routine calculation to verify that the map in question is compatible with the DG $A$-module structures. It thus defines for each $X \in \mathsf{D}(A)$ a natural morphism $\eta_X \colon U \otimes_k X \to \mathrm{Hom}_k(X^*, U)$. It is easy to see that those $X$ for which $\eta_X$ is an isomorphism form a thick subcategory of $\mathsf{D}(A)$. It contains $k$ and hence contains every DG $A$-module $V$ with $\mathrm{rank}_k\, \mathrm{H}(V)$ finite; see 1.4(1). $\qquad\square$

The next result extends the equality $\mathrm{cx}_A U = \mathrm{inj}\, \mathrm{cx}_A U$ from Theorem 4.7.

**Theorem 4.9.** *If $A$ is a DG Hopf $k$-algebra with $\mathrm{Ext}_A(k, k)$ finitely generated as $k$-algebra, and $U$ and $V$ are DG $A$-modules with $\mathrm{H}(U)$ and $\mathrm{H}(V)$ finite over $k$, then*

$$\mathrm{cx}_A(U, V) = \mathrm{cx}_A(V, U).$$

*Proof.* The desired assertion results from the following chain of equalities:

$$
\begin{aligned}
\mathrm{cx}_A(U, V) &= \mathrm{cx}_A(U, V^{**}) \\
&= \mathrm{cx}_A(U \otimes_k V^*, k) \\
&= \mathrm{cx}_A(k, U \otimes_k V^*) \\
&= \mathrm{cx}_A(k, \mathrm{Hom}_k(V, U)) \\
&= \mathrm{cx}_A(V, U).
\end{aligned}
$$

The first and fourth ones come, respectively, from isomorphisms $V \simeq V^{**}$ and $U \otimes_k V^* \simeq \mathrm{Hom}_k(V, U)$ given by Lemma 4.8. The second and fifth ones one follow from the adjunction isomorphism (4.1.3). Theorem 4.7 supplies the middle link. $\qquad\square$

We have not yet provided any non-trivial example of DG Hopf $k$-algebra $A$ with finitely generated cohomology algebra. To state a general result in this direction, recall that $A$ is said to be *cocommutative* if its comultiplication satisfies the equality $\Delta = \tau\Delta$, where $\tau\colon A \otimes_k A \to A \otimes_k A$ is defined by $\tau(a \otimes b) = (-1)^{|a||b|} b \otimes a$.

The following result is Wilkerson's main theorem in [17]:

**4.10.** If $A$ is a cocommutative DG Hopf $k$-algebra with zero differential and with $\mathrm{rank}_k A$ finite, then the graded $k$-algebra $S = \mathrm{Ext}_A(k, k)$ is finitely generated.

In positive characteristic the proof of the theorem depends on the action of the Steenrod algebra on $S$; the existence of such an action is another non-trivial result, see [11]. The situation is completely different in characteristic zero, where a celebrated theorem of Hopf shows that every non-negative cocommutative graded Hopf algebra is isomorphic, as an algebra, to the exterior algebra on a vector space of finite rank; in this case the cohomology is well known, and is computed next.

## 5. Exterior algebras I.

In this section $k$ is a field and $\Lambda$ a DG algebra with $\partial^\Lambda = 0$ and with $\Lambda$ an exterior $k$-algebra on alternating indeterminates $\xi_1, \ldots, \xi_c$ of positive odd degrees.

**Proposition 5.1.** *The graded $k$-algebra $S = \mathrm{Ext}_\Lambda^*(k, k)$ is a polynomial ring over $k$ on commuting indeterminates $\chi_1, \ldots, \chi_c$ with $|\chi_i| = -(|\xi_i| + 1)$.*

*Proof.* One has $\Lambda \cong k\langle \xi_1 \rangle \otimes_k B$, where $k\langle \xi_j \rangle$ denotes the exterior algebra on the graded vector space $k\xi_i$ and $B = k\langle \xi_2 \rangle \otimes_k \cdots \otimes_k k\langle \xi_c \rangle$. By Lemma 3.3 and induction, it suffices to treat the case $\Lambda = k\langle \xi \rangle$. We use the notation from Construction 3.1.

Set $d = |\xi|$, and let $F$ be a DG $\Lambda$-module, whose underlying graded $\Lambda$-module has a basis $\{e_n\}_{n \geqslant 0}$ with $|e_n| = (d + 1)n$, and with differential defined by the formulas $\partial(e_0) = 0$ and $\partial(e_n) = \xi e_{n-1}$ for $n \geq 1$. An elementary verification shows that the $\Lambda$-linear map $\varepsilon \colon F \to k$ with $\varepsilon(e_0) = 1$ and $\varepsilon(e_n) = 0$ for $n \geq 1$ is a quasi-isomorphism of DG $\Lambda$-modules, and thus a semifree resolution of $k$. It yields

$$\mathrm{Ext}_\Lambda^n(k, k) = \mathrm{H}^n(\mathrm{Hom}_\Lambda(F, k)) = \mathrm{Hom}_\Lambda(F, k)^n$$
$$= \begin{cases} k\chi^{(i)} & \text{for } n = (d + 1)i \geq 0; \\ 0 & \text{otherwise}, \end{cases}$$

where $\chi^{(i)} \colon F \to k$ is the chain map of DG $\Lambda$-modules defined by $\chi^{(i)}(e_n) = 0$ for $n \neq i$ and $\chi^{(i)}(e_i) = 1$. Setting $\widetilde{\chi}^{(i)}(e_n) = e_{n-i}$ for $n \geq i$ and $\widetilde{\chi}^{(i)}(e_n) = 0$ for $n < i$ one obtains a chain map $\widetilde{\chi}^{(i)} \colon F \to F$, such that $\varepsilon \widetilde{\chi}^{(i)} = \chi^{(i)}$. The definition of composition products yields equalities

$$[\chi^{(j)}][\chi^{(i)}] = [\chi^{(j)} \circ \widetilde{\chi}^{(i)}] = [\chi^{(i+j)}].$$

for all $i, j \geq 0$. They show that the isomorphism of graded $k$-vector spaces $k[\chi] \to S$, which sends $\chi^i$ to $[\chi^{(i)}]$ for each $i \geq 0$, is a homomorphism of graded $k$-algebras. $\qquad\square$

*Remark* 5.2. The universal property of exterior algebras guarantees that there is a unique homomorphism of graded $k$-algebras $\Delta \colon \Lambda \to \Lambda \otimes_k \Lambda$ with

$$\Delta(\xi_i) = \xi_i \otimes 1 + 1 \otimes \xi_i \quad \text{for } 1 \leq i \leq c.$$

It evidently satisfies $\Delta(\varepsilon^A \otimes A) = \mathrm{id}^A = \Delta(A \otimes_k \varepsilon^A)$, and so is the comultiplication of a graded Hopf algebra structure on $\Lambda$.

In view of the preceding proposition and remark, Theorems 4.7 and 4.9 yield:

**Theorem 5.3.** *For DG $\Lambda$-modules $U$, $V$ with $\mathrm{H}(U)$ and $\mathrm{H}(V)$ finite over $k$ one has*

$$\mathrm{cx}_A(V, U) = \mathrm{cx}_A(U, V) \leq \mathrm{cx}_\Lambda(U, U) = \mathrm{inj\,cx}_\Lambda U = \mathrm{cx}_\Lambda U \leq \mathrm{cx}_\Lambda k = c.$$

$$\square$$

# 6. Complete intersection local rings I.

In this section $(R, \mathfrak{m}, k)$ denotes a local ring. When $R$ is complete intersection (the definition is recalled below) we open a path to an exterior algebra, and use it to transport results from Section 5.

The *embedding dimension* of $(R, \mathfrak{m}, k)$ is the number $\mathrm{edim}\, R = \mathrm{rank}_k(\mathfrak{m}/\mathfrak{m}^2)$; by Nakayama's Lemma, it is equal to the minimal number of generators of $\mathfrak{m}$.

**Construction 6.1.** Choose a minimal set of generators $r_1, \ldots, r_e$ of $\mathfrak{m}$, and let $K$ denote the Koszul complex on $r_1, \ldots, r_e$. The functor $- \otimes_R K$ preserves quasi-isomorphisms of complexes of $R$-modules, so it defines an exact functor

$$\mathsf{k} \colon \mathsf{D}(R) \to \mathsf{D}(K).$$

**Lemma 6.2.** *For each homologically finite complex of $R$-modules $M$ one has:*

$$\mathrm{cx}_R M = \mathrm{cx}_K \mathsf{k}(M) \quad \text{and} \quad \mathrm{inj\,cx}_R M = \mathrm{inj\,cx}_K \mathsf{k}(M).$$

*Proof.* Let $G \to M$ be a semifree resolution over $R$, and note that the induced morphism $G \otimes_R K \to M \otimes_R K$ then is a semifree resolution over $K$. The expression for $\mathrm{cx}_R M$ now follows from the isomorphisms of $k$-vector spaces

$$\mathrm{Ext}_K^n(M \otimes_R K, k) = \mathrm{H}^n(\mathrm{Hom}_K(G \otimes_R K, k))$$
$$\cong \mathrm{H}^n(\mathrm{Hom}_R(G, k))$$
$$= \mathrm{Ext}_R^n(M, k) .$$

For the second equality, choose a semifree resolution $F \to k$ over $K$. One has an isomorphism $kM \simeq \mathrm{Hom}_R(K, \Sigma^e M)$ of DG $K$-modules, where $e = \mathrm{edim}\, R$. For each $n \in \mathbb{Z}$ adjunction gives the isomorphism of $k$-vector spaces below, and the last equality holds because $F$ is semifree over $R$:

$$\mathrm{Ext}_K^n(k, M \otimes_R K) = \mathrm{H}^n(\mathrm{Hom}_K(F, \mathrm{Hom}_R(K, \Sigma^e M)))$$
$$\cong \mathrm{H}^n(\mathrm{Hom}_R(F, \Sigma^e M))$$
$$= \mathrm{Ext}_R^{n-e}(k, M) . \qquad \square$$

Krull's Principal Ideal Theorem implies $\mathrm{edim}\, R \geq \mathrm{dim}\, R$, where $\dim R$ denotes the Krull dimension of $R$. The *codimension* of $R$ is the number $\mathrm{codim}\, R = \mathrm{edim}\, R - \dim R$. Rings of codimension zero are called *regular*. Cohen's Structure Theorem shows that the $\mathfrak{m}$-adic completion $\widehat{R}$ of $R$ admits a surjective homomorphism of rings $\varkappa \colon (Q, \mathfrak{q}, k) \to \widehat{R}$, with $Q$ a regular local ring; see [13, §29]. Such a homomorphism is called a *Cohen presentation* of $\widehat{R}$. One has $\dim Q \geq \mathrm{edim}\, R$, and equality is equivalent to $\mathrm{Ker}(\varkappa) \subseteq \mathfrak{q}^2$; such a presentation is said to be *minimal*.

Any Cohen presentation $\varkappa$ can be refined to a minimal one. Indeed, choose elements $q_1, \ldots, q_n$ in $\mathfrak{q}$ mapping to a $k$-basis of $(\mathrm{Ker}(\varkappa) + \mathfrak{q}^2)/\mathfrak{q}^2$. One then has

$$\mathrm{codim}\, Q/(\boldsymbol{q}) = (\mathrm{edim}\, Q - n) - (\dim Q - n) = \mathrm{codim}\, Q = 0 ,$$

so the local ring $Q/(\boldsymbol{q})$ is regular along with $Q$. The map $\varkappa$ factors through a homomorphism $Q/(\boldsymbol{q}) \to \widehat{R}$, and the latter is a minimal Cohen presentation.

**Construction 6.3.** Let $\iota \colon R \to \widehat{R}$ be the completion map and choose a minimal Cohen presentation $\varkappa \colon (Q, \mathfrak{q}, k) \to \widehat{R}$.

Choose a minimal set of generators $r_1, \ldots, r_e$ of $\mathfrak{m}$ and then pick in $Q$ elements $q_1, \ldots, q_e$ with $\varkappa(q_i) = \iota(r_i)$ for $i = 1, \ldots, e$. Let $K$ denote the Koszul complex on $r_1, \ldots, r_e$, and $E$ be the one on $q_1, \ldots, q_e$. The definition of Koszul complexes allows one to identify the DG algebras $\widehat{R} \otimes_R K$ and $\widehat{R} \otimes_Q E$.

Choose a minimal set of generators $\{f_1, \ldots, f_c\}$ of the ideal $\operatorname{Ker} \varkappa$, let $A$ be the Koszul complex on this set, and let $\pi \colon A \to \widehat{R}$ denote the canonical projection.

Set $\Lambda = A \otimes_Q k$, and note that this is a DG algebra with zero differential, and its underlying graded algebra is the exterior algebra $\bigwedge_k (A_1 \otimes_Q k)$.

The ring $R$ is said to be *complete intersection* if in some Cohen presentation $\varkappa \colon Q \to \widehat{R}$ the ideal $\operatorname{Ker} \varkappa$ can be generated by a $Q$-regular sequence. When this is the case, the kernel of *every* Cohen presentation is generated by a regular sequence, and for each minimal presentation such a sequence consists of codim $R$ elements.

In the next three lemmas $(R, \mathfrak{m}, k)$ *denotes a complete intersection local ring.*

**Lemma 6.4.** *The following maps are quasi-isomorphisms of DG algebras:*

$$K = R \otimes_R K \xrightarrow{\iota \otimes_R K} \widehat{R} \otimes_R K = \widehat{R} \otimes_Q E \xleftarrow{\pi \otimes_Q E} A \otimes_Q E \xrightarrow{A \otimes_Q \varepsilon^E} A \otimes_Q k = \Lambda .$$

*Proof.* As $\widehat{R}$ is flat over $R$, one can identify $\operatorname{H}(\iota \otimes_R K)$ with the map

$$\iota \otimes_R \operatorname{H}(K) \colon R \otimes_R \operatorname{H}(K) \to \widehat{R} \otimes_R \operatorname{H}(K) ,$$

which is bijective because $\operatorname{H}(K)$ is a direct sum of shifts of $k$.

The sequence $f_1, \ldots, f_c$ is $Q$-regular because $R$ is complete intersection, hence $\pi$ is a quasi-isomorphism, and then so is $\pi \otimes_Q E$.

Since the ring $Q$ is regular and the elements $q_1, \ldots, q_e$ minimally generate $\mathfrak{q}$, they form a $Q$-regular sequence, so $\varepsilon^E$ is a quasi-isomorphism, and then so is $A \otimes_Q \varepsilon^E$. $\qquad\square$

In view of 2.4, the quasi-isomorphisms in Lemma 6.4 define an equivalence of categories $\mathsf{j} \colon \mathsf{D}(K) \to \mathsf{D}(\Lambda)$. Lemmas 6.2 and 2.5 then give:

**Lemma 6.5.** *For every homologically finite complex of $R$-modules $M$ one has*

$$\operatorname{cx}_R M = \operatorname{cx}_\Lambda \mathsf{jk}(M) \quad and \quad \operatorname{inj} \operatorname{cx}_R M = \operatorname{inj} \operatorname{cx}_\Lambda \mathsf{jk}(M) . \qquad\square$$

**Lemma 6.6.** *In $\mathsf{D}(\Lambda)$ there is an isomorphism $\mathsf{jk}(k) \simeq \bigoplus_{i \geq 0} \Sigma^i k^{\binom{c}{i}}$.*

*Proof.* Choose bases $x_1, \ldots, x_c$ of $A_1$ over $Q$ with $\partial(x_i) = f_i$ for $1 \leq i \leq c$ and $y_1, \ldots, y_e$ of $E_1$ over $Q$ with $\partial(y_j) = q_j$ for $1 \leq j \leq e$; Thus $f_i = \sum_{j=1}^{d} b_{ij} q_j$ with $b_{ij} \in Q$. There is a unique homomorphism $\alpha \colon A \to E$ of DG algebras over $Q$ with

$$\alpha(x_i) = \sum_{j=1}^{e} b_{ij} y_j \quad \text{for} \quad 1 \leq i \leq c.$$

It appears in a commutative diagram of DG algebras

$$K = R \otimes_R K \xrightarrow{\iota \otimes_R K} \widehat{R} \otimes_R K = \widehat{R} \otimes_Q E \xleftarrow{\pi \otimes_Q E} A \otimes_Q E \xrightarrow{A \otimes_Q \varepsilon^E} A \otimes_Q k = \Lambda$$

$$\downarrow{\varepsilon^R \otimes_R K} \qquad \downarrow{\varepsilon^R \otimes_Q E} \qquad \downarrow{\varepsilon^{\widehat{R}} \otimes_Q E} \qquad \downarrow{\alpha \otimes_Q E} \qquad \downarrow{\alpha \otimes_Q k}$$

$$k(k) = k \otimes_R K =\!\!=\!\!= k \otimes_R K = k \otimes_Q E \xleftarrow[\simeq]{\varepsilon^E \otimes_Q E} E \otimes_Q E \xrightarrow[\simeq]{E \otimes_Q \varepsilon^E} E \otimes_Q k$$

where $\varepsilon^E \otimes_Q E$ and $E \otimes_Q \varepsilon^E$ are quasi-isomorphisms because $\varepsilon^E$ is one.

The map $\epsilon^E \otimes_R K$ defines the action of $K$ on $k(k)$, and $\alpha \otimes_Q k$ that of $\Lambda$ on $E \otimes_Q k$. The commutativity of the diagram implies and $jk(k) \simeq E \otimes_Q k$ in $\mathsf{D}(\Lambda)$, see 2.4. The minimality of the Cohen presentation $\varkappa$ means that each $b_{ij}$ is in $\mathfrak{q}$, so one has $k(k) \cong \bigoplus_{i \geq 0} \Sigma^i k^{\binom{c}{i}}$ as DG $K$-modules, and hence in $\mathsf{D}(\Lambda)$ one gets

$$jk(k) \simeq j\left( \bigoplus_{i \geq 0} \Sigma^i k^{\binom{c}{i}} \right) \simeq \bigoplus_{i \geq 0} \Sigma^i j(k)^{\binom{c}{i}} \simeq \bigoplus_{i \geq 0} \Sigma^i k^{\binom{c}{i}} ,$$

where the last isomorphism holds because one has $j(k) \simeq k$ by Lemma 2.5. $\qquad \square$

We come to the main result of this section. All its assertions are known: the first equality is proved in [2, 5.3], the inequality follows from [8, 4.2], and the second equality from [16, Thm. 6]. The point here is that they are deduced, in a uniform way, from the corresponding relations for DG modules over exterior algebras, established in Theorem 5.3, which ultimately are much simpler to prove.

**Theorem 6.7.** *If $(R, \mathfrak{m}, k)$ is complete intersection, then for every complex of $R$-modules $M$ with $\mathrm{H}(M)$ finitely generated the following inequalities hold:*

$$\mathrm{inj}\,\mathrm{cx}_R\, M = \mathrm{cx}_R\, M \leq \mathrm{cx}_R\, k = \mathrm{codim}\, R \,.$$

*Proof.* The isomorphism of Lemma 6.6 implies $\mathrm{cx}_\Lambda\, k = \mathrm{cx}_\Lambda\, jk(k)$. Now Lemma 6.5 translates part of Theorem 5.3 into the desired statement. $\qquad \square$

The equalities in the theorem may fail when $R$ is not complete intersection:

*Remark* 6.8. Gulliksen [9, 2.3] proved that the condition $\mathrm{cx}_R\, k < \infty$ characterizes local complete intersection rings among all local rings.

Jorgensen and Şega [10, 1.2] construct Gorenstein local $k$-algebras $R$ with $\mathrm{rank}_k\, R$ finite and modules $M$ with $\{\mathrm{cx}_R\, M, \mathrm{inj}\,\mathrm{cx}_R\, M\} = \{1, \infty\}$, in one order or the other.

*Remark* 6.9. The unused portion of Theorem 5.3 suggests that the relations

$$\mathrm{cx}_R\,(N, M) = \mathrm{cx}_R\,(M, N) \leq \mathrm{cx}_R\,(M, M) = \mathrm{cx}_R\, M$$

might hold also over complete intersections. They do, see [4, 5.7], but we know of no simple way to deduce them from Theorem 5.3.


# 7. Projective levels

In this section $A$ is a DG algebra. We collect some results on $A$-levels of DG modules, which are reminiscent of theorems on projective dimension for modules.

**7.1.** For any DG $A$-module $U$ the following hold.

(1) One has $\operatorname{level}_A^A(U) \leq l + 1$ if and only if $U$ is isomorphic in $\mathsf{D}(A)$ to a direct summand of some semifree DG module $F$ with a semifree filtration having $F^l = F$ and every $F^n/F^{n-1}$ of finite rank over $A^\natural$; see 2.1.
(2) When $A$ is non-negative, $\mathsf{H}_0(A)$ is a field, and $\mathsf{H}_i(U) = 0$ for all $i \ll 0$, one has $\operatorname{level}_A^A(U) < \infty$ if and only if $\operatorname{Tor}^A(k, U)$ is finite over $k$.
(3) When $A$ has zero differential and is noetherian, and the graded $A$-module $\mathsf{H}(U)$ is finitely generated, the following inequalities hold:

$$\operatorname{level}_A^A(U) \leq \operatorname{proj\,dim}_A \mathsf{H}(U) + 1 \leq \operatorname{gl\,dim} A + 1 \,.$$

Indeed, (1) is proved in [5, 4.2], (2) in [5, 4.8], and (3) in [5, 5.5].

The preceding results pertain to homological algebra, in the sense that they do not depend on the structure of the ring $A^\flat = \bigoplus_{n \in \mathbb{Z}} A_n$. For algebras over fields the next result contains as a special case the New Intersection Theorem, see [14], a central result in commutative algebra, and effectively belongs to the latter subject.

**7.2.** Let $A$ be a DG algebra with zero differential and $U$ a DG $A$-module.
  If the ring $A^\flat$ is commutative, noetherian, and is an algebra over a field, then

$$\operatorname{level}_A^A(U) \geq \operatorname{height} I + 1 \,,$$

where $I$ is the annihilator of the $A^\flat$-module $\bigoplus_{n \in \mathbb{Z}} \mathsf{H}_n(U)$; see [5, 5.1].


# 8. Exterior algebras II.

In this section $k$ is a field, and $\Lambda$ a DG algebra with $\partial^\Lambda = 0$ and $\Lambda^\natural$ an exterior $k$-algebra on alternating indeterminates of positive odd degrees $d_1, \ldots, d_c$.
  We recall a version of the Bernstein-Gelfand-Gelfand equivalence from [5, 7.4]:

**8.1.** Let $S$ be a DG algebra with $\partial^S = 0$ and $S^\natural$ a polynomial ring over $k$ on commuting indeterminates of degrees $-(d_1 + 1), \ldots, -(d_c + 1)$; set $d = d_1 + \cdots + d_c$.

There exist exact functors $\mathsf{D}(\Lambda) \xrightarrow{\mathsf{h}} \mathsf{D}(S) \xrightarrow{\mathsf{t}} \mathsf{D}(\Lambda)$ inducing inverse equivalences

$$
\begin{array}{ccc}
\mathsf{thick}_\Lambda(k) & \underset{\underset{\mathsf{t}}{\longleftarrow}}{\overset{\overset{\mathsf{h}}{\longrightarrow}}{\equiv}} & \mathsf{thick}_S(S) \\
\cup| & & \cup| \\
\mathsf{thick}_\Lambda(\Lambda) & \underset{\underset{\mathsf{t}}{\longleftarrow}}{\overset{\overset{\mathsf{h}}{\longrightarrow}}{\equiv}} & \mathsf{thick}_S(k)
\end{array}
\tag{8.1.1}
$$

such that in $\mathsf{D}(S)$ and $\mathsf{D}(\Lambda)$, respectively, there are isomorphisms

$$
\begin{array}{ll}
\mathsf{h}(\Lambda) \simeq \Sigma^d k & \qquad \mathsf{t}(k) \simeq \Sigma^{-d}\Lambda \\
& \text{and} \\
\mathsf{h}(k) \simeq S & \qquad \mathsf{t}(S) \simeq k\,.
\end{array}
\tag{8.1.2}
$$

Let $\mathsf{D}^\mathsf{f}(\Lambda)$ be the full subcategory of $\mathsf{D}(\Lambda)$ whose objects are the DG modules $U$ with $\mathrm{H}(U)$ finite over $\Lambda$, and $\mathsf{D}^\mathsf{f}(S)$ the corresponding subcategory of $\mathsf{D}(S)$. The next proposition refines the equivalences in (8.1.1); see Remark 8.3.

**Proposition 8.2.** *For each $U \in \mathsf{D}^\mathsf{f}(\Lambda)$ and each $M \in \mathsf{D}^\mathsf{f}(S)$ one has*

$$
\mathrm{cx}_\Lambda U = \dim_S \mathrm{H}(\mathsf{h}U) \quad \text{and} \quad \dim_S M = \mathrm{cx}_\Lambda \mathrm{H}(\mathsf{t}M)\,,
$$

*and for each $n \in \mathbb{N}$ the functors $\mathsf{h}$ and $\mathsf{t}$ induce inverse equivalences*

$$
\begin{array}{ccc}
\mathsf{D}^\mathsf{f}(\Lambda) & \underset{\underset{\mathsf{t}}{\longleftarrow}}{\overset{\overset{\mathsf{h}}{\longrightarrow}}{\equiv}} & \mathsf{D}^\mathsf{f}(S) \\
\cup| & & \cup| \\
\{U \in \mathsf{D}^\mathsf{f}(\Lambda) \mid \mathrm{cx}_\Lambda U \le n\} & \underset{\underset{\mathsf{t}}{\longleftarrow}}{\overset{\overset{\mathsf{h}}{\longrightarrow}}{\equiv}} & \{M \in \mathsf{D}^\mathsf{f}(S) \mid \dim_S \mathrm{H}(M) \le n\}\,.
\end{array}
\tag{8.2.1}
$$

*Proof.* One has $\mathsf{D}^\mathsf{f}(\Lambda) = \mathsf{thick}_\Lambda(k)$ by 1.4(1). As the polynomial ring $S$ has finite global dimension, 7.1(3) gives $\mathsf{D}^\mathsf{f}(S) = \mathsf{thick}_S(S)$. Thus, the top row of (8.1.1) gives inverse equivalences between $\mathsf{D}^\mathsf{f}(\Lambda)$ and $\mathsf{D}^\mathsf{f}(S)$. It is easy to verify that the subcategories in the lower row are thick, so it suffices to compute $\mathrm{cx}_\Lambda U$ and $\dim_S \mathrm{H}(M)$.

The equivalence $\mathsf{h}$ gives the first isomorphism in the chain

$$
\mathrm{RHom}_\Lambda(k, U) \simeq \mathrm{RHom}_S(\mathsf{h}(k), \mathsf{h}(U)) \simeq \mathrm{RHom}_S(S, \mathsf{h}(U)) \simeq \mathsf{h}(U)\,;
$$

the second one comes from (8.1.2), the third is clear. In homology, one obtains

$$
\mathrm{Ext}^n_\Lambda(k, U) \cong \mathrm{H}^n(\mathsf{h}(U))
$$

for every $n \in \mathbb{Z}$, whence the second equality in the following sequence; the first one comes from Theorem 5.3, the third from the Hilbert-Serre Theorem, see 4.6:

$$\mathrm{cx}_\Lambda U = \mathrm{inj}\,\mathrm{cx}_\Lambda U = \mathrm{cx}\,(\,\mathrm{rank}_k\,\mathrm{H}_n(\mathsf{h}(U)))_{n \in \mathbb{Z}} = \dim_S \mathrm{H}(\mathsf{h}(U))\,.$$

Finally, from the already proved assertions one obtains

$$\dim_S \mathrm{H}(M) = \dim_S \mathrm{H}(\mathsf{ht}(M)) = \mathrm{cx}_\Lambda \mathsf{t}(M)\,. \qquad \square$$

*Remark* 8.3. For $U \in \mathsf{D}^\mathsf{f}(\Lambda)$ the definition of complexity shows that $\mathrm{cx}_\Lambda U \le 0$ is equivalent to the finiteness of the $k$-vector space $\mathrm{Ext}_\Lambda(U, k)$. It is the graded $k$-dual of $\mathrm{Tor}^\Lambda(U, k)$, so by 7.1(2) its finiteness is equivalent to $U \in \mathsf{thick}_\Lambda(\Lambda)$. On the other hand, $M \in \mathsf{D}^\mathsf{f}(S)$ has $\dim_S \mathrm{H}(M) \le 0$ if and only if $M$ is in $\mathsf{thick}_S(k)$; see 1.4(1). Thus, for $n = 0$ diagram (8.2.1) reduces to (8.1.1).

**Theorem 8.4.** *If $U$ is a DG $\Lambda$-module with $0 < \mathrm{rank}_k \mathrm{H}(U) < \infty$, then one has*

$$\mathrm{card}\{n \in \mathbb{Z} \mid \mathrm{H}_n(U) \ne 0\} \ge \mathrm{level}_\Lambda^k(U) \ge c - \mathrm{cx}_\Lambda U + 1\,.$$

*Proof.* The inequality on the left comes from 1.4(3). The one on the right results from the following chain of (in)equalities

$$\begin{aligned}
\mathrm{level}_\Lambda^k(U) &= \mathrm{level}_S^{\mathsf{h}(k)}(\mathsf{h}(U)) \\
&= \mathrm{level}_S^S(\mathsf{h}(U)) \\
&\ge \mathrm{height}(\mathrm{Ann}_S\,\mathrm{H}(\mathsf{h}(U))) + 1 \\
&= \dim S - \dim_S \mathrm{H}(\mathsf{h}(U)) + 1 \\
&= \dim S - \mathrm{cx}_\Lambda U + 1 \\
&= c - \mathrm{cx}_\Lambda V + 1\,.
\end{aligned}$$

The first one holds because $\mathsf{h}$ is an equivalence, see 1.2(2), the second comes from (8.1.2), and the third from 7.2; the remaining equalities hold by [13, Exercise 5.1][1], by Lemma 8.2, and because $\dim S = c$ holds. $\qquad \square$

The inequalities in Theorem 8.4 are tight.

**Example 8.5.** For each integer $i$ with $0 \le i \le c$ and $\Lambda^{(i)} = \Lambda/(\xi_1, \ldots, \xi_i)$ one has

$$\mathrm{card}\{n \in \mathbb{Z} \mid \mathrm{H}_n(\Lambda^{(i)}) \ne 0\} = c - i + 1 \quad \text{and} \quad \mathrm{cx}_R \Lambda^{(i)} = i\,.$$

---

[1] solved on page 288

Indeed, the first one is clear. For $\Gamma^{(i)} = \Lambda/(\xi_{i+1}, \ldots, \xi_n)$ one has

$$\operatorname{Ext}^n_\Lambda(\Lambda^{(i)}, k) \cong \operatorname{Ext}^n_{\Gamma^{(i)} \otimes_k \Lambda^{(i)}}(\Lambda^{(i)}, k) \cong \operatorname{Ext}^n_{\Gamma^{(i)}}(k, k),$$

because $\Lambda$ is isomorphic to $\Gamma^{(i)} \otimes_k \Lambda^{(i)}$ as $k$-algebras. Theorem 5.3 gives $\operatorname{cx}_{\Gamma^{(i)}} k = i$.

## 9. Complete intersection local rings II.

We finish the paper with a new proof of [5, 11.3]. The original argument uses a reduction, constructed in [2], to a complete intersection ring of smaller codimension and a bounded complex of free modules over it, then refers to the main theorem of [5], which applies to such complexes over arbitrary local rings. Here we describe a direct reduction to results on exterior algebras, established in Section 8.

**Theorem 9.1.** *If $(R, \mathfrak{m}, k)$ is a complete intersection local ring and $M$ a complex of $R$-modules with $\mathrm{H}(M)$ finite and nonzero, then one has inequalities*

$$\sum_{n \in \mathbb{Z}} \ell\ell_R \, \mathrm{H}_n(M) \geq \operatorname{level}^k_R(M) \geq \operatorname{codim} R - \operatorname{cx}_R M + 1.$$

*Proof.* The inequality on the left is a special case of 1.4(2).

Set $c = \operatorname{codim} R$. Let $K$ be the Koszul complex on a minimal set of generators for $\mathfrak{m}$ and $\Lambda$ the DG algebra with zero differential, and with underlying graded algebra an exterior algebra over $k$ on $c$ generators of degree one. Construction 6.1 and Lemma 6.4 provide exact functors of triangulated categories

$$\mathsf{D}(R) \xrightarrow{\;\mathsf{k}\;} \mathsf{D}(K) \xrightarrow[\equiv]{\;\mathsf{j}\;} \mathsf{D}(\Lambda)$$

the second of which is an equivalence. One then has the following relations

$$\begin{aligned}
\operatorname{level}^k_R(M) &\geq \operatorname{level}^{\mathsf{jk}(k)}_\Lambda(\mathsf{jk}(M)) \\
&= \operatorname{level}^k_\Lambda(\mathsf{jk}(M)) \\
&\geq c - \operatorname{cx}_\Lambda(\mathsf{jk}(M)) + 1 \\
&= \operatorname{codim} R - \operatorname{cx}_R M + 1
\end{aligned}$$

where the inequalities are given by 1.2(2) and by Theorem 8.4, while the equalities come from 1.2(1) and Lemmas 6.6, and from Lemma 6.5. $\qquad\square$

As in Theorem 8.4 the inequalities in Theorem 9.1 are tight.

**Example 9.2.** Let $k$ be a field and set $R = k[x_1, \ldots, x_c]/(x_1^2, \ldots, x_c^2)$. For every integer $i$ with $0 \le i \le c$, set $R^{(i)} = R/(x_1, \ldots, x_i)$. One then has equalities

$$\ell\ell_R R_i = c - i + 1, \quad \operatorname{codim} R = c, \quad \text{and} \quad \operatorname{cx}_R R^{(i)} = i.$$

Indeed, the first two are clear. For $Q^{(i)} = R/(x_{i+1}, \ldots, x_n)$ one has

$$\operatorname{Ext}_R^n(R^{(i)}, k) \cong \operatorname{Ext}_{Q^{(i)} \otimes_k R^{(i)}}^n(R^{(i)}, k) \cong \operatorname{Ext}_{Q^{(i)}}^n(k, k),$$

because $R$ is isomorphic to $Q^{(i)} \otimes_k R^{(i)}$ as $k$-algebras. Theorem 6.7 gives $\operatorname{cx}_{Q^{(i)}} k = i$.

The number on the right hand side of the formula in Theorem 9.1 is finite whenever the module $M$ has finite complexity. However, this condition alone does not imply the inequalities in the theorem, even when the ring $R$ is Gorenstein.

**Example 9.3.** Let $k$ be a field and $c$ an integer with $c \ge 3$. The ring

$$R = \frac{k[x_1, \ldots, x_c]}{\sum_{h=1}^{c-1}(x_h^2 - x_{h+1}^2) + \sum_{1 \le i < j \le c}(x_i x_j)}$$

is Gorenstein, but not complete intersection, and its module $M = R$ has

$$\ell\ell_R M = \operatorname{level}_R^k(M) = 3 < c - 0 + 1 = \operatorname{codim} R - \operatorname{cx}_R M + 1.$$

It is worth noting that a very special case of Theorem 9.1 was initially discovered when studying actions of finite elementary abelian groups on finite CW complexes:

*Remark* 9.4. Let $k$ be a field of positive characteristic $p$ and $G$ an elementary abelian $p$-group of rank $c$. The group algebra $kG$ is isomorphic to $k[x_1, \ldots, x_c]/(x_1^p, \ldots, x_c^p)$, which is a complete intersection of codimension $c$. Over $kG$ Theorem 9.1 was proved by Carlsson [7] for $M^\natural$ of finite rank over $A^\natural$ and $p$ equal to 2, and for general $M$ and $p$ by Allday, Baumgartner, and Puppe; see [1, 4.6.42].

# References

[1] C. Allday, V. Puppe, *Cohomological methods in transf ormation groups*, Cambridge Stud. Adv. Math. **32**, Cambridge Univ. Press, Cambridge, 1993.

[2] L. L. Avramov, *Modules of finite virtual projective dimension*, Invent. Math. **96** (1989), 71–101.

[3] L. L. Avramov, *Homological asymptotics of modules over local rings*, Commutative algebra (Berkeley, CA, 1987), Math. Sci. Res. Inst. Publ., **15**, Springer, New York, 1989; 33–62.

[4] L. L. Avramov, R.-O. Buchweitz, *Support varieties and cohomology over complete intersections*, Invent. Math. **142** (2000), 285–318.

[5] L. L. Avramov, R.-O. Buchweitz, S. B. Iyengar, C. Miller, *Homology of finite free complexes*, Adv. Math. **223** (2010), 1731–1781.

[6] L. L. Avramov, S. Halperin, *Through the looking glass: a dictionary between rational homotopy theory and local algebra*, Algebra, algebraic topology and their interactions (Stockholm, 1983), Lecture Notes in Math. **1183**, Springer, Berlin, 1986; 1–27

[7] G. Carlsson, *On the homology of finite free $(\mathbb{Z}/2)^k$-complexes*, Invent. Math. **74** (1983), 139–147.

[8] T. H. Gulliksen, *A change of ring theorem with applications to Poincaré series and intersection multiplicity*, Math. Scand. **34** (1974), 167–183.

[9] T. H. Gulliksen, *On the deviations of a local ring*, Math. Scand. **47** (1980), 5–20.

[10] D. A. Jorgensen, L. M. Şega, *Asymmetric complete resolutions and vanishing of Ext over Gorenstein rings*, Int. Math. Res. Not. **2005**, 3459–3477.

[11] A. Liulevicius, *The factorization of cyclic reduced powers by secondary cohomology operations*, Mem. Amer. Math. Soc. **42**, Amer. Math. Soc., Providence, RI, 1962.

[12] S. Maclane, *Homology*, Grundlehren math. Wissenschaften, **114**, Springer, Berlin, 1963.

[13] H. Matsumura, *Commutative ring theory*, Cambridge Stud. Adv. Math. **8** Cambridge Univ. Press, Cambridge, 1986.

[14] P. C. Roberts, *Multiplicities and Chern classes in local algebra*, Cambridge Tracts Math. **133**, Cambridge Univ. Press, Cambridge, 1998.

[15] W. Smoke, *Dimension and multiplicity over graded algebras*, J. Algebra **21** (1972), 149–173.

[16] J. Tate, *Homology of noetherian rings and local rings*, Illinois J. Math. **1** (1957), 14–25

[17] C. Wilkerson, *The cohomology algebras of finite-dimensional Hopf algebras*, Trans. Amer. Math. Soc. **264** (1981), 137–150.

DEPARTMENT OF MATHEMATICS, UNIVERSITY OF NEBRASKA, LINCOLN, NE 68588, U.S.A.
*E-mail address*: avramov@math.unl.edu

DEPARTMENT OF MATHEMATICS, UNIVERSITY OF NEBRASKA, LINCOLN, NE 68588, U.S.A.
*E-mail address*: iyengar@math.unl.edu

# Cluster algebras, quiver representations and triangulated categories

BERNHARD KELLER

ABSTRACT. This is an introduction to some aspects of Fomin-Zelevinsky's cluster algebras and their links with the representation theory of quivers and with Calabi-Yau triangulated categories. It is based on lectures given by the author at summer schools held in 2006 (Bavaria) and 2008 (Jerusalem). In addition to by now classical material, we present the outline of a proof of the periodicity conjecture for pairs of Dynkin diagrams (details will appear elsewhere) and recent results on the interpretation of mutations as derived equivalences.

## Contents



## 1. Introduction

### 1.1. Context

Cluster algebras were invented by S. Fomin and A. Zelevinsky [50] in the spring of the year 2000 in a project whose aim it was to develop a combinatorial approach to the results obtained by G. Lusztig concerning total positivity in algebraic groups [103] on the one hand and canonical bases in quantum groups [102] on the other hand (let us stress that canonical bases were discovered

*Date*: July 2008, last modified on May 13, 2010.

independently and simultaneously by M. Kashiwara [83]). Despite great progress during the last few years [52] [17] [55], we are still relatively far from these initial aims. Presently, the best results on the link between cluster algebras and canonical bases are probably those of C. Geiss, B. Leclerc and J. Schröer [64] [65] [62] [61] [63] but even they cannot construct canonical bases from cluster variables for the moment. Despite these difficulties, the theory of cluster algebras has witnessed spectacular growth thanks notably to the many links that have been discovered with a wide range of subjects including

- Poisson geometry [69] [70] ... ,
- integrable systems [54] ... ,
- higher Teichmüller spaces [43] [44] [45] [46] ... ,
- combinatorics and the study of combinatorial polyhedra like the Stasheff associahedra [34] [33] [99] [48] [108] [49] ... ,
- commutative and non commutative algebraic geometry, in particular the study of stability conditions in the sense of Bridgeland [23] [21] [24], Calabi-Yau algebras [71] [35], Donaldson-Thomas invariants [123] [95] [96] [98] ... ,
- and last not least the representation theory of quivers and finite-dimensional algebras, cf. for example the surveys [9] [114] [116] .

We refer to the introductory papers [53] [132] [134] [135] [136] and to the cluster algebras portal [47] for more information on cluster algebras and their links with other parts of mathematics.

The link between cluster algebras and quiver representations follows the spirit of categorification: One tries to interpret cluster algebras as combinatorial (perhaps $K$-theoretic) invariants associated with categories of representations. Thanks to the rich structure of these categories, one can then hope to prove results on cluster algebras which seem beyond the scope of the purely combinatorial methods. It turns out that the link becomes especially beautiful if we use *triangulated categories* constructed from categories of quiver representations.

## 1.2.  Contents

We start with an informal presentation of Fomin-Zelevinsky's classification theorem and of the cluster algebras (without coefficients) associated with Dynkin diagrams. Then we successively introduce quiver mutations, the cluster algebra associated with a quiver, and the cluster algebra with coefficients associated with an 'ice quiver' (a quiver some of whose vertices are frozen). We illustrate cluster algebras with coefficients on a number of examples appearing as coordinate algebras of homogeneous varieties.

Sections 5, 6 and 7 are devoted to the (additive) categorification of cluster algebras. We start by recalling basic notions from the representation theory of quivers. Then we present a fundamental link between indecomposable representations and cluster variables: the Caldero-Chapoton formula. After a brief reminder on derived categories in general, we give the canonical presentation in terms of generators and relations of the derived category of a Dynkin quiver. This yields in particular a presentation for the module category, which we use to sketch Caldero-Chapoton's proof of their formula. Then we introduce the cluster category and survey its many links to the cluster algebra in the finite case. Most of these links are still valid, mutatis mutandis, in the acyclic case, as we see in section 6. Surprisingly enough, one can go even further and categorify interesting classes of cluster algebras using generalizations of the cluster category, which are still triangulated categories and Calabi-Yau of dimension 2. We present this relatively recent theory in section 7. In section 8, we apply it to sketch a proof of the periodicity conjecture for pairs of Dynkin diagrams (details will appear elsewhere [86]). In the final section 9, we give an interpretation of quiver mutation in terms of derived equivalences. We use this framework to establish links between various ways of lifting the mutation operation from combinatorics to linear or homological algebra: mutation of cluster-tilting objects, spherical collections and decorated representations.

## Acknowledgments

These notes are based on lectures given at the IRTG-Summerschool 2006 (Schloss Reisensburg, Bavaria) and at the Midrasha Mathematicae 2008 (Hebrew University, Jerusalem). I thank the organizers of these events for their generous invitations and for providing stimulating working conditions. I am grateful to Thorsten Holm, Peter Jørgensen and Raphael Rouquier for their encouragment and for accepting to include these notes in the proceedings of the 'Workshop on triangulated categories' they organized at Leeds in 2006. It is a pleasure to thank to Carles Casacuberta, André Joyal, Joachim Kock, Amnon Neeman and Frank Neumann for an invitation to the Centre de Recerca Matemàtica, Barcelona, where most of this text was written down. I thank Lingyan Guo and Sefi Ladkani for kindly pointing out misprints and inaccuracies. I am indebted to Tom Bridgeland, Bernard Leclerc, David Kazhdan, Tomoki Nakanishi, Raphaël Rouquier and Michel Van den Bergh for helpful conversations.

# 2. An informal introduction to cluster-finite cluster algebras

## 2.1. The classification theorem

Let us start with a remark on terminology: a *cluster* is a group of similar things or people positioned or occurring closely together [121], as in the combination 'star cluster'. In French, 'star cluster' is translated as 'amas d'étoiles', whence the term 'algèbre amassée' for cluster algebra.

We postpone the precise definition of a cluster algebra to section 3. For the moment, the following description will suffice: A cluster algebra is a commutative $\mathbb{Q}$-algebra endowed with a family of distinguished generators (the *cluster variables*) grouped into overlapping subsets (the *clusters*) of fixed finite cardinality, which are constructed recursively using *mutations*.

The set of cluster variables in a cluster algebra may be finite or infinite. The first important result of the theory is the classification of those cluster algebras where it is finite: the *cluster-finite* cluster algebras. This is the

**Classification Theorem 2.1** (Fomin-Zelevinsky [52]). *The cluster-finite cluster algebras are parametrized by the finite root systems (like semisimple complex Lie algebras).*

It follows that for each Dynkin diagram $\Delta$, there is a canonical cluster algebra $\mathcal{A}_\Delta$. It turns out that $\mathcal{A}_\Delta$ occurs naturally as a subalgebra of the field of rational functions $\mathbb{Q}(x_1, \ldots, x_n)$, where $n$ is the number of vertices of $\Delta$. Since $\mathcal{A}_\Delta$ is generated by its cluster variables (like any cluster algebra), it suffices to produce the (finite) list of these variables in order to describe $\mathcal{A}_\Delta$. Now for the algebras $\mathcal{A}_\Delta$, the recursive construction via mutations mentioned above simplifies considerably. In fact, it turns out that one can *directly construct the cluster variables* without first constructing the clusters. This is made possible by

## 2.2. The knitting algorithm

The general algorithm will become clear from the following three examples. We start with the simplest non trivial Dynkin diagram:

$$\Delta = A_2 : \quad \circ \!\!-\!\!\!-\!\!\!-\!\! \circ \, .$$

We first choose a numbering of its vertices and an orientation of its edges:

$$\vec{\Delta} = \vec{A}_2 : \quad 1 \longrightarrow 2 \, .$$

Now we draw the so-called *repetition* (or *Bratteli diagram*) $\mathbb{Z}\vec{\Delta}$ associated with $\Delta$: We first draw the product $\mathbb{Z} \times \vec{\Delta}$ made up of a countable number of copies of $\Delta$ (drawn slanted upwards); then for each arrow $\alpha : i \to j$ of $\Delta$, we add a new family of arrows $(n, \alpha^*) : (n, j) \to (n + 1, i)$, $n \in \mathbb{Z}$ (drawn slanted downwards). We refer to section 5.5 for the formal definition. Here is the result for $\vec{\Delta} = \vec{A}_2$:

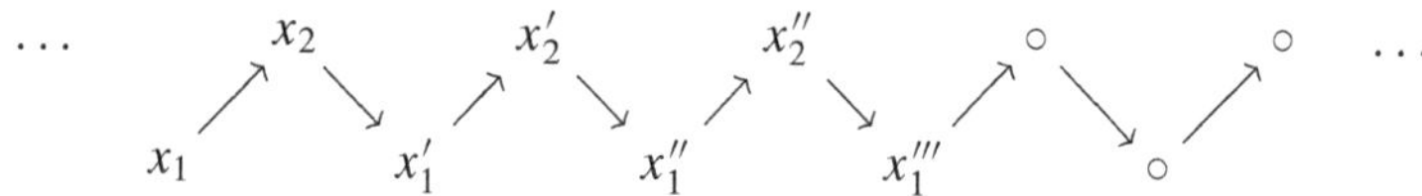

We will now assign a cluster variable to each vertex of the repetition. We start by assigning $x_1$ and $x_2$ to the vertices of the zeroth copy of $\vec{\Delta}$. Next, we construct new variables $x_1'$, $x_2'$, $x_1''$, … by 'knitting' from the left to the right (an analogous procedure can be used to go from the right to the left).

To compute $x_1'$, we consider its immediate predecessor $x_2$, add 1 to it and divide the result by the left translate of $x_1'$, to wit the variable $x_1$. This yields

$$x_1' = \frac{1 + x_2}{x_1}.$$

Similarly, we compute $x_2'$ by adding 1 to its predecessor $x_1'$ and dividing the result by the left translate $x_2$:

$$x_2' = \frac{1 + x_1'}{x_2} = \frac{x_1 + 1 + x_2}{x_1 x_2}.$$

Using the same rule for $x_1''$ we obtain

$$x_1'' = \frac{1 + x_2'}{x_1'} = \left( \frac{x_1 x_2 + x_1 + 1 + x_2}{x_1 x_2} \right) / \left( \frac{1 + x_2}{x_1} \right) = \frac{1 + x_1}{x_2}.$$

Here something remarkable has happened: The numerator $x_1 x_2 + x_1 + 1 + x_2$ is actually divisible by $1 + x_2$ so that the denominator remains a monomial (contrary to what one might expect). We continue with

$$x_2'' = \frac{1 + x_1''}{x_2'} = \left( \frac{x_2 + 1 + x_1}{x_2} \right) / \left( \frac{x_1 + 1 + x_2}{x_1 x_2} \right) = x_1 ,$$

a result which is perhaps even more surprising. Finally, we get

$$x_1''' = \frac{1 + x_2''}{x_1''} = (1 + x_1)/ \left( \frac{1 + x_1}{x_2} \right) = x_2.$$

Clearly, from here on, the pattern will repeat. We could have computed 'towards the left' and would have found the same repeating pattern. In conclusion, there are the 5 cluster variables $x_1$, $x_2$, $x_1'$, $x_2'$ and $x_1''$ and the cluster algebra $\mathcal{A}_{A_2}$ is the $\mathbb{Q}$-subalgebra (not the subfield!) of $\mathbb{Q}(x_1, x_2)$ generated by these 5 variables.

Before going on to a more complicated example, let us record the remarkable phenomena we have observed:

(1) All denominators of all cluster variables are monomials. In other words, the cluster variables are Laurent polynomials. This *Laurent phenomenon* holds for all cluster variables in all cluster algebras, as shown by Fomin and Zelevinsky [51].

(2) The computation is *periodic* and thus only yields finitely many cluster variables. Of course, this was to be expected by the classification theorem above. In fact, the procedure generalizes easily from Dynkin diagrams to arbitrary trees, and then periodicity characterizes Dynkin diagrams among trees.

(3) *Numerology*: We have obtained 5 cluster variables. Now we have $5 = 2 + 3$ and this decomposition does correspond to a natural partition of the set of cluster variables into the two *initial* cluster variables $x_1$ and $x_2$ and the three non initial ones $x_1'$, $x_2'$ and $x_1''$. The latter are in natural bijection with the positive roots $\alpha_1$, $\alpha_1 + \alpha_2$ and $\alpha_2$ of the root system of type $A_2$ with simple roots $\alpha_1$ and $\alpha_2$. To see this, it suffices to look at the denominators of the three variables: The denominator $x_1^{d_1} x_2^{d_2}$ corresponds to the root $d_1\alpha_1 + d_2\alpha_2$. It was proved by Fomin-Zelevinsky [52] that this generalizes to arbitrary Dynkin diagrams. In particular, the number of cluster variables in the cluster algebra $\mathcal{A}_\Delta$ always equals the sum of the rank and the number of positive roots of $\Delta$.

Let us now consider the example $A_3$: We choose the following linear orientation:

$$1 \longrightarrow 2 \longrightarrow 3 .$$

The associated repetition looks as follows:

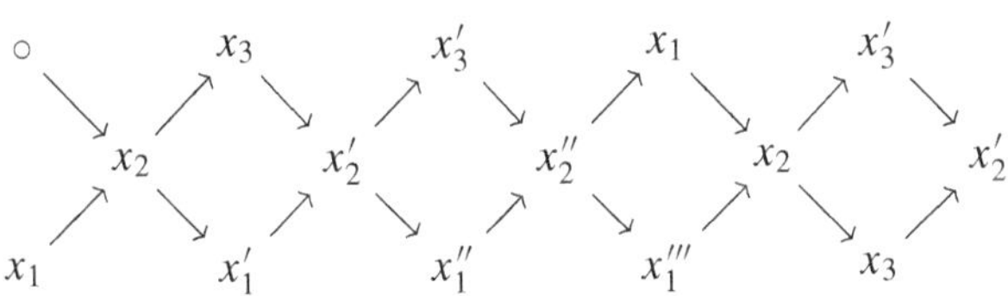

The computation of $x_1'$ is as before:

$$x_1' = \frac{1 + x_2}{x_1} .$$

However, to compute $x_2'$, we have to modify the rule, since $x_2'$ has *two* immediate predecessors with associated variables $x_1'$ and $x_3$. In the formula, we simply take the *product* over all immediate predecessors:

$$x_2' = \frac{1 + x_1' x_3}{x_2} = \frac{x_1 + x_3 + x_2 x_3}{x_2 x_3}.$$

Similarly, for the following variables $x_3', x_1'', \ldots$. We obtain the periodic pattern shown in the diagram above. In total, we find $9 = 3 + 6$ cluster variables, namely

$$x_1, x_2, x_3, \frac{1 + x_2}{x_1}, \frac{x_1 + x_3 + x_2 x_3}{x_1 x_2}, \frac{x_1 + x_1 x_2 + x_3 + x_2 x_3}{x_1 x_2 x_3},$$

$$\frac{x_1 + x_3}{x_2}, \frac{x_1 + x_1 x_2 + x_3}{x_2 x_3}, \frac{1 + x_2}{x_3}.$$

The cluster algebra $\mathcal{A}_{A_3}$ is the subalgebra of the field $\mathbb{Q}(x_1, x_2, x_3)$ generated by these variables. Again we observe that all denominators are monomials. Notice also that $9 = 3 + 6$ and that 3 is the rank of the root system associated with $A_3$ and 6 its number of positive roots. Moreover, if we look at the denominators of the non initial cluster variables (those other than $x_1, x_2, x_3$), we see a natural bijection with the positive roots

$$\alpha_1, \alpha_1 + \alpha_2, \alpha_1 + \alpha_2 + \alpha_3, \alpha_2, \alpha_2 + \alpha_3, \alpha_3$$

of the root system of $A_3$, where $\alpha_1, \alpha_2, \alpha_3$ denote the three simple roots.

Finally, let us consider the non simply laced Dynkin diagram $\Delta = G_2$:

$$\circ \overset{(3,1)}{\rule{2cm}{0.4pt}} \circ.$$

The associated Cartan matrix is

$$\begin{pmatrix} 2 & -3 \\ -1 & 2 \end{pmatrix}$$

and the associated root system of rank 2 looks as follows:

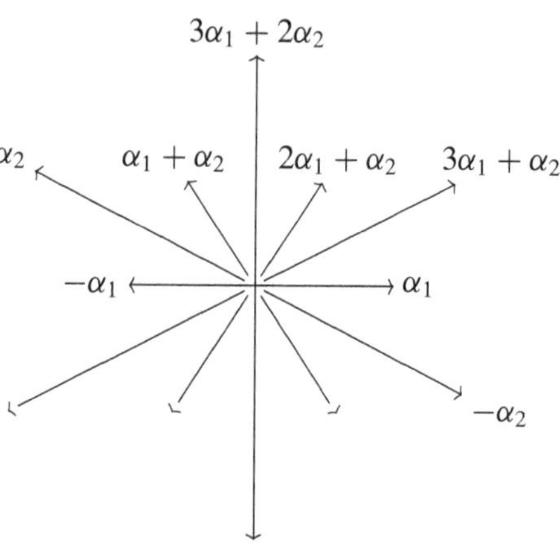

We choose an orientation of the valued edge of $G_2$ to obtain the following valued oriented graph:

$$\vec{\Delta} : 1 \xrightarrow{(3,1)} 2.$$

Now the repetition also becomes a valued oriented graph

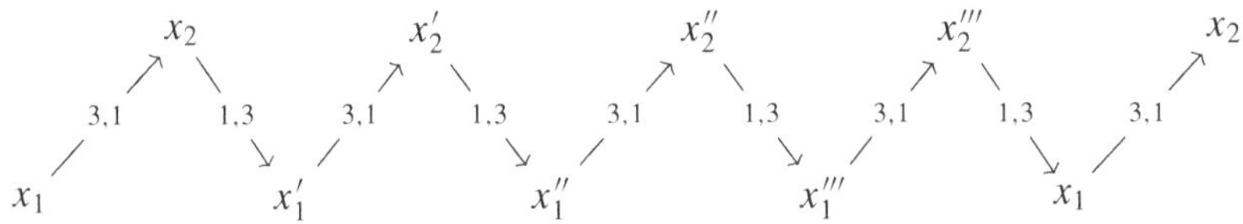

The mutation rule is a variation on the one we are already familiar with: In the recursion formula, each predecessor $p$ of a cluster variable $x$ has to be raised to the power indicated by the valuation 'closest' to $p$. Thus, we have for example

$$x_1' = \frac{1+x_2}{x_1}, \quad x_2' = \frac{1+(x_1')^3}{x_2} = \frac{1+x_1^3+3x_2+3x_2^2+x_2^3}{x_1^3 x_2},$$

$$x_1' = \frac{1+x_2'}{x_1'} = \frac{\cdots}{x_1^2 x_2},$$

where we can read off the denominators from the decompositions of the positive roots as linear combinations of simple roots given above. We find $8 = 2 + 6$ cluster variables, which together generate the cluster algebra $\mathcal{A}_{G_2}$ as a subalgebra of $\mathbb{Q}(x_1, x_2)$.

## 3. Symmetric cluster algebras without coefficients

In this section, we will construct the cluster algebra associated with an antisymmetric matrix with integer coefficients. Instead of using matrices, we will use quivers (without loops or 2-cycles), since they are easy to visualize and well-suited to our later purposes.

### 3.1. Quivers

Let us recall that a *quiver* $Q$ is an oriented graph. Thus, it is a quadruple given by a set $Q_0$ (the set of vertices), a set $Q_1$ (the set of arrows) and two maps $s : Q_1 \to Q_0$ and $t : Q_1 \to Q_0$ which take an arrow to its source respectively its target. Our quivers are 'abstract graphs' but in practice we draw them as in this example:

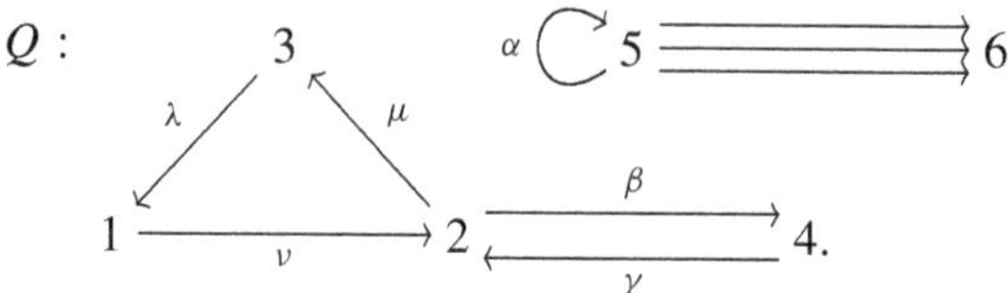

A *loop* in a quiver $Q$ is an arrow $\alpha$ whose source coincides with its target; a *2-cycle* is a pair of distinct arrows $\beta \neq \gamma$ such that the source of $\beta$ equals the target of $\gamma$ and vice versa. It is clear how to define *3-cycles, connected components*. ... A quiver is *finite* if both, its set of vertices and its set of arrows, are finite.

## 3.2. Seeds and mutations

Fix an integer $n \geq 1$. A *seed* is a pair $(R, u)$, where

a) $R$ is a finite quiver without loops or 2-cycles with vertex set $\{1, \ldots, n\}$;
b) $u$ is a free generating set $\{u_1, \ldots, u_n\}$ of the field $\mathbb{Q}(x_1, \ldots, x_n)$ of fractions of the polynomial ring $\mathbb{Q}[x_1, \ldots, x_n]$ in $n$ indeterminates.

Notice that in the quiver $R$ of a seed, all arrows between any two given vertices point in the same direction (since $R$ does not have 2-cycles). Let $(R, u)$ be a seed and $k$ a vertex of $R$. The *mutation* $\mu_k(R, u)$ of $(R, u)$ at $k$ is the seed $(R', u')$, where

a) $R'$ is obtained from $R$ as follows:
  1) reverse all arrows incident with $k$;
  2) for all vertices $i \neq j$ distinct from $k$, modify the number of arrows between $i$ and $j$ as follows:

| $R$ | $R'$ |
|---|---|
| $i \xrightarrow{\;r\;} j$ <br> $s \searrow \quad \nearrow t$ <br> $k$ | $i \xrightarrow{\;r+st\;} j$ <br> $s \nwarrow \quad \swarrow t$ <br> $k$ |
| $i \xrightarrow{\;r\;} j$ <br> $s \nwarrow \quad \swarrow t$ <br> $k$ | $i \xrightarrow{\;r-st\;} j$ <br> $s \searrow \quad \nearrow t$ <br> $k$ |

where $r, s, t$ are non negative integers, an arrow $i \xrightarrow{l} j$ with $l \geq 0$ means that $l$ arrows go from $i$ to $j$ and an arrow $i \xrightarrow{l} j$ with $l \leq 0$ means that $-l$ arrows go from $j$ to $i$.

b) $u'$ is obtained from $u$ by replacing the element $u_k$ with

$$u'_k = \frac{1}{u_k} \left( \prod_{\text{arrows } i \to k} u_i + \prod_{\text{arrows } k \to j} u_j \right). \tag{3.2.1}$$

In the *exchange relation* (3.2.1), if there are no arrows from $i$ with target $k$, the product is taken over the empty set and equals 1. It is not hard to see that $\mu_k(R, u)$ is indeed a seed and that $\mu_k$ is an involution: we have $\mu_k(\mu_k(R, u)) = (R, u)$. Notice that the expression given in (3.2.1) for $u'_k$ is *subtraction-free*.

To a quiver $R$ without loops or 2-cycles with vertex set $\{1, \ldots, n\}$ there corresponds the $n \times n$ antisymmetric integer matrix $B$ whose entry $b_{ij}$ is the number of arrows $i \to j$ minus the number of arrows $j \to i$ in $R$ (notice that at least one of these numbers is zero since $R$ does not have 2-cycles). Clearly, this correspondence yields a bijection. Under this bijection, the matrix $B'$ corresponding to the mutation $\mu_k(R)$ has the entries

$$b'_{ij} = \begin{cases} -b_{ij} & \text{if } i = k \text{ or } j = k; \\ b_{ij} + \text{sgn}(b_{ik})[b_{ik}b_{kj}]_+ & \text{else,} \end{cases}$$

where $[x]_+ = \max(x, 0)$. This is matrix mutation as it was defined by Fomin-Zelevinsky in their seminal paper [50], *cf.* also [55].

### 3.3. Examples of seed and quiver mutations

Let $R$ be the cyclic quiver

$$\tag{3.3.1}$$

and $u = \{x_1, x_2, x_3\}$. If we mutate at $k = 1$, we obtain the quiver

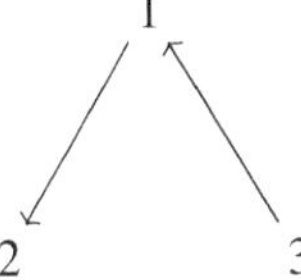

and the set of fractions given by $u'_1 = (x_2 + x_3)/x_1$, $u'_2 = u_2 = x_2$ and $u'_3 = u_3 = x_3$. Now, if we mutate again at 1, we obtain the original seed. This is a general fact: Mutation at $k$ is an involution. If, on the other hand, we mutate

$(R', u')$ at 2, we obtain the quiver

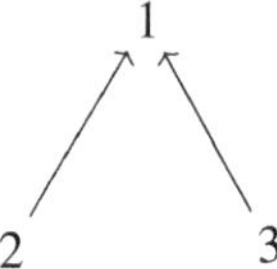

and the set $u''$ given by $u_1'' = u_1' = (x_2 + x_3)/x_1$, $u_2' = \frac{x_1+x_2+x_3}{x_1x_2}$ and $u_3'' = u_3' = x_3$.

An important special case of quiver mutation is the mutation at a source (a vertex without incoming arrows) or a sink (a vertex without outgoing arrows). In this case, the mutation only reverses the arrows incident with the mutating vertex. It is easy to see that all orientations of a tree are mutation equivalent and that only sink and source mutations are needed to pass from one orientation to any other.

Let us consider the following, more complicated quiver glued together from four 3-cycles:

$$(3.3.2)$$

If we successively perform mutations at the vertices 5, 3, 1 and 6, we obtain the sequence of quivers (we use [87])

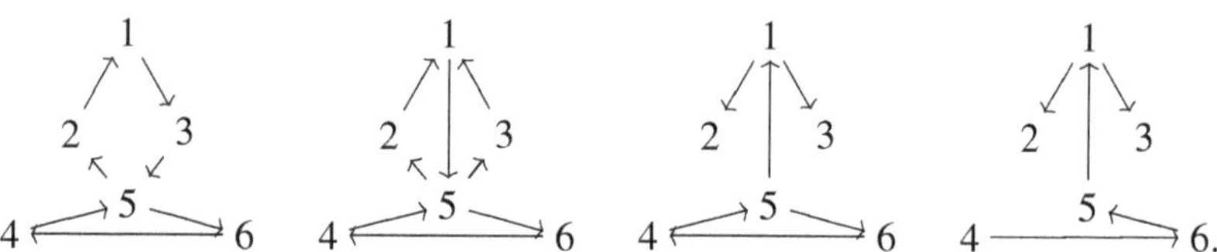

Notice that the last quiver no longer has any oriented cycles and is in fact an orientation of the Dynkin diagram of type $D_6$. The sequence of new fractions appearing in these steps is

$$u_5' = \frac{x_3x_4 + x_2x_6}{x_5}, \quad u_3' = \frac{x_3x_4 + x_1x_5 + x_2x_6}{x_3x_5},$$

$$u_1' = \frac{x_2x_3x_4 + x_3^2x_4 + x_1x_2x_5 + x_2^2x_6 + x_2x_3x_6}{x_1x_3x_5}, \quad u_6' = \frac{x_3x_4 + x_4x_5 + x_2x_6}{x_5x_6}.$$

It is remarkable that all the denominators appearing here are monomials and that all the coefficients in the numerators are positive.

Finally, let us consider the quiver

$$(3.3.3)$$

One can show [91] that it is impossible to transform it into a quiver without oriented cycles by a finite sequence of mutations. However, its mutation class (the set of all quivers obtained from it by iterated mutations) contains many quivers with just one oriented cycle, for example

In fact, in this example, the mutation class is finite and it can be completely computed using, for example, [87]: It consists of 5739 quivers up to isomorphism. The above quivers are members of the mutation class containing relatively few arrows. The initial quiver is the unique member of its mutation class with the largest number of arrows. Here are some other quivers in the mutation class with a relatively large number of arrows:

Only 84 among the 5739 quivers in the mutation class contain double arrows (and none contain arrows of multiplicity $\geq 3$). Here is a typical example

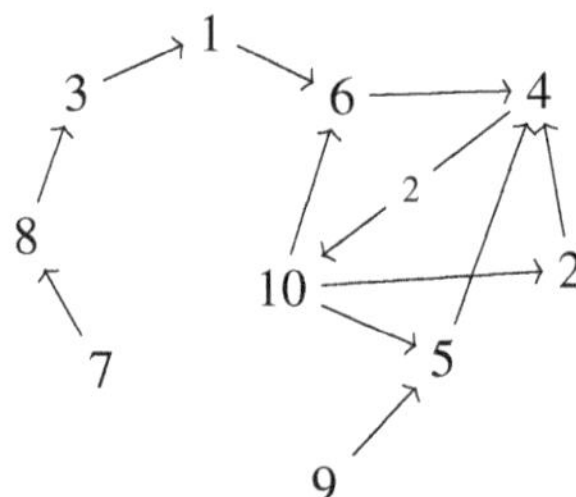

The classification of the quivers with a finite mutation class is still open. Many examples are given in [49] and [38].

The quivers (3.3.1), (3.3.2) and (3.3.3) are part of a family which appears in the study of the cluster algebra structure on the coordinate algebra of the subgroup of upper unitriangular matrices in $SL(n, \mathbb{C})$, *cf.* section 4.6. The quiver (3.3.3) is associated with the elliptic root system $E_8^{(1,1)}$ in the notations of Saito [118], *cf.* Remark 19.4 in [64]. The study of coordinate algebras on varieties associated with reductive algebraic groups (in particular, double Bruhat cells) has provided a major impetus for the development of cluster algebras, *cf.* [17].

## 3.4. Definition of cluster algebras

Let $Q$ be a finite quiver without loops or 2-cycles with vertex set $\{1, \ldots, n\}$. Consider the *initial seed* $(Q, x)$ consisting of $Q$ and the set $x$ formed by the variables $x_1, \ldots, x_n$. Following [50] we define

- the *clusters with respect to* $Q$ to be the sets $u$ appearing in seeds $(R, u)$ obtained from $(Q, x)$ by iterated mutation,
- the *cluster variables* for $Q$ to be the elements of all clusters,
- the *cluster algebra* $\mathcal{A}_Q$ to be the $\mathbb{Q}$-subalgebra of the field $\mathbb{Q}(x_1, \ldots, x_n)$ generated by all the cluster variables.

Thus, the cluster algebra consists of all $\mathbb{Q}$-linear combinations of monomials in the cluster variables. It is useful to define yet another combinatorial object associated with this recursive construction: The *exchange graph* associated with $Q$ is the graph whose vertices are the seeds modulo simultaneous renumbering of the vertices and the associated cluster variables and whose edges correspond to mutations.

A remarkable theorem due to Gekhtman-Shapiro-Vainshtein states that each cluster $u$ occurs in a unique seed $(R, u)$, *cf.* [68].

Notice that the knitting algorithm only produced the cluster variables whereas this definition yields additional structure: the clusters.

## 3.5. The example $A_2$

Here the computation of the exchange graph is essentially equivalent to performing the knitting algorithm. If we denote the cluster variables by $x_1, x_2, x_1'$, $x_2'$ and $x_1''$ as in section 2.2, then the exchange graph is the pentagon

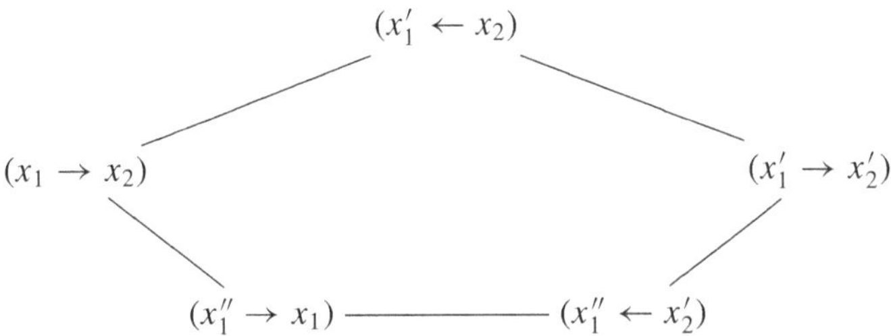

where we have written $x_1 \to x_2$ for the seed $(1 \to 2, \{x_1, x_2\})$. Notice that it is not possible to find a consistent labeling of the edges by 1's and 2's. The reason for this is that the vertices of the exchange graph are not the seeds but the seeds up to renumbering of vertices and variables. Here the clusters are precisely the pairs of consecutive variables in the cyclic ordering of $x_1, \ldots, x_1''$.

## 3.6. The example $A_3$

Let us consider the quiver

$$Q : 1 \longrightarrow 2 \longrightarrow 3$$

obtained by endowing the Dynkin diagram $A_3$ with a linear orientation. By applying the recursive construction to the initial seed $(Q, x)$ one finds exactly fourteen seeds (modulo simultaneous renumbering of vertices and cluster variables). These are the vertices of the exchange graph, which is isomorphic to the third Stasheff associahedron [122] [34]:

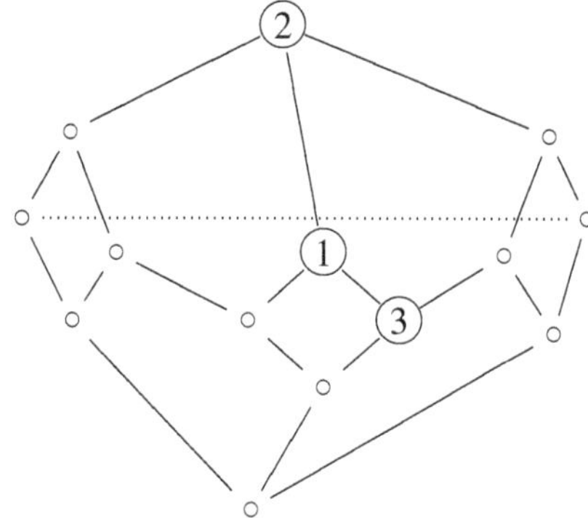

The vertex labeled 1 corresponds to $(Q, x)$, the vertex 2 to $\mu_2(Q, x)$, which is given by

$$1 \xleftarrow{\qquad\qquad} 2 \xleftarrow{\qquad} 3\,, \left\{ x_1, \frac{x_1 + x_3}{x_2}, x_3 \right\}\,,$$

and the vertex 3 to $\mu_1(Q, x)$, which is given by

$$1 \xleftarrow{\qquad} 2 \xrightarrow{\qquad} 3\,, \left\{ \frac{1 + x_2}{x_1}, x_2, x_3 \right\}.$$

As expected (section 2.2), we find a total of $3 + 6 = 9$ cluster variables, which correspond bijectively to the faces of the exchange graph. The clusters $x_1, x_2, x_3$ and $x_1', x_2, x_3$ also appear naturally as slices of the repetition, where by a *slice*, we mean a full connected subquiver containing a representative of each orbit under the horizontal translation (a subquiver is *full* if, with any two vertices, it contains all the arrows between them).

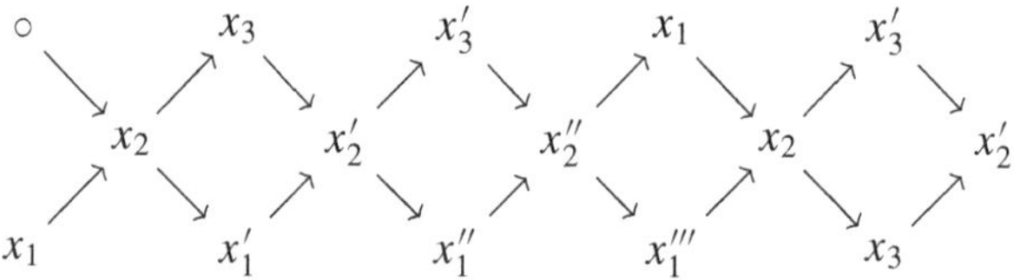

In fact, as it is easy to check, each slice yields a cluster. However, some clusters do not come from slices, for example the cluster $x_1, x_3, x_1''$ associated with the seed $\mu_2(Q, x)$.

## 3.7. Cluster algebras with finitely many cluster variables

The phenomena observed in the above examples are explained by the following key theorem:

**Theorem 3.1** (Fomin-Zelevinsky [52]). *Let $Q$ be a finite connected quiver without loops or 2-cycles with vertex set $\{1, \ldots, n\}$. Let $\mathcal{A}_Q$ be the associated cluster algebra.*

a) *All cluster variables are Laurent polynomials, i.e. their denominators are monomials.*

b) *The number of cluster variables is finite iff $Q$ is mutation equivalent to an orientation of a simply laced Dynkin diagram $\Delta$. In this case, $\Delta$ is unique and the non initial cluster variables are in bijection with the positive roots of $\Delta$; namely, if we denote the simple roots by $\alpha_1, \ldots, \alpha_n$, then for each positive root $\sum d_i \alpha_i$, there is a unique non initial cluster variable whose denominator is $\prod_i x_i^{d_i}$.*

c) *The knitting algorithm yields all cluster variables iff the quiver $Q$ has two vertices or is an orientation of a simply laced Dynkin diagram $\Delta$.*

The theorem can be extended to the non simply laced case if we work with valued quivers as in the example of $G_2$ in section 2.2.

It is not hard to check that the knitting algorithm yields exactly the cluster variables obtained by iterated mutations at sinks and sources. Remarkably, in the Dynkin case, all cluster variables can be obtained in this way.

The construction of the cluster algebra shows that if the quiver $Q$ is mutation-equivalent to $Q'$, then we have an isomorphism

$$\mathcal{A}_{Q'} \xrightarrow{\sim} \mathcal{A}_Q$$

preserving clusters and cluster variables. Thus, to prove that the condition in b) is sufficient, it suffices to show that $\mathcal{A}_Q$ is cluster-finite if the underlying graph of $Q$ is a Dynkin diagram.

No normal form for mutation-equivalence is known in general and it is unkown how to decide whether two given quivers are mutation-equivalent. However, for certain restricted classes, the answer to this problem is known: Trivially, two quivers with two vertices are mutation-equivalent iff they are isomorphic. But it is already a non-trivial problem to decide when a quiver

$$1 \xrightarrow{\phantom{xx}r\phantom{xx}} 2 \; , \\ {}_s\nwarrow \quad \swarrow_t \\ 3$$

where $r$, $s$ and $t$ are non negative integers, is mutation-equivalent to a quiver without a 3-cycle: As shown in [15], this is the case iff the 'Markoff inequality'

$$r^2 + s^2 + t^2 - rst > 4$$

holds or one among $r$, $s$ and $t$ is $< 2$.

For a general quiver $Q$, a criterion for $\mathcal{A}_Q$ to be cluster-finite in terms of quadratic forms was given in [14]. In practice, the quickest way to decide whether a concretely given quiver is cluster-finite and to determine its cluster-type is to compute its mutation-class using [87]. For example, the reader can easily check that for $3 \leq n \leq 8$, the following quiver glued together from $n-2$ triangles

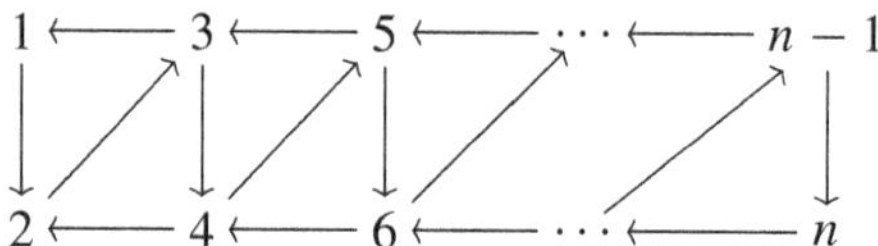

is cluster-finite of respective cluster-type $A_3$, $D_4$, $D_5$, $E_6$, $E_7$ and $E_8$ and that it is not cluster-finite if $n > 8$.

# 4. Cluster algebras with coefficients

In their combinatorial properties, cluster algebras with coefficients are very similar to those without coefficients which we have considered up to now. The great virtue of cluster algebras with coefficients is that they proliferate in nature as algebras of coordinates on homogeneous varieties. We will define cluster algebras with coefficients and illustrate their ubiquity on several examples.

## 4.1. Definition

Let $1 \leq n \leq m$ be integers. An *ice quiver of type* $(n, m)$ is a quiver $\tilde{Q}$ with vertex set

$$\{1, \ldots, m\} = \{1, \ldots, n\} \cup \{n+1, \ldots, m\}$$

such that there are no arrows between any vertices $i$, $j$ which are strictly greater than $n$. The *principal part* of $\tilde{Q}$ is the full subquiver $Q$ of $\tilde{Q}$ whose vertex set is $\{1, \ldots, n\}$ (a subquiver is *full* if, with any two vertices, it contains all the arrows between them). The vertices $n+1, \ldots, m$ are often called *frozen vertices*. The cluster algebra

$$\mathcal{A}_{\tilde{Q}} \subset \mathbb{Q}(x_1, \ldots, x_m)$$

is defined as before but

– only mutations with respect to vertices in the principal part are allowed and no arrows are drawn between the vertices $> n$,

– in a cluster

$$u = \{u_1, \ldots, u_n, x_{n+1}, \ldots, x_m\}$$

only $u_1, \ldots, u_n$ are called cluster variables; the elements $x_{n+1}, \ldots, x_m$ are called *coefficients*; to make things clear, the set $u$ is often called an *extended cluster*;

– the *cluster type of* $\widetilde{Q}$ is that of $Q$ if it is defined.

Often, one also considers localizations of $\mathcal{A}_{\widetilde{Q}}$ obtained by inverting certain coefficients. Notice that the datum of $\widetilde{Q}$ corresponds to that of the integer $m \times n$-matrix $\widetilde{B}$ whose top $n \times n$-submatrix $B$ is antisymmetric and whose entry $b_{ij}$ equals the number of arrows $i \to j$ or the opposite of the number of arrows $j \to i$. The matrix $B$ is called the *principal part* of $\widetilde{B}$. One can also consider valued ice quivers, which will correspond to $m \times n$-matrices whose principal part is antisymmetrizable.

## 4.2. Example: $SL(2, \mathbb{C})$

Let us consider the algebra of regular functions on the algebraic group $SL(2, \mathbb{C})$, *i.e.* the algebra

$$\mathbb{C}[a, b, c, d]/(ad - bc - 1).$$

We claim that this algebra *has a cluster algebra structure*, namely that it is isomorphic to the complexification of the cluster algebra with coefficients associated with the following ice quiver

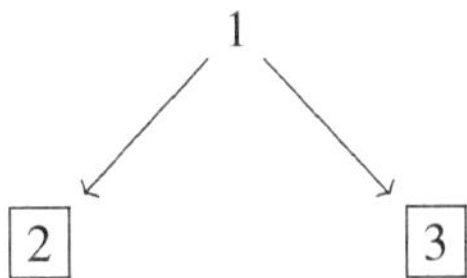


where we have framed the frozen vertices. Indeed, here the principal part $Q$ only consists of the vertex 1 and we can only perform one mutation, whose associated exchange relation reads

$$x_1 x_1' = 1 + x_2 x_3 \text{ or } x_1 x_1' - x_2 x_3 = 1.$$

We obtain an isomorphism as required by sending $x_1$ to $a$, $x_1'$ to $d$, $x_2$ to $b$ and $x_3$ to $c$. We describe this situation by saying that the quiver

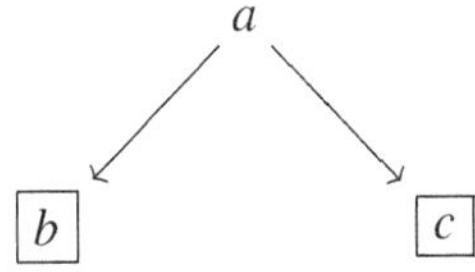

whose vertices are labeled by the images of the corresponding variables, is an initial seed for a cluster structure on the algebra $A$. Notice that this cluster structure is not unique.

## 4.3. Example: Planes in affine space

As a second example, let us consider the algebra $A$ of polynomial functions on the cone over the Grassmannian of planes in $\mathbb{C}^{n+3}$. This algebra is the $\mathbb{C}$-algebra generated by the Plücker coordinates $x_{ij}$, $1 \leq i < j \leq n + 3$, subject to the Plücker relations: for each quadruple of integers $i < j < k < l$, we have

$$x_{ik}x_{jl} = x_{ij}x_{kl} + x_{jk}x_{il}.$$

Each plane $P$ in $\mathbb{C}^{n+3}$ gives rise to a straight line in this cone, namely the one generated by the $2 \times 2$-minors $x_{ij}$ of any $(n + 3) \times 2$-matrix whose columns generate $P$. Notice that the monomials in the Plücker relation are naturally associated with the sides and the diagonals of the square

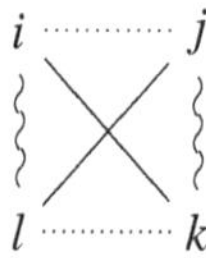


The relation expresses the product of the variables associated with the diagonals as the sum of the monomials associated with the two pairs of opposite sides.

Now the idea is that the Plücker relations are exactly the exchange relations for a suitable structure of cluster algebra with coefficients on the coordinate ring. To formulate this more precisely, let us consider a regular $(n + 3)$-gon in the plane with vertices numbered $1, \ldots, n + 2$, and consider the variable $x_{ij}$ as associated with the segment $[ij]$ joining the vertices $i$ and $j$.

**Proposition 4.1** ([52, Example 12.6]). *The algebra $A$ has a structure of cluster algebra with coefficients such that*

- *the coefficients are the variables $x_{ij}$ associated with the sides of the $(n + 3)$-gon;*
- *the cluster variables are the variables $x_{ij}$ associated with the diagonals of the $(n + 3)$-gon;*
- *the clusters are the n-tuples of variables whose associated diagonals form a triangulation of the $(n + 3)$-gon.*

*Moreover, the exchange relations are exactly the Plücker relations and the cluster type is $A_n$.*

Thus, a triangulation of the $(n + 3)$-gon determines an initial seed for the cluster algebra and hence an ice quiver $\widetilde{Q}$ whose frozen vertices correspond to the sides of the $(n + 3)$-gon and whose non frozen variables to the diagonals in the triangulation. The arrows of the quiver are determined by the exchange relations which appear when we wish to replace one diagonal $[ik]$ of the triangulation by its *flip*, i.e. the unique diagonal $[jl]$ different from $[ik]$ which does not cross any other diagonal of the triangulation. It is not hard to see that this means that the underlying graph of $\widetilde{Q}$ is the graph dual to the triangulation and that the orientation of the edges of this graph is induced by the choice of an orientation of the plane. Here is an example of a triangulation and the associated ice quiver:

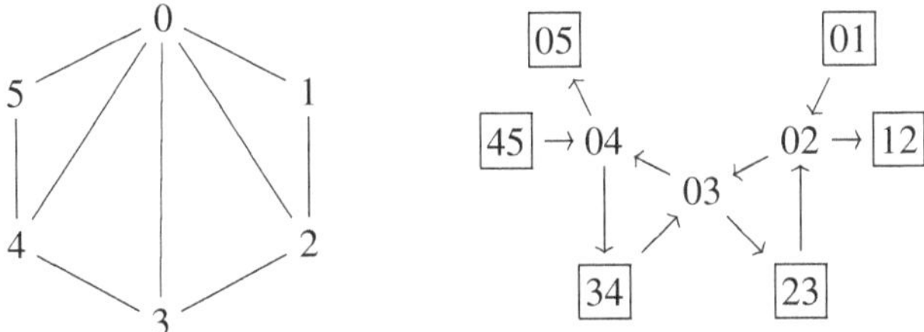

## 4.4. Example: The big cell of the Grassmannian

We consider the cone over the big cell in the Grassmannian of $k$-dimensional subspaces of the space of rows $\mathbb{C}^n$, where $1 \leq k \leq n$ are fixed integers such that $l = n - k$ is greater or equal to 2. In more detail, let $G$ be the group $SL(n, \mathbb{C})$ and $P$ the subgroup of $G$ formed by the block lower triangular matrices with diagonal blocks of sizes $k \times k$ and $l \times l$. The quotient $P \backslash G$ identifies with our Grassmannian. The big cell is the image under $\pi : G \to P \backslash G$ of the space of block upper triangular matrices whose diagonal is the identity matrix and whose upper right block is an arbitrary $k \times l$-matrix $Y$. The projection $\pi$ induces an isomorphism between the space $M_{k \times l}(\mathbb{C})$ of these matrices and the big cell. In particular, the algebra $A$ of regular functions on the big cell is the algebra of polynomials in the coefficients $y_{ij}$, $1 \leq i \leq k$, $1 \leq j \leq l$, of $Y$. Now for $1 \leq i \leq k$ and $1 \leq j \leq l$, let $F_{ij}$ be the largest square submatrix of $Y$ whose lower left corner is $(i, j)$ and let $s(i, j)$ be its size. Put

$$f_{ij} = (-1)^{(k-i)(s(i,j)-1)} \det(F_{ij}).$$

**Theorem 4.2** ([69]). *The algebra A has the structure of a cluster algebra with coefficients whose initial seed is given by*

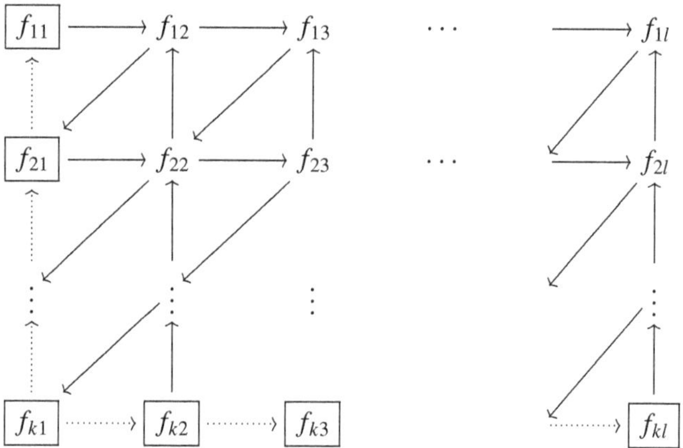

The following table indicates when these algebras are cluster-finite and what their cluster-type is:

| $k \setminus n$ | 2 | 3 | 4 | 5 | 6 | 7 |
|---|---|---|---|---|---|---|
| 2 | $A_1$ | $A_2$ | $A_3$ | $A_4$ | $A_5$ | $A_6$ |
| 3 | $A_2$ | $D_4$ | $E_6$ | $E_8$ | | |
| 4 | $A_3$ | $E_6$ | | | | |
| 5 | $A_4$ | $E_8$ | | | | |

The homogeneous coordinate ring of the Grassmannian $Gr(k, n)$ itself also has a cluster algebra structure [120] and so have partial flag varieties, double Bruhat cells, Schubert varieties ..., *cf.* [62] [17].

## 4.5. Compatible Poisson structures

Recall that the group $SL(n, \mathbb{C})$ has a canonical Poisson structure given by the *Sklyanin bracket*, which is defined by

$$\omega(x_{ij}, x_{\alpha\beta}) = (\mathrm{sign}(\alpha - i) - \mathrm{sign}(\beta - j))x_{i\beta}x_{\alpha j}$$

where the $x_{ij}$ are the coordinate functions on $SL(n, \mathbb{C})$. This bracket makes $G = SL(n, \mathbb{C})$ into a Poisson-Lie group and $P \setminus G$ into a Poisson $G$-variety for each subgroup $P$ of $G$ containing the subgroup $B$ of lower triangular matrices. In particular, the big cell of the Grassmannian considered above inherits a Poisson bracket.

**Theorem 4.3** ([69]). *This bracket is compatible with the cluster algebra structure in the sense that each extended cluster is a* log-canonical *coordinate system,* *i.e. we have*

$$\omega(u_i, u_j) = \omega_{ij}^{(u)} u_i u_j$$

*for certain (integer) constants $\omega_{ij}^{(u)}$ depending on the extended cluster $u$. Moreover, the coefficients are central for $\omega$.*

This theorem admits the following generalization: Let $\widetilde{Q}$ be an ice quiver. Define the *cluster variety* $\mathcal{X}(\widetilde{Q})$ to be obtained by glueing the complex tori indexed by the clusters $u$

$$T^{(u)} = (\mathbb{C}^*)^n = \mathrm{Spec}(\mathbb{C}[u_1, u_1^{-1}, \ldots, u_m, u_m^{-1}])$$

using the exchange relations as glueing maps, where $m$ is the number of vertices of $\widetilde{Q}$.

**Theorem 4.4** ([69]). *Suppose that the principal part $Q$ of $\widetilde{Q}$ is connected and that the matrix $\widetilde{B}$ associated with $\widetilde{Q}$ is of maximal rank. Then the vector space of Poisson structures on $\mathcal{X}(\widetilde{Q})$ compatible with the cluster algebra structure is of dimension*

$$1 + \binom{m - n}{2}.$$

Notice that in general, the cluster variety $\mathcal{X}(\widetilde{Q})$ is an open subset of the spectrum of the (complexified) cluster algebra. For example, for the cluster algebra associated with $SL(2, \mathbb{C})$ which we have considered above, the cluster variety is the union of the elements

$$\begin{pmatrix} a & b \\ c & d \end{pmatrix}$$

of $SL(2, \mathbb{C})$ such that we have $abc \neq 0$ or $bcd \neq 0$. The cluster variety is always regular, but the spectrum of the cluster algebra may be singular. For example, the spectrum of the cluster algebra associated with the ice quiver

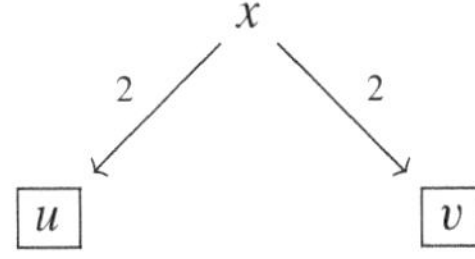


is the hypersurface in $\mathbb{C}^4$ defined by the equation $xx' = u^2 + v^2$, which is singular at the origin. The corresponding cluster variety is obtained by removing the points with $x = x' = u^2 + v^2 = 0$ and is regular.

In the above theorem, the assumption that $\widetilde{B}$ be of full rank is essential. Otherwise, there may not exist any Poisson bracket compatible with the cluster algebra structure. However, as shown in [70], for any cluster algebra with coefficients, there are 'dual Poisson structures', namely certain 2-forms, which are compatible with the cluster algebra structure.

## 4.6. Example: The maximal unipotent subgroup of $SL(n + 1, \mathbb{C})$

Let $n$ be a non negative integer and $N$ the subgroup of $SL(n + 1, \mathbb{C})$ formed by the upper triangular matrices with all diagonal coefficients equal to 1. For $1 \leq i, j \leq n + 1$ and $g \in N$, let $F_{ij}(g)$ be the maximal square submatrix of $g$ whose lower left corner is $(i, j)$. Let $f_{ij}(g)$ be the determinant of $F_{ij}(g)$. We consider the functions $f_{ij}$ for $1 \leq i \leq n$ and $i + j \leq n + 1$.

**Theorem 4.5** ([17]). *The coordinate algebra* $\mathbb{C}[N]$ *has an upper cluster algebra structure whose initial seed is given by*

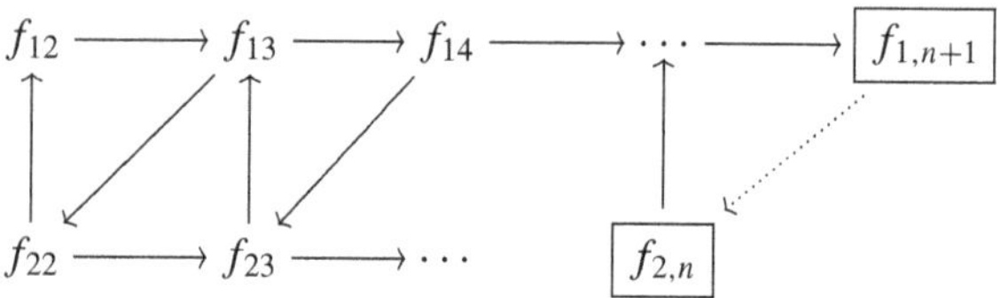

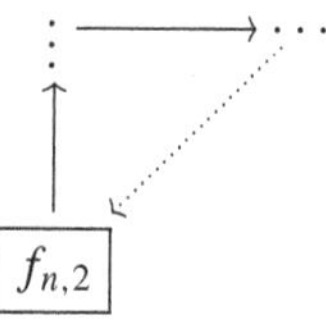

We refer to [17] for the notion of 'upper' cluster algebra structure. It is not hard to check that this structure is of cluster type $A_3$ for $n = 3$, $D_6$ for $n = 4$ and cluster-infinite for $n \geq 5$. For $n = 5$, this cluster algebra is related to the elliptic root system $E_8^{(1,1)}$ in the notations of Saito [118], *cf.* [64].

A theorem of Fekete [42] generalized in [16] claims that a square matrix of order $n + 1$ is *totally positive* (*i.e.* all its minors are $> 0$) if and only if the following $(n + 1)^2$ minors of g are positive: all minors occupying several initial rows and several consecutive columns, and all minors occupying several initial columns and several consecutive rows. It follows that an element $g$ of $N$ is totally positive if $f_{ij}(g) > 0$ for the $f_{ij}$ belonging to the initial seed above. The same holds for the $u_1, \ldots, u_m$ in place of these $f_{ij}$ for any cluster $u$ of this cluster algebra because each exchange relation expresses the new variable *subtraction-free* in the old variables.

Geiss-Leclerc-Schröer have shown [65] that each monomial in the variables of an arbitrary cluster belongs to Lusztig's dual semicanonical basis of $\mathbb{C}[N]$ [104]. They also show that the dual semicanonical basis of $\mathbb{C}[N]$ is different from the dual canonical basis of Lusztig and Kashiwara except in types $A_2$, $A_3$ and $A_4$ [64].

## 5. Categorification via cluster categories: the finite case

### 5.1. Quiver representations and Gabriel's theorem

We refer to the books [117] [59] [5] and [4] for a wealth of information on the representation theory of quivers and finite-dimensional algebras. Here, we will only need very basic notions.

Let $Q$ be a finite quiver without oriented cycles. For example, $Q$ can be an orientation of a simply laced Dynkin diagram or the quiver

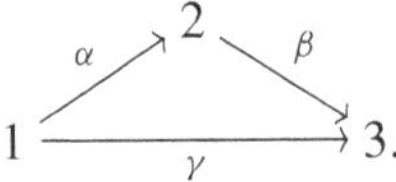

Let $k$ be an algebraically closed field. A *representation of $Q$* is a diagram of finite-dimensional vector spaces of the shape given by $Q$. More formally, a representation of $Q$ is the datum $V$ of

- a finite-dimensional vector space $V_i$ for each vertex $i$ or $Q$,
- a linear map $V_\alpha : V_i \rightarrow V_j$ for each arrow $\alpha : i \rightarrow j$ of $Q$.

Thus, in the above example, a representation of $Q$ is a (not necessarily commutative) diagram

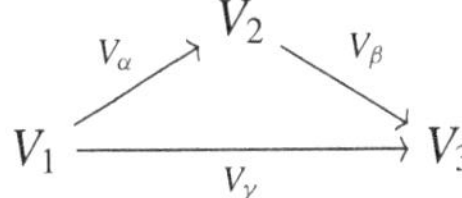

formed by three finite-dimensional vector spaces and three linear maps. A *morphism of representations* is a morphism of diagrams. More formally, a morphism of representations $f : V \rightarrow W$ is the datum of a linear map $f_i :

$V_i \to W_i$ for each vertex $i$ of $Q$ such that the square

$$
\begin{array}{ccc}
V_i & \xrightarrow{\;V_\alpha\;} & V_j \\
\downarrow{\scriptstyle f_i} & & \downarrow{\scriptstyle f_j} \\
W_i & \xrightarrow[\;W_\alpha\;]{} & W_j
\end{array}
$$

commutes for all arrows $\alpha : i \to j$ of $Q$. The *composition* of morphisms is defined in the natural way. We thus obtain the *category of representations* $\mathsf{rep}(Q)$. A morphism $f : V \to W$ of this category is an isomorphism iff its components $f_i$ are invertible for all vertices $i$ of $Q_0$.

For example, let $Q$ be the quiver

$$ 1 \longrightarrow 2 \, , $$

and

$$ V : \quad V_1 \xrightarrow{\;V_\alpha\;} V_2 $$

a representation of $Q$. By choosing basis in the spaces $V_1$ and $V_2$ we find an isomorphism of representations

$$
\begin{array}{ccc}
V_1 & \xrightarrow{\;V_\alpha\;} & V_2 \\
\uparrow & & \uparrow \\
k^n & \xrightarrow[\;A\;]{} & k^p \, ,
\end{array}
$$

where, by abuse of notation, we denote by $A$ the multiplication by a $p \times n$-matrix $A$. We know that we have

$$ PAQ = \begin{bmatrix} I_r & 0 \\ 0 & 0 \end{bmatrix} $$

for invertible matrices $P$ and $Q$, where $r$ is the rank of $A$. Let us denote the right hand side by $I_r \oplus 0$. Then we have an isomorphism of representations

$$
\begin{array}{ccc}
k^n & \xrightarrow{\;A\;} & k^p \\
\uparrow{\scriptstyle Q} & & \uparrow{\scriptstyle P^{-1}} \\
k^n & \xrightarrow[\;I_r \oplus 0\;]{} & k^p
\end{array}
$$

We thus obtain a normal form for the representations of this quiver.

Now the category $\mathsf{rep}_k(Q)$ is in fact an *abelian category*: Its direct sums, kernels and cokernels are computed componentwise. Thus, if $V$ and $W$ are two representations, then the *direct sum* $V \oplus W$ is the representation given by

$$(V \oplus W)_i = V_i \oplus W_i \text{ and } (V \oplus W)_\alpha = V_\alpha \oplus W_\alpha ,$$

for all vertices $i$ and all arrows $\alpha$ of $Q$. For example, the above representation in normal form is isomorphic to the direct sum

$$(k \xrightarrow{1} k)^r \oplus (k \longrightarrow 0)^{n-r} \oplus (0 \longrightarrow k)^{p-r}.$$

The *kernel* of a morphism of representations $f : V \to W$ is given by

$$\ker(f)_i = \ker(f_i : V_i \to W_i)$$

endowed with the maps induced by the $V_\alpha$ and similarly for the cokernel. A *subrepresentation* $V'$ of a representation $V$ is given by a family of subspaces $V_i' \subset V_i$, $i \in Q_0$, such that the image of $V_i'$ under $V_\alpha$ is contained in $V_j'$ for each arrow $\alpha : i \to j$ of $Q$. A sequence

$$0 \longrightarrow U \longrightarrow V \longrightarrow W \longrightarrow 0$$

of representations is a *short exact sequence* if the sequence

$$0 \longrightarrow U_i \longrightarrow V_i \longrightarrow W_i \longrightarrow 0$$

is exact for each vertex $i$ of $Q$.

A representation $V$ is *simple* if it is non zero and if for each subrepresentation $V'$ of $V$ we have $V' = 0$ or $V/V' = 0$. Equivalently, a representation is simple if it has exactly two subrepresentations. A representation $V$ is *indecomposable* if it is non zero and in each decomposition $V = V' \oplus V''$, we have $V' = 0$ or $V'' = 0$. Equivalently, a representation is indecomposable if it has exactly two direct factors.

In the above example, the representations

$$k \longrightarrow 0 \text{ and } 0 \longrightarrow k$$

are simple. The representation

$$V = (k \xrightarrow{1} k)$$

is not simple: It has the non trivial subrepresentation $0 \longrightarrow k$. However, it is indecomposable. Indeed, each endomorphism $f : V \to V$ is given by two equal components $f_1 = f_2$ so that the endomorphism algebra of $V$ is one-dimensional. If $V$ was a direct sum $V' \oplus V''$ for two non-zero subspaces, the endomorphism algebra of $V$ would contain the product of the endomorphism algebras of $V'$ and $V''$ and thus would have to be at least of dimension 2. Since

$V$ is indecomposable, the exact sequence

$$0 \to (0 \xrightarrow{\quad} k) \to (k \xrightarrow{\ 1\ } k) \to (k \xrightarrow{\quad} 0) \to 0$$

is not a split exact sequence.

If $Q$ is an arbitrary quiver, for each vertex $i$, we define the representation $S_i$ by

$$(S_i)_j = \begin{cases} k & i = j \\ 0 & \text{else.} \end{cases}$$

Then clearly the representations $S_i$ are simple and pairwise non isomorphic. As an exercise, the reader may show that if $Q$ does not have oriented cycles, then each representation admits a finite filtration whose subquotients are among the $S_i$. Thus, in this case, each simple representation is isomorphic to one of the representations $S_i$.

Recall that a (possibly non commutative) ring is *local* if its non invertible elements form an ideal.

**Decomposition Theorem 5.1** (Azumaya-Fitting-Krull-Remak-Schmidt).

a) *A representation is indecomposable iff its endomorphism algebra is local.*
b) *Each representation decomposes into a finite sum of indecomposable representations, unique up to isomorphism and permutation.*

As we have seen above, for quivers without oriented cycles, the classification of the simple representations is trivial. On the other hand, the problem of classifying the indecomposable representations is non trivial. Let us examine this problem in a few examples: For the quiver $1 \to 2$, we have checked the existence in part b) directly. The uniqueness in b) then implies that each indecomposable representation is isomorphic to exactly one of the representations $S_1$, $S_2$ and

$$k \xrightarrow{\ 1\ } k \,.$$

Similarly, using elementary linear algebra it is not hard to check that each indecomposable representation of the quiver

$$\vec{A}_n : 1 \longrightarrow 2 \longrightarrow \cdots \longrightarrow n$$

is isomorphic to a representation $I[p, q]$, $1 \le p < q \le n$, which takes the vertices $i$ in the interval $[p, q]$ to $k$, the arrows linking them to the identity and all other vertices to zero. In particular, the number of isomorphism classes of indecomposable representations of $\vec{A}_n$ is $n(n + 1)/2$.

The representations of the quiver

$$1 \circlearrowright \alpha$$

are the pairs $(V_1, V_\alpha)$ consisting of a finite-dimensional vector space and an endomorphism and the morphisms of representations are the 'intertwining operators'. It follows from the existence and uniqueness of the Jordan normal form that a system of representatives of the isomorphism classes of indecomposable representations is formed by the representations $(k^n, J_{n,\lambda})$, where $n \geq 1$ is an integer, $\lambda$ a scalar and $J_{n,\lambda}$ the Jordan block of size $n$ with eigenvalue $\lambda$.

The *Kronecker quiver*

$$1 \rightrightarrows 2$$

admits the following infinite family of pairwise non isomorphic representations:

$$k \underset{\mu}{\overset{\lambda}{\rightrightarrows}} k \ ,$$

where $(\lambda : \mu)$ runs through the projective line.

**Question 5.2.** *For which quivers are there only finitely many isomorphism classes of indecomposable representations?*

To answer this question, we define the *dimension vector* of a representation $V$ to be the sequence $\underline{\dim}\, V$ of the dimensions $\dim V_i$, $i \in Q_0$. For example, the dimension vectors of the indecomposable representations of $\vec{A}_2$ are the pairs

$$\underline{\dim}\, S_1 = [10] \, , \quad \underline{\dim}\, S_2 = [01] \, , \quad \underline{\dim}\, (k \to k) = [11].$$

We define the *Tits form*

$$q_Q : \mathbb{Z}^{Q_0} \to \mathbb{Z}$$

by

$$q_Q(v) = \sum_{i \in Q_0} v_i^2 - \sum_{\alpha \in Q_1} v_{s(\alpha)} v_{t(\alpha)}.$$

Notice that the Tits form does not depend on the orientation of the arrows of $Q$ but only on its underlying graph. We say that the quiver $Q$ is *representation-finite* if, up to isomorphism, it has only finitely many indecomposable representations. We say that a vector $v \in \mathbb{Z}^{Q_0}$ is a *root* of $q_Q$ if $q_Q(v) = 1$ and that it is *positive* if its components are $\geq 0$.

**Theorem 5.3** (Gabriel [58]). *Let $Q$ be a connected quiver and assume that $k$ is algebraically closed. The following are equivalent:*

(i)  *Q is representation-finite;*

(ii)  $q_Q$ *is positive definite;*

(iii)  *The underlying graph of Q is a simply laced Dynkin diagram $\Delta$.*

*Moreover, in this case, the map taking a representation to its dimension vector yields a bijection from the set of isomorphism classes of indecomposable representations to the set of positive roots of the Tits form $q_Q$.*

It is not hard to check that if the conditions hold, the positive roots of $q_Q$ are in turn in bijection with the positive roots of the root system $\Phi$ associated with $\Delta$, via the map taking a positive root $v$ of $q_Q$ to the element

$$\sum_{i \in Q_0} v_i \alpha_i$$

of the root lattice of $\Phi$.

Let us consider the example of the quiver $Q = \vec{A}_2$. In this case, the Tits form is given by

$$q_Q(v) = v_1^2 + v_2^2 - v_1 v_2.$$

It is positive definite and its positive roots are indeed precisely the dimension vectors

$$[01]\,,\ [10]\,,\ [11]$$

of the indecomposable representations.

Gabriel's theorem has been generalized to non algebraically closed ground fields by Dlab and Ringel [40]. Let us illustrate the main idea on one simple example: Consider the category of diagrams

$$V :\ V_1 \xrightarrow{\ f\ } V_2$$

where $V_1$ is a finite-dimensional real vector space, $V_2$ a finite-dimensional complex vector space and $f$ an $\mathbb{R}$-linear map. Morphisms are given in the natural way. Then we have the following complete list of representatives of the isomorphism classes of indecomposables:

$$\mathbb{R} \to 0\,,\ \mathbb{R}^2 \to \mathbb{C}\,,\ \mathbb{R} \to \mathbb{C}\,,\ 0 \to \mathbb{C}.$$

The corresponding dimension vectors are

$$[10]\,,\ [21]\,,\ [11]\,,\ [01].$$

They correspond bijectively to the positive roots of the root system $B_2$.

## 5.2. Tame and wild quivers

The quivers with infinitely many isomorphism classes of indecomposables can be further subdivided into two important classes: A quiver is *tame* if it has infinitely many isomorphism classes of indecomposables but these occur in 'families of at most one parameter' (we refer to [117] [4] for the precise definition). The Kronecker quiver is a typical example. A quiver is *wild* if there are 'families of indecomposables of $\geq 2$ parameters'. One can show that in this case, there are families of an arbitrary number of parameters and that the classification of the indecomposables over any fixed wild algebra would entail the classification of the the indecomposables over all finite-dimensional algebras. The following three quivers are representation-finite, tame and wild respectively:

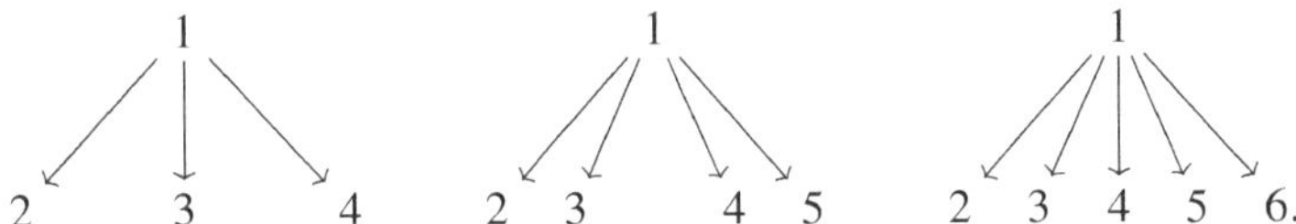

**Theorem 5.4** (Donovan-Freislich [41], Nazarova [110]). *Let $Q$ be a connected quiver and assume that $k$ is algebraically closed. Then $Q$ is tame iff the underlying graph of $Q$ is a simply laced extended Dynkin diagram.*

Let us recall the list of simply laced extended Dynkin quivers. In each case, the number of vertices of the diagram $\widetilde{D}_n$ equals $n + 1$.

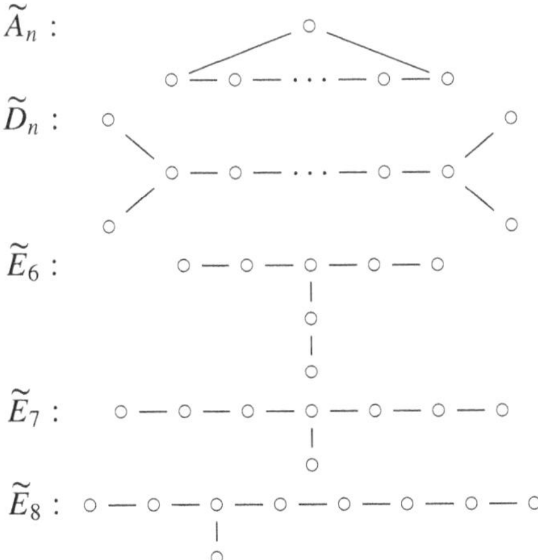

The following theorem is a first illustration of the close connection between cluster algebras and the representation theory of quivers. Let $Q$ be a finite quiver without oriented cycles and let $v(Q)$ be the supremum of the multiplicities of the arrows occurring in all quivers mutation-equivalent to $Q$.

**Theorem 5.5.**
a) *$Q$ is representation-finite iff $v(Q)$ equals 1.*
b) *$Q$ is tame iff $v(Q)$ equals 2.*
c) *$Q$ is wild iff $v(Q) \geq 3$ iff $v(Q) = \infty$.*
d) *The mutation class of $Q$ is finite iff $Q$ has two vertices, is representation-finite or tame.*

Here, part a) follows from Gabriel's theorem and part (iii) of Theorem 1.8 in [52]. Part b) follows from parts a) and c) by exclusion of the third. For part c), let us first assume that $Q$ is wild. Then it is proved at the end of the proof of theorem 3.1 in [13] that $v(Q) = \infty$. Conversely, let us assume that $v(Q) \geq 3$. Then using Theorem 5 of [29] we obtain that $Q$ is wild. Part d) is proved in [13].

## 5.3. The Caldero-Chapoton formula

Let $\Delta$ be a simply laced Dynkin diagram and $Q$ a quiver with underlying graph $\Delta$. Suppose that the set of vertices of $\Delta$ and $Q$ is the set of the natural numbers $1, 2, \ldots, n$. We already know from part b) of theorem 3.1 that for each positive root

$$\alpha = \sum_{i=1}^{n} d_i \alpha_i$$

of the corresponding root system, there is a unique non initial cluster variable $X_\alpha$ with denominator

$$x_1^{d_1} \ldots x_n^{d_n}.$$

By combining this with Gabriel's theorem, we get the

**Corollary 5.6.** *The map taking an indecomposable representation $V$ with dimension vector $(d_i)$ of $Q$ to the unique non initial cluster variable $X_V$ whose denominator is $x_1^{d_1} \ldots x_n^{d_n}$ induces a bijection from the set of ismorphism classes of indecomposable representations to the set of non initial cluster variables.*

Let us consider this bijection for $Q = \vec{A}_2$:

$$S_2 = (0 \to k) \qquad\qquad P_1 = (k \to k) \qquad\qquad S_1 = (k \to 0)$$

$$X_{S_2} = \frac{1 + x_1}{x_2} \qquad X_{P_1} = \frac{x_1 + 1 + x_2}{x_1 x_2} \qquad S_{S_1} = \frac{1 + x_2}{x_1}$$

We observe that for the two simple representations, the numerator contains exactly two terms: the number of subrepresentations of the simple representation! Moreover, the representation $P_1$ has exactly three subrepresentations and the numerator of $X_{P_1}$ contains three terms. In fact, it turns out that this phenomenon is general in type $A$. But now let us consider the following quiver with underlying graph $D_4$

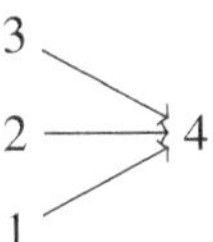

and the dimension vector $d$ with $d_1 = d_2 = d_3 = 1$ and $d_4 = 2$. The unique (up to isomorphism) indecomposable representation $V$ with dimension vector $d$ consists of a plane $V_4$ together with three lines in general position $V_i \subset V_4$, $i = 1, 2, 3$. The corresponding cluster variable is

$$X_4 = \frac{1}{x_1 x_2 x_3 x_4^2} (1 + 3x_4 + 3x_4^2 + x_4^3 + 2x_1 x_2 x_3 + 3x_1 x_2 x_3 x_4 + x_1^2 x_2^2 x_3^2).$$

Its numerator contains a total of 14 monomials. On the other hand, it is easy to see that $V_4$ has only 13 types of submodules: twelve submodules are determined by their dimension vectors but for the dimension vector $e = (0, 0, 0, 1)$, we have a family of submodules: Each submodule of this dimension vector corresponds to the choice of a line in $V_4$. Thus for this dimension vector $e$, the family of submodules is parametrized by a projective line. Notice that the Euler characteristic of the projective line is 2 (since it is a sphere: the Riemann sphere). So if we attribute weight 1 to the submodules determined by their dimension vector and weight 2 to this $\mathbb{P}^1$-family, we find a 'total submodule weight' equal to the number of monomials in the numerator. These considerations led Caldero-Chapoton [27] to the following definition, whose ingredients we describe below: Let $Q$ be a finite quiver with vertices $1, \ldots, n$, and $V$ a finite-dimensional representation of $Q$. Let $d$ be the dimension vector of $V$. Define

$$CC(V) = \frac{1}{x_1^{d_1} x_2^{d_2} \ldots x_n^{d_n}} \left( \sum_{0 \leq e \leq d} \chi(\mathrm{Gr}_e(V)) \prod_{i=1}^{n} x_i^{\sum_{j \to i} e_j + \sum_{i \to j}(d_j - e_j)} \right).$$

Here the sum is taken over all vectors $e \in \mathbb{N}^n$ such that $0 \leq e_i \leq d_i$ for all $i$. For each such vector $e$, the *quiver Grassmannian* $\mathrm{Gr}_e(V)$ is the variety of $n$-tuples of subspaces $U_i \subset V_i$ such that $\dim U_i = e_i$ and the $U_i$ form a subrepresentation of $V$. By taking such a subrepresentation to the family of the $U_i$, we obtain a map

$$\mathrm{Gr}_e(V) \to \prod_{i=1}^{n} \mathrm{Gr}_{e_i}(V_i)\,,$$

where $\mathrm{Gr}_{e_i}(V_i)$ denotes the ordinary Grassmannian of $e_i$-dimensional subspaces of $V_i$. Recall that the Grassmannian carries a canonical structure of projective variety. It is not hard to see that for a family of subspaces $(U_i)$ the condition of being a subrepresentation is a closed condition so that the quiver Grassmannian identifies with a projective subvariety of the product of ordinary Grassmannians. If $k$ is the field of complex numbers, the Euler characteristic $\chi$ is taken with respect to singular cohomology with coefficients in $\mathbb{Q}$ (or any other field). If $k$ is an arbitrary algebraically closed field, we use étale cohomology to define $\chi$. The most important properties of $\chi$ are (*cf. e.g.* section 7.4 in [67])

(1) $\chi$ is additive with respect to disjoint unions;
(2) if $p : E \to X$ is a morphism of algebraic varieties such that the Euler characteristic of the fiber over a point $x \in X$ does not depend on $x$, then $\chi(E)$ is the product of $\chi(X)$ by the Euler characteristic of the fiber over any point $x \in X$.

**Theorem 5.7** (Caldero-Chapoton [27]). *Let $Q$ be a Dynkin quiver and $V$ an indecomposable representation. Then we have $CC(V) = X_V$, the cluster variable obtained from $V$ by composing Fomin-Zelevinsky's bijection with Gabriel's.*

Caldero-Chapoton's proof of the theorem was by induction. One of the aims of the following sections is to explain 'on what' they did the induction.

## 5.4. The derived category

Let $k$ be an algebraically closed field and $Q$ a (possibly infinite) quiver without oriented cycles (we will impose more restrictive conditions on $Q$ later). For example, $Q$ could be the quiver

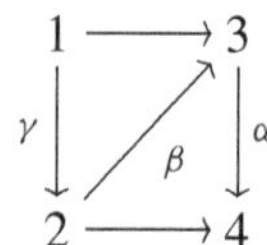

A *path* of $Q$ is a formal composition of $\geq 0$ arrows. For example, the sequence $(4|\alpha|\beta|\gamma|1)$ is a path of length 3 in the above example (notice that we include the source and target vertices of the path in the notation). For each vertex $i$ of

$Q$, we have the *lazy path* $e_i = (i|i)$, the unique path of length 0 which starts at $i$ and stops at $i$ and does nothing in between. The *path category* has set of objects $Q_0$ (the set of vertices of $Q$) and, for any vertices $i$, $j$, the morphism space from $i$ to $j$ is the vector space whose basis consists of all paths from $i$ to $j$. Composition is induced by composition of paths and the unit morphisms are the lazy paths. If $Q$ is finite, we define the *path algebra* to be the matrix algebra

$$kQ = \bigoplus_{i,j \in Q_0} \mathrm{Hom}(i, j)$$

where multiplication is matrix multiplication. Equivalently, the path algebra has as a basis all paths and its product is given by concatenating composable paths and equating the product of non composable paths to zero. The path algebra has the sum of the lazy paths as its unit element

$$1 = \sum_{i \in Q_0} e_i.$$

The idempotent $e_i$ yields the *projective right module*

$$P_i = e_i kQ.$$

The modules $P_i$ generate the *category of k-finite-dimensional right modules* $\mathrm{mod}\, kQ$. Each arrow $\alpha$ from $i$ to $j$ yields a map $P_i \to P_j$ given by left multiplication by $\alpha$. (If we were to consider – heaven forbid – left modules, the analogous map would be given by right multiplication by $\alpha$ and it would go in the direction opposite to that of $\alpha$. Whence our preference for right modules).

Notice that we have an equivalence of categories

$$\mathrm{rep}_k(Q^{op}) \to \mathrm{mod}\, kQ$$

sending a representation $V$ of the opposite quiver $Q^{op}$ to the sum

$$\bigoplus_{i \in Q_0} V_i$$

endowed with the natural right action of the path algebra. Conversely, a $kQ$-module $M$ gives rise to the representation $V$ with $V_i = Me_i$ for each vertex $i$ of $Q$ and $V_\alpha$ given by right multiplication by $\alpha$ for each arrow $\alpha$ of $Q$. The category $\mathrm{mod}\, kQ$ is *abelian*, *i.e.* it is additive, has kernels and cokernels and for each morphism $f$ the cokernel of its kernel is canonically isomorphic to the kernel of its cokernel.

The category $\mathrm{mod}\, kQ$ is *hereditary*. Recall from [32] that this means that submodules of projective modules are projective; equivalently, that all extension groups in degrees $i \geq 2$ vanish:

$$\mathsf{Ext}^i_{kQ}(L, M) = 0;$$

equivalently, that $kQ$ is of global dimension $\leq 1$; ... Thus, in the spirit of noncommutative algebraic geometry approached via abelian categories, we should think of $\mathsf{mod}\, kQ$ as a 'non commutative curve'.

We define $\mathcal{D}_Q$ to be the *bounded derived category* $\mathcal{D}^b(\mathsf{mod}\, kQ)$ of the abelian category $\mathsf{mod}\, kQ$. Thus, the objects of $\mathcal{D}_Q$ are the bounded complexes of (right) $kQ$-modules

$$\ldots \to 0 \to \ldots \to M^p \xrightarrow{d^p} M^{p+1} \to \ldots \to 0 \to \ldots .$$

Its morphisms are obtained from morphisms of complexes by formally inverting all quasi-isomorphisms. We refer to [125] [84] ... for in depth treatments of the fundamentals of this construction. Below, we will give a complete and elementary description of the category $\mathcal{D}_Q$ if $Q$ is a Dynkin quiver. We have the following general facts: The functor

$$\mathsf{mod}\, kQ \to \mathcal{D}_Q$$

taking a module $M$ to the complex concentrated in degree $0$

$$\ldots \to 0 \to M \to 0 \to \ldots$$

is a fully faithful embedding. From now on, we will identify modules with complexes concentrated in degree $0$. If $L$ and $M$ are two modules, then we have a canonical isomorphism

$$\mathsf{Ext}^i_{kQ}(L, M) \xrightarrow{\sim} \mathsf{Hom}_{\mathcal{D}_Q}(L, M[i])$$

for all $i \in \mathbb{Z}$, where $M[i]$ denotes the complex $M$ shifted by $i$ degrees to the left: $M[i]^p = M^{p+i}$, $p \in \mathbb{Z}$, and endowed with the differential $d_{M[i]} = (-1)^i d_M$. The category $\mathcal{D}_Q$ has all finite direct sums (and they are given by direct sums of complexes) and the decomposition theorem 5.1 holds. Moreover, each object is isomorphic to a direct sum of shifted copies of modules (this holds more generally in the derived category of any hereditary abelian category, for example the derived category of coherent sheaves on an algebraic curve). The category $\mathcal{D}_Q$ is abelian if and only if the quiver $Q$ does not have any arrows. However, it is always *triangulated*. This means that it is $k$-linear (it is additive, and the morphism sets are endowed with $k$-vector space structures so that the composition is bilinear) and endowed with the following extra structure:

a) a *suspension (or shift) functor* $\Sigma : \mathcal{D}_Q \to \mathcal{D}_Q$, namely the functor taking a complex $M$ to $M[1]$;

b) a class of *triangles* (sometimes called 'distinguished triangles'), namely the
sequences

$$L \to M \to N \to \Sigma L$$

which are 'induced' by short exact sequences of complexes.

The class of triangles satisfies certain axioms, *cf. e.g.* [125]. The most important
consequence of these axioms is that the triangles induce long exact sequences
in the functors $\mathsf{Hom}(X, ?)$ and $\mathsf{Hom}(?, X)$, *i.e.* for each object $X$ of $\mathcal{D}_Q$, the
sequences

$$\ldots (X, \Sigma^{-1} N) \to (X, L) \to (X, M) \to (X, N) \to (X, \Sigma L) \to \ldots$$

and

$$\ldots (\Sigma^{-1} N, X) \leftarrow (L, X) \leftarrow (M, X) \leftarrow (N, X) \leftarrow (\Sigma L, X) \leftarrow \ldots$$

are exact.

## 5.5. Presentation of the derived category of a Dynkin quiver

From now on, we assume that $Q$ is a Dynkin quiver. Let $\mathbb{Z}Q$ be its repetition
(*cf.* section 2.2). So the vertices of $\mathbb{Z}Q$ are the pairs $(p, i)$, where $p$ is an integer
and $i$ a vertex of $Q$ and the arrows of $\mathbb{Z}Q$ are obtained as follows: each arrow
$\alpha : i \to j$ of $Q$ yields the arrows

$$(p, \alpha) : (p, i) \to (p, j) , \quad p \in \mathbb{Z} ,$$

and the arrows

$$\sigma(p, \alpha) : (p - 1, j) \to (p, i) , \quad p \in \mathbb{Z}.$$

We extend $\sigma$ to a map defined on all arrows of $\mathbb{Z}Q$ by defining

$$\sigma(\sigma(p, \alpha)) = (p - 1, \alpha).$$

We endow $\mathbb{Z}Q$ with the map $\sigma$ and with the automorphism $\tau : \mathbb{Z}Q \to \mathbb{Z}Q$
taking $(p, i)$ to $(p - 1, i)$ and $(p, \alpha)$ to $(p - 1, \alpha)$ for all vertices $i$ of $Q$, all
arrows $\alpha$ of $Q$ and all integers $p$.

For a vertex $v$ of $\mathbb{Z}Q$ the *mesh ending at* $v$ is the full subquiver

(5.5.1)

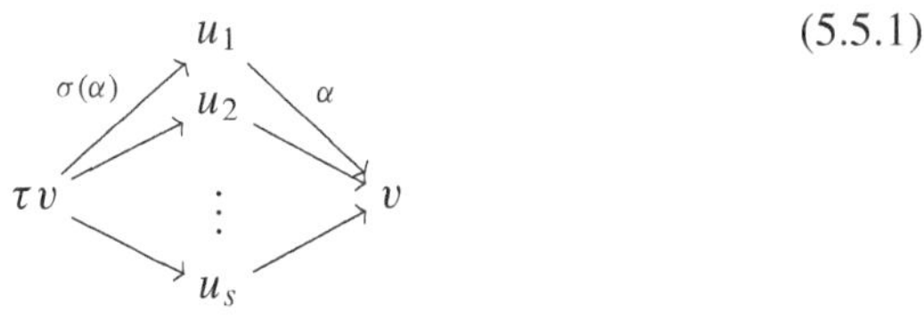

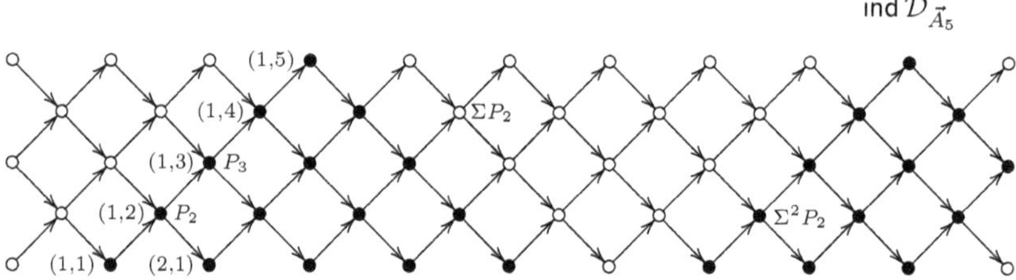

Figure 1.  The repetition of type $A_n$

formed by $v$, $\tau(v)$ and all sources $u$ of arrows $\alpha : u \to v$ of $\mathbb{Z}Q$ ending in $v$. We define the *mesh ideal* to be the (two-sided) ideal of the path category of $\mathbb{Z}Q$ which is generated by all *mesh relators*

$$r_v = \sum_{\text{arrows } \alpha:u\to v} \alpha\sigma(\alpha)\,,$$

where $v$ runs through the vertices of $\mathbb{Z}Q$. The *mesh category* is the quotient of the path category of $\mathbb{Z}Q$ by the mesh ideal.

**Theorem 5.8** (Happel [73]).

a) *There is a canonical bijection $v \mapsto M_v$ from the set of vertices of $\mathbb{Z}Q$ to the set of isomorphism classes of indecomposables of $\mathcal{D}_Q$ which takes the vertex $(1, i)$ to the indecomposable projective $P_i$.*

b) *Let* $\mathrm{ind}\,\mathcal{D}_Q$ *be the full subcategory of indecomposables of $\mathcal{D}Q$. The bijection of a) lifts to an equivalence of categories from the mesh category of $\mathbb{Z}Q$ to the category* $\mathrm{ind}\,\mathcal{D}_Q$.

In figure 1, we see the repetition for $Q = \vec{A}_5$ and the map taking its vertices to the indecomposable objects of the derived category. The vertices marked $\bullet$ belonging to the left triangle are mapped to indecomposable modules. The vertex $(1, i)$ corresponds to the indecomposable projective $P_i$. The arrow $(1, i) \to (1, i + 1)$, $1 \le i \le 5$, is mapped to the left multiplication by the arrow $i \to i + 1$. The functor takes a mesh (5.5.1) to a triangle

$$M_{\tau v} \longrightarrow \bigoplus_{i=1}^{s} M_{u_i} \longrightarrow M_v \longrightarrow \Sigma M_{\tau v} \qquad (5.5.2)$$

called an *Auslander-Reiten triangle* or *almost split triangle*, cf. [74]. If $M_v$ and $M_{\tau v}$ are modules, then so is the middle term and the triangle comes from an exact sequence of modules

$$0 \longrightarrow M_{\tau v} \longrightarrow \bigoplus_{i=1}^{s} M_{u_i} \longrightarrow M_v \longrightarrow 0$$

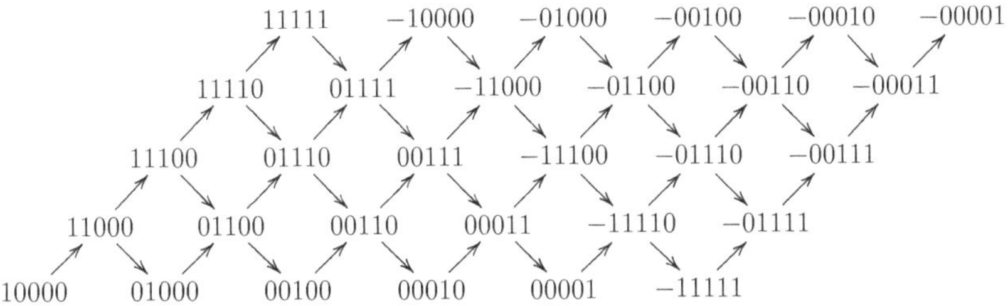

Figure 2. Some dimension vectors of indecomposables in $\mathcal{D}_{\tilde{A}_5}$

called an *Auslander-Reiten sequence* or *almost split sequence, cf.* [5]. These almost split triangles respectively sequences can be characterized intrinsically in $\mathcal{D}_Q$ respectively $\mathsf{mod}\,kQ$.

Recall that the Grothendieck group $K_0(\mathcal{T})$ of a triangulated category is the quotient of the free abelian group on the isomorphism classes $[X]$ of objects $X$ of $\mathcal{T}$ by the subgroup generated by all elements

$$[X] - [Y] + [Z]$$

arising from triangles $(X, Y, Z)$ of $\mathcal{T}$. In the case of $\mathcal{D}_Q$, the natural map

$$K_0(\mathsf{mod}\,kQ) \to K_0(\mathcal{D}_Q)$$

is an isomorphism (its inverse sends a complex to the alternating sum of the classes of its homologies). Since $K_0(\mathsf{mod}\,kQ)$ is free on the classes $[S_i]$ associated with the simple modules, the same holds for $K_0(\mathcal{D}_Q)$ so that its elements are given by $n$-tuples of integers. We write $\underline{\dim}\,M$ for the image in $K_0(\mathcal{D}_Q)$ of an object $M$ of $K_0(\mathcal{D}_Q)$ and call $\underline{\dim}\,M$ the *dimension vector of M*. Then each triangle (5.5.2) yields an equality

$$\underline{\dim}\,M_v = \sum_{i=1}^{s} \underline{\dim}\,M_{u_i} - \underline{\dim}\,M_{\tau v}.$$

Using these equalities, we can easily determine $\underline{\dim}\,M$ for each indecomposable $M$ starting from the known dimension vectors $\underline{\dim}\,P_i$, $1 \leq i \leq n$. In the above example, we find the dimension vectors listed in figure (2).

Thanks to the theorem, the automorphism $\tau$ of the repetition yields a $k$-linear automorphism, still denoted by $\tau$, of the derived category $\mathcal{D}_Q$. This automorphism has several intrinsic descriptions:

1) As shown in [60], it is the right derived functor of the left exact Coxeter functor $\mathsf{rep}(Q^{op}) \to \mathsf{rep}(Q^{op})$ introduced by Bernstein-Gelfand-Ponomarev [19] in their proof of Gabriel's theorem. If we identify $K_0(\mathcal{D}_Q)$ with the root

lattice via Gabriel's theorem, then the automorphism induced by $\tau^{-1}$ equals the the Coxeter transformation $c$. As shown by Gabriel [60], the identity $c^h = \mathbf{1}$, where $h$ is the Coxeter number, lifts to an isomorphism of functors

$$\tau^{-h} \xrightarrow{\sim} \Sigma^2. \tag{5.5.3}$$

2) It can be expressed in terms of the Serre functor of $\mathcal{D}_Q$: Recall that for a $k$-linear triangulated category $\mathcal{T}$ with finite-dimensional morphism spaces, a Serre functor is an autoequivalence $S : \mathcal{T} \to \mathcal{T}$ such that the Serre duality formula holds: We have bifunctorial isomorphisms

$$D\,\mathsf{Hom}(X, Y) \xrightarrow{\sim} \mathsf{Hom}(Y, SX)\,, \quad X, Y \in \mathcal{T}\,,$$

where $D$ is the duality $\mathsf{Hom}_k(?, k)$ over the ground field. Notice that this determines the functor $S$ uniquely up to isomorphism. In the case of $\mathcal{D}_Q = \mathcal{D}^b(\mathsf{mod}\,kQ)$, it is not hard to prove that a Serre functor exists (it is given by the left derived functor of the tensor product by the bimodule $D(kQ)$). Now the autoequivalence $\tau$, the suspension functor $\Sigma$ and the Serre functor $S$ are linked by the fundamental isomorphism

$$\tau \Sigma \xrightarrow{\sim} S. \tag{5.5.4}$$

## 5.6. Caldero-Chapoton's proof

The above description of the derived category yields in particular a description of the module category, which is a full subcategory of the derived category. This description was used by Caldero-Chapoton [27] to prove their formula. Let us sketch the main steps in their proof: Recall that we have defined a surjective map $v \mapsto X_v$ from the set of vertices of the repetition to the set of cluster variables such that

a) we have $X_{(0,i)} = x_i$ for $1 \le i \le n$ and
b) we have

$$X_{\tau v} X_v = 1 + \prod_{\text{arrows } w \to v} X_w$$

for all vertices $v$ of the repetition.

We wish to show that we have

$$X_v = CC(M_v)$$

for all vertices $v$ such that $M_v$ is an indecomposable module. This is done by induction on the distance of $v$ from the vertices $(1, i)$ in the quiver $\mathbb{Z}Q$. More precisely, one shows the following

a) We have $CC(P_i) = X_{(1,i)}$ for each indecomposable projective $P_i$. Here we use the fact that submodules of projectives are projective in order to explicitly compute $CC(P_i)$.

b1) For each split exact sequence

$$0 \to L \to E \to M \to 0 ,$$

we have

$$CC(L)CC(M) = CC(E).$$

Thus, if $E = E_1 \oplus \ldots E_s$ is a decomposition into indecomposables, then

$$CC(E) = \prod_{i=1}^{s} CC(E_i).$$

b2) If

$$0 \to L \to E \to M \to 0$$

is an almost split exact sequence, then we have

$$CC(E) + 1 = CC(L)CC(M).$$

It is now clear how to prove the equality $X_v = CC(M_v)$ by induction by proceeding from the projective indecomposables to the right.

### 5.7. The cluster category

The *cluster category*

$$\mathcal{C}_Q = \mathcal{D}_Q/(\tau^{-1}\Sigma)^{\mathbb{Z}} = \mathcal{D}_Q/(S^{-1}\Sigma^2)^{\mathbb{Z}}$$

is the orbit category of the derived category under the action of the cyclic group generated by the autoequivalence $\tau^{-1}\Sigma = S^{-1}\Sigma^2$. This means that the objects of $\mathcal{C}_Q$ are the same as those of the derived category $\mathcal{D}_Q$ and that for two objects $X$ and $Y$, the morphism space from $X$ to $Y$ in $\mathcal{C}_Q$ is

$$\mathcal{C}_Q(X, Y) = \bigoplus_{p\in\mathbb{Z}} \mathcal{D}_Q(X, (S^{-1}\Sigma^2)^p Y).$$

Morphisms are composed in the natural way. This definition is due to Buan-Marsh-Reineke-Reiten-Todorov [10], who were trying to obtain a better understanding of the 'decorated quiver representations' introduced by Reineke-Marsh-Zelevinsky [106]. For quivers of type $A$, an equivalent category was defined independently by Caldero-Chapoton-Schiffler [28] using an entirely different description. Clearly the category $\mathcal{C}_Q$ is $k$-linear. It is not hard to check that its morphism spaces are finite-dimensional.

One can show [90] that $\mathcal{C}_Q$ admits a canonical structure of triangulated category such that the projection functor $\pi : \mathcal{D}_Q \to \mathcal{C}_Q$ becomes a triangle functor (in general, orbit categories of triangulated categories are no longer triangulated). The Serre functor $S$ of $\mathcal{D}_Q$ clearly induces a Serre functor in $\mathcal{C}_Q$, which we still denote by $S$. Now, by the definition of $\mathcal{C}_Q$ (and its triangulated structure), we have an isomorphism of triangle functors

$$S \xrightarrow{\sim} \Sigma^2.$$

This means that $\mathcal{C}_Q$ is 2-*Calabi-Yau*. Indeed, for an integer $d \in \mathbb{Z}$, a triangulated category $\mathcal{T}$ with finite-dimensional morphism spaces is $d$-Calabi-Yau if it admits a Serre functor isomorphic as a triangle functor to the $d$th power of its suspension functor.

## 5.8. From cluster categories to cluster algebras

We keep the notations and hypotheses of the previous section. The suspension functor $\Sigma$ and the Serre functor $S$ induce automorphisms of the repetition $\mathbb{Z}Q$ which we still denote by $\Sigma$ and $S$ respectively. The orbit quiver $\mathbb{Z}Q/(\tau^{-1}\Sigma)^{\mathbb{Z}}$ inherits the automorphism $\tau$ and the map $\sigma$ (defined on arrows only) and thus has a well-defined mesh category. Recall that we write $\mathsf{Ext}^1(X, Y)$ for $\mathsf{Hom}(X, \Sigma Y)$ in any triangulated category.

**Theorem 5.9** ([10] [11]).

a) *The decomposition theorem holds for the cluster category and the mesh category of $\mathbb{Z}Q/(\tau^{-1}\Sigma)^{\mathbb{Z}}$ is canonically equivalent to the full subcategory* $\mathsf{ind}\,\mathcal{C}_Q$ *of the indecomposables of $\mathcal{C}_Q$. Thus, we have an induced bijection $L \mapsto X_L$ from the set of isomorphism classes of indecomposables of $\mathcal{C}_Q$ to the set of all cluster variables of $\mathcal{A}_Q$ which takes the shifted projective $\Sigma P_i$ to the initial variable $x_i$, $1 \leq i \leq n$.*

b) *Under this bijection, the clusters correspond to the* cluster-tilting *sets, i.e. the sets of pairwise non isomorphic indecomposables $T_1, \ldots, T_n$ such that we have*

$$Ext^1(T_i, T_j) = 0$$

   *for all $i, j$.*

c) *If $T_1, \ldots, T_n$ is cluster-tilting, then the quiver (cf. below) of the endomorphism algebra of the sum $T = \bigoplus_{i=1}^{n} T_i$ does not have loops nor 2-cycles and* *the associated antisymmetric matrix is the exchange matrix of the unique seed containing the cluster $X_{T_1}, \ldots, X_{T_n}$.*

In part b), the condition implies in particular that $\mathsf{Ext}^1(T_i, T_i)$ vanishes. However, for a Dynkin quiver $Q$, we have $\mathsf{Ext}^1(L, L) = 0$ for each indecomposable $L$ of $\mathcal{C}_Q$. A *cluster-tilting object of* $\mathcal{C}_Q$ is the direct sum of the objects $T_1, \ldots, T_n$ of a cluster-tilting set. Since these are pairwise non-isomorphic indecomposables, the datum of $T$ is equivalent to that of the $T_i$. A *cluster-tilted algebra of type $Q$* is the endomorphism algebra of a cluster-tilting object of $\mathcal{C}_Q$. In part c), the most subtle point is that the quiver does not have loops or 2-cycles [11]. Let us recall what one means by the *quiver of a finite-dimensional algebra over an algebraically closed field*:

**Proposition-Definition 5.10** (Gabriel). *Let $B$ be a finite-dimensional algebra over the algebraically closed ground field $k$.*

a) *There exists a quiver $Q_B$, unique up to isomorphism, such that $B$ is Morita equivalent to the algebra $kQ_B/I$, where $I$ is an ideal of $kQ_B$ contained in the square of the ideal generated by the arrows of $Q_B$.*
b) *The ideal $I$ is not unique in general but we have $I = 0$ iff $B$ is hereditary.*
c) *There is a bijection $i \mapsto S_i$ between the vertices of $Q_B$ and the isomorphism classes of simple $B$-modules. The number of arrows from a vertex $i$ to a vertex $j$ equals the dimension of $\mathsf{Ext}^1_B(S_j, S_i)$.*

In our case, the algebra $B$ is the endomorphism algebra of the sum $T$ of the cluster-tilting set $T_1, \ldots, T_n$ in $\mathcal{C}_Q$. In this case, the Morita equivalence of a) even becomes an isomorphism (because the $T_i$ are pairwise non isomorphic). For a suitable choice of this isomorphism, the idempotent $e_i$ associated with the vertex $i$ is sent to the identity of $T_i$ and the images of the arrows from $i$ to $j$ yield a basis of the space of *irreducible morphisms*

$$\mathsf{irr}_T(T_i, T_j) = \mathsf{rad}_T(T_i, T_j)/\mathsf{rad}^2_T(T_i, T_j) \,,$$

where $\mathsf{rad}_T(T_i, T_j)$ denotes the vector space of non isomorphisms from $T_i$ to $T_j$ (thanks to the locality of the endomorphism rings, this set is indeed closed under addition) and $\mathsf{rad}^2_T$ the subspace of non isomorphisms admitting a non trivial factorization:

$$\mathsf{rad}^2_T(T_i, T_j) = \sum_{r=1}^{n} \mathsf{rad}_T(T_r, T_j)\,\mathsf{rad}_T(T_i, T_r).$$

As an illustration of theorem 5.9, we consider the cluster-tilting set $T_1, \ldots, T_5$ in $\mathcal{C}_{\tilde{A}_5}$ depicted in figure 3. Here the vertices labeled $0, 1, \ldots,$ 4 have to be identified with the vertices labeled $20, 21, \ldots, 24$ (in this order) to

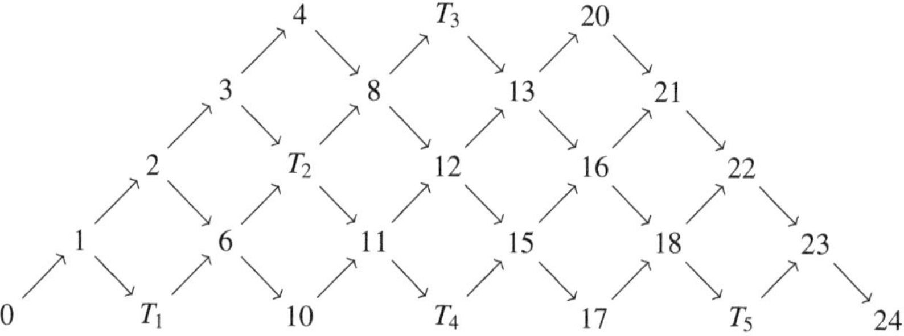

Figure 3. A cluster-tilting set in $A_5$

obtain the orbit quiver $\mathbb{Z}Q/(\tau^{-1}\Sigma)^{\mathbb{Z}}$. In the orbit category, we have $\tau \xrightarrow{\sim} \Sigma$ so that $\Sigma T_1$ is the indecomposable associated to vertex 0, for example. Using this and the description of the morphisms in the mesh category, it is easy to check that we do have

$$\mathsf{Ext}^1(T_i, T_j) = 0$$

for all $i$, $j$. It is also easy to determine the spaces of morphisms

$$\mathsf{Hom}_{\mathcal{C}_Q}(T_i, T_j)$$

and the compositions of morphisms. Determining these is equivalent to determining the endomorphism algebra

$$\mathsf{End}(T) = \mathsf{Hom}(T, T) = \bigoplus_{i,j} \mathsf{Hom}(T_i, T_j).$$

This algebra is easily seen to be *isomorphic* to the algebra given by the following quiver $Q'$

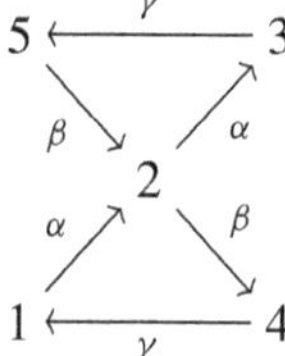

with the relations

$$\alpha\beta = 0\,, \ \beta\gamma = 0\,, \ \gamma\alpha = 0.$$

Thus the quiver of $\mathsf{End}(T)$ is $Q'$. It encodes the exchange matrix of the associated cluster

$$X_{T_1} = \frac{1 + x_2}{x_1}$$

$$X_{T_2} = \frac{x_1 x_2 + x_1 x_4 + x_3 x_4 + x_2 x_3 x_4}{x_1 x_2 x_3}$$

$$X_{T_3} = \frac{x_1 x_2 x_3 + x_1 x_2 x_3 x_4 + x_1 x_2 x_5 + x_1 x_4 x_5 + x_3 x_4 x_5 + x_2 x_3 x_4 x_5}{x_1 x_2 x_3 x_4 x_5}$$

$$X_{T_4} = \frac{x_2 + x_4}{x_3}$$

$$X_{T_5} = \frac{1 + x_4}{x_5}.$$

## 5.9. A $K$-theoretic interpretation of the exchange matrix

Keep the notations and hypotheses of the preceding section. Let $T_1, \ldots, T_n$ be a cluster-tilting set, $T$ the sum of the $T_i$ and $B$ its endomorphism algebra. For two finite-dimensional right $B$-modules $L$ and $M$ put

$$\langle L, M \rangle_a = \dim \mathsf{Hom}(L, M) - \dim \mathsf{Ext}^1(L, M)$$
$$- \dim \mathsf{Hom}(M, L) + \dim \mathsf{Ext}^1(M, L).$$

This is the antisymmetrization of a truncated Euler form. A priori it is defined on the split Grothendieck group of the category $\mathsf{mod}\, B$ (*i.e.* the quotient of the free abelian group on the isomorphism classes divided by the subgroup generated by all relations obtained from direct sums in $\mathsf{mod}\, B$).

**Proposition 5.11** (Palu). *The form* $\langle, \rangle_a$ *descends to an antisymmetric form on* $K_0(\mathsf{mod}\, B)$. *Its matrix in the basis of the simples is the exchange matrix associated with the cluster corresponding to* $T_1, \ldots, T_n$.

## 5.10. Mutation of cluster-tilting sets

Let us recall two axioms of triangulated categories:

TR1  For each morphism $u : X \to Y$, there exists a triangle

$$X \xrightarrow{u} Y \to Z \to \Sigma X.$$

TR2  A sequence

$$X \xrightarrow{u} Y \xrightarrow{v} Z \xrightarrow{w} \Sigma X$$

is a triangle if and only if the sequence

$$Y \xrightarrow{v} Z \xrightarrow{w} \Sigma X \xrightarrow{-u} \Sigma Y$$

is a triangle.

One can show that in TR1, the triangle is unique up to (non unique) isomorphism. In particular, up to isomorphism, the object $Z$ is uniquely determined by $u$. Notice the sign in TR2. It follows from TR1 and TR2 that a given morphism also occurs as the second (respectively third) morphism in a triangle.

Now, with the notations and hypotheses of the preceding section, suppose that $T_1, \ldots, T_n$ is a cluster-tilting set and $Q'$ the quiver of the endomorphism algebra $B$ of the sum of the $T_i$. As explained after proposition-definition 5.10, we have a surjective algebra morphism

$$kQ' \to \bigoplus_{i,j} \mathsf{Hom}(T_i, T_j)$$

which takes the idempotent $e_i$ to the identity of $T_i$ and the arrows $i \to j$ to irreducible morphisms $T_i \to T_j$, for all vertices $i, j$ of $Q'$ (*cf.* the above example computation of $B$ and $Q' = Q_B$).

Now let $k$ be a vertex of $Q'$ (the mutating vertex). We choose triangles

$$T_k \xrightarrow{u} \bigoplus_{\substack{\text{arrows} \\ k \to i}} T_i \to T_k^* \to \Sigma T_k$$

and

$${}^*T_k \to \bigoplus_{\substack{\text{arrows} \\ j \to k}} T_j \xrightarrow{v} T_k \to \Sigma {}^*T_k \,,$$

where the component of $u$ (respectively $v$) corresponding to an arrow $\alpha : k \to i$ (respectively $j \to k$) is the corresponding morphism $T_k \to T_i$ (respectively $T_j \to T_k$). These triangles are unique up to isomorphism and called the *exchange triangles* associated with $k$ and $T_1, \ldots, T_n$.

**Theorem 5.12** ([10]).
a) *The objects $T_k^*$ and ${}^*T_k$ are isomorphic.*
b) *The set obtained from $T_1, \ldots, T_n$ by replacing $T_k$ with $T_k^*$ is cluster-tilting and its associated cluster is the mutation at $k$ of the cluster associated with $T_1, \ldots, T_n$.*
c) *Two indecomposables $L$ and $M$ appear as the the pair $(T_k, T_k^*)$ associated with an exchange if and only if the space $\mathsf{Ext}^1(L, M)$ is one-dimensional. In this case, the exchange triangles are the unique (up to isomorphism) non split triangles*

$$L \to E \to M \to \Sigma L \text{ and } M \to E' \to L \to \Sigma M.$$

Let us extend the map $L \mapsto X_L$ from indecomposable to decomposable objects of $\mathcal{C}_Q$ by requiring that we have

$$X_N = X_{N_1} X_{N_2}$$

whenever $N = N_1 \oplus N_2$ (this is compatible with the muliplicativity of the Caldero-Chapoton map). We know that if $u_1, \ldots, u_n$ is a cluster and $B = (b_{ij})$ the associated exchange matrix, then the mutation at $k$ yields the variable $u'_k$ such that

$$u_k u'_k = \prod_{\substack{\text{arrows} \\ k \to i}} u_i + \prod_{\substack{\text{arrows} \\ j \to k}} u_j.$$

By combining this with the exchange triangles, we see that in the situation of c), we have

$$X_L X_M = X_E + X_{E'}.$$

We would like to generalize this identity to the case where the space $\mathsf{Ext}^1(L, M)$ is of higher dimension. For three objects $L$, $M$ and $N$ of $\mathcal{C}_Q$, let $Ext^1(L, M)_N$ be the subset of $\mathsf{Ext}^1(L, M)$ formed by those morphisms $\varepsilon : L \to \Sigma M$ such that in the triangle

$$M \to E \to L \xrightarrow{\varepsilon} \Sigma M,$$

the object $E$ is isomorphic to $N$ (we do not fix an isomorphism). Notice that this subset is a cone (*i.e.* stable under multiplication by non zero scalars) in the vector space $\mathsf{Ext}^1(L, M)$.

**Proposition 5.13** ([30]). *The subset* $\mathsf{Ext}^1(L, M)_N$ *is constructible in* $\mathsf{Ext}^1(L, M)$. *In particular, it is a union of algebraic subvarieties. It is empty for all but finitely isomorphism classes of objects* N.

If $k$ is the field of complex numbers, we denote by $\chi$ the Euler characteristic with respect to singular cohomology with coefficients in a field. If $k$ is an arbitrary algebraically closed field, we denote by $\chi$ the Euler characteristic with respect to étale cohomology with proper support.

**Theorem 5.14** ([30]). *Suppose that* L *and* M *are objects of* $\mathcal{C}_Q$ *such that* $\mathsf{Ext}^1(L, M) \neq 0$. *Then we have*

$$X_L X_M = \sum_N \frac{\chi(\mathbb{P}\,\mathsf{Ext}^1(L, M)_N) + \chi(\mathbb{P}\,\mathsf{Ext}^1(M, L)_N)}{\chi(\mathbb{P}\,\mathsf{Ext}^1(L, M))} X_N,$$

*where the sum is taken over all isomorphism classes of objects* N *of* $\mathcal{C}_Q$.

Notice that in the theorem, the objects $L$ and $M$ may be decomposable so that $X_L$ and $X_M$ will not be cluster variables in general and the $X_N$ do not form a linearly independent set in the cluster algebra. Thus, the formula should be considered as a relation rather than as an alternative definition for the multiplication of the cluster algebra. Notice that it nevertheless bears a close resemblance to the product formula in a dual Hall algebra: For two objects $L$ and $M$ in a finitary abelian category of finite global dimension, we have

$$[L] * [M] = \sum_{[N]} \frac{|\mathrm{Ext}^1(L, M)_N|}{|\mathrm{Ext}^1(L, M)|} [N] \, ,$$

where the brackets denote isomorphism classes and the vertical bars the cardinalities of the underlying sets, *cf.* Proposition 1.5 of [119].

## 6. Categorification via cluster categories: the acyclic case

### 6.1. Categorification

Let $Q$ be a connected finite quiver without oriented cycles with vertex set $\{1, \ldots, n\}$. Let $k$ be an algebraically closed field. We have seen in section 5.4 how to define the bounded derived category $\mathcal{D}_Q$. We still have a fully faithful functor from the mesh category of $\mathbb{Z}Q$ to the category of indecomposables of $\mathcal{D}_Q$ but this functor is very far from being essentially surjective. In fact, its image does not even contain the injective indecomposable $kQ$-modules. The methods of the preceding section therefore do not generalize but most of the results continue to hold. The derived category $\mathcal{D}_Q$ still has a Serre functor (the total left derived functor of the tensor product functor $? \otimes_B D(kQ)$). We can form the cluster category

$$\mathcal{C}_Q = \mathcal{D}_Q / (S^{-1} \Sigma^2)^{\mathbb{Z}}$$

as before and it is still a triangulated category in a canonical way such that the projection $\pi : \mathcal{D}_Q \to \mathcal{C}_Q$ becomes a triangle functor [90]. Moreover, the decomposition theorem 5.1 holds for $\mathcal{C}_Q$ and each object $L$ of $\mathcal{C}_Q$ decomposes into a direct sum

$$L = \pi(M) \oplus \bigoplus_{i=1}^{n} \pi(\Sigma P_i)^{m_i}$$

for some module $M$ and certain multiplicities $m_i$, $1 \leq i \leq n$, *cf.* [10]. We put

$$X_L = CC(M) \prod_{i=1}^{n} x_i^{m_i} \, ,$$

where $CC(M)$ is defined as in section 5.3 Notice that in general, $X_L$ can only be expected to be an element of the fraction field $\mathbb{Q}(x_1, \ldots, x_n)$, not of the cluster algebra $\mathcal{A}_Q$ inside this field. (The exponents in the formula for $X_L$ are perhaps more transparent in equation 7.5.1 below).

**Theorem 6.1.** *Let $Q$ be a finite quiver without oriented cycles with vertex set $\{1, \ldots, n\}$.*

a) *The map $L \mapsto X_L$ induces a bijection from the set of isomorphism classes of rigid indecomposables of the cluster category $\mathcal{C}_Q$ onto the set of cluster variables of the cluster algebra $\mathcal{A}_Q$.*
b) *Under this bijection, the clusters correspond exactly to the cluster-tilting sets, i.e. the sets $T_1, \ldots, T_n$ of rigid indecomposables such that*

$$Ext^1(T_i, T_j) = 0$$

*for all $i$, $j$.*
c) *For a cluster-tilting set $T_1, \ldots, T_n$, the quiver of the endomorphism algebra of the sum of the $T_i$ does not have loops nor 2-cycles and encodes the exchange matrix of the [68] seed containing the corresponding cluster.*
d) *If $L$ and $M$ are rigid indecomposables such that the space $\mathsf{Ext}^1(L, M)$ is one-dimensional, then we have the generalized exchange relation*

$$X_L X_M = X_B + X_{B'} \tag{6.1.1}$$

*where $B$ and $B'$ are the middle terms of 'the' non split triangles*

$$L \longrightarrow B \longrightarrow M \longrightarrow \Sigma L \;\; and \;\; M \longrightarrow B' \longrightarrow L \longrightarrow \Sigma M \,.$$

Parts a), b) and d) of the theorem are proved in [29] and part c) in [11]. The proofs build on work by many authors notably Buan-Marsh-Reiten-Todorov [12] Buan-Marsh-Reiten [11], Buan-Marsh-Reineke-Reiten-Todorov [10], Marsh-Reineke-Zelevinsky [106], . . . and especially on Caldero-Chapoton's explicit formula for $X_L$ proved in [27] for orientations of simply laced Dynkin diagrams. Another crucial ingredient of the proof is the Calabi-Yau property of the cluster category. An alternative proof of part c) was given by A. Hubery [79] for quivers whose underlying graph is an extended simply laced Dynkin diagram.

We describe the main steps of the proof of a). The mutation of cluster-tilting sets is defined using the construction of section 5.10.

1) If $T$ is a cluster-tilting object, then the quiver $Q_T$ of its endomorphism algebra does not have loops or 2-cycles. If $T'$ is obtained from $T$ by mutation

at the summand $T_1$, then the quiver $Q_{T'}$ of the endomorphism algebra of $T'$ is the mutation at the vertex 1 of the quiver $Q_T$, *cf.* [11].

2) Each rigid indecomposable is contained in a cluster-tilting set. Any two cluster-tilting sets are linked by a finite sequence of mutations. This is deduced in [10] from the work of Happel-Unger [76].

3) If $(T_1, T_1^*)$ is an exchange pair and

$$T_1^* \to E \to T_1 \to \Sigma T_1^* \text{ and } T_1 \to E' \to T_1^* \to \Sigma T_1$$

are the exchange triangles, then we have

$$X_{T_1} X_{T_1^*} = X_E + X_{E'}.$$

This is shown in [29].

It follows from 1)-3) that the map $L \to X_L$ does take rigid indecomposables to cluster variables and that each cluster variable is obtained in this way. It remains to be shown that a rigid indecomposable $L$ is determined up to isomorphism by $X_L$. This follows from

4) If $M$ is a rigid indecomposable module, the denominator of $X_M$ is $x_1^{d_1} \ldots x_n^{d_n}$, *cf.* [29].

Indeed, a rigid indecomposable module $M$ is determined, up to isomorphism, by its dimension vector.

We sum up the relations between the cluster algebra and the cluster category in the following table

| cluster algebra | cluster category |
|---|---|
| multiplication | direct sum |
| addition | ? |
| cluster variables | rigid indecomposables |
| clusters | cluster-tilting sets |
| mutation | mutation |
| exchange relation | exchange triangles |
| $xx^* = m + m'$ | $T_k \to M \to T_k^* \to \Sigma T_k$ |
| | $T_k^* \to M' \to T_k \to \Sigma T_k^*$ |

## 6.2. Two applications

Theorem 6.1 does shed new light on cluster algebras. In particular, thanks to the theorem, Caldero and Reineke [31] have made significant progress towards the

**Conjecture 6.2.** *Suppose that $Q$ does not have oriented cycles. Then all cluster variables of $\mathcal{A}_Q$ belong to $\mathbb{N}[x_1^{\pm}, \ldots, x_n^{\pm}]$.*

This conjecture is a consequence of a general conjecture of Fomin-Zelevinsky [50], which here is specialized to the case of cluster algebras associated with acyclic quivers, for cluster expansions in the initial cluster. Caldero-Reineke's work in [31] is based on Lusztig's [105] and in this sense it does not quite live up to the hopes that cluster theory ought to explain Lusztig's results. Notice that in [31], the above conjecture is stated as a theorem. However, a gap in the proof was found by Nakajima [109]: the authors incorrectly identify their parameter $q$ with Lusztig's parameter $v$, whereas the correct identification is $v = -\sqrt{q}$.

Here are two applications to the exchange graph of the cluster algebra associated with an acyclic quiver $Q$:

**Corollary 6.3** ([29]).

a) *For any cluster variable $x$, the set of seeds whose clusters contain $x$ form a connected subgraph of the exchange graph.*

b) *The set of seeds whose quiver does not have oriented cycles form a connected subgraph (possibly empty) of the exchange graph.*

For acyclic cluster algebras, parts a) and b) confirm conjecture 4.14 parts (3) and (4) by Fomin-Zelevinsky in [53]. By b), the cluster algebra associated with a quiver without oriented cycles has a well-defined cluster-type.

## 6.3. Cluster categories and singularities

The construction of cluster categories may seem a bit artificial. Nevertheless, cluster categories do occur 'in nature'. In particular, certain triangulated categories associated with singularities are equivalent to cluster categories. We illustrate this on the following example: Let the cyclic group $G$ of order 3 act on a three-dimensional complex vector space $V$ by scalar multiplication with a primitive third root of unity. Let $S$ be the completion at the origin of the coordinate algebra of $V$ and let $R = S^G$ the fixed point algebra, corresponding to the completion of the singularity at the origin of the quotient $V /\!/ G$. The algebra $R$ is a Gorenstein ring, *cf. e.g.* [128], and an isolated singularity of dimension 3, *cf. e.g.* Corollary 8.2 of [81]. The category $CM(R)$ of maximal Cohen-Macaulay modules is an exact Frobenius category and its stable category $\underline{CM}(R)$ is a triangulated category. By Auslander's results [6], *cf.* Lemma 3.10 of [130], it is 2-Calabi Yau. One can show that it is equivalent to the cluster category $\mathcal{C}_Q$

for the quiver

$$Q : 1 \rightrightarrows 2$$

by an equivalence which takes the cluster-tilting object $T = kQ$ to $S$ considered as an $R$-module. This example can be found in [91], where it is deduced from an abstract characterization of cluster categories. A number of similar examples can be found in [25] and [93].

# 7. Categorification via 2-Calabi-Yau categories

The extension of the results of the preceding sections to quivers containing oriented cycles is the subject of ongoing research, *cf.* for example [39] [61] [8]. Here we present an approach based on the fact that many arguments developed for cluster categories apply more generally to suitable triangulated categories, whose most important property it is to be Calabi-Yau of dimension 2. As an application, we will sketch a proof of the periodicity conjecture in section 8 (details will appear elsewhere [86]).

## 7.1. Definition and main examples

Let $k$ be an algebraically closed field and let $\mathcal{C}$ a triangulated category with suspension functor $\Sigma$ where all idempotents split (*i.e.* each idempotent endomorphism $e$ of an object $M$ is the projection onto $M_1$ along $M_2$ in a decomposition $M = M_1 \oplus M_2$). We assume that

1) $\mathcal{C}$ is Hom-finite (*i.e.* we have $\dim \mathcal{C}(L, M) < \infty$ for all $L, M$ in $\mathcal{C}$) and the decomposition theorem 5.1 holds for $\mathcal{C}$;
2) $\mathcal{C}$ is 2-Calabi-Yau, *i.e.* we are given bifunctorial isomorphisms

$$D\mathcal{C}(L, M) \xrightarrow{\sim} \mathcal{C}(M, \Sigma^2 L) , \quad L, M \in \mathcal{C};$$

3) $\mathcal{C}$ admits a cluster-tilting object $T$, *i.e.*
   a) $T$ is the sum of pairwise non-isomorphic indecomposables,
   b) $T$ is rigid and
   c) for each object $L$ of $\mathcal{C}$, if $\mathsf{Ext}^1(T, L)$ vanishes, then $L$ belongs to the category $\mathsf{add}(T)$ of direct factors of finite direct sums of copies of $T$.

If all these assumptions hold, we say that $(\mathcal{C}, T)$ *is a 2-Calabi-Yau category with cluster tilting object*. If $Q$ is a finite quiver, we say that a 2-Calabi-Yau category $\mathcal{C}$ with cluster-tilting object $T$ is a *2-Calabi-Yau realization of $Q$* if $Q$ is the quiver of the endomorphism algebra of $T$.

For example, if $\mathcal{C}$ is the cluster category of a finite quiver $Q$ without oriented cycles, conditions 1) and 2) hold and an object $T$ is cluster-tilting in the above sense iff it is the direct sum of a cluster-tilting set $T_1, \ldots, T_n$, where $n$ is the number of vertices of $Q$, *cf.* [10]. The 'initial' cluster-tilting object in this case is $T = kQ$ (the image in $\mathcal{C}_Q$ of the free module of rank one) and $(\mathcal{C}, kQ)$ is the *canonical 2-Calabi-Yau realization* of the quiver $Q$ (without oriented cycles).

The second main class of examples comes from the work of Geiss-Leclerc-Schröer: Let $\Delta$ be a simply laced Dynkin diagram, $\vec{\Delta}$ a quiver with underlying graph $\Delta$ and $\overline{\Delta}$ the *doubled quiver* obtained from $\vec{\Delta}$ by adjoining an arrow $\alpha^* : j \to i$ for each arrow $\alpha : i \to j$. The *preprojective algebra* $\Lambda = \Lambda(\Delta)$ is the quotient of the path algebra of $\overline{\Delta}$ by the ideal generated by the relator

$$\sum_{\alpha \in Q_1} \alpha\alpha^* - \alpha^*\alpha.$$

For example, if $\vec{\Delta}$ is the quiver

$$1 \xrightarrow{\ \alpha\ } 2 \xrightarrow{\ \beta\ } 3 \ ,$$

then $\overline{\Delta}$ is the quiver

$$1 \underset{\alpha^*}{\overset{\alpha}{\rightleftarrows}} 2 \underset{\beta^*}{\overset{\beta}{\rightleftarrows}} 3$$

and the ideal generated by the above sum of commutators is also generated by the elements

$$\alpha^*\alpha, \ \alpha\alpha^* - \beta^*\beta, \ \beta\beta^*.$$

It is classical, *cf. e.g.* [115], that the algebra $\Lambda = \Lambda(\Delta)$ is a finite-dimensional (!) selfinjective algebra (*i.e.* $\Lambda$ is injective as a right $\Lambda$-module over itself). Let $\mathsf{mod}\,\Lambda$ denote the category of $k$-finite-dimensional right $\Lambda$-modules. The *stable module category* $\underline{\mathsf{mod}}\Lambda$ is the quotient of $\mathsf{mod}\,\Lambda$ by the ideal of all morphisms factoring through a projective module. This category carries a canonical triangulated structure (like any stable module category of a self-injective algebra): The suspension is constructed by choosing exact sequences of modules

$$0 \to L \to IL \to \Sigma L \to 0$$

where $IL$ is injective (but not necessarily functorial in $L$; the object $\Sigma L$ becomes functorial in $L$ when we pass to the stable category). The triangles are by definition isomorphic to standard triangles obtained from exact sequences of modules as follows: Let

$$0 \to L \xrightarrow{i} M \xrightarrow{p} N \to 0$$

be a short exact sequence of mod $\Lambda$. Choose a commutative diagram

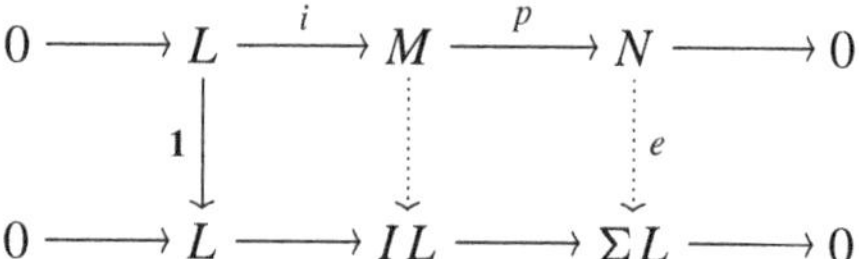

Then the image of $(i, p, e)$ is a standard triangle in the stable module category. As shown in [36], the stable module category $\mathcal{C} = \underline{\mathrm{mod}}\Lambda$ is 2-Calabi-Yau and it is easy to check that assumption 1) holds.

**Theorem 7.1** (Geiss-Leclerc-Schröer). *The category $\mathcal{C} = \underline{\mathrm{mod}}\Lambda$ admits a cluster-tilting object $T$ such that the quiver of $\mathrm{End}(T)$ is obtained from that of the category of indecomposable $k\Delta$-modules by deleting the injective vertices and adding an arrow $v \to \tau v$ for each non projective vertex $v$.*

Here, by the quiver of the category of indecomposable $k\Delta$-modules, we mean the full subquiver of the repetition $\mathbb{Z}Q$ which is formed by the vertices corresponding to modules (complexes concentrated in degree 0). Thus for $\Delta = \vec{A}_5$, this quiver is as follows:

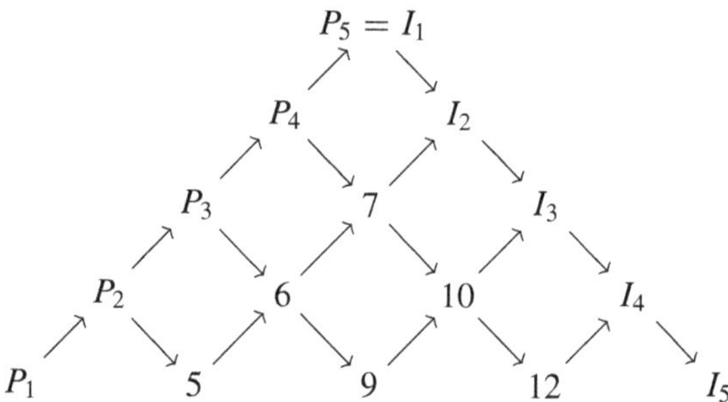

where we have marked the indecomposable projectives $P_i$ and the indecomposable injectives $I_j$. If we remove the vertices corresponding to the indecomposable injectives (and all the arrows incident with them) and add an arrow $v \to \tau v$ for each vertex not corresponding to an indecomposable projective, we obtain the following quiver

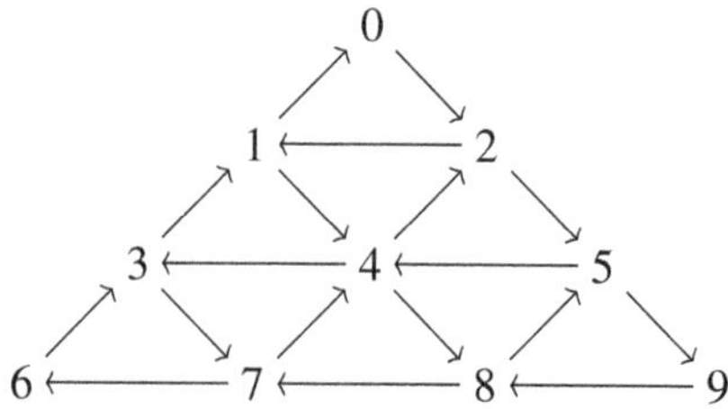

In a series of papers [64] [66] [65] [62] [61] [67] [63], Geiss-Leclerc-Schröer have obtained remarkable results for a class of quivers which are important in the study of (dual semi-)canonical bases. They use an analogue [67] of the Caldero-Chapoton map due ultimately to Lusztig [104]. The class they consider has been further enlarged by Buan-Iyama-Reiten-Scott [7]. Thanks to their results, an analogue of Caldero-Chapoton's formula and a weakened version of theorem 6.1 was proved in [57] for an even larger class.

## 7.2. Calabi-Yau reduction

Suppose that $(\mathcal{C}, T)$ is a 2-Calabi-Yau realization of a quiver $Q$ so that for $1 \leq i \leq n$, the vertex $i$ of $Q$ corresponds to the indecomposable summand $T_i$ of $T$. Let $J$ be a subset of the set of vertices of $Q$ and let $Q'$ be the quiver obtained from $Q$ by deleting all vertices in $J$ and all arrows incident with one of these vertices. Let $\mathcal{U}$ be the full subcategory of $\mathcal{C}$ formed by the objects $U$ such that $\mathrm{Ext}^1(T_j, U) = 0$ for all $j \in J$. Note that all $T_i$, $1 \leq i \leq n$, belong to $\mathcal{U}$. Let $\langle T_j | j \in J \rangle$ denote the ideal of $\mathcal{U}$ generated by the identities of the objects $T_j$, $j \in J$. By imitating the construction of the triangulated structure on a stable category, one can endow the quotient

$$\mathcal{C}' = \mathcal{U} / \langle T_j | j \in J \rangle$$

with a canonical structure of triangulated category, *cf.* [81].

**Theorem 7.2** (Iyama-Yoshino [81]). *The pair $(\mathcal{C}', T)$ is a 2-Calabi-Yau realization of the quiver $Q'$. Moreover, the projection $\mathcal{U} \to \mathcal{C}'$ induces a bijection between the cluster-tilting sets of $\mathcal{C}$ containing the $T_j$, $j \in J$, and the cluster-tilting sets of $\mathcal{C}'$.*

## 7.3. Mutation

Let $(\mathcal{C}, T)$ be a 2-Calabi-Yau category with cluster-tilting object. Let $T_1$ be an indecomposable direct factor of $T$.

**Theorem 7.3** (Iyama-Yoshino [81]). *Up to isomorphism, there is a unique indecomposable object $T_1^*$ not isomorphic to $T_1$ such that the object $\mu_1(T)$ obtained from $T$ by replacing the indecomposable summand $T_1$ with $T_1^*$ is cluster-tilting.*

We call $\mu_1(T)$ the *mutation* of $T$ at $T_1$. If $\mathcal{C}$ is the cluster category of a finite quiver without oriented cycles, this operation specializes of course to the one defined in section 5.10. However, in general, the quiver $Q$ of the endomorphism algebra of $T$ may contain loops and 2-cycles and then the quiver of the endomorphism algebra of $\mu_1(T)$ is not determined by $Q$. Let us illustrate this phenomenon on the following example (taken from Proposition 2.6 of [25]): Let $\mathcal{C}$ be the orbit category of the bounded derived category $\mathcal{D}_{\tilde{D}_6}$ under the action of the autoequivalence $\tau^2$. Then $\mathcal{C}$ satisfies the assumptions 1) and 2) of section 7.1. Its category of indecomposables is equivalent to the mesh category of the quiver

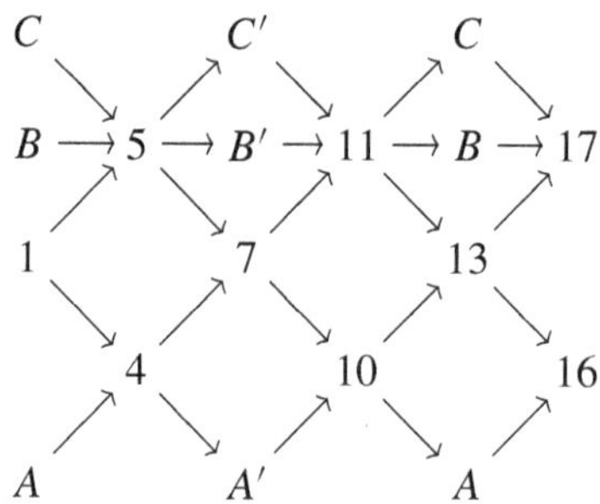

where the vertices labeled $A$, 1, $B$, $C$, 4, 5 on the left have to be identified with the vertices labeled $A$, 13, $B$, $C$, 16, 17 on the right. In this case, there are exactly 6 indecomposable rigid objects, namely $A$, $B$, $C$, $A'$, $B'$ and $C'$. There are exactly 6 cluster-tilting sets. The following is the exchange graph: Its vertices are cluster-tilting sets (we write $AC$ instead of $\{A, C\}$) and its edges represent mutations.

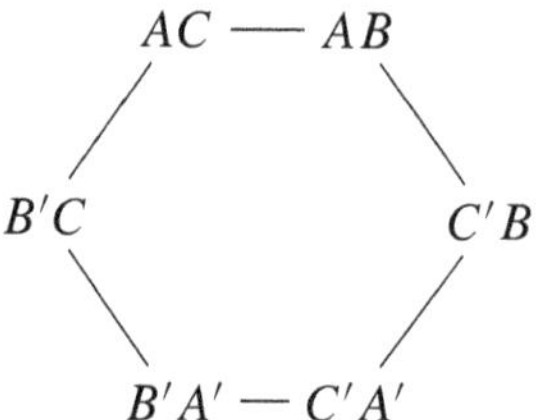

The quivers of the endomorphism algebras are as follows:

$$AC, AB: \quad \circ \rightleftarrows \circ \;\circlearrowright$$
$$B'C, C'B: \quad \circ \rightleftarrows \circ$$
$$B'A', C'A': \quad \circlearrowleft\circ \rightleftarrows \circ$$

In the setting of the above theorem, there are still exchange triangles as in theorem 5.12 but their description is different: Let $\mathcal{T}'$ be the full subcategory of $\mathcal{C}$ formed by the direct sums of indecomposables $T_i$, where $i$ is different from $k$. A *left $\mathcal{T}'$-approximation* of $T_k$ is a morphism $f : T_k \to T'$ with $T'$

in $\mathcal{T}'$ such that any morphism from $T_k$ to an object of $\mathcal{T}'$ factors through $f$. A left $\mathcal{T}'$-approximation $f$ is *minimal* if for each endomorphism $g$ of $T'$, the equality $gf = f$ implies that $g$ is invertible. Dually, one defines (minimal) right $\mathcal{T}'$-approximations. It is not hard to show that they always exist.

**Theorem 7.4** (Iyama-Yoshino [81]). *If $(T_1, T_1^*)$ is an exchange pair, there are non split triangles, unique up to isomorphism,*

$$T_1 \xrightarrow{f} E' \to T_1^* \to \Sigma T_1 \quad and \quad T_1^* \to E \xrightarrow{g} T_1 \to \Sigma T_1^*$$

*such that $f$ is a minimal left $\mathcal{T}'$-approximation and $g$ a minimal right $\mathcal{T}'$-approximation.*

## 7.4. Simple mutations, reachable cluster-tilting objects

Let $(\mathcal{C}, T)$ be a 2-CY category with cluster-tilting object and $T_1$ an indecomposable direct summand of $T$. Let $(T_1, T_1^*)$ be the corresponding exchange pair and

$$T_1^* \to E \to T_1 \to \Sigma T_1^*$$

the exchange triangle. The long exact sequence induced in $\mathcal{C}(T_1, ?)$ by this triangle yields a short exact sequence

$$\mathcal{C}'_T(T_1, T_1) \to \mathcal{C}(T_1, T_1) \to \mathsf{Ext}^1(T_1, T_1^*) \to 0 \, ,$$

where the leftmost term is the space of those endomorphisms of $T_1$ which factor through a sum of copies of $T/T_1$. Now the algebra $\mathcal{C}(T_1, T_1)$ is local and its residue field is $k$ (since $k$ is algebraically closed). We deduce the following lemma.

**Lemma 7.5.** *The quiver of the endomorphism algebra of $T$ does not have a loop at the vertex corresponding to $T_1$ iff we have $\dim \mathsf{Ext}^1(T_1, T_1^*) = 1$ iff $\mathsf{Ext}^1(T_1, T_1^*)$ is a simple module over $\mathcal{C}(T_1, T_1)$. In this case, in the exchange triangles*

$$T_1^* \to E \to T_1 \to \Sigma T_1^* \text{ and } T_1 \to E' \to T_1^* \to \Sigma T_1 \, ,$$

*we have*

$$E = \bigoplus_{\substack{arrows \\ i \to 1}} T_i \text{ and } E' = \bigoplus_{\substack{arrows \\ 1 \to j}} T_j \, .$$

We say that the mutation at $T_1$ is *simple* if the conditions of the lemma hold. If $\mathcal{C}$ is the cluster category of a finite quiver without oriented cycles, all mutations in $\mathcal{C}$ are simple, by part c) of theorem 6.1. By a theorem of Geiss-Leclerc-Schröer, if $\mathcal{C}$ is the stable module category of a preprojective algebra of Dynkin type, the quiver of any cluster-tilting object in $\mathcal{C}$ does not have loops nor 2-cycles. So again, all mutations are simple in this case.

**Theorem 7.6** (Buan-Iyama-Reiten-Scott [7]). *Suppose that the quivers $Q$ and $Q'$ of the endomorphism algebras of $T$ and $T' = \mu_1(T)$ do not have loops nor 2-cycles. Then $Q'$ is the mutation of $Q$ at the vertex 1.*

We define a cluster-tilting object $T'$ to be *reachable from* $T$ if there is a sequence of mutations

$$T = T^{(0)} \rightsquigarrow T^{(1)} \rightsquigarrow \ldots \rightsquigarrow T^{(N)} = T'$$

such that the quiver of $\mathsf{End}(T^{(i)})$ does not have loops nor 2-cycles for all $1 \leq i \leq N$. We define a rigid indecomposable of $\mathcal{C}$ to be *reachable from* $T$ if it is a direct summand of a reachable cluster-tilting object.

**Corollary 7.7.** *If a cluster-tilting object $T'$ is reachable from $T$, then the quiver of the endomorphism algebra of $T'$ is mutation-equivalent to the quiver of the endomorphism algebra of $T$. If $\mathcal{C}$ is a cluster-category or the stable module category of the preprojective algebra of a Dynkin diagram, then all quivers mutation-equivalent to $Q$ are obtained in this way.*

## 7.5. Combinatorial invariants

Let $(\mathcal{C}, T)$ be a 2-Calabi-Yau category with cluster-tilting object. Let $\mathcal{T}$ be the full subcategory whose objects are all direct factors of finite direct sums of copies of $T$. Notice that $\mathcal{T}$ is equivalent to the category of finitely generated projective modules over the endomorphism algebra of $T$. Let $K_0(\mathcal{T})$ be the Grothendieck group of the additive category $\mathcal{T}$. Thus, the group $K_0(\mathcal{T})$ is free abelian on the isomorphism classes of the indecomposable summands of $T$.

**Lemma 7.8** (Keller-Reiten [92]). *For each object $L$ of $\mathcal{C}$, there is a triangle*

$$T_1 \rightarrow T_0 \rightarrow L \rightarrow \Sigma T_1$$

*such that $T_0$ and $T_1$ belong to $\mathcal{T}$. The difference*

$$[T_0] - [T_1]$$

*considered as an element of $K_0(\mathcal{T})$ does not depend on the choice of this triangle.*

In the situation of the lemma, we define the *index* $\mathrm{ind}(L)$ *of* $L$ as the element $[T_0] - [T_1]$ of $K_0(\mathcal{T})$.

**Theorem 7.9** (Dehy-Keller [37]).

a) *Two rigid objects are isomorphic iff their indices are equal.*

b) *The indices of the indecomposable summands of a cluster-tilting object form a basis of $K_0(\mathcal{T})$. In particular, all cluster-tilting objects have the same number of pairwise non isomorphic indecomposable summands.*

Let $B$ be the endomorphism algebra of $T$. For two finite-dimensional right $B$-modules $L$ and $M$ put

$$\langle L, M \rangle_a = \dim \mathrm{Hom}(L, M) - \dim \mathrm{Ext}^1(L, M)$$
$$- \dim \mathrm{Hom}(M, L) + \dim \mathrm{Ext}^1(M, L).$$

This is the antisymmetrization of a truncated Euler form. A priori it is defined on the split Grothendieck group of the category $\mathrm{mod}\,B$ (*i.e.* the quotient of the free abelian group on the isomorphism classes divided by the subgroup generated by all relations obtained from direct sums in $\mathrm{mod}\,B$).

**Proposition 7.10** (Palu [112]). *The form $\langle, \rangle_a$ descends to an antisymmetric form on $K_0(\mathrm{mod}\,B)$. Its matrix in the basis of the simples is the antisymmetric matrix associated with the quiver of $B$ (loops and 2-cycles do not contribute to this matrix).*

Let $T_1, \ldots, T_n$ be the pairwise non isomorphic indecomposable direct summands of $T$. For $L \in \mathcal{C}$, we define the integer $g_i(L)$ to be the multiplicity of $[T_i]$ in the index $\mathrm{ind}(L)$, $1 \leq i \leq n$, and we define the element $X'_L$ of the field $\mathbb{Q}(x_1, \ldots, x_n)$ by

$$X'_L = \prod_{i=1}^{n} x_i^{g_i(L)} \sum_e \chi(\mathrm{Gr}_e(\mathrm{Ext}^1(T, L))) \prod_{i=1}^{n} x_i^{\langle S_i, e \rangle_a}, \tag{7.5.1}$$

where $S_i$ is the simple quotient of the indecomposable projective $B$-module $P_i = \mathrm{Hom}(T, T_i)$. Notice that we have $X'_{T_i} = x_i$, $1 \leq i \leq n$. If $\mathcal{C}$ is the cluster-category of a finite quiver $Q$ without oriented cycles and $T = kQ$, then we have $X'_L = X_{\Sigma L}$ in the notations of section 6 and the formula for $X'_L$ is essentially another expression for the Caldero-Chapoton formula.

Now let $Q$ be the quiver of the endomorphism algebra of $T$ in $\mathcal{C}$ and let $\mathcal{A}_Q$ be the associated cluster algebra.

134                        *Bernhard Keller*

**Theorem 7.11** (Palu [112]). *If $L$ and $M$ are objects of $C$ such that $\mathsf{Ext}^1(L, M)$ is one-dimensional, then we have*

$$X'_L X'_M = X'_E + X'_{E'},$$

*where*

$$L \to E \to M \to \Sigma L \text{ and } M \to E' \to L \to \Sigma M$$

*are 'the' two non split triangles. Thus, if $L$ is a rigid indecomposable reachable from $T$, then $X'_L$ is a cluster variable of $\mathcal{A}_Q$.*

**Corollary 7.12.** *Suppose that $C$ is the cluster category $C_Q$ of a finite quiver $Q$ without oriented cycles. Let $\mathcal{A}_Q \subset \mathbb{Q}(x_1, \ldots, x_n)$ be the associated cluster algebra and $x \in \mathcal{A}_Q$ a cluster variable. Let $L \in C_Q$ be the unique (up to isomorphism) indecomposable rigid object such that $x = X_{\Sigma L}$. Let $u_1, \ldots, u_n$ be an arbitrary cluster of $\mathcal{A}_Q$ and $T_1, \ldots, T_n$ the cluster-tilting set such that $u_i = X_{T_i}$, $1 \le i \le n$. Then the expression of $x$ as a Laurent polynomial in $u_1, \ldots, u_n$ is given by*

$$x = X'_L(u_1, \ldots, u_n).$$

The expression for $X'_L$ makes it natural to define the polynomial $F'_L \in \mathbb{Z}[y_1, \ldots y_n]$ by

$$F'_L = \sum_e \chi(\mathsf{Gr}_e(\mathsf{Ext}^1(T, L))) \prod_{j=1}^n y_j^{e_j}. \tag{7.5.2}$$

We then have

$$X'_L = \prod_{i=1}^n x_i^{g_i(L)} F'_L \left( \prod_{i=1}^n x_i^{b_{i1}}, \ldots, \prod_{i=1}^n x_i^{b_{in}} \right).$$

The polynomial $F'_L$ is related to Fomin-Zelevinsky's $F$-polynomials [54] [55], as we will see below.

## 7.6. More mutants categorified

Let $Q$ be a finite quiver without loops nor 2-cycles with vertex set $\{1, \ldots, n\}$. Let $\mathbb{T}_n$ be the regular $n$-ary tree: Its edges are labeled by the integers $1, \ldots, n$ such that the $n$ edges emanating from each vertex carry different labels. Let $t_0$ be a vertex of $\mathbf{T}_n$. To each vertex $t$ of $\mathbb{T}_n$ we associate a seed $(Q_t, \mathbf{x}_t)$ (*cf.* section 3.2) such that at $t = t_0$, we have $Q_t = Q$ and $\mathbf{x}_t = \{x_1, \ldots, x_n\}$ and whenever $t$ is linked to $t'$ by an edge labeled $i$, we have $(Q_{t'}, \mathbf{x}_{t'}) = \mu_i(Q_t, \mathbf{x}_t)$.

Now assume that $Q$ admits a 2-Calabi-Yau realization $(\mathcal{C}, T)$. According to the mutation theorem 7.6, we can associate a cluster-tilting object $T_t$ to each vertex $t$ of $\mathbb{T}_n$ such that at $t = t_0$, we have $T_t = T$ and that whenever $t$ is linked to $t'$ by an edge labeled $i$, we have $T_{t'} = \mu_i(T_t)$.

Now let

$$t_0 \xrightarrow{\ i_1\ } t_1 \xrightarrow{\ i_2\ } \cdots \xrightarrow{\ i_N\ } t_N$$

be a path in $\mathbb{T}_n$ and suppose that for each $1 \leq i \leq N$, the quiver of the endomorphism algebra of $T_{t_i}$ does not have loops nor 2-cycles. Then it follows by induction from theorem 7.6 that the quiver of the endomorphism algebra of $T_{t_i}$ is $Q_{t_i}$, $1 \leq i \leq N$, and from theorem 7.11 that the cluster $\mathbf{x}_{t_i}$ equals the image under $L \mapsto X'_L$ of the set of indecomposable direct factors of $T_{t_i}$.

Following [55], let us consider three other pieces of data associated with each vertex $t$ of $\mathbb{T}_n$:

- the tropical $Y$-variables $y_{1,t}, \ldots, y_{n,t}$,
- the $F$-polynomials $F_{1,t}, \ldots, F_{n,t}$,
- the (non tropical) $Y$-variables $Y_{1,t}, \ldots, Y_{n,t}$.

Here the tropical $Y$-variables are monomials in the indeterminates $y_1, \ldots, y_n$ and their inverses. At $t = t_0$, we have $y_{i,t} = y_i$, $1 \leq i \leq n$. If $t$ is linked to $t'$ by an edge labelled $i$, then

$$y_{j,t'} = \begin{cases} y_{i,t}^{-1} & \text{if } j = i \\ y_{j,t}\, y_{i,t}^{[b_{ij}]_+} (y_{i,t} \oplus 1)^{-b_{ij}} & \text{if } j \neq i. \end{cases}$$

Here, $(b_{ij})$ is the antisymmetric matrix associated with $Q_t$, for an integer $a$, we write $[a]_+$ for $\max(a, 0)$, and, for a monomial $m$ which is the product of powers $y_j^{e_j}$, $1 \leq j \leq n$, we write $m \oplus 1$ for the product of the factors $y_j^{e_j}$ with negative exponents $e_j$. Notice that $[b_{ij}]_+$ is the number of arrows from $i$ to $j$ in $Q_t$.

The $F$-polynomials lie in $\mathbb{Z}[y_1, \ldots, y_n]$. At $t = t_0$, they all equal 1. If $t$ is linked to $t'$ by an edge labeled $i$, then

$$F_{j,t'} = F_{j,t} \quad \text{if } j \neq i,$$

$$F_{i,t'} = \frac{1}{F_{i,t}} \left( \prod_{c_{li} > 0} y_l^{c_{li}} \prod_{j=1}^{n} F_{j,t}^{[b_{ij}]_+} + \prod_{c_{li} < 0} y_l^{-c_{li}} \prod_{j=1}^{n} F_{j,t}^{[-b_{ij}]_+} \right),$$

where the $c_{li}$ are the exponents in the tropical $Y$-variables $y_{i,t} = \prod_{l=1}^{n} y_l^{c_{li}}$ and $B = (b_{ij})$ is the antisymmetric matrix associated with $Q_t$.

Finally, the non tropical $Y$-variables $Y_{j,t}$ lie in the field $\mathbb{Q}(y_1, \ldots, y_n)$. At $t = t_0$, we have $Y_{j,t} = y_j$, $1 \leq j \leq n$, and if $t$ and $t'$ are linked by an edge labeled $i$, then

$$Y_{j,t'} = \begin{cases} Y_{i,t}^{-1} & \text{if } j = i \\ Y_{j,t}Y_{i,t}^{[b_{ij}]_+}(Y_{i,t}+1)^{-b_{ij}} & \text{if } j \neq i. \end{cases} \qquad (7.6.1)$$

The *principal extension* $\tilde{Q}$ of $Q$ is the quiver obtained from $Q$ by adding, for each vertex $i$ of $Q$, a new vertex $i'$ and a new arrow $i' \to i$. We assume that we are given a 2-Calabi-Yau realization $(\tilde{\mathcal{C}}, \tilde{T})$ of $\tilde{Q}$. Let $\mathcal{U} \subset \tilde{\mathcal{C}}$ be the full subcategory of the objects $U$ such that $\mathsf{Ext}^1(T_{i'}, U) = 0$ for each new vertex $i'$. Then, according to the theorem on Calabi-Yau reductions (*cf.* section 7.2), the quotient category

$$\mathcal{U}/\langle T_{i'} | i \in Q_0 \rangle$$

together with the image of $\tilde{T}$ is a 2-Calabi-Yau realization of the quiver $Q$. We assume that $(\mathcal{C}, T)$ is this realization. Now let

$$t_0 \xrightarrow{\phantom{xx} i_1 \phantom{xx}} t_1 \xrightarrow{\phantom{xx} i_2 \phantom{xx}} \cdots \xrightarrow{\phantom{xx} i_N \phantom{xx}} t_N$$

be a path in $\mathbb{T}_n$ and suppose that for each $1 \le i \le N$, the quiver of the endomorphism algebra of $T_{t_i}$ does not have loops nor 2-cycles. Let $t = t_N$ and let $T_1', \ldots, T_n'$ be the indecomposable summands of $T' = T_t$. Recall from part b) of theorem 7.9 that the indices $\mathsf{ind}(T_l')$, $1 \le l \le n$, form a basis of the group $K_0(\mathcal{T})$, where $\mathcal{T}$ is the full subcategory of $\mathcal{C}$ formed by the direct summands of finite direct sums of copies of $T$.

**Theorem 7.13.**

a) *The exchange matrix $B_t = (b_{ij})$ associated with $t$ is the antisymmetric matrix associated with the quiver of the endomorphism algebra of $T_t$.*

b) *We have $y_{j,t} = \prod_{l=1}^n y_l^{c_{lj}}$, $1 \le j \le n$, where $c_{lj}$ is defined by*

$$[T_j] = \sum_{l=1}^n c_{lj}\, \mathsf{ind}(T_l').$$

c) *We have $F_{j,t} = F_{T_j'}'$, $1 \le j \le n$, where $F_{T_j'}'$ is defined by equation 7.5.2.*

d) *We have*

$$Y_{j,t} = y_{j,t} \prod_{i=1}^n F_{i,t}^{b_{ij}}.$$

Here part a) follows by induction from theorem 7.6, part b) is proved by applying theorem 7.6 and Theorem 12 of [111] to the category $\tilde{\mathcal{C}}$, part c) is Theorem 5.3 of [57] and part d) is Proposition 3.12 of [55].

## 7.7. 2-CY categories from algebras of global dimension 2

The results of the preceding sections only become interesting if we are able to construct 2-Calabi-Yau realizations for large classes of quivers. In the case of a quiver without oriented cycles, this problem is solved by the cluster category. The results of Geiss-Leclerc-Schröer provide another large class of quivers admitting 2-Calabi-Yau realizations. Here, we will exhibit a construction which generalizes both cluster categories of quivers without oriented cycles and the stable categories of preprojective algebras of Dynkin diagrams.

Let $k$ be an algebraically closed field and $A$ a finite-dimensional $k$-algebra of global dimension $\leq 2$. For example, $A$ can be the path algebra of a finite quiver without oriented cycles. Let $\mathcal{D}_A$ be the bounded derived category of the category $\mathsf{mod}\, A$ of $k$-finite-dimensional right $A$-modules. It admits a Serre functor, namely the total derived functor of the tensor product $? \otimes_A DA$ with the $k$-dual bimodule of $A$ considered as a bimodule over itself. We can form the orbit category

$$\mathcal{D}_A/(S^{-1}\Sigma^2)^{\mathbb{Z}}.$$

This is a $k$-linear category endowed with a suspension functor (induced by $\Sigma$) but in general it is no longer triangulated. Nevertheless, one can show that it embeds fully faithfully into a 'smallest triangulated overcategory' [90]. We denote this overcategory by $\mathcal{C}_A$ and call it the *generalized cluster category of $A$*.

**Theorem 7.14** (Amiot [1]). *If the functor*

$$\mathsf{Tor}_2^A(?, DA) : \mathsf{mod}\, A \to \mathsf{mod}\, A$$

*is nilpotent (i.e. some power of it vanishes), then $\mathcal{C}_A$ is $\mathsf{Hom}$-finite and 2-Calabi-Yau. Moreover, the image $T$ of $A$ in $\mathcal{C}_A$ is a cluster-tilting object. The quiver of its endomorphism algebra is obtained from that of $A$ by adding, for each pair of vertices $(i, j)$, a number of arrows equal to*

$$\dim \mathsf{Tor}_2^A(S_j, S_i^{op})$$

*from $i$ to $j$, where $S_j$ is the simple right module associated with $j$ and $S_i^{op}$ the simple left module associated with $i$.*

We consider two classes of examples obtained from this theorem: First, let $\Delta$ be a simply laced Dynkin diagram and $k\vec{\Delta}$ the path algebra of a quiver with underlying graph $\Delta$. Let $A$ the *Auslander algebra of $k\vec{\Delta}$, i.e.* the endomorphism algebra of the direct sum of a system of representatives of the indecomposable $B$-modules modulo isomorphism. Then it is not hard to check that the

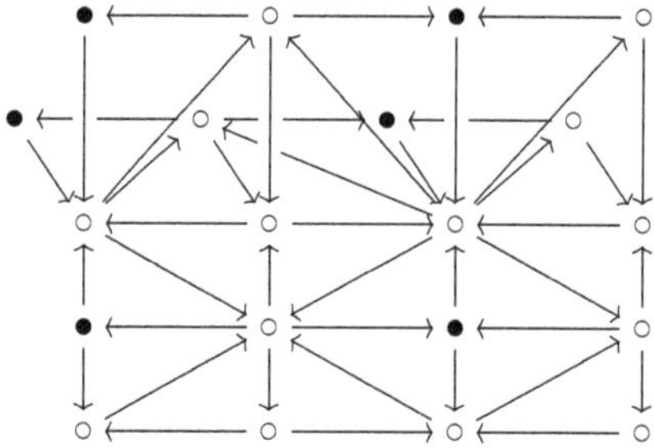

Figure 4.  The quiver $\vec{A}_4 \otimes \vec{D}_5$

assumptions of the theorem hold. The quiver of $A$ is simply the quiver of
the category of indecomposables of $k\vec{\Delta}$ and the minimal relations correspond
to its meshes. For example, if we choose $\Delta = \vec{A}_4$ with the linear orientation, the quiver of the endomorphism algebra of the image $T$ of $A$ in $\mathcal{C}_A$ is
the quiver obtained from Geiss-Leclerc-Schröer's construction at the end of
section 7.1.

As a second class of examples, we consider an algebra $A$ which is the tensor
product $kQ \otimes_k kQ'$ of two path algebras of quivers $Q$ and $Q'$ without oriented
cycles. Such an algebra is clearly of global dimension $\leq 2$. Let us assume that $Q$
and $Q'$ are moreover Dynkin quivers. Then it is not hard to check that the functor
$\mathrm{Tor}_2(?, DA)$ is indeed nilpotent. Thus, the theorem applies. Another elementary
exercise in homological algebra shows that the space $\mathrm{Tor}_2^A(S_{j,j'}, S_{i,i'}^{op})$ is at most
one-dimensional and that it is non zero if and only if there is an arrow $i \to j$ in
$Q$ and an arrow $i' \to j'$ in $Q'$. This immediately yields the shape of the quiver
of the endomorphism algebra of the image $T$ of $A$ in $\mathcal{C}_A$: It is the *tensor product*
$Q \otimes Q'$ obtained from the product $Q \times Q'$ by adding an arrow $(j, j') \to (i, i')$
for each pair of arrows $i \to j$ of $Q$ and $i' \to j'$ of $Q'$. For example, for
suitable orientations of $A_4$ and $D_5$, we obtain the quiver of figure 4. Notice
that if we perform mutations at the six vertices of the form $(i, i')$, where $i$
is a sink of $Q = \vec{A}_4$ and $i'$ a source of $Q' = \vec{D}_5$ (they are marked by $\bullet$), we
obtain the quiver of figure 5 related to the periodicity conjecture for $(A_4, D_5)$,
*cf.* below.

## 8. Application: The periodicity conjecture

Let $\Delta$ and $\Delta'$ be two Dynkin diagrams with vertex sets $I$ and $I'$. Let $A$ and $A'$
be the incidence matrices of $\Delta$ and $\Delta'$, *i.e.* if $C$ is the Cartan matrix of $\Delta$ and $J$

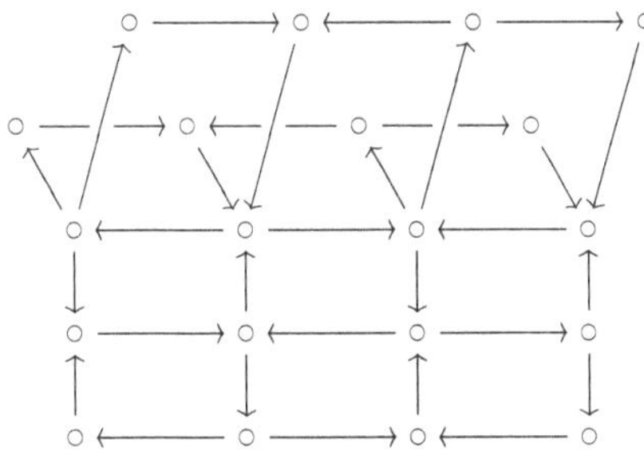

Figure 5. The quiver $\vec{A}_4 \square \vec{D}_5$

the identity matrix of the same format, then $A = 2J - C$. Let $h$ and $h'$ be the Coxeter numbers of $\Delta$ and $\Delta'$.

The *Y-system of algebraic equations* associated with the pair of Dynkin diagrams $(\Delta, \Delta')$ is a system of countably many recurrence relations in the variables $Y_{i,i',t}$, where $(i, i')$ is a vertex of $\Delta \times \Delta'$ and $t$ an integer. The system reads as follows:

$$Y_{i,i',t-1} Y_{i,i',t+1} = \frac{\prod_{j \in I}(1 + Y_{j,i',t})^{a_{ij}}}{\prod_{j' \in I'}(1 + Y_{i,j',t}^{-1})^{a'_{i'j'}}}. \tag{8.0.1}$$

**Periodicity Conjecture 8.1.** *All solutions to this system are periodic in $t$ of period dividing $2(h + h')$.*

Here is an algebraic reformulation: Let $K$ be the fraction field of the ring of integer polynomials in the variables $Y_{ii'}$, where $i$ runs through the set of vertices $I$ of $\Delta$ and $i'$ through the set of vertices $I'$ of $\Delta'$. Since $\Delta$ is a tree, the set $I$ is the disjoint union of two subsets $I_+$ and $I_-$ such that there are no edges between any two vertices of $I_+$ and no edges between any two vertices of $I_-$. Analogously, $I'$ is the disjoint union of two sets of vertices $I'_+$ and $I'_-$. For a vertex $(i, i')$ of the product $I \times I'$, define $\varepsilon(i, i')$ to be 1 if $(i, i')$ lies in $I_+ \times I'_+ \cup I_- \times I'_-$ and $-1$ otherwise. For $\varepsilon = \pm 1$, define an automorphism $\tau_\varepsilon$ of $K$ by $\tau_\varepsilon(Y_{i,i'}) = Y_{i,i'}$ if $\varepsilon(i, i') \neq \varepsilon$ and

$$\tau_\varepsilon(Y_{ii'}) = Y_{ii'}^{-1} \prod_j (1 + Y_{ji'})^{a_{ij}} \prod_{j'} (1 + Y_{ij'}^{-1})^{-a'_{i'j'}} \tag{8.0.2}$$

if $\varepsilon(i, i') = \varepsilon$. Finally, define an automorphism $\varphi$ of $K$ by

$$\varphi = \tau_+ \tau_-. \tag{8.0.3}$$

Then, as in [54], it is not hard to check that the periodicity conjecture holds iff the order of the automorphism $\varphi$ is finite and divides $h + h'$.

The conjecture was formulated

- by Al. B. Zamolodchikov for $(\Delta, A_1)$, where $\Delta$ is simply laced [131, (12)];
- by Ravanini-Valleriani-Tateo for $(\Delta, \Delta')$, where $\Delta$ and $\Delta'$ are simply laced or 'tadpoles' [113, (6.2)];
- by Kuniba-Nakanishi for $(\Delta, A_n)$, where $\Delta$ is not necessarily simply laced [100, (2a)], see also Kuniba-Nakanishi-Suzuki [101, B.6]; notice however that in the non simply laced case, the $Y$-system given by Kuniba-Nakanishi is different from the one above and that they conjecture periodicity with period dividing twice the sum of the *dual* Coxeter numbers;

It was proved

- for $(A_n, A_1)$ by Frenkel-Szenes [56] (who produced explicit solutions) and by Gliozzi-Tateo [72] (via volumes of threefolds computed using triangulations),
- by Fomin-Zelevinsky [54] for $(\Delta, A_1)$, where $\Delta$ is not necessarily simply laced (via the cluster approach and a computer check for the exceptional types; a uniform proof can be given using [129]),
- for $(A_n, A_m)$ by Volkov [127], who exhibited explicit solutions using cross ratios, and by Szenes [124], who interpreted the system as a system of flat connections on a graph; an equivalent statement was proved by Henriques [77];
- by Hernandez-Leclerc for $(A_n, A_1)$ using representations of quantum affine algebras (which yield formulas for solutions in terms of $q$-characters). They expect to treat $(A_n, \Delta)$ similarly, *cf.* [78].

**Theorem 8.2** ([86]). *The periodicity conjecture 8.1 is true.*

Let us sketch a proof based on 2-Calabi-Yau categories (details will appear elsewhere [86]): First, using the folding technique of Fomin-Zelevinsky's [54] one reduces the conjecture to the case where both $\Delta$ and $\Delta'$ are simply laced. Thus, from now on, we assume that $\Delta$ and $\Delta'$ are simply laced.

*First step:* We choose quivers $Q$ and $Q'$ whose underlying graphs are $\Delta$ and $\Delta'$. We assume that $Q$ and $Q'$ are *alternating*, *i.e.* that each vertex is either a source or a sink. If $i$ is a vertex of $Q$ or $Q'$, we put $\varepsilon(i) = 1$ if $i$ is a source and $\varepsilon(i) = -1$ if $i$ is a sink. For example, we can consider the following quivers

$$\vec{A}_4 : \quad 1 \longleftarrow 2 \longrightarrow 3 \longleftarrow 4 \ ,$$

$$\vec{D}_5 : \quad 1 \longleftarrow 2 \longrightarrow 3 \ \begin{matrix} \nearrow 4 \\ \searrow 5. \end{matrix}$$

We define the *square product* $Q \square Q'$ to be the quiver obtained from $Q \times Q'$ by reversing all the arrows in the full subquivers of the form $\{i\} \times Q'$ and $Q \times \{i'\}$, where $i$ is a sink of $Q$ and $i'$ a source of $Q'$. The square product of the above quivers $\vec{A}_4$ and $\vec{D}_5$ is depicted in figure 5. The *initial $Y$-seed* associated with $Q \square Q'$ is the pair $\mathbf{y}_0$ formed by the quiver $Q \square Q'$ and the family of variables $Y_{i,i'}, (i, i') \in I \times I'$. We can apply mutations to it using the quiver mutation rule and the mutation rule for (non tropical) $Y$-variables given in equation 7.6.1.

*A general construction.* Let $R$ be a quiver and $v$ a sequence of vertices $v_1, \ldots, v_N$ of $R$. We assume that the composed mutation

$$\mu_v = \mu_{v_N} \ldots \mu_{v_2} \mu_{v_1}$$

transforms $R$ into itself. Then clearly the same holds for the inverse sequence

$$\mu_v^{-1} = \mu_{v_1} \mu_{v_2} \ldots \mu_{v_N}.$$

Now the *restricted $Y$-pattern* associated with $R$ and $\mu_v$ is the sequence of $Y$-seeds obtained from the initial $Y$-seed $\mathbf{y}_0$ associated with $R$ by applying all integral powers of $\mu_v$. Thus this pattern is given by a sequence of seeds $\mathbf{y}_t$, $t \in \mathbb{Z}$, such that $\mathbf{y}_0$ is the initial $Y$-seed associated with $R$ and, for all $t \in \mathbb{Z}$, $\mathbf{y}_{t+1}$ is obtained from $\mathbf{y}_t$ by the sequence of mutations $\mu_v$.

*Second step.* For two elements $\sigma, \sigma'$ of $\{+, -\}$ define the following composed mutation of $Q \square Q'$:

$$\mu_{\sigma,\sigma'} = \prod_{\varepsilon(i)=\sigma, \varepsilon(i')=\sigma'} \mu_{(i,i')}.$$

Notice that there are no arrows between any two vertices of the index set so that the order in the product does not matter. Then it is easy to check that $\mu_{+,+}\mu_{-,-}$ and $\mu_{-,+}\mu_{+,-}$ both transform $Q \square Q'$ into $(Q \square Q')^{op}$ and vice versa. Thus the composed sequence of mutations

$$\mu_\square = \mu_{-,-}\mu_{+,+}\mu_{-,+}\mu_{+,-}$$

transforms $Q \square Q'$ into itself. We define the *$Y$-system* $\mathbf{y}_\square$ associated with $Q \square Q'$ to be the restricted $Y$-pattern associated with $Q \square Q'$ and $\mu_\square$. As in section 8 of [55], one checks that the identity $\varphi^{h+h'} = \mathbf{1}$ follows if one shows that the system $\mathbf{y}_\square$ is periodic of period dividing $h + h'$.

*Third step.* One checks easily that we have

$$\mu_{+-}(Q \square Q') = Q \otimes Q',$$

where the tensor product $Q \otimes Q'$ is defined at the end of section 7.7. For the above quivers $\vec{A}_4$ and $\vec{D}_5$, the tensor product is depicted in figure 4. Therefore, the periodicity of the restricted $Y$-system associated with $Q \square Q'$ and $\mu_\square$ is equivalent to that of the restricted $Y$-system associated with $Q \otimes Q'$ and

$$\mu_\otimes = \mu_{+,-}\mu_{-,-}\mu_{+,+}\mu_{-,+}.$$

*Fourth step.* As we have seen in section 7.7, the quiver $Q \otimes Q'$ admits a 2-Calabi-Yau realization given by the cluster category $\mathcal{C}_{kQ\otimes kQ'}$ associated with the tensor product of the path algebras $kQ \otimes kQ'$. Let $T$ be the initial cluster tilting object. By theorem 7.3, we can define its iterated mutations

$$T_t = \mu_\otimes^t(T)$$

for all $t \in \mathbb{Z}$. Now we use the

**Proposition 8.3.** *None of the endomorphism quivers occurring in the sequence of mutated cluster-tilting objects joining $T$ to $T_t$ contains loops nor 2-cycles.*

It is not hard to see that $(\mathcal{C}_{kQ\otimes kQ'}, T)$ is the Calabi-Yau reduction of a 2-Calabi-Yau realization of the principal extension of $Q \square Q'$. Therefore, it follows from theorem 7.13 that the quiver of the endomorphism algebra of $T_t$ is $Q \otimes Q'$ and that the $Y$-variables in the $Y$-seed $\mathbf{y}_{\square,t}$ can be expressed in terms of the triangulated category $\mathcal{C}_{kQ\otimes kQ'}$ and the objects $T$ and $T_t$. Thus, it suffices to show that $T_t$ is isomorphic to $T$ whenever $t$ is an integer multiple of $h + h'$.

*Fifth step.* Let $\tau \otimes \mathbf{1}$ denote the auto-equivalence of the bounded derived category of $kQ \otimes kQ'$ given by the total left derived functor of the tensor product with the bimodule complex $(\Sigma^{-1}D(kQ)) \otimes kQ'$, where $D$ is duality over $k$. It induces an autoequivalence of $\mathcal{C}_{kQ\otimes kQ'}$ which we still denote by $\tau \otimes \mathbf{1}$. Define $\Phi$ to be the autoequivalence $\tau^{-1} \otimes \mathbf{1}$ of $\mathcal{C}_{kQ\otimes kQ'}$.

**Proposition 8.4.** *For each integer $t$, the image $\Phi^t(T)$ is isomorphic to $T_t$.*

The proposition is proved by showing that the indices of the two objects are equal. This suffices by theorem 7.9. Now we conclude thanks to the following categorical periodicity result:

**Proposition 8.5.** *The power $\Phi^{h+h'}$ is isomorphic to the identity functor.*

Let us sketch the proof of this proposition: With the natural abuse of notation, the Serre functor $S$ of the bounded derived category of $kQ \otimes kQ'$ is given by the 'tensor product' $S \otimes S$ of the Serre functors for $kQ$ and $kQ'$. In the generalized cluster category, the Serre functor becomes isomorphic to the square of the suspension functor. So we have

$$S = S \otimes S = \Sigma^2 = \Sigma \otimes \Sigma$$

as autoequivalences of $\mathcal{C}_{kQ\otimes kQ'}$. Now recall from equation 5.5.4 that $\tau = \Sigma^{-1}S$. Thus we get the isomorphism

$$\tau \otimes \tau = 1$$

of autoequivalences of $\mathcal{C}_{kQ\otimes kQ'}$. So we have

$$\tau^{-1} \otimes 1 = 1 \otimes \tau.$$

Now we compute $\Phi^{h+h'}$ by using the left hand side for the first $h$ factors and the right hand side for the last $h'$ factors:

$$\Phi^{h+h'} = (\tau^{-1} \otimes 1)^{h}(1 \otimes \tau)^{h'}.$$

But we know from equation 5.5.3 that $\tau^{-h} = \Sigma^2$. So we find

$$\Phi^{h+h'} = (\Sigma^2 \otimes 1)(1 \otimes \Sigma^{-2}) = \Sigma^2 \Sigma^{-2} = 1$$

as required.

# 9. Quiver mutation and derived equivalence

Let $Q$ be a finite quiver and $i$ a *source* of $Q$, *i.e.* no arrows have target $i$. Then the mutation $Q'$ of $Q$ at $i$ is simply obtained by reversing all the arrows starting at $i$. In this case, the categories of representations of $Q$ and $Q'$ are related by the Bernstein-Gelfand-Ponomarev reflection functors [19] and these induce equivalences in the derived categories [73]. We would like to present a similar categorical interpretation for mutation at arbitrary vertices. For this, we use recent work by Derksen-Weyman-Zelevinsky [39] and a construction due to Ginzburg [71]. We first recall the classical reflection functors in a form which generalizes well.

## 9.1. A reminder on reflection functors

We keep the above notations. In the sequel, $k$ is a field and $kQ$ is the path algebra of $Q$ over $k$. For each vertex $j$ of $Q$, we write $e_j$ for the lazy path at $j$, the idempotent of $kQ$ associated with $j$. We write $P_j = e_j kQ$ for the corresponding indecomposable right $kQ$-module, $\mathsf{Mod}\,kQ$ for the category of (all) right $kQ$-modules and $\mathcal{D}(kQ)$ for its derived category.

The module categories over $kQ$ and $kQ'$ are linked by a pair of adjoint functors [19]

$$\mathsf{Mod}\,kQ'$$

$$F_0 \left\downarrow \quad \right\uparrow G_0$$

$$\mathsf{Mod}\,kQ.$$

The right adjoint $G_0$ takes a representation $V$ of $Q^{op}$ to the representation $V'$ with $V'_j = V_j$ for $j \neq i$ and where $V'_i$ is the kernel of the map

$$\bigoplus_{\substack{\text{arrows of } Q \\ i \to j}} V_j \longrightarrow V_i$$

whose components are the images under $V$ of the arrows $i \to j$.

Then the left derived functor of $F_0$ and the right derived functor of $G_0$ are quasi-inverse equivalences [73]

$$\mathcal{D}kQ'$$

$$F \left\downarrow \quad \right\uparrow G$$

$$\mathcal{D}kQ.$$

The functor $F$ sends the indecomposable projective $P'_j$, $j \neq i$, to $P_j$ and the indecomposable projective $P'_i$ to the cone over the morphism

$$P_i \to \bigoplus_{\substack{\text{arrows of } Q \\ i \to j}} P_j.$$

In fact, the sum of the images of all the $P'_j$ has a natural structure of complex of $kQ'$-$kQ$-bimodules and $F$ can also be described as the derived tensor product over $kQ'$ with this complex of bimodules.

The example of the following quiver $Q$

$$
\begin{array}{ccc}
& 1 & \\
b \nearrow & & \searrow a \\
2 & \xleftarrow{\;c\;} & 3
\end{array}
$$

and its mutation $Q'$ at 1

$$Q' : 2 \leftarrow 1 \leftarrow 3$$

shows that if we mutate at a vertex which is neither a sink nor a source, then the derived categories of representations of $Q$ and $Q'$ are not equivalent in general. The cluster-tilted algebra associated with the mutation of $Q'$ at 1 (*cf.* section 5.8) is the quotient of $kQ$ by the ideal generated by $ab$, $bc$ and $ca$. It is of infinite global dimension and therefore not derived equivalent to $kQ$, either.

Clearly, in order to understand mutation from a representation-theoretic point of view, more structure is needed. Now quiver mutation has been independently invented and investigated in the physics literature, *cf.* for example equation (12.2) on page 70 in [26] (I thank to S. Fomin for this reference). In the physics context, a crucial role is played by the so-called superpotentials, *cf.* for example [18]. This lead Derksen-Weyman-Zelevinsky to their systematic study of quivers with potentials and their mutations in [39]. We now sketch their main result.

## 9.2. Mutation of quivers with potentials

Let $Q$ be a finite quiver. Let $\widehat{kQ}$ be the *completed path algebra, i.e.* the completion of the path algebra at the ideal generated by the arrows of $Q$. Thus, $\widehat{kQ}$ is a topological algebra and the paths of $Q$ form a topologial basis so that the underlying vector space of $\widehat{kQ}$ is

$$\prod_{p \text{ path}} kp.$$

The *continuous Hochschild homology* of $\widehat{kQ}$ is the vector space $\mathsf{HH}_0$ obtained as the quotient of $\widehat{kQ}$ by the closure of the subspace generated by all commutators. It admits a topological basis formed by the *cycles* of $Q$, *i.e.* the orbits of paths $p = (i|\alpha_n|\ldots|\alpha_1|i)$ of length $n \geq 0$ with identical source and target under the action of the cyclic group of order $n$. In particular, the space $\mathsf{HH}_0$ is a product of copies of $k$ indexed by the vertices if $Q$ does not have oriented cycles. For each arrow $a$ of $Q$, the *cyclic derivative with respect to $a$* is the unique linear map

$$\partial_a : \mathsf{HH}_0 \to \widehat{kQ}$$

which takes the class of a path $p$ to the sum

$$\sum_{p=uav} vu$$

taken over all decompositions of $p$ as a concatenation of paths $u$, $a$, $v$, where $u$ and $v$ are of length $\geq 0$. A *potential* on $Q$ is an element $W$ of $\mathsf{HH}_0$ whose expansion in the basis of cycles does not involve cycles of length 0.

Now assume that $Q$ does not have loops or 2-cycles.

**Theorem 9.1** (Derksen-Weyman-Zelevinsky [39]). *The mutation operation*

$$Q \mapsto \mu_i(Q)$$

*admits a good extension to quivers with potentials*

$$(Q, W) \mapsto \mu_i(Q, W) = (Q', W'),$$

i.e. *the quiver $Q'$ is isomorphic to $\mu_i(Q)$ if $W$ is generic.*

Here, 'generic' means that $W$ avoids a certain countable union of hypersurfaces in the space of potentials. The main ingredient of the proof of this theorem is the construction of a 'minimal model', which, in a different language, was also obtained in [82] and [97].

If $W$ is not generic, then $\mu_i(Q)$ is not necessarily isomorphic to $Q'$ but still isomorphic to its 2-reduction, *i.e.* the quiver obtained from $Q'$ by removing the arrows of a maximal set of pairwise disjoint 2-cycles. For example, the mutation of the quiver

$$Q: \quad \begin{array}{c} 1 \\ b \nearrow \ \searrow a \\ 2 \xleftarrow{\ c\ } 3 \end{array} \qquad (9.2.1)$$

with the potential $W = abc$ at the vertex 1 is the quiver with potential

$$Q' : 2 \leftarrow 1 \leftarrow 3, \qquad W = 0.$$

On the other hand, the mutation of the above cyclic quiver $Q$ with the potential $W = abcabc$ at the vertex 1 is the quiver

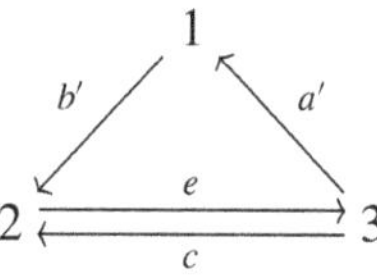

with the potential $ecec + eb'a'$.

Two quivers with potentials $(Q, W)$ and $(Q', W')$ are *right equivalent* if there is an isomorphism $\varphi : \widehat{kQ} \to \widehat{kQ'}$ taking $W$ to $W'$. It is shown in [39] (*cf.* also [133]) that the mutation $\mu_i$ induces an involution on the set of right equivalence classes of quivers with potentials, where the quiver does not have loops and does not have a 2-cycle passing through $i$.

## 9.3. Derived equivalence of Ginzburg dg algebras

We will associate a derived equivalence of differential graded (=dg) algebras with each mutation of a quiver with potential. The dg algebra in question is the algebra $\Gamma = \Gamma(Q, W)$ associated by Ginzburg with an arbitrary quiver $Q$ with potential $W$, *cf.* section 4.3 of [71]. The underlying graded algebra of $\Gamma(Q, W)$ is the (completed) graded path algebra of a *graded quiver, i.e.* a quiver where each arrow has an associated integer degree. We first describe this graded quiver $\widetilde{Q}$: It has the same vertices as $Q$. Its arrows are

- the arrows of $Q$ (they all have degree 0),
- an arrow $a^* : j \to i$ of degree $-1$ for each arrow $a : i \to j$ of $Q$,
- loops $t_i : i \to i$ of degree $-2$ associated with each vertex $i$ of $Q$.

We now consider the completion $\Gamma = \widehat{k\widetilde{Q}}$ (formed in the category of graded algebras) and endow it with the unique continuous differential of degree 1 such that on the generators we have:

- $da = 0$ for each arrow $a$ of $Q$,
- $d(a^*) = \partial_a W$ for each arrow $a$ of $Q$,
- $d(t_i) = e_i(\sum_a [a, a^*])e_i$ for each vertex $i$ of $Q$, where $e_i$ is the idempotent associated with $i$ and the sum runs over the set of arrows of $Q$.

One checks that we do have $d^2 = 0$ and that the homology in degree 0 of $\Gamma$ is the *Jacobi algebra* as defined in [39]:

$$\mathcal{P}(Q, W) = \widehat{kQ}/(\partial_\alpha W \mid \alpha \in Q_1).$$

Let us consider two typical examples: Let $Q$ be the quiver with one vertex and three loops labeled $X$, $Y$ and $Z$. Let $W = XYZ - XZY$. Then the cyclic derivatives of $W$ yield the commutativity relations between $X$, $Y$ and $Z$ and the Jacobi algebra is canonically isomorphic to the power series algebra $k[[X, Y, Z]]$. It is not hard to check that in this example, the homology of $\Gamma$ is concentrated in degree 0 so that we have a quasi-isomorphism $\Gamma \to \mathcal{P}(Q, W)$. Using theorem 5.3.1 of Ginzburg's [71], one can show that this is the case if and only if the full subcategory of the derived category of the Jacobi algebra formed by the complexes whose homology is of finite total dimension is 3-Calabi-Yau as a triangulated category (*cf.* also below).

As a second example, we consider the Ginzburg dg algebra associated with the cyclic quiver 9.2.1 with the potential $W = abc$. Here is the graded quiver $\tilde{Q}$:

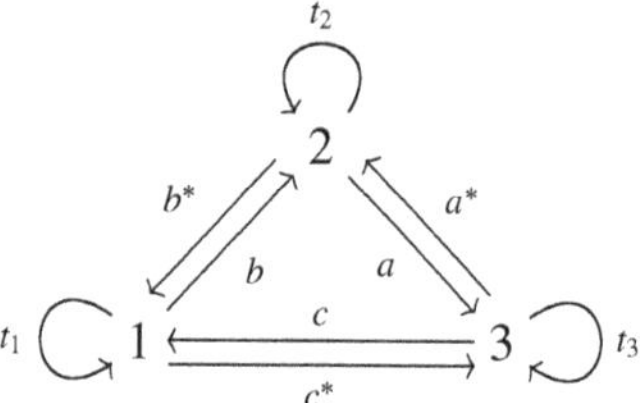


The differential is given by

$$d(a^*) = bc \,, \ d(b^*) = ca \,, \ d(c^*) = ab \,, \ d(t_1) = cc^* - b^*b \,, \ \dots$$

It is not hard to show that for each $i \leq 0$, we have a canonical isomorphism

$$H^i(\Gamma) = C_{\vec{A}_3}(T, \Sigma^i T)$$

where $\vec{A}_3$ is the quiver $1 \to 2 \to 3$, $C_{\vec{A}_3}$ its cluster category and $T$ the sum of the images of the modules $P_1$, $P_3$ and $P_3/P_2$. Since $\Sigma$ is an autoequivalence of finite order of the cluster category $C_{\vec{A}_3}$, we see that $\Gamma$ has non vanishing homology in infinitely many degrees $< 0$. In particular, $\Gamma$ is not quasi-isomorphic to the Jacobi algebra.

Let us denote by $\mathcal{D}\Gamma$ the derived category of $\Gamma$. Its objects are the differential $\mathbb{Z}$-graded right $\Gamma$-modules and its morphisms obtained from morphisms of dg $\Gamma$-modules (homogeneous of degree 0 and which commute with the differential) by formally inverting all quasi-isomorphisms, *cf.* [89]. Let us denote by $\mathsf{per}\,\Gamma$ the *perfect derived category, i.e.* the full subcategory of $\mathcal{D}\Gamma$ which is the closure of the free right $\Gamma$-module $\Gamma_\Gamma$. under shifts, extensions and passage to direct factors. Finally, we denote by $\mathcal{D}_{fd}(\Gamma)$ the *finite-dimensional derived category, i.e.* the full subcategory of $\mathcal{D}\Gamma$ formed by the dg modules whose homology is of finite total dimension (!). We recall that the objects of $\mathsf{per}(\Gamma)$ can be intrinsically characterized as the *compact* ones, *i.e.* those whose associated covariant $\mathsf{Hom}$-functor commutes with arbitrary coproducts. The objects $M$ of the bounded derived category are characterized by the fact that $\mathsf{Hom}(P, M)$ is finite-dimensional for each object $P$ of $\mathsf{per}(\Gamma)$.

The following facts will be shown in [85]. Notice that they hold for arbitrary $Q$ and $W$.

1) The dg algebra $\Gamma$ is *homologically smooth, i.e.* it is perfect as an object of the derived category of $\Gamma^e = \Gamma \widehat{\otimes} \Gamma^{op}$. This implies that the bounded derived category is contained in the perfect derived category, *cf. e.g.* [88].

2) The dg algebra $\Gamma$ is 3-*Calabi-Yau as a bimodule*, *i.e.* there is an isomorphism in the derived category of $\Gamma^e$

$$\mathrm{RHom}_{\Gamma^e}(\Gamma, \Gamma^e) \xrightarrow{\sim} \Gamma[-3].$$

As a consequence, the finite-dimensional derived category $\mathcal{D}_{fd}(\Gamma)$ is 3-Calabi-Yau as a triangulated category, *cf. e.g.* lemma 4.1 of [88].

3) The triangulated category $\mathcal{D}(\Gamma)$ admits a $t$-structure whose left aisle $\mathcal{D}_{\leq 0}$ consists of the dg modules $M$ such that $H^i(M) = 0$ for all $i > 0$. This $t$-structure induces a $t$-structure on $\mathcal{D}_{fd}(\Gamma)$ whose heart is equivalent to the category of finite-dimensional (and hence nilpotent) modules over the Jacobi algebra.

The *generalized cluster category* [2] is the idempotent completion $\mathcal{C}_{(Q,W)}$ of the quotient

$$\mathrm{per}(\Gamma)/\mathcal{D}_{fd}(\Gamma).$$

The name is justified by the fact that if $Q$ is a quiver without oriented cycles (and so $W = 0$), then $\mathcal{C}_{(Q,W)}$ is triangle equivalent to $\mathcal{C}_Q$. In general, the category $\mathcal{C}_{(Q,W)}$ has infinite-dimensional $\mathsf{Hom}$-spaces. However, if $H^0(\Gamma)$ is finite-dimensional, then $\mathcal{C}_{(Q,W)}$ is a 2-Calabi-Yau category with cluster-tilting object $T = \Gamma$, in the sense of section 7.1, *cf.* [2] (the present version of [2] uses non completed Ginzburg algebras; the complete case will be included in a future version).

From now on, we suppose that $W$ does not involve cycles of length $\leq 1$. Then the simple $\widetilde{Q}$-modules $S_i$ associated with the vertices of $Q$ yield a basis of the Grothendieck group $K_0(\mathcal{D}_{fd}(\Gamma))$. Thanks to the Euler form

$$\langle P, M \rangle = \chi(\mathrm{RHom}(P, M)), \ \ P \in \mathrm{per}(\Gamma), \ \ M \in \mathcal{D}_{fd}(\Gamma),$$

this Grothendieck group is dual to $K_0(\mathrm{per}(\Gamma))$ and the dg modules $P_i = e_i \Gamma$ form the basis dual to the $S_i$. The Euler form also induces an antisymmetric bilinear form on $K_0(\mathcal{D}_{fd}(\Gamma))$ ('Poisson form'). If $W$ does not involve cycles of length $\leq 2$, then the quiver of the Jacobi algebra is $Q$ and the matrix of the form on $\mathcal{D}_{fd}(\Gamma)$ in the basis of the $S_i$ is given by

$$\langle S_i, S_j \rangle = |\{\text{arrows } j \to i \text{ of } Q\}| - |\{\text{arrows } i \to j \text{ of } Q\}|.$$

The dual of this form is given by the map ('symplectic form')

$$K_0(k) \to K_0(\mathrm{per}(\Gamma \widehat{\otimes} \Gamma^{op}) = K_0(\mathrm{per}(\Gamma)) \otimes_{\mathbb{Z}} K_0(\mathrm{per}(\Gamma))$$

associated with the $k$-$\Gamma^e$-bimodule $\Gamma$.

Now suppose that $W$ does not involve cycles of length $\leq 2$ and that $Q$ does not have loops nor 2-cycles. Then each simple $S_i$ is a (3-)spherical object in the

3-Calabi-Yau category $\mathcal{D}_{fd}(\Gamma)$, *i.e.* we have an isomorphism of graded algebras

$$\mathsf{Ext}^*(S_i, S_i) \xrightarrow{\sim} H^*(S^3, k).$$

Moreover, the quiver $Q$ encodes the dimensions of the $\mathsf{Ext}^1$-groups between the $S_i$: We have

$$\dim \mathsf{Ext}^1(S_i, S_j) = |\{\text{arrows } j \to i \text{ in } Q\}|.$$

In particular, for $i \neq j$, we have either $\mathsf{Ext}^1(S_i, S_j) = 0$ or $\mathsf{Ext}^1(S_j, S_i) = 0$ because $Q$ does not contain 2-cycles. It follows that the $S_i$, $i \in Q_0$, form a *spherical collection* in $\mathcal{D}_{fd}(\Gamma)$ in the sense of [98]. Conversely, each spherical collection in an ($A_\infty$-) 3-Calabi-Yau category in the sense of [98] can be obtained in this way from the Ginzburg algebra associated to a quiver with potential.

The categories we have considered so far are summed up in the sequence of triangulated categories

$$0 \longrightarrow \mathcal{D}_{fd}(\Gamma) \longrightarrow \mathsf{per}(\Gamma) \longrightarrow \mathcal{C}_{(Q,W)} \longrightarrow 0 . \tag{9.3.1}$$

This sequence is 'exact up to factors', *i.e.* the left hand term is a thick subcategory of the middle term and the right hand term is the idempotent closure of the quotient. Notice that the left hand category is 3-Calabi-Yau, the middle one does not have a Serre functor and the right hand one is 2-Calabi-Yau if the Jacobi algebra is finite-dimensional.

We keep the last hypotheses: $Q$ does not have loops or 2-cycles and $W$ does not involve cycles of length $\leq 2$. Let $i$ be a vertex of $Q$ and $\Gamma'$ the Ginzburg algebra of the mutated quiver with potential $\mu_i(Q, W)$. The following theorem improves on Vitória's [126], *cf.* also [107] [18].

**Theorem 9.2** ([94]). *There is an equivalence of derived categories*

$$F : \mathcal{D}(\Gamma') \to \mathcal{D}(\Gamma)$$

*which takes the dg modules $P'_j$ to $P_j$ for $j \neq i$ and $P'_i$ to the cone on the morphism*

$$P_i \longrightarrow \bigoplus_{\substack{arrows \\ i \to j}} P_j .$$

*It induces triangle equivalences* $\mathsf{per}(\Gamma') \xrightarrow{\sim} \mathsf{per}(\Gamma)$ *and* $\mathcal{D}_{fd}(\Gamma') \to \mathcal{D}_{fd}(\Gamma)$.

In fact, the functor $F$ is given by the left derived functor of the tensor product by a suitable $\Gamma'$-$\Gamma$-bimodule.

By transport of the canonical $t$-structure on $\mathcal{D}(\Gamma')$, we obtain new $t$-structures on $\mathcal{D}(\Gamma)$ and $\mathcal{D}_{fd}(\Gamma)$. They are related to the canonical ones by a tilt (in the sense of [75]) at the simple object corresponding to the vertex $i$. By iterating mutation (as far as possible), one obtains many new $t$-structures on $\mathcal{D}_{fd}(\Gamma)$.

In the context of Iyama-Reiten's [80], the dg algebras $\Gamma$ and $\Gamma'$ have their homologies concentrated in degree $0$ and the equivalence of the theorem is given by the tilting module constructed in [loc. cit.].

## 9.4. A geometric illustration

To illustrate the sequence 9.3.1, let us consider the example of the quiver

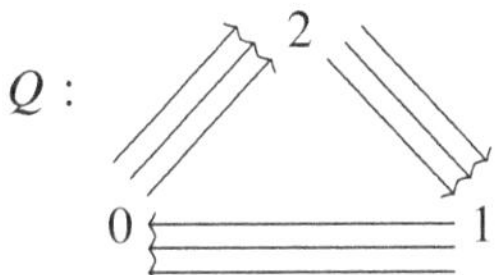

where the arrows going out from $i$ are labeled $x_i$, $y_i$, $z_i$, $0 \le i \le 2$, endowed with the potential

$$W = \sum_{i=0}^{2}(x_i y_i z_i - x_i z_i y_i).$$

Let $p : \omega \to \mathbb{P}^2$ be the canonical bundle on $\mathbb{P}^2$. Then we have a triangle equivalence

$$\mathcal{D}^b(\mathsf{coh}(\omega)) \xrightarrow{\sim} \mathsf{per}(\Gamma)$$

which sends $p^*(O(-i))$ to the dg module $P_i$, $0 \le i \le 2$. Under this equivalence, the subcategory $\mathcal{D}_{fd}\Gamma$ corresponds to the subcategory $\mathcal{D}^b_Z(\mathsf{coh}(\omega))$ of complexes of coherent sheaves whose homology is supported on the zero section $Z$ of the bundle $\omega$. This subcategory is indeed Calabi-Yau of dimension 3. Bridgeland has studied its $t$-structures [22] by linking them to $t$-structures and mutations (in the sense of Rudakov's school) in the derived category of coherent sheaves on the projective plane. In this example, the Ginzburg algebra has its homology concentrated in degree 0. So it is quasi-isomorphic to the Jacobi algebra. The Jacobi algebra is infinite-dimensional so that the generalized cluster category $\mathcal{C}_{(Q,W)}$ is not $\mathsf{Hom}$-finite. The category $\mathcal{C}_{(Q,W)}$ identifies with the quotient

$$\mathcal{D}^b(\mathsf{coh}(\omega))/\mathcal{D}^b_Z(\mathsf{coh}(\omega))$$

and thus with the bounded derived category $\mathcal{D}^b(\mathsf{coh}(\omega \setminus Z))$ of coherent sheaves on the complement of the zero section of $\omega$. In order to understand why this category is 'close to being' Calabi-Yau of dimension 2, we consider the scheme $C$ obtained from $\omega$ by contracting the zero section $Z$ to a point $P_0$. The projection $\omega \to C$ induces an isomorphism from $\omega \setminus Z$ onto $C \setminus \{P_0\}$ and so we have an equivalence

$$\mathcal{C}_{(Q,W)} \xrightarrow{\sim} \mathcal{D}^b(\mathrm{coh}(\omega \setminus Z)) \xrightarrow{\sim} \mathcal{D}^b(\mathrm{coh}(C \setminus \{P_0\})).$$

Let $\widehat{C}$ denote the completion of $C$ at the singular point $P_0$. Then $\widehat{C}$ is of dimension 3 and has $P_0$ as its unique closed point. Thus the subscheme $\widehat{C} \setminus \{P_0\}$ is of dimension 2. The induced functor

$$\mathcal{C}_{(Q,W)} \xrightarrow{\sim} \mathcal{D}^b(\mathrm{coh}(C \setminus \{P_0\})) \to \mathcal{D}^b(\mathrm{coh}(\widehat{C} \setminus \{P_0\}))$$

yields a 'completion' of $\mathcal{C}_{(Q,W)}$ which is of 'dimension 2' and Calabi-Yau in a generalized sense, *cf.* [35].

## 9.5. Ginzburg algebras from algebras of global dimension 2

Let us link the cluster categories obtained from algebras of global dimension 2 in section 7.7 to the generalized cluster categories $\mathcal{C}_{(Q,W)}$ constructed above.

Let $A$ be an algebra given as the quotient $kQ'/I$ of the path algebra of a finite quiver $Q'$ by an ideal $I$ contained in the square of the ideal $J$ generated by the arrows of $Q'$. Assume that $A$ is of global dimension 2 (but not necessarily of finite dimension over $k$). We construct a quiver with potential $(Q, W)$ as follows: Let $R$ be the union over all pairs of vertices $(i, j)$ of a set of representatives of the vectors belonging to a basis of

$$\mathrm{Tor}_2^A(S_j, S_i^{op}) = e_j(I/(IJ + JI))e_i.$$

We think of these representatives as 'minimal relations' from $i$ to $j$, *cf.* [20]. For each such representative $r$, let $\rho_r$ be a new arrow from $j$ to $i$. We define $Q$ to be obtained from $Q'$ by adding all the arrows $\rho_r$ and the potential is given by

$$W = \sum_{r \in R} r\rho_r.$$

Recall that a *tilting module* over an algebra $B$ is a $B$-module $T$ such that the total derived functor of the tensor product by $T$ over the endomorphism algebra $\mathrm{End}_B(T)$ is an equivalence

$$\mathcal{D}(\mathrm{End}_B(T)) \xrightarrow{\sim} \mathcal{D}(B).$$

The second assertion of part a) of the following theorem generalizes a result by Assem-Brüstle-Schiffler [3].

**Theorem 9.3** ([85]).
a) *The category $\mathcal{C}_{(Q,W)}$ is triangle equivalent to the cluster category $\mathcal{C}_A$. This equivalence takes $\Gamma$ to the image $\pi(A)$ of $A$ in $\mathcal{C}_A$ and thus induces an isomorphism from the Jacobi algebra $\mathcal{P}(Q, W)$ onto the endomorphism algebra of the image of $A$ in $\mathcal{C}_A$.*
b) *If $T$ is a tilting module over $kQ''$ for a quiver without oriented cycles $Q''$ and $A$ is the endomorphism algebra of $T$, then $\mathcal{C}_{(Q,W)}$ is triangle equivalent to $\mathcal{C}_{Q''}$.*

## 9.6. Cluster-tilting objects, spherical collections, decorated representations

We consider the setup of theorem 9.2. The equivalence $F$ induces equivalences $\mathrm{per}(\Gamma') \to \mathrm{per}(\Gamma)$ and $\mathcal{D}_{fd}(\Gamma') \to \mathcal{D}_{fd}(\Gamma)$. The last equivalence takes $S_i'$ to $\Sigma S_i$ and, for $j \neq i$, the module $S_j'$ to $S_j$ if $\mathrm{Ext}^1(S_j, S_i) = 0$ and, more generally, to the middle term $F S_j'$ of the universal extension

$$S_i^d \to F S_j' \to S_j \to \Sigma S_i^d ,$$

where the components of the third morphism form a basis of $\mathrm{Ext}^1(S_j, S_i)$. This means that the images $F S_j'$, $j \in Q_0$, form the *left mutated spherical collection* in $\mathcal{D}_{fd}(\Gamma)$ in the sense of [98].

If $H^0(\Gamma)$ is finite-dimensional so that $\mathcal{C}_{(Q,W)}$ is a 2-Calabi-Yau category with cluster-tilting object, then the images of the $F P_j$, $j \in Q_0$, form the mutated cluster-tilting object, as it follows from the description in lemma 7.5.

Now suppose that $H^0(\Gamma)$ is finite-dimensional. Then we can establish a connection with decorated representations and their mutations in the sense of [39]: Recall from [loc. cit.] that a decorated representation of $(Q, W)$ is given by a finite-dimensional (hence nilpotent) module $M$ over the Jacobi algebra and a collection of vector spaces $V_j$ indexed by the vertices $j$ of $Q$. Given an object $L$ of $\mathcal{C}_{(Q,W)}$, we put $M = \mathcal{C}(\Gamma, L)$ and, for each vertex $j$, we choose a vector space $V_j$ of maximal dimension such that the triangle

$$T_1 \to T_0 \to L \to \Sigma T_1 \tag{9.6.1}$$

of lemma 7.8 admits a direct factor

$$V_j \otimes P_j \to 0 \to \Sigma V_j \otimes P_j \to \Sigma V_j \otimes P_j.$$

Let us write $GL$ for the decorated representation thus constructed. The assignment $L \mapsto GL$ defines a bijection between isomorphism classes of objects $L$ in $\mathcal{C}_{(Q,W)}$ and right equivalence classes of decorated representations. It is compatible with mutations: If $(M', V')$ is the image $GL'$ of an object $L'$ of $\mathcal{C}_{(Q',W')}$, then the mutation $(Q, W, M, V)$ of $(Q', W', M', V')$ at $i$ in the sense of [39] is right equivalent to $(Q, W, M'', V'')$ where $(M'', V'') = GFL'$ and $F$ is the equivalence of theorem 9.2.

# References


[1] Claire Amiot, *Cluster categories for algebras of global dimension 2 and quivers with potential*, Annales de l'institut Fourier **59** (2009), no. 6, 2525–2590.

[2] Claire Amiot, *Sur les petites catégories triangulées*, Ph. D. thesis, Université Paris Diderot - Paris 7, Juillet 2008.

[3] Ibrahim Assem, Thomas Brüstle, and Ralf Schiffler, *Cluster-tilted algebras as trivial extensions*, Bull. London Math. Soc. **40** (2008), 151–162.

[4] Ibrahim Assem, Daniel Simson, and Andrzej Skowroński, *Elements of the representation theory of associative algebras. Vol. 1*, London Mathematical Society Student Texts, vol. 65, Cambridge University Press, Cambridge, 2006, Techniques of representation theory.

[5] M. Auslander, I. Reiten, and S. Smalø, *Representation theory of Artin algebras*, Cambridge Studies in Advanced Mathematics, vol. 36, Cambridge University Press, 1995 (English).

[6] Maurice Auslander, *Functors and morphisms determined by objects*, Representation theory of algebras (Proc. Conf., Temple Univ., Philadelphia, Pa., 1976), Dekker, New York, 1978, pp. 1–244. Lecture Notes in Pure Appl. Math., Vol. 37.

[7] Aslak Bakke Buan, Osamu Iyama, Idun Reiten, and Jeanne Scott, *Cluster structures for 2-Calabi-Yau categories and unipotent groups*, arXiv:math.RT/0701557. Compos. Math. **145** (2009), no. 4, 1035–1079.

[8] Aslak Bakke Buan, Osamu Iyama, Idun Reiten, and David Smith, *Mutation of cluster-tilting objects and potentials*, arXiv:0804.3813. to appear in Amer. J. Math.

[9] Aslak Bakke Buan and Robert Marsh, *Cluster-tilting theory*, Trends in representation theory of algebras and related topics, Contemp. Math., vol. 406, Amer. Math. Soc., Providence, RI, 2006, pp. 1–30.

[10] Aslak Bakke Buan, Robert J. Marsh, Markus Reineke, Idun Reiten, and Gordana Todorov, *Tilting theory and cluster combinatorics*, Advances in Mathematics **204** **(2)** (2006), 572–618.

[11] Aslak Bakke Buan, Robert J. Marsh, and Idun Reiten, *Cluster mutation via quiver representations*, Comm. Math. Helv. **83** (2008), 143–177.

[12] Aslak Bakke Buan, Robert J. Marsh, Idun Reiten, and Gordana Todorov, *Clusters and seeds in acyclic cluster algebras*, Proc. Amer. Math. Soc. **135** (2007), no. 10, 3049–3060 (electronic), With an appendix coauthored in addition by P. Caldero and B. Keller.

[13] Aslak Bakke Buan and Idun Reiten, *Acyclic quivers of finite mutation type*, Int. Math. Res. Not. (2006), Art. ID 12804, 10.

[14] Michael Barot, Christof Geiss, and Andrei Zelevinsky, *Cluster algebras of finite type and positive symmetrizable matrices*, J. London Math. Soc. (2) **73** (2006), no. 3, 545–564.

[15] Andre Beineke, Thomas Bruestle, and Lutz Hille, *Cluster-cyclic quivers with three vertices and the Markov equation*, arXiv:math/0612213. to appear in Algebr. Represent. Theory.

[16] Arkady Berenstein, Sergey Fomin, and Andrei Zelevinsky, *Parametrizations of canonical bases and totally positive matrices*, Adv. Math. **122** (1996), no. 1, 49–149.

[17] ———, *Cluster algebras. III. Upper bounds and double Bruhat cells*, Duke Math. J. **126** (2005), no. 1, 1–52.

[18] David Berenstein and Michael Douglas, *Seiberg duality for quiver gauge theories*, arXiv:hep-th/0207027v1.

[19] I. N. Bernšteĭn, I. M. Gel′fand, and V. A. Ponomarev, *Coxeter functors, and Gabriel's theorem*, Uspehi Mat. Nauk **28** (1973), no. 2(170), 19–33.

[20] Klaus Bongartz, *Algebras and quadratic forms*, J. London Math. Soc. (2) **28** (1983), no. 3, 461–469.

[21] Tom Bridgeland, *Stability conditions and Hall algebras*, Talk at the meeting 'Recent developments in Hall algebras', Luminy, November 2006.

[22] ———, *t-structures on some local Calabi-Yau varieties*, J. Algebra **289** (2005), no. 2, 453–483.

[23] ———, *Derived categories of coherent sheaves*, International Congress of Mathematicians. Vol. II, Eur. Math. Soc., Zürich, 2006, pp. 563–582.

[24] ———, *Stability conditions on triangulated categories*, Ann. of Math. (2) **166** (2007), no. 2, 317–345.

[25] Igor Burban, Osamu Iyama, Bernhard Keller, and Idun Reiten, *Cluster tilting for one-dimensional hypersurface singularities*, Adv. Math. **217** (2008), no. 6, 2443–2484.

[26] F. Cachazo, B. Fiol, K. Intriligator, S. Katz, and C. Vafa, *A geometric unification of dualities*, Nucl. Phys. B **628** (2002), 3–78.

[27] Philippe Caldero and Frédéric Chapoton, *Cluster algebras as Hall algebras of quiver representations*, Comment. Math. Helv. **81** (2006), no. 3, 595–616.

[28] Philippe Caldero, Frédéric Chapoton, and Ralf Schiffler, *Quivers with relations arising from clusters ($A_n$ case)*, Trans. Amer. Math. Soc. **358** (2006), no. 5, 1347–1364.

[29] Philippe Caldero and Bernhard Keller, *From triangulated categories to cluster algebras. II*, Ann. Sci. École Norm. Sup. (4) **39** (2006), no. 6, 983–1009.

[30] Philippe Caldero and Bernhard Keller, *From triangulated categories to cluster algebras*, Inv. Math. **172** (2008), 169–211.

[31] Philippe Caldero and Markus Reineke, *On the quiver Grassmannian in the acyclic case*, Journal of Pure and Applied Algebra. **212** (2008), no. 11, 2369–2380.

[32] Henri Cartan and Samuel Eilenberg, *Homological algebra*, Princeton University Press, Princeton, N. J., 1956.

[33] Frédéric Chapoton, *Enumerative properties of generalized associahedra*, Sém. Lothar. Combin. **51** (2004/05), Art. B51b, 16 pp. (electronic).

[34] Frédéric Chapoton, Sergey Fomin, and Andrei Zelevinsky, *Polytopal realizations of generalized associahedra*, Canad. Math. Bull. **45** (2002), no. 4, 537–566, Dedicated to Robert V. Moody.

[35] Joe Chuang and Raphaël Rouquier, book in preparation.

[36] William Crawley-Boevey, *On the exceptional fibres of Kleinian singularities*, Amer. J. Math. **122** (2000), no. 5, 1027–1037.

[37] Raika Dehy and Bernhard Keller, *On the combinatorics of rigid objects in 2-Calabi-Yau categories*, International Mathematics Research Notices **2008** (2008), rnn029–17.

[38] Harm Derksen and Theodore Owen, *New graphs of finite mutation type*, Electron. J. Combin. **15** (2008), no. 1. Research Paper 139, 15pp.

[39] Harm Derksen, Jerzy Weyman, and Andrei Zelevinsky, *Quivers with potentials and their representations I: Mutations*, Selecta Mathematica. **14** (2008), 59–119.

[40] Vlastimil Dlab and Claus Michael Ringel, *Indecomposable representations of graphs and algebras*, Mem. Amer. Math. Soc. **6** (1976), no. 173, v+57.

[41] Peter Donovan and Mary Ruth Freislich, *The representation theory of finite graphs and associated algebras*, Carleton University, Ottawa, Ont., 1973, Carleton Mathematical Lecture Notes, No. 5.

[42] M. Fekete, *Ueber ein Problem von Laguerre*, Rend. Circ. Mat. Palermo **34** (1912), 89–100, 110–120.

[43] V. V. Fock and A. B. Goncharov, *Cluster ensembles, quantization and the dilogarithm*, Annales scientifiques de l'ENS **42** (2009), no. 6, 865–930.

[44] ______, *Cluster ensembles, quantization and the dilogarithm II: The intertwiner*, arXiv:math/0702398.

[45] ______, *The quantum dilogarithm and representations of quantized cluster varieties*, arXiv:math/0702397. to appear in Jnuent. Math.

[46] V. V. Fock and A. B. Goncharov, *Cluster $\mathcal{X}$-varieties, amalgamation, and Poisson-Lie groups*, Algebraic geometry and number theory, Progr. Math., vol. 253, Birkhäuser Boston, Boston, MA, 2006, pp. 27–68.

[47] Sergey Fomin, *Cluster algebras portal*, `www.math.lsa.umich.edu/~fomin/cluster.html`.

[48] Sergey Fomin and Nathan Reading, *Generalized cluster complexes and Coxeter combinatorics*, Int. Math. Res. Not. (2005), no. 44, 2709–2757.

[49] Sergey Fomin, Michael Shapiro, and Dylan Thurston, *Cluster algebras and triangulated surfaces. Part I: Cluster complexes*, Acta Mathematica. **201** (2008), no. 1, 83–146.

[50] Sergey Fomin and Andrei Zelevinsky, *Cluster algebras. I. Foundations*, J. Amer. Math. Soc. **15** (2002), no. 2, 497–529 (electronic).

[51] ______, *The Laurent phenomenon*, Adv. in Appl. Math. **28** (2002), no. 2, 119–144.

[52] ______, *Cluster algebras. II. Finite type classification*, Invent. Math. **154** (2003), no. 1, 63–121.

[53] ______, *Cluster algebras: notes for the CDM-03 conference*, Current developments in mathematics, 2003, Int. Press, Somerville, MA, 2003, pp. 1–34.

[54] ______, *Y-systems and generalized associahedra*, Ann. of Math. (2) **158** (2003), no. 3, 977–1018.

[55] ______, *Cluster algebras IV: Coefficients*, Compositio Mathematica **143** (2007), 112–164.

[56] Edward Frenkel and András Szenes, *Thermodynamic Bethe ansatz and dilogarithm identities. I*, Math. Res. Lett. **2** (1995), no. 6, 677–693.

[57] Changjian Fu and Bernhard Keller, *On cluster algebras with coefficients and 2-Calabi-Yau categories*, Trans. Amer. Math. Soc. **362** (2010), 859–895.

[58] P. Gabriel, *Unzerlegbare Darstellungen I*, Manuscripta Math. **6** (1972), 71–103.

[59] P. Gabriel and A. V. Roiter, *Representations of finite-dimensional algebras*, Encyclopaedia Math. Sci., vol. 73, Springer–Verlag, 1992.

[60] Peter Gabriel, *Auslander-Reiten sequences and representation-finite algebras*, Representation theory, I (Proc. Workshop, Carleton Univ., Ottawa, Ont., 1979), Springer, Berlin, 1980, pp. 1–71.

[61] Christof Geiß, Bernard Leclerc, and Jan Schröer, *Cluster algebra structures and semicanonical bases for unipotent groups*, arXiv:math/0703039.

[62] ———, *Partial flag varieties and preprojective algebras*, Ann. Jnst. Fourier. (Grenoble) **58** (2008), no. 3, 825–876.

[63] ———, *Preprojective algebras and cluster algebras*, Trends in representation theory and related topics, EMS Ser. Congr. Rep., Eur. Math. Soc., Zürich 2008, pp. 253–283.

[64] ———, *Semicanonical bases and preprojective algebras*, Ann. Sci. École Norm. Sup. (4) **38** (2005), no. 2, 193–253.

[65] ———, *Rigid modules over preprojective algebras*, Invent. Math. **165** (2006), no. 3, 589–632.

[66] ———, *Auslander algebras and initial seeds for cluster algebras*, J. Lond. Math. Soc. (2) **75** (2007), no. 3, 718–740.

[67] ———, *Semicanonical bases and preprojective algebras. II. A multiplication formula*, Compos. Math. **143** (2007), no. 5, 1313–1334.

[68] Michael Gekhtman, Michael Shapiro, and Alek Vainshtein, *On the properties of the exchange graph of a cluster algebra*, Math. Res. Lett. **15** (2008), no. 2, 321–330.

[69] ———, *Cluster algebras and Poisson geometry*, Mosc. Math. J. **3** (2003), no. 3, 899–934, 1199, {Dedicated to Vladimir Igorevich Arnold on the occasion of his 65th birthday}.

[70] ———, *Cluster algebras and Weil-Petersson forms*, Duke Math. J. **127** (2005), no. 2, 291–311.

[71] Victor Ginzburg, *Calabi-Yau algebras*, arXiv:math/0612139v3 [math.AG].

[72] F. Gliozzi and R. Tateo, *Thermodynamic Bethe ansatz and three-fold triangulations*, Internat. J. Modern Phys. A **11** (1996), no. 22, 4051–4064.

[73] Dieter Happel, *On the derived category of a finite-dimensional algebra*, Comment. Math. Helv. **62** (1987), no. 3, 339–389.

[74] ———, *Triangulated categories in the representation theory of finite-dimensional algebras*, Cambridge University Press, Cambridge, 1988.

[75] Dieter Happel, Idun Reiten, and Sverre O. Smalø, *Tilting in abelian categories and quasitilted algebras*, Mem. Amer. Math. Soc. **120** (1996), no. 575, viii+ 88.

[76] Dieter Happel and Luise Unger, *On a partial order of tilting modules*, Algebr. Represent. Theory **8** (2005), no. 2, 147–156.

[77] André Henriques, *A periodicity theorem for the octahedron recurrence*, J. Algebraic Combin. **26** (2007), no. 1, 1–26.

[78] David Hernandez and Bernard Leclerc, Cluster algebras and quantum affine algebras, arXiv:0903.1452v1 [math.Q4].

[79] Andrew Hubery, *Acyclic cluster algebras via Ringel-Hall algebras*, Preprint available at the author's home page.

[80] Osamu Iyama and Idun Reiten, *Fomin-Zelevinsky mutation and tilting modules over Calabi-Yau algebras*, Amer. J. Math. **130** (2008), no. 4, 1087–1149.

[81] Osamu Iyama and Yuji Yoshino, *Mutations in triangulated categories and rigid Cohen-Macaulay modules*, Inv. Math. **172** (2008), 117–168.

[82] Hiroshige Kajiura, *Noncommutative homotopy algebras associated with open strings*, Rev. Math. Phys. **19** (2007), no. 1, 1–99.

[83] Masaki Kashiwara, *Bases cristallines*, C. R. Acad. Sci. Paris Sér. I Math. **311** (1990), no. 6, 277–280.

[84] Masaki Kashiwara and Pierre Schapira, *Sheaves on manifolds*, Grundlehren der Mathematischen Wissenschaften [Fundamental Principles of Mathematical Sciences], vol. 292, Springer-Verlag, Berlin, 1994, With a chapter in French by Christian Houzel, Corrected reprint of the 1990 original.

[85] Bernhard Keller, *Deformed CY-completions*, arXiv:0908.3499 [math.RT].

[86] ______, *The periodicity conjecture for pairs of Dynkin diagrams*, arXiv:1001:1531.

[87] ______, *Quiver mutation in Java*, Java applet available at the author's home page.

[88] ______, *Triangulated Calabi-Yau categories*, Trends in Representation Theory of Algebras (A. Skowronski, ed.), Eur. Math. Soc., 2008, pp. 467–489.

[89] ______, *Deriving DG categories*, Ann. Sci. École Norm. Sup. (4) **27** (1994), no. 1, 63–102.

[90] ______, *On triangulated orbit categories*, Doc. Math. **10** (2005), 551–581.

[91] Bernhard Keller and Idun Reiten, *Acyclic Calabi-Yau categories are cluster categories*, preprint, 2006, with an appendix by Michel Van den Bergh, arXiv:math.RT/0610594, to appear in Compos. Math.

[92] ______, *Cluster-tilted algebras are Gorenstein and stably Calabi-Yau*, Advances in Mathematics **211** (2007), 123–151.

[93] Bernhard Keller and Michel Van den Bergh, *On two examples of Iyama and Yoshino*, arXiv:0803.0720.

[94] Bernhard Keller and Dong Yang, Derived equivalences from mutations of quivers with potential, arXiv:0906:0761.

[95] Maxim Kontsevich, *Donaldson-Thomas invariants*, Mathematische Arbeitstagung June 22-28, 2007, MPI Bonn, www.mpim-bonn.mpg.de/preprints/.

[96] ______, *Donaldson-Thomas invariants, stability conditions and cluster transformations*, Report on joint work with Y. Soibelman, talks at the IHES, October 11 and 12, 2007.

[97] Maxim Kontsevich and Yan Soibelman, *Notes on A-infinity algebras, A-infinity categories and non-commutative geometry. I*, arXiv:math.RA/0606241.

[98] ______, *Stability structures, Donaldson-Thomas invariants and cluster transformations*, arXiv:0811.2435.

[99] Christian Krattenthaler, *The F-triangle of the generalised cluster complex*, Topics in discrete mathematics, Algorithms Combin., vol. 26, Springer, Berlin, 2006, pp. 93–126.

[100] A. Kuniba and T. Nakanishi, *Spectra in conformal field theories from the Rogers dilogarithm*, Modern Phys. Lett. A **7** (1992), no. 37, 3487–3494.

[101] Atsuo Kuniba, Tomoki Nakanishi, and Junji Suzuki, *Functional relations in solvable lattice models. I. Functional relations and representation theory*, Internat. J. Modern Phys. A **9** (1994), no. 30, 5215–5266.

[102] G. Lusztig, *Canonical bases arising from quantized enveloping algebras*, J. Amer. Math. Soc. **3** (1990), no. 2, 447–498.

[103] ______, *Total positivity in reductive groups*, Lie theory and geometry, Progr. Math., vol. 123, Birkhäuser Boston, Boston, MA, 1994, pp. 531–568.

[104] ______, *Semicanonical bases arising from enveloping algebras*, Adv. Math. **151** (2000), no. 2, 129–139.

[105] George Lusztig, *Canonical bases and Hall algebras*, Representation theories and algebraic geometry (Montreal, PQ, 1997), NATO Adv. Sci. Inst. Ser. C Math. Phys. Sci., vol. 514, Kluwer Acad. Publ., Dordrecht, 1998, pp. 365–399.

[106] Robert Marsh, Markus Reineke, and Andrei Zelevinsky, *Generalized associahedra via quiver representations*, Trans. Amer. Math. Soc. **355** (2003), no. 10, 4171–4186 (electronic).

[107] Subir Mukhopadhyay and Koushik Ray, *Seiberg duality as derived equivalence for some quiver gauge theories*, J. High Energy Phys. (2004), no. 2, 070, 22 pp. (electronic).

[108] Gregg Musiker, *A graph theoretic expansion formula for cluster algebras of type $B_n$ and $D_n$*, arXiv:0710.3574v1 [math.CO]. to appear in the Annals of Combinatorics.

[109] Hiraku Nakajima, *Quiver varieties and cluster algebras*, arXiv:0905002 [math.QA].

[110] L. A. Nazarova, *Representations of quivers of infinite type*, Izv. Akad. Nauk SSSR Ser. Mat. **37** (1973), 752–791.

[111] Yann Palu, *Grothendieck group and generalized mutation rule for 2-calabi-yau triangulated categories*, Journal of Pure and Applied Algebra. **213** (2009), no. 7, 1438–1449.

[112] ______, *Cluster characters for 2-calabi-yau triangulated categories*, Annales de l'institut Fourier **58** (2008), no. 6, 2221–2248.

[113] F. Ravanini, A. Valleriani, and R. Tateo, *Dynkin TBAs*, Internat. J. Modern Phys. A **8** (1993), no. 10, 1707–1727.

[114] Idun Reiten, *Tilting theory and cluster algebras*, preprint available at www.institut.math.jussieu.fr/~keller/ictp2006/lecturenotes/reiten.pdf.

[115] Claus Michael Ringel, *The preprojective algebra of a quiver*, Algebras and modules, II (Geiranger, 1996), CMS Conf. Proc., vol. 24, Amer. Math. Soc., Providence, RI, 1998, pp. 467–480.

[116] ______, *Some remarks concerning tilting modules and tilted algebras. Origin. Relevance. Future.*, Handbook of Tilting Theory, LMS Lecture Note Series, vol. 332, Cambridge Univ. Press, Cambridge, 2007, pp. 49–104.

[117] C.M. Ringel, *Tame algebras and integral quadratic forms*, Lecture Notes in Mathematics, vol. 1099, Springer Verlag, 1984.

[118] Kyoji Saito, *Extended affine root systems. I. Coxeter transformations*, Publ. Res. Inst. Math. Sci. **21** (1985), no. 1, 75–179.

[119] Olivier Schiffmann, *Lectures on Hall algebras*, math.RT/0611617.

[120] Joshua S. Scott, *Grassmannians and cluster algebras*, Proc. London Math. Soc. (3) **92** (2006), no. 2, 345–380.

[121] Catherine Soanes and Angus Stevenson, *Concise Oxford English Dictionary*, Oxford University Press, Oxford, 2004.

[122] James Dillon Stasheff, *Homotopy associativity of H-spaces. I, II*, Trans. Amer. Math. Soc. 108 (1963), 275–292; ibid. **108** (1963), 293–312.

[123] Balázs Szendrői, *Non-commutative Donaldson-Thomas theory and the conifold*, Geom. Toplo. **12** (2008), no. 2, 1171–1202.

[124] András Szenes, *Periodicity of Y-systems and flat connections*, Lett. Math. Phys. **89** (2009), no. 3, 217–230.

[125] Jean-Louis Verdier, *Des catégories dérivées des catégories abéliennes*, Astérisque, vol. 239, Société Mathématique de France, 1996 (French).

[126] Jorge Vitória, *Mutations vs. Seiberg duality*, J. Algebra **321** (2009), no. 3, 816–828.

[127] Alexandre Yu. Volkov, *On the periodicity conjecture for Y-systems*, Comm. Math. Phys. **276** (2007), no. 2, 509–517.

[128] Keiichi Watanabe, *Certain invariant subrings are Gorenstein. I, II*, Osaka J. Math. **11** (1974), 1–8; ibid. 11 (1974), 379–388.

[129] Shih-Wei Yang and Andrei Zelevinsky, *Cluster algebras of finite type via Coxeter elements and principal minors*, Transform. Groups **13** (2008), no. 3–4, 855–895.

[130] Yuji Yoshino, *Cohen-Macaulay modules over Cohen-Macaulay rings*, London Mathematical Society Lecture Note Series, vol. 146, Cambridge University Press, Cambridge, 1990.

[131] Al. B. Zamolodchikov, *On the thermodynamic Bethe ansatz equations for reflectionless ADE scattering theories*, Phys. Lett. B **253** (1991), no. 3–4, 391–394.

[132] Andrei Zelevinsky, *Cluster algebras: notes for 2004 IMCC (Chonju, Korea, August 2004)*, arXiv:math.RT/0407414.

[133] Andrei Zelevinsky, *Mutations for quivers with potentials: Oberwolfach talk, april 2007*, arXiv:0706.0822.

[134] ———, *From Littlewood-Richardson coefficients to cluster algebras in three lectures*, Symmetric functions 2001: surveys of developments and perspectives, NATO Sci. Ser. II Math. Phys. Chem., vol. 74, Kluwer Acad. Publ., Dordrecht, 2002, pp. 253–273.

[135] ———, *Cluster algebras: origins, results and conjectures*, Advances in algebra towards millennium problems, SAS Int. Publ., Delhi, 2005, pp. 85–105.

[136] ———, *What is a cluster algebra?*, Notices of the A.M.S. **54** (2007), no. 11, 1494–1495.

Université Paris Diderot – Paris 7, UFR de Mathématiques, Institut de Mathématiques de Jussieu, UMR 7586 du CNRS, Case 7012, 2, place Jussieu, 75251 Paris Cedex 05, France

*E-mail address*: keller@math.jussieu.fr

# Localization theory for triangulated categories

HENNING KRAUSE

## Contents



## 1. Introduction

These notes provide an introduction to the theory of localization for triangulated categories. Localization is a machinery to formally invert morphisms in a category. We explain this formalism in some detail and we show how it is applied to triangulated categories.

There are basically two ways to approach the localization theory for triangulated categories and both are closely related to each other. To explain this, let us fix a triangulated category $\mathcal{T}$. The first approach is *Verdier localization*. For this one chooses a full triangulated subcategory $\mathcal{S}$ of $\mathcal{T}$ and constructs a universal exact functor $\mathcal{T} \to \mathcal{T}/\mathcal{S}$ which annihilates the objects belonging to $\mathcal{S}$. In fact, the quotient category $\mathcal{T}/\mathcal{S}$ is obtained by formally inverting all morphisms $\sigma$ in $\mathcal{T}$ such that the cone of $\sigma$ belongs to $\mathcal{S}$.

On the other hand, there is *Bousfield localization*. In this case one considers an exact functor $L \colon \mathcal{T} \to \mathcal{T}$ together with a natural morphism $\eta X \colon X \to LX$ for all $X$ in $\mathcal{T}$ such that $L(\eta X) = \eta(LX)$ is invertible. There are two full triangulated subcategories arising from such a localization functor $L$. We have

"

the subcategory Ker $L$ formed by all $L$-acyclic objects, and we have the essential image Im $L$ which coincides with the subcategory formed by all $L$-local objects. Note that $L$, Ker $L$, and Im $L$ determine each other. Moreover, $L$ induces an equivalence $\mathcal{T}/\operatorname{Ker} L \xrightarrow{\sim} \operatorname{Im} L$. Thus a Bousfield localization functor $\mathcal{T} \to \mathcal{T}$ is nothing but the composite of a Verdier quotient functor $\mathcal{T} \to \mathcal{T}/\mathcal{S}$ with a fully faithful right adjoint $\mathcal{T}/\mathcal{S} \to \mathcal{T}$.

Having introduced these basic objects, there are a number of immediate questions. For example, given a triangulated subcategory $\mathcal{S}$ of $\mathcal{T}$, can we find a localization functor $L\colon \mathcal{T} \to \mathcal{T}$ satisfying Ker $L = \mathcal{S}$ or Im $L = \mathcal{S}$? On the other hand, if we start with $L$, which properties of Ker $L$ and Im $L$ are inherited from $\mathcal{T}$? It turns out that well generated triangulated categories in the sense of Neeman [33] provide an excellent setting for studying these questions.

Let us discuss briefly the relevance of well generated categories. The concept generalizes that of a compactly generated triangulated category. For example, the derived category of unbounded chain complexes of modules over some fixed ring is compactly generated. Also, the stable homotopy category of CW-spectra is compactly generated. Given any localization functor $L$ on a compactly generated triangulated category, it is rare that Ker $L$ or Im $L$ are compactly generated. However, in all known examples Ker $L$ and Im $L$ are well generated. The following theorem provides a conceptual explanation; it combines several results from Section 7.

**Theorem.** *Let $\mathcal{T}$ be a well generated triangulated category and $\mathcal{S}$ a full triangulated subcategory which is closed under small coproducts. Then the following are equivalent.*

(1) *The triangulated category $\mathcal{S}$ is well generated.*
(2) *The triangulated category $\mathcal{T}/\mathcal{S}$ is well generated.*
(3) *There exists a cohomological functor $H\colon \mathcal{T} \to \mathcal{A}$ into a locally presentable abelian category such that $H$ preserves small coproducts and $\mathcal{S} = \operatorname{Ker} H$.*
(4) *There exists a small set $\mathcal{S}_0$ of objects in $\mathcal{S}$ such that $\mathcal{S}$ admits no proper full triangulated subcategory closed under small coproducts and containing $\mathcal{S}_0$.*

*Moreover, in this case there exists a localization functor $L\colon \mathcal{T} \to \mathcal{T}$ such that* Ker $L = \mathcal{S}$.

Note that every abelian Grothendieck category is locally presentable; in particular every module category is locally presentable.

Our approach for studying localization functors on well generated triangulated categories is based on the interplay between triangulated and abelian structure. A well known construction due to Freyd provides for any triangulated category $\mathcal{T}$ an abelian category $A(\mathcal{T})$ together with a universal cohomological functor $\mathcal{T} \to A(\mathcal{T})$. However, the category $A(\mathcal{T})$ is usually far too big and therefore not manageable. If $\mathcal{T}$ is well generated, then we have a canonical filtration

$$A(\mathcal{T}) = \bigcup_\alpha A_\alpha(\mathcal{T})$$

indexed by all regular cardinals, such that for each $\alpha$ the category $A_\alpha(\mathcal{T})$ is abelian and locally $\alpha$-presentable in the sense of Gabriel and Ulmer [17]. Moreover, each inclusion $A_\alpha(\mathcal{T}) \to A(\mathcal{T})$ admits an exact right adjoint and the composite

$$H_\alpha : \mathcal{T} \longrightarrow A(\mathcal{T}) \longrightarrow A_\alpha(\mathcal{T})$$

is the universal cohomological functor into a locally $\alpha$-presentable abelian category. Thus we may think of the functors $\mathcal{T} \to A_\alpha(\mathcal{T})$ as successive approximations of $\mathcal{T}$ by locally presentable abelian categories. For instance, there exists for each object $X$ in $\mathcal{T}$ some cardinal $\alpha(X)$ such that the induced map $\mathcal{T}(X, Y) \to A_\beta(\mathcal{T})(H_\beta X, H_\beta Y)$ is bijective for all $Y$ in $\mathcal{T}$ and all $\beta \geq \alpha(X)$.

These notes are organized as follows. We start off with an introduction to categories of fractions and localization functors for arbitrary categories. Then we apply this to triangulated categories. First we treat arbitrary triangulated categories and explain the localization in the sense of Verdier and Bousfield. Then we pass to compactly and well generated triangulated categories where Brown representability provides an indispensable tool for constructing localization functors. Module categories and their derived categories are used to illustrate most of the concepts; see [12] for complementary material from topology. The results on well generated categories are based on facts from the theory of locally presentable categories; we have collected these in a separate appendix.

## Acknowledgement

The plan to write an introduction to the theory of triangulated localization took shape during the "Workshop on Triangulated Categories" in Leeds 2006. I wish to thank the organizers Thorsten Holm, Peter Jørgensen, and Raphaël Rouquier for their skill and diligence in organizing this meeting. Most of these notes were then written during a three months stay in 2007 at the Centre de Recerca Matemàtica in Barcelona as a participant of the special program "Homotopy Theory and Higher Categories". I am grateful to the organizers Carles Casacuberta, Joachim Kock, and Amnon Neeman for creating a stimulating atmosphere and for several helpful discussions. Finally, I would like to thank Xiao-Wu Chen, Daniel Murfet, and Jan Šťovíček for their helpful comments on a preliminary version of these notes.

# 2. Categories of fractions and localization functors

## 2.1. Categories

Throughout we fix a universe of sets in the sense of Grothendieck [19]. The members of this universe will be called *small sets*.

Let $\mathcal{C}$ be a category. We denote by $\mathrm{Ob}\,\mathcal{C}$ the set of objects and by $\mathrm{Mor}\,\mathcal{C}$ the set of morphisms in $\mathcal{C}$. Given objects $X, Y$ in $\mathcal{C}$, the set of morphisms $X \to Y$ will be denoted by $\mathcal{C}(X, Y)$. The identity morphism of an object $X$ is denoted by $\mathrm{id}_{\mathcal{C}}\,X$ or just $\mathrm{id}\,X$. If not stated otherwise, we always assume that the morphisms between two fixed objects of a category form a small set.

A category $\mathcal{C}$ is called *small* if the isomorphism classes of objects in $\mathcal{C}$ form a small set. In that case we define the *cardinality* of $\mathcal{C}$ as $\mathrm{card}\,\mathcal{C} = \sum_{X,Y \in \mathcal{C}_0} \mathrm{card}\,\mathcal{C}(X, Y)$ where $\mathcal{C}_0$ denotes a representative set of objects of $\mathcal{C}$, meeting each isomorphism class exactly once.

Let $F : \mathcal{I} \to \mathcal{C}$ be a functor from a small (indexing) category $\mathcal{I}$ to a category $\mathcal{C}$. Then we write $\varinjlim_{i \in \mathcal{I}} Fi$ for the colimit of $F$, provided it exists. Given a cardinal $\alpha$, the colimit of $F$ is called $\alpha$-*colimit* if $\mathrm{card}\,\mathcal{I} < \alpha$. An example of a colimit is the coproduct $\coprod_{i \in I} X_i$ of a family $(X_i)_{i \in I}$ of objects in $\mathcal{C}$ where the indexing set $I$ is always assumed to be small. We say that a category $\mathcal{C}$ *admits small coproducts* if for every family $(X_i)_{i \in I}$ of objects in $\mathcal{C}$ which is indexed by a small set $I$ the coproduct $\coprod_{i \in I} X_i$ exists in $\mathcal{C}$. Analogous terminology is used for limits and products.

## 2.2. Categories of fractions

Let $F : \mathcal{C} \to \mathcal{D}$ be a functor. We say that $F$ makes a morphism $\sigma$ of $\mathcal{C}$ invertible if $F\sigma$ is invertible. The set of all those morphisms which $F$ inverts is denoted by $\Sigma(F)$.

Given a category $\mathcal{C}$ and any set $\Sigma$ of morphisms of $\mathcal{C}$, we consider the *category of fractions* $\mathcal{C}[\Sigma^{-1}]$ together with a canonical *quotient functor*

$$Q_\Sigma : \mathcal{C} \longrightarrow \mathcal{C}[\Sigma^{-1}]$$

having the following properties.

(Q1) $Q_\Sigma$ makes the morphisms in $\Sigma$ invertible.

(Q2) If a functor $F : \mathcal{C} \to \mathcal{D}$ makes the morphisms in $\Sigma$ invertible, then there is a unique functor $\bar{F} : \mathcal{C}[\Sigma^{-1}] \to \mathcal{D}$ such that $F = \bar{F} \circ Q_\Sigma$.

Note that $\mathcal{C}[\Sigma^{-1}]$ and $Q_\Sigma$ are essentially unique if they exists. Now let us sketch the construction of $\mathcal{C}[\Sigma^{-1}]$ and $Q_\Sigma$. At this stage, we ignore set-theoretic issues, that is, the morphisms between two objects of $\mathcal{C}[\Sigma^{-1}]$ need not to form a small set. We put $\operatorname{Ob}\mathcal{C}[\Sigma^{-1}] = \operatorname{Ob}\mathcal{C}$. To define the morphisms of $\mathcal{C}[\Sigma^{-1}]$, consider the quiver (i.e. oriented graph) with set of vertices $\operatorname{Ob}\mathcal{C}$ and with set of arrows the disjoint union $(\operatorname{Mor}\mathcal{C}) \amalg \Sigma^{-1}$, where $\Sigma^{-1} = \{\sigma^{-1}\colon Y \to X \mid \Sigma \ni \sigma\colon X \to Y\}$. Let $\mathcal{P}$ be the set of paths in this quiver (i.e. finite sequences of composable arrows), together with the obvious composition which is the concatenation operation and denoted by $\circ_\mathcal{P}$. We define $\operatorname{Mor}\mathcal{C}[\Sigma^{-1}]$ as the quotient of $\mathcal{P}$ modulo the following relations:

(1) $\beta \circ_\mathcal{P} \alpha = \beta \circ \alpha$ for all composable morphisms $\alpha, \beta \in \operatorname{Mor}\mathcal{C}$.
(2) $\operatorname{id}_\mathcal{P} X = \operatorname{id}_\mathcal{C} X$ for all $X \in \operatorname{Ob}\mathcal{C}$.
(3) $\sigma^{-1} \circ_\mathcal{P} \sigma = \operatorname{id}_\mathcal{P} X$ and $\sigma \circ_\mathcal{P} \sigma^{-1} = \operatorname{id}_\mathcal{P} Y$ for all $\sigma\colon X \to Y$ in $\Sigma$.

The composition in $\mathcal{P}$ induces the composition of morphisms in $\mathcal{C}[\Sigma^{-1}]$. The functor $Q_\Sigma$ is the identity on objects and on $\operatorname{Mor}\mathcal{C}$ the composite

$$\operatorname{Mor}\mathcal{C} \xrightarrow{\text{inc}} (\operatorname{Mor}\mathcal{C}) \amalg \Sigma^{-1} \xrightarrow{\text{inc}} \mathcal{P} \xrightarrow{\text{can}} \operatorname{Mor}\mathcal{C}[\Sigma^{-1}].$$

Having completed the construction of the category of fractions $\mathcal{C}[\Sigma^{-1}]$, let us mention that it is also called *quotient category* or *localization* of $\mathcal{C}$ with respect to $\Sigma$.

## 2.3. Adjoint functors

Let $F\colon \mathcal{C} \to \mathcal{D}$ and $G\colon \mathcal{D} \to \mathcal{C}$ be a pair of functors and assume that $F$ is left adjoint to $G$. We denote by

$$\theta\colon F \circ G \to \operatorname{Id}\mathcal{D} \quad \text{and} \quad \eta\colon \operatorname{Id}\mathcal{C} \to G \circ F$$

the corresponding adjunction morphisms. Let $\Sigma = \Sigma(F)$ denote the set of morphisms $\sigma$ of $\mathcal{C}$ such that $F\sigma$ is invertible. Recall that a morphism $\mu\colon F \to F'$ between two functors is invertible if for each object $X$ the morphism $\mu X\colon FX \to F'X$ is invertible.

**Proposition 2.3.1.** *The following statements are equivalent.*

(1) *The functor $G$ is fully faithful.*
(2) *The morphism $\theta\colon F \circ G \to \operatorname{Id}\mathcal{D}$ is invertible.*
(3) *The functor $\bar{F}\colon \mathcal{C}[\Sigma^{-1}] \to \mathcal{D}$ satisfying $F = \bar{F} \circ Q_\Sigma$ is an equivalence.*

*Proof.* See [18, I.1.3]. $\qquad\qquad\qquad\qquad\qquad\qquad\qquad\qquad\qquad\square$

## 2.4. Localization functors

A functor $L\colon \mathcal{C} \to \mathcal{C}$ is called a *localization functor* if there exists a morphism $\eta\colon \operatorname{Id}\mathcal{C} \to L$ such that $L\eta\colon L \to L^2$ is invertible and $L\eta = \eta L$. Note that we only require the existence of $\eta$; the actual morphism is not part of the definition of $L$. However, we will see that $\eta$ is determined by $L$, up to a unique isomorphism $L \to L$.

**Proposition 2.4.1.** *Let $L\colon \mathcal{C} \to \mathcal{C}$ be a functor and $\eta\colon \operatorname{Id}\mathcal{C} \to L$ be a morphism. Then the following are equivalent.*

(1) *$L\eta\colon L \to L^2$ is invertible and $L\eta = \eta L$.*
(2) *There exists a functor $F\colon \mathcal{C} \to \mathcal{D}$ and a fully faithful right adjoint $G\colon \mathcal{D} \to \mathcal{C}$ such that $L = G \circ F$ and $\eta\colon \operatorname{Id}\mathcal{C} \to G \circ F$ is the adjunction morphism.*

*Proof.* (1) $\Rightarrow$ (2): Let $\mathcal{D}$ denote the full subcategory of $\mathcal{C}$ formed by all objects $X$ such that $\eta X$ is invertible. For each $X \in \mathcal{D}$, let $\theta X\colon LX \to X$ be the inverse of $\eta X$. Define $F\colon \mathcal{C} \to \mathcal{D}$ by $FX = LX$ and let $G\colon \mathcal{D} \to \mathcal{C}$ be the inclusion. We claim that $F$ and $G$ form an adjoint pair. In fact, it is straightforward to check that the maps

$$\mathcal{D}(FX, Y) \longrightarrow \mathcal{C}(X, GY), \quad \alpha \mapsto G\alpha \circ \eta X,$$

and

$$\mathcal{C}(X, GY) \longrightarrow \mathcal{D}(FX, Y), \quad \beta \mapsto \theta Y \circ F\beta,$$

are mutually inverse bijections.

(2) $\Rightarrow$ (1): Let $\theta\colon FG \to \operatorname{Id}\mathcal{D}$ denote the second adjunction morphism. Then the composites

$$F \xrightarrow{F\eta} FGF \xrightarrow{\theta F} F \quad \text{and} \quad G \xrightarrow{\eta G} GFG \xrightarrow{G\theta} G$$

are identity morphisms; see [27, IV.1]. We know from Proposition 2.3.1 that $\theta$ is invertible because $G$ is fully faithful. Therefore $L\eta = GF\eta$ is invertible. Moreover, we have

$$L\eta = GF\eta = (G\theta F)^{-1} = \eta GF = \eta L. \qquad \square$$

**Corollary 2.4.2.** *A functor $L\colon \mathcal{C} \to \mathcal{C}$ is a localization functor if and only if there exists a functor $F\colon \mathcal{C} \to \mathcal{D}$ and a fully faithful right adjoint $G\colon \mathcal{D} \to \mathcal{C}$ such that $L = G \circ F$. In that case there exist a unique equivalence $\mathcal{C}[\Sigma^{-1}] \to \mathcal{D}$ making the following diagram commutative*

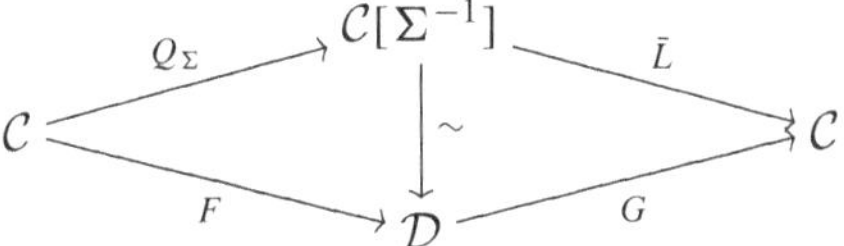

*where $\Sigma$ denotes the set of morphisms $\sigma$ in $C$ such that $L\sigma$ is invertible.*

*Proof.* The characterization of a localization functor follows from Proposition 2.4.1. Now observe that $\Sigma$ equals the set of morphisms $\sigma$ in $C$ such that $F\sigma$ is invertible since $G$ is fully faithful. Thus we can apply Proposition 2.3.1 to obtain the equivalence $C[\Sigma^{-1}] \to \mathcal{D}$ making the diagram commutative. $\square$

## 2.5. Local objects

Given a localization functor $L\colon C \to C$, we wish to describe those objects $X$ in $C$ such that $X \xrightarrow{\sim} LX$. To this end, it is convenient to make the following definition. An object $X$ in a category $C$ is called *local* with respect to a set $\Sigma$ of morphisms if for every morphism $W \to W'$ in $\Sigma$ the induced map $C(W', X) \to C(W, X)$ is bijective. Now let $F\colon C \to \mathcal{D}$ be a functor and let $\Sigma(F)$ denote the set of morphisms $\sigma$ of $C$ such that $F\sigma$ is invertible. An object $X$ in $C$ is called *F*-local if it is local with respect to $\Sigma(F)$.

**Lemma 2.5.1.** *Let $F\colon C \to \mathcal{D}$ be a functor and $X$ an object of $C$. Suppose there are two morphisms $\eta_1\colon X \to Y_1$ and $\eta_2\colon X \to Y_2$ such that $F\eta_i$ is invertible and $Y_i$ is $F$-local for $i = 1, 2$. Then there exists a unique isomorphism $\phi\colon Y_1 \to Y_2$ such that $\eta_2 = \phi \circ \eta_1$.*

*Proof.* The morphism $\eta_1$ induces a bijection $C(Y_1, Y_2) \to C(X, Y_2)$ and we take for $\phi$ the unique morphism which is sent to $\eta_2$. Exchanging the roles of $\eta_1$ and $\eta_2$, we obtain the inverse for $\phi$. $\square$

**Proposition 2.5.2.** *Let $L\colon C \to C$ be a localization functor and $\eta\colon \mathrm{Id}\,C \to L$ a morphism such that $L\eta$ is invertible. Then the following are equivalent for an object $X$ in $C$.*

(1) *The object $X$ is $L$-local.*
(2) *The map $C(LW, X) \to C(W, X)$ induced by $\eta W$ is bijective for all $W$ in $C$.*
(3) *The morphism $\eta X\colon X \to LX$ is invertible.*
(4) *The map $C(W, X) \to C(LW, LX)$ induced by $L$ is bijective for all $W$ in $C$.*
(5) *The object $X$ is isomorphic to $LX'$ for some object $X'$ in $C$.*

*Proof.* (1) $\Rightarrow$ (2): The morphism $\eta W$ belongs to $\Sigma(L)$ and therefore $C(\eta W, X)$ is bijective if $X$ is $L$-local.

(2) $\Rightarrow$ (3): Put $W = X$. We obtain a morphism $\phi: LX \to X$ which is an inverse for $\eta X$. More precisely, we have $\phi \circ \eta X = \text{id}\, X$. On the other hand,

$$\eta X \circ \phi = L\phi \circ \eta LX = L\phi \circ L\eta X = L(\phi \circ \eta X) = \text{id}\, LX.$$

Thus $\eta X$ is invertible.

(3) $\Leftrightarrow$ (4): We use the factorization $\mathcal{C} \xrightarrow{F} \mathcal{D} \xrightarrow{G} \mathcal{C}$ of $L$ from Proposition 2.4.1. Then we obtain for each $W$ in $\mathcal{C}$ a factorization

$$\mathcal{C}(W, X) \longrightarrow \mathcal{C}(W, LX) \xrightarrow{\sim} \mathcal{C}(FW, FX) \xrightarrow{\sim} \mathcal{C}(LW, LX)$$

of the map $f_W: \mathcal{C}(W, X) \to \mathcal{C}(LW, LX)$ induced by $L$. Here, the first map is induced by $\eta X$, the second follows from the adjunction, and the third is induced by $G$. Thus $f_W$ is bijective for all $W$ iff the first map is bijective for all $W$ iff $\eta X$ is invertible.

(3) $\Rightarrow$ (5): Take $X' = X$.

(5) $\Rightarrow$ (1): We use again the factorization $\mathcal{C} \xrightarrow{F} \mathcal{D} \xrightarrow{G} \mathcal{C}$ of $L$ from Proposition 2.4.1. Fix $\sigma$ in $\Sigma(L)$ and observe that $F\sigma$ is invertible. Then we have $\mathcal{C}(\sigma, X) \cong \mathcal{C}(\sigma, G(FX')) \cong \mathcal{D}(F\sigma, FX')$ and this implies that $\mathcal{C}(\sigma, X)$ is bijective since $F\sigma$ is invertible. $\qquad\square$

Given a functor $F: \mathcal{C} \to \mathcal{D}$, we denote by $\text{Im}\, F$ the *essential image* of $F$, that is, the full subcategory of $\mathcal{D}$ which is formed by all objects isomorphic to $FX$ for some object $X$ in $\mathcal{C}$.

**Corollary 2.5.3.** *Let $L: \mathcal{C} \to \mathcal{C}$ be a localization functor. Then $L$ induces an equivalence $\mathcal{C}[\Sigma(L)^{-1}] \xrightarrow{\sim} \text{Im}\, L$ and $\text{Im}\, L$ is the full subcategory of $\mathcal{C}$ consisting of all $L$-local subobjects.*

*Proof.* Write $L$ as composite $\mathcal{C} \xrightarrow{F} \text{Im}\, L \xrightarrow{G} \mathcal{C}$ of two functors, where $FX = LX$ for all $X$ in $\mathcal{C}$ and $G$ is the inclusion functor. Then it follows from Corollary 2.4.2 that $F$ induces an equivalence $\mathcal{C}[\Sigma(L)^{-1}] \xrightarrow{\sim} \text{Im}\, L$. The second assertion is an immediate consequence of Proposition 2.5.2. $\qquad\square$

Given a localization functor $L: \mathcal{C} \to \mathcal{C}$ and an object $X$ in $\mathcal{C}$, the morphism $X \to LX$ is initial among all morphisms to an object in $\text{Im}\, \mathcal{D}$ and terminal among all morphisms in $\Sigma(L)$. The following statement makes this precise.

**Corollary 2.5.4.** *Let $L: \mathcal{C} \to \mathcal{C}$ be a localization functor and $\eta: \text{Id}\, \mathcal{C} \to L$ a morphism such that $L\eta$ is invertible. Then for each morphism $\eta X: X \to LX$ the following holds.*

(1) *The object $LX$ belongs to $\text{Im}\, L$ and every morphism $X \to Y$ with $Y$ in $\text{Im}\, L$ factors uniquely through $\eta X$.*

(2) *The morphism $\eta X$ belongs to $\Sigma(L)$ and factors uniquely through every morphism $X \to Y$ in $\Sigma(L)$.*

*Proof.* Apply Proposition 2.5.2. $\qquad\qquad\qquad\qquad\qquad\qquad\qquad\quad\square$

**Remark 2.5.5.** (1) Let $L \colon C \to C$ be a localization functor and suppose there are two morphisms $\eta_i \colon \mathrm{Id}\,C \to L$ such that $L\eta_i$ is invertible for $i = 1, 2$. Then there exists a unique isomorphism $\phi \colon L \xrightarrow{\sim} L$ such that $\eta_2 = \phi \circ \eta_1$. This follows from Lemma 2.5.1.

(2) Given any functor $F \colon C \to \mathcal{D}$, the full subcategory of $F$-local objects is closed under taking all limits which exist in $C$.

## 2.6. Existence of localization functors

We provide a criterion for the existence of a localization functor $L$; it explains how $L$ is determined by the category of $L$-local objects.

**Proposition 2.6.1.** *Let $C$ be a category and $\mathcal{D}$ a full subcategory. Suppose that every object in $C$ isomorphic to one in $\mathcal{D}$ belongs to $\mathcal{D}$. Then the following are equivalent.*

(1) *There exists a localization functor $L \colon C \to C$ with $\mathrm{Im}\,L = \mathcal{D}$.*
(2) *For every object $X$ in $C$ there exists a morphism $\eta X \colon X \to X'$ with $X'$ in $\mathcal{D}$ such that every morphism $X \to Y$ with $Y$ in $\mathcal{D}$ factors uniquely through $\eta X$.*
(3) *The inclusion functor $\mathcal{D} \to C$ admits a left adjoint.*

*Proof.* (1) $\Rightarrow$ (2): Suppose there exists a localization functor $L \colon C \to C$ with $\mathrm{Im}\,L = \mathcal{D}$ and let $\eta \colon \mathrm{Id}\,C \to L$ be a morphism such that $L\eta$ is invertible. Then Proposition 2.5.2 shows that $C(\eta X, Y)$ is bijective for all $Y$ in $\mathcal{D}$.

(2) $\Rightarrow$ (3): The morphisms $\eta X$ provide a functor $F \colon C \to \mathcal{D}$ by sending each $X$ in $C$ to $X'$. It is straightforward to check that $F$ is a left adjoint for the inclusion $\mathcal{D} \to C$.

(3) $\Rightarrow$ (1): Let $G \colon \mathcal{D} \to C$ denote the inclusion and $F$ its right adjoint. Then $L = G \circ F$ is a localization functor with $\mathrm{Im}\,L = \mathcal{D}$ by Proposition 2.4.1. $\quad\square$

## 2.7. Localization functors preserving coproducts

We characterize the fact that a localization functor preserves small coproducts.

**Proposition 2.7.1.** *Let $L \colon C \to C$ be a localization functor and suppose the category $C$ admits small coproducts. Then the following are equivalent.*

(1) *The functor L preserves small coproducts.*
(2) *The L-local objects are closed under taking small coproducts in $\mathcal{C}$.*
(3) *The right adjoint of the quotient functor $\mathcal{C} \to \mathcal{C}[\Sigma(L)^{-1}]$ preserves small coproducts.*

*Proof.* $(1) \Rightarrow (2)$: Let $(X_i)_{i \in I}$ be a family of $L$-local objects. Thus the natural morphisms $X_i \to LX_i$ are invertible by Proposition 2.5.2 and they induce an isomorphism

$$\coprod_i X_i \xrightarrow{\sim} \coprod_i LX_i \xrightarrow{\sim} L(\coprod_i X_i).$$

It follows that $\coprod_i X_i$ is $L$-local.

$(2) \Leftrightarrow (3)$: We can identify $\mathcal{C}[\Sigma(L)^{-1}] = \operatorname{Im} L$ by Corollary 2.5.3 and then the right adjoint of the quotient functor identifies with the inclusion $\operatorname{Im} L \to \mathcal{C}$. Thus the right adjoint preserves small coproducts if and only if the inclusion $\operatorname{Im} L \to \mathcal{C}$ preserves small coproducts.

$(3) \Rightarrow (1)$: Write $L$ as composite $\mathcal{C} \to \mathcal{C}[\Sigma(L)^{-1}] \to \mathcal{C}$ of the quotient functor $Q$ with its right adjoint $\bar{L}$. Then $Q$ preserves small coproducts since it is a left adjoint. It follows that $L$ preserves small coproducts if $\bar{L}$ preserves small coproducts. $\qquad\square$

## 2.8. Colocalization functors

A functor $\Gamma \colon \mathcal{C} \to \mathcal{C}$ is called *colocalization functor* if its opposite functor $\Gamma^{\mathrm{op}} \colon \mathcal{C}^{\mathrm{op}} \to \mathcal{C}^{\mathrm{op}}$ is a localization functor. We call an object $X$ in $\mathcal{C}$ $\Gamma$-*colocal* if it is $\Gamma^{\mathrm{op}}$-local when viewed as an object of $\mathcal{C}^{\mathrm{op}}$. Note that a colocalization functor $\Gamma \colon \mathcal{C} \to \mathcal{C}$ induces an equivalence

$$\mathcal{C}[\Sigma(\Gamma)^{-1}] \xrightarrow{\sim} \operatorname{Im} \Gamma$$

and the essential image $\operatorname{Im} \Gamma$ equals the full subcategory of $\mathcal{C}$ consisting of all $\Gamma$-colocal objects.

**Remark 2.8.1.** We think of $\Gamma$ as $L$ turned upside down; this explains our notation. Another reason for the use of $\Gamma$ is the interpretation of local cohomology as colocalization.

## 2.9. Example: Localization of modules

Let $A$ be an associative ring and denote by $\operatorname{Mod} A$ the category of (right) $A$-modules. Suppose that $A$ is commutative and let $S \subseteq A$ be a multiplicatively closed subset, that is, $1 \in S$ and $st \in S$ for all $s, t \in S$. We denote by

$$S^{-1}A = \{x/s \mid x \in A \text{ and } s \in S\}$$

the ring of fractions. For each $A$-module $M$, let

$$S^{-1}M = \{x/s \mid x \in M \text{ and } s \in S\}$$

be the localized module. An $S^{-1}A$-module $N$ becomes an $A$-module via restriction of scalars along the canonical ring homomorpism $A \to S^{-1}A$. We obtain a pair of functors

$$F: \operatorname{Mod} A \longrightarrow \operatorname{Mod} S^{-1}A, \quad M \mapsto S^{-1}M \cong M \otimes_A S^{-1}A,$$

$$G: \operatorname{Mod} S^{-1}A \longrightarrow \operatorname{Mod} A, \quad N \mapsto N \cong \operatorname{Hom}_{S^{-1}A}(S^{-1}A, N).$$

Moreover, for each pair of modules $M$ over $A$ and $N$ over $S^{-1}A$, we have natural morphisms

$$\eta M: M \longrightarrow (G \circ F)M = S^{-1}M, \quad x \mapsto x/1,$$

$$\theta N: S^{-1}N = (F \circ G)N \longrightarrow N, \quad x/s \mapsto xs^{-1}.$$

These natural morphisms induce mutually inverse bijections as follows:

$$\operatorname{Hom}_A(M, GN) \xrightarrow{\sim} \operatorname{Hom}_{S^{-1}A}(FM, N), \quad \alpha \mapsto \theta N \circ F\alpha,$$

$$\operatorname{Hom}_{S^{-1}A}(FM, N) \xrightarrow{\sim} \operatorname{Hom}_A(M, GN), \quad \beta \mapsto G\beta \circ \eta M.$$

It is clear that the functors $F$ and $G$ form an adjoint pair, that is, $F$ is a left adjoint of $G$ and $G$ is a right adjoint of $F$. Moreover, the adjunction morphism $\theta: F \circ G \to \operatorname{Id}$ is invertible. Therefore the composite $L = G \circ F$ is a localization functor.

Let us formulate this slightly more generally. Fix a ring homomorphism $f: A \to B$. Then it is well known that the restriction functor $\operatorname{Mod} B \to \operatorname{Mod} A$ is fully faithful if and only if $f$ is an epimorphism; see [45, Proposition XI.1.2]. Thus the functor $\operatorname{Mod} A \to \operatorname{Mod} A$ taking a module $M$ to $M \otimes_A B$ is a localization functor provided that $f$ is an epimorphism.

## 2.10. Example: Localization of spectra

A *spectrum* $E$ is a sequence of based topological spaces $E_n$ and based homeomorphisms $E_n \to \Omega E_{n+1}$. A morphism of spectra $E \to F$ is a sequence of based continuous maps $E_n \to F_n$ strictly compatible with the given structural homeomorphisms. The homotopy groups of a spectrum $E$ are the groups $\pi_n E = \pi_{n+i}(E_i)$ for $i \geq 0$ and $n + i \geq 0$. A morphism between spectra is a *weak equivalence* if it induces an isomorphism on homotopy groups. The *stable homotopy category* $\mathrm{Ho}\,\mathcal{S}$ is obtained from the category $\mathcal{S}$ of spectra by formally inverting the weak equivalences. Thus $\mathrm{Ho}\,\mathcal{S} = \mathcal{S}[\Sigma^{-1}]$ where $\Sigma$ denotes the set of weak equivalences. We refer to [2, 39] for details.

## 2.11. Notes

The category of fractions is introduced by Gabriel and Zisman in [18], but the idea of formally inverting elements can be traced back much further; see for instance [36]. The appropriate context for localization functors is the theory of monads; see [27].

# 3. Calculus of fractions

## 3.1. Calculus of fractions

Let $\mathcal{C}$ be a category and $\Sigma$ a set of morphisms in $\mathcal{C}$. The category of fractions $\mathcal{C}[\Sigma^{-1}]$ admits an elementary description if some extra assumptions on $\Sigma$ are satisfied. We say that $\Sigma$ *admits a calculus of left fractions* if the following holds.

(LF1) If $\sigma, \tau$ are composable morphisms in $\Sigma$, then $\tau \circ \sigma$ is in $\Sigma$. The identity morphism id $X$ is in $\Sigma$ for all $X$ in $\mathcal{C}$.

(LF2) Each pair of morphisms $X' \xleftarrow{\sigma} X \xrightarrow{\alpha} Y$ with $\sigma$ in $\Sigma$ can be completed to a commutative square

$$\begin{array}{ccc} X & \xrightarrow{\ \alpha\ } & Y \\ \big\downarrow{\sigma} & & \big\downarrow{\sigma'} \\ X' & \xrightarrow{\ \alpha'\ } & Y' \end{array}$$

such that $\sigma'$ is in $\Sigma$.

(LF3) Let $\alpha, \beta : X \to Y$ be morphisms in $\mathcal{C}$. If there is a morphism $\sigma : X' \to X$ in $\Sigma$ with $\alpha \circ \sigma = \beta \circ \sigma$, then there exists a morphism $\tau : Y \to Y'$ in $\Sigma$ with $\tau \circ \alpha = \tau \circ \beta$.

Now assume that $\Sigma$ admits a calculus of left fractions. Then one obtains a new category $\Sigma^{-1}\mathcal{C}$ as follows. The objects are those of $\mathcal{C}$. Given objects $X$ and $Y$, we call a pair $(\alpha, \sigma)$ of morphisms

$$X \xrightarrow{\ \alpha\ } Y' \xleftarrow{\ \sigma\ } Y$$

in $\mathcal{C}$ with $\sigma$ in $\Sigma$ a *left fraction*. The morphisms $X \to Y$ in $\Sigma^{-1}\mathcal{C}$ are equivalence classes $[\alpha, \sigma]$ of such left fractions, where two diagrams $(\alpha_1, \sigma_1)$ and $(\alpha_2, \sigma_2)$ are equivalent if there exists a commutative diagram

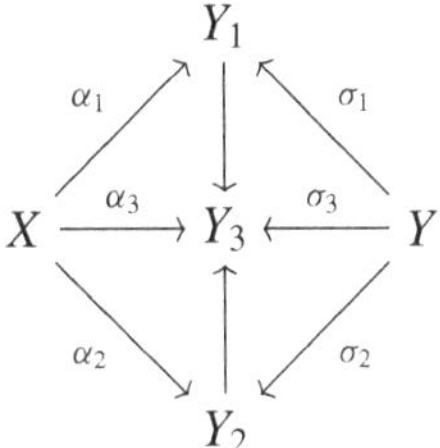

with $\sigma_3$ in $\Sigma$. The composition of two equivalence classes $[\alpha, \sigma]$ and $[\beta, \tau]$ is by definition the equivalene class $[\beta' \circ \alpha, \sigma' \circ \tau]$ where $\sigma'$ and $\beta'$ are obtained from condition (LF2) as in the following commutative diagram.

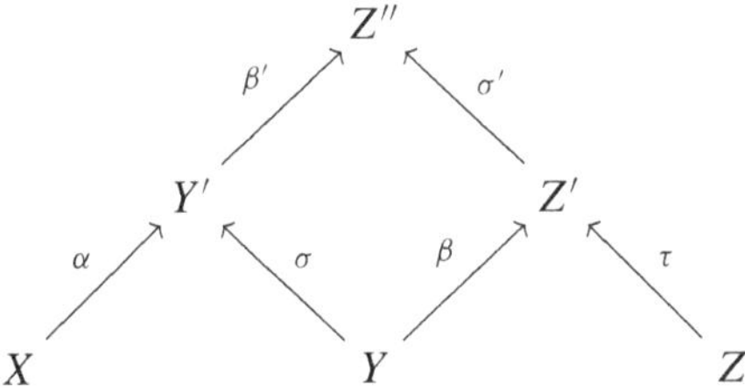

We obtain a canonical functor

$$P_\Sigma : \mathcal{C} \longrightarrow \Sigma^{-1}\mathcal{C}$$

by taking the identity map on objects and by sending a morphism $\alpha : X \to Y$ to the equivalence class $[\alpha, \mathrm{id}\, Y]$. Let us compare $P_\Sigma$ with the quotient functor $Q_\Sigma : \mathcal{C} \to \mathcal{C}[\Sigma^{-1}]$.

**Proposition 3.1.1.** *The functor* $F : \Sigma^{-1}\mathcal{C} \to \mathcal{C}[\Sigma^{-1}]$ *which is the identity map on objects and which takes a morphism* $[\alpha, \sigma]$ *to* $(Q_\Sigma \sigma)^{-1} \circ Q_\Sigma \alpha$ *is an isomorphism.*

*Proof.* The functor $P_\Sigma$ inverts all morphisms in $\Sigma$ and factors therefore through $Q_\Sigma$ via a functor $G : \mathcal{C}[\Sigma^{-1}] \to \Sigma^{-1}\mathcal{C}$. It is straightforward to check that $F \circ G = \mathrm{Id}$ and $G \circ F = \mathrm{Id}$. $\qquad\square$

From now on, we will identify $\Sigma^{-1}\mathcal{C}$ with $\mathcal{C}[\Sigma^{-1}]$ whenever $\Sigma$ admits a calculus of left fractions. A set of morphisms $\Sigma$ in $\mathcal{C}$ *admits a calculus of right fractions* if the dual conditions of (LF1) – (LF3) are satisfied. Moreover, $\Sigma$ is called a *multiplicative system* if it admits both, a calculus of left fractions and a calculus of right fractions. Note that all results about sets of morphisms admitting a calculus of left fractions have a dual version for sets of morphisms admitting a calculus of right fractions.

## 3.2. Calculus of fractions and adjoint functors

Given a category $\mathcal{C}$ and a set of morphisms $\Sigma$, it is an interesting question to ask when the quotient functor $\mathcal{C} \to \mathcal{C}[\Sigma^{-1}]$ admits a right adjoint. It turns out that this problem is closely related to the property of $\Sigma$ to admit a calculus of left fractions.

**Lemma 3.2.1.** *Let $F: \mathcal{C} \to \mathcal{D}$ and $G: \mathcal{D} \to \mathcal{C}$ be a pair of adjoint functors. Assume that the right adjoint $G$ is fully faithful and let $\Sigma$ be the set of morphisms $\sigma$ in $\mathcal{C}$ such that $F\sigma$ is invertible. Then $\Sigma$ admits a calculus of left fractions.*

*Proof.* We need to check the conditions (LF1) – (LF3). Observe first that $L = G \circ F$ is a localization functor so that we can apply Proposition 2.5.2.

(LF1): This condition is clear because $F$ is a functor.

(LF2): Let $X' \xleftarrow{\sigma} X \xrightarrow{\alpha} Y$ be a pair of morphisms with $\sigma$ in $\Sigma$. This can be completed to a commutative square

$$
\begin{array}{ccc}
X & \xrightarrow{\ \alpha\ } & Y \\
\downarrow{\sigma} & & \downarrow{\sigma'} \\
X' & \xrightarrow{\ \alpha'\ } & Y'
\end{array}
$$

if we take for $\sigma'$ the morphism $\eta Y: Y \to LY$ in $\Sigma$, because the map $\mathcal{C}(\sigma, LY)$ is surjective by Proposition 2.5.2.

(LF3): Let $\alpha, \beta: X \to Y$ be morphisms in $\mathcal{C}$ and suppose there is a morphism $\sigma: X' \to X$ in $\Sigma$ with $\alpha \circ \sigma = \beta \circ \sigma$. Then we take $\tau = \eta Y$ in $\Sigma$ and have $\tau \circ \alpha = \tau \circ \beta$, because the map $\mathcal{C}(\sigma, LY)$ is injective by Proposition 2.5.2. $\square$

**Lemma 3.2.2.** *Let $\mathcal{C}$ be a category and $\Sigma$ a set of morphisms admitting a calculus of left fractions. Then the following are equivalent for an object $X$ in $\mathcal{C}$.*

(1) *$X$ is local with respect to $\Sigma$.*
(2) *The quotient functor induces a bijection $\mathcal{C}(W, X) \to \mathcal{C}[\Sigma^{-1}](W, X)$ for all $W$.*

*Proof.* (1) $\Rightarrow$ (2): To show that $f_W: \mathcal{C}(W, X) \to \mathcal{C}[\Sigma^{-1}](W, X)$ is surjective, choose a left fraction $W \xrightarrow{\alpha} X' \xleftarrow{\sigma} X$ with $\sigma$ in $\Sigma$. Then there exists $\tau: X' \to X$ with $\tau \circ \sigma = \mathrm{id}\, X$ since $X$ is local. Thus $f_W(\tau \circ \alpha) = [\alpha, \sigma]$. To show that $f_W$ is injective, suppose that $f_W(\alpha) = f_W(\beta)$. Then we have $\sigma \circ \alpha = \sigma \circ \beta$ for some $\sigma: X \to X'$ in $\Sigma$. The morphism $\sigma$ is a section because $X$ is local, and therefore $\alpha = \beta$.

$(2) \Rightarrow (1)$: Let $\sigma: W \to W'$ be a morphism in $\Sigma$. Then we have $\mathcal{C}(\sigma, X) \cong \mathcal{C}[\Sigma^{-1}]([\sigma, \mathrm{id}\, W'], X)$. Thus $\mathcal{C}(\sigma, X)$ is bijective since $[\sigma, \mathrm{id}\, W']$ is invertible. $\qquad\square$

**Proposition 3.2.3.** *Let $\mathcal{C}$ be a category, $\Sigma$ a set of morphisms admitting a calculus of left fractions, and $Q: \mathcal{C} \to \mathcal{C}[\Sigma^{-1}]$ the quotient functor. Then the following are equivalent.*

(1) *The functor $Q$ has a right adjoint (which is then fully faithful).*
(2) *For each object $X$ in $\mathcal{C}$, there exist a morphism $\eta X: X \to X'$ such that $X'$ is local with respect to $\Sigma$ and $Q(\eta X)$ is invertible.*

*Proof.* $(1) \Rightarrow (2)$: Denote by $Q_\rho$ the right adjoint of $Q$ and by $\eta: \mathrm{Id}\,\mathcal{C} \to Q_\rho Q$ the adjunction morphism. We take for each object $X$ in $\mathcal{C}$ the morphism $\eta X: X \to Q_\rho Q X$. Note that $Q_\rho Q X$ is local by Proposition 2.5.2.

$(2) \Rightarrow (1)$: We fix objects $X$ and $Y$. Then we have two natural bijections

$$\mathcal{C}[\Sigma^{-1}](X, Y) \xrightarrow{\sim} \mathcal{C}[\Sigma^{-1}](X, Y') \xleftarrow{\sim} \mathcal{C}(X, Y').$$

The first is induced by $\eta Y: Y \to Y'$ and is bijective since $Q(\eta Y)$ is invertible. The second map is bijective by Lemma 3.2.2, since $Y'$ is local with respect to $\Sigma$. Thus we obtain a right adjoint for $Q$ by sending each object $Y$ of $\mathcal{C}[\Sigma^{-1}]$ to $Y'$. $\qquad\square$

## 3.3. A criterion for the fractions to form a small set

Let $\mathcal{C}$ be a category and $\Sigma$ a set of morphisms in $\mathcal{C}$. Suppose that $\Sigma$ admits a calculus of left fractions. From the construction of $\mathcal{C}[\Sigma^{-1}]$ we cannot expect that for any given pair of objects $X$ and $Y$ the equivalence classes of fractions in $\mathcal{C}[\Sigma^{-1}](X, Y)$ form a small set. The situation is different if the category $\mathcal{C}$ is small. Then it is clear that $\mathcal{C}[\Sigma^{-1}](X, Y)$ is a small set for all objects $X, Y$. The following criterion generalizes this simple observation.

**Lemma 3.3.1.** *Let $\mathcal{C}$ be a category and $\Sigma$ a set of morphisms in $\mathcal{C}$ which admits a calculus of left fractions. Let $Y$ be an object in $\mathcal{C}$ and suppose that there exists a small set $S = S(Y, \Sigma)$ of objects in $\mathcal{C}$ such that for every morphism $\sigma: Y \to Y'$ in $\Sigma$ there is a morphism $\tau: Y' \to Y''$ with $\tau \circ \sigma$ in $\Sigma$ and $Y''$ in $S$. Then $\mathcal{C}[\Sigma^{-1}](X, Y)$ is a small set for every object $X$ in $\mathcal{C}$.*

*Proof.* The condition on $Y$ implies that every fraction $X \xrightarrow{\alpha} Y' \xleftarrow{\sigma} Y$ is equivalent to one of the form $X \xrightarrow{\alpha'} Y'' \xleftarrow{\sigma'} Y$ with $Y''$ in $S$. Clearly, the fractions of the form $(\alpha', \sigma')$ with $\sigma' \in C(Y, Y'')$ and $Y'' \in S$ form a small set. $\qquad\square$

## 3.4. Calculus of fractions for subcategories

We provide a criterion such that the calculus of fractions for a set of morphisms in a category $C$ is compatible with the passage to a subcategory of $C$.

**Lemma 3.4.1.** *Let $C$ be a category and $\Sigma$ a set of morphisms admitting a calculus of left fractions. Suppose $D$ is a full subcategory of $C$ such that for every morphism $\sigma \colon Y \to Y'$ in $\Sigma$ with $Y$ in $D$ there is a morphism $\tau \colon Y' \to Y''$ with $\tau \circ \sigma$ in $\Sigma \cap D$. Then $\Sigma \cap D$ admits a calculus of left fractions and the induced functor $D[(\Sigma \cap D)^{-1}] \to C[\Sigma^{-1}]$ is fully faithful.*

*Proof.* It is straightforward to check (LF1) – (LF3) for $\Sigma \cap D$. Now let $X, Y$ be objects in $D$. Then we need to show that the induced map

$$f \colon D[(\Sigma \cap D)^{-1}](X, Y) \longrightarrow C[\Sigma^{-1}](X, Y)$$

is bijective. The map sends the equivalence class of a fraction to the equivalence class of the same fraction. If $[\alpha, \sigma]$ belongs to $C[\Sigma^{-1}](X, Y)$ and $\tau$ is a morphism with $\tau \circ \sigma$ in $\Sigma \cap D$, then $[\tau \circ \alpha, \tau \circ \sigma]$ belongs to $D[(\Sigma \cap D)^{-1}](X, Y)$ and $f$ sends it to $[\alpha, \sigma]$. Thus $f$ is surjective. A similar argument shows that $f$ is injective. $\qquad\square$

**Example 3.4.2.** Let $A$ be a commutative noetherian ring and $S \subseteq A$ a multiplicatively closed subset. Denote by $\Sigma$ the set of morphisms $\sigma$ in Mod $A$ such that $S^{-1}\sigma$ is invertible. Then $\Sigma$ is a multiplicative system and one can show directly that for the subcategory mod $A$ of finitely generated $A$-modules and $T = \Sigma \cap \bmod A$ the dual of the condition in Lemma 3.4.1 holds. Thus the induced functor

$$(\bmod A)[T^{-1}] \longrightarrow (\mathrm{Mod}\, A)[\Sigma^{-1}]$$

is fully faithful.

## 3.5. Calculus of fractions and coproducts

We provide a criterion for the quotient functor $C \to C[\Sigma^{-1}]$ to preserve small coproducts.

**Proposition 3.5.1.** *Let $C$ be a category which admits small coproducts. Suppose that $\Sigma$ is a set of morphisms in $C$ which admits a calculus of left fractions. If $\coprod_i \sigma_i$ belongs to $\Sigma$ for every family $(\sigma_i)_{i\in I}$ in $\Sigma$, then the category $C[\Sigma^{-1}]$ admits small coproducts and the quotient functor $C \to C[\Sigma^{-1}]$ preserves small coproducts.*

*Proof.* Let $(X_i)_{i\in I}$ be a family of objects in $C[\Sigma^{-1}]$ which is indexed by a small set $I$. We claim that the coproduct $\coprod_i X_i$ in $C$ is also a coproduct in $C[\Sigma^{-1}]$. Thus we need to show that for every object $Y$, the canonical map

$$C[\Sigma^{-1}](\coprod_i X_i, Y) \longrightarrow \prod_i C[\Sigma^{-1}](X_i, Y) \tag{3.5.1}$$

is bijective.

To check surjectivity of (3.5.1), let $(X_i \xrightarrow{\alpha_i} Z_i \xleftarrow{\sigma_i} Y)_{i\in I}$ be a family of left fractions. Using (LF2), we obtain a commutative diagram

$$\begin{array}{ccccc}
\coprod_i X_i & \xrightarrow{\coprod_i \alpha_i} & \coprod_i Z_i & \xleftarrow{\coprod_i \sigma_i} & \coprod_i Y \\
& & \downarrow & & \downarrow{\scriptstyle \pi_Y} \\
& & Z & \xleftarrow{\ \sigma\ } & Y
\end{array}$$

where $\pi_Y : \coprod_i Y \to Y$ is the summation morphism and $\sigma \in \Sigma$. It is easily checked that

$$(X_i \to Z \xleftarrow{\sigma} Y) \sim (X_i \xrightarrow{\alpha_i} Z_i \xleftarrow{\sigma_i} Y)$$

for all $i \in I$, and therefore (3.5.1) sends $\coprod_i X_i \to Z \xleftarrow{\sigma} Y$ to the family $(X_i \xrightarrow{\alpha_i} Z_i \xleftarrow{\sigma_i} Y_i)_{i\in I}$.

To check injectivity of (3.5.1), let $\coprod_i X_i \xrightarrow{\alpha'} Z' \xleftarrow{\sigma'} Y$ and $\coprod_i X_i \xrightarrow{\alpha''} Z'' \xleftarrow{\sigma''} Y$ be left fraction such that

$$(X_i \xrightarrow{\alpha'_i} Z' \xleftarrow{\sigma'} Y) \sim (X_i \xrightarrow{\alpha''_i} Z'' \xleftarrow{\sigma''} Y)$$

for all $i$. We may assume that $Z' = Z = Z''$ and $\sigma' = \sigma = \sigma''$ since we can choose morphisms $\tau' \colon Z' \to Z$ and $\tau'' \colon Z'' \to Z$ with $\tau' \circ \sigma' = \tau'' \circ \sigma'' \in \Sigma$. Thus there are morphisms $\beta_i \colon Z \to Z_i$ with $\beta_i \circ \alpha'_i = \beta_i \circ \alpha''_i$ and $\beta_i \circ \sigma \in \Sigma$ for all $i$. Each $\beta_i$ belongs to the *saturation* $\bar\Sigma$ of $\Sigma$ which is the set of all morphisms in $C$ which become invertible in $C[\Sigma^{-1}]$. Note that a morphism $\phi$ in $C$ belongs to $\bar\Sigma$ if and only if there are morphisms $\phi'$ and $\phi''$ such that $\phi \circ \phi'$ and $\phi'' \circ \phi$ belong to $\Sigma$. Therefore $\bar\Sigma$ is also closed under taking coproducts. Moreover, $\bar\Sigma$ admits a calculus of left fractions, and we obtain therefore a commutative diagram

$$\begin{array}{ccc}
\coprod_i X_i & \longrightarrow \coprod_i Z \xrightarrow{\ \pi_Z\ } & Z \\
& \downarrow{\coprod_i \beta_i} & \downarrow{\tau} \\
& \coprod_i Z_i \longrightarrow & Z^*
\end{array}$$

with $\tau \in \bar{\Sigma}$. Thus $\tau \circ \sigma \in \bar{\Sigma}$, and we have

$$\Big(\coprod_i X_i \xrightarrow{\alpha'} Z \xleftarrow{\sigma} Y\Big) \sim \Big(\coprod_i X_i \xrightarrow{\alpha''} Z \xleftarrow{\sigma} Y\Big)$$

since $\pi_Z \circ \coprod_i \alpha_i' = \alpha'$ and $\pi_Z \circ \coprod_i \alpha_i'' = \alpha'$. Therefore the map (3.5.1) is also injective, and this completes the proof. $\qquad\square$

**Example 3.5.2.** Let $C$ be a category which admits small coproducts and $L\colon C \to C$ be a localization functor. Then a morphism $\sigma$ in $C$ belongs to $\Sigma = \Sigma(L)$ if and only if the induced map $C(\sigma, LX)$ is invertible for every object $X$ in $C$. Thus $\Sigma$ is closed under taking small coproducts and therefore the quotient functor $C \to C[\Sigma^{-1}]$ preserves small coproducts.

### 3.6. Notes

The calculus of fractions for categories has been developed by Gabriel and Zisman in [18] as a tool for homotopy theory.

# 4. Localization for triangulated categories

## 4.1. Triangulated categories

Let $T$ be an additive category with an equivalence $S\colon T \to T$. A *triangle* in $T$ is a sequence $(\alpha, \beta, \gamma)$ of morphisms

$$X \xrightarrow{\ \alpha\ } Y \xrightarrow{\ \beta\ } Z \xrightarrow{\ \gamma\ } SX,$$

and a morphism between two triangles $(\alpha, \beta, \gamma)$ and $(\alpha', \beta', \gamma')$ is a triple $(\phi_1, \phi_2, \phi_3)$ of morphisms in $T$ making the following diagram commutative.

$$\begin{array}{ccccccc}
X & \xrightarrow{\ \alpha\ } & Y & \xrightarrow{\ \beta\ } & Z & \xrightarrow{\ \gamma\ } & SX \\
\downarrow{\phi_1} & & \downarrow{\phi_2} & & \downarrow{\phi_3} & & \downarrow{S\phi_1} \\
X' & \xrightarrow{\ \alpha'\ } & Y' & \xrightarrow{\ \beta'\ } & Z' & \xrightarrow{\ \gamma'\ } & SX'
\end{array}$$

The category $\mathcal{T}$ is called *triangulated* if it is equipped with a set of distinguished triangles (called *exact triangles*) satisfying the following conditions.

(TR1) A triangle isomorphic to an exact triangle is exact. For each object $X$, the triangle $0 \to X \xrightarrow{\text{id}} X \to 0$ is exact. Each morphism $\alpha$ fits into an exact triangle $(\alpha, \beta, \gamma)$.

(TR2) A triangle $(\alpha, \beta, \gamma)$ is exact if and only if $(\beta, \gamma, -S\alpha)$ is exact.

(TR3) Given two exact triangles $(\alpha, \beta, \gamma)$ and $(\alpha', \beta', \gamma')$, each pair of morphisms $\phi_1$ and $\phi_2$ satisfying $\phi_2 \circ \alpha = \alpha' \circ \phi_1$ can be completed to a morphism

$$
\begin{array}{ccccccc}
X & \xrightarrow{\alpha} & Y & \xrightarrow{\beta} & Z & \xrightarrow{\gamma} & SX \\
\downarrow{\scriptstyle\phi_1} & & \downarrow{\scriptstyle\phi_2} & & \downarrow{\scriptstyle\phi_3} & & \downarrow{\scriptstyle S\phi_1} \\
X' & \xrightarrow{\alpha'} & Y' & \xrightarrow{\beta'} & Z' & \xrightarrow{\gamma'} & SX'
\end{array}
$$

of triangles.

(TR4) Given exact triangles $(\alpha_1, \alpha_2, \alpha_3)$, $(\beta_1, \beta_2, \beta_3)$, and $(\gamma_1, \gamma_2, \gamma_3)$ with $\gamma_1 = \beta_1 \circ \alpha_1$, there exists an exact triangle $(\delta_1, \delta_2, \delta_3)$ making the following diagram commutative.

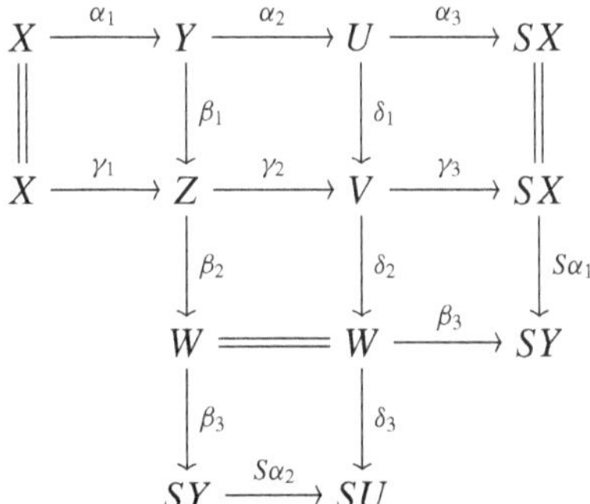

Recall that an idempotent endomorphism $\phi = \phi^2$ of an object $X$ in an additive category *splits* if there exists a factorization $X \xrightarrow{\pi} Y \xrightarrow{\iota} X$ of $\phi$ with $\pi \circ \iota = \text{id}\, Y$.

**Remark 4.1.1.** Suppose a triangulated category $\mathcal{T}$ admits countable coproducts. Then every idempotent endomorphism splits. More precisely, let $\phi \colon X \to X$ be an idempotent morphism in $\mathcal{T}$, and denote by $Y$ a homotopy colimit of the sequence

$$
X \xrightarrow{\phi} X \xrightarrow{\phi} X \xrightarrow{\phi} \cdots .
$$

The morphism $\phi$ factors through the canonical morphism $\pi\colon X \to Y$ via a morphism $\iota\colon Y \to X$, and we have $\pi \circ \iota = \mathrm{id}\, Y$. Thus $\phi$ splits; see [33, Proposition 1.6.8] for details.

## 4.2. Exact functors

An *exact* functor $\mathcal{T} \to \mathcal{U}$ between triangulated categories is a pair $(F, \mu)$ consisting of a functor $F\colon \mathcal{T} \to \mathcal{U}$ and an isomorphism $\mu\colon F \circ S_{\mathcal{T}} \to S_{\mathcal{U}} \circ F$ such that for every exact triangle $X \xrightarrow{\alpha} Y \xrightarrow{\beta} Z \xrightarrow{\gamma} S_{\mathcal{T}} X$ in $\mathcal{T}$ the triangle

$$FX \xrightarrow{F\alpha} FY \xrightarrow{F\beta} FZ \xrightarrow{\mu X \circ F\gamma} S_{\mathcal{U}}(FX)$$

is exact in $\mathcal{U}$.

We have the following useful lemma.

**Lemma 4.2.1.** *Let $F\colon \mathcal{T} \to \mathcal{U}$ and $G\colon \mathcal{U} \to \mathcal{T}$ be an adjoint pair of functors between triangulated categories. If one of both functors is exact, then also the other is exact.*

*Proof.* See [33, Lemma 5.3.6]. $\qquad\qquad\square$

## 4.3. Multiplicative systems

Let $\mathcal{T}$ be a triangulated category and $\Sigma$ a set of morphisms which is a multiplicative system. Recall this means that $\Sigma$ admits a calculus of left and right fractions. Then we say that $\Sigma$ is *compatible with the triangulation* if

(1) given $\sigma$ in $\Sigma$, the morphism $S^n\sigma$ belongs to $\Sigma$ for all $n \in \mathbb{Z}$, and
(2) given a morphism $(\phi_1, \phi_2, \phi_3)$ between exact triangles with $\phi_1$ and $\phi_2$ in $\Sigma$, there is also a morphism $(\phi_1, \phi_2, \phi_3')$ with $\phi_3'$ in $\Sigma$.

**Lemma 4.3.1.** *Let $\mathcal{T}$ be a triangulated category and $\Sigma$ a multiplicative system of morphisms which is compatible with the triangulation. Then the quotient category $\mathcal{T}[\Sigma^{-1}]$ carries a unique triangulated structure such that the quotient functor $\mathcal{T} \to \mathcal{T}[\Sigma^{-1}]$ is exact.*

*Proof.* The equivalence $S\colon \mathcal{T} \to \mathcal{T}$ induces a unique equivalence $\mathcal{T}[\Sigma^{-1}] \to \mathcal{T}[\Sigma^{-1}]$ which commutes with the quotient functor $Q\colon \mathcal{T} \to \mathcal{T}[\Sigma^{-1}]$. This follows from the fact that $S\Sigma = \Sigma$. Now take as exact triangles in $\mathcal{T}[\Sigma^{-1}]$ all those isomorphic to images of exact triangles in $\mathcal{T}$. It is straightforward to verify the axioms (TR1) – (TR4); see [48, II.2.2.6]. The functor $Q$ is exact by construction. In particular, we have $Q \circ S_{\mathcal{T}} = S_{\mathcal{T}[\Sigma^{-1}]} \circ Q$. $\qquad\square$

## 4.4. Cohomological functors

A functor $H\colon \mathcal{T} \to \mathcal{A}$ from a triangulated category $\mathcal{T}$ to an abelian category $\mathcal{A}$ is *cohomological* if $H$ sends every exact triangle in $\mathcal{T}$ to an exact sequence in $\mathcal{A}$.

**Example 4.4.1.** For each object $X$ in $\mathcal{T}$, the representable functors $\mathcal{T}(X, -)\colon \mathcal{T} \to \mathrm{Ab}$ and $\mathcal{T}(-, X)\colon \mathcal{T}^{\mathrm{op}} \to \mathrm{Ab}$ into the category $\mathrm{Ab}$ of abelian groups are cohomological functors.

**Lemma 4.4.2.** *Let* $H\colon \mathcal{T} \to \mathcal{A}$ *be a cohomological functor. Then the set* $\Sigma$ *of morphisms* $\sigma$ *in* $\mathcal{T}$ *such that* $H(S^n \sigma)$ *is invertible for all* $n \in \mathbb{Z}$ *forms a multiplicative system which is compatible with the triangulation of* $\mathcal{T}$.

*Proof.* We need to verify that $\Sigma$ admits a calculus of left and right fractions. In fact, it is sufficient to check conditions (LF1) – (LF3), because then the dual conditions are satisfied as well since the definition of $\Sigma$ is self-dual.

(LF1): This condition is clear because $H$ is a functor.

(LF2): Let $\alpha\colon X \to Y$ and $\sigma\colon X \to X'$ be morphisms with $\sigma$ in $\Sigma$. We complete $\alpha$ to an exact triangle and apply (TR3) to obtain the following morphism between exact triangles.

$$
\begin{array}{ccccccc}
W & \longrightarrow & X & \xrightarrow{\ \alpha\ } & Y & \longrightarrow & SW \\
\| & & \downarrow{\scriptstyle \sigma} & & \downarrow{\scriptstyle \sigma'} & & \| \\
W & \longrightarrow & X' & \xrightarrow{\ \alpha'\ } & Y' & \longrightarrow & SW
\end{array}
$$

Then the 5-lemma shows that $\sigma'$ belongs to $\Sigma$.

(LF3): Let $\alpha, \beta\colon X \to Y$ be morphisms in $\mathcal{T}$ and $\sigma\colon X' \to X$ in $\Sigma$ such that $\alpha \circ \sigma = \beta \circ \sigma$. Complete $\sigma$ to an exact triangle $X' \xrightarrow{\sigma} X \xrightarrow{\phi} X'' \to SX'$. Then $\alpha - \beta$ factors through $\phi$ via some morphism $\psi\colon X'' \to Y$. Now complete $\psi$ to an exact triangle $X'' \xrightarrow{\psi} Y \xrightarrow{\tau} Y' \to SX''$. Then $\tau$ belongs to $\Sigma$ and $\tau \circ \alpha = \tau \circ \beta$.

It remains to check that $\Sigma$ is compatible with the triangulation. Condition (1) is clear from the definition of $\Sigma$. For condition (2), observe that given any morphism $(\phi_1, \phi_2, \phi_3)$ between exact triangles with $\phi_1$ and $\phi_2$ in $\Sigma$, we have that $\phi_3$ belongs to $\Sigma$. This is an immediate consequence of the 5-lemma. $\qquad\square$

## 4.5. Triangulated and thick subcategories

Let $\mathcal{T}$ be a triangulated category. A non-empty full subcategory $\mathcal{S}$ is a *triangulated subcategory* if the following conditions hold.

(TS1) $S^n X \in \mathcal{S}$ for all $X \in \mathcal{S}$ and $n \in \mathbb{Z}$.

(TS2) Let $X \to Y \to Z \to SX$ be an exact triangle in $\mathcal{T}$. If two objects from $\{X, Y, Z\}$ belong to $\mathcal{S}$, then also the third.

A triangulated subcategory $\mathcal{S}$ is *thick* if in addition the following condition holds.

(TS3) Let $X \xrightarrow{\pi} Y \xrightarrow{\iota} X$ be morphisms in $\mathcal{T}$ such that $\mathrm{id}\, Y = \pi \circ \iota$. If $X$ belongs to $\mathcal{S}$, then also $Y$.

Note that a triangulated subcategory $\mathcal{S}$ of $\mathcal{T}$ inherits a canonical triangulated structure from $\mathcal{T}$.

Next observe that a triangulated subcategory $\mathcal{S}$ of $\mathcal{T}$ is thick provided that $\mathcal{S}$ admits countable coproducts. This follows from the fact that in a triangulated category with countable coproducts all idempotent endomorphisms split.

Let $\mathcal{T}$ be a triangulated category and let $F\colon \mathcal{T} \to \mathcal{U}$ be an additive functor. The *kernel* $\mathrm{Ker}\, F$ of $F$ is by definition the full subcategory of $\mathcal{T}$ which is formed by all objects $X$ such that $FX = 0$. If $F$ is an exact functor into a triangulated category, then $\mathrm{Ker}\, F$ is a thick subcategory of $\mathcal{T}$. Also, if $F$ is a cohomological functor into an abelian category, then $\bigcap_{n \in \mathbb{Z}} S^n(\mathrm{Ker}\, F)$ is a thick subcategory of $\mathcal{T}$.

## 4.6. Verdier localization

Let $\mathcal{T}$ be a triangulated category. Given a triangulated subcategory $\mathcal{S}$, we denote by $\Sigma(\mathcal{S})$ the set of morphisms $X \to Y$ in $\mathcal{T}$ which fit into an exact triangle $X \to Y \to Z \to SX$ with $Z$ in $\mathcal{S}$.

**Lemma 4.6.1.** *Let $\mathcal{T}$ be a triangulated category and $\mathcal{S}$ a triangulated subcategory. Then $\Sigma(\mathcal{S})$ is a multiplicative system which is compatible with the triangulation of $\mathcal{T}$.*

*Proof.* The proof is similar to that of Lemma 4.4.2; see [48, II.2.1.8] for details. $\qquad\square$

The *localization* of $\mathcal{T}$ with respect to a triangulated subcategory $\mathcal{S}$ is by definition the quotient category

$$\mathcal{T}/\mathcal{S} := \mathcal{T}[\Sigma(\mathcal{S})^{-1}]$$

together with the quotient functor $\mathcal{T} \to \mathcal{T}/\mathcal{S}$.

**Proposition 4.6.2.** *Let $\mathcal{T}$ be a triangulated category and $\mathcal{S}$ a full triangulated subcategory. Then the category $\mathcal{T}/\mathcal{S}$ and the quotient functor $Q\colon \mathcal{T} \to \mathcal{T}/\mathcal{S}$ have the following properties.*

(1) *The category $\mathcal{T}/\mathcal{S}$ carries a unique triangulated structure such that $Q$ is exact.*
(2) *A morphism in $\mathcal{T}$ is annihilated by $Q$ if and only if it factors through an object in $\mathcal{S}$.*
(3) *The kernel $\operatorname{Ker} Q$ is the smallest thick subcategory containing $\mathcal{S}$.*
(4) *Every exact functor $\mathcal{T} \to \mathcal{U}$ annihilating $\mathcal{S}$ factors uniquely through $Q$ via an exact functor $\mathcal{T}/\mathcal{S} \to \mathcal{U}$.*
(5) *Every cohomological functor $\mathcal{T} \to \mathcal{A}$ annihilating $\mathcal{S}$ factors uniquely through $Q$ via a cohomological functor $\mathcal{T}/\mathcal{S} \to \mathcal{A}$.*

*Proof.* (1) follows from Lemma 4.3.1.

(2) Let $\phi$ be a morphism in $\mathcal{T}$. We have $Q\phi = 0$ iff $\sigma \circ \phi = 0$ for some $\sigma \in \Sigma(\mathcal{S})$ iff $\phi$ factors through some object in $\mathcal{S}$.

(3) Let $X$ be an object in $\mathcal{T}$. Then $QX = 0$ if and only if $Q(\operatorname{id} X) = 0$. Thus part (2) implies that the kernel of $Q$ conists of all direct factors of objects in $\mathcal{S}$.

(4) An exact functor $F\colon \mathcal{T} \to \mathcal{U}$ annihilating $\mathcal{S}$ inverts every morphism in $\Sigma(\mathcal{S})$. Thus there exists a unique functor $\bar{F}\colon \mathcal{T}/\mathcal{S} \to \mathcal{U}$ such that $F = \bar{F} \circ Q$. The functor $\bar{F}$ is exact because an exact triangle $\Delta$ in $\mathcal{T}/\mathcal{S}$ is up to isomorphism of the form $Q\Delta'$ for some exact triangle $\Delta'$ in $\mathcal{T}$. Thus $\bar{F}\Delta \cong F\Delta'$ is exact.

(5) Analogous to (4). $\qquad\square$

## 4.7. Localization of subcategories

Let $\mathcal{T}$ be a triangulated category with two full triangulated subcategories $\mathcal{T}'$ and $\mathcal{S}$. Then we put $\mathcal{S}' = \mathcal{S} \cap \mathcal{T}'$ and have $\Sigma_{\mathcal{T}'}(\mathcal{S}') = \Sigma_{\mathcal{T}}(\mathcal{S}) \cap \mathcal{T}'$. Thus we can form the following commutative diagram of exact functors

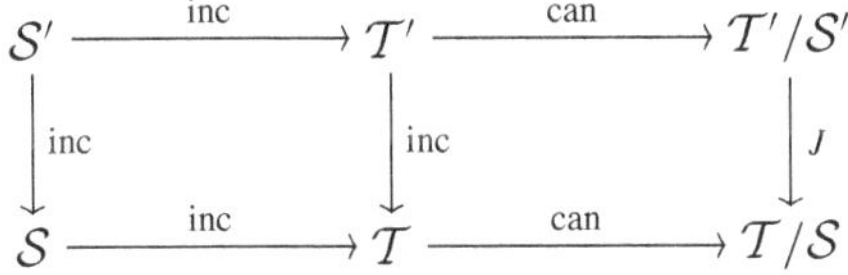

and ask when the functor $J$ is fully faithful. We have the following criterion.

**Lemma 4.7.1.** *Let $\mathcal{T}$, $\mathcal{T}'$, $\mathcal{S}$, $\mathcal{S}'$ be as above. Suppose that either*

(1) *every morphism from an object in $\mathcal{S}$ to an object in $\mathcal{T}'$ factors through some object in $\mathcal{S}'$, or*

(2) *every morphism from an object in $\mathcal{T}'$ to an object in $\mathcal{S}$ factors through some object in $\mathcal{S}'$.*

*Then the induced functor $J : \mathcal{T}'/\mathcal{S}' \to \mathcal{T}/\mathcal{S}$ is fully faithful.*

*Proof.* Suppose that condition (1) holds. We apply the criterion from Lemma 3.4.1. Thus we take a morphism $\sigma : Y \to Y'$ from $\Sigma(\mathcal{S})$ with $Y$ in $\mathcal{T}'$ and need to find $\tau : Y' \to Y''$ such that $\tau \circ \sigma$ belongs to $\Sigma(\mathcal{S}) \cap \mathcal{T}'$. To this end complete $\sigma$ to an exact triangle $X \xrightarrow{\phi} Y \xrightarrow{\sigma} Y' \to SX$. Then $X$ belongs to $\mathcal{S}$ and by our assumption we have a factorization $X \xrightarrow{\phi'} Z \xrightarrow{\phi''} Y$ of $\phi$ with $Z$ in $\mathcal{S}'$. Complete $\phi''$ to an exact triangle $Z \xrightarrow{\phi''} Y \xrightarrow{\psi} Y'' \to SZ$. Then (TR3) yields a morphism $\tau : Y' \to Y''$ satisfying $\psi = \tau \circ \sigma$. In particular, $\tau \circ \sigma$ lies in $\Sigma(\mathcal{S}) \cap \mathcal{T}'$ since $Z$ belongs to $\mathcal{S}'$. The proof using condition (2) is dual. $\quad\square$

## 4.8. Orthogonal subcategories

Let $\mathcal{T}$ be a triangulated category and $\mathcal{S}$ a triangulated subcategory. Then we define two full subcategories

$$\mathcal{S}^{\perp} = \{Y \in \mathcal{T} \mid \mathcal{T}(X, Y) = 0 \text{ for all } X \in \mathcal{S}\}$$
$$^{\perp}\mathcal{S} = \{X \in \mathcal{T} \mid \mathcal{T}(X, Y) = 0 \text{ for all } Y \in \mathcal{S}\}$$

and call them *orthogonal subcategories* with respect to $\mathcal{S}$. Note that $\mathcal{S}^{\perp}$ and $^{\perp}\mathcal{S}$ are thick subcategories of $\mathcal{T}$.

**Lemma 4.8.1.** *Let $\mathcal{T}$ be a triangulated category and $\mathcal{S}$ a triangulated subcategory. Then the following are equivalent for an object $Y$ in $\mathcal{T}$.*

(1) *$Y$ belongs to $\mathcal{S}^{\perp}$.*
(2) *$Y$ is $\Sigma(\mathcal{S})$-local, that is, $\mathcal{T}(\sigma, Y)$ is bijective for all $\sigma$ in $\Sigma(\mathcal{S})$.*
(3) *The quotient functor induces a bijection $\mathcal{T}(X, Y) \to \mathcal{T}/\mathcal{S}(X, Y)$ for all $X$ in $\mathcal{T}$.*

*Proof.* (1) $\Rightarrow$ (2): Suppose $\mathcal{T}(X, Y) = 0$ for all $X$ in $\mathcal{S}$. Then every $\sigma$ in $\Sigma(\mathcal{S})$ induces a bijection $\mathcal{T}(\sigma, Y)$ because $\mathcal{T}(-, Y)$ is cohomological. Thus $Y$ is $\Sigma(\mathcal{S})$-local.

(2) $\Rightarrow$ (1): Suppose that $Y$ is $\Sigma(\mathcal{S})$-local. If $X$ belongs to $\mathcal{S}$, then the morphism $\sigma : X \to 0$ belongs to $\Sigma(\mathcal{S})$ and induces therefore a bijection $\mathcal{C}(\sigma, Y)$. Thus $Y$ belongs to $\mathcal{S}^{\perp}$.

(2) $\Leftrightarrow$ (3): Apply Lemma 3.2.2. $\quad\square$

### 4.9. Bousfield localization

Let $\mathcal{T}$ be a triangulated category. We wish to study exact localization functors $L\colon \mathcal{T} \to \mathcal{T}$. To be more precise, we assume that $L$ is an exact functor and that $L$ is a localization functor in the sense that there exists a morphism $\eta\colon \operatorname{Id}\mathcal{C} \to L$ with $L\eta\colon L \to L^2$ being invertible and $L\eta = \eta L$. Note that there is an isomorphism $\mu\colon L \circ S \xrightarrow{\sim} S \circ L$ since $L$ is exact, and there exists a unique choice such that $\mu X \circ \eta S X = S\eta X$ for all $X$ in $\mathcal{T}$. This follows from Lemma 2.5.1.

We observe that the kernel of an exact localization functor is a thick subcategory of $\mathcal{T}$. The following fundamental result characterizes the thick subcategories of $\mathcal{T}$ which are of this form.

**Proposition 4.9.1.** *Let $\mathcal{T}$ be a triangulated category and $\mathcal{S}$ a thick subcategory. Then the following are equivalent.*

(1) *There exists an exact localization functor $L\colon \mathcal{T} \to \mathcal{T}$ with $\operatorname{Ker} L = \mathcal{S}$.*
(2) *The inclusion functor $\mathcal{S} \to \mathcal{T}$ admits a right adjoint.*
(3) *For each $X$ in $\mathcal{T}$ there exists an exact triangle $X' \to X \to X'' \to SX''$ with $X'$ in $\mathcal{S}$ and $X''$ in $\mathcal{S}^{\perp}$.*
(4) *The quotient functor $\mathcal{T} \to \mathcal{T}/\mathcal{S}$ admits a right adjoint.*
(5) *The composite $\mathcal{S}^{\perp} \xrightarrow{\mathrm{inc}} \mathcal{T} \xrightarrow{\mathrm{can}} \mathcal{T}/\mathcal{S}$ is an equivalence.*
(6) *The inclusion functor $\mathcal{S}^{\perp} \to \mathcal{T}$ admits a left adjoint and $^{\perp}(\mathcal{S}^{\perp}) = \mathcal{S}$.*

*Proof.* Let $I\colon \mathcal{S} \to \mathcal{T}$ and $J\colon \mathcal{S}^{\perp} \to \mathcal{T}$ denote the inclusions and $Q\colon \mathcal{T} \to \mathcal{T}/\mathcal{S}$ the quotient functor.

(1) $\Rightarrow$ (2): Suppose that $L\colon \mathcal{T} \to \mathcal{T}$ is an exact localization functor with $\operatorname{Ker} L = \mathcal{S}$ and let $\eta\colon \operatorname{Id}\mathcal{T} \to L$ be a morphism such that $L\eta$ is invertible. We obtain a right adjoint $I_\rho\colon \mathcal{T} \to \mathcal{S}$ for the inclusion $I$ by completing for each $X$ in $\mathcal{T}$ the morphism $\eta X$ to an exact triangle $I_\rho X \xrightarrow{\theta X} X \xrightarrow{\eta X} LX \to S(I_\rho X)$. Note that $I_\rho X$ belongs to $\mathcal{S}$ since $L\eta X$ is invertible. Moreover, $\mathcal{T}(W, \theta X)$ is bijective for all $W$ in $\mathcal{S}$ since $\mathcal{T}(W, LX) = 0$ by Lemma 4.8.1. Here we use that $LX$ is $\Sigma(L)$-local by Proposition 2.5.2 and that $\Sigma(L) = \Sigma(\mathcal{S})$. Thus $I_\rho$ provides a right adjoint for $I$ since $\mathcal{T}(W, I_\rho X) \cong \mathcal{T}(IW, X)$ for all $W$ in $\mathcal{S}$ and $X$ in $\mathcal{T}$. In particular, we see that the exact triangle defining $I_\rho X$ is, up to a unique isomorphism, uniquely determined by $X$. Therefore $I_\rho$ is well defined.

(2) $\Rightarrow$ (3): Suppose that $I_\rho\colon \mathcal{T} \to \mathcal{S}$ is a right adjoint of the inclusion $I$. We fix an object $X$ in $\mathcal{T}$ and complete the adjunction morphism $\theta X\colon I_\rho X \to X$ to an exact triangle $I_\rho X \xrightarrow{\theta X} X \to X'' \to S(I_\rho X)$. Clearly, $I_\rho X$ belongs to $\mathcal{S}$. We have $\mathcal{T}(W, X'') = 0$ for all $W$ in $\mathcal{S}$ since $\mathcal{T}(W, \theta X)$ is bijective. Thus $X''$ belongs to $\mathcal{S}^{\perp}$.

                    *Henning Krause*

(3) $\Rightarrow$ (4): We apply Proposition 3.2.3 to obtain a right adjoint for the quotient functor $Q$. To this end fix an object $X$ in $\mathcal{T}$ and an exact triangle $X' \to X \xrightarrow{\eta} X'' \to SX''$ with $X'$ in $\mathcal{S}$ and $X''$ in $\mathcal{S}^\perp$. The morphism $\eta$ belongs to $\Sigma(\mathcal{S})$ by definition, and the object $X''$ is $\Sigma(\mathcal{S})$-local by Lemma 4.8.1. Now it follows from Proposition 3.2.3 that $Q$ admits a right adjoint.

(4) $\Rightarrow$ (1): Let $Q_\rho \colon \mathcal{T}/\mathcal{S} \to \mathcal{T}$ denote a right adjoint of $Q$. This functor is fully faithful by Proposition 2.3.1 and exact by Lemma 4.2.1. Thus $L = Q_\rho \circ Q$ is an exact functor with $\operatorname{Ker} L = \operatorname{Ker} Q = \mathcal{S}$. Moreover, $L$ is a localization functor by Corollary 2.4.2.

(4) $\Rightarrow$ (5): Let $Q_\rho \colon \mathcal{T}/\mathcal{S} \to \mathcal{T}$ denote a right adjoint of $Q$. The composite $Q \circ J \colon \mathcal{S}^\perp \to \mathcal{T}/\mathcal{S}$ is fully faithful by Lemma 4.8.1. Given an object $X$ in $\mathcal{T}/\mathcal{S}$, we have $Q(Q_\rho X) \cong X$ by Proposition 2.3.1, and $Q_\rho X$ belongs to $\mathcal{S}^\perp$, since $\mathcal{T}(W, Q_\rho X) \cong \mathcal{T}/\mathcal{S}(QW, X) = 0$ for all $W$ in $\mathcal{S}$. Thus $Q \circ J$ is dense and therefore an equivalence.

(5) $\Rightarrow$ (6): Suppose $Q \circ J \colon \mathcal{S}^\perp \to \mathcal{T}/\mathcal{S}$ is an equivalence and let $F \colon \mathcal{T}/\mathcal{S} \to \mathcal{S}^\perp$ be a quasi-inverse. We have for all $X$ in $\mathcal{T}$ and $Y$ in $\mathcal{S}^\perp$

$$\mathcal{T}(X, JY) \xrightarrow{\sim} \mathcal{T}/\mathcal{S}(QX, QJY) \xrightarrow{\sim} \mathcal{S}^\perp(FQX, FQJY) \xrightarrow{\sim} \mathcal{S}^\perp(FQX, Y),$$

where the first bijection follows from Lemma 4.8.1 and the others are clear from the choice of $F$. Thus $F \circ Q$ is a left adjoint for the inclusion $J$.

It remains to show that $^\perp(\mathcal{S}^\perp) = \mathcal{S}$. The inclusion $^\perp(\mathcal{S}^\perp) \supseteq \mathcal{S}$ is clear. Now let $X$ be an object of $^\perp(\mathcal{S}^\perp)$. Then we have

$$\mathcal{T}/\mathcal{S}(QX, QX) \cong \mathcal{S}^\perp(FQX, FQX) \cong \mathcal{T}(X, J(FQX)) = 0.$$

Thus $QX = 0$ and therefore $X$ belongs to $\mathcal{S}$.

(6) $\Rightarrow$ (3): Suppose that $J_\lambda \colon \mathcal{T} \to \mathcal{S}^\perp$ is a left adjoint of the inclusion $J$. We fix an object $X$ in $\mathcal{T}$ and complete the adjunction morphism $\mu X \colon X \to J_\lambda X$ to an exact triangle $X' \to X \xrightarrow{\mu X} J_\lambda X \to SX'$. Clearly, $J_\lambda X$ belongs to $\mathcal{S}^\perp$. We have $\mathcal{T}(X', Y) = 0$ for all $Y$ in $\mathcal{S}^\perp$ since $\mathcal{T}(\mu X, Y)$ is bijective. Thus $X'$ belongs to $^\perp(\mathcal{S}^\perp) = \mathcal{S}$. $\qquad\square$

The following diagram displays the functors which arise from a localization functor $L \colon \mathcal{T} \to \mathcal{T}$. We use the convention that $F_\rho$ denotes a right adjoint of a functor $F$.

$$\mathcal{S} \underset{I=\mathrm{inc}}{\overset{I_\rho}{\rightleftarrows}} \mathcal{T} \underset{Q=\mathrm{can}}{\overset{Q_\rho}{\rightleftarrows}} \mathcal{T}/\mathcal{S} \qquad (L = Q_\rho \circ Q \quad \text{and} \quad \Gamma = I \circ I_\rho)$$

## 4.10. Acyclic and local objects

Let $\mathcal{T}$ be a triangulated category and $L : \mathcal{T} \to \mathcal{T}$ an exact localization functor. An object $X$ in $\mathcal{T}$ is by definition *L-acyclic* if $LX = 0$. Recall that an object in $\mathcal{T}$ is *L*-local if and only if it belongs to the essential image $\operatorname{Im} L$ of $L$; see Proposition 2.5.2. The exactness of $L$ implies that $\mathcal{S} := \operatorname{Ker} L$ is a thick subcategory and that $\Sigma(L) = \Sigma(\mathcal{S})$. Therefore *L*-local and $\Sigma(\mathcal{S})$-local objects coincide.

The following result says that acyclic and local objects form an orthogonal pair.

**Proposition 4.10.1.** *Let* $L : \mathcal{T} \to \mathcal{T}$ *be an exact localization functor. Then we have*

$$\operatorname{Ker} L = {}^{\perp}(\operatorname{Im} L) \quad and \quad (\operatorname{Ker} L)^{\perp} = \operatorname{Im} L.$$

*More explictly, the following holds.*

(1) $X \in \mathcal{T}$ *is L-acyclic if and only if* $\mathcal{T}(X, Y) = 0$ *for every L-local object Y.*
(2) $Y \in \mathcal{T}$ *is L-local if and only if* $\mathcal{T}(X, Y) = 0$ *for every L-acyclic object X.*

*Proof.* (1) We write $L = G \circ F$ where $F$ is a functor and $G$ a fully faithful right adjoint; see Corollary 2.4.2. Suppose first we have given objects $X$, $Y$ such that $X$ is *L*-acyclic and $Y$ is *L*-local. Observe that $FX = 0$ since $G$ is faithful. Thus

$$\mathcal{T}(X, Y) \cong \mathcal{T}(X, GFY) \cong \mathcal{T}(FX, FY) = 0.$$

Now suppose that $X$ is an object with $\mathcal{T}(X, Y) = 0$ for all *L*-local $Y$. Then

$$\mathcal{T}(FX, FX) \cong \mathcal{T}(X, GFX) = 0$$

and therefore $FX = 0$. Thus $X$ is *L*-acyclic.
(2) This is a reformulation of Lemma 4.8.1. $\qquad\square$

## 4.11. A functorial triangle

Let $\mathcal{T}$ be a triangulated category and $L : \mathcal{T} \to \mathcal{T}$ an exact localization functor. We denote by $\eta : \operatorname{Id} \mathcal{T} \to L$ a morphism such that $L\eta$ is invertible. It follows from Proposition 4.9.1 and its proof that we obtain an exact functor $\Gamma : \mathcal{T} \to \mathcal{T}$ by completing for each $X$ in $\mathcal{T}$ the morphism $\eta X$ to an exact triangle

$$\Gamma X \xrightarrow{\theta X} X \xrightarrow{\eta X} LX \longrightarrow S(\Gamma X). \tag{4.11.1}$$

The exactness of $\Gamma$ follows from Lemma 4.2.1. Observe that $\Gamma X$ is $L$-acyclic and that $LX$ is $L$-local. In fact, the exact triangle (4.11.1) is essentially determined by these properties. This is a consequence of the following basic properties of $L$ and $\Gamma$.

**Proposition 4.11.1.** *The functors $L, \Gamma \colon \mathcal{T} \to \mathcal{T}$ have the following properties.*

(1) $L$ *induces an equivalence* $\mathcal{T}/\operatorname{Ker} L \xrightarrow{\sim} \operatorname{Im} L$.
(2) $L$ *induces a left adjoint for the inclusion* $\operatorname{Im} L \to \mathcal{T}$.
(3) $\Gamma$ *induces a right adjoint for the inclusion* $\operatorname{Ker} L \to \mathcal{T}$.

*Proof.* (1) is a reformulation of Corollary 2.5.3, and (2) follows from Corollary 2.4.2. (3) is an immediate consequence of the construction of $\Gamma$ via Proposition 4.9.1. $\qquad\square$

**Proposition 4.11.2.** *Let $L \colon \mathcal{T} \to \mathcal{T}$ be an exact localization functor and $X$ an object in $\mathcal{C}$. Given any exact triangle $X' \to X \to X'' \to SX'$ with $X'$ $L$-acyclic and $X''$ $L$-local, there are unique isomorphisms $\alpha$ and $\beta$ making the following diagram commutative.*

$$
\begin{array}{ccccccc}
X' & \longrightarrow & X & \longrightarrow & X'' & \longrightarrow & SX' \\
\downarrow{\scriptstyle \alpha} & & \| & & \downarrow{\scriptstyle \beta} & & \downarrow{\scriptstyle S\alpha} \\
\Gamma X & \xrightarrow{\ \theta X\ } & X & \xrightarrow{\ \eta X\ } & LX & \longrightarrow & S(\Gamma X)
\end{array}
\qquad (4.11.2)
$$

*Proof.* The morphism $\theta X$ induces a bijection $\mathcal{T}(X', \theta X)$ since $X'$ is acyclic. Thus $X' \to X$ factors uniquely through $\theta X$ via a morphism $\alpha \colon X' \to \Gamma X$. An application of (TR3) gives a morphism $\beta \colon X'' \to LX$ making the diagram (4.11.2) commutative. Now apply $L$ to this diagram. Then $L\beta$ is an isomorphism since $LX' = 0 = L\Gamma X$, and $L\beta$ is isomorphic to $\beta$ since $X''$ and $LX$ are $L$-local. Thus $\beta$ is an isomorphism, and therefore $\alpha$ is an isomorphism. $\qquad\square$

## 4.12. Localization versus colocalization

For exact functors on triangulated categories, we have the following symmetry principle relating localization and colocalization.

**Proposition 4.12.1.** *Let $\mathcal{T}$ be a triangulated category.*

(1) *Suppose $L \colon \mathcal{T} \to \mathcal{T}$ is an exact localization functor and $\Gamma \colon \mathcal{T} \to \mathcal{T}$ the functor which is defined in terms of the exact triangle (4.11.1). Then $\Gamma$ is an exact colocalization functor with $\operatorname{Ker} \Gamma = \operatorname{Im} L$ and $\operatorname{Im} \Gamma = \operatorname{Ker} L$.*

(2) *Suppose $\Gamma\colon \mathcal{T} \to \mathcal{T}$ is an exact colocalization functor and $L\colon \mathcal{T} \to \mathcal{T}$ the functor which is defined in terms of the exact triangle (4.11.1). Then $L$ is an exact localization functor with $\operatorname{Ker} L = \operatorname{Im} \Gamma$ and $\operatorname{Im} L = \operatorname{Ker} \Gamma$.*

*Proof.* It suffices to prove (1) because (2) is the dual statement. So let $L\colon \mathcal{T} \to \mathcal{T}$ be an exact localization functor. It follows from the construction of $\Gamma$ that it is of the form $\Gamma = I \circ I_\rho$ where $I_\rho$ denotes a right adjoint of the fully faithful inclusion $I\colon \operatorname{Ker} L \to \mathcal{T}$. Thus $\Gamma$ is a colocalization functor by Corollary 2.4.2. The exactness of $\Gamma$ follows from Lemma 4.2.1, and the identities $\operatorname{Ker} \Gamma = \operatorname{Im} L$ and $\operatorname{Im} \Gamma = \operatorname{Ker} L$ are easily derived from the exact triangle (4.11.1). $\qquad\square$

## 4.13. Recollements

A *recollement* is by definition a diagram of exact functors

$$\mathcal{T}' \underset{I_\lambda}{\overset{I_\rho}{\rightleftarrows}} \mathcal{T} \underset{Q_\lambda}{\overset{Q_\rho}{\rightleftarrows}} \mathcal{T}'' \tag{4.13.1}$$

satisfying the following conditions.

(1) $I_\lambda$ is a left adjoint and $I_\rho$ a right adjoint of $I$.
(2) $Q_\lambda$ is a left adjoint and $Q_\rho$ a right adjoint of $Q$.
(3) $I_\lambda I \cong \operatorname{Id} \mathcal{T}' \cong I_\rho I$ and $Q Q_\rho \cong \operatorname{Id} \mathcal{T}'' \cong Q Q_\lambda$.
(4) $\operatorname{Im} I = \operatorname{Ker} Q$.

Note that the isomorphisms in (3) are supposed to be the adjunction morphisms resulting from (1) and (2).

A recollement gives rise to various localization and colocalization functors for $\mathcal{T}$. First observe that the functors $I$, $Q_\lambda$, and $Q_\rho$ are fully faithful; see Proposition 2.3.1. Therefore $Q_\rho Q$ and $I I_\lambda$ are localization functors and $Q_\lambda Q$ and $I I_\rho$ are colocalization functors. This follows from Corollary 2.4.2. Note that the localization functor $L = Q_\rho Q$ has the additional property that the inclusion $\operatorname{Ker} L \to \mathcal{T}$ admits a left adjoint. Moreover, $L$ determines the recollement up to an equivalence.

**Proposition 4.13.1.** *Let $L\colon \mathcal{T} \to \mathcal{T}$ be an exact localization functor and suppose that the inclusion $\operatorname{Ker} L \to \mathcal{T}$ admits a left adjoint. Then $L$ induces a recollement of the following form.*

$$\operatorname{Ker} L \underset{\longleftarrow}{\overset{\overset{\longleftarrow}{\xrightarrow{\;\text{inc}\;}}}{}} \mathcal{T} \underset{\longleftarrow}{\overset{\overset{\longleftarrow}{\longrightarrow}}{}} \operatorname{Im} L$$

*Moreover, any recollement for $\mathcal{T}$ is, up to equivalences, of this form for some exact localization functor $L\colon \mathcal{T} \to \mathcal{T}$.*

*Proof.* We apply Proposition 4.9.1 and its dual assertion. Observe first that any localization functor $L: \mathcal{T} \to \mathcal{T}$ induces the following diagram.

$$\mathrm{Ker}\, L \;\underset{I=\mathrm{inc}}{\overset{I_\rho=\Gamma}{\rightleftarrows}}\; \mathcal{T} \;\underset{Q=L}{\overset{Q_\rho=\mathrm{inc}}{\rightleftarrows}}\; \mathrm{Im}\, L$$

The functor $I$ admits a left adjoint if and only if $Q$ admits a left adjoint. Thus the diagram can be completed to a recollement if and only if the inclusion $I$ admits a left adjoint.

Suppose now there is given a recollement of the form (4.13.1). Then $L = Q_\rho Q$ is a localization functor and the inclusion $\mathrm{Ker}\, L \to \mathcal{T}$ admits a left adjoint. The functor $I$ induces an equivalence $\mathcal{T}' \xrightarrow{\sim} \mathrm{Ker}\, L$ and $Q_\rho$ induces an equivalence $\mathcal{T}'' \xrightarrow{\sim} \mathrm{Im}\, L$. It is straightforward to formulate and check the various compatibilities of these equivalences. $\qquad\square$

As a final remark, let us mention that for any recollement of the form (4.13.1), the functors $Q_\lambda$ and $Q_\rho$ provide two (in general different) embeddings of $\mathcal{T}''$ into $\mathcal{T}$. If we identify $\mathcal{T}' = \mathrm{Im}\, I$, then $Q_\rho$ identifies $\mathcal{T}''$ with $(\mathcal{T}')^{\perp}$ and $Q_\lambda$ identifies $\mathcal{T}''$ with $^{\perp}(\mathcal{T}')$; see Proposition 4.10.1.

### 4.14. Example: The derived category of a module category

Let $A$ be an associative ring. We denote by $\mathbf{K}(\mathrm{Mod}\, A)$ the category of chain complexes of $A$-modules whose morphisms are the homotopy classes of chain maps. The functor $H^n: \mathbf{K}(\mathrm{Mod}\, A) \to \mathrm{Mod}\, A$ taking the cohomology of a complex in degree $n$ is cohomological. A morphism $\phi$ is called *quasi-isomorphism* if $H^n\phi$ is an isomorphism for all $n \in \mathbb{Z}$, and we denote the set of all quasi-isomorphisms by qis. Then

$$\mathbf{D}(A) := \mathbf{D}(\mathrm{Mod}\, A) := \mathbf{K}(\mathrm{Mod}\, A)[\mathrm{qis}^{-1}]$$

is by definition the *derived category* of $\mathrm{Mod}\, A$. The kernel of the quotient functor $Q: \mathbf{K}(\mathrm{Mod}\, A) \to \mathbf{D}(\mathrm{Mod})$ is the full subcategory $\mathbf{K}_{\mathrm{ac}}(\mathrm{Mod}\, A)$ which is formed by all acyclic complexes. Note that $Q$ admits a left adjoint $Q_\lambda$ taking each complex to its K-projective resolution and a right adjoint $Q_\rho$ taking each complex to its K-injective resolution. Thus we obtain the following recollement.

$$\mathbf{K}_{\mathrm{ac}}(\mathrm{Mod}\, A) \;\overset{\mathrm{inc}}{\rightleftarrows}\; \mathbf{K}(\mathrm{Mod}\, A) \;\underset{Q_\lambda}{\overset{Q_\rho}{\rightleftarrows\!\!\!\overset{Q}{\longrightarrow}}}\; \mathbf{D}(\mathrm{Mod}\, A) \qquad (4.14.1)$$

It follows that for each pair of chain complexes $X, Y$ the set of morphisms $\mathbf{D}(\mathrm{Mod}\, A)(X, Y)$ is small, since $Q_\lambda$ induces a bijection with

$\mathbf{K}(\operatorname{Mod} A)(Q_\lambda X, Q_\lambda Y)$. The adjoints of $Q$ are discussed in more detail in Section 5.8.

## 4.15. Example: A derived category without small morphism sets

For any abelian category $\mathcal{A}$, the *derived category* $\mathbf{D}(\mathcal{A})$ is by definition $\mathbf{K}(\mathcal{A})[\mathrm{qis}^{-1}]$. Here, $\mathbf{K}(\mathcal{A})$ denotes the category of chain complexes in $\mathcal{A}$ whose morphisms are the homotopy classes of chain maps, and qis denotes the set of quasi-isomorphisms. Let us identify objects in $\mathcal{A}$ with chain complexes concentrated in degree zero.

We give an example of an abelian category $\mathcal{A}$ and an object $X$ in $\mathcal{A}$ such that the set $\operatorname{Ext}^1_{\mathcal{A}}(X, X) \cong \mathbf{D}(\mathcal{A})(X, SX)$ is not small. This example is taken from Freyd [14, pp. 131] and has been pointed out to me by Neeman.

Let $U$ denote the set of all cardinals of small sets. This set is not small. Consider the free associative $\mathbb{Z}$-algebra $\mathbb{Z}\langle U \rangle$ which is generated by the elements of $U$. Now let $\mathcal{A} = \operatorname{Mod} A$ denote the category of $A$-modules, where it is assumed that the underlying set of each module is small. Let $\mathbb{Z}$ denote the trivial $A$-module, that is, $zu = 0$ for all $z \in \mathbb{Z}$ and $u \in U$. We claim that the set $\operatorname{Ext}^1_{\mathcal{A}}(\mathbb{Z}, \mathbb{Z})$ is not small. To see this, define for each $u \in U$ an $A$-module $E_u = \mathbb{Z} \oplus \mathbb{Z}$ by

$$(z_1, z_2)x = \begin{cases} (z_2, 0) & \text{if } x = u, \\ (0, 0) & \text{if } x \neq u, \end{cases}$$

where $(z_1, z_2) \in E_u$ and $x \in U$. Then we have short exact sequences $0 \to \mathbb{Z} \xrightarrow{\left[\begin{smallmatrix}1\\0\end{smallmatrix}\right]} E_u \xrightarrow{[0\ 1]} \mathbb{Z} \to 0$ which yield pairwise different elements of $\operatorname{Ext}^1_{\mathcal{A}}(\mathbb{Z}, \mathbb{Z})$ as $u$ runs though the elements in $U$.

## 4.16. Example: The recollement induced by an idempotent

Recollements can be defined for abelian categories in the same way as for triangulated categories. A typical example arises for any module category from an idempotent element of the underlying ring.

Let $A$ be an associative ring and $e^2 = e \in A$ an idempotent. Then the functor $F \colon \operatorname{Mod} A \to \operatorname{Mod} eAe$ taking a module $M$ to $Me$ and restriction along $p \colon A \to A/AeA$ induce the following recollement.

$$\operatorname{Mod} A/AeA \underset{\xleftarrow{\hspace{2em}}}{\overset{\xleftarrow{\hspace{2em}}}{\xrightarrow{\ \ p_* \ \ }}} \operatorname{Mod} A \underset{\xrightarrow{-\otimes_{eAe} eA}}{\overset{\xleftarrow{\operatorname{Hom}_{eAe}(Ae,-)}}{\xrightarrow{\ \ F \ \ }}} \operatorname{Mod} eAe$$

Note that we can describe adjoints of $F$ since

$$F = \operatorname{Hom}_A(eA, -) = - \otimes_A Ae.$$

The recollement for Mod $A$ induces the following recollement of triangulated categories for $\mathbf{D}(A)$.

$$\operatorname{Ker}\mathbf{D}(F) \underset{\longleftarrow}{\overset{\longleftarrow}{\xrightarrow{\hspace{1.2cm}\text{inc}\hspace{1.2cm}}}} \mathbf{D}(A) \underset{-\otimes^{\mathbf{L}}_{eAe}eA}{\overset{\mathbf{R}\operatorname{Hom}_{eAe}(Ae,-)}{\underset{\longleftarrow}{\overset{\longleftarrow}{\xrightarrow{\hspace{0.8cm}\mathbf{D}(F)\hspace{0.8cm}}}}}} \mathbf{D}(eAe)$$

The functor $F$ is exact and $\mathbf{D}(F)$ takes by definition a complex $X$ to $FX$. The functor $\mathbf{D}(p_*)\colon \mathbf{D}(A/AeA) \to \mathbf{D}(A)$ identifies $\mathbf{D}(A/AeA)$ with $\operatorname{Ker}\mathbf{D}(F)$ if and only if $\operatorname{Tor}^A_i(A/AeA, A/AeA) = 0$ for all $i > 0$.

## 4.17. Notes

Triangulated categories were introduced independently in algebraic geometry by Verdier in his thèse [48], and in algebraic topology by Puppe [38]. Grothendieck and his school used the formalism of triangulated and derived categories for studying homological properties of abelian categories. Early examples are Grothendieck duality and local cohomology for categories of sheaves. The basic example of a triangulated category from topology is the stable homotopy category.

Localizations of triangulated categories are discussed in Verdier's thèse [48]. In particular, he introduced the localization (or *Verdier quotient*) of a triangulated category with respect to a triangulated subcategory. In the context of stable homotopy theory, it is more common to think of localization functors as endofunctors; see for instance the work of Bousfield [8], which explains the term *Bousfield localization*. The standard reference for recollements is [6]. Resolutions of unbounded complexes were first studied by Spaltenstein in [44]; see also [5].

# 5. Localization via Brown representability

## 5.1. Brown representatbility

Let $\mathcal{T}$ be a triangulated category and suppose that $\mathcal{T}$ has small coproducts. A *localizing subcategory* of $\mathcal{T}$ is by definition a thick subcategory which is closed under taking small coproducts. A localizing subcategory of $\mathcal{T}$ is *generated* by a fixed set of objects if it is the smallest localizing subcategory of $\mathcal{T}$ which contains this set.

We say that $\mathcal{T}$ is *perfectly generated* by some small set $\mathcal{S}$ of objects of $\mathcal{T}$ provided the following holds.

(PG1) There is no proper localizing subcategory of $\mathcal{T}$ which contains $\mathcal{S}$.
(PG2) Given a family $(X_i \to Y_i)_{i \in I}$ of morphisms in $\mathcal{T}$ such that the induced map $\mathcal{T}(C, X_i) \to \mathcal{T}(C, Y_i)$ is surjective for all $C \in \mathcal{S}$ and $i \in I$, the induced map

$$\mathcal{T}(C, \coprod_i X_i) \longrightarrow \mathcal{T}(C, \coprod_i Y_i)$$

is surjective.

We say that a triangulated category $\mathcal{T}$ with small products is *perfectly cogenerated* if $\mathcal{T}^{\mathrm{op}}$ is perfectly generated.

**Theorem 5.1.1.** *Let $\mathcal{T}$ be a triangulated category with small coproducts and suppose $\mathcal{T}$ is perfectly generated.*

(1) *A functor $F : \mathcal{T}^{\mathrm{op}} \to \mathrm{Ab}$ is cohomological and sends small coproducts in $\mathcal{T}$ to products if and only if $F \cong \mathcal{T}(-, X)$ for some object $X$ in $\mathcal{T}$.*
(2) *An exact functor $\mathcal{T} \to \mathcal{U}$ between triangulated categories preserves small coproducts if and only if it has a right adjoint.*

*Proof.* For a proof of (1) see [24, Theorem A]. To prove (2), suppose that $F$ preserves small coproducts. Then one defines the right adjoint $G : \mathcal{U} \to \mathcal{T}$ by sending an object $X$ in $\mathcal{U}$ to the object in $\mathcal{T}$ representing $\mathcal{U}(F-, X)$. Thus $\mathcal{U}(F-, X) \cong \mathcal{T}(-, GX)$. Conversely, given a right adjoint of $F$, it is automatic that $F$ preserves small coproducts. $\qquad\square$

**Remark 5.1.2.** (1) In the presence of (PG2), condition (PG1) is equivalent to the following: For an object $X$ in $\mathcal{T}$, we have $X = 0$ if $\mathcal{T}(S^n C, X) = 0$ for all $C \in \mathcal{S}$ and $n \in \mathbb{Z}$.

(2) A perfectly generated triangulated category $\mathcal{T}$ has small products. In fact, Brown representability implies that for any family of objects $X_i$ in $\mathcal{T}$ the functor $\prod_i \mathcal{T}(-, X_i)$ is represented by an object in $\mathcal{T}$.


## 5.2. Localization functors via Brown representability

The existence of localization functors is basically equivalent to the existence of certain right adjoints; see Proposition 4.9.1. We combine this observation with Brown's representability theorem and obtain the following.

**Proposition 5.2.1.** *Let $\mathcal{T}$ be a triangulated category which admits small coproducts and fix a localizing subcategory $\mathcal{S}$.*

(1) *Suppose $\mathcal{S}$ is perfectly generated. Then there exists an exact localization functor $L\colon \mathcal{T} \to \mathcal{T}$ with $\operatorname{Ker} L = \mathcal{S}$.*

(2) *Suppose $\mathcal{T}$ is perfectly generated. Then there exists an exact localization functor $L\colon \mathcal{T} \to \mathcal{T}$ with $\operatorname{Ker} L = \mathcal{S}$ if and only if the morphisms between any two objects in $\mathcal{T}/\mathcal{S}$ form a small set.*

*Proof.* The existence of a localization functor $L$ with $\operatorname{Ker} L = \mathcal{S}$ is equivalent to the existence of a right adjoint for the inclusion $\mathcal{S} \to \mathcal{T}$, and it is equivalent to the existence of a right adjoint for the quotient functor $\mathcal{T} \to \mathcal{T}/\mathcal{S}$. Both functors preserve small coproducts since $\mathcal{S}$ is closed under taking small coproducts; see Proposition 3.5.1. Now apply Theorem 5.1.1 for the existence of right adjoints. $\qquad\qquad\square$

## 5.3. Compactly generated triangulated categories

Let $\mathcal{T}$ be a triangulated category with small coproducts. An object $X$ in $\mathcal{T}$ is called *compact* (or *small*) if every morphism $X \to \coprod_{i \in I} Y_i$ in $\mathcal{T}$ factors through $\coprod_{i \in J} Y_i$ for some finite subset $J \subseteq I$. Note that $X$ is compact if and only if the representable functor $\mathcal{T}(X, -)\colon \mathcal{T} \to \mathrm{Ab}$ preserves small coproducts. The compact objects in $\mathcal{T}$ form a thick subcategory which we denote by $\mathcal{T}^c$.

The triangulated category $\mathcal{T}$ is called *compactly generated* if it is perfectly generated by a small set of compact objects. Observe that condition (PG2) is automatically satisfied if every object in $\mathcal{S}$ is compact.

A compactly generated triangulated category $\mathcal{T}$ is perfectly cogenerated. To see this, let $\mathcal{S}$ be a set of compact generators. Then the objects representing $\operatorname{Hom}_{\mathbb{Z}}(\mathcal{T}(C, -), \mathbb{Q}/\mathbb{Z})$, where $C$ runs through all objects in $\mathcal{S}$, form a set of perfect cogenerators for $\mathcal{T}$.

The following proposition is a reformulation of Brown representability for compactly generated triangulated categories.

**Proposition 5.3.1.** *Let $F\colon \mathcal{T} \to \mathcal{U}$ be an exact functor between triangulated categories. Suppose that $\mathcal{T}$ has small coproducts and that $\mathcal{T}$ is compactly generated.*

(1) *The functor $F$ admits a right adjoint if and only if $F$ preserves small coproducts.*

(2) *The functor $F$ admits a left adjoint if and only if $F$ preserves small products.*

## 5.4. Right adjoint functors preserving coproducts

The following lemma provides a characterization of the fact that a right adjoint functor preserves small coproducts. This will be useful in the context of compactly generated categories.

**Lemma 5.4.1.** *Let $F: \mathcal{T} \to \mathcal{U}$ be an exact functor between triangulated categories which has a right adjoint $G$.*

(1) *If $G$ preserves small coproducts, then $F$ preserves compactness.*
(2) *If $F$ preserves compactness and $\mathcal{T}$ is generated by compact objects, then $G$ preserves small coproducts.*

*Proof.* Let $X$ be an object in $\mathcal{T}$ and $(Y_i)_{i \in I}$ a family of objects in $\mathcal{U}$.

(1) We have

$$\mathcal{U}(FX, \coprod_i Y_i) \cong \mathcal{T}(X, G(\coprod_i Y_i)) \cong \mathcal{T}(X, \coprod_i GY_i). \tag{5.4.1}$$

If $X$ is compact, then the isomorphism shows that a morphism $FX \to \coprod_i Y_i$ factors through a finite coproduct. It follows that $FX$ is compact.

(2) Let $X$ be compact. Then the canonical morphism $\phi: \coprod_i GY_i \to G(\coprod_i Y_i)$ induces an isomorphism

$$\mathcal{T}(X, \coprod_i GY_i) \cong \coprod_i \mathcal{T}(X, GY_i) \cong \coprod_i \mathcal{U}(FX, Y_i) \cong \mathcal{U}(FX, \coprod_i Y_i)$$

$$\cong \mathcal{T}(X, G(\coprod_i Y_i)),$$

where the last isomorphism uses that $FX$ is compact. It is easily checked that the objects $X'$ in $\mathcal{T}$ such that $\mathcal{T}(X', \phi)$ is an isomorphism form a localizing subcategory of $\mathcal{T}$. Thus $\phi$ is an isomorphism because the compact objects generate $\mathcal{T}$. $\qquad\square$

## 5.5. Localization functors preserving coproducts

The following result provides a characterization of the fact that an exact localization functor $L$ preserves small coproducts; in that case one calls $L$ *smashing*. The example given below explains this terminology.

**Proposition 5.5.1.** *Let $\mathcal{T}$ be a category with small coproducts and $L: \mathcal{T} \to \mathcal{T}$ an exact localization functor. Then the following are equivalent.*

(1) *The functor $L: \mathcal{T} \to \mathcal{T}$ preserves small coproducts.*
(2) *The colocalization functor $\Gamma: \mathcal{T} \to \mathcal{T}$ with $\operatorname{Ker} \Gamma = \operatorname{Im} L$ preserves small coproducts.*

(3) *The right adjoint of the inclusion functor $\operatorname{Ker} L \to \mathcal{T}$ preserves small coproducts.*

(4) *The right adjoint of the quotient functor $\mathcal{T} \to \mathcal{T}/\operatorname{Ker} L$ preserves small coproducts.*

(5) *The subcategory $\operatorname{Im} L$ of all $L$-local objects is closed under taking small coproducts.*

*If $\mathcal{T}$ is perfectly generated, in addition the following is equivalent.*

(6) *There exists a recollement of the following form.*

$$\operatorname{Im} L \underset{\longleftarrow}{\overset{\longleftarrow}{\xrightarrow{\quad\text{inc}\quad}}} \mathcal{T} \underset{\longleftarrow}{\overset{\longleftarrow}{\xrightarrow{\hspace{2cm}}}} \operatorname{Ker} L \qquad (5.5.1)$$

*Proof.* (1) $\Leftrightarrow$ (4) $\Leftrightarrow$ (5) follows from Proposition 2.7.1.

(1) $\Leftrightarrow$ (2) $\Leftrightarrow$ (3) is easily deduced from the functorial triangle (4.11.1) relating $L$ and $\Gamma$.

(5) $\Leftrightarrow$ (6): Assume that $\mathcal{T}$ is perfectly generated. Then we can apply Brown's representability theorem and consider the sequence

$$\operatorname{Im} L \xrightarrow{\ I\ } \mathcal{T} \xrightarrow{\ Q\ } \operatorname{Ker} L$$

where $I$ denotes the inclusion and $Q$ a right adjoint of the inclusion $\operatorname{Ker} L \to \mathcal{T}$. Note that $Q$ induces an equivalence $\mathcal{T}/\operatorname{Im} L \xrightarrow{\sim} \operatorname{Ker} L$; see Propositions 4.11.1 and 4.12.1. The functors $I$ and $Q$ have left adjoints. Thus the pair $(I, Q)$ gives rise to a recollement if and only if $I$ and $Q$ both admit right adjoints. It follows from Proposition 4.9.1 that this happens if and only if $Q$ admits a right adjoint. Now Brown's representability theorem implies that this is equivalent to the fact that $Q$ preserves small coproducts. And Proposition 3.5.1 shows that $Q$ preserves small coproducts if and only if $\operatorname{Im} L$ is closed under taking small coproducts. This finishes the proof. $\qquad\square$

**Remark 5.5.2.** (1) The implication (6) $\Rightarrow$ (5) holds without any extra assumption on $\mathcal{T}$.

(2) Suppose an exact localization functor $L \colon \mathcal{T} \to \mathcal{T}$ preserves small coproducts. If $\mathcal{T}$ is compactly generated, then $\operatorname{Im} L$ is compactly generated. This follows from Lemma 5.4.1, because the left adjoint of the inclusion $\operatorname{Im} L \to \mathcal{T}$ sends the compact generators of $\mathcal{T}$ to compact generators for $\operatorname{Im} L$. A similar argument shows that $\operatorname{Im} L$ is perfectly generated provided that $\mathcal{T}$ is perfectly generated.

**Example 5.5.3.** Let $\mathcal{S}$ be the stable homotopy category of spectra and $\wedge$ its smash product. Then an exact localization functor $L \colon \mathcal{S} \to \mathcal{S}$ preserves small coproducts if and only if $L$ is of the form $L = - \wedge E$ for some spectrum $E$. We sketch the argument. Let $S$ denote the sphere spectrum. There exists a natural morphism $\eta X : X \wedge LS \to LX$ for each $X$ in $\mathcal{S}$. Suppose that $L$ preserves small coproducts. Then the subcategory of objects $X$ in $\mathcal{S}$ such that $\eta X$ is invertible contains $S$ and is closed under forming small coproduts and exact triangles. Thus $L = - \wedge E$ for $E = LS$.

Let $L : \mathcal{T} \to \mathcal{T}$ be an exact localization functor which induces a recollement of the form (5.5.1). Then the sequence $\operatorname{Ker} L \to \mathcal{T} \to \operatorname{Im} L$ of left adjoint functors induces a sequence

$$(\operatorname{Ker} L)^c \longrightarrow \mathcal{T}^c \longrightarrow (\operatorname{Im} L)^c$$

of exact functors, by Lemma 5.4.1. This sequence is of some interest. The functor $(\operatorname{Ker} L)^c \to \mathcal{T}^c$ is fully faithful and identifies $(\operatorname{Ker} L)^c$ with a thick subcategory of $\mathcal{T}^c$, whereas the functor $\mathcal{T}^c \to (\operatorname{Im} L)^c$ shares some formal properties with a quotient functor. A typical example arises from finite localization; see Theorem 5.6.1. However, there are examples where $\mathcal{T}$ is compactly generated but $0 = (\operatorname{Ker} L)^c \subseteq \operatorname{Ker} L \neq 0$; see [25] for details.

## 5.6. Finite localization

A common type of localization for triangulated categories is finite localization. Here, we explain the basic result and refer to our discussion of well generated categories for a more general approach and further details.

Let $\mathcal{T}$ be a compactly generated triangulated category and suppose we have given a subcategory $\mathcal{S}' \subseteq \mathcal{T}^c$. Let $\mathcal{S}$ denote the localizing subcategory generated by $\mathcal{S}'$. Then $\mathcal{S}$ is compactly generated and therefore the inclusion functor $\mathcal{S} \to \mathcal{T}$ admits a right adjoint by Brown's representability theorem. In particular, we have a localization functor $L : \mathcal{T} \to \mathcal{T}$ with $\operatorname{Ker} L = \mathcal{S}$ and the morphisms between any pair of objects in $\mathcal{T}/\mathcal{S}$ form a small set; see Proposition 5.2.1. We observe that the compact objects of $\mathcal{S}$ identify with the smallest thick subcategory of $\mathcal{T}^c$ containing $\mathcal{S}'$. This follows from Corollary 7.2.2. Thus we obtain the following commutative diagram of exact functors.

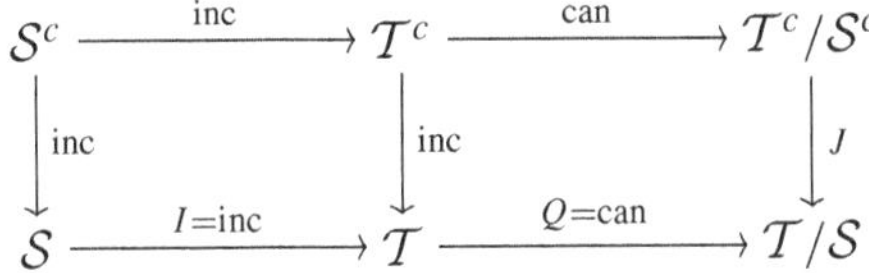

**Theorem 5.6.1.** *Let $\mathcal{T}$ and $\mathcal{S}$ be as above. Then the quotient category $\mathcal{T}/\mathcal{S}$ is compactly generated. The induced exact functor $J : \mathcal{T}^c/\mathcal{S}^c \to \mathcal{T}/\mathcal{S}$ is fully faithful and the category $(\mathcal{T}/\mathcal{S})^c$ of compact objects equals the full subcategory*

*consisting of all direct factors of objects in the image of J. Moreover, the
inclusion $\mathcal{S}^{\perp} \to \mathcal{T}$ induces the following recollement.*

$$\mathcal{S}^{\perp} \underset{\longleftarrow}{\overset{\longleftarrow}{\xrightarrow{\quad\text{inc}\quad}}} \mathcal{T} \underset{\longleftarrow}{\overset{\longleftarrow}{\longrightarrow}} \mathcal{S}$$

*Proof.* The inclusion $I$ preserves compactness and therefore the right adjoint $I_{\rho}$
preserves small coproducts by Lemma 5.4.1. Thus $Q_{\rho}$ preserve small coproducts by Proposition 5.5.1, and therefore $Q$ preserves compactness, again by
Lemma 5.4.1. It follows that $J$ induces a functor $\mathcal{T}^{c}/\mathcal{S}^{c} \to (\mathcal{T}/\mathcal{S})^{c}$. In particular, $Q$ sends a set of compact generators of $\mathcal{T}$ to a set of compact generators
for $\mathcal{T}/\mathcal{S}$.

Next we apply Lemma 4.7.1 to show that $J$ is fully faithful. For this, one
needs to check that every morphism from a compact object in $\mathcal{T}$ to an object
in $\mathcal{S}$ factors through some object in $\mathcal{S}^{c}$. This follows from Theorem 7.2.1.
The image of $J$ is a full triangulated subcategory of $\mathcal{T}^{c}$ which generates $\mathcal{T}/\mathcal{S}$.
Another application of Corollary 7.2.2 shows that every compact object of $\mathcal{T}/\mathcal{S}$
is a direct factor of some object in the image of $J$.

Let $L \colon \mathcal{T} \to \mathcal{T}$ denote the localization functor with $\operatorname{Ker} L = \mathcal{S}$. Then $\mathcal{S}^{\perp}$
equals the full subcategory of $L$-local objects. This subcategory is closed under
small coproducts since $\mathcal{S}$ is generated by compact objects. Thus the existence
of the recollement follows from Proposition 5.5.1. $\qquad\qquad\qquad\square$


## 5.7. Cohomological localization via localization of graded modules

Let $\mathcal{T}$ be a triangulated category which admits small coproducts. Suppose that
$\mathcal{T}$ is generated by a small set of compact objects. We fix a graded[1] ring $\Lambda$ and
a graded cohomological functor

$$H^{*} \colon \mathcal{T} \longrightarrow \mathcal{A}$$

into the category $\mathcal{A}$ of graded $\Lambda$-modules. Thus $H^{*}$ is a functor which sends
each exact triangle in $\mathcal{T}$ to an exact sequence in $\mathcal{A}$, and we have an isomorphism
$H^{*} \circ S \cong T \circ H^{*}$ where $T$ denotes the shift functor for $\mathcal{A}$. In addition, we
assume that $H^{*}$ preserves small products and coproducts.

**Theorem 5.7.1.** *Let $L \colon \mathcal{A} \to \mathcal{A}$ be an exact localization functor for the category $\mathcal{A}$ of graded $\Lambda$-modules. Then there exists an exact localization functor*

---

[1] All graded rings and modules are graded over $\mathbb{Z}$. Morphisms between graded modules are
degree zero maps.

$\tilde{L}: \mathcal{T} \to \mathcal{T}$ *such that the following square commutes up to a natural isomorphism.*

$$\begin{array}{ccc} \mathcal{T} & \xrightarrow{\ \tilde{L}\ } & \mathcal{T} \\ \downarrow{\scriptstyle H^*} & & \downarrow{\scriptstyle H^*} \\ \mathcal{A} & \xrightarrow{\ L\ } & \mathcal{A} \end{array}$$

*More precisely, the adjunction morphisms* $\mathrm{Id}\,\mathcal{A} \to L$ *and* $\mathrm{Id}\,\mathcal{T} \to \tilde{L}$ *induce for each* $X$ *in* $\mathcal{T}$ *the following isomorphisms.*

$$H^*\tilde{L}X \xrightarrow{\sim} L(H^*\tilde{L}X) = LH^*(\tilde{L}X) \xleftarrow{\sim} LH^*(X)$$

*An object* $X$ *in* $\mathcal{T}$ *is* $\tilde{L}$*-acyclic if and only if* $H^*X$ *is* $L$*-acyclic. If an object* $X$ *in* $\mathcal{T}$ *is* $\tilde{L}$*-local, then* $H^*X$ *is* $L$*-local. The converse holds, provided that* $H^*$ *reflects isomorphisms.*

*Proof.* We recall that $\mathcal{T}$ is perfectly cogenerated because it is compactly generated. Thus Brown's representability theorem provides a compact object $C$ in $\mathcal{T}$ such that

$$H^*X \cong \mathcal{T}(C, X)^* := \coprod_{i \in \mathbb{Z}} \mathcal{T}(C, S^i Y) \quad \text{for all} \quad X \in \mathcal{T}.$$

Now consider the essential image $\mathrm{Im}\,L$ of $L$ which equals the full subcategory formed by all $L$-local objects in $\mathcal{A}$. Because $L$ is exact, this subcategory is *coherent*, that is, for any exact sequence $X_1 \to X_2 \to X_3 \to X_4 \to X_5$ with $X_1, X_2, X_4, X_5 \in \mathcal{A}$, we have $X_3 \in \mathcal{A}$. This is an immediate consequence of the 5-lemma. In addition, $\mathrm{Im}\,L$ is closed under taking small products. The $L$-local objects form an abelian Grothendieck category and therefore $\mathrm{Im}\,L$ admits an injective cogenerator, say $I$; see [16]. Using again Brown's representability theorem, there exists $\tilde{I}$ in $\mathcal{T}$ such that

$$\mathcal{A}(H^* -, I) \cong \mathcal{T}(-, \tilde{I}) \quad \text{and therefore} \quad \mathcal{A}(H^* -, I)^* \cong \mathcal{T}(-, \tilde{I})^*. \quad (5.7.1)$$

Now consider the subcategory $\mathcal{V}$ of $\mathcal{T}$ which is formed by all objects $X$ in $\mathcal{T}$ such that $H^*X$ is $L$-local. This is a triangulated subcategory which is closed under taking small products. Observe that $\tilde{I}$ belongs to $\mathcal{V}$. To prove this, take a free presentation

$$F_1 \longrightarrow F_0 \longrightarrow H^*C \longrightarrow 0$$

over $\Lambda$ and apply $\mathcal{A}(-, I)^*$ to it. Using the isomorphism (5.7.1), we see that $H^*\tilde{I}$ belongs to $\mathrm{Im}\,L$ because $\mathrm{Im}\,L$ is coherent and closed under taking small products.

Now let $\mathcal{U}$ denote the smallest triangulated subcategory of $\mathcal{T}$ containing $\tilde{I}$ and closed under taking small products. Observe that $\mathcal{U} \subseteq \mathcal{V}$. We claim that $\mathcal{U}$ is perfectly cogenerated by $\tilde{I}$. Thus, given a family of morphisms $\phi_i \colon X_i \to Y_i$ in $\mathcal{U}$ such that $\mathcal{T}(Y_i, \tilde{I}) \to \mathcal{T}(X_i, \tilde{I})$ is surjective for all $i$, we need to show that $\mathcal{T}(\prod_i Y_i, \tilde{I}) \to \mathcal{T}(\prod_i X_i, \tilde{I})$ is surjective. We argue as follows. If $\mathcal{T}(Y_i, \tilde{I}) \to \mathcal{T}(X_i, \tilde{I})$ is surjective, then the isomorphism (5.7.1) implies that $H^*\phi_i$ is a monomorphism since $I$ is an injective cogenerator for $\operatorname{Im} L$. Thus the product $\prod_i \phi_i \colon \prod_i X_i \to \prod_i Y_i$ induces a monomorphism $H^* \prod_i \phi_i = \prod_i H^*\phi_i$ and therefore $\mathcal{T}(\prod_i \phi_i, \tilde{I})$ is surjective. We conclude from Brown's representability theorem that the inclusion functor $G \colon \mathcal{U} \to \mathcal{T}$ has a left adjoint $F \colon \mathcal{T} \to \mathcal{U}$. Thus $\tilde{L} = G \circ F$ is a localization functor by Corollary 2.4.2.

Next we show that an object $X \in \mathcal{T}$ is $\tilde{L}$-acyclic if and only if $H^*X$ is $L$-acyclic. This follows from Proposition 4.10.1 and the isomorphism (5.7.1), because we have

$$\tilde{L}X = 0 \iff \mathcal{T}(X, \tilde{I}) = 0 \iff \mathcal{A}(H^*X, I) = 0 \iff LH^*X = 0.$$

Now denote by $\eta \colon \operatorname{Id} \mathcal{A} \to L$ and $\tilde{\eta} \colon \operatorname{Id} \mathcal{T} \to \tilde{L}$ the adjunction morphisms and consider the following commutative square.

$$
\begin{array}{ccc}
H^*X & \xrightarrow{\;\eta H^*X\;} & LH^*X \\
{\scriptstyle H^*\tilde{\eta}X}\big\downarrow & & \big\downarrow{\scriptstyle LH^*\tilde{\eta}X} \\
H^*\tilde{L}X & \xrightarrow{\;\eta H^*\tilde{L}X\;} & LH^*\tilde{L}X
\end{array}
\tag{5.7.2}
$$

We claim that $LH^*\tilde{\eta}X$ and $\eta H^*\tilde{L}X$ are invertible for each $X$ in $\mathcal{T}$. The morphism $\tilde{\eta}X$ induces an exact triangle

$$X' \to X \xrightarrow{\tilde{\eta}X} \tilde{L}X \to SX'$$

with $\tilde{L}X' = 0 = \tilde{L}SX'$. Applying the cohomological functor $LH^*$, we see that $LH^*\tilde{\eta}X$ is an isomorphism, since $LH^*X' = 0 = LH^*SX'$. Thus $LH^*\tilde{\eta}$ is invertible. The morphism $\eta H^*\tilde{L}X$ is invertible because $H^*\tilde{L}X$ is $L$-local. This follows from the fact that $\tilde{L}X$ belongs to $\mathcal{U}$.

The commutative square (5.7.2) implies that $H^*\tilde{\eta}X$ is invertible if and only if $\eta H^*X$ is invertible. Thus if $X$ is $\tilde{L}$-local, then $H^*X$ is $L$-local. The converse holds if $H^*$ reflects isomorphisms. $\qquad\square$

**Remark 5.7.2.** (1) The localization functor $\tilde{L}$ is essentially uniquely determined by $H^*$ and $L$, because $\operatorname{Ker}\tilde{L} = \operatorname{Ker} LH^*$.

(2) Suppose that $C$ is a generator of $\mathcal{T}$. If $L$ preserves small coproducts, then it follows that $\tilde{L}$ preserves small coproducts. In fact, the assumption

implies that $H^*\tilde{L}$ preserves small coproducts, since $LH^* \cong H^*\tilde{L}$. But $H^*$ reflects isomorphisms because $C$ is a generator of $\mathcal{T}$. Thus $\tilde{L}$ preserves small coproducts.

## 5.8. Example: Resolutions of chain complexes

Let $A$ be an associative ring. Then the derived category $\mathbf{D}(A)$ of unbounded chain complexes of modules over $A$ is compactly generated. A compact generator is the ring $A$, viewed as a complex concentrated in degree zero. Let us be more precise, because we want to give an explicit construction of $\mathbf{D}(A)$ which implies that the morphisms between any two objects in $\mathbf{D}(A)$ form a small set. Moreover, we combine Brown representability with Proposition 4.9.1 to provide descriptions of the adjoints $Q_\lambda$ and $Q_\rho$ of the quotient functor $Q\colon \mathbf{K}(\mathrm{Mod}\,A) \to \mathbf{D}(A)$ which appear in the recollement (4.14.1).

Denote by $\mathrm{Loc}\,A$ the localizing subcategory of $\mathbf{K}(\mathrm{Mod}\,A)$ which is generated by $A$. Then $\mathrm{Loc}\,A$ is a compactly generated triangulated category and $(\mathrm{Loc}\,A)^\perp = \mathbf{K}_{\mathrm{ac}}(\mathrm{Mod}\,A)$ since

$$\mathbf{K}(\mathrm{Mod}\,A)(A, S^n X) \cong H^n X.$$

Brown representability provides a right adjoint for the inclusion $\mathrm{Loc}\,A \to \mathbf{K}(\mathrm{Mod}\,A)$ and therefore the composite $F\colon \mathrm{Loc}\,A \xrightarrow{\mathrm{inc}} \mathbf{K}(\mathrm{Mod}\,A) \xrightarrow{\mathrm{can}} \mathbf{D}(A)$ is an equivalence by Proposition 4.9.1. The right adjoint of the inclusion $\mathrm{Loc}\,A \to \mathbf{K}(\mathrm{Mod}\,A)$ annihilates the acyclic complexes and induces therefore a functor $\mathbf{D}(A) \to \mathrm{Loc}\,A$ (which is a quasi-inverse for $F$). The composite with the inclusion $\mathrm{Loc}\,A \to \mathbf{K}(\mathrm{Mod}\,A)$ is the left adjoint $Q_\lambda$ of $Q$ and takes a complex to its *K-projective resolution*.

Now fix an injective cogenerator $I$ for the category of $A$-modules, for instance $I = \mathrm{Hom}_{\mathbb{Z}}(A, \mathbb{Q}/\mathbb{Z})$. We denote by $\mathrm{Coloc}\,I$ the smallest thick subcategory of $\mathbf{K}(\mathrm{Mod}\,A)$ closed under small products and containing $I$. Then $I$ is a perfect cogenerator for $\mathrm{Coloc}\,I$ and $^\perp(\mathrm{Coloc}\,I) = \mathbf{K}_{\mathrm{ac}}(\mathrm{Mod}\,A)$ since

$$\mathbf{K}(\mathrm{Mod}\,A)(S^n X, I) \cong \mathrm{Hom}_A(H^n X, I).$$

Brown representability provides a left adjoint for the inclusion $\mathrm{Coloc}\,I \to \mathbf{K}(\mathrm{Mod}\,A)$ and therefore the composite $G\colon \mathrm{Coloc}\,I \xrightarrow{\mathrm{inc}} \mathbf{K}(\mathrm{Mod}\,A) \xrightarrow{\mathrm{can}} \mathbf{D}(A)$ is an equivalence by Proposition 4.9.1. The left adjoint of the inclusion $\mathrm{Coloc}\,I \to \mathbf{K}(\mathrm{Mod}\,A)$ annihilates the acyclic complexes and induces therefore a functor $\mathbf{D}(A) \to \mathrm{Coloc}\,I$ (which is a quasi-inverse for $G$). The composition with the inclusion Coloc $I \to \mathbf{K}(\mathrm{Mod}\, A)$ is the right adjoint $Q_\rho$ of $Q$ and takes a complex to its *K-injective resolution*.

## 5.9. Example: Homological epimorphisms

Let $f\colon A \to B$ be a ring homomorphism and $f_*\colon \mathbf{D}(B) \to \mathbf{D}(A)$ the functor given by restriction of scalars. Clearly, $f_*$ preserves small products and coproducts. Thus Brown representability implies the existence of left and right adjoints for $f_*$ since $\mathbf{D}(A)$ is compactly generated. For instance, the left adjoint is the derived tensor functor $- \otimes^{\mathbf{L}}_A B \colon \mathbf{D}(A) \to \mathbf{D}(B)$ which preserves compactness.

The functor $f_*$ is fully faithful if and only if $(f_*-) \otimes^{\mathbf{L}}_A B \cong \mathrm{Id}\,\mathbf{D}(B)$ iff $B \otimes_A B \cong B$ and $\mathrm{Tor}^A_i(B, B) = 0$ for all $i > 0$. In that case $f$ is called *homological epimorphism* and the exact functor $L\colon \mathbf{D}(A) \to \mathbf{D}(A)$ sending $X$ to $f_*(X \otimes^{\mathbf{L}}_A B)$ is a localization functor.

Take for instance a commutative ring $A$ and let $f\colon A \to S^{-1}A = B$ be the localization with respect to a multiplicatively closed subset $S \subseteq A$. Then the induced exact localization functor $L\colon \mathbf{D}(A) \to \mathbf{D}(A)$ takes a chain complex $X$ to $S^{-1}X$. Note that $L$ preserves small coproducts. In particular, $L$ gives rise to the following recollement.

$$\mathbf{D}(B) \xrightleftharpoons[{-\otimes^{\mathbf{L}}_A B}]{\overset{\mathbf{RHom}_A(B,-)}{\longleftarrow}\ \xrightarrow{\ f_*\ }\ \longleftarrow} \mathbf{D}(A) \rightleftarrows \mathcal{U}$$

The triangulated category $\mathcal{U}$ is equivalent to the kernel of $- \otimes^{\mathbf{L}}_A B$, and one can show that $\mathrm{Ker}(- \otimes^{\mathbf{L}}_A B)$ is the localizing subcategory of $\mathbf{D}(A)$ generated by the complexes of the form

$$\cdots \to 0 \to A \xrightarrow{x} A \to 0 \to \cdots \qquad (x \in S).$$

## 5.10. Notes

The Brown representability theorem in homotopy theory is due to Brown [9]. Generalizations of the Brown representability theorem for triangulated categories can be found in work of Franke [13], Keller [21], and Neeman [31, 33]. The version used here is taken from [24]. The finite localization theorem for compactly generated triangulated categories is due to Neeman [30]; it is based on previous work of Bousfield, Ravenel, Thomason-Trobaugh, Yao, and others. The cohomological localization functors commuting with localization functors of graded modules have been used to set up local cohomology functors in [7].

# 6. Well generated triangulated categories

## 6.1. Regular cardinals

A cardinal $\alpha$ is called *regular* if $\alpha$ is not the sum of fewer than $\alpha$ cardinals, all smaller than $\alpha$. For example, $\aleph_0$ is regular because the sum of finitely many finite cardinals is finite. Also, the successor $\kappa^+$ of every infinite cardinal $\kappa$ is regular. In particular, there are arbitrarily large regular cardinals. For more details on regular cardinals, see for instance [26].

## 6.2. Localizing subcategories

Let $\mathcal{T}$ be a triangulated category and $\alpha$ a regular cardinal. A coproduct in $\mathcal{T}$ is called $\alpha$-*coproduct* if it has less than $\alpha$ factors. A full subcategory of $\mathcal{T}$ is called $\alpha$-*localizing* if it is a thick subcategory and closed under taking $\alpha$-coproducts. Given a subcategory $\mathcal{S} \subseteq \mathcal{T}$, we denote by $\mathrm{Loc}_\alpha \, \mathcal{S}$ the smallest $\alpha$-localizing subcategory of $\mathcal{T}$ which contains $\mathcal{S}$. Note that $\mathrm{Loc}_\alpha \, \mathcal{S}$ is small provided that $\mathcal{S}$ is small.

A full subcategory of $\mathcal{T}$ is called *localizing* if it is a thick subcategory and closed under taking small coproducts. The smallest localizing subcategory containing a subcategory $\mathcal{S} \subseteq \mathcal{T}$ is $\mathrm{Loc}\,\mathcal{S} = \bigcup_\alpha \mathrm{Loc}_\alpha \, \mathcal{S}$ where $\alpha$ runs through all regular cardinals. We call $\mathrm{Loc}\,\mathcal{S}$ the localizing subcategory generated by $\mathcal{S}$.

## 6.3. Well generated triangulated categories

Let $\mathcal{T}$ be a triangulated category which admits small coproducts and fix a regular cardinal $\alpha$. An object $X$ in $\mathcal{T}$ is called $\alpha$-*small* if every morphism $X \to \coprod_{i \in I} Y_i$ in $\mathcal{T}$ factors through $\coprod_{i \in J} Y_i$ for some subset $J \subseteq I$ with card $J < \alpha$. The triangulated category $\mathcal{T}$ is called $\alpha$-*well generated* if it is perfectly generated by a small set of $\alpha$-small objects, and $\mathcal{T}$ is called *well generated* if it is $\beta$-well generated for some regular cardinal $\beta$.

Suppose $\mathcal{T}$ is $\alpha$-well generated by a small set $\mathcal{S}$ of $\alpha$-small objects. Given any regular cardinal $\beta \geq \alpha$, we denote by $\mathcal{T}^\beta$ the $\beta$-localizing subcategory $\mathrm{Loc}_\beta \, \mathcal{S}$ generated by $\mathcal{S}$ and call the objects of $\mathcal{T}^\beta$ $\beta$-*compact*. Choosing a representative for each isomorphism class, one can show that the $\beta$-compact objects form a small set of $\beta$-small perfect generators for $\mathcal{T}$. Moreover, $\mathcal{T}^\beta$ does not depend on the choice of $\mathcal{S}$. For a proof we refer to [23, Lemma 5]; see also Proposition 6.8.1 and Remark 6.10.2. Note that $\mathcal{T} = \bigcup_\beta \mathcal{T}^\beta$, where $\beta$ runs through all regular cardinals greater or equal than $\alpha$, because $\bigcup_\beta \mathcal{T}^\beta$ is a triangulated subcategory containing $\mathcal{S}$ and closed under small coproducts.

**Remark 6.3.1.** The $\alpha$-small objects of $\mathcal{T}$ form an $\alpha$-localizing subcategory.

**Example 6.3.2.** A triangulated category $\mathcal{T}$ is $\aleph_0$-well generated if and only if $\mathcal{T}$ is compactly generated. In that case $\mathcal{T}^{\aleph_0} = \mathcal{T}^c$.

**Example 6.3.3.** Let $\mathcal{A}$ be the category of sheaves of abelian groups on a non-compact, connected manifold of dimension at least 1. Then the derived category $\mathbf{D}(\mathcal{A})$ of unbounded chain complexes is well generated, but the only compact object in $\mathbf{D}(\mathcal{A})$ is the zero object; see [34]. For more examples of well generated but not compactly generated triangulated categories, see [32].

## 6.4. Filtered categories

Let $\alpha$ be a regular cardinal. A category $\mathcal{C}$ is called *$\alpha$-filtered* if the following holds.

(FIL1) There exists an object in $\mathcal{C}$.

(FIL2) For every family $(X_i)_{i \in I}$ of fewer than $\alpha$ objects there exists an object $X$ with morphisms $X_i \to X$ for all $i$.

(FIL3) For every family $(\phi_i \colon X \to Y)_{i \in I}$ of fewer than $\alpha$ morphisms there exists a morphism $\psi \colon Y \to Z$ with $\psi \phi_i = \psi \phi_j$ for all $i$ and $j$.

One drops the cardinal $\alpha$ and calls $\mathcal{C}$ *filtered* in case it is $\aleph_0$-filtered.

Given a functor $F \colon \mathcal{C} \to \mathcal{D}$, we use the term *$\alpha$-filtered colimit* for the colimit $\varinjlim_{X \in \mathcal{C}} FX$ provided that $\mathcal{C}$ is a small $\alpha$-filtered category.

**Lemma 6.4.1.** *Let $i \colon \mathcal{C}' \to \mathcal{C}$ be a fully faithful functor with $\mathcal{C}$ a small $\alpha$-filtered category. Suppose that $i$ is cofinal in the sense that for any $X \in \mathcal{C}$ there is an object $Y \in \mathcal{C}'$ and a morphism $X \to iY$. Then $\mathcal{C}'$ is a small $\alpha$-filtered category, and for any functor $F \colon \mathcal{C} \to \mathcal{D}$ into a category which admits $\alpha$-filtered colimits, the natural morphism*

$$\varinjlim_{Y \in \mathcal{C}'} F(iY) \longrightarrow \varinjlim_{X \in \mathcal{C}} FX$$

*is an isomorphism.*

*Proof.* See [19, Proposition 8.1.3]. $\qquad\qquad\square$

A full subcategory $\mathcal{C}'$ of a small $\alpha$-filtered category $\mathcal{C}$ is called *cofinal* if for any $X \in \mathcal{C}$ there is an object $Y \in \mathcal{C}'$ and a morphism $X \to Y$.

## 6.5. Comma categories

Let $T$ be a triangulated category which admits small coproducts and fix a full subcategory $S$. Given an object $X$ in $T$, let $S/X$ denote the category whose objects are pairs $(C, \mu)$ with $C \in S$ and $\mu \in T(C, X)$. The morphisms $(C, \mu) \to (C', \mu')$ are the morphisms $\gamma : C \to C'$ in $T$ making the following diagram commutative.

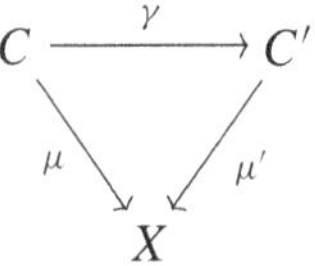

Analogously, one defines for a morphism $\phi : X \to X'$ in $T$ the category $S/\phi$ whose objects are commuting squares in $T$ of the form

$$\begin{array}{ccc} C & \xrightarrow{\gamma} & C' \\ \mu \downarrow & & \downarrow \mu' \\ X & \xrightarrow{\phi} & X' \end{array}$$

with $C, C' \in S$.

**Lemma 6.5.1.** *Let $\alpha$ be a regular cardinal and $S$ an $\alpha$-localizing subcategory of $T$. Then the categories $S/X$ and $S/\phi$ are $\alpha$-filtered for each object $X$ and each morphism $\phi$ in $T$.*

*Proof.* Straightforward. □

## 6.6. The comma category of an exact triangle

Let $T$ be a triangulated category. We consider the category of pairs $(\phi_1, \phi_2)$ of composable morphisms $X_1 \xrightarrow{\phi_1} X_2 \xrightarrow{\phi_2} X_3$ in $T$. A morphism $\mu : (\phi_1, \phi_2) \to (\phi_1', \phi_2')$ is a triple $\mu = (\mu_1, \mu_2, \mu_3)$ of morphisms in $T$ making the following diagram commutative.

$$\begin{array}{ccccc} X_1 & \xrightarrow{\phi_1} & X_2 & \xrightarrow{\phi_2} & X_3 \\ \mu_1 \downarrow & & \mu_2 \downarrow & & \mu_3 \downarrow \\ X_1' & \xrightarrow{\phi_1'} & X_2' & \xrightarrow{\phi_2'} & X_3' \end{array}$$

A pair $(\phi_1, \phi_2)$ of composable morphisms is called *exact* if it fits into an exact triangle $X_1 \xrightarrow{\phi_1} X_2 \xrightarrow{\phi_2} X_3 \xrightarrow{\phi_3} SX_1$.

**Lemma 6.6.1.** *Let* $\mu\colon (\gamma_1, \gamma_2) \to (\phi_1, \phi_2)$ *be a morphism between pairs of composable morphisms and suppose that* $(\phi_1, \phi_2)$ *is exact. Then* $\mu$ *factors through an exact pair of composable morphisms which belong to the smallest full triangulated subcategory containing* $\gamma_1$ *and* $\gamma_2$.

*Proof.* We proceed in two steps. The first step provides a factorization of $\mu$ through a pair $(\gamma_1', \gamma_2')$ of composable morphisms such that $\gamma_2'\gamma_1' = 0$. To achieve this, complete $\gamma_1$ to an exact triangle $C_1 \xrightarrow{\gamma_1} C_2 \xrightarrow{\bar{\gamma}_2} \bar{C}_3 \to SC_1$. Note that $\phi_2\mu_2$ factors through $\bar{\gamma}_2$. Now complete $\left[\begin{smallmatrix}\gamma_2 \\ \bar{\gamma}_2\end{smallmatrix}\right]$ to an exact triangle $C_2 \xrightarrow{\left[\begin{smallmatrix}\gamma_2 \\ \bar{\gamma}_2\end{smallmatrix}\right]} C_3 \amalg \bar{C}_3 \xrightarrow{[\delta\ \bar{\delta}]} C_3' \to SC_2$ and observe that $\mu_3$ factors through $\delta$ via a morphism $\mu_3'\colon C_3' \to X_3$. Thus we obtain the following factorization of $\mu$ with $(\delta\gamma_2)\gamma_1 = -\bar{\delta}\bar{\gamma}_2\gamma_1 = 0$.

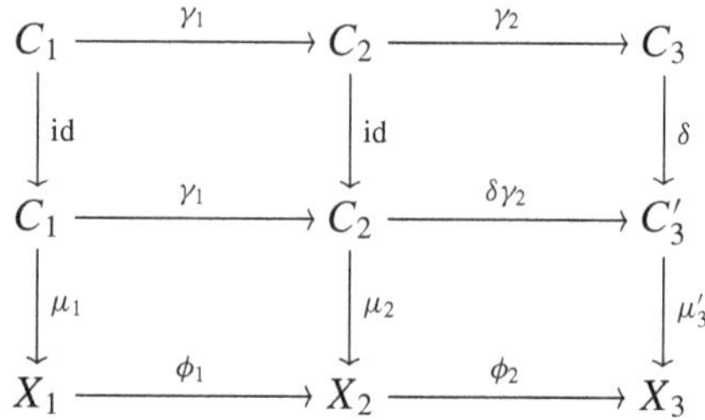

For the second step we may assume that $\gamma_2\gamma_1 = 0$. We complete $\gamma_2$ to an exact triangle $\bar{C}_1 \xrightarrow{\bar{\gamma}_1} C_2 \xrightarrow{\gamma_2} C_3 \to S\bar{C}_1$. Clearly, $\gamma_1$ factors through $\bar{\gamma}_1$ via a morphism $\rho\colon C_1 \to \bar{C}_1$ and $\mu_2\bar{\gamma}_1$ factors through $\phi_1$ via a morphism $\sigma\colon \bar{C}_1 \to X_1$. Thus we obtain the following factorization of $\mu$

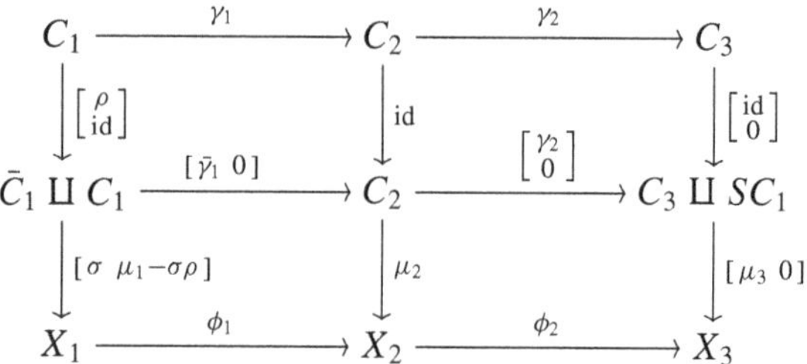

where the middle row fits into an exact triangle. $\qquad\square$

The following statement is a reformulation of the previous one in terms of cofinal subcategories.

**Proposition 6.6.2.** *Let* $\mathcal{T}$ *be a triangulated category and* $\mathcal{S}$ *a full triangulated subcategory. Suppose that* $X_1 \xrightarrow{\phi_1} X_2 \xrightarrow{\phi_2} X_3 \xrightarrow{\phi_3} SX_1$ *is an exact triangle in*

*$T$ and denote by $S/(\phi_1, \phi_2)$ the category whose objects are the commutative diagrams in $T$ of the following form.*

$$
\begin{array}{ccccc}
C_1 & \xrightarrow{\gamma_1} & C_2 & \xrightarrow{\gamma_2} & C_3 \\
\downarrow & & \downarrow & & \downarrow \\
X_1 & \xrightarrow{\phi_1} & X_2 & \xrightarrow{\phi_2} & X_3
\end{array}
$$

*such each $C_i$ belongs to $S$. Then the full subcategory formed by the diagrams such that there exists an exact triangle $C_1 \xrightarrow{\gamma_1} C_2 \xrightarrow{\gamma_2} C_3 \xrightarrow{\gamma_3} SC_1$ is a cofinal subcategory of $S/(\phi_1, \phi_2)$.*

## 6.7. A Kan extension

Let $T$ be a triangulated category with small coproducts and $S$ a small full subcategory. Suppose that the objects of $S$ are $\alpha$-small and that $S$ is closed under $\alpha$-coproducts. We denote by $\mathrm{Add}_\alpha(S^{\mathrm{op}}, \mathrm{Ab})$ the category of $\alpha$-product preserving functors $S^{\mathrm{op}} \to \mathrm{Ab}$. This is a locally presentable abelian category in the sense of [17] and we refer to the Appendix B for basic facts on locally presentable categories. Depending on the choice of $S$, we can think of $\mathrm{Add}_\alpha(S^{\mathrm{op}}, \mathrm{Ab})$ as a locally presentable approximation of the triangulated category $T$. In order to make this precise, we need to introduce various functors.

Let $h_T \colon T \to A(T)$ denote the abelianization of $T$; see Appendix A. Sometimes we write $\widehat{T}$ instead of $A(T)$. The inclusion functor $f \colon S \to T$ induces a functor

$$
f_* \colon A(T) \longrightarrow \mathrm{Add}_\alpha(S^{\mathrm{op}}, \mathrm{Ab}), \quad X \mapsto A(T)((h_T \circ f)-, X),
$$

and we observe that the composite

$$
T \xrightarrow{h_T} A(T) \xrightarrow{f_*} \mathrm{Add}_\alpha(S^{\mathrm{op}}, \mathrm{Ab})
$$

is the restricted Yoneda functor sending each $X \in T$ to $T(-, X)|_S$.

The next proposition discusses a left adjoint for $f_*$. To this end, we denote for any category $C$ by $h_C$ the Yoneda functor sending $X$ in $C$ to $C(-, X)$.

**Proposition 6.7.1.** *The functor $f_*$ admits a left adjoint $f^*$ which makes the following diagram commutative.*

$$
\begin{array}{ccc}
S & \xrightarrow{h_S} & \mathrm{Add}_\alpha(S^{\mathrm{op}}, \mathrm{Ab}) \\
\downarrow{\scriptstyle f=\mathrm{inc}} & & \downarrow{\scriptstyle f^*} \\
T & \xrightarrow{h_T} & A(T)
\end{array}
$$

*Moreover, the functor $f^*$ has the following properties.*

(1)  *$f^*$ is fully faithful and identifies $\mathrm{Add}_\alpha(\mathcal{S}^{\mathrm{op}}, \mathrm{Ab})$ with the full subcategory formed by all colimits of objects in $\{\mathcal{T}(-, X) \mid X \in \mathcal{S}\}$.*
(2)  *$f_*$ preserves small coproducts if and only if (PG2) holds for $\mathcal{S}$.*
(3)  *Suppose that $\mathcal{S}$ is a triangulated subcategory of $\mathcal{T}$. Then for $X$ in $\mathcal{T}$, the adjunction morphism $f^* f_*(h_\mathcal{T} X) \to h_\mathcal{T} X$ identifies with the canonical morphism*

$$\operatorname*{colim}_{(C,\mu)\in\mathcal{S}/X} h_\mathcal{T} C \longrightarrow h_\mathcal{T} X.$$

*Proof.* The functor $f^*$ is constructed as a left Kan extension. To explain this, it is convenient to identify $\mathrm{Add}_\alpha(\mathcal{S}^{\mathrm{op}}, \mathrm{Ab})$ with the category $\mathrm{Lex}_\alpha(\widehat{\mathcal{S}}^{\mathrm{op}}, \mathrm{Ab})$ of left exact functors $\widehat{\mathcal{S}}^{\mathrm{op}} \to \mathrm{Ab}$ which preserve $\alpha$-products. To be more precise, the Yoneda functor $h\colon \mathcal{S} \to \widehat{\mathcal{S}}$ induces an equivalence

$$\mathrm{Lex}_\alpha(\widehat{\mathcal{S}}^{\mathrm{op}}, \mathrm{Ab}) \xrightarrow{\sim} \mathrm{Add}_\alpha(\mathcal{S}^{\mathrm{op}}, \mathrm{Ab}), \quad F \mapsto F \circ h,$$

because every additive functor $\mathcal{S}^{\mathrm{op}} \to \mathrm{Ab}$ extends uniquely to a left exact functor $\widehat{\mathcal{S}}^{\mathrm{op}} \to \mathrm{Ab}$; see Lemma A.1.

Using this identification, the existence of a fully faithful left adjoint $\mathrm{Lex}_\alpha(\widehat{\mathcal{S}}^{\mathrm{op}}, \mathrm{Ab}) \to A(\mathcal{T})$ for $f_*$ and its basic properties follow from Lemma B.6, because the inclusion $f\colon \mathcal{S} \to \mathcal{T}$ induces a fully faithful and right exact functor $\widehat{f}\colon \widehat{\mathcal{S}} \to \widehat{\mathcal{T}} = A(\mathcal{T})$. This functor preserves $\alpha$-coproducts and identifies $\widehat{\mathcal{S}}$ with a full subcategory of $\alpha$-presentable objects, since the objects from $\mathcal{S}$ are $\alpha$-small in $\mathcal{T}$.

(2) Let $\Sigma = \Sigma(f_*)$ denote the set of morphisms of $A(\mathcal{T})$ which $f_*$ makes invertible. It follows from Proposition 2.3.1 that $f_*$ induces an equivalence

$$A(\mathcal{T})[\Sigma^{-1}] \xrightarrow{\sim} \mathrm{Lex}_\alpha(\widehat{\mathcal{S}}^{\mathrm{op}}, \mathrm{Ab}),$$

and therefore $f_*$ preserves small coproducts if and only if $\Sigma$ is closed under taking small coproducts, by Proposition 3.5.1. It is not hard to check that $f_*$ is exact, and therefore a morphism in $A(\mathcal{T})$ belongs to $\Sigma$ if and only if its kernel and cokernel are annihilated by $f_*$. Now observe that an object $F$ in $A(\mathcal{T})$ with presentation $\mathcal{T}(-, X) \to \mathcal{T}(-, Y) \to F \to 0$ is annihilated by $f_*$ if and only if $\mathcal{T}(C, X) \to \mathcal{T}(C, Y)$ is surjective for all $C \in \mathcal{S}$. It follows hat $f_*$ preserves small coproducts if and only if (PG2) holds for $\mathcal{S}$.

(3) Let $F = f_*(h_\mathcal{T} X) = \mathcal{T}(-, X)|_\mathcal{S}$. Then Lemma B.7 implies that $F = \operatorname*{colim}_{(C,\mu)\in\mathcal{S}/X} h_\mathcal{S} C$, since $\mathcal{S}/F = \mathcal{S}/X$. Thus $f^* F = \operatorname*{colim}_{(C,\mu)\in\mathcal{S}/X} h_\mathcal{T} C$.  $\square$

**Corollary 6.7.2.** *Let $\mathcal{T}$ be a triangulated category with small coproducts. Suppose $\mathcal{T}$ is $\alpha$-well generated and denote by $\mathcal{T}^\alpha$ the full subcategory formed by all $\alpha$-compact objects. Then the functor $\mathcal{T} \to A(\mathcal{T})$ taking an object $X$ to*

$$\operatorname*{colim}_{(C,\mu)\in\mathcal{T}^\alpha/X} \mathcal{T}(-,C)$$

*preserves small coproducts.*

## 6.8. A criterion for well generatedness

Let $\mathcal{T}$ be a triangulated category which admits small coproducts. The following result provides a useful criterion for $\mathcal{T}$ to be well generated in terms of cohomological functors into locally presentable abelian categories.

**Proposition 6.8.1.** *Let $\mathcal{T}$ be a triangulated category with small coproducts and $\alpha$ a regular cardinal. Let $\mathcal{S}_0$ be a small set of objects and denote by $\mathcal{S}$ the full subcategory formed by all $\alpha$-coproducts of objects in $\mathcal{S}_0$. Then the following are equivalent.*

(1) *The objects of $\mathcal{S}_0$ are $\alpha$-small and (PG2) holds for $\mathcal{S}_0$.*
(2) *The objects of $\mathcal{S}$ are $\alpha$-small and (PG2) holds for $\mathcal{S}$.*
(3) *The functor $H\colon \mathcal{T} \to \operatorname{Add}_\alpha(\mathcal{S}^{\mathrm{op}}, \mathrm{Ab})$ taking $X$ to $\mathcal{T}(-,X)|_{\mathcal{S}}$ preserves small coproducts.*

*Proof.* It is clear that (1) and (2) are equivalent, and it follows from Proposition 6.7.1 that (2) implies (3). To prove that (3) implies (2), assume that $H$ preserves small coproducts. Let $\phi\colon X \to \coprod_{i\in I} Y_i$ be a morphism in $\mathcal{T}$ with $X \in \mathcal{S}$. Write $\coprod_{i\in I} Y_i = \operatorname*{colim}_{J\subseteq I} Y_J$ as $\alpha$-filtered colimit of coproducts $Y_J = \coprod_{i\in J} Y_i$ with card $J < \alpha$. Then we have

$$\operatorname*{colim}_{J\subseteq I} \mathcal{T}(X, Y_J) \cong \operatorname*{colim}_{J\subseteq I} \operatorname{Hom}_{\mathcal{S}}(\mathcal{S}(-, X), HY_J)$$

$$\cong \operatorname{Hom}_{\mathcal{S}}(\mathcal{S}(-, X), \operatorname*{colim}_{J\subseteq I} HY_J)$$

$$\cong \operatorname{Hom}_{\mathcal{S}}(\mathcal{S}(-, X), \coprod_{i\in I} HY_i)$$

$$\cong \operatorname{Hom}_{\mathcal{S}}(\mathcal{S}(-, X), H(\coprod_{i\in I} Y_i))$$

$$\cong \mathcal{T}(X, \coprod_{i\in I} Y_i).$$

Thus $\phi$ factors through some $Y_J$, and it follows that $X$ is $\alpha$-small. Now Proposition 6.7.1 implies that (PG2) holds for $\mathcal{S}$. $\qquad\square$

## 6.9. Cohomological functors via filtered colimits

The following theorem shows that cohomological functors on well generated triangulated categories can be computed via filtered colimits. This generalizes a fact which is well known for compactly generated triangulated categories. We say that an abelian category has *exact $\alpha$-filtered colimits* provided that every $\alpha$-filtered colimit of exact sequences is exact.

**Theorem 6.9.1.** *Let $\mathcal{T}$ be a triangulated category with small coproducts. Suppose $\mathcal{T}$ is $\alpha$-well generated and denote by $\mathcal{T}^\alpha$ the full subcategory formed by all $\alpha$-compact objects. Let $\mathcal{A}$ be an abelian category which has small coproducts and exact $\alpha$-filtered colimits. If $H : \mathcal{T} \to \mathcal{A}$ is a cohomological functor which preserves small coproducts, then we have for $X$ in $\mathcal{T}$ a natural isomorphism*

$$\operatorname*{colim}_{(C,\mu)\in\mathcal{T}^\alpha/X} HC \xrightarrow{\sim} HX. \tag{6.9.1}$$

*Proof.* The left hand term of (6.9.1) defines a functor $\tilde{H} : \mathcal{T} \to \mathcal{A}$ and we need to show that the canonical morphism $\tilde{H} \to H$ is invertible.

First observe that $\tilde{H}$ is cohomological. This is a consequence of Proposition 6.6.2 and Lemma 6.4.1, because for any exact triangle $X_1 \to X_2 \to X_3 \to SX_1$ in $\mathcal{T}$, the sequence $\tilde{H}X_1 \to \tilde{H}X_2 \to \tilde{H}X_3$ can be written as $\alpha$-filtered colimit of exact sequences in $\mathcal{A}$.

Next we claim that $\tilde{H}$ preserves small coproducts. To this end consider the exact functor $\bar{H} : A(\mathcal{T}) \to \mathcal{A}$ which extends $H$; see Lemma A.2. Note that $\bar{H}$ preserve small coproducts because $H$ has this property. We have for $X$ in $\mathcal{T}$

$$\tilde{H}X = \operatorname*{colim}_{(C,\mu)\in\mathcal{T}^\alpha/X} \bar{H}\big(\mathcal{T}(-,C)\big) \cong \bar{H}\big( \operatorname*{colim}_{(C,\mu)\in\mathcal{T}^\alpha/X} \mathcal{T}(-,C)\big).$$

Now the assertion follows from Corollary 6.7.2.

To complete the proof, consider the full subcategory $\mathcal{T}'$ consisting of those objects $X$ in $\mathcal{T}$ such that the morphism $\tilde{H}X \to HX$ is an isomorphism. Clearly, $\mathcal{T}'$ is a triangulated subcategory since both functors are cohomological, it is closed under taking small coproducts since they are preserved by both functors, and it contains $\mathcal{T}^\alpha$. Thus $\mathcal{T}' = \mathcal{T}$. $\qquad\square$

**Remark 6.9.2.** For an alternative proof of the fact that $\tilde{H}$ is cohomological, one uses Lemma B.5.

## 6.10.  A universal property

Let $T$ be a triangulated category which admits small coproducts and is $\alpha$-well generated. We denote by $A_\alpha(T)$ the full subcategory of $A(T)$ which is formed by all colimits of objects $T(-, X)$ with $X$ in $T^\alpha$. Observe that $A_\alpha(T)$ is a locally presentable abelian category with exact $\alpha$-filtered colimits. This follows from Proposition 6.7.1 and the discussion in Appendix B, because $A_\alpha(T)$ can be identified with a category of left exact functors.

We have two functors

$$H_\alpha : T \longrightarrow A_\alpha(T), \quad X \mapsto \operatorname*{colim}_{(C,\mu)\in T^\alpha/X} T(-, C),$$

$$h_\alpha : T \longrightarrow \mathrm{Add}_\alpha((T^\alpha)^{\mathrm{op}}, \mathrm{Ab}), \quad X \mapsto T(-, X)|_{T^\alpha},$$

which are related by an equivalence as follows.

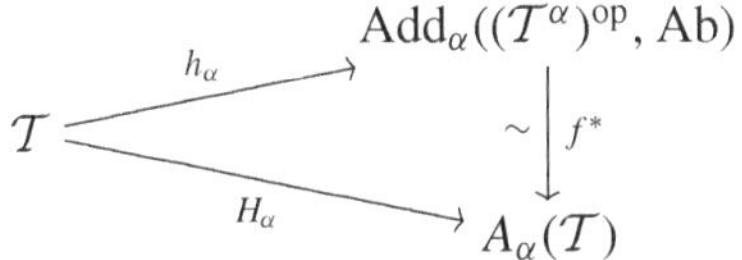

The functor $f^*$ is induced by the inclusion $f : T^\alpha \to T$ and discussed in Proposition 6.7.1. In particular, there it is shown that $f^*(h_\alpha X) = f^* f_*(h_T X) = H_\alpha X$ for all $X$ in $T$.

**Proposition 6.10.1.** *The functor $H_\alpha : T \to A_\alpha(T)$ has the following universal property.*

(1) *The functor $H_\alpha$ is a cohomological functor to an abelian category with small coproducts and exact $\alpha$-filtered colimits and $H_\alpha$ preserves small coproducts.*

(2) *Given a cohomological functor $H : T \to A$ to an abelian category with small coproducts and exact $\alpha$-filtered colimits such that $H$ preserves small coproducts, there exists an essentially unique exact functor $\bar{H} : A_\alpha(T) \to A$ which preserves small coproducts and satisfies $H = \bar{H} \circ H_\alpha$.*

*Proof.* (1) It is clear that $h_\alpha$ is cohomological and it follows from Proposition 6.7.1 that $h_\alpha$ preserves small coproducts.

(2) Given $H : T \to A$, we denote by $\tilde{H} : A(T) \to A$ the exact functor which extends $H$, and we define $\bar{H} : A_\alpha(T) \to A$ by sending each $X$ to $\tilde{H} X$.

The following commutative diagram illustrates this construction.

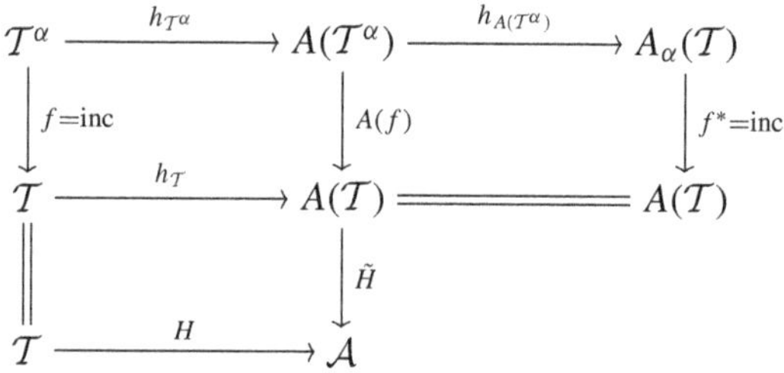

Let us check the properties of $\bar{H}$. The functor $\bar{H}$ preserves small coproducts since $\tilde{H}$ has this property. The functor $\bar{H}$ is exact when restricted to $A(\mathcal{T}^\alpha)$. Thus it follows from Lemma B.5 that $\bar{H}$ is exact. The equality $H = \bar{H} \circ H_\alpha$ is a consequence of Theorem 6.9.1 since both functors coincide on $\mathcal{T}^\alpha$. Suppose now there is a second functor $A_\alpha(\mathcal{T}) \to \mathcal{A}$ having the properties of $\bar{H}$. Then both functors agree on $\mathcal{P} = \{\mathcal{T}(-, X) \mid X \in \mathcal{T}^\alpha\}$ and therefore on all of $A_\alpha(\mathcal{T})$ since each object in $A_\alpha(\mathcal{T})$ is a colimit of objects in $\mathcal{P}$ and both functors preserve colimits. $\qquad\square$

**Remark 6.10.2.** The universal property can be used to show that the category $\mathcal{T}^\alpha$ of $\alpha$-compact objects does not depend on the choice of a perfectly generating set for $\mathcal{T}$. More precisely, if $\mathcal{T}$ is $\alpha$-well generated, then two $\alpha$-localizing subcategories coincide if each contains a small set of $\alpha$-small perfect generators. This follows from the fact that the functor $H_\alpha$ identifies the $\alpha$-compact objects with the $\alpha$-presentable projective objects of $A_\alpha(\mathcal{T})$.

## 6.11. Notes

Well generated triangulated were introduced and studied by Neeman in his book [33] as a natural generalization of compactly generated triangulated categories. For an alternative approach which simplifies the definition, see [23]. More recently, well generated categories with specific models have been studied; see [37, 47] for work involving algebraic models via differential graded categories, and [20] for topological models. In [43], Rosický used combinatorial models and showed that there exist universal cohomological functors into locally presentable categories which are full. Interesting consequences of this fact are discussed in [35]. The description of the universal cohomological functors in terms of filtered colimits seems to be new.

# 7. Localization for well generated categories

## 7.1. Cohomological localization

The following theorem shows that cohomological functors on well generated triangulated categories induce localization functors. This generalizes a fact which is well known for compactly generated triangulated categories.

**Theorem 7.1.1.** *Let $\mathcal{T}$ be a triangulated category with small coproducts which is well generated. Let $H\colon \mathcal{T} \to \mathcal{A}$ be a cohomological functor into an abelian category which has small coproducts and exact $\alpha$-filtered colimits for some regular cardinal $\alpha$. Suppose also that $H$ preserves small coproducts. Then there exists an exact localization functor $L\colon \mathcal{T} \to \mathcal{T}$ such that for each object $X$ we have $LX = 0$ if and only if $H(S^n X) = 0$ for all $n \in \mathbb{Z}$.*

*Proof.* We may assume that $\mathcal{T}$ is $\alpha$-well generated. Let $\Sigma = \Sigma(H)$ denote the set of morphisms $\sigma$ in $\mathcal{T}$ such $H\sigma$ is invertible. Next we assume that $S\Sigma = \Sigma$. Otherwise, we replace $\mathcal{A}$ by a countable product $\mathcal{A}^{\mathbb{Z}}$ of copies of $\mathcal{A}$ and $H$ by $(HS^n)_{n\in\mathbb{Z}}$. Then $\Sigma$ admits a calculus of right fractions by Lemma 4.4.2, and we apply the criterion of Lemma 3.3.1 to show that the morphisms between any two objects in $\mathcal{T}[\Sigma^{-1}]$ form a small set. The existence of a localization functor $L\colon \mathcal{T} \to \mathcal{T}$ with $\operatorname{Ker} L = \operatorname{Ker} H$ then follows from Proposition 5.2.1.

Thus we need to specify for each object $Y$ of $\mathcal{T}$ a small set of objects $S(Y, \Sigma)$ such that for every morphism $X \to Y$ in $\Sigma$, there exist a morphism $X' \to X$ in $\Sigma$ with $X'$ in $S(Y, \Sigma)$. Suppose that $Y$ belongs to $\mathcal{T}^\kappa$. We define by induction $\kappa_{-1} = \kappa + \alpha$ and

$$\kappa_n = \sup\{\operatorname{card} \mathcal{T}^\alpha/U \mid U \in \mathcal{T}^{\kappa_{n-1}}\}^+ + \kappa_{n-1} \quad \text{for} \quad n \geq 0.$$

Then we put $S(Y, \Sigma) = \mathcal{T}^{\bar\kappa}$ with $\bar\kappa = (\sum_{n\geq 0} \kappa_n)^+$.

Now fix $\sigma\colon X \to Y$ in $\Sigma$. The morphism $X' \to X$ in $\Sigma$ with $X'$ in $S(Y, \Sigma)$ is constructed as follows. The canonical morphism $\pi\colon \coprod_{(C,\mu)\in\mathcal{T}^\alpha/X} C \to X$ induces an epimorphism $H\pi$ by Theorem 6.9.1. We can choose $\mathcal{C} \subseteq \mathcal{T}^\alpha/X$ with $\operatorname{card}\mathcal{C} \leq \operatorname{card}\mathcal{T}^\alpha/Y$ such that $\pi_0\colon X_0 = \coprod_{(C,\mu)\in\mathcal{C}} C \to X$ induces an epimorphism $H\pi_0$ since $H\sigma$ is invertible. More precisely, we call two objects $(C, \mu)$ and $(C', \mu')$ of $\mathcal{T}^\alpha/X$ equivalent if $\sigma\mu = \sigma\mu'$, and we choose as objects of $\mathcal{C}$ precisely one representative for each equivalence class.

Suppose we have already constructed $\pi_i\colon X_i \to X$ with $X_i$ in $\mathcal{T}^{\kappa_i}$ for some $i \geq 0$. Then we form the following commutative diagram with exact rows.

$$U_i \xrightarrow{\ \iota_i\ } X_i \xrightarrow{\ \pi_i\ } X \longrightarrow SU_i$$

with vertical maps $\sigma_i$, $\|$, $\sigma$, $S\sigma_i$ to

$$V_i \longrightarrow X_i \longrightarrow Y \longrightarrow SV_i$$

Note that $H\sigma_i$ is invertible. Thus we can choose $\mathcal{C}_i \subseteq T^\alpha/U_i$ with

$$\operatorname{card}\mathcal{C}_i \leq \operatorname{card} T^\alpha/V_i \leq \operatorname{card} T^\alpha/X_i + \operatorname{card} T^\alpha/Y < \kappa_{i+1}$$

such that $\xi_i \colon \coprod_{(C,\mu)\in\mathcal{C}_i} C \to U_i$ induces an epimorphism $H\xi_i$. Now complete $\iota_i \circ \xi_i$ to an exact triangle and define $\pi_{i+1} \colon X_{i+1} \to X$ by the commutativity of the following diagram.

$$\coprod_{(C,\mu)\in\mathcal{C}_i} C \xrightarrow{\ \iota_i\circ\xi_i\ } X_i \xrightarrow{\ \phi_i\ } X_{i+1} \longrightarrow S\Big(\coprod_{(C,\mu)\in\mathcal{C}_i} C\Big)$$

with $\pi_i$, $\pi_{i+1}$ to $X$.

Observe that $X_{i+1}$ belongs to $T^{\kappa_{i+1}}$ and that $\operatorname{Ker} H\pi_i = \operatorname{Ker} H\phi_i$. The $\phi_i$ induce an exact triangle

$$\coprod_{i\in\mathbb{N}} X_i \xrightarrow{(\mathrm{id}-\phi_i)} \coprod_{i\in\mathbb{N}} X_i \xrightarrow{\ \psi\ } X' \longrightarrow S\Big(\coprod_{i\in\mathbb{N}} X_i\Big) \qquad (7.1.1)$$

such that $X'$ belongs to $S(Y, \Sigma)$ and the morphism $(\pi_i)\colon \coprod_{i\in\mathbb{N}} X_i \to X$ factors through $\psi$ via a morphism $\tau\colon X' \to X$. We claim that $H\tau$ is invertible. In fact, the lemma below implies that the $\pi_i$ induce the following exact sequence

$$0 \longrightarrow \coprod_{i\in\mathbb{N}} HX_i \xrightarrow{(\mathrm{id}-H\phi_i)} \coprod_{i\in\mathbb{N}} HX_i \xrightarrow{(H\pi_i)} HX \longrightarrow 0.$$

On the other hand, the exact triangle (7.1.1) induces the exact sequence

$$H\Big(\coprod_{i\in\mathbb{N}} X_i\Big) \xrightarrow{H(\mathrm{id}-\phi_i)} H\Big(\coprod_{i\in\mathbb{N}} X_i\Big) \xrightarrow{H\psi} HX' \longrightarrow HS\Big(\coprod_{i\in\mathbb{N}} X_i\Big),$$

and a comparison shows that $H\tau$ is invertible. Here, we use again that $H$ preserves small coproducts, and this completes the proof. $\qquad\square$

**Lemma 7.1.2.** *Let $\mathcal{A}$ be an abelian category which admits countable coproducts. Then a sequence of epimorphisms $(\pi_i)_{i\in\mathbb{N}}$*

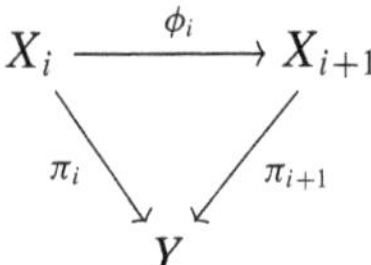

$$X_i \xrightarrow{\ \phi_i\ } X_{i+1}$$

with $\pi_i$, $\pi_{i+1}$ to $Y$.

*satisfying* $\pi_i = \pi_{i+1} \circ \phi_i$ *and* $\operatorname{Ker} \pi_i = \operatorname{Ker} \phi_i$ *for all* $i$ *induces an exact sequence*

$$0 \longrightarrow \coprod_{i \in \mathbb{N}} X_i \xrightarrow{(\mathrm{id} - \phi_i)} \coprod_{i \in \mathbb{N}} X_i \xrightarrow{(\pi_i)} Y \longrightarrow 0.$$

*Proof.* The assumption $U_i := \operatorname{Ker} \pi_i = \operatorname{Ker} \phi_i$ implies that there exists a morphism $\pi_i' \colon Y \to X_i$ with $\pi_i \pi_i' = \operatorname{id} Y$ and $\phi_i \pi_i' = \pi_{i+1}'$ for all $i \geq 1$. Thus we have a sequence of commuting squares

$$
\begin{array}{ccc}
U_i \amalg Y & \xrightarrow{\ [\,\mathrm{inc}\ \pi_i'\,]\ } & X_i \\[2pt]
\left\downarrow{\scriptstyle\begin{bmatrix} 0 & 0 \\ 0 & \mathrm{id} \end{bmatrix}}\right. & & \left\downarrow{\scriptstyle \phi_i}\right. \\[2pt]
U_{i+1} \amalg Y & \xrightarrow{\ [\,\mathrm{inc}\ \pi_{i+1}'\,]\ } & X_{i+1}
\end{array}
$$

where the horizontal maps are isomorphisms. Taking colimits on both sides, the assertion follows. $\qquad\square$

## 7.2. Localization with respect to a small set of objects

Let $T$ be a well generated triangulated category and $S$ a localizing subcategory which is generated by a small set of objects. The following result says that $S$ and $T/S$ are both well generated and that the filtration $T = \bigcup_\alpha T^\alpha$ via $\alpha$-compact objects induces canonical filtrations

$$S = \bigcup_\alpha (S \cap T^\alpha) \quad \text{and} \quad T/S = \bigcup_\alpha T^\alpha / (S \cap T^\alpha).$$

**Theorem 7.2.1.** *Let* $T$ *be a well generated triangulated category and* $S$ *a localizing subcategory which is generated by a small set of objects. Fix a regular cardinal* $\alpha$ *such that* $T$ *is* $\alpha$-*well generated and* $S$ *is generated by* $\alpha$-*compact objects.*

(1) *An object $X$ in $T$ belongs to $S$ if and only if every morphism $C \to X$ from an object $C$ in $T^\alpha$ factors through some object in $S \cap T^\alpha$.*
(2) *The localizing subcategory $S$ and the quotient category $T/S$ are $\alpha$-well generated.*
(3) *We have $S^\alpha = S \cap T^\alpha$ and a commutative diagram of exact functors*

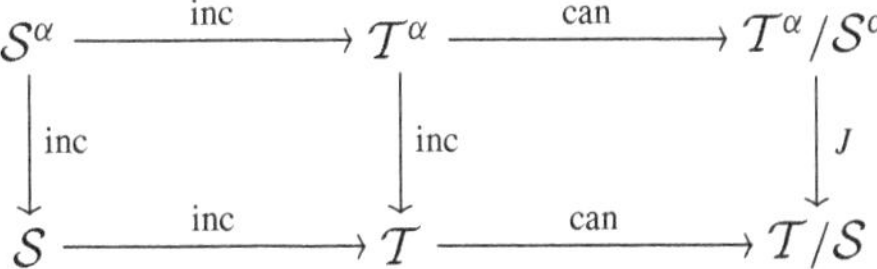

*such that $J$ is fully faithful. Moreover, $J$ induces a functor $T^\alpha/S^\alpha \to (T/S)^\alpha$ such that every object of $(T/S)^\alpha$ is a direct factor of an object in the image of $J$. This functor is an equivalence if $\alpha > \aleph_0$.*

*Proof.* Let $C = S \cap T^\alpha$. Then the inclusion $i: C \to T^\alpha$ induces a fully faithful and exact functor $i^*: \mathrm{Add}_\alpha(C^{\mathrm{op}}, \mathrm{Ab}) \to \mathrm{Add}_\alpha((T^\alpha)^{\mathrm{op}}, \mathrm{Ab})$ which is left adjoint to the functor $i_*$ taking $F$ to $F \circ i$; see Lemma B.8. Note that the image $\mathrm{Im}\, i^*$ of $i^*$ is closed under small coproducts. We consider the restricted Yoneda functor $h_\alpha: T \to \mathrm{Add}_\alpha((T^\alpha)^{\mathrm{op}}, \mathrm{Ab})$ taking $X$ to $T(-, X)|_{T^\alpha}$ and observe that $h_\alpha^{-1}(\mathrm{Im}\, i^*)$ is a localizing subcategory of $T$ containing $C$. Thus we obtain a functor $H$ making the following diagram commutative.

$$
\begin{array}{ccc}
S & \xrightarrow{\quad\mathrm{inc}\quad} & T \\
\big\downarrow{\scriptstyle H} & & \big\downarrow{\scriptstyle h_\alpha} \\
\mathrm{Add}_\alpha(C^{\mathrm{op}}, \mathrm{Ab}) & \xrightarrow{\quad i^*\quad} & \mathrm{Add}_\alpha((T^\alpha)^{\mathrm{op}}, \mathrm{Ab})
\end{array}
$$

Let us compare $H$ with the restricted Yoneda functor

$$H': S \longrightarrow \mathrm{Add}_\alpha(C^{\mathrm{op}}, \mathrm{Ab}), \quad X \mapsto S(-, X)|_C.$$

In fact, we have an isomorphism

$$H \xrightarrow{\sim} i_* \circ i^* \circ H = i_* \circ h_\alpha|_S = H'$$

and $H'$ preserves small coproducts since $h_\alpha$ does. It follows from Proposition 6.8.1 that $C$ provides a small set of $\alpha$-small perfect generators for $S$. Thus $S$ is $\alpha$-well generated and $S^\alpha = \mathrm{Loc}_\alpha C = S \cap T^\alpha$.

Next we apply Proposition 5.2.1 and obtain a localization functor $L: T \to T$ with $\mathrm{Ker}\, L = S$. We use $L$ to show that $S = h_\alpha^{-1}(\mathrm{Im}\, i^*)$. We know already that $S \subseteq h_\alpha^{-1}(\mathrm{Im}\, i^*)$. Now let $X$ be an object in $h_\alpha^{-1}(\mathrm{Im}\, i^*)$ and consider the exact triangle $\Gamma X \to X \to LX \to S(\Gamma X)$. Then $T(C, LX) = 0$ for all $C \in C$ and therefore $i_* h_\alpha LX = 0$. On the other hand, $h_\alpha LX = i^* F$ for some functor $F$ and therefore $0 = i_* h_\alpha LX = i_* i^* F \cong F$. Thus $LX = 0$ and therefore $X$ belongs to $S$. This shows $S = h_\alpha^{-1}(\mathrm{Im}\, i^*)$.

Now we prove (1) and use the description of the essential image of $i^*$ from Lemma B.8. We have for an object $X$ in $T$ that $X$ belongs to $S$ iff $h_\alpha X$ belongs to $\mathrm{Im}\, i^*$ iff every morphism $T^\alpha(-, C) \to h_\alpha X$ with $C \in T^\alpha$ factors through $T^\alpha(-, C')$ for some $C' \in C$ iff every morphism $C \to X$ with $C \in T^\alpha$ factors through some $C' \in C$.

An immediate consequence of (1) is the fact that $J$ is fully faithful. This follows from Lemma 4.7.1.

Now consider the quotient functor $q : \mathcal{T}^\alpha \to \mathcal{T}^\alpha/\mathcal{S}^\alpha$. This induces an exact functor $q^* : \mathrm{Add}_\alpha((\mathcal{T}^\alpha)^{\mathrm{op}}, \mathrm{Ab}) \to \mathrm{Add}_\alpha((\mathcal{T}^\alpha/\mathcal{S}^\alpha)^{\mathrm{op}}, \mathrm{Ab})$ which is left adjoint to the fully faithful functor $q_*$ taking $F$ to $F \circ q$; see Lemma B.8. Clearly, $q^* \circ h_\alpha$ annihilates $\mathcal{S}$ and induces therefore a functor $K$ making the following diagram commutative.

$$
\begin{array}{ccc}
\mathcal{T} & \xrightarrow{\;\;Q=\mathrm{can}\;\;} & \mathcal{T}/\mathcal{S} \\
\Big\downarrow{\scriptstyle h_\alpha} & & \Big\downarrow{\scriptstyle K} \\
\mathrm{Add}_\alpha((\mathcal{T}^\alpha)^{\mathrm{op}}, \mathrm{Ab}) & \xrightarrow{\;\;q^*\;\;} & \mathrm{Add}_\alpha((\mathcal{T}^\alpha/\mathcal{S}^\alpha)^{\mathrm{op}}, \mathrm{Ab})
\end{array}
$$

Note that $Q$ admits a right adjoint which we denote by $Q_\rho$. We identify $\mathcal{T}^\alpha/\mathcal{S}^\alpha$ via $J$ with a full triangulated subcategory of $\mathcal{T}/\mathcal{S}$ and consider the restricted Yoneda functor

$$
K' : \mathcal{T}/\mathcal{S} \longrightarrow \mathrm{Add}_\alpha((\mathcal{T}^\alpha/\mathcal{S}^\alpha)^{\mathrm{op}}, \mathrm{Ab}), \quad X \mapsto \mathcal{T}/\mathcal{S}(-, X)|_{\mathcal{T}^\alpha/\mathcal{S}^\alpha}.
$$

Adjointness gives the following isomorphism

$$
\mathcal{T}/\mathcal{S}(JqC, X) = \mathcal{T}/\mathcal{S}(QC, X) \cong \mathcal{T}(C, Q_\rho X)
$$

for all $C \in \mathcal{T}^\alpha$ and $X \in \mathcal{T}/\mathcal{S}$. Thus we have an isomorphism

$$
K \xleftarrow{\sim} K \circ Q \circ Q_\rho = q^* \circ h_\alpha \circ Q_\rho \cong q^* \circ q_* \circ K' \xrightarrow{\sim} K'
$$

and $K'$ preserves small coproducts since $h_\alpha$ does. It follows from Proposition 6.8.1 that $\mathcal{T}^\alpha/\mathcal{S}^\alpha$ provides a small set of $\alpha$-small perfect generators for $\mathcal{T}/\mathcal{S}$. Thus $\mathcal{T}/\mathcal{S}$ is $\alpha$-well generated and $(\mathcal{T}/\mathcal{S})^\alpha = \mathrm{Loc}_\alpha(\mathcal{T}^\alpha/\mathcal{S}^\alpha)$. $\qquad\square$

**Corollary 7.2.2.** *Let $\mathcal{T}$ be an $\alpha$-well generated triangulated category and $\mathcal{S}$ a localizing subcategory generated by a small set $\mathcal{S}_0$ of $\alpha$-compact objects. Then $\mathcal{S}$ is $\alpha$-well generated and $\mathcal{S}^\alpha$ equals the $\alpha$-localizing subcategory generated by $\mathcal{S}_0$.*

*Proof.* In the preceding proof of Theorem 7.2.1, we can choose for $\mathcal{C}$ instead of $\mathcal{S} \cap \mathcal{T}^\alpha$ the $\alpha$-localizing subcategory of $\mathcal{T}$ which is generated by $\mathcal{S}_0$. Then the proof shows that $\mathcal{C}$ provides a small set of $\alpha$-small perfect generators for $\mathcal{S}$. Thus we have $\mathcal{S}^\alpha = \mathcal{C}$ by definition. $\qquad\square$

The localization with respect to a localizing subcategory generated by a small set of objects can be interpreted in various ways. The following remark provides some indication.

**Remark 7.2.3.** (1) Let $\mathcal{T}$ be a well generated triangulated category and $\phi$ a morphism in $\mathcal{T}$. Then there exists a universal exact localization functor

$L\colon \mathcal{T} \to \mathcal{T}$ inverting $\phi$. To see this, complete $\phi$ to an exact triangle $X \xrightarrow{\phi} Y \to Z \to SX$ and let $L$ be the localization functor such that $\operatorname{Ker} L$ equals the localizing subcategory generated by $Z$. Conversely, any exact localization functor $L\colon \mathcal{T} \to \mathcal{T}$ is the universal exact localization functor inverting some morphism $\phi$ provided that $\operatorname{Ker} L$ is generated by a small set $\mathcal{S}_0$ of objects. To see this, take $\phi\colon 0 \to \coprod_{X \in \mathcal{S}_0} X$.

(2) Let $\mathcal{T}$ be a triangulated category and $L\colon \mathcal{T} \to \mathcal{T}$ an exact localization functor such that $\mathcal{S} = \operatorname{Ker} L$ is generated by a single object $W$. Then the first morphism $\Gamma X \to X$ from the functorial triangle $\Gamma X \to X \to LX \to S(\Gamma X)$ is called *cellularization* and the second morphism $X \to LX$ is called *nullification* with respect to $W$. The objects in $\mathcal{S}$ are *built from $W$*.

## 7.3. Functors between well generated categories

We consider functors between well generated triangulated categories which are exact and preserve small coproducts. The following result shows that such functors are controlled by their restriction to the subcategory of $\alpha$-compact objects for some regular cardinal $\alpha$.

**Proposition 7.3.1.** *Let $F\colon \mathcal{T} \to \mathcal{U}$ be an exact functor between $\alpha$-well generated triangulated categories. Suppose that $F$ preserves small coproducts and let $G$ be a right adjoint.*

(1) *There exists a regular cardinal $\beta_0 \geq \alpha$ such that $F$ preserves $\beta_0$-compactness. In that case $F$ preserves $\beta$-compactness for all regular $\beta \geq \beta_0$.*

(2) *Given a regular cardinal $\beta \geq \beta_0$, the restriction $f\colon \mathcal{T}^\beta \to \mathcal{U}^\beta$ of $F$ induces the following diagram of functors which commute up to natural isomorphisms.*

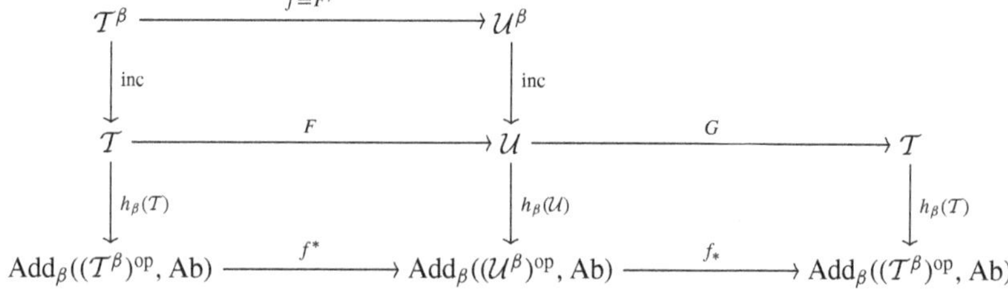

*Proof.* (1) Choose $\beta_0 \geq \alpha$ such that $F(\mathcal{T}^\alpha) \subseteq \mathcal{U}^{\beta_0}$. Then we get for $\beta \geq \beta_0$

$$F(\mathcal{T}^\beta) = F(\operatorname{Loc}_\beta \mathcal{T}^\alpha) \subseteq \operatorname{Loc}_\beta F(\mathcal{T}^\alpha) \subseteq \operatorname{Loc}_\beta \mathcal{U}^{\beta_0} = \mathcal{U}^\beta.$$

(2) We apply Theorem 6.9.1 to show that $h_\beta(\mathcal{U}) \circ F \cong f^* \circ h_\beta(\mathcal{T})$. In fact, it follows from Proposition 6.10.1 and Lemma B.8 that both composites are cohomological functors, preserve small coproducts, and agree on $\mathcal{T}^\beta$.

The isomorphism $h_\beta(\mathcal{T}) \circ G \cong f_* \circ h_\beta(\mathcal{U})$ follows from the adjointness of $F$ and $G$, since $\mathcal{T}(C, GX) \cong \mathcal{U}(fC, X)$ for every $C \in \mathcal{T}^\beta$ and $X \in \mathcal{U}$.  $\square$

## 7.4. The kernel of a functor between well generated categories

We show that the class of well generated triangulated categories is closed under taking kernels of exact functors which preserve small coproducts.

**Theorem 7.4.1.** *Let $F : \mathcal{T} \to \mathcal{U}$ be an exact functor between $\alpha$-well generated triangulated categories and suppose that $F$ preserves small coproducts. Let $\mathcal{S} = \mathrm{Ker}\, F$ and choose a regular cardinal $\beta \geq \alpha$ such that $F$ preserves $\beta$-compactness.*

(1) *An object $X$ in $\mathcal{T}$ belongs to $\mathcal{S}$ if and only if every morphism $C \to X$ with $C \in \mathcal{T}^\beta$ factors through a morphism $\gamma : C \to C'$ in $\mathcal{T}^\beta$ satisfying $F\gamma = 0$.*
(2) *Suppose $\beta > \aleph_0$. Then $\mathcal{S}$ is $\beta$-well generated and $\mathcal{S}^\beta = \mathcal{S} \cap \mathcal{T}^\beta$.*

*Proof.* Let $f : \mathcal{T}^\beta \to \mathcal{U}^\beta$ be the restriction of $F$ and denote by $\mathfrak{I}$ the set of morphisms in $\mathcal{T}^\beta$ which are annihilated by $F$.

(1) Let $X$ be an object in $\mathcal{T}$. Then it follows from Proposition 7.3.1 that $FX = 0$ if and only $f^* h_\beta X = 0$. Now Lemma B.8 implies that $f^* h_\beta X = 0$ iff each morphism $C \to X$ with $C \in \mathcal{T}^\beta$ factors through some morphism $C \to C'$ in $\mathfrak{I}$.

(2) Let $\mathcal{S}'$ denote the localizing subcategory of $\mathcal{T}$ which is generated by all homotopy colimits of sequences

$$C_0 \longrightarrow C_1 \longrightarrow C_2 \longrightarrow \cdots$$

of morphisms in $\mathfrak{I}$. We claim that $\mathcal{S}' = \mathrm{Ker}\, F$. Clearly, we have $\mathcal{S}' \subseteq \mathrm{Ker}\, F$. Now fix an object $X \in \mathrm{Ker}\, F$. We have seen in (1) that each morphism $\mu : C \to X$ with $C \in \mathcal{T}^\beta$ factors through some morphism $C \to C'$ in $\mathfrak{I}$. We obtain by induction a sequence

$$C = C_0 \xrightarrow{\gamma_0} C_1 \xrightarrow{\gamma_1} C_2 \xrightarrow{\gamma_2} \cdots$$

of morphisms in $\mathfrak{I}$ such that $\mu$ factors through each finite composite $\gamma_i \ldots \gamma_0$. Thus $\mu$ factors through the homotopy colimit of this sequence and therefore through an object of $\mathcal{S}' \cap \mathcal{T}^\beta$. Here one uses that $\beta > \aleph_0$. We conclude from Theorem 7.2.1 that $X$ belongs to $\mathcal{S}'$. Moreover, we conclude from this theorem that $\mathcal{S}'$ is $\beta$-well generated.  $\square$

**Remark 7.4.2.** It is necessary to assume in part (2) of the preceding theorem that $\beta > \aleph_0$. For example, there exists a ring $A$ with Jacobson radical $\mathfrak{r}$ such that the functor $F = - \otimes_A^{\mathbf{L}} A/\mathfrak{r} \colon \mathbf{D}(A) \to \mathbf{D}(A/\mathfrak{r})$ satisfies $\mathcal{S} = \mathrm{Ker}\, F \neq 0$ but $\mathcal{S} \cap \mathbf{D}(A)^c = 0$; see [22].

Observe that Theorem 7.4.1 provides a partial answer to the telescope conjecture for compactly generated categories. This conjecture claims that the kernel of a localization functor $L \colon \mathcal{T} \to \mathcal{T}$ is generated by compact objects provided that $L$ preserves small coproducts. Part (1) implies that $\mathcal{S} = \mathrm{Ker}\, L$ is generated by morphisms between compact objects, and part (2) says that $\mathcal{S}$ is generated by $\aleph_1$-compact objects. I am grateful to Amnon Neeman for explaining to me how to deduce (2) from (1). The following corollary makes the connection with the telescope conjecture more precise; just put $\alpha = \aleph_0$.

**Corollary 7.4.3.** *Let $L \colon \mathcal{T} \to \mathcal{T}$ be an exact localization functor which preserves small coproducts. Suppose that $\mathcal{T}$ is $\alpha$-well generated and let $\beta \geq \max(\alpha, \aleph_1)$. Then $\mathcal{S} = \mathrm{Ker}\, L$ is $\beta$-well generated and $\mathcal{S}^\beta = \mathcal{S} \cap \mathcal{T}^\beta$.*

*Proof.* Let $L \colon \mathcal{T} \to \mathcal{T}$ be a localization functor which preserves small coproducts. Write $L = G \circ F$ as the composite of a quotient functor $F \colon \mathcal{T} \to \mathcal{U}$ and a fully faithful right adjoint $G$, where $\mathcal{U} = \mathcal{T}/\mathcal{S}$ and $\mathcal{S} = \mathrm{Ker}\, L$. Then $G$ preserves small coproducts by Proposition 5.5.1. The isomorphism (5.4.1) from the proof of Lemma 5.4.1 shows that $F$ preserves $\alpha$-smallness and sends a set of perfect generators of $\mathcal{T}$ to a set of perfect generators of $\mathcal{U}$. In particular, $F$ preserves $\beta$-compactness for all regular $\beta \geq \alpha$. Now apply Theorem 7.4.1. $\qquad\square$

## 7.5. The kernel of a cohomological functor on a well generated category

The following result says that kernels of cohomological functors from well generated triangulated categories into locally presentable abelian categories are well generated. The argument is basically the same as that for kernels of exact functors between well generated triangulated categories.

**Theorem 7.5.1.** *Let $H \colon \mathcal{T} \to \mathcal{A}$ be a cohomological functor from a well generated triangulated category into a locally presentable abelian category and suppose that $H$ preserves small coproducts. Let $\mathcal{S}$ denote the localizing subcategory of $\mathcal{T}$ consisting of all objects $X$ such that $H(S^n X) = 0$ for all $n \in \mathbb{Z}$. Then $\mathcal{S}$ is a well generated triangulated category.*

*Proof.* Replacing $H$ by $(HS^n)_{n\in\mathbb{Z}}$, we may assume that $S = \operatorname{Ker} H$. Choose a regular cardinal $\alpha$ such that $\mathcal{T}$ is $\alpha$-well generated and $\mathcal{A}$ is locally $\alpha$-presentable. Then we have $H(\mathcal{T}^\alpha) \subseteq \mathcal{A}^\beta$ for some regular cardinal $\beta$ and we assume $\beta \geq \max(\alpha, \aleph_1)$. The description of $H$ in Theorem 6.9.1 shows that $H$ restricts to a functor $h\colon \mathcal{T}^\beta \to \mathcal{A}^\beta$, and we denote by $\bar{h}\colon A(\mathcal{T}^\beta) \to \mathcal{A}^\beta$ the induced exact functor. Then we obtain the following functor

$$h^*\colon \operatorname{Add}_\beta((\mathcal{T}^\beta)^{\mathrm{op}}, \mathrm{Ab}) \xrightarrow{\sim} \operatorname{Lex}_\beta(A(\mathcal{T}^\beta)^{\mathrm{op}}, \mathrm{Ab}) \xrightarrow{\bar{h}^*} \operatorname{Lex}_\beta((\mathcal{A}^\beta)^{\mathrm{op}}, \mathrm{Ab}) \xrightarrow{\sim} \mathcal{A}$$

where the first equivalence follows from Lemma B.1 and the second equivalence follows from Lemma B.6. The functor $\bar{h}^*$ is a left Kan extension; it takes a filtered colimit

$$F = \operatorname*{colim}_{(C,\mu)\in A(\mathcal{T}^\beta)/F} A(\mathcal{T}^\beta)(-, C) \quad \text{to} \quad \operatorname*{colim}_{(C,\mu)\in A(\mathcal{T}^\beta)/F} \mathcal{A}^\beta(-, \bar{h}C).$$

Note that $h^*$ is exact and preserves small coproducts. This follows from Lemma B.5 and the fact that $\bar{h}^*$ is left adjoint to the restriction functor $\bar{h}_*$.

The composite $h^* \circ h_\beta \colon \mathcal{T} \to \mathcal{A}$ coincides with $H$ on $\mathcal{T}^\beta$ and therefore $h^* \circ h_\beta \cong H$ by Theorem 6.9.1. In particular, we have for each $X$ in $\mathcal{T}$ that $HX = 0$ if and only if $h^*(h_\beta X) = 0$. Now we use the same argument as in the proof of Theorem 7.4.1 and show that $\operatorname{Ker} H$ is generated by all homotopy colimits of countable sequences of morphisms in $\mathcal{T}^\beta$ which are annihilated by $H$. $\qquad\square$

## 7.6. Localization of well generated categories versus abelian localization

We demonstrate the interplay between triangulated and abelian localization. To this end recall from Proposition 6.10.1 that we have for each well generated category $\mathcal{T}$ a universal cohomological functor $H_\alpha \colon \mathcal{T} \to A_\alpha(\mathcal{T})$ into a locally $\alpha$-presentable abelian category. We show that each exact localization functor for $\mathcal{T}$ can be extended to an exact localization functor for $A_\alpha(\mathcal{T})$ for some regular cardinal $\alpha$.

**Theorem 7.6.1.** *Let $\mathcal{T}$ be a well generated triangulated category and $L\colon \mathcal{T} \to \mathcal{T}$ an exact localization functor. Suppose that $\operatorname{Ker} L$ is well generated. Then there exists a regular cardinal $\alpha$ and an exact localization functor*

$L'\colon A_\alpha(\mathcal{T}) \to A_\alpha(\mathcal{T})$ *such that the following square commutes up to a natural isomorphism.*

$$
\begin{array}{ccc}
\mathcal{T} & \xrightarrow{\ \ L\ \ } & \mathcal{T} \\
\ \downarrow{\scriptstyle H_\alpha} & & \ \downarrow{\scriptstyle H_\alpha} \\
A_\alpha(\mathcal{T}) & \xrightarrow{\ \ L'\ \ } & A_\alpha(\mathcal{T})
\end{array}
$$

*More precisely, the adjunction morphisms* $\operatorname{Id}\mathcal{T} \to L$ *and* $\operatorname{Id}A_\alpha(\mathcal{T}) \to L'$ *induce for each $X$ in $\mathcal{T}$ the following isomorphisms.*

$$
H_\alpha L X \xrightarrow{\ \sim\ } L'(H_\alpha L X) = L' H_\alpha(L X) \xleftarrow{\ \sim\ } L' H_\alpha(X)
$$

*An object $X$ in $\mathcal{T}$ is $L$-acyclic if and only if $H_\alpha X$ is $L'$-acyclic, and $X$ is $L$-local if and only if $H_\alpha X$ is $L'$-local.*

*Proof.* Choose a regular cardinal $\alpha > \aleph_0$ such that $\mathcal{T}$ is $\alpha$-well generated and $\mathcal{S} = \operatorname{Ker} L$ is generated by $\alpha$-compact objects. Let $\mathcal{U} = \mathcal{T}/\mathcal{S}$ and write $L = G \circ F$ as the composite of the quotient functor $F\colon \mathcal{T} \to \mathcal{U}$ with its right adjoint $G\colon \mathcal{U} \to \mathcal{T}$.

Now identify $A_\alpha(\mathcal{T}) = \operatorname{Add}_\alpha((\mathcal{T}^\alpha)^{\mathrm{op}}, \mathrm{Ab})$ and $A_\alpha(\mathcal{U}) = \operatorname{Add}_\alpha((\mathcal{U}^\alpha)^{\mathrm{op}}, \mathrm{Ab})$. The induced functor $f\colon \mathcal{T}^\alpha \to \mathcal{U}^\alpha$ equals, up to an equivalence, the quotient functor $\mathcal{T}^\alpha \to \mathcal{T}^\alpha/\mathcal{S}^\alpha$, by Theorem 7.2.1. From $f$ we obtain a pair of adjoint functors $f^*$ and $f_*$ by Lemma B.8. Both functors are exact and the right adjoint $f_*$ is fully faithful. Thus we obtain an exact localization functor $L' = f_* \circ f^*$ for $A_\alpha(\mathcal{T})$ by Corollary 2.4.2. The commutativity $H_\alpha \circ L \cong L' \circ H_\alpha$ and the assertions about acyclic and local objects then follow from Proposition 7.3.1. $\qquad\square$

## 7.7. Example: The derived category of an abelian Grothendieck category

Let $\mathcal{A}$ be an abelian Grothendieck category. Then the derived category $\mathbf{D}(\mathcal{A})$ of unbounded chain complexes is a well generated triangulated category. Let us sketch an argument.

The Popescu-Gabriel theorem says that for each generator $G$ of $\mathcal{A}$, the functor $T = \mathcal{A}(G, -)\colon \mathcal{A} \to \operatorname{Mod} A$ (where $A = \mathcal{A}(G, G)$ denotes the endomorphism ring of $G$) is fully faithful and admits an exact left adjoint, say $Q$; see [45, Theorem X.4.1]. Consider the cohomological functor $H\colon \mathbf{D}(A) \to \mathcal{A}$ taking a complex $X$ to $Q(\coprod_{n \in \mathbb{Z}} H^n X)$. Then an application of Theorem 7.5.1 shows that $\mathcal{S} = \operatorname{Ker} H$ is well generated, and therefore $\mathbf{D}(A)/\mathcal{S}$ is well generated by Theorem 7.2.1.

Next observe that $\mathbf{K}(Q)$ induces an equivalence

$$\mathbf{K}(\mathrm{Mod}\,A)/(\mathrm{Ker}\,\mathbf{K}(Q)) \xrightarrow{\sim} \mathbf{K}(\mathcal{A})$$

since $\mathbf{K}(Q)$ has $\mathbf{K}(T)$ as a fully faithful right adjoint. Moreover, the cohomology of each object in the kernel of $\mathbf{K}(Q)$ lies in the kernel of $Q$. Thus we obtain the following commutative diagram.

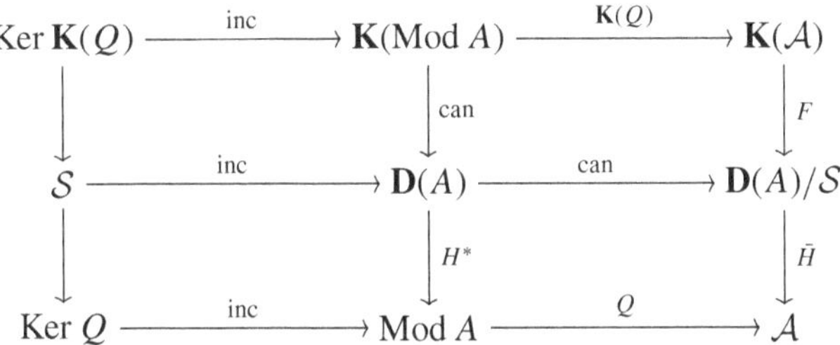

It is easily checked that the kernel of $F$ consists of all acyclic complexes. Thus $F$ induces an equivalence $\mathbf{D}(\mathcal{A}) \xrightarrow{\sim} \mathbf{D}(A)/\mathcal{S}$.

## 7.8. Notes

Given a triangulated category $\mathcal{T}$, there are two basic questions when one studies exact localization functors $\mathcal{T} \to \mathcal{T}$. One can ask for the existence of a localization functor with some prescribed kernel, and one can ask for a classification, or at least some structural results, for the set of all localization functors on $\mathcal{T}$. Well generated categories provide a suitable setting for some partial answers.

The fact that cohomological functors induce localization functors is well known for compactly generated triangulated categories [28], but the result seems to be new for well generated categories. The localization theorem which describes the localization with respect to a small set of objects is due to Neeman [33]. The example of the derived category of an abelian Grothendieck category is discussed in [3, 34]. The description of the kernel of an exact functor between well generated categories seems to be new. A motivation for this is the telescope conjecture which is due to Bousfield and Ravenel [8, 40] and originally formulated for the stable homotopy category of CW-spectra.

It is interesting to note that the existence of localization functors depends to some extent on axioms from set theory; see for instance [11, 10].

# 8. Epilogue: Beyond well generatedness

Well generated triangulated categories were introduced by Neeman as a class of triangulated categories which includes all compactly generated categories and behaves well with respect to localization. We have discussed in Sections 6 and 7 most of the basic properties of well generated categories but the picture is still not complete because some important questions remain open. For instance, given a well generated triangulated category $\mathcal{T}$, we do not know when a localizing subcategory arises as the kernel of a localization functor and when it is generated by a small set of objects. Also, one might ask when the set of all localizing subcategories is small. Another aspect is Brown representability. We do know that every cohomological functor $\mathcal{T}^{\mathrm{op}} \to \mathrm{Ab}$ preserving small products is representable, but what about covariant functors $\mathcal{T} \to \mathrm{Ab}$? It seems that one obtains more insight by studying the universal cohomological functors $\mathcal{T} \to A_\alpha(\mathcal{T})$; in particular we need to know when they are full; see [43, 35] for some recent work in this direction.

Instead of answering these open questions, let us be adventurous and move a little bit beyond the class of well generated categories. In fact, there are natural examples of triangulated categories which are not well generated. Such examples arise from additive categories by taking their homotopy category of chain complexes. More precisely, let $\mathcal{A}$ be an additive category and suppose that $\mathcal{A}$ admits small coproducts. We denote by $\mathbf{K}(\mathcal{A})$ the category of chain complexes in $\mathcal{A}$ whose morphisms are the homotopy classes of chain maps. Take for instance the category $\mathcal{A} = \mathrm{Ab}$ of abelian groups. Then one can show that $\mathbf{K}(\mathrm{Ab})$ is not well generated; see [33]. In fact, more is true. The category $\mathbf{K}(\mathrm{Ab})$ admits no small set of generators, that is, any localizing subcategory generated by a small set of objects is a proper subcategory. However, it is not difficult to show that any localizing subcategory generated by a small set of objects is well generated. So we may think of $\mathbf{K}(\mathrm{Ab})$ as *locally well generated*. In fact, discussions with Jan Šťovíček suggest that $\mathbf{K}(\mathcal{A})$ is locally well generated whenever $\mathcal{A}$ is locally finitely presented; see [46]. Recall that $\mathcal{A}$ is *locally finitely presented* if $\mathcal{A}$ has filtered colimits and there exists a small set of finitely presented objects $\mathcal{A}_0$ such that every object can be written as the filtered colimit of objects in $\mathcal{A}_0$. On the other hand, $\mathbf{K}(\mathcal{A})$ is only generated by a small set of objects if $\mathcal{A} = \mathrm{Add}\,\mathcal{A}_0$ for some small set of objects $\mathcal{A}_0$. Here, $\mathrm{Add}\,\mathcal{A}_0$ denotes the smallest subcategory of $\mathcal{A}$ which contains $\mathcal{A}_0$ and is closed under taking small coproducts and direct summands. We refer to [46] for further details.

# Appendix A.  The abelianization of a triangulated category

Let $C$ be an additive category. We consider functors $F \colon C^{\mathrm{op}} \to \mathrm{Ab}$ into the category of abelian groups and call a sequence $F' \to F \to F''$ of functors *exact* if the induced sequence $F'X \to FX \to F''X$ of abelian groups is exact for all $X$ in $C$. A functor $F$ is said to be *coherent* if there exists an exact sequence (called *presentation*)

$$C(-, X) \longrightarrow C(-, Y) \longrightarrow F \longrightarrow 0.$$

The morphisms between two coherent functors form a small set by Yoneda's lemma, and the coherent functors $C^{\mathrm{op}} \to \mathrm{Ab}$ form an additive category with cokernels. We denote this category by $\widehat{C}$.

A basic tool is the fully faithful *Yoneda functor* $h_C \colon C \to \widehat{C}$ which sends an object $X$ to $C(-, X)$. One might think of this functor as the completion of $C$ with respect to the formation of finite colimits. To formulate some further properties, we recall that a morphism $X \to Y$ is a *weak kernel* for a morphism $Y \to Z$ if the induced sequence $C(-, X) \to C(-, Y) \to C(-, Z)$ is exact.

**Lemma A.1.** *Let $C$ be an additive category.*

(1) *Given an additive functor $H \colon C \to A$ to an additive category which admits cokernels, there is (up to a unique isomorphism) a unique right exact functor $\bar{H} \colon \widehat{C} \to A$ such that $H = \bar{H} \circ h_C$.*
(2) *If $C$ has weak kernels, then $\widehat{C}$ is an abelian category.*
(3) *If $C$ has small coproducts, then $\widehat{C}$ has small coproducts and the Yoneda functor preserves small coproducts.*

*Proof.* (1) Extend $H$ to $\bar{H}$ by sending $F$ in $\widehat{C}$ with presentation

$$C(-, X) \xrightarrow{\;(-,\phi)\;} C(-, Y) \longrightarrow F \longrightarrow 0$$

to the cokernel of $H\phi$.

(2) The category $\widehat{C}$ has cokernels, and it is therefore sufficient to show that $\widehat{C}$ has kernels. To this end fix a morphism $F_1 \to F_2$ with the following presentation.

$$
\begin{array}{ccccccc}
C(-, X_1) & \longrightarrow & C(-, Y_1) & \longrightarrow & F_1 & \longrightarrow & 0 \\
\downarrow & & \downarrow & & \downarrow & & \\
C(-, X_2) & \longrightarrow & C(-, Y_2) & \longrightarrow & F_2 & \longrightarrow & 0
\end{array}
$$

We construct the kernel $F_0 \to F_1$ by specifying the following presentation.

$$\begin{array}{ccccccc}
\mathcal{C}(-, X_0) & \longrightarrow & \mathcal{C}(-, Y_0) & \longrightarrow & F_0 & \longrightarrow & 0 \\
\downarrow & & \downarrow & & \downarrow & & \\
\mathcal{C}(-, X_1) & \longrightarrow & \mathcal{C}(-, Y_1) & \longrightarrow & F_1 & \longrightarrow & 0
\end{array}$$

First the morphism $Y_0 \to Y_1$ is obtained from the weak kernel sequence

$$Y_0 \longrightarrow X_2 \amalg Y_1 \longrightarrow Y_2.$$

Then the morphisms $X_0 \to X_1$ and $X_0 \to Y_0$ are obtained from the weak kernel sequence

$$X_0 \longrightarrow X_1 \amalg Y_0 \longrightarrow Y_1.$$

(3) For every family of functors $F_i$ having a presentation

$$\mathcal{C}(-, X_i) \xrightarrow{(-, \phi_i)} \mathcal{C}(-, Y_i) \longrightarrow F_i \longrightarrow 0,$$

the coproduct $F = \coprod_i F_i$ has a presentation

$$\mathcal{C}(-, \coprod_i X_i) \xrightarrow{(-, \amalg \phi_i)} \mathcal{C}(-, \coprod_i Y_i) \longrightarrow F \longrightarrow 0.$$

Thus coproducts in $\widehat{\mathcal{C}}$ are not computed pointwise. $\qquad\qquad\square$

The assigment $\mathcal{C} \mapsto \widehat{\mathcal{C}}$ is functorial in the following weak sense. Given a functor $F \colon \mathcal{C} \to \mathcal{D}$, there is (up to a unique isomorphism) a unique right exact functor $\widehat{F} \colon \widehat{\mathcal{C}} \to \widehat{\mathcal{D}}$ extending the composite $h_{\mathcal{D}} \circ F \colon \mathcal{C} \to \widehat{\mathcal{D}}$.

Now let $\mathcal{T}$ be a triangulated category. Then we write $A(\mathcal{T}) = \widehat{\mathcal{T}}$ and call this category the *abelianization* of $\mathcal{T}$, because the Yoneda functor $\mathcal{T} \to A(\mathcal{T})$ is the universal cohomological functor for $\mathcal{T}$.

**Lemma A.2.** *Let $\mathcal{T}$ be a triangulated category. Then the category $A(\mathcal{T})$ is abelian and the Yoneda functor $h_{\mathcal{T}} \colon \mathcal{T} \to A(\mathcal{T})$ is cohomological.*

(1) *Given a cohomological functor $H \colon \mathcal{T} \to \mathcal{A}$ to an abelian category, there is (up to a unique isomorphism) a unique exact functor $\bar{H} \colon A(\mathcal{T}) \to \mathcal{A}$ such that $H = \bar{H} \circ h_{\mathcal{T}}$.*

(2) *Given an exact functor $F \colon \mathcal{T} \to \mathcal{T}'$ between triangulated categories, there is (up to a unique isomorphism) a unique exact functor $A(F) \colon A(\mathcal{T}) \to A(\mathcal{T}')$ such that $h_{\mathcal{T}'} \circ F = A(F) \circ h_{\mathcal{T}}$.*

*Proof.* The category $\mathcal{T}$ has weak kernels and therefore $A(\mathcal{T})$ is abelian. Note that the weak kernel of a morphism $Y \to Z$ is obtained by completing the morphism to an exact triangle $X \to Y \to Z \to SX$.

(1) Let $H\colon \mathcal{T} \to \mathcal{A}$ be a cohomological functor and let $\bar{H}\colon A(\mathcal{T}) \to \mathcal{A}$ be the right exact functor extending $H$ which exists by Lemma A.1. Then $\bar{H}$ is exact because $H$ is cohomological.

(2) Let $F\colon \mathcal{T} \to \mathcal{T}'$ be exact. Then $H = h_{\mathcal{T}'} \circ F$ is a cohomological functor and we let $A(F) = \bar{H}$ be the exact functor which extends $H$. $\qquad\square$

The assignment $\mathcal{T} \mapsto A(\mathcal{T})$ from triangulated categories to abelian categories preserves various properties of exact functors between triangulated categories. Let us mention some of them.

**Lemma A.3.** *Let $F\colon \mathcal{T} \to \mathcal{T}'$ and $G\colon \mathcal{T}' \to \mathcal{T}$ be exact functors between triangulated categories.*

(1) *$F$ is fully faithful if and only if $A(F)$ is fully faithful.*
(2) *If $F$ induces an equivalence $\mathcal{T}/\operatorname{Ker} F \xrightarrow{\sim} \mathcal{T}'$, then $A(F)$ induces an equivalence $A(\mathcal{T})/(\operatorname{Ker} A(F)) \xrightarrow{\sim} A(\mathcal{T}')$.*
(3) *$F$ preserves small (co)products if and only if $A(F)$ preserves small (co)products.*
(4) *$F$ is left adjoint to $G$ if and only if $A(F)$ is left adjoint to $A(G)$.*

*Proof.* Straightforward. $\qquad\square$

## Notes

The abelianization of a triangulated category appears in Verdier's thèse [48] and in Freyd's work on the stable homotopy category [15]. Note that their construction is slightly different from the one given here, which is based on coherent functors in the sense of Auslander [4].

## Appendix B. Locally presentable abelian categories

Fix a regular cardinal $\alpha$ and a small additive category $\mathcal{C}$ which admits $\alpha$-coproducts. We denote by $\operatorname{Add}(\mathcal{C}^{\mathrm{op}}, \mathrm{Ab})$ the category of additive functors $\mathcal{C}^{\mathrm{op}} \to \mathrm{Ab}$ into the category of abelian groups. This is an abelian category which admits small (co)products. In fact, (co)kernels and (co)products are computed pointwise in Ab. Given functors $F$ and $G$ in $\operatorname{Add}(\mathcal{C}^{\mathrm{op}}, \mathrm{Ab})$, we write $\operatorname{Hom}_{\mathcal{C}}(F, G)$ for the set of morphisms $F \to G$. The most important objects in $\operatorname{Add}(\mathcal{C}^{\mathrm{op}}, \mathrm{Ab})$ are the *representable functors* $\mathcal{C}(-, X)$ with $X \in \mathcal{C}$. Recall that Yoneda's lemma provides a bijection

$$\operatorname{Hom}_{\mathcal{C}}(\mathcal{C}(-, X), F) \xrightarrow{\sim} FX$$

for all $F\colon \mathcal{C}^{\mathrm{op}} \to \mathrm{Ab}$ and $X \in \mathcal{C}$.

We denote by $\mathrm{Add}_\alpha(\mathcal{C}^{\mathrm{op}}, \mathrm{Ab})$ the full subcategory of $\mathrm{Add}(\mathcal{C}^{\mathrm{op}}, \mathrm{Ab})$ which is formed by all functors preserving $\alpha$-products. This is an exact abelian subcategory, because kernels and cokernels of morphism between $\alpha$-product preserving functors preserve $\alpha$-products. In particular, $\mathrm{Add}_\alpha(\mathcal{C}^{\mathrm{op}}, \mathrm{Ab})$ is an abelian category.

Now suppose that $\mathcal{C}$ admits cokernels. Then $\mathrm{Lex}_\alpha(\mathcal{C}^{\mathrm{op}}, \mathrm{Ab})$ denotes the full subcategory of $\mathrm{Add}(\mathcal{C}^{\mathrm{op}}, \mathrm{Ab})$ which is formed by all left exact functors preserving $\alpha$-products. This category is locally presentable in the sense of Gabriel and Ulmer and we refer to [17, §5] for an extensive treatment. In this appendix we collect some basic facts.

First observe that $\alpha$-filtered colimits in $\mathrm{Lex}_\alpha(\mathcal{C}^{\mathrm{op}}, \mathrm{Ab})$ are computed pointwise. This follows from the fact that in $\mathrm{Ab}$ taking $\alpha$-filtered colimits commutes with taking $\alpha$-limits; see [17, Satz 5.12]. In particular, $\mathrm{Lex}_\alpha(\mathcal{C}^{\mathrm{op}}, \mathrm{Ab})$ has small coproducts because every small coproduct is the $\alpha$-filtered colimit of its sub-coproducts with less than $\alpha$ factors.

Next we show that one can identify $\mathrm{Add}_\alpha(\mathcal{C}^{\mathrm{op}}, \mathrm{Ab})$ with a category of left exact functors. To this end consider the Yoneda functor $h_\mathcal{C}\colon \mathcal{C} \to \widehat{\mathcal{C}}$ taking $X$ to $\mathcal{C}(-, X)$.

**Lemma B.1.** *Let $\mathcal{C}$ be a small additive category with $\alpha$-coproducts. Then the Yoneda functor induces an equivalence*

$$\mathrm{Lex}_\alpha(\widehat{\mathcal{C}}^{\mathrm{op}}, \mathrm{Ab}) \xrightarrow{\sim} \mathrm{Add}_\alpha(\mathcal{C}^{\mathrm{op}}, \mathrm{Ab})$$

*by taking a functor $F$ to $F \circ h_\mathcal{C}$.*

*Proof.* Use that every additive functor $\mathcal{C}^{\mathrm{op}} \to \mathrm{Ab}$ extends uniquely to a left exact functor $\widehat{\mathcal{C}}^{\mathrm{op}} \to \mathrm{Ab}$; see Lemma A.1. $\qquad\square$

From now on we assume that $\mathcal{C}$ admits $\alpha$-coproducts and cokernels. Given any additive functor $F\colon \mathcal{C}^{\mathrm{op}} \to \mathrm{Ab}$, we consider the category $\mathcal{C}/F$ whose objects are pairs $(C, \mu)$ consisting of an object $C \in \mathcal{C}$ and an element $\mu \in FC$. A morphism $(C, \mu) \to (C', \mu')$ is a morphism $\phi\colon C \to C'$ such that $F\phi(\mu') = \mu$.

**Lemma B.2.** *Let $F\colon \mathcal{C}^{\mathrm{op}} \to \mathrm{Ab}$ be an additive functor.*

(1) *The canonical morphism*

$$\operatorname*{colim}_{(C,\mu)\in\mathcal{C}/F} \mathcal{C}(-, C) \longrightarrow F$$

*in $\mathrm{Add}(\mathcal{C}^{\mathrm{op}}, \mathrm{Ab})$ is an isomorphism.*

(2) *The functor $F$ belongs to $\mathrm{Lex}_\alpha(\mathcal{C}^{\mathrm{op}}, \mathrm{Ab})$ if and only if the category $\mathcal{C}/F$ is $\alpha$-filtered.*

*Proof.* (1) is easy. For (2), see [17, Satz 5.3]. $\qquad\square$

The representable functors in $\mathrm{Lex}_\alpha(\mathcal{C}^{\mathrm{op}}, \mathrm{Ab})$ share the following finiteness property. Recall that an object $X$ from an additive category $\mathcal{A}$ with $\alpha$-filtered colimits is $\alpha$-*presentable* if the representable functor $\mathcal{A}(X, -)\colon \mathcal{A} \to \mathrm{Ab}$ preserves $\alpha$-filtered colimits. Next observe that the inclusion $\mathrm{Lex}_\alpha(\mathcal{C}^{\mathrm{op}}, \mathrm{Ab}) \to \mathrm{Add}(\mathcal{C}^{\mathrm{op}}, \mathrm{Ab})$ preserves $\alpha$-filtered colimits. This follows from the fact that in $\mathrm{Ab}$ taking $\alpha$-filtered colimits commutes with taking $\alpha$-limits. This has the following consequence.

**Lemma B.3.** *For each $X$ in $\mathcal{C}$, the representable functor $\mathcal{C}(-, X)$ is an $\alpha$-presentable object of $\mathrm{Lex}_\alpha(\mathcal{C}^{\mathrm{op}}, \mathrm{Ab})$.*

*Proof.* Combine Yoneda's lemma with the fact that the inclusion $\mathrm{Lex}_\alpha(\mathcal{C}^{\mathrm{op}}, \mathrm{Ab}) \to \mathrm{Add}(\mathcal{C}^{\mathrm{op}}, \mathrm{Ab})$ preserves $\alpha$-filtered colimits. $\qquad\square$

There is a general result for the category $\mathrm{Lex}_\alpha(\mathcal{C}^{\mathrm{op}}, \mathrm{Ab})$ which says that taking $\alpha$-filtered colimits commutes with taking $\alpha$-limits; see [17, Korollar 7.12]. Here we need the following special case.

**Lemma B.4.** *Suppose the category $\mathrm{Lex}_\alpha(\mathcal{C}^{\mathrm{op}}, \mathrm{Ab})$ is abelian. Then an $\alpha$-filtered colimit of exact sequences is again exact.*

*Proof.* We need to show that taking $\alpha$-filtered colimits commutes with taking kernels and cokernels. A cokernel is nothing but a colimit and therefore taking colimits and cokernels commute. The statement about kernels follows from the fact that the inclusion $\mathrm{Lex}_\alpha(\mathcal{C}^{\mathrm{op}}, \mathrm{Ab}) \to \mathrm{Add}(\mathcal{C}^{\mathrm{op}}, \mathrm{Ab})$ preserves kernels and $\alpha$-filtered colimits. Thus we can compute kernels and $\alpha$-filtered colimits in $\mathrm{Add}(\mathcal{C}^{\mathrm{op}}, \mathrm{Ab})$ and therefore in the category $\mathrm{Ab}$ of abelian groups. In $\mathrm{Ab}$ it is well known that taking kernels and filtered colimits commute. $\qquad\square$

**Lemma B.5.** *Suppose that $\mathcal{C}$ is abelian. Then $\mathrm{Lex}_\alpha(\mathcal{C}^{\mathrm{op}}, \mathrm{Ab})$ is abelian and the Yoneda functor $h_{\mathcal{C}}\colon \mathcal{C} \to \mathrm{Lex}_\alpha(\mathcal{C}^{\mathrm{op}}, \mathrm{Ab})$ is exact. Given an abelian category $\mathcal{A}$ which admits small coproducts and exact $\alpha$-filtered colimits, and given a functor $F\colon \mathrm{Lex}_\alpha(\mathcal{C}^{\mathrm{op}}, \mathrm{Ab}) \to \mathcal{A}$ preserving $\alpha$-filtered colimits, we have that $F$ is exact if and only if $F \circ h_{\mathcal{C}}$ is exact.*

*Proof.* We use the analogue of Lemma B.2 for morphisms which says that each morphism $\phi$ in $\mathrm{Lex}_\alpha(\mathcal{C}^{\mathrm{op}}, \mathrm{Ab})$ can be written as $\alpha$-filtered colimit $\phi = \underrightarrow{\mathrm{colim}}_{i \in \mathcal{C}/\phi} \phi_i$

of morphisms between representable functors. Thus one computes

$$\operatorname{Coker}\phi = \underset{i\in\mathcal{C}/\phi}{\operatorname{colim}}\operatorname{Coker}\phi_i \quad\text{and}\quad \operatorname{Ker}\phi = \underset{i\in\mathcal{C}/\phi}{\operatorname{colim}}\operatorname{Ker}\phi_i,$$

and we see that $\operatorname{Lex}_\alpha(\mathcal{C}^{\mathrm{op}},\operatorname{Ab})$ is abelian; see Lemma B.4. The formula for kernels and cokernels shows that each exact sequence can be written as $\alpha$-filtered colimit of exact sequences in the image of the Yoneda embedding. The criterion for the exactness of a functor $\operatorname{Lex}_\alpha(\mathcal{C}^{\mathrm{op}},\operatorname{Ab})\to\mathcal{A}$ is an immediate consequence. $\qquad\square$

Let $\mathcal{A}$ be a cocomplete additive category. We denote by $\mathcal{A}^\alpha$ the full subcategory which is formed by all $\alpha$-presentable objects. Following [17], the category $\mathcal{A}$ is called *locally $\alpha$-presentable* if $\mathcal{A}^\alpha$ is small and each object is an $\alpha$-filtered colimit of $\alpha$-presentable objects. We call $\mathcal{A}$ *locally presentable* if it is locally $\beta$-presentable for some cardinal $\beta$. Note that we have for each locally presentable category $\mathcal{A}$ a filtration $\mathcal{A}=\bigcup_\beta\mathcal{A}^\beta$ where $\beta$ runs through all regular cardinals. We have already seen that $\operatorname{Lex}_\alpha(\mathcal{C}^{\mathrm{op}},\operatorname{Ab})$ is locally $\alpha$-presentable, and the next lemma implies that, up to an equivalence, all locally $\alpha$-presentable categories are of this form.

Let $f\colon\mathcal{C}\to\mathcal{A}$ be a fully faithful and right exact functor into a cocomplete additive category. Suppose that $f$ preserves $\alpha$-coproducts and that each object in the image of $f$ is $\alpha$-presentable. Then $f$ induces the functor

$$f_*\colon\mathcal{A}\longrightarrow\operatorname{Lex}_\alpha(\mathcal{C}^{\mathrm{op}},\operatorname{Ab}),\quad X\mapsto\mathcal{A}(f-,X),$$

and the following lemma discusses its left adjoint.

**Lemma B.6.** *There is a fully faithful functor $f^*\colon\operatorname{Lex}_\alpha(\mathcal{C}^{\mathrm{op}},\operatorname{Ab})\to\mathcal{A}$ which sends each representable functor $\mathcal{C}(-,X)$ to $fX$ and identifies $\operatorname{Lex}_\alpha(\mathcal{C}^{\mathrm{op}},\operatorname{Ab})$ with the full subcategory of $\mathcal{A}$ formed by all colimits of objects in the image of $f$. The functor $f^*$ is a left adjoint of $f_*$.*

*Proof.* The functor is the left Kan extension of $f$; it takes $F=\underset{(C,\mu)\in\mathcal{C}/F}{\operatorname{colim}}\mathcal{C}(-,C)$ in $\operatorname{Lex}_\alpha(\mathcal{C}^{\mathrm{op}},\operatorname{Ab})$ to $\underset{(C,\mu)\in\mathcal{C}/F}{\operatorname{colim}}fC$ in $\mathcal{A}$. We refer to [17, Satz 7.8] for details. $\quad\square$

Suppose now that $\mathcal{C}$ is a triangulated category. The following lemma characterizes the cohomological functors $\mathcal{C}^{\mathrm{op}}\to\operatorname{Ab}$.

**Lemma B.7.** *Let $\mathcal{C}$ be a small triangulated category and suppose $\mathcal{C}$ admits $\alpha$-coproducts. For a functor $F$ in $\operatorname{Add}_\alpha(\mathcal{C}^{\mathrm{op}},\operatorname{Ab})$ the following are equivalent.*

(1) *The category $C/F$ is $\alpha$-filtered.*
(2) *$F$ is an $\alpha$-filtered colimit of representable functors.*
(3) *$F$ is a cohomological functor.*

*Proof.* The implications (1) $\Rightarrow$ (2) $\Rightarrow$ (3) are clear. So we prove (3) $\Rightarrow$ (1). It is convenient to identify $\mathrm{Add}_\alpha(C^{\mathrm{op}}, \mathrm{Ab})$ with $\mathrm{Lex}_\alpha(\widehat{C}^{\mathrm{op}}, \mathrm{Ab})$ and this identifies $F$ with the left exact functor $\bar{F} \colon \widehat{C} \to \mathrm{Ab}$ which extends $F$. In fact, $\bar{F}$ is exact since $F$ is cohomological, by Lemma A.2. Now write $\bar{F}$ as $\alpha$-filtered colimit of representable functors $\bar{F} = \operatorname*{colim}_{(M,\nu)\in\widehat{C}/\bar{F}} \widehat{C}(-, M)$; see Lemma B.2. The exactness of $\bar{F}$ implies that the representable functors $C(-, C)$ with $C \in C$ form a full subcategory of $\widehat{C}/\bar{F}$ which is cofinal. We identify this subcategory with $C/F$ and conclude from Lemma 6.4.1 that $C/F$ is $\alpha$-filtered. $\qquad\square$

Next we discuss the functoriality of the assignment $C \mapsto \mathrm{Add}_\alpha(C^{\mathrm{op}}, \mathrm{Ab})$.

**Lemma B.8.** *Let $f \colon C \to D$ be an exact functor between small triangulated categories which admit $\alpha$-coproducts. Suppose that $f$ preserves $\alpha$-coproducts. Then the restriction functor*

$$f_* \colon \mathrm{Add}_\alpha(D^{\mathrm{op}}, \mathrm{Ab}) \longrightarrow \mathrm{Add}_\alpha(C^{\mathrm{op}}, \mathrm{Ab}), \quad F \mapsto F \circ f,$$

*has a left adjoint $f^*$ which sends $C(-, X)$ to $D(-, fX)$ for all $X$ in $C$. Moreover, the following holds.*

(1) *The functors $f_*$ and $f^*$ are exact.*
(2) *Suppose $f$ induces an equivalence $C/\operatorname{Ker} f \xrightarrow{\sim} D$. Then $f_*$ is fully faithful.*
(3) *Suppose $f$ is fully faithful. Then $f^*$ is fully faithful. Moreover, a cohomological functor $F \colon D^{\mathrm{op}} \to \mathrm{Ab}$ is in the essential image of $f^*$ if and only if every morphism $D(-, D) \to F$ factors through $D(-, fC)$ for some object $C$ in $C$.*
(4) *A cohomological functor $F \colon C^{\mathrm{op}} \to \mathrm{Ab}$ belongs to the kernel of $f^*$ if and only if every morphism $C(-, C) \to F$ factors through a morphism $C(-, \gamma) \colon C(-, C) \to C(-, C')$ such that $f\gamma = 0$.*

*Proof.* The left adjoint of $f_*$ is the left Kan extension. We can describe it explicitly if we identify $\mathrm{Add}_\alpha(C^{\mathrm{op}}, \mathrm{Ab})$ with $\mathrm{Lex}_\alpha(\widehat{C}^{\mathrm{op}}, \mathrm{Ab})$; see Lemma B.1. Given a functor $F$ in $\mathrm{Lex}_\alpha(\widehat{C}^{\mathrm{op}}, \mathrm{Ab})$ written as $\alpha$-filtered colimit $F = \operatorname*{colim}_{(C,\mu)\in\widehat{C}/F} \widehat{C}(-, C)$ of representable functors, we put

$$f^* F = \operatorname*{colim}_{(C,\mu)\in\widehat{C}/F} \widehat{D}(-, \widehat{f}C).$$

Thus $f^*$ makes the following diagram commutative.

$$
\begin{array}{ccccccc}
\mathcal{C} & \xrightarrow{h_{\mathcal{C}}} & \widehat{\mathcal{C}} & \xrightarrow{h_{\widehat{\mathcal{C}}}} & \mathrm{Lex}_\alpha(\widehat{\mathcal{C}}^{\mathrm{op}}, \mathrm{Ab}) & \xrightarrow{=} & \mathrm{Add}_\alpha(\mathcal{C}^{\mathrm{op}}, \mathrm{Ab}) \\
\downarrow{\scriptstyle f} & & \downarrow{\scriptstyle \widehat{f}} & & \downarrow{\scriptstyle f^*} & & \downarrow{\scriptstyle f^*} \\
\mathcal{D} & \xrightarrow{h_{\mathcal{D}}} & \widehat{\mathcal{D}} & \xrightarrow{h_{\widehat{\mathcal{D}}}} & \mathrm{Lex}_\alpha(\widehat{\mathcal{D}}^{\mathrm{op}}, \mathrm{Ab}) & \xrightarrow{=} & \mathrm{Add}_\alpha(\mathcal{D}^{\mathrm{op}}, \mathrm{Ab})
\end{array}
$$

We check that $f^*$ is a left adjoint for $f_*$. For a representable functor $F = \widehat{\mathcal{C}}(-, X)$ we have

$$
\begin{aligned}
\mathrm{Hom}_{\widehat{\mathcal{D}}}(f^*\widehat{\mathcal{C}}(-, X), G) = \mathrm{Hom}_{\widehat{\mathcal{D}}}(\widehat{\mathcal{D}}(-, \widehat{f}X), G) &\cong G(\widehat{f}X) \\
&= f_* G(X) \cong \mathrm{Hom}_{\widehat{\mathcal{C}}}(\widehat{\mathcal{C}}(-, X), f_* G)
\end{aligned}
$$

for all $G$ in $\mathrm{Lex}_\alpha(\widehat{\mathcal{D}}^{\mathrm{op}}, \mathrm{Ab})$. Clearly, this isomorphism extends to every colimit of representable functors.

(1) The exactness of $f_*$ is clear because a sequence $F' \to F \to F''$ in $\mathrm{Add}_\alpha(\mathcal{C}^{\mathrm{op}}, \mathrm{Ab})$ is exact if and only if $F'X \to FX \to F''X$ is exact for all $X$ in $\mathcal{C}$. For the exactness of $f^*$ we identify again $\mathrm{Add}(\mathcal{C}^{\mathrm{op}}, \mathrm{Ab})$ with $\mathrm{Lex}_\alpha(\widehat{\mathcal{C}}^{\mathrm{op}}, \mathrm{Ab})$ and apply Lemma B.5. Thus we need to check that the composition of $f^*$ with the Yoneda functor $h_{\widehat{\mathcal{C}}}$ is exact. But we have that $f^* \circ h_{\widehat{\mathcal{C}}} = h_{\widehat{\mathcal{D}}} \circ \widehat{f}$, and now the exactness follows from that of $f$. Finally, we use the fact that taking $\alpha$-filtered colimits in $\mathrm{Add}_\alpha(\mathcal{D}^{\mathrm{op}}, \mathrm{Ab})$ is exact by Lemma B.4.

(2) It is well known that for any epimorphism $f \colon \mathcal{C} \to \mathcal{D}$ of additive categories inducing a bijection $\mathrm{Ob}\,\mathcal{C} \to \mathrm{Ob}\,\mathcal{D}$, the restriction functor $\mathrm{Add}(\mathcal{D}^{\mathrm{op}}, \mathrm{Ab}) \to \mathrm{Add}(\mathcal{C}^{\mathrm{op}}, \mathrm{Ab})$ is fully faithful; see [29, Corollary 5.2]. Given a triangulated subcategory $\mathcal{C}' \subseteq \mathcal{C}$, the quotient functor $\mathcal{C} \to \mathcal{C}/\mathcal{C}'$ is an epimorphism. Thus the assertion follows since $\mathrm{Add}_\alpha(\mathcal{C}^{\mathrm{op}}, \mathrm{Ab})$ is a full subcategory of $\mathrm{Add}(\mathcal{C}^{\mathrm{op}}, \mathrm{Ab})$.

(3) We keep our identification $\mathrm{Add}_\alpha(\mathcal{C}^{\mathrm{op}}, \mathrm{Ab}) = \mathrm{Lex}_\alpha(\widehat{\mathcal{C}}^{\mathrm{op}}, \mathrm{Ab})$ and consider the adjunction morphism $\eta \colon \mathrm{Id} \to f_* \circ f^*$. We claim that $\eta$ is an isomorphism. Because $f$ is fully faithful, $\eta F$ is an isomorphism for each representable functor $F = \widehat{\mathcal{C}}(-, X)$. It follows that $\eta F$ is an isomorphism for all $F$ since $f^*$ and $f_*$ both preserve $\alpha$-filtered colimits and each $F$ can be expressed as $\alpha$-filtered colimit of representable functors. Now Proposition 2.3.1 implies that $f^*$ is fully faithful.

Let $F$ be a cohomological functor in $\mathrm{Add}_\alpha(\mathcal{D}^{\mathrm{op}}, \mathrm{Ab})$ and apply Lemma B.7 to write the functor as $\alpha$-filtered colimit $F = \underset{(D,\mu)\in \mathcal{D}/F}{\mathrm{colim}}\ \mathcal{D}(-, D)$ of representable functors. Suppose first that every morphism $\mathcal{D}(-, D) \to F$ factors through $\mathcal{D}(-, fC)$ for some $C \in \mathcal{C}$. Then $\mathrm{Im}\,f/F$ is a cofinal subcategory of $\mathcal{D}/F$

and therefore $F = \underset{(D,\mu)\in\mathrm{Im}\, f/F}{\mathrm{colim}}\ \mathcal{D}(-, D)$ by Lemma 6.4.1. Thus $F$ belongs to the essential image of $f^*$ since $\mathcal{D}(-, fC) = f^*\mathcal{C}(-, C)$ for all $C \in \mathcal{C}$ and the essential image is closed under taking colimits. Now suppose that $F$ belongs to the essential image of $f^*$. Then $F = f^*G \cong f^*f_*f^*G = f^*f_*F$ for some $G$. The functor $f_*F$ is cohomological and therefore $f_*F = \underset{(C,\mu)\in\mathcal{C}/f_*F}{\mathrm{colim}}\ \mathcal{C}(-, C)$, again by Lemma B.7. Thus $F \cong \underset{(C,\mu)\in\mathcal{C}/f_*F}{\mathrm{colim}}\ \mathcal{D}(-, fC)$ and we use Lemma B.3 to conclude that each morphism $\mathcal{D}(-, D) \to F$ factors through $\mathcal{D}(-, fC)$ for some $(C, \mu) \in \mathcal{C}/f_*F$.

(4) Let $F$ be a cohomological functor in $\mathrm{Add}_\alpha(\mathcal{C}^{\mathrm{op}}, \mathrm{Ab})$ and apply Lemma B.7 to write the functor as $\alpha$-filtered colimit $F = \underset{(C,\mu)\in\mathcal{C}/F}{\mathrm{colim}}\ \mathcal{C}(-, C)$ of representable functors. Now $f^*F = \underset{(C,\mu)\in\mathcal{C}/F}{\mathrm{colim}}\ \mathcal{D}(-, fC) = 0$ if and only if for each $D \in \mathcal{D}$, we have $\underset{(C,\mu)\in\mathcal{C}/F}{\mathrm{colim}}\ \mathcal{D}(D, fC) = 0$. This happens iff for each $(C, \mu) \in \mathcal{C}/F$, we find a morphism $\gamma\colon C \to C'$ in $\mathcal{C}/F$ inducing a map $\mathcal{D}(fC, fC) \to \mathcal{D}(fC, fC')$ which annihilates the identity morphism. But this means that $f\gamma = 0$ and that $\mu\colon \mathcal{C}(-, C) \to F$ factors through $\mathcal{C}(-, \gamma)$. $\square$

## Notes

Locally presentable categories were introduced and studied by Gabriel and Ulmer in [17]; see [1] for a modern treatment. In [33], Neeman initiated the use of locally presentable abelian categories for studying triangulated categories.

## References

[1] J. Adámek and J. Rosický, *Locally presentable and accessible categories*, Cambridge Univ. Press, Cambridge, 1994.

[2] J. F. Adams, *Stable homotopy and generalised homology*, Univ. Chicago Press, Chicago, Ill., 1974.

[3] L. Alonso Tarrío, A. Jeremías López and M. J. Souto Salorio, Localization in categories of complexes and unbounded resolutions, Canad. J. Math. **52** (2000), no. 2, 225–247.

[4] M. Auslander, Coherent functors, in *Proc. Conf. Categorical Algebra (La Jolla, Calif., 1965)*, 189–231, Springer, New York, 1966.

[5] L. Avramov and S. Halperin, Through the looking glass: a dictionary between rational homotopy theory and local algebra, in *Algebra, algebraic topology and their interactions (Stockholm, 1983)*, 1–27, Lecture Notes in Math., 1183, Springer, Berlin, 1986.

[6] A. A. Beĭlinson, J. Bernstein and P. Deligne, Faisceaux pervers, in *Analysis and topology on singular spaces, I (Luminy, 1981)*, 5–171, Astérisque, 100, Soc. Math. France, Paris, 1982.

[7] D. Benson, S. B. Iyengar and H. Krause, Local cohomology and support for triangulated categories, Ann. Sci. École Norm. Sup. (4) **41** (2008), no. 4, 575–621.

[8] A. K. Bousfield, The localization of spectra with respect to homology, Topology **18** (1979), no. 4, 257–281.

[9] E. H. Brown, Jr., Cohomology theories, Ann. of Math. (2) **75** (1962), 467–484.

[10] C. Casacuberta, J. J. Gutiérrez and J. Rosický, Are all localizing subcategories of stable homotopy categories coreflective?, preprint (2007).

[11] C. Casacuberta, D. Scevenels and J. H. Smith, Implications of large-cardinal principles in homotopical localization, Adv. Math. **197** (2005), no. 1, 120–139.

[12] W. G. Dwyer, Localizations, in *Axiomatic, enriched and motivic homotopy theory*, 3–28, Kluwer Acad. Publ., Dordrecht, 2004.

[13] J. Franke, On the Brown representability theorem for triangulated categories, Topology **40** (2001), no. 4, 667–680.

[14] P. Freyd, *Abelian categories. An introduction to the theory of functors*, Harper & Row, New York, 1964.

[15] P. Freyd, Stable homotopy, in *Proc. Conf. Categorical Algebra (La Jolla, Calif., 1965)*, 121–172, Springer, New York, 1966.

[16] P. Gabriel, Des catégories abéliennes, Bull. Soc. Math. France **90** (1962), 323–448.

[17] P. Gabriel and F. Ulmer, *Lokal präsentierbare Kategorien*, Lecture Notes in Math., 221, Springer, Berlin, 1971.

[18] P. Gabriel and M. Zisman, *Calculus of fractions and homotopy theory*, Springer-Verlag New York, Inc., New York, 1967.

[19] A. Grothendieck and J. L. Verdier, Préfaisceaux, in *SGA 4, Théorie des Topos et Cohomologie Etale des Schémas, Tome 1. Thórie des Topos*, Lect. Notes in Math., vol. 269, Springer, Heidelberg, 1972-1973, pp. 1–184.

[20] A. Heider, Two results from Morita theory of stable model categories, arXiv:0707.0707v1.

[21] B. Keller, Deriving DG categories, Ann. Sci. École Norm. Sup. (4) **27** (1994), no. 1, 63–102.

[22] B. Keller, A remark on the generalized smashing conjecture, Manuscripta Math. **84** (1994), no. 2, 193–198.

[23] H. Krause, On Neeman's well generated triangulated categories, Doc. Math. **6** (2001), 121–126 (electronic).

[24] H. Krause, A Brown representability theorem via coherent functors, Topology **41** (2002), no. 4, 853–861.

[25] H. Krause, Cohomological quotients and smashing localizations, Amer. J. Math. **127** (2005), no. 6, 1191–1246.

[26] A. Lévy, *Basic set theory*, Springer, Berlin, 1979.

[27] S. MacLane, *Categories for the working mathematician*, Springer, New York, 1971.

[28] H. R. Margolis, *Spectra and the Steenrod algebra*, North-Holland, Amsterdam, 1983.

[29] B. Mitchell, The dominion of Isbell, Trans. Amer. Math. Soc. **167** (1972), 319–331.

[30] A. Neeman, The connection between the $K$-theory localization theorem of Thomason, Trobaugh and Yao and the smashing subcategories of Bousfield and Ravenel, Ann. Sci. École Norm. Sup. (4) **25** (1992), no. 5, 547–566.

[31] A. Neeman, The Grothendieck duality theorem via Bousfield's techniques and Brown representability, J. Amer. Math. Soc. **9** (1996), no. 1, 205–236.

[32] A. Neeman, Non-compactly generated categories, Topology **37** (1998), no. 5, 981–987.

[33] A. Neeman, *Triangulated categories*, Ann. of Math. Stud., 148, Princeton Univ. Press, Princeton, NJ, 2001.

[34] A. Neeman, On the derived category of sheaves on a manifold, Doc. Math. **6** (2001), 483–488 (electronic).

[35] A. Neeman, Brown representability follows from Rosický, preprint (2007).

[36] O. Ore, Linear equations in non-commutative fields, Ann. of Math. (2) **32** (1931), no. 3, 463–477.

[37] M. Porta, The Popescu-Gabriel theorem for triangulated categories, arXiv:0706.4458v1.

[38] D. Puppe, On the structure of stable homotopy theory, in *Colloquium on algebraic topology*, Aarhus Universitet Matematisk Institut (1962), 65–71.

[39] D. G. Quillen, *Homotopical algebra*, Lecture Notes in Math., 43, Springer, Berlin, 1967.

[40] D. C. Ravenel, Localization with respect to certain periodic homology theories, Amer. J. Math. **106** (1984), no. 2, 351–414.

[41] J. Rickard, Idempotent modules in the stable category, J. London Math. Soc. (2) **56** (1997), no. 1, 149–170.

[42] J. Rickard, Bousfield localization for representation theorists, in *Infinite length modules (Bielefeld, 1998)*, 273–283, Birkhäuser, Basel.

[43] J. Rosický, Generalized Brown representability in homotopy categories, Theory Appl. Categ. **14** (2005), no. 19, 451–479 (electronic).

[44] N. Spaltenstein, Resolutions of unbounded complexes, Compositio Math. **65** (1988), no. 2, 121–154.

[45] B. Stenström, *Rings of quotients*, Springer, New York, 1975.

[46] J. Šťovíček, Locally well generated homotopy categories of complexes, arXiv:0810.5684v1.

[47] G. Tabuada, Homotopy theory of well-generated algebraic triangulated categories, J. K-theory, in press.

[48] J.-L. Verdier, Des catégories dérivées des catégories abéliennes, Astérisque No. 239 (1996), xii+253 pp. (1997).

HENNING KRAUSE, INSTITUT FÜR MATHEMATIK, UNIVERSITÄT PADERBORN, 33095 PADERBORN, GERMANY.

*E-mail address*: hkrause@math.upb.de

# Homological algebra in bivariant K-theory and other triangulated categories. I

RALF MEYER AND RYSZARD NEST

ABSTRACT. Bivariant (equivariant) K-theory is the standard setting for non-commutative topology. We may carry over various techniques from homotopy theory and homological algebra to this setting. Here we do this for some basic notions from homological algebra: phantom maps, exact chain complexes, projective resolutions, and derived functors. We introduce these notions and apply them to examples from bivariant K-theory.

An important observation of Beligiannis is that we can approximate our category by an Abelian category in a canonical way, such that our homological concepts reduce to the corresponding ones in this Abelian category. We compute this Abelian approximation in several interesting examples, where it turns out to be very concrete and tractable.

The derived functors comprise the second page of a spectral sequence that, in favourable cases, converges towards Kasparov groups and other interesting objects. This mechanism is the common basis for many different spectral sequences. Here we only discuss the very simple 1-dimensional case, where the spectral sequences reduce to short exact sequences.

## 1. Introduction

It is well-known that many basic constructions from homotopy theory extend to categories of $C^*$-algebras. As we argued in [17], the framework of *triangulated categories* is ideal for this purpose. The notion of triangulated category was introduced by Jean-Louis Verdier to formalise the properties of the derived category of an Abelian category. Stable homotopy theory provides further classical examples of triangulated categories. The triangulated category structure encodes basic information about manipulations with long exact sequences and (total) derived functors. The main point of [17] is that the domain of the Baum–Connes assembly map is the total left derived functor of the functor that maps a $G$-$C^*$-algebra $A$ to $K_*(G \ltimes_r A)$.

2000 *Mathematics Subject Classification.* 18E30, 19K35, 46L80, 55U35.
This research was supported by the EU-Network *Quantum Spaces and Noncommutative Geometry* (Contract HPRN-CT-2002-00280).

The relevant triangulated categories in non-commutative topology come from Kasparov's bivariant K-theory. This bivariant version of K-theory carries a composition product that turns it into a category. The formal properties of this and related categories are surveyed in [15], with an audience of homotopy theorists in mind.

*Projective resolutions* are among the most fundamental concepts in homological algebra; several others like derived functors are based on it. Projective resolutions seem to live in the underlying Abelian category and not in its derived category. This is why *total* derived functor make more sense in triangulated categories than the derived functors themselves. Nevertheless, we can define derived functors in triangulated categories and far more general categories. This goes back to Samuel Eilenberg and John C. Moore ([9]). We learned about this theory in articles by Apostolos Beligiannis ([4]) and J. Daniel Christensen ([8]).

Homological algebra in non-Abelian categories is always *relative*, that is, we need additional structure to get started. This is useful because we may fit the additional data to our needs. In a triangulated category $\mathfrak{T}$, there are several kinds of additional data that yield equivalent theories; following [8], we use an *ideal* in $\mathfrak{T}$. We only consider ideals $\mathfrak{I}$ in $\mathfrak{T}$ of the form

$$\mathfrak{I}(A, B) := \{x \in \mathfrak{T}(A, B) \mid F(x) = 0\}$$

for a stable homological functor $F \colon \mathfrak{T} \to \mathfrak{C}$ into a stable Abelian category $\mathfrak{C}$. Here *stable* means that $\mathfrak{C}$ carries a suspension automorphism and that $F$ intertwines the suspension automorphisms on $\mathfrak{T}$ and $\mathfrak{C}$, and homological means that exact triangles yield exact sequences. Ideals of this form are called *homological ideals*.

A basic example is the ideal in the Kasparov category KK defined by

$$\mathfrak{I}_{\mathrm{K}}(A, B) := \{f \in \mathrm{KK}(A, B) \mid 0 = \mathrm{K}_*(f) \colon \mathrm{K}_*(A) \to \mathrm{K}_*(B)\}. \tag{1.1}$$

For a compact (quantum) group $G$, we define two ideals $\mathfrak{I}_{\ltimes} \subseteq \mathfrak{I}_{\ltimes,\mathrm{K}} \subseteq \mathrm{KK}^G$ in the equivariant Kasparov category $\mathrm{KK}^G$ by

$$\mathfrak{I}_{\ltimes}(A, B) := \{f \in \mathrm{KK}^G(A, B) \mid G \ltimes f = 0 \text{ in } \mathrm{KK}(G \ltimes A, G \ltimes B)\}, \tag{1.2}$$

$$\mathfrak{I}_{\ltimes,\mathrm{K}}(A, B) := \{f \in \mathrm{KK}^G(A, B) \mid \mathrm{K}_*(G \ltimes f) = 0\}, \tag{1.3}$$

where $\mathrm{K}_*(G \ltimes f)$ denotes the map $\mathrm{K}_*(G \ltimes A) \to \mathrm{K}_*(G \ltimes B)$ induced by $f$.

For a locally compact group $G$ and a (suitable) family of subgroups $\mathcal{F}$, we define the homological ideal

$$\mathcal{VC}_{\mathcal{F}}(A, B) := \{f \in \mathrm{KK}^G(A, B) \mid \mathrm{Res}_G^H(f) = 0 \text{ in } \mathrm{KK}^H(A, B) \text{ for all } H \in \mathcal{F}\}. \tag{1.4}$$

If $\mathcal{F}$ is the family of compact subgroups, then $\mathcal{VC}_{\mathcal{F}}$ is related to the Baum–Connes assembly map ([17]). Of course, there are analogous ideals in more classical categories of (spectra of) $G$-CW-complexes.

All these examples can be analysed using the machinery we explain. We carry this out in some cases in Sections 4 and 5.

We use an ideal $\mathfrak{J}$ to carry over various notions from homological algebra to our triangulated category $\mathfrak{T}$. In order to see what they mean in examples, we characterise them using a stable homological functor $F\colon \mathfrak{T} \to \mathfrak{C}$ with $\ker F = \mathfrak{J}$. This is often easy. For instance, a chain complex with entries in $\mathfrak{T}$ is $\mathfrak{J}$-*exact* if and only if $F$ maps it to an exact chain complex in the Abelian category $\mathfrak{C}$ (see Lemma 28), and a morphism in $\mathfrak{T}$ is an $\mathfrak{J}$-*epimorphism* if and only if $F$ maps it to an epimorphism. Here we may take any functor $F$ with $\ker F = \mathfrak{J}$.

But the most crucial notions like projective objects and resolutions require a more careful choice of the functor $F$. Here we need the *universal $\mathfrak{J}$-exact functor*, which is a stable homological functor $F$ with $\ker F = \mathfrak{J}$ such that any other such functor factors uniquely through $F$ (up to natural equivalence). The universal $\mathfrak{J}$-exact functor and its applications are due to Apostolos Beligiannis ([4]).

If $F\colon \mathfrak{T} \to \mathfrak{C}$ is universal, then $F$ detects $\mathfrak{J}$-projective objects, and it iden-tifies $\mathfrak{J}$-derived functors with derived functors in the Abelian category $\mathfrak{C}$ (see Theorem 59). Thus all our homological notions reduce to their counterparts in the Abelian category $\mathfrak{C}$.

In order to apply this, we need to know when a functor $F$ with $\ker F = \mathfrak{J}$ is the universal one. We develop a new, useful criterion for this purpose here, which uses partially defined adjoint functors (Theorem 57).

Our criterion shows that the universal $\mathfrak{J}_{\mathrm{K}}$-exact functor for the ideal $\mathfrak{J}_{\mathrm{K}}$ in KK in (1.1) is the K-theory functor $\mathrm{K}_*$, considered as a functor from KK to the category $\mathfrak{Ab}_{\mathrm{c}}^{\mathbb{Z}/2}$ of countable $\mathbb{Z}/2$-graded Abelian groups (see Theorem 63). Hence the derived functors for $\mathfrak{J}_{\mathrm{K}}$ only involve Ext and Tor for Abelian groups.

For the ideal $\mathfrak{J}_{\ltimes,\mathrm{K}}$ in $\mathrm{KK}^G$ in (1.3), we get the functor

$$\mathrm{KK}^G \to \mathfrak{Mod}(\mathrm{Rep}\,G)_{\mathrm{c}}^{\mathbb{Z}/2}, \qquad A \mapsto \mathrm{K}_*(G \ltimes A), \tag{1.5}$$

where $\mathfrak{Mod}(\mathrm{Rep}\,G)_{\mathrm{c}}^{\mathbb{Z}/2}$ denotes the Abelian category of countable $\mathbb{Z}/2$-graded modules over the representation ring $\mathrm{Rep}\,G$ of the compact (quantum) group $G$ (see Theorem 72); here we use a certain canonical $\mathrm{Rep}\,G$-module structure on $\mathrm{K}_*(G \ltimes A)$. Hence derived functors with respect to $\mathfrak{J}_{\ltimes,\mathrm{K}}$ involve Ext and Tor for $\mathrm{Rep}\,G$-modules.

We do not need the $\mathrm{Rep}\,G$-module structure on $\mathrm{K}_*(G \ltimes A)$ to define $\mathfrak{J}_{\ltimes,\mathrm{K}}$: our machinery notices automatically that such a module structure is missing. The universality of the functor in (1.5) clarifies in what sense homological algebra with Rep $G$-modules is a *linearisation* of algebraic topology with $G$-C$^*$-algebras.

The universal homological functor for the ideal $\mathfrak{I}_{\ltimes}$ is quite similar to the one for $\mathfrak{I}_{\ltimes,K}$ (see Theorem 73). There is a canonical Rep $G$-module structure on $G \ltimes A$ as an object of KK, and the universal $\mathfrak{I}_{\ltimes}$-exact functor is essentially the functor $A \mapsto G \ltimes A$, viewed as an object of a suitable Abelian category that encodes this Rep $G$-module structure; it also involves a fully faithful embedding of KK in an Abelian category due to Peter Freyd ([10]).

The derived functors that we have discussed above appear in a spectral sequence which – in favourable cases – computes morphism spaces in $\mathfrak{T}$ (like $\mathrm{KK}^G(A, B)$) and other homological functors. This spectral sequence is a generalisation of the *Adams spectral sequence* in stable homotopy theory and is the main motivation for [8]. Much earlier, such spectral sequences were studied by Hans-Berndt Brinkmann in [7]. In [16], this spectral sequence is applied to our bivariant K-theory examples. Here we only consider the much easier case where this spectral sequence degenerates to an exact sequence (see Theorem 66). This generalises the familiar Universal Coefficient Theorem for $\mathrm{KK}_*(A, B)$.

## 2. Homological ideals in triangulated categories

After fixing some basic notation, we introduce several interesting ideals in bivariant Kasparov categories; we are going to discuss these ideals throughout this article. Then we briefly recall what a triangulated category is and introduce homological ideals. Before we begin, we should point out that the choice of ideal is important because all our homological notions depend on it. It seems to be a matter of experimentation and experience to find the right ideal for a given purpose.

### 2.1. Generalities about ideals in additive categories

All categories we consider will be *additive*, that is, they have a zero object and finite direct products and coproducts which agree, and the morphism spaces carry Abelian group structures such that the composition is additive in each variable ([13]).

**Notation 1.** Let $\mathfrak{C}$ be an additive category. We write $\mathfrak{C}(A, B)$ for the group of morphisms $A \to B$ in $\mathfrak{C}$, and $A \in\in \mathfrak{C}$ to denote that $A$ is an object of the category $\mathfrak{C}$.

**Definition 2.** An *ideal* $\mathfrak{I}$ in $\mathfrak{C}$ is a family of subgroups $\mathfrak{I}(A, B) \subseteq \mathfrak{C}(A, B)$ for all $A, B \in\in \mathfrak{C}$ such that

$$\mathfrak{C}(C, D) \circ \mathfrak{I}(B, C) \circ \mathfrak{C}(A, B) \subseteq \mathfrak{I}(A, D) \qquad \text{for all } A, B, C, D \in\in \mathfrak{C}.$$

We write $\mathfrak{I}_1 \subseteq \mathfrak{I}_2$ if $\mathfrak{I}_1(A, B) \subseteq \mathfrak{I}_2(A, B)$ for all $A, B \in\in \mathfrak{C}$. Clearly, the ideals in $\mathfrak{T}$ form a complete lattice. The largest ideal $\mathfrak{C}$ consists of all morphisms in $\mathfrak{C}$; the smallest ideal $0$ contains only zero morphisms.

**Definition 3.** Let $\mathfrak{C}$ and $\mathfrak{C}'$ be additive categories and let $F\colon \mathfrak{C} \to \mathfrak{C}'$ be an additive functor. Its *kernel* $\ker F$ is the ideal in $\mathfrak{C}$ defined by

$$\ker F(A, B) := \{ f \in \mathfrak{C}(A, B) \mid F(f) = 0 \}.$$

This should be distinguished from the *kernel on objects*, consisting of all objects with $F(A) \cong 0$, which is used much more frequently. The kernel is the class of $\ker F$-contractible objects that we introduce below.

**Definition 4.** Let $\mathfrak{I} \subseteq \mathfrak{T}$ be an ideal. Its *quotient category* $\mathfrak{C}/\mathfrak{I}$ has the same objects as $\mathfrak{C}$ and morphism groups $\mathfrak{C}(A, B)/\mathfrak{I}(A, B)$.

The quotient category is again additive, and the obvious functor $F\colon \mathfrak{C} \to \mathfrak{C}/\mathfrak{I}$ is additive and satisfies $\ker F = \mathfrak{I}$. Thus any ideal $\mathfrak{I}$ in $\mathfrak{C}$ is of the form $\ker F$ for a canonical additive functor $F$.

The additivity of $\mathfrak{C}/\mathfrak{I}$ and $F$ depends on the fact that any ideal $\mathfrak{I}$ is compatible with *finite* products in the following sense: the natural isomorphisms

$$\mathfrak{C}(A, B_1 \times B_2) \xrightarrow{\cong} \mathfrak{C}(A, B_1) \times \mathfrak{C}(A, B_2),$$
$$\mathfrak{C}(A_1 \times A_2, B) \xrightarrow{\cong} \mathfrak{C}(A_1, B) \times \mathfrak{C}(A_2, B)$$

restrict to isomorphisms

$$\mathfrak{I}(A, B_1 \times B_2) \xrightarrow{\cong} \mathfrak{I}(A, B_1) \times \mathfrak{I}(A, B_2),$$
$$\mathfrak{I}(A_1 \times A_2, B) \xrightarrow{\cong} \mathfrak{I}(A_1, B) \times \mathfrak{I}(A_2, B).$$

## 2.2. Examples of ideals

**Example 5.** Let KK be the *Kasparov category*, whose objects are the separable C*-algebras and whose morphism spaces are the Kasparov groups $KK_0(A, B)$, with the Kasparov product as composition. Let $\mathfrak{Ab}^{\mathbb{Z}/2}$ be the category of $\mathbb{Z}/2$-graded Abelian groups. Both categories are evidently additive.

K-theory is an additive functor $K_* \colon KK \to \mathfrak{Ab}^{\mathbb{Z}/2}$. We let $\mathfrak{I}_K := \ker K_*$ (as in (1.1)). Thus $\mathfrak{I}_K(A, B) \subseteq KK(A, B)$ is the kernel of the natural map

$$\gamma \colon KK(A, B) \to \mathrm{Hom}\big(K_*(A), K_*(B)\big) := \prod_{n \in \mathbb{Z}/2} \mathrm{Hom}\big(K_n(A), K_n(B)\big).$$

There is another interesting ideal in KK, namely, the kernel of a natural map

$$\kappa \colon \mathfrak{I}_K(A, B) \to \mathrm{Ext}\big(K_*(A), K_{*+1}(B)\big) := \prod_{n \in \mathbb{Z}/2} \mathrm{Ext}\big(K_n(A), K_{n+1}(B)\big)$$

due to Lawrence Brown (see [23]), whose definition we now recall. We represent $f \in KK(A, B) \cong \mathrm{Ext}\big(A, C_0(\mathbb{R}, B)\big)$ by a C*-algebra extension $C_0(\mathbb{R}, B) \otimes \mathbb{K} \rightarrowtail E \twoheadrightarrow A$. This yields an exact sequence

$$\begin{array}{ccccc}
K_1(B) & \longrightarrow & K_0(E) & \longrightarrow & K_0(A) \\
\uparrow{\scriptstyle f_*} & & & & \downarrow{\scriptstyle f_*} \\
K_1(A) & \longleftarrow & K_1(E) & \longleftarrow & K_0(B).
\end{array} \qquad (2.1)$$

The vertical maps in (2.1) are the two components of $\gamma(f)$. If $f \in \mathfrak{I}_K(A, B)$, then (2.1) splits into two extensions of Abelian groups, which yield an element $\kappa(f)$ in $\mathrm{Ext}\big(K_*(A), K_{*+1}(B)\big)$.

**Example 6.** Let $G$ be a second countable, locally compact group. Let $KK^G$ be the associated *equivariant Kasparov category*; its objects are the separable $G$-C*-algebras and its morphism spaces are the groups $KK^G(A, B)$, with the Kasparov product as composition. If $H \subseteq G$ is a closed subgroup, then there is a *restriction functor* $\mathrm{Res}_G^H \colon KK^G \to KK^H$, which simply forgets part of the equivariance.

If $\mathcal{F}$ is a set of closed subgroups of $G$, we define an ideal $\mathcal{VC}_{\mathcal{F}}$ in $KK^G$ by

$$\mathcal{VC}_{\mathcal{F}}(A, B) := \{f \in KK^G(A, B) \mid \mathrm{Res}_G^H(f) = 0 \text{ for all } H \in \mathcal{F}\}$$

as in (1.4). Of course, the condition $\mathrm{Res}_G^H(f) = 0$ is supposed to hold in $KK^H(A, B)$. We are mainly interested in the case where $\mathcal{F}$ is the family of all compact subgroups of $G$ and simply denote the ideal by $\mathcal{VC}$ in this case.

This ideal arises if we try to compute $G$-equivariant homology theories in terms of $H$-equivariant homology theories for $H \in \mathcal{F}$. The ideal $\mathcal{VC}$ is closely related to the approach to the Baum–Connes assembly map in [17].

The authors feel more at home with Kasparov theory than with spectra. Many readers will prefer to work in categories of spectra of, say, $G$-CW-complexes.

We do not introduce these categories here; but it shoud be clear enough that they support similar restriction functors, which provide analogues of the ideals $\mathcal{VC}_{\mathcal{F}}$.

**Example 7.** Let $G$ and $KK^G$ be as in Example 6. Using the crossed product functor (also called *descent* functor)

$$G \ltimes {\llcorner}: KK^G \to KK, \qquad A \mapsto G \ltimes A,$$

we define ideals $\mathfrak{I}_{\ltimes} \subseteq \mathfrak{I}_{\ltimes,K} \subseteq KK^G$ as in (1.2) and (1.3) by

$$\mathfrak{I}_{\ltimes}(A, B) := \{ f \in KK^G(A, B) \mid G \ltimes f = 0 \text{ in } KK(G \ltimes A, G \ltimes B)\},$$
$$\mathfrak{I}_{\ltimes,K}(A, B) := \{ f \in KK^G(A, B) \mid K_*(G \ltimes f) = 0 \colon K_*(G \ltimes A) \to K_*(G \ltimes B)\}.$$

We only study these ideals for compact $G$. In this case, the *Green–Julg Theorem* identifies $K_*(G \ltimes A)$ with the $G$-equivariant K-theory $K_*^G(A)$ (see [11]). Hence the ideal $\mathfrak{I}_{\ltimes,K} \subseteq KK^G$ is a good equivariant analogue of the ideal $\mathfrak{I}_K$ in KK.

Literally the same definition as above provides ideals $\mathfrak{I}_{\ltimes} \subseteq \mathfrak{I}_{\ltimes,K} \subseteq KK^G$ if $G$ is a compact *quantum* group. We will always allow this more general situation below, but readers unfamiliar with quantum groups may ignore this.

**Remark 8.** We emphasise quantum groups here because Examples 6 and 7 become closely related in this context. This requires a quantum group analogue of the ideals $\mathcal{VC}_{\mathcal{F}}$ in $KK^G$ of Example 6. If $G$ is a locally compact quantum group, then Saad Baaj and Georges Skandalis construct a $G$-equivariant Kasparov category $KK^G$ in [2]. There is a forgetful functor $\mathrm{Res}_G^H \colon KK^G \to KK^H$ for each closed quantum subgroup $H \subseteq G$. Therefore, a family $\mathcal{F}$ of closed quantum subgroups yields an ideal $\mathcal{VC}_{\mathcal{F}}$ in $KK^G$ as in Example 6.

Let $G$ be a compact group as in Example 7. Any crossed product $G \ltimes A$ carries a canonical *coaction* of $G$, that is, a coaction of the discrete quantum group $C^*(G)$. *Baaj–Skandalis duality* asserts that this yields an equivalence of categories $KK^G \cong KK^{C^*(G)}$ (see [2]). We get back the crossed product functor $KK^G \to KK$ by composing this equivalence with the restriction functor $KK^{C^*(G)} \to KK$ for the trivial quantum subgroup. Hence $\mathfrak{I}_{\ltimes} \subseteq KK^G$ corresponds by Baaj–Skandalis duality to $\mathcal{VC}_{\mathcal{F}} \subseteq KK^{C^*(G)}$, where $\mathcal{F}$ consists only of the trivial quantum subgroup.

Thus the constructions in Examples 6 and 7 are both special cases of a more general construction for locally compact quantum groups.

Finally, we consider a classical example from homological algebra.

**Example 9.** Let $\mathfrak{C}$ be an Abelian category. Let $\mathfrak{Ho}(A)$ be the homotopy category of unbounded chain complexes

$$\cdots \to C_n \xrightarrow{\delta_n} C_{n-1} \xrightarrow{\delta_{n-1}} C_{n-2} \xrightarrow{\delta_{n-2}} C_{n-3} \to \cdots$$

over $\mathfrak{C}$. The space of morphisms $A \to B$ in $\mathfrak{Ho}(\mathfrak{C})$ is the space $[A, B]$ of *homotopy classes* of chain maps from $A$ to $B$.

Taking homology defines functors $H_n \colon \mathfrak{Ho}(\mathfrak{C}) \to \mathfrak{C}$ for $n \in \mathbb{Z}$, which we combine to a single functor $\mathrm{H}_* \colon \mathfrak{Ho}(\mathfrak{C}) \to \mathfrak{C}^{\mathbb{Z}}$. We let $\mathfrak{I}_{\mathrm{H}} \subseteq \mathfrak{Ho}(\mathfrak{C})$ be its kernel:

$$\mathfrak{I}_{\mathrm{H}}(A, B) := \{ f \in [A, B] \mid \mathrm{H}_*(f) = 0 \}. \tag{2.2}$$

We also consider the category $\mathfrak{Ho}(\mathfrak{C}; \mathbb{Z}/p)$ of *p-periodic* chain complexes over $\mathfrak{C}$ for $p \in \mathbb{N}_{\geq 1}$; its objects satisfy $C_n = C_{n+p}$ and $\delta_n = \delta_{n+p}$ for all $n \in \mathbb{Z}$, and chain maps and homotopies are required $p$-periodic as well. The category $\mathfrak{Ho}(\mathfrak{C}; \mathbb{Z}/2)$ plays a role in connection with cyclic cohomology, especially with local cyclic cohomology ([14, 21]). The category $\mathfrak{Ho}(\mathfrak{C}; \mathbb{Z}/1)$ is isomorphic to the category of chain complexes without grading. By convention, we let $\mathbb{Z}/0 = \mathbb{Z}$, so that $\mathfrak{Ho}(\mathfrak{C}; \mathbb{Z}/0) = \mathfrak{Ho}(\mathfrak{C})$.

The homology of a periodic chain complex is, of course, periodic, so that we get a homological functor $\mathrm{H}_* \colon \mathfrak{Ho}(\mathfrak{C}; \mathbb{Z}/p) \to \mathfrak{C}^{\mathbb{Z}/p}$; here $\mathfrak{C}^{\mathbb{Z}/p}$ denotes the category of $\mathbb{Z}/p$-graded objects of $\mathfrak{C}$. We let $\mathfrak{I}_{\mathrm{H}} \subseteq \mathfrak{Ho}(\mathfrak{C}; \mathbb{Z}/p)$ be the kernel of $\mathrm{H}_*$ as in (2.2).

### 2.3. What is a triangulated category?

A *triangulated category* is a category $\mathfrak{T}$ with a *suspension automorphism* $\Sigma \colon \mathfrak{T} \to \mathfrak{T}$ and a class of *exact triangles*, subject to various axioms (see [17, 19, 25]). An exact triangle is a diagram in $\mathfrak{T}$ of the form

$$A \to B \to C \to \Sigma A \qquad \text{or} \qquad$$

$$
\begin{array}{ccc}
A & \longrightarrow & B \\
 & \nwarrow_{[1]} & \swarrow \\
 & C, &
\end{array}
$$

where the [1] in the arrow $C \to A$ warns us that this map has *degree* 1. A *morphism* of triangles is a triple of maps $\alpha, \beta, \gamma$ making the obvious diagram commute.

A typical example is the homotopy category $\mathfrak{Ho}(\mathfrak{C}; \mathbb{Z}/p))$ of $\mathbb{Z}/p$-graded chain complexes. Here the suspension functor is the (signed) *translation* functor

$$\Sigma\big((C_n, d_n)\big) := (C_{n-1}, -d_{n-1}) \qquad \text{on objects,}$$
$$\Sigma\big((f_n)\big) := (f_{n-1}) \qquad \text{on morphisms;}$$

a triangle is exact if it is isomorphic to a *mapping cone triangle*

$$A \xrightarrow{f} B \to \mathrm{cone}(f) \to \Sigma A$$

for some chain map $f$; the maps $B \to \mathrm{cone}(f) \to \Sigma A$ are the canonical ones. It is well-known that this defines a triangulated category for $p = 0$; the arguments for $p \geq 1$ are essentially the same.

Another classical example is the stable homotopy category, say, of compactly generated pointed topological spaces (it is not particularly relevant which category of spaces or spectra we use). The suspension is $\Sigma(A) := \mathbb{S}^1 \wedge A$; a triangle is exact if it is isomorphic to a *mapping cone triangle*

$$A \xrightarrow{f} B \to \mathrm{cone}(f) \to \Sigma A$$

for some map $f$; the maps $B \to \mathrm{cone}(f) \to \Sigma A$ are the canonical ones.

We are mainly interested in the categories $\mathrm{KK}$ and $\mathrm{KK}^G$ introduced in §2.2. Their triangulated category structure is discussed in detail in [17]. We are facing a notational problem because the functor $X \mapsto C_0(X)$ from pointed compact spaces to C*-algebras is *contravariant*, so that *mapping cone triangles* now have the form

$$A \xleftarrow{f} B \leftarrow \mathrm{cone}(f) \leftarrow C_0(\mathbb{R}, A)$$

for a $*$-homomorphism $f \colon B \to A$; here

$$\mathrm{cone}(f) = \big\{(a, b) \in C_0\big((0, \infty], A\big) \times B \mid a(\infty) = f(b)\big\}$$

and the maps $C_0(\mathbb{R}, A) \to \mathrm{cone}(f) \to B$ are the obvious ones, $a \mapsto (a, 0)$ and $(a, b) \mapsto b$.

It is reasonable to view a $*$-homomorphism from $A$ to $B$ as a morphism from $B$ to $A$. Nevertheless, we prefer the convention that an algebra homomorphism $A \to B$ is a morphism $A \to B$. But then the most natural triangulated category structure lives on the opposite category $\mathrm{KK}^{\mathrm{op}}$. This creates only notational difficulties because the opposite category of a triangulated category inherits a canonical triangulated category structure, which has "the same" exact triangles. However, the passage to opposite categories exchanges suspensions and desuspensions and modifies some sign conventions. Thus the functor $A \mapsto C_0(\mathbb{R}, A)$, which is the suspension functor in $\mathrm{KK}^{\mathrm{op}}$, becomes the *desuspension* functor in $\mathrm{KK}$. Fortunately, Bott periodicity implies that $\Sigma^2 \cong \mathrm{id}$, so that $\Sigma$ and $\Sigma^{-1}$ agree.

Depending on your definition of a triangulated category, you may want the suspension to be an equivalence or isomorphism of categories. In the latter case, you must replace $\mathrm{KK}^{(G)}$ by an equivalent category (see [17]); since this is not important here, we do not bother about this issue.

A triangle in $\mathrm{KK}^{(G)}$ is called *exact* if it is isomorphic to a mapping cone triangle

$$C_0(\mathbb{R}, B) \to \mathrm{cone}(f) \to A \xrightarrow{f} B$$

for some (equivariant) $*$-homomorphism $f$.

An important source of exact triangles in $\mathrm{KK}^G$ are *extensions*. If $A \rightarrowtail B \twoheadrightarrow C$ is an extension of $G$-$\mathrm{C}^*$-algebras with an equivariant completely positive contractive section, then it yields a class in $\mathrm{Ext}(C, A) \cong \mathrm{KK}(\Sigma^{-1}C, A)$; the resulting triangle

$$\Sigma^{-1}C \to A \to B \to C$$

in $\mathrm{KK}^G$ is exact and called an *extension triangle*. It is easy to see that any exact triangle is isomorphic to an extension triangle.

It is shown in [17] that KK and $\mathrm{KK}^G$ for a locally compact group $G$ are triangulated categories with this extra structure. The same holds for the equivariant Kasparov theory $\mathrm{KK}^S$ with respect to any $\mathrm{C}^*$-bialgebra $S$; this theory was defined by Baaj and Skandalis in [2].

The triangulated category axioms are discussed in greater detail in [17, 19, 25]. They encode some standard machinery for manipulating long exact sequences. Most of them amount to formal properties of mapping cones and mapping cylinders, which we can prove as in classical topology. The only axiom that requires more care is that any morphism $f \colon A \to B$ should be part of an exact triangle.

Unlike in [17], we prefer to construct this triangle as an extension triangle because this works in greater generality; we have taken this idea from Radu Popescu and Alexander Bonkat ([6, 20]). Any element in $\mathrm{KK}_0^S(A, B) \cong \mathrm{KK}_1^S\big(A, C_0(\mathbb{R}, B)\big)$ can be represented by an extension $\mathbb{K}(\mathcal{H}) \rightarrowtail E \twoheadrightarrow A$ with an equivariant completely positive contractive section, where $\mathcal{H}$ is a full $S$-equivariant Hilbert $C_0(\mathbb{R}, B)$-module, so that $\mathbb{K}(\mathcal{H})$ is $\mathrm{KK}^S$-equivalent to $C_0(\mathbb{R}, B)$. Hence the resulting extension triangle in $\mathrm{KK}^S$ is isomorphic to one of the form

$$C_0(\mathbb{R}, A) \to C_0(\mathbb{R}, B) \to E \to A;$$

by construction, it contains the suspension of the given class in $\mathrm{KK}_0^S(A, B)$; it is easy to remove the suspension.

**Definition 10.** Let $\mathfrak{T}$ be a triangulated and $\mathfrak{C}$ an Abelian category. A covariant functor $F \colon \mathfrak{T} \to \mathfrak{C}$ is called *homological* if $F(A) \to F(B) \to F(C)$ is exact at $F(B)$ for all exact triangles $A \to B \to C \to \Sigma A$. A contravariant functor with the analogous exactness property is called *cohomological*.

Let $A \to B \to C \to \Sigma A$ be an exact triangle. Then a homological functor $F : \mathfrak{T} \to \mathfrak{C}$ yields a natural long exact sequence

$$\cdots \to F_{n+1}(C) \to F_n(A) \to F_n(B) \to F_n(C) \to F_{n-1}(A) \to F_{n-1}(B) \to \cdots$$

with $F_n(A) := F(\Sigma^{-n} A)$ for $n \in \mathbb{Z}$, and a cohomological functor $F : \mathfrak{T}^{\mathrm{op}} \to \mathfrak{C}$ yields a natural long exact sequence

$$\cdots \leftarrow F^{n+1}(C) \leftarrow F^n(A) \leftarrow F^n(B) \leftarrow F^n(C) \leftarrow F^{n-1}(A) \leftarrow F^{n-1}(B) \leftarrow \cdots$$

with $F^n(A) := F(\Sigma^{-n} A)$.

**Proposition 11.** *Let $\mathfrak{T}$ be a triangulated category. The functors*

$$\mathfrak{T}(A, \_) : \mathfrak{T} \to \mathfrak{Ab}, \qquad B \mapsto \mathfrak{T}(A, B)$$

*are homological for all $A \in\in \mathfrak{T}$. Dually, the functors*

$$\mathfrak{T}(\_, B) : \mathfrak{T}^{\mathrm{op}} \to \mathfrak{Ab}, \qquad A \mapsto \mathfrak{T}(A, B)$$

*are cohomological for all $B \in\in \mathfrak{T}$.*

Observe that

$$\mathfrak{T}^n(A, B) = \mathfrak{T}(\Sigma^{-n} A, B) \cong \mathfrak{T}(A, \Sigma^n B) \cong \mathfrak{T}_{-n}(A, B).$$

**Definition 12.** A *stable additive category* is an additive category equipped with an (additive) automorphism $\Sigma$, called *suspension*.

A *stable homological functor* is a homological functor $F : \mathfrak{T} \to \mathfrak{C}$ into a stable Abelian category $\mathfrak{C}$ together with natural isomorphisms $F\big(\Sigma_{\mathfrak{T}}(A)\big) \cong \Sigma_{\mathfrak{C}}\big(F(A)\big)$ for all $A \in\in \mathfrak{T}$.

**Example 13.** The category $\mathfrak{C}^{\mathbb{Z}/p}$ of $\mathbb{Z}/p$-graded objects of an Abelian category $\mathfrak{C}$ is stable for any $p \in \mathbb{N}$; the suspension automorphism merely shifts the grading. The functors $K_* : KK \to \mathfrak{Ab}^{\mathbb{Z}/2}$ and $H_* : \mathfrak{Ho}(\mathfrak{C}; \mathbb{Z}/p) \to \mathfrak{C}^{\mathbb{Z}/p}$ introduced in Examples 5 and 9 are stable homological functors.

If $F : \mathfrak{T} \to \mathfrak{C}$ is any homological functor, then

$$F_* : \mathfrak{T} \to \mathfrak{C}^{\mathbb{Z}}, \qquad A \mapsto \big(F_n(A)\big)_{n \in \mathbb{Z}}$$

is a stable homological functor. Many of our examples satisfy *Bott periodicity*, that is, there is a natural isomorphism $F_2(A) \cong F(A)$. Then we get a stable homological functor $F_* : \mathfrak{T} \to \mathfrak{C}^{\mathbb{Z}/2}$. A typical example for this is the functor $K_*$.

**Definition 14.** A functor $F\colon \mathfrak{T} \to \mathfrak{T}'$ between two triangulated categories is called *exact* if it intertwines the suspension automorphisms (up to specified natural isomorphisms) and maps exact triangles in $\mathfrak{T}$ again to exact triangles in $\mathfrak{T}'$.

**Example 15.** The restriction functor $\mathrm{Res}_G^H\colon \mathrm{KK}^G \to \mathrm{KK}^H$ for a closed quantum subgroup $H$ of a locally compact quantum group $G$ and the crossed product functors $G \ltimes \lrcorner,\, G \ltimes_\mathrm{r} \lrcorner\colon \mathrm{KK}^G \to \mathrm{KK}$ are exact because they preserve mapping cone triangles.

Let $F\colon \mathfrak{T}_1 \to \mathfrak{T}_2$ be an exact functor. If $G\colon \mathfrak{T}_2 \to?$ is exact, homological, or cohomological, then so is $G \circ F$.

Using Examples 13 and 15, we see that the functors that define the ideals $\ker \gamma$ in Example 5, $\mathcal{VC}_{\mathcal{F}}$ in Example 6, $\mathfrak{I}_\ltimes$, and $\mathfrak{I}_{\ltimes,\mathrm{K}}$ in Example 7, and $\mathfrak{I}_\mathrm{H}$ in Example 9 are all stable and either homological or exact.

## 2.4. The universal homological functor

The following general construction of Peter Freyd ([10]) plays an important role in [4]. For an additive category $\mathfrak{C}$, let $\mathfrak{Fun}(\mathfrak{C}^{\mathrm{op}}, \mathfrak{Ab})$ be the category of contravariant additive functors $\mathfrak{C} \to \mathfrak{Ab}$, with natural transformations as morphisms. Unless $\mathfrak{C}$ is essentially small, this is not quite a category because the morphisms may form classes instead of sets. We may ignore this set-theoretic problem because the bivariant Kasparov categories that we are interested in are essentially small, and the subcategory $\mathfrak{Coh}(\mathfrak{C})$ of $\mathfrak{Fun}(\mathfrak{C}^{\mathrm{op}}, \mathfrak{Ab})$ that we are going to use later on is an honest category for any $\mathfrak{C}$.

The category $\mathfrak{Fun}(\mathfrak{C}^{\mathrm{op}}, \mathfrak{Ab})$ is Abelian: if $f\colon F_1 \to F_2$ is a natural transformation, then its kernel, cokernel, image, and co-image are computed pointwise on the objects of $\mathfrak{C}$, so that they boil down to the corresponding constructions with Abelian groups.

The *Yoneda embedding* is an additive functor

$$\mathbb{Y}\colon \mathfrak{C} \to \mathfrak{Fun}(\mathfrak{C}^{\mathrm{op}}, \mathfrak{Ab}), \qquad B \mapsto \mathfrak{T}(\lrcorner, B).$$

This functor is fully faithful, and there are natural isomorphisms

$$\mathrm{Hom}(\mathbb{Y}(B), F) \cong F(B) \qquad \text{for all } F \in\in \mathfrak{Fun}(\mathfrak{C}^{\mathrm{op}}, \mathfrak{Ab}),\ B \in\in \mathfrak{T}$$

by the *Yoneda Lemma*. A functor $F \in\in \mathfrak{Fun}(\mathfrak{C}^{\mathrm{op}}, \mathfrak{Ab})$ is called *representable* if it is isomorphic to $\mathbb{Y}(B)$ for some $B \in\in \mathfrak{C}$. Hence $\mathbb{Y}$ yields an equivalence of categories between $\mathfrak{C}$ and the subcategory of representable functors in $\mathfrak{Fun}(\mathfrak{C}^{\mathrm{op}}, \mathfrak{Ab})$.

A functor $F \in\in \mathfrak{Fun}(\mathfrak{C}^{\mathrm{op}}, \mathfrak{Ab})$ is called *finitely presented* if there is an exact sequence $\mathbb{Y}(B_1) \to \mathbb{Y}(B_2) \to F \to 0$ with $B_1, B_2 \in\in \mathfrak{T}$. Since $\mathbb{Y}$ is fully faithful, this means that $F$ is the cokernel of $\mathbb{Y}(f)$ for a morphism $f$ in $\mathfrak{C}$. We let $\mathfrak{Coh}(\mathfrak{C})$ be the full subcategory of finitely presented functors in $\mathfrak{Fun}(\mathfrak{C}^{\mathrm{op}}, \mathfrak{Ab})$. Since representable functors belong to $\mathfrak{Coh}(\mathfrak{C})$, we still have a Yoneda embedding $\mathbb{Y} \colon \mathfrak{C} \to \mathfrak{Coh}(\mathfrak{C})$. Although the category $\mathfrak{Coh}(\mathfrak{T})$ tends to be very big and therefore unwieldy, it plays an important theoretical role.

**Theorem 16** (Freyd's Theorem). *Let $\mathfrak{T}$ be a triangulated category.*

*Then $\mathfrak{Coh}(\mathfrak{T})$ is a stable Abelian category that has enough projective and enough injective objects, and the projective and injective objects coincide.*

*The functor $\mathbb{Y} \colon \mathfrak{T} \to \mathfrak{Coh}(\mathfrak{T})$ is fully faithful, stable, and homological. Its essential range $\mathbb{Y}(\mathfrak{T})$ consists of projective-injective objects. Conversely, an object of $\mathfrak{Coh}(\mathfrak{T})$ is projective-injective if and only if it is a retract of an object of $\mathbb{Y}(\mathfrak{T})$.*

*The functor $\mathbb{Y}$ is the universal (stable) homological functor in the following sense: any (stable) homological functor $F \colon \mathfrak{T} \to \mathfrak{C}'$ to a (stable) Abelian category $\mathfrak{C}'$ factors uniquely as $F = \bar{F} \circ \mathbb{Y}$ for a (stable) exact functor $F \colon \mathfrak{Coh}(\mathfrak{T}) \to \mathfrak{C}'$.*

If idempotents in $\mathfrak{T}$ split – as in all our examples – then $\mathbb{Y}(\mathfrak{T})$ is closed under retracts, so that $\mathbb{Y}(\mathfrak{T})$ is equal to the class of projective-injective objects in $\mathfrak{Coh}(\mathfrak{T})$.

## 2.5. Homological ideals in triangulated categories

Let $\mathfrak{T}$ be a triangulated category, let $\mathfrak{C}$ be a stable additive category, and let $F \colon \mathfrak{T} \to \mathfrak{C}$ be a stable homological functor. Then $\ker F$ is a stable ideal in the following sense:

**Definition 17.** An ideal $\mathfrak{I}$ in $\mathfrak{T}$ is called *stable* if the suspension isomorphisms $\Sigma \colon \mathfrak{T}(A, B) \overset{\cong}{\to} \mathfrak{T}(\Sigma A, \Sigma B)$ for $A, B \in\in \mathfrak{T}$ restrict to isomorphisms

$$\Sigma \colon \mathfrak{I}(A, B) \overset{\cong}{\to} \mathfrak{I}(\Sigma A, \Sigma B).$$

If $\mathfrak{I}$ is stable, then there is a unique suspension automorphism on $\mathfrak{T}/\mathfrak{I}$ for which the canonical functor $\mathfrak{T} \to \mathfrak{T}/\mathfrak{I}$ is stable. Thus the stable ideals are exactly the kernels of stable additive functors.

**Definition 18.** An ideal $\mathfrak{I}$ in a triangulated category $\mathfrak{T}$ is called *homological* if it is the kernel of a stable homological functor.

**Remark 19.** Freyd's Theorem shows that $\mathbb{Y}$ induces a bijection between (stable) exact functors $\mathfrak{Coh}(\mathfrak{T}) \to \mathfrak{C}'$ and (stable) homological functors $\mathfrak{T} \to \mathfrak{C}'$ because $\bar{F} \circ \mathbb{Y}$ is homological if $\bar{F}\colon \mathfrak{Coh}(\mathfrak{T}) \to \mathfrak{C}'$ is exact. Hence the notion of homological functor is independent of the triangulated category structure on $\mathfrak{T}$ because the Yoneda embedding $\mathbb{Y}\colon \mathfrak{T} \to \mathfrak{Coh}(\mathfrak{T})$ does not involve any additional structure. Hence the notion of homological ideal only uses the suspension automorphism, not the class of exact triangles.

All the ideals considered in §2.2 except for $\ker \kappa$ in Example 5 are kernels of stable homological functors or exact functors. Those of the first kind are homological by definition. If $F\colon \mathfrak{T} \to \mathfrak{T}'$ is an exact functor between two triangulated categories, then $\mathbb{Y} \circ F\colon \mathfrak{T} \to \mathfrak{Coh}(\mathfrak{T}')$ is a stable homological functor with $\ker \mathbb{Y} \circ F = \ker F$ by Freyd's Theorem 16. Hence kernels of exact functors are homological as well.

Is any homological ideal the kernel of an exact functor? This is *not* the case:

**Proposition 20.** *Let $\mathfrak{Der}(\mathfrak{Ab})$ be the derived category of the category $\mathfrak{Ab}$ of Abelian groups. Define the ideal $\mathfrak{I}_{\mathrm{H}}$ in $\mathfrak{Der}(\mathfrak{Ab})$ as in Example 9. This ideal is not the kernel of an exact functor.*

We postpone the proof to the end of §3.1 because it uses the machinery of §3.1.

It takes some effort to characterise homological ideals because $\mathfrak{T}/\mathfrak{I}$ is almost never Abelian. The results in [4, §2–3] show that an ideal is homological if and only if it is *saturated* in the notation of [4]. We do not discuss this notion here because most ideals that we consider are obviously homological. The only example where we could profit from an abstract characterisation is the ideal $\ker \kappa$ in Example 5.

There is no obvious homological functor whose kernel is $\ker \kappa$ because $\kappa$ is not a functor on KK. Nevertheless, $\ker \kappa$ is the kernel of an exact functor; the relevant functor is the functor KK $\to$ UCT, where UCT is the variant of KK that satisfies the Universal Coefficient Theorem in complete generality. This functor can be constructed as a localisation of KK (see [17]). The Universal Coefficient Theorem implies that its kernel is exactly $\ker \kappa$.

## 3. From homological ideals to derived functors

Once we have a stable homological functor $F\colon \mathfrak{T} \to \mathfrak{C}$, it is not surprising that we can do a certain amount of homological algebra in $\mathfrak{T}$. For instance, we may call a chain complex of objects of $\mathfrak{T}$ *$F$-exact* if $F$ maps it to an exact chain complex in $\mathfrak{C}$; and we may call an object $F$-*projective* if $F$ maps it to a projective object in $\mathfrak{C}$. But are these definitions reasonable?

We propose that a reasonable homological notion should depend only on the ideal ker $F$. We will see that the notion of $F$-exact chain complex is reasonable and only depends on ker $F$. In contrast, the notion of projectivity above depends on $F$ and is only reasonable in special cases. There is another, more useful, notion of projective object that depends only on the ideal ker $F$.

Various notions from homological algebra still make sense in the context of homological ideals in triangulated categories. Our discussion mostly follows [1, 4, 8, 9]. All our definitions involve only the ideal, not a stable homological functor that defines it. We reformulate them in terms of an exact or a stable homological functor defining the ideal in order to understand what they mean in concrete cases. Following [9], we construct projective objects using adjoint functors.

The most sophisticated concept in this section is the *universal $\mathfrak{I}$-exact functor*, which gives us complete control over projective resolutions and derived functors. We can usually describe such functors very concretely.

### 3.1. Basic notions

We introduce some useful terminology related to an ideal:

**Definition 21.** Let $\mathfrak{I}$ be a homological ideal in a triangulated category $\mathfrak{T}$.

– Let $f : A \to B$ be a morphism in $\mathfrak{T}$; embed it in an exact triangle $A \xrightarrow{f} B \xrightarrow{g} C \xrightarrow{h} \Sigma A$. We call $f$
  * $\mathfrak{I}$-*monic* if $h \in \mathfrak{I}$;
  * $\mathfrak{I}$-*epic* if $g \in \mathfrak{I}$;
  * an $\mathfrak{I}$-*equivalence* if it is both $\mathfrak{I}$-monic and $\mathfrak{I}$-epic, that is, $g, h \in \mathfrak{I}$;
  * an $\mathfrak{I}$-*phantom map* if $f \in \mathfrak{I}$.
– An object $A \in \mathfrak{T}$ is called $\mathfrak{I}$-*contractible* if $\mathrm{id}_A \in \mathfrak{I}(A, A)$.
– An exact triangle $A \xrightarrow{f} B \xrightarrow{g} C \xrightarrow{h} \Sigma A$ in $\mathfrak{T}$ is called $\mathfrak{I}$-*exact* if $h \in \mathfrak{I}$.

The notions of monomorphism (or monic morphism) and epimorphism (or epic morphism) – which can be found in any book on category theory such as [13] – are categorical ways to express injectivity or surjectivity of maps. A morphism in an Abelian category that is both monic and epic is invertible.

The classes of $\mathfrak{I}$-phantom maps, $\mathfrak{I}$-monics, $\mathfrak{I}$-epics, and of $\mathfrak{I}$-exact triangles determine each other uniquely because we can embed any morphism in an exact triangle in any position. It is a matter of taste which of these is considered most fundamental. Following Daniel Christensen ([8]), we favour the phantom maps. Other authors prefer exact triangles instead ([1,4,9]). Of course, the notion of an $\mathfrak{I}$-phantom map is redundant; it becomes more relevant if we consider, say, the class of $\mathfrak{I}$-exact triangles as our basic notion.

Notice that $f$ is $\mathfrak{I}$-epic or $\mathfrak{I}$-monic if and only if $-f$ is. If $f$ is $\mathfrak{I}$-epic or $\mathfrak{I}$-monic, then so are $\Sigma^n(f)$ for all $n \in \mathbb{Z}$ because $\mathfrak{I}$ is stable. Similarly, (signed) suspensions of $\mathfrak{I}$-exact triangles remain $\mathfrak{I}$-exact triangles.

**Lemma 22.** *Let* $F \colon \mathfrak{T} \to \mathfrak{C}$ *be a stable homological functor into a stable Abelian category* $\mathfrak{C}$.

- *A morphism $f$ in $\mathfrak{T}$ is*
  - *$*$ a* ker *$F$-phantom map if and only if $F(f) = 0$;*
  - *$*$* ker *$F$-monic if and only if $F(f)$ is monic;*
  - *$*$* ker *$F$-epic if and only if $F(f)$ is epic;*
  - *$*$ a* ker *$F$-equivalence if and only if $F(f)$ is invertible.*
- *An object $A \in\in \mathfrak{T}$ is* ker *$F$-contractible if and only if $F(A) = 0$.*
- *An exact triangle $A \to B \to C \to \Sigma A$ is* ker *$F$-exact if and only if*

$$0 \to F(A) \to F(B) \to F(C) \to 0$$

*is a short exact sequence in* $\mathfrak{C}$.

*Proof.* Sequences in $\mathfrak{C}$ of the form $X \xrightarrow{0} Y \xrightarrow{f} Z$ or $X \xrightarrow{f} Y \xrightarrow{0} Z$ are exact at $Y$ if and only if $f$ is monic or epic, respectively. Moreover, a sequence of the form $X \xrightarrow{0} Y \to Z \to U \xrightarrow{0} W$ is exact if and only if $0 \to Y \to Z \to U \to 0$ is exact.

Combined with the long exact homology sequences for $F$ and suitable exact triangles, these observations yield the assertions about monomorphisms, epimorphisms, and exact triangles. The description of equivalences and contractible objects follows, and phantom maps are trivial, anyway. $\qquad\square$

Now we specialise these notions to the ideal $\mathfrak{I}_K \subseteq KK$ of Example 5, replacing $\mathfrak{I}_K$ by $K$ in our notation to avoid clutter.

- Let $f \in KK(A, B)$ and let $K_*(f) \colon K_*(A) \to K_*(B)$ be the induced map. Then $f$ is
  - $*$ a K-*phantom map* if and only if $K_*(f) = 0$;
  - $*$ K-*monic* if and only if $K_*(f)$ is injective;
  - $*$ K-*epic* if and only if $K_*(f)$ is surjective;
  - $*$ a K-*equivalence* if and only if $K_*(f)$ is invertible.

- A C*-algebra $A \in\in \mathrm{KK}$ is K-*contractible* if and only if $\mathrm{K}_*(A) = 0$.
- An exact triangle $A \to B \to C \to \Sigma A$ in $\mathrm{KK}$ is K-*exact* if and only if

$$0 \to \mathrm{K}_*(A) \to \mathrm{K}_*(B) \to \mathrm{K}_*(C) \to 0$$

is a short exact sequence (of $\mathbb{Z}/2$-graded Abelian groups).

Similar things happen for the other ideals in §2.2 that are *naturally* defined as kernels of stable homological functors.

**Remark 23.** It is crucial for the above theory that we consider functors that are both *stable* and *homological*. Everything fails if we drop either assumption and consider functors such as $\mathrm{K}_0(A)$ or $\mathrm{Hom}\big(\mathbb{Z}/4, \mathrm{K}_*(A)\big)$.

**Lemma 24.** *An object $A \in\in \mathfrak{T}$ is $\mathfrak{I}$-contractible if and only if $0\colon 0 \to A$ is an $\mathfrak{I}$-equivalence. A morphism $f$ in $\mathfrak{T}$ is an $\mathfrak{I}$-equivalence if and only if its generalised mapping cone is $\mathfrak{I}$-contractible.*

Thus the classes of $\mathfrak{I}$-equivalences and of $\mathfrak{I}$-contractible objects determine each other. But they do not allow us to recover the ideal itself. For instance, the ideals $\mathfrak{I}_\mathrm{K}$ and $\ker \kappa$ in Example 5 have the same contractible objects and equivalences.

*Proof.* Recall that the generalised mapping cone of $f$ is the object $C$ that fits in an exact triangle $A \xrightarrow{f} B \to C \to \Sigma A$. The long exact sequence for this triangle yields that $F(f)$ is invertible if and only if $F(C) = 0$, where $F$ is some stable homological functor $F$ with $\ker F = \mathfrak{I}$. Now the second assertion follows from Lemma 22. Since the generalised mapping cone of $0 \to A$ is $A$, the first assertion is a special case of the second one. $\qquad\square$

Many ideals are defined as $\ker F$ for an exact functor $F\colon \mathfrak{T} \to \mathfrak{T}'$ between triangulated categories. We can also use such a functor to describe the above notions:

**Lemma 25.** *Let $\mathfrak{T}$ and $\mathfrak{T}'$ be triangulated categories and let $F\colon \mathfrak{T} \to \mathfrak{T}'$ be an exact functor.*

- *A morphism $f \in \mathfrak{T}(A, B)$ is*
  - *a $\ker F$-phantom map if and only if $F(f) = 0$;*
  - *$\ker F$-monic if and only if $F(f)$ is (split) monic.*
  - *$\ker F$-epic if and only if $F(f)$ is (split) epic;*
  - *a $\ker F$-equivalence if and only if $F(f)$ is invertible.*
- *An object $A \in\in \mathfrak{T}$ is $\ker F$-contractible if and only if $F(A) = 0$.*
- *An exact triangle $A \to B \to C \to \Sigma A$ is $\ker F$-exact if and only if the exact triangle $F(A) \to F(B) \to F(C) \to F(\Sigma A)$ in $\mathfrak{T}'$ splits.*

We will explain the notation during the proof.

*Proof.* A morphism $f\colon X \to Y$ in $\mathfrak{T}'$ is called *split epic* (*split monic*) if there is $g\colon Y \to X$ with $f \circ g = \mathrm{id}_Y$ ($g \circ f = \mathrm{id}_X$). An exact triangle $X \xrightarrow{f} Y \xrightarrow{g} Z \xrightarrow{h} \Sigma X$ is said to *split* if $h = 0$. This immediately yields the characterisation of $\ker F$-exact triangles. Any split triangle is isomorphic to a direct sum triangle, so that $f$ is split monic and $g$ is split epic ([19, Corollary 1.2.7]). Conversely, either of these conditions implies that the triangle is split.

Since the $\ker F$-exact triangles determine the $\ker F$-epimorphisms and $\ker F$-monomorphisms, the latter are detected by $F(f)$ being split epic or split monic, respectively. It is clear that split epimorphisms and split monomorphisms are epimorphisms and monomorphisms, respectively. The converse holds in a triangulated category because if we embed a monomorphism or epimorphism in an exact triangle, then one of the maps is forced to vanish, so that the exact triangle splits.

Finally, a morphism is invertible if and only if it is both split monic and split epic, and the zero map $F(A) \to F(A)$ is invertible if and only if $F(A) = 0$. $\quad\square$

Alternatively, we may prove Lemma 22 using the Yoneda embedding $\mathbb{Y}\colon \mathfrak{T}' \to \mathfrak{Coh}(\mathfrak{T}')$. The assertions about phantom maps, equivalences, and contractibility boil down to the observation that $\mathbb{Y}$ is fully faithful. The assertions about monomorphisms and epimorphisms follow because a map $f\colon A \to B$ in $\mathfrak{T}'$ becomes epic (monic) in $\mathfrak{Coh}(\mathfrak{T}')$ if and only if it is split epic (monic) in $\mathfrak{T}'$.

There is a similar description for $\bigcap \ker F_i$ for a set $\{F_i\}$ of exact functors. This applies to the ideal $\mathcal{VC}_{\mathcal{F}}$ for a family of (quantum) subgroups $\mathcal{F}$ in a locally compact (quantum) group $G$ (Example 6). Replacing $\mathcal{VC}_{\mathcal{F}}$ by $\mathcal{F}$ in our notation to avoid clutter, we get:

- A morphism $f \in \mathrm{KK}^G(A, B)$ is
  * an $\mathcal{F}$-*phantom map* if and only if $\mathrm{Res}^H_G(f) = 0$ in $\mathrm{KK}^H$ for all $H \in \mathcal{F}$;
  * $\mathcal{F}$-*epic* if and only if $\mathrm{Res}^H_G(f)$ is (split) epic in $\mathrm{KK}^H$ for all $H \in \mathcal{F}$;
  * $\mathcal{F}$-*monic* if and only if $\mathrm{Res}^H_G(f)$ is (split) monic in $\mathrm{KK}^H$ for all $H \in \mathcal{F}$;
  * an $\mathcal{F}$-*equivalence* if and only if $\mathrm{Res}^H_G(f)$ is a $\mathrm{KK}^H$-equivalence for all $H \in \mathcal{F}$.
- A $G$-$C^*$-algebra $A \in\in \mathrm{KK}^G$ is $\mathcal{F}$-*contractible* if and only if $\mathrm{Res}^H_G(A) \cong 0$ in $\mathrm{KK}^H$ for all $H \in \mathcal{F}$.
- An exact triangle $A \to B \to C \to \Sigma A$ in $\mathrm{KK}^G$ is $\mathcal{F}$-*exact* if and only if

$$\mathrm{Res}^H_G(A) \to \mathrm{Res}^H_G(B) \to \mathrm{Res}^H_G(C) \to \Sigma\,\mathrm{Res}^H_G(A)$$

is a split exact triangle in $\mathrm{KK}^H$ for all $H \in \mathcal{F}$.

You may write down a similar list for the ideal $\mathfrak{I}_\ltimes \subseteq \mathrm{KK}^G$ of Example 7.

Lemma 25 allows us to prove that the ideal $\mathfrak{I}_\mathrm{H}$ in $\mathfrak{Der}(\mathfrak{Ab})$ cannot be the kernel of an exact functor:

*Proof of Proposition* 20. We embed $\mathfrak{Ab} \to \mathfrak{Der}(\mathfrak{Ab})$ as chain complexes concentrated in degree 0. The generator $\tau \in \mathrm{Ext}(\mathbb{Z}/2, \mathbb{Z}/2)$ corresponds to the extension of Abelian groups $\mathbb{Z}/2 \rightarrowtail \mathbb{Z}/4 \twoheadrightarrow \mathbb{Z}/2$, where the first map is multiplication by 2 and the second map is the natural projection. We get an exact triangle

$$\mathbb{Z}/2 \to \mathbb{Z}/4 \to \mathbb{Z}/2 \xrightarrow{\tau} \mathbb{Z}/2[1]$$

in $\mathfrak{Der}(\mathfrak{Ab})$. This triangle is $\mathfrak{I}_\mathrm{H}$-exact because the map $\mathbb{Z}/2 \to \mathbb{Z}/4$ is injective as a group homomorphism and hence $\mathfrak{I}_\mathrm{H}$-monic in $\mathfrak{Der}(\mathfrak{Ab})$.

Assume there were an exact functor $F : \mathfrak{Der}(\mathfrak{Ab}) \to \mathfrak{T}'$ with $\ker F = \mathfrak{I}_\mathrm{H}$. Then $F(\tau) = 0$, so that $F$ maps our triangle to a split triangle and $F(\mathbb{Z}/4) \cong F(\mathbb{Z}/2) \oplus F(\mathbb{Z}/2)$ by Lemma 25. It follows that $F(2 \cdot \mathrm{id}_{\mathbb{Z}/4}) = 2 \cdot \mathrm{id}_{F(\mathbb{Z}/4)} = 0$ because $2 \cdot \mathrm{id}_{F(\mathbb{Z}/2)} = F(2 \cdot \mathrm{id}_{\mathbb{Z}/2}) = 0$. Hence $2 \cdot \mathrm{id}_{\mathbb{Z}/4} \in \ker F = \mathfrak{I}_\mathrm{H}$, which is false. This contradiction shows that there is no exact functor $F$ with $\ker F = \mathfrak{I}_\mathrm{H}$. $\qquad\square$

One of the most interesting questions about an ideal is whether all $\mathfrak{I}$-contractible objects vanish or, equivalently, whether all $\mathfrak{I}$-equivalences are invertible. These two questions are equivalent by Lemma 24. The answer is negative for the ideal $\mathfrak{I}_\mathrm{K}$ in KK because the Universal Coefficient Theorem does not hold for arbitrary separable C*-algebras. Therefore, we also get counterexamples for the ideal $\mathfrak{I}_{\ltimes,\mathrm{K}}$ in $\mathrm{KK}^G$ for a compact quantum group. In contrast, if $G$ is a connected Lie group with torsion-free fundamental group, then $\mathfrak{I}_\ltimes$-equivalences in $\mathrm{KK}^G$ are invertible (see [18]). If $G$ is an amenable group, then $\mathcal{VC}$-equivalences in $\mathrm{KK}^G$ are invertible; this follows from the proof of the Baum–Connes Conjecture for these groups by Nigel Higson and Gennadi Kasparov (see [17]). These examples show that this question is subtle and may involve difficult analysis.

## 3.2. Exact chain complexes

The notion of $\mathfrak{I}$-exactness, which we have only defined for exact triangles so far, will now be extended to chain complexes. Our definition differs from Beligiannis' one ([1,4]), which we recall first.

Let $\mathfrak{T}$ be a triangulated category and let $\mathfrak{I}$ be a homological ideal in $\mathfrak{T}$.

**Definition 26.** A chain complex

$$C_\bullet := (\cdots \to C_{n+1} \xrightarrow{d_{n+1}} C_n \xrightarrow{d_n} C_{n-1} \xrightarrow{d_{n-1}} C_{n-2} \to \cdots)$$

in $\mathfrak{T}$ is called $\mathfrak{I}$-*decomposable* if there is a sequence of $\mathfrak{I}$-exact triangles

$$K_{n+1} \xrightarrow{g_n} C_n \xrightarrow{f_n} K_n \xrightarrow{h_n} \Sigma K_{n+1}$$

with $d_n = g_{n-1} \circ f_n \colon C_n \to C_{n-1}$.

Such complexes are called $\mathfrak{I}$-exact in [1,4]. This definition is inspired by the following well-known fact: a chain complex over an Abelian category is exact if and only if it splits into short exact sequences of the form $K_n \rightarrowtail C_n \twoheadrightarrow K_{n-1}$ as in Definition 26.

We prefer another definition of exactness because we have not found a general explicit criterion for a chain complex to be $\mathfrak{I}$-decomposable.

**Definition 27.** Let $C_\bullet = (C_n, d_n)$ be a chain complex over $\mathfrak{T}$. For each $n \in \mathbb{N}$, embed $d_n$ in an exact triangle

$$C_n \xrightarrow{d_n} C_{n-1} \xrightarrow{f_n} X_n \xrightarrow{g_n} \Sigma C_n. \tag{3.1}$$

We call $C_\bullet$ $\mathfrak{I}$-*exact in degree n* if the map $X_n \xrightarrow{g_n} \Sigma C_n \xrightarrow{\Sigma f_{n+1}} \Sigma X_{n+1}$ belongs to $\mathfrak{I}(X_n, \Sigma X_{n+1})$. This does not depend on auxiliary choices because the exact triangles in (3.1) are unique up to (non-canonical) isomorphism.

We call $C_\bullet$ $\mathfrak{I}$-*exact* if it is $\mathfrak{I}$-exact in degree $n$ for all $n \in \mathbb{Z}$.

This definition is designed to make the following lemma true:

**Lemma 28.** *Let $F \colon \mathfrak{T} \to \mathfrak{C}$ be a stable homological functor into a stable Abelian category $\mathfrak{C}$ with* ker $F = \mathfrak{I}$. *A chain complex $C_\bullet$ over $\mathfrak{T}$ is $\mathfrak{I}$-exact in degree $n$ if and only if*

$$F(C_{n+1}) \xrightarrow{F(d_{n+1})} F(C_n) \xrightarrow{F(d_n)} F(C_{n-1})$$

*is exact at $F(C_n)$.*

*Proof.* The complex $C_\bullet$ is $\mathfrak{I}$-exact in degree $n$ if and only if the map

$$\Sigma^{-1} F(X_n) \xrightarrow{\Sigma^{-1} F(g_n)} F(C_n) \xrightarrow{F(f_{n+1})} F(X_{n+1})$$

vanishes. Equivalently, the range of $\Sigma^{-1}F(g_n)$ is contained in the kernel of $F(f_{n+1})$. The long exact sequences

$$\cdots \to \Sigma^{-1}F(X_n) \xrightarrow{\Sigma^{-1}F(g_n)} F(C_n) \xrightarrow{F(d_n)} F(C_{n-1}) \to \cdots,$$

$$\cdots \to F(C_{n+1}) \xrightarrow{F(d_{n+1})} F(C_n) \xrightarrow{F(f_{n+1})} F(X_{n+1}) \to \cdots$$

show that the range of $\Sigma^{-1}F(g_n)$ and the kernel of $F(f_{n+1})$ are equal to the kernel of $F(d_n)$ and the range of $F(d_{n+1})$, respectively. Hence $C_\bullet$ is $\mathfrak{I}$-exact in degree $n$ if and only if $\ker F(d_n) \subseteq \operatorname{range} F(d_{n+1})$. Since $d_n \circ d_{n+1} = 0$, this is equivalent to $\ker F(d_n) = \operatorname{range} F(d_{n+1})$. $\qquad\square$

**Corollary 29.** *$\mathfrak{I}$-decomposable chain complexes are $\mathfrak{I}$-exact.*

*Proof.* Let $F\colon \mathfrak{T} \to \mathfrak{C}$ be a stable homological functor with $\ker F = \mathfrak{I}$. If $C_\bullet$ is $\mathfrak{I}$-decomposable, then $F(C_\bullet)$ is obtained by splicing short exact sequences in $\mathfrak{C}$. This implies that $F(C_\bullet)$ is exact, so that $C_\bullet$ is $\mathfrak{I}$-exact by Lemma 28. $\quad\square$

The converse implication in Corollary 29 fails in general (see Example 37).

**Example 30.** For the ideal $\mathfrak{I}_K$ in KK, Lemma 28 yields that a chain complex $C_\bullet$ over KK is K-*exact* (in degree $n$) if and only if the chain complex

$$\cdots \to K_*(C_{n+1}) \to K_*(C_n) \to K_*(C_{n-1}) \to \cdots$$

of $\mathbb{Z}/2$-graded Abelian groups is exact (in degree $n$). Similar remarks apply to the other ideals in §2.2 that are defined as kernels of stable homological functors.

As a trivial example, we consider the largest possible ideal $\mathfrak{I} = \mathfrak{T}$. This ideal is defined by the zero functor. Lemma 28 or the definition yield that *all* chain complexes are $\mathfrak{T}$-exact. In contrast, it seems hard to characterise the $\mathfrak{I}$-decomposable chain complexes, already for $\mathfrak{I} = \mathfrak{T}$.

**Lemma 31.** *A chain complex of length* 3

$$\cdots \to 0 \to A \xrightarrow{f} B \xrightarrow{g} C \to 0 \to \cdots$$

*is $\mathfrak{I}$-exact if and only if there are an $\mathfrak{I}$-exact exact triangle $A' \xrightarrow{f'} B' \xrightarrow{g'} C' \to \Sigma A'$ and a commuting diagram*

$$\begin{array}{ccccc}
A' & \xrightarrow{f'} & B' & \xrightarrow{g'} & C' \\
\sim\downarrow{\alpha} & & \sim\downarrow{\beta} & & \sim\downarrow{\gamma} \\
A & \xrightarrow{f} & B & \xrightarrow{g} & C
\end{array} \qquad (3.2)$$

*where the vertical maps* $\alpha, \beta, \gamma$ *are* $\mathfrak{I}$-*equivalences. Furthermore, we can achieve that* $\alpha$ *and* $\beta$ *are identity maps.*

*Proof.* Let $F$ be a stable homological functor with $\mathfrak{I} = \ker F$.

Suppose first that we are in the situation of (3.2). Lemma 22 yields that $F(\alpha)$, $F(\beta)$, and $F(\gamma)$ are invertible and that $0 \to F(A') \to F(B') \to F(C') \to 0$ is a short exact sequence. Hence so is $0 \to F(A) \to F(B) \to F(C) \to 0$. Now Lemma 28 yields that our given chain complex is $\mathfrak{I}$-exact.

Conversely, suppose that we have an $\mathfrak{I}$-exact chain complex. By Lemma 28, this means that $0 \to F(A) \to F(B) \to F(C) \to 0$ is a short exact sequence. Hence $f : A \to B$ is $\mathfrak{I}$-monic. Embed $f$ in an exact triangle $A \to B \to C' \to \Sigma A$. Since $f$ is $\mathfrak{I}$-monic, this triangle is $\mathfrak{I}$-exact. Let $\alpha = \mathrm{id}_A$ and $\beta = \mathrm{id}_B$. Since the functor $\mathfrak{T}(\llcorner, C)$ is cohomological and $g \circ f = 0$, we can find a map $\gamma : C' \to C$ making (3.2) commute. The functor $F$ maps the rows of (3.2) to short exact sequences by Lemmas 28 and 22. Now the Five Lemma yields that $F(\gamma)$ is invertible, so that $\gamma$ is an $\mathfrak{I}$-equivalence. $\qquad\square$

**Remark 32.** Lemma 31 implies that $\mathfrak{I}$-exact chain complexes of length 3 are $\mathfrak{I}$-decomposable. We do not expect this for chain complexes of length 4. But we have not searched for a counterexample.

Which chain complexes over $\mathfrak{T}$ are $\mathfrak{I}$-exact for $\mathfrak{I} = 0$ and hence for any homological ideal? The next definition provides the answer.

**Definition 33.** A chain complex $C_\bullet$ over a triangulated category is called *homologically exact* if $F(C_\bullet)$ is exact for any homological functor $F : \mathfrak{T} \to \mathfrak{C}$.

**Example 34.** If $A \to B \to C \to \Sigma A$ is an exact triangle, then the chain complex

$$\cdots \to \Sigma^{-1}A \to \Sigma^{-1}B \to \Sigma^{-1}C \to A \to B \to C \to \Sigma A \to \Sigma B \to \Sigma C \to \cdots$$

is homologically exact by the definition of a homological functor.

**Lemma 35.** *Let* $F : \mathfrak{T} \to \mathfrak{T}'$ *be an exact functor between two triangulated categories. Let* $C_\bullet$ *be a chain complex over* $\mathfrak{T}$. *The following are equivalent:*

(1) $C_\bullet$ *is* $\ker F$-*exact in degree* $n$;
(2) $F(C_\bullet)$ *is* $\mathfrak{I}$-*exact in degree* $n$ *with respect to* $\mathfrak{I} = 0$;
(3) *the chain complex* $\mathbb{Y} \circ F(C_\bullet)$ *in* $\mathfrak{Coh}(\mathfrak{T}')$ *is exact in degree* $n$;
(4) $F(C_\bullet)$ *is homologically exact in degree* $n$;
(5) *the chain complexes of Abelian groups* $\mathfrak{T}'\big(A, F(C_\bullet)\big)$ *are exact in degree* $n$ *for all* $A \in\in \mathfrak{T}'$.

*Proof.* By Freyd's Theorem 16, $\mathbb{Y} \circ F \colon \mathfrak{T} \to \mathfrak{Coh}(\mathfrak{T}')$ is a stable homological functor with $\ker F = \ker(\mathbb{Y} \circ F)$. Hence Lemma 28 yields (1) $\Longleftrightarrow$ (3). Similarly, we have (2) $\Longleftrightarrow$ (3) because $\mathbb{Y} \colon \mathfrak{T}' \to \mathfrak{Coh}(\mathfrak{T}')$ is a stable homological functor with $\ker \mathbb{Y} = 0$. Freyd's Theorem 16 also asserts that any homological functor $F \colon \mathfrak{T}' \to \mathfrak{C}'$ factors as $\bar{F} \circ \mathbb{Y}$ for an exact functor $\bar{F}$. Hence (3)$\Longrightarrow$(4). Proposition 11 yields (4)$\Longrightarrow$(5). Finally, (5) $\Longleftrightarrow$ (3) because kernels and cokernels in $\mathfrak{Coh}(\mathfrak{T}')$ are computed pointwise on objects of $\mathfrak{T}'$. $\qquad\square$

**Remark 36.** More generally, consider a set of exact functors $F_i \colon \mathfrak{T} \to \mathfrak{T}'_i$. As in the proof of the equivalence (1) $\Longleftrightarrow$ (2) in Lemma 35, we see that a chain complex $C_\bullet$ is $\bigcap \ker F_i$-exact (in degree $n$) if and only if the chain complexes $F_i(C_\bullet)$ are exact (in degree $n$) for all $i$.

As a consequence, a chain complex $C_\bullet$ over $\mathrm{KK}^G$ for a locally compact quantum group $G$ is $\mathcal{F}$-exact if and only if $\mathrm{Res}^H_G(C_\bullet)$ is homologically exact for all $H \in \mathcal{F}$. A chain complex $C_\bullet$ over $\mathrm{KK}^G$ for a compact quantum group $G$ is $\mathfrak{I}_\ltimes$-exact if and only if the chain complex $G \ltimes C_\bullet$ over $\mathrm{KK}$ is homologically exact.

**Example 37.** We exhibit an $\mathfrak{I}$-exact chain complex that is not $\mathfrak{I}$-decomposable for the ideal $\mathfrak{I} = 0$. By Lemma 25, any 0-exact triangle is split. Therefore, a chain complex is 0-decomposable if and only if it is a direct sum of chain complexes of the form $0 \to K_n \xrightarrow{\mathrm{id}} K_n \to 0$. Hence any decomposable chain complex is contractible and therefore mapped by any homological functor to a contractible chain complex. By the way, if idempotents in $\mathfrak{T}$ split then a chain complex is 0-decomposable if and only if it is contractible.

As we have remarked in Example 34, the chain complex

$$\cdots \to \Sigma^{-1} C \to A \to B \to C \to \Sigma A \to \Sigma B \to \Sigma C \to \Sigma^2 A \to \cdots$$

is homologically exact for any exact triangle $A \to B \to C \to \Sigma A$. But such chain complexes need not be contractible. A counterexample is the exact triangle $\mathbb{Z}/2 \to \mathbb{Z}/4 \to \mathbb{Z}/2 \to \Sigma \mathbb{Z}/2$ in $\mathfrak{Der}(\mathfrak{Ab})$, which we have already used in the proof of Proposition 20. The resulting chain complex over $\mathfrak{Der}(\mathfrak{Ab})$ cannot be contractible because $\mathrm{H}_*$ maps it to a non-contractible chain complex.

### 3.2.1. More homological algebra with chain complexes

Using our notion of exactness for chain complexes, we can do homological algebra in the homotopy category $\mathfrak{Ho}(\mathfrak{T})$. We briefly sketch some results in this direction, assuming some familiarity with more advanced notions from homological algebra. We will not use this later.

The $\mathfrak{I}$-exact chain complexes form a thick subcategory of $\mathfrak{Ho}(\mathfrak{T})$ because of Lemma 28. We let $\mathfrak{Der} := \mathfrak{Der}(\mathfrak{T}, \mathfrak{I})$ be the localisation of $\mathfrak{Ho}(\mathfrak{T})$ at this subcategory and call it the *derived category of $\mathfrak{T}$ with respect to $\mathfrak{I}$*.

We let $\mathfrak{Der}^{\geq n}$ and $\mathfrak{Der}^{\leq n}$ be the full subcategories of $\mathfrak{Der}$ consisting of chain complexes that are $\mathfrak{I}$-exact in degrees less than $n$ and greater than $n$, respectively.

**Theorem 38.** *The pair of subcategories $\mathfrak{Der}^{\geq 0}$, $\mathfrak{Der}^{\leq 0}$ forms a* truncation *structure (t-structure) on $\mathfrak{Der}$ in the sense of* [3].

*Proof.* The main issue here is the truncation of chain complexes. Let $C_\bullet$ be a chain complex over $\mathfrak{T}$. We embed the map $d_0$ in an exact triangle $C_0 \to C_{-1} \to X \to \Sigma C_0$ and let $C_\bullet^{\geq 0}$ be the chain complex

$$\cdots \to C_2 \to C_1 \to C_0 \to C_{-1} \to X \to \Sigma C_0 \to \Sigma C_{-1} \to \Sigma X$$
$$\to \Sigma^2 C_0 \to \cdots .$$

This chain complex is $\mathfrak{I}$-exact – even homologically exact – in negative degrees, that is, $C_\bullet^{\geq 0} \in \mathfrak{Der}^{\geq 0}$. The triangulated category structure allows us to construct a chain map $C_\bullet^{\geq 0} \to C_\bullet$ that is an isomorphism on $C_n$ for $n \geq -1$. Hence its mapping cone $C_\bullet^{\leq -1}$ is $\mathfrak{I}$-exact – even contractible – in degrees $\geq 0$, that is, $C_\bullet^{\leq -1} \in\in \mathfrak{Der}^{\leq -1}$. By construction, we have an exact triangle

$$C_\bullet^{\geq 0} \to C_\bullet \to C_\bullet^{\leq -1} \to \Sigma C_\bullet^{\geq 0}$$

in $\mathfrak{Der}$.

We also have to check that there is no non-zero morphism $C_\bullet \to D_\bullet$ in $\mathfrak{Der}$ if $C_\bullet \in\in \mathfrak{Der}^{\geq 0}$ and $D_\bullet \in\in \mathfrak{Der}^{\leq -1}$. Recall that morphisms in $\mathfrak{Der}$ are represented by diagrams $C_\bullet \xleftarrow{\sim} \tilde{C}_\bullet \to D_\bullet$ in $\mathfrak{Ho}(\mathfrak{T})$, where the first map is an $\mathfrak{I}$-equivalence. Hence $\tilde{C}_\bullet \in\in \mathfrak{Der}^{\geq 0}$ as well. We claim that any chain map $f \colon \tilde{C}_\bullet^{\geq 0} \to D_\bullet^{\leq -1}$ is homotopic to 0. Since the maps $\tilde{C}_\bullet^{\geq 0} \to C_\bullet$ and $D_\bullet \to D_\bullet^{\leq -1}$ are $\mathfrak{I}$-equivalences, any morphism $C_\bullet \to D_\bullet$ vanishes in $\mathfrak{Der}$.

It remains to prove the claim. In a first step, we use that $D_\bullet^{\leq -1}$ is contractible in degrees $\geq 0$ to replace $f$ by a homotopic chain map supported in degrees $< 0$. In a second step, we use that $\tilde{C}_\bullet^{\geq 0}$ is homologically exact in the relevant degrees to recursively construct a chain homotopy between $f$ and 0. $\qquad\square$

Any truncation structure gives rise to an Abelian category, its *core*. The core of $\mathfrak{Der}$ is the full subcategory $\mathfrak{C}$ of all chain complexes that are $\mathfrak{I}$-exact except in degree 0. This is a stable Abelian category, and the standard embedding $\mathfrak{T} \to \mathfrak{Ho}(\mathfrak{T})$ yields a stable homological functor $F \colon \mathfrak{T} \to \mathfrak{C}$ with $\ker F = \mathfrak{I}$.

This functor is characterised uniquely by the following universal property: any (stable) homological functor $H \colon \mathfrak{T} \to \mathfrak{C}'$ with $\mathfrak{I} \subseteq \ker H$ factors uniquely as $H = \bar{H} \circ F$ for an exact functor $\bar{H} \colon \mathfrak{C} \to \mathfrak{C}'$. We construct $\bar{H}$ in three steps.

First, we lift $H$ to an exact functor $\mathfrak{Ho}(H) \colon \mathfrak{Ho}(\mathfrak{T}, \mathfrak{I}) \to \mathfrak{Ho}(\mathfrak{C}')$. Secondly, $\mathfrak{Ho}(H)$ descends to a functor $\mathfrak{Der}(H) \colon \mathfrak{Der}(\mathfrak{T}, \mathfrak{I}) \to \mathfrak{Der}(\mathfrak{C}')$. Finally, $\mathfrak{Der}(H)$ restricts to a functor $\bar{H} \colon \mathfrak{C} \to \mathfrak{C}'$ between the cores. Since $\mathfrak{I} \subseteq \ker H$, an $\mathfrak{I}$-exact chain complex is also $\ker H$-exact. Hence $\mathfrak{Ho}(H)$ preserves exactness of chain complexes by Lemma 28. This allows us to construct $\mathfrak{Der}(H)$ and shows that $\mathfrak{Der}(H)$ is compatible with truncation structures. This allows us to restrict it to an exact functor between the cores. Finally, we use that the core of the standard truncation structure on $\mathfrak{Der}(\mathfrak{C})$ is $\mathfrak{C}$. It is easy to see that we have $\bar{H} \circ F = H$.

Especially, we get an exact functor $\mathfrak{Der}(F) \colon \mathfrak{Der}(\mathfrak{T}, \mathfrak{I}) \to \mathfrak{Der}(\mathfrak{C})$, which restricts to the identity functor $\mathrm{id}_{\mathfrak{C}}$ between the cores. Hence $\mathfrak{Der}(F)$ is fully faithful on the thick subcategory generated by $\mathfrak{C} \subseteq \mathfrak{Der}(\mathfrak{T}, \mathfrak{I})$. It seems plausible that $\mathfrak{Der}(F)$ should be an equivalence of categories under some mild conditions on $\mathfrak{I}$ and $\mathfrak{T}$.

We will continue our study of the functor $F \colon \mathfrak{T} \to \mathfrak{C}$ in §3.7. The universal property determines it uniquely. Beligiannis ([4]) has another, simpler construction.

## 3.3. Projective objects

Let $\mathfrak{I}$ be a homological ideal in a triangulated category $\mathfrak{T}$.

**Definition 39.** A homological functor $F \colon \mathfrak{T} \to \mathfrak{C}$ is called $\mathfrak{I}$-*exact* if $F(f) = 0$ for all $\mathfrak{I}$-phantom maps $f$ or, equivalently, $\mathfrak{I} \subseteq \ker F$. An object $A \in\in \mathfrak{T}$ is called $\mathfrak{I}$-*projective* if the functor $\mathfrak{T}(A, \llcorner) \colon \mathfrak{T} \to \mathfrak{Ab}$ is $\mathfrak{I}$-exact. Dually, an object $B \in\in \mathfrak{T}$ is called $\mathfrak{I}$-*injective* if the functor $\mathfrak{T}(\llcorner, B) \colon \mathfrak{T} \to \mathfrak{Ab}^{\mathrm{op}}$ is $\mathfrak{I}$-exact.

We write $\mathfrak{P}_{\mathfrak{I}}$ for the class of $\mathfrak{I}$-projective objects in $\mathfrak{T}$.

The notions of projective and injective object are dual to each other: if we pass to the opposite category $\mathfrak{T}^{\mathrm{op}}$ with the canonical triangulated category structure and use the same ideal $\mathfrak{I}^{\mathrm{op}}$, then this exchanges the roles of projective and injective objects. Therefore, it suffices to discuss one of these two notions in the following. We will only treat projective objects because all the ideals in §2.2 have enough projective objects, but most of them do not have enough injective objects.

Notice that the functor $F$ is $\mathfrak{I}$-exact if and only if the associated stable functor $F_* \colon \mathfrak{T} \to \mathfrak{C}^{\mathbb{Z}}$ is $\mathfrak{I}$-exact because $\mathfrak{I}$ is stable.

Since we require $F$ to be homological, the long exact homology sequence and Lemma 28 yield that the following conditions are all equivalent to $F$ being $\mathfrak{I}$-exact:

– $F$ maps $\mathfrak{I}$-epimorphisms to epimorphisms in $\mathfrak{C}$;
– $F$ maps $\mathfrak{I}$-monomorphisms to monomorphisms in $\mathfrak{C}$;
– $0 \to F(A) \to F(B) \to F(C) \to 0$ is a short exact sequence in $\mathfrak{C}$ for any $\mathfrak{I}$-exact triangle $A \to B \to C \to \Sigma A$;
– $F$ maps $\mathfrak{I}$-exact chain complexes to exact chain complexes in $\mathfrak{C}$.

This specialises to equivalent definitions of $\mathfrak{I}$-projective objects.

**Lemma 40.** *An object $A \in\in \mathfrak{T}$ is $\mathfrak{I}$-projective if and only if $\mathfrak{I}(A, B) = 0$ for all $B \in\in \mathfrak{T}$.*

*Proof.* If $f \in \mathfrak{I}(A, B)$, then $f = f_*(\mathrm{id}_A)$. This has to vanish if $A$ is $\mathfrak{I}$-projective. Suppose, conversely, that $\mathfrak{I}(A, B) = 0$ for all $B \in\in \mathfrak{T}$. If $f \in \mathfrak{I}(B, B')$, then $\mathfrak{T}(A, f)$ maps $\mathfrak{T}(A, B)$ to $\mathfrak{I}(A, B') = 0$, so that $\mathfrak{T}(A, f) = 0$. Hence $A$ is $\mathfrak{I}$-projective. $\qquad\square$

An $\mathfrak{I}$-exact functor also has the following properties (which are strictly weaker than being $\mathfrak{I}$-exact):

– $F$ maps $\mathfrak{I}$-equivalences to isomorphisms in $\mathfrak{C}$;
– $F$ maps $\mathfrak{I}$-contractible objects to 0 in $\mathfrak{C}$.

Again we may specialise this to $\mathfrak{I}$-projective objects.

**Lemma 41.** *The class of $\mathfrak{I}$-exact homological functors $\mathfrak{T} \to \mathfrak{Ab}$ or $\mathfrak{T} \to \mathfrak{Ab}^{\mathrm{op}}$ is closed under composition with $\Sigma^{\pm 1} \colon \mathfrak{T} \to \mathfrak{T}$, retracts, direct sums, and direct products. The class $\mathfrak{P}_{\mathfrak{I}}$ of $\mathfrak{I}$-projective objects is closed under (de)suspensions, retracts, and possibly infinite direct sums (as far as they exist in $\mathfrak{T}$).*

*Proof.* The first assertion follows because direct sums and products of Abelian groups are exact; the second one is a special case. $\qquad\square$

**Notation 42.** Let $\mathfrak{P}$ be a set of objects of $\mathfrak{T}$. We let $(\mathfrak{P})_\oplus$ be the smallest class of objects of $\mathfrak{T}$ that contains $\mathfrak{P}$ and is closed under retracts and direct sums (as far as they exist in $\mathfrak{T}$).

By Lemma 41, $(\mathfrak{P})_\oplus$ consists of $\mathfrak{I}$-projective objects if $\mathfrak{P}$ does. We say that $\mathfrak{P}$ *generates all $\mathfrak{I}$-projective objects* if $(\mathfrak{P})_\oplus = \mathfrak{P}_{\mathfrak{I}}$. In examples, it is usually easier to describe a class of generators in this sense.

## 3.4. Projective resolutions

**Definition 43.** Let $\mathfrak{I} \subseteq \mathfrak{T}$ be a homological ideal in a triangulated category and let $A \in\in \mathfrak{T}$. A *one-step $\mathfrak{I}$-projective resolution* is an $\mathfrak{I}$-epimorphism $\pi : P \to A$ with $P \in\in \mathfrak{P}_{\mathfrak{I}}$. An *$\mathfrak{I}$-projective resolution* of $A$ is an $\mathfrak{I}$-exact chain complex

$$\cdots \xrightarrow{\delta_{n+1}} P_n \xrightarrow{\delta_n} P_{n-1} \xrightarrow{\delta_{n-1}} \cdots \xrightarrow{\delta_1} P_0 \xrightarrow{\delta_0} A$$

with $P_n \in\in \mathfrak{P}_{\mathfrak{I}}$ for all $n \in \mathbb{N}$.

We say that $\mathfrak{I}$ has *enough projective objects* if each $A \in\in \mathfrak{T}$ has a one-step $\mathfrak{I}$-projective resolution.

The following proposition contains the basic properties of projective resolutions, which are familiar from the similar situation for Abelian categories.

**Proposition 44.** *If $\mathfrak{I}$ has enough projective objects, then any object of $\mathfrak{T}$ has an $\mathfrak{I}$-projective resolution (and vice versa).*

*Let $P_\bullet \to A$ and $P'_\bullet \to A'$ be $\mathfrak{I}$-projective resolutions. Then any map $A \to A'$ may be lifted to a chain map $P_\bullet \to P'_\bullet$, and this lifting is unique up to chain homotopy. Two $\mathfrak{I}$-projective resolutions of the same object are chain homotopy equivalent. As a result, the construction of projective resolutions provides a functor*

$$P : \mathfrak{T} \to \mathfrak{Ho}(\mathfrak{T}).$$

*Let $A \xrightarrow{f} B \xrightarrow{g} C \xrightarrow{h} \Sigma A$ be an $\mathfrak{I}$-exact triangle. Then there exists a canonical map $\eta : P(C) \to P(A)[1]$ in $\mathfrak{Ho}(\mathfrak{T})$ such that the triangle*

$$P(A) \xrightarrow{P(f)} P(B) \xrightarrow{P(g)} P(C) \xrightarrow{\eta} P(A)[1]$$

*in $\mathfrak{Ho}(\mathfrak{T})$ is exact; here [1] denotes the translation functor in $\mathfrak{Ho}(\mathfrak{T})$, which has nothing to do with the suspension in $\mathfrak{T}$.*

*Proof.* Let $A \in\in \mathfrak{T}$. By assumption, there is a one-step $\mathfrak{I}$-projective resolution $\delta_0 : P_0 \to A$, which we embed in an exact triangle $A_1 \to P_0 \to A \to \Sigma A_1$. Since $\delta_0$ is $\mathfrak{I}$-epic, this triangle is $\mathfrak{I}$-exact. By induction, we construct a sequence of such $\mathfrak{I}$-exact triangles $A_{n+1} \to P_n \to A_n \to \Sigma A_{n+1}$ for $n \in \mathbb{N}$ with $P_n \in\in \mathfrak{P}$ and $A_0 = A$. By composition, we obtain maps $\delta_n : P_n \to P_{n-1}$ for $n \geq 1$, which satisfy $\delta_n \circ \delta_{n+1} = 0$ for all $n \geq 0$. The resulting chain complex

$$\cdots \to P_n \xrightarrow{\delta_n} P_{n-1} \xrightarrow{\delta_{n-1}} P_{n-2} \to \cdots \to P_1 \xrightarrow{\delta_1} P_0 \xrightarrow{\delta_0} A \to 0$$

is $\mathfrak{I}$-decomposable by construction and therefore $\mathfrak{I}$-exact by Corollary 29.

The remaining assertions are proved exactly as their classical counterparts in homological algebra. We briefly sketch the arguments. Let $P_\bullet \to A$ and $P'_\bullet \to A'$ be $\mathfrak{I}$-projective resolutions and let $f \in \mathfrak{T}(A, A')$. We construct $f_n \in \mathfrak{T}(P_n, P'_n)$ by induction on $n$ such that the diagrams

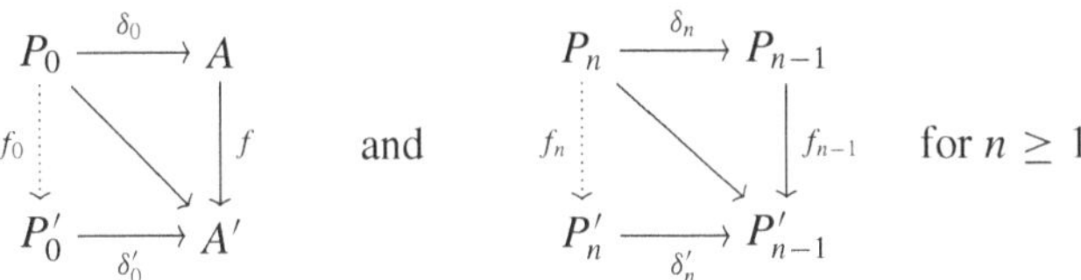

commute. We must check that this is possible. Since the chain complex $P'_\bullet \to A$ is $\mathfrak{I}$-exact and $P_n$ is $\mathfrak{I}$-projective for all $n \geq 0$, the chain complexes

$$\cdots \to \mathfrak{T}(P_n, P'_m) \xrightarrow{(\delta'_m)_*} \mathfrak{T}(P_n, P'_{m-1}) \to \cdots \to \mathfrak{T}(P_n, P'_0) \xrightarrow{(\delta'_0)_*} \mathfrak{T}(P_n, A) \to 0$$

are exact for all $n \in \mathbb{N}$. This allows us to find maps $f_n$ as above. By construction, these maps form a chain map lifting $f \colon A \to A'$. Its uniqueness up to chain homotopy is proved similarly. If we apply this unique lifting result to two $\mathfrak{I}$-projective resolutions of the same object, we get the uniqueness of $\mathfrak{I}$-projective resolutions up to chain homotopy equivalence. Hence we get a well-defined functor $P \colon \mathfrak{T} \to \mathfrak{Ho}(\mathfrak{T})$.

Now consider an $\mathfrak{I}$-exact triangle $A \to B \to C \to \Sigma A$ as in the third paragraph of the lemma. Let $X_\bullet$ be the mapping cone of some chain map $P(A) \to P(B)$ in the homotopy class $P(f)$. This chain complex is supported in degrees $\geq 0$ and has $\mathfrak{I}$-projective entries because $X_n = P(A)_{n-1} \oplus P(B)_n$. The map $X_0 = 0 \oplus P(B)_0 \to B \to C$ yields a chain map $X_\bullet \to C$, that is, the composite map $X_1 \to X_0 \to C$ vanishes. By construction, this chain map lifts the given map $B \to C$ and we have an exact triangle $P(A) \to P(B) \to X_\bullet \to P(A)[1]$ in $\mathfrak{Ho}(\mathfrak{T})$. It remains to observe that $X_\bullet \to C$ is $\mathfrak{I}$-exact. Then $X_\bullet$ is an $\mathfrak{I}$-projective resolution of $C$. Since such resolutions are unique up to chain homotopy equivalence, we get a canonical isomorphism $X_\bullet \cong P(C)$ in $\mathfrak{Ho}(\mathfrak{T})$ and hence the assertion in the third paragraph.

Let $F$ be a stable homological functor with $\mathfrak{I} = \ker F$. We have to check that $F(X_\bullet) \to F(C)$ is a resolution. This reduces to a well-known diagram chase in Abelian categories, using that $F\big(P(A)\big) \to F(A)$ and $F\big(P(B)\big) \to F(B)$ are resolutions and that $F(A) \rightarrowtail F(B) \twoheadrightarrow F(C)$ is exact. $\qquad\square$

## 3.5. Derived functors

We only define derived functors if there are enough projective objects because this case is rather easy and suffices for our applications. The general case can be reduced to the familiar case of Abelian categories using the results of §3.2.1.

**Definition 45.** Let $\mathfrak{I}$ be a homological ideal in a triangulated category $\mathfrak{T}$ with enough projective objects. Let $F\colon \mathfrak{T} \to \mathfrak{C}$ be an additive functor with values in an Abelian category $\mathfrak{C}$. It induces a functor $\mathfrak{Ho}(F)\colon \mathfrak{Ho}(\mathfrak{T}) \to \mathfrak{Ho}(\mathfrak{C})$, applying $F$ pointwise to chain complexes. Let $P\colon \mathfrak{T} \to \mathfrak{Ho}(\mathfrak{T})$ be the projective resolution functor constructed in Proposition 44. Let $H_n\colon \mathfrak{Ho}(\mathfrak{C}) \to \mathfrak{C}$ be the $n$th homology functor for some $n \in \mathbb{N}$. The composite functor

$$\mathbb{L}_n F\colon \mathfrak{T} \xrightarrow{P} \mathfrak{Ho}(\mathfrak{T}) \xrightarrow{\mathfrak{Ho}(F)} \mathfrak{Ho}(\mathfrak{C}) \xrightarrow{H_n} \mathfrak{C}$$

is called the $n$th *left derived functor* of $F$. If $F\colon \mathfrak{T}^{\mathrm{op}} \to \mathfrak{C}$ is an additive functor, then the corresponding functor $H^n \circ \mathfrak{Ho}(F) \circ P\colon \mathfrak{T}^{\mathrm{op}} \to \mathfrak{C}$ is denoted by $\mathbb{R}^n F$ and called the $n$th *right derived functor* of $F$.

More concretely, let $A \in\in \mathfrak{T}$ and let $(P_\bullet, \delta_\bullet)$ be an $\mathfrak{I}$-projective resolution of $A$. If $F$ is covariant, then $\mathbb{L}_n F(A)$ is the homology at $F(P_n)$ of the chain complex

$$\cdots \to F(P_{n+1}) \xrightarrow{F(\delta_{n+1})} F(P_n) \xrightarrow{F(\delta_n)} F(P_{n-1}) \to \cdots \to F(P_0) \to 0.$$

If $F$ is contravariant, then $\mathbb{R}^n F(A)$ is the cohomology at $F(P_n)$ of the cochain complex

$$\cdots \leftarrow F(P_{n+1}) \xleftarrow{F(\delta_{n+1})} F(P_n) \xleftarrow{F(\delta_n)} F(P_{n-1}) \leftarrow \cdots \leftarrow F(P_0) \leftarrow 0.$$

**Lemma 46.** *Let $A \to B \to C \to \Sigma A$ be an $\mathfrak{I}$-exact triangle. If $F\colon \mathfrak{T} \to \mathfrak{C}$ is a covariant additive functor, then there is a long exact sequence*

$$\cdots \to \mathbb{L}_n F(A) \to \mathbb{L}_n F(B) \to \mathbb{L}_n F(C) \to \mathbb{L}_{n-1} F(A)$$
$$\to \cdots \to \mathbb{L}_1 F(C) \to \mathbb{L}_0 F(A) \to \mathbb{L}_0 F(B) \to \mathbb{L}_0 F(C) \to 0.$$

*If $F$ is contravariant instead, then there is a long exact sequence*

$$\cdots \leftarrow \mathbb{R}^n F(A) \leftarrow \mathbb{R}^n F(B) \leftarrow \mathbb{R}^n F(C) \leftarrow \mathbb{R}^{n-1} F(A)$$
$$\leftarrow \cdots \leftarrow \mathbb{R}^1 F(C) \leftarrow \mathbb{R}^0 F(A) \leftarrow \mathbb{R}^0 F(B) \leftarrow \mathbb{R}^0 F(C) \leftarrow 0.$$

*Proof.* This follows from the third assertion of Proposition 44 together with the well-known long exact homology sequence for exact triangles in $\mathfrak{Ho}(\mathfrak{C})$. $\quad\square$

**Lemma 47.** *Let $F\colon \mathfrak{T} \to \mathfrak{C}$ be a homological functor. The following assertions are equivalent:*

(1) *$F$ is $\mathfrak{I}$-exact;*
(2) *$\mathbb{L}_0 F(A) \cong F(A)$ and $\mathbb{L}_p F(A) = 0$ for all $p > 0$, $A \in\in \mathfrak{T}$;*
(3) *$\mathbb{L}_0 F(A) \cong F(A)$ for all $A \in\in \mathfrak{T}$.*

*The analogous assertions for contravariant functors are equivalent as well.*

*Proof.* If $F$ is $\mathfrak{I}$-exact, then $F$ maps $\mathfrak{I}$-exact chain complexes in $\mathfrak{T}$ to exact chain complexes in $\mathfrak{C}$. This applies to $\mathfrak{I}$-projective resolutions, so that $(1)\Longrightarrow(2)\Longrightarrow(3)$. It follows from (3) and Lemma 46 that $F$ maps $\mathfrak{I}$-epimorphisms to epimorphisms. Since this characterises $\mathfrak{I}$-exact functors, we get $(3)\Longrightarrow(1)$. $\qquad\square$

It can happen that $\mathbb{L}_p F = 0$ for all $p > 0$ although $F$ is not $\mathfrak{I}$-exact.

We have a natural transformation $\mathbb{L}_0 F(A) \to F(A)$ (or $F(A) \to \mathbb{R}^0 F(A)$), which is induced by the augmentation map $P_\bullet \to A$ for an $\mathfrak{I}$-projective resolution. Lemma 47 shows that these maps are usually not bijective, although this happens frequently for derived functors on Abelian categories.

**Definition 48.** We let $\mathrm{Ext}^n_{\mathfrak{T},\mathfrak{I}}(A, B)$ be the $n$th right derived functor with respect to $\mathfrak{I}$ of the contravariant functor $A \mapsto \mathfrak{T}(A, B)$.

We have natural maps $\mathfrak{T}(A, B) \to \mathrm{Ext}^0_{\mathfrak{T},\mathfrak{I}}(A, B)$, which usually are not invertible. Lemma 46 yields long exact sequences

$$\cdots \leftarrow \mathrm{Ext}^n_{\mathfrak{T},\mathfrak{I}}(A, D) \leftarrow \mathrm{Ext}^n_{\mathfrak{T},\mathfrak{I}}(B, D) \leftarrow \mathrm{Ext}^n_{\mathfrak{T},\mathfrak{I}}(C, D) \leftarrow \mathrm{Ext}^{n-1}_{\mathfrak{T},\mathfrak{I}}(A, D) \leftarrow$$
$$\cdots \leftarrow \mathrm{Ext}^1_{\mathfrak{T},\mathfrak{I}}(C, D) \leftarrow \mathrm{Ext}^0_{\mathfrak{T},\mathfrak{I}}(A, D) \leftarrow \mathrm{Ext}^0_{\mathfrak{T},\mathfrak{I}}(B, D) \leftarrow \mathrm{Ext}^0_{\mathfrak{T},\mathfrak{I}}(C, D) \leftarrow 0$$

for any $\mathfrak{I}$-exact exact triangle $A \to B \to C \to \Sigma A$ and any $D \in\in \mathfrak{T}$.

We claim that there are similar long exact sequences

$$0 \to \mathrm{Ext}^0_{\mathfrak{T},\mathfrak{I}}(D, A) \to \mathrm{Ext}^0_{\mathfrak{T},\mathfrak{I}}(D, B) \to \mathrm{Ext}^0_{\mathfrak{T},\mathfrak{I}}(D, C) \to \mathrm{Ext}^1_{\mathfrak{T},\mathfrak{I}}(D, A) \to \cdots$$
$$\to \mathrm{Ext}^{n-1}_{\mathfrak{T},\mathfrak{I}}(D, C) \to \mathrm{Ext}^n_{\mathfrak{T},\mathfrak{I}}(D, A) \to \mathrm{Ext}^n_{\mathfrak{T},\mathfrak{I}}(D, B) \to \mathrm{Ext}^n_{\mathfrak{T},\mathfrak{I}}(D, C) \to \cdots$$

in the second variable. Since $P(D)_n$ is $\mathfrak{I}$-projective, the sequences

$$0 \to \mathfrak{T}(P(D)_n, A) \to \mathfrak{T}(P(D)_n, B) \to \mathfrak{T}(P(D)_n, C) \to 0$$

are exact for all $n \in \mathbb{N}$. This extension of chain complexes yields the desired long exact sequence.

We list a few more elementary properties of derived functors. We only spell things out for the left derived functors $\mathbb{L}_n F : \mathfrak{T} \to \mathfrak{C}$ of a covariant functor $F : \mathfrak{T} \to \mathfrak{C}$. Similar assertions hold for right derived functors of contravariant functors.

The derived functors $\mathbb{L}_n F$ satisfy $\mathfrak{I} \subseteq \ker \mathbb{L}_n F$ and hence descend to functors $\mathbb{L}_n F : \mathfrak{T}/\mathfrak{I} \to \mathfrak{C}$ because the zero map $P(A) \to P(B)$ is a chain map lifting of $f$ if $f \in \mathfrak{I}(A, B)$. As a consequence, $\mathbb{L}_n F(A) \cong 0$ if $A$ is $\mathfrak{I}$-contractible. The long exact homology sequences of Lemma 46 show that $\mathbb{L}_n F(f) : \mathbb{L}_n F(A) \to \mathbb{L}_n F(B)$ is invertible if $f \in \mathfrak{T}(A, B)$ is an $\mathfrak{I}$-equivalence.

**Warning 49.** The derived functors $\mathbb{L}_n F$ are *not homological* and therefore do not deserve to be called $\mathfrak{I}$-exact even though they vanish on $\mathfrak{I}$-phantom maps. Lemma 46 shows that these functors are only half-exact on $\mathfrak{I}$-exact triangles. Thus $\mathbb{L}_n F(f)$ need not be monic (or epic) if $f$ is $\mathfrak{I}$-monic (or $\mathfrak{I}$-epic). The problem is that the $\mathfrak{I}$-projective resolution functor $P \colon \mathfrak{T} \to \mathfrak{Ho}(\mathfrak{T})$ is not exact – it even fails to be stable.

The following remarks require a more advanced background in homological algebra and are not going to be used in the sequel.

**Remark 50.** The derived functors introduced above, especially the Ext functors, can be interpreted in terms of *derived categories*.

We have already observed in §3.2.1 that the $\mathfrak{I}$-exact chain complexes form a thick subcategory of $\mathfrak{Ho}(\mathfrak{T})$. The augmentation map $P(A) \to A$ of an $\mathfrak{I}$-projective resolution of $A \in\in \mathfrak{T}$ is a quasi-isomorphism with respect to this thick subcategory. The chain complex $P(A)$ is projective (see [12]), that is, for any chain complex $C_\bullet$, the space of morphisms $A \to C_\bullet$ in the derived category $\mathfrak{Der}(\mathfrak{T}, \mathfrak{I})$ agrees with $[P(A), C_\bullet]$. Especially, $\mathrm{Ext}^n_{\mathfrak{T},\mathfrak{I}}(A, B)$ is the space of morphisms $A \to B[n]$ in $\mathfrak{Der}(\mathfrak{T}, \mathfrak{I})$.

Now let $F \colon \mathfrak{T} \to \mathfrak{C}$ be an additive covariant functor. Extend it to an exact functor $\bar{F} \colon \mathfrak{Ho}(\mathfrak{T}) \to \mathfrak{Ho}(\mathfrak{C})$. It has a total left derived functor

$$\mathbb{L}\bar{F} \colon \mathfrak{Der}(\mathfrak{T}, \mathfrak{I}) \to \mathfrak{Der}(\mathfrak{C}), \qquad A \mapsto \bar{F}\big(P(A)\big).$$

By definition, we have $\mathbb{L}_n F(A) := H_n\big(\mathbb{L}\bar{F}(A)\big)$.

**Remark 51.** In classical Abelian categories, the Ext groups form a graded ring, and the derived functors form graded modules over this graded ring. The same happens in our context. The most conceptual construction of these products uses the description of derived functors sketched in Remark 50.

Recall that we may view elements of $\mathrm{Ext}^n_{\mathfrak{T},\mathfrak{I}}(A, B)$ as morphisms $A \to B[n]$ in the derived category $\mathfrak{Der}(\mathfrak{T}, \mathfrak{I})$. Taking translations, we can also view them as morphisms $A[m] \to B[n + m]$ for any $m \in \mathbb{Z}$. The usual composition in the category $\mathfrak{Der}(\mathfrak{T}, \mathfrak{I})$ therefore yields an associative product

$$\mathrm{Ext}^n_{\mathfrak{T},\mathfrak{I}}(B, C) \otimes \mathrm{Ext}^m_{\mathfrak{T},\mathfrak{I}}(A, B) \to \mathrm{Ext}^{n+m}_{\mathfrak{T},\mathfrak{I}}(A, C).$$

Thus we get a graded additive category with morphism spaces $\big(\mathrm{Ext}^n_{\mathfrak{T},\mathfrak{I}}(A, B)\big)_{n\in\mathbb{N}}$.

Similarly, if $F \colon \mathfrak{T} \to \mathfrak{C}$ is an additive functor and $\mathbb{L}\bar{F} \colon \mathfrak{Der}(\mathfrak{T}, \mathfrak{I}) \to \mathfrak{Der}(\mathfrak{C})$ is as in Remark 51, then a morphism $A \to B[n]$ in $\mathfrak{Der}(\mathfrak{T}, \mathfrak{I})$ induces a morphism $\mathbb{L}\bar{F}(A) \to \mathbb{L}\bar{F}(B)[n]$ in $\mathfrak{Der}(\mathfrak{C})$. Passing to homology, we get canonical maps

$$\mathrm{Ext}^n_{\mathfrak{T},\mathfrak{J}}(A, B) \to \mathrm{Hom}_{\mathfrak{C}}\big(\mathbb{L}F_m(A), \mathbb{L}F_{m-n}(B)\big) \qquad \forall m \geq n,$$

which satisfy an appropriate associativity condition. For a contravariant functor, we get canonical maps

$$\mathrm{Ext}^n_{\mathfrak{T},\mathfrak{J}}(A, B) \to \mathrm{Hom}_{\mathfrak{C}}\big(\mathbb{R}F^m(B), \mathbb{R}F^{m+n}(A)\big) \qquad \forall m \geq 0.$$

### 3.6. Projective objects via adjointness

We develop a method for constructing enough projective objects. Let $\mathfrak{T}$ and $\mathfrak{C}$ be stable additive categories, let $F: \mathfrak{T} \to \mathfrak{C}$ be a stable additive functor, and let $\mathfrak{J} := \ker F$. In our applications, $\mathfrak{T}$ is triangulated and the functor $F$ is either exact or stable and homological.

Recall that a covariant functor $R: \mathfrak{T} \to \mathfrak{Ab}$ is *(co)representable* if it is naturally isomorphic to $\mathfrak{T}(A, \_)$ for some $A \in\in \mathfrak{T}$, which is then unique. If the functor $B \mapsto \mathfrak{C}(A, F(B))$ on $\mathfrak{T}$ is representable, we write $F^{\vdash}(A)$ for the representing object. By construction, we have natural isomorphisms

$$\mathfrak{T}(F^{\vdash}(A), B) \cong \mathfrak{C}\big(A, F(B)\big)$$

for all $B \in \mathfrak{T}$. Let $\mathfrak{C}'$ be the full subcategory of all objects $A \in\in \mathfrak{C}$ for which $F^{\vdash}(A)$ is defined. Then $F^{\vdash}$ is a functor $\mathfrak{C}' \to \mathfrak{T}$, which we call the *(partially defined) left adjoint* of $F$. Although one usually assumes $\mathfrak{C} = \mathfrak{C}'$, we shall also need $F^{\vdash}$ in cases where it is not defined everywhere.

The functor $B \mapsto \mathfrak{C}(A, F(B))$ for $A \in\in \mathfrak{C}'$ vanishes on $\mathfrak{J} = \ker F$ for trivial reasons. Hence $F^{\vdash}(A) \in\in \mathfrak{T}$ is $\mathfrak{J}$-projective. This simple observation is surprisingly powerful: as we shall see, it often yields all $\mathfrak{J}$-projective objects.

**Remark 52.** We have $F^{\vdash}(\Sigma A) \cong \Sigma F^{\vdash}(A)$ for all $A \in\in \mathfrak{C}'$, so that $\Sigma(\mathfrak{C}') = \mathfrak{C}'$. Moreover, $F^{\vdash}$ commutes with infinite direct sums (as far as they exist in $\mathfrak{T}$) because

$$\mathfrak{T}\left(\bigoplus F^{\vdash}(A_i), B\right) \cong \prod \mathfrak{T}(F^{\vdash}(A_i), B) \cong \prod \mathfrak{C}(A_i, F(B))$$
$$\cong \mathfrak{C}\left(\bigoplus A_i, F(B)\right).$$

**Example 53.** Consider the functor $K_*: KK \to \mathfrak{Ab}^{\mathbb{Z}/2}$. Let $\mathbb{Z} \in\in \mathfrak{Ab}^{\mathbb{Z}/2}$ denote the trivially graded Abelian group $\mathbb{Z}$. Notice that

$$\mathrm{Hom}\big(\mathbb{Z}, K_*(A)\big) \cong K_0(A) \cong KK(\mathbb{C}, A),$$
$$\mathrm{Hom}\big(\mathbb{Z}[1], K_*(A)\big) \cong K_1(A) \cong KK(C_0(\mathbb{R}), A),$$

where $\mathbb{Z}[1]$ means $\mathbb{Z}$ in odd degree. Hence $\mathrm{K}_*^\vdash(\mathbb{Z}) = \mathbb{C}$ and $\mathrm{K}_*^\vdash(\mathbb{Z}[1]) = C_0(\mathbb{R})$. More generally, Remark 52 shows that $\mathrm{K}_*^\vdash(A)$ is defined if both the even and odd parts of $A \in\in \mathfrak{Ab}^{\mathbb{Z}/2}$ are countable free Abelian groups: it is a direct sum of at most countably many copies of $\mathbb{C}$ and $C_0(\mathbb{R})$. Hence all such countable direct sums are $\mathfrak{I}_\mathrm{K}$-projective (we briefly say K-*projective*). As we shall see, $\mathrm{K}_*^\vdash$ is not defined on all of $\mathfrak{Ab}^{\mathbb{Z}/2}$; this is typical of homological functors.

**Example 54.** Consider the functor $\mathrm{H}_* \colon \mathfrak{Ho}(\mathfrak{C}; \mathbb{Z}/p) \to \mathfrak{C}^{\mathbb{Z}/p}$ of Example 9. Let $j \colon \mathfrak{C}^{\mathbb{Z}/p} \to \mathfrak{Ho}(\mathfrak{C}; \mathbb{Z}/p)$ be the functor that views an object of $\mathfrak{C}^{\mathbb{Z}/p}$ as a $p$-periodic chain complex whose boundary map vanishes.

A chain map $j(A) \to B_\bullet$ for $A \in\in \mathfrak{C}^{\mathbb{Z}/p}$ and $B_\bullet \in\in \mathfrak{Ho}(\mathfrak{C}; \mathbb{Z}/p)$ is a family of maps $\varphi_n \colon A_n \to \ker(d_n \colon B_n \to B_{n-1})$. Such a family is chain homotopic to $0$ if and only if each $\varphi_n$ lifts to a map $A_n \to B_{n+1}$. Suppose that $A_n$ is projective for all $n \in \mathbb{Z}/p$. Then such a lifting exists if and only if $\varphi_n(A_n) \subseteq d_{n+1}(B_{n+1})$. Hence

$$[j(A), B_\bullet] \cong \prod_{n\in\mathbb{Z}/p} \mathfrak{C}\big(A_n, \mathrm{H}_n(B_\bullet)\big) \cong \mathfrak{C}^{\mathbb{Z}/p}\big(A, \mathrm{H}_*(B_\bullet)\big).$$

As a result, the left adjoint of $\mathrm{H}_*$ is defined on the subcategory of projective objects $\mathfrak{P}(\mathfrak{C})^{\mathbb{Z}/p} \subseteq \mathfrak{C}^{\mathbb{Z}/p}$ and agrees there with the restriction of $j$. We will show in §3.8 that all $\mathfrak{I}_\mathrm{H}$-projective objects are of the form $\mathrm{H}_*^\vdash(A)$ for some $A \in\in \mathfrak{P}(\mathfrak{C})^{\mathbb{Z}/p}$ (provided $\mathfrak{C}$ has enough projective objects).

By duality, analogous results hold for injective objects: the domain of the *right* adjoint of $\mathrm{H}_*$ is the subcategory of injective objects of $\mathfrak{C}^{\mathbb{Z}/p}$, the right adjoint is equal to $j$ on this subcategory, and this provides all $\mathrm{H}_*$-injective objects of $\mathfrak{Ho}(\mathfrak{C}; \mathbb{Z}/p)$.

These examples show that $F^\vdash$ yields many $\ker F$-projective objects. We want to get *enough* $\ker F$-projective objects in this fashion, assuming that $F^\vdash$ is defined on enough of $\mathfrak{C}$. In order to treat ideals of the form $\bigcap F_i$, we now consider a more complicated setup. Let $\{\mathfrak{C}_i \mid i \in I\}$ be a set of stable homological or triangulated categories together with full subcategories $\mathfrak{P}\mathfrak{C}_i \subseteq \mathfrak{C}_i$ and stable homological or exact functors $F_i \colon \mathfrak{T} \to \mathfrak{C}_i$ for all $i \in I$. Assume that

- the left adjoint $F_i^\vdash$ is defined on $\mathfrak{P}\mathfrak{C}_i$ for all $i \in I$;
- there is an epimorphism $P \to F_i(A)$ in $\mathfrak{C}_i$ with $P \in\in \mathfrak{P}\mathfrak{C}_i$ for any $i \in I$, $A \in\in \mathfrak{T}$;
- the set of functors $F_i^\vdash \colon \mathfrak{P}\mathfrak{C}_i \to \mathfrak{T}$ is *cointegrable*, that is, $\bigoplus_{i\in I} F_i^\vdash(B_i)$ exists for all families of objects $B_i \in \mathfrak{P}\mathfrak{C}_i, i \in I$.

The reason for the notation $\mathfrak{PC}_i$ is that for a homological functor $F_i$ we usually take $\mathfrak{PC}_i$ to be the class of projective objects of $\mathfrak{C}_i$; if $F_i$ is exact, then we often take $\mathfrak{PC}_i = \mathfrak{C}_i$. But it may be useful to choose a smaller category, as long as it satisfies the second condition above.

**Proposition 55.** *In this situation, there are enough $\mathfrak{I}$-projective objects, and $\mathfrak{P}_\mathfrak{I}$ is generated by $\bigcup_{i \in I}\{F_i^{\vdash}(B) \mid B \in \mathfrak{PC}_i\}$. More precisely, an object of $\mathfrak{T}$ is $\mathfrak{I}$-projective if and only if it is a retract of $\bigoplus_{i \in I} F_i^{\vdash}(B_i)$ for a family of objects $B_i \in \mathfrak{PC}_i$.*

*Proof.* Let $\tilde{\mathfrak{P}}_0 := \bigcup_{i \in I}\{F_i^{\vdash}(B) \mid B \in \mathfrak{PC}_i\}$ and $\mathfrak{P}_0 := (\tilde{\mathfrak{P}}_0)_\oplus$. To begin with, we observe that any object of the form $F_i^{\vdash}(B)$ with $B \in\in \mathfrak{PC}_i$ is $\ker F_i$-projective and hence $\mathfrak{I}$-projective because $\mathfrak{I} \subseteq \ker F_i$. Hence $\mathfrak{P}_0$ consists of $\mathfrak{I}$-projective objects.

Let $A \in\in \mathfrak{T}$. For each $i \in I$, there is an epimorphism $p_i \colon B_i \to F_i(A)$ with $B_i \in \mathfrak{PC}_i$. The direct sum $B := \bigoplus_{i \in I} F_i^{\vdash}(B_i)$ exists. We have $B \in\in \mathfrak{P}_0$ by construction. We are going to construct an $\mathfrak{I}$-epimorphism $p \colon B \to A$. This shows that there are enough $\mathfrak{I}$-projective objects.

The maps $p_i \colon B_i \to F_i(A)$ provide maps $\hat{p}_i \colon F_i^{\vdash}(B_i) \to A$ via the adjointness isomorphisms $\mathfrak{T}(F_i^{\vdash}(B_i), A) \cong \mathfrak{C}_i\big(B_i, F_i(A)\big)$. We let $p := \sum \hat{p}_i \colon \bigoplus F_i^{\vdash}(B_i) \to A$. We must check that $p$ is an $\mathfrak{I}$-epimorphism. Equivalently, $p$ is $\ker F_i$-epic for all $i \in I$; this is, in turn equivalent to $F_i(p)$ being an epimorphism in $\mathfrak{C}_i$ for all $i \in I$, because of Lemma 22 or 25. This is what we are going to prove.

The identity map on $F_i^{\vdash}(B_i)$ yields a map $\alpha_i \colon B_i \to F_i F_i^{\vdash}(B_i)$ via the adjointness isomorphism $\mathfrak{T}\big(F_i^{\vdash}(B_i), F_i^{\vdash}(B_i)\big) \cong \mathfrak{C}_i\big(B_i, F_i F_i^{\vdash}(B_i)\big)$. Composing with the map

$$F_i F_i^{\vdash}(B_i) \to F_i \left(\bigoplus F_i^{\vdash}(B_i)\right) = F_i(B)$$

induced by the coordinate embedding $F_i^{\vdash}(B_i) \to B$, we get a map $\alpha_i' \colon B_i \to F_i(B)$. The naturality of the adjointness isomorphisms yields $F_i(\hat{p}_i) \circ \alpha_i = p_i$ and hence $F_i(p) \circ \alpha_i' = p_i$. The map $p_i$ is an epimorphism by assumption. Now we use a cancellation result for epimorphisms: if $f \circ g$ is an epimorphism, then so is $f$. Thus $F_i(p)$ is an epimorphism as desired.

If $A$ is $\mathfrak{I}$-projective, then the $\mathfrak{I}$-epimorphism $p \colon B \to A$ splits; to see this, embed $p$ in an exact triangle $N \to B \to A \to \Sigma N$ and observe that the map $A \to \Sigma N$ belongs to $\mathfrak{I}(A, \Sigma N) = 0$. Therefore, $A$ is a retract of $B$. Since $\mathfrak{P}_0$ is closed under retracts and $B \in\in \mathfrak{P}_0$, we get $A \in\in \mathfrak{P}_0$. Hence $\tilde{\mathfrak{P}}_0$ generates all $\mathfrak{I}$-projective objects. $\qquad\square$

### 3.7. The universal exact homological functor

For the following results, it is essential to define an ideal by a single functor $F$ instead of a family of functors as in Proposition 55.

**Definition 56.** Let $\mathfrak{I}$ be a homological ideal in a triangulated category $\mathfrak{T}$. An $\mathfrak{I}$-exact stable homological functor $F: \mathfrak{T} \to \mathfrak{C}$ is called *universal* if any other $\mathfrak{I}$-exact stable homological functor $G: \mathfrak{T} \to \mathfrak{C}'$ factors as $\bar{G} = G \circ F$ for a stable exact functor $\bar{G}: \mathfrak{C} \to \mathfrak{C}'$ that is unique up to natural isomorphism.

This universal property characterises $F$ uniquely up to natural isomorphism. We have constructed such a functor in §3.2.1. Beligiannis constructs it in [4, §3] using a localisation of the Abelian category $\mathfrak{Coh}(\mathfrak{T})$ at a suitable Serre subcategory; he calls this functor *projectivisation functor* and its target category *Steenrod category*. This notation is motivated by the special case of the Adams spectral sequence. The following theorem allows us to check whether a given functor is universal:

**Theorem 57.** *Let $\mathfrak{T}$ be a triangulated category, let $\mathfrak{I} \subseteq \mathfrak{T}$ be a homological ideal, and let $F: \mathfrak{T} \to \mathfrak{C}$ be an $\mathfrak{I}$-exact stable homological functor into a stable Abelian category $\mathfrak{C}$; let $\mathfrak{PC}$ be the class of projective objects in $\mathfrak{C}$. Suppose that idempotent morphisms in $\mathfrak{T}$ split.*

*The functor $F$ is the universal $\mathfrak{I}$-exact stable homological functor and there are enough $\mathfrak{I}$-projective objects in $\mathfrak{T}$ if and only if*

- *$\mathfrak{C}$ has enough projective objects;*
- *the adjoint functor $F^{\vdash}$ is defined on $\mathfrak{PC}$;*
- *$F \circ F^{\vdash}(A) \cong A$ for all $A \in\in \mathfrak{PC}$.*

*Proof.* Suppose first that $F$ is universal and that there are enough $\mathfrak{I}$-projective objects. Then $F$ is equivalent to the projectivisation functor of [4]. The various properties of this functor listed in [4, Proposition 4.19] include the following:

- there are enough projective objects in $\mathfrak{C}$;
- $F$ induces an equivalence of categories $\mathfrak{P}_{\mathfrak{I}} \cong \mathfrak{PC}$ ($\mathfrak{P}_{\mathfrak{I}}$ is the class of projective objects in $\mathfrak{T}$);
- $\mathfrak{C}\big(F(A), F(B)\big) \cong \mathfrak{T}(A, B)$ for all $A \in\in \mathfrak{P}_{\mathfrak{I}}$, $B \in\in \mathfrak{T}$.

Here we use the assumption that idempotents in $\mathfrak{T}$ split. The last property is equivalent to $F^{\vdash} \circ F(A) \cong A$ for all $A \in\in \mathfrak{P}_{\mathfrak{I}}$. Since $\mathfrak{P}_{\mathfrak{I}} \cong \mathfrak{PC}$ via $F$, this implies that $F^{\vdash}$ is defined on $\mathfrak{PC}$ and that $F \circ F^{\vdash}(A) \cong A$ for all $A \in\in \mathfrak{PC}$. Thus $F$ has the properties listed in the statement of the theorem.

Now suppose conversely that $F$ has these properties. Let $\mathfrak{P}'_{\mathfrak{I}} \subseteq \mathfrak{T}$ be the essential range of $F^{\vdash}: \mathfrak{PC} \to \mathfrak{T}$. We claim that $\mathfrak{P}'_{\mathfrak{I}}$ is the class of all

ℑ-projective objects in $\mathfrak{T}$. Since $F \circ F^{\vdash}$ is equivalent to the identity functor on $\mathfrak{PC}$ by assumption, $F|_{\mathfrak{P}'_{\mathfrak{I}}}$ and $F^{\vdash}$ provide an equivalence of categories $\mathfrak{P}'_{\mathfrak{I}} \cong \mathfrak{PC}$. Since $\mathfrak{C}$ is assumed to have enough projectives, the hypotheses of Proposition 55 are satisfied. Hence there are enough ℑ-projective objects in $\mathfrak{T}$, and any object of $\mathfrak{P}_{\mathfrak{I}}$ is a retract of an object of $\mathfrak{P}'_{\mathfrak{I}}$. Idempotent morphisms in the category $\mathfrak{P}'_{\mathfrak{I}} \cong \mathfrak{PC}$ split because $\mathfrak{C}$ is Abelian and retracts of projective objects are again projective. Hence $\mathfrak{P}'_{\mathfrak{I}}$ is closed under retracts in $\mathfrak{T}$, so that $\mathfrak{P}'_{\mathfrak{I}} = \mathfrak{P}_{\mathfrak{I}}$. It also follows that $F$ and $F^{\vdash}$ provide an equivalence of categories $\mathfrak{P}_{\mathfrak{I}} \cong \mathfrak{PC}$. Hence $F^{\vdash} \circ F(A) \cong A$ for all $A \in\in \mathfrak{P}_{\mathfrak{I}}$, so that we get $\mathfrak{C}\big(F(A), F(B)\big) \cong \mathfrak{T}(F^{\vdash} \circ F(A), B) \cong \mathfrak{T}(A, B)$ for all $A \in\in \mathfrak{P}_{\mathfrak{I}}$, $B \in\in \mathfrak{T}$.

Now let $G \colon \mathfrak{T} \to \mathfrak{C}'$ be a stable homological functor. We will later assume $G$ to be ℑ-exact, but the first part of the following argument works in general. Since $F$ provides an equivalence of categories $\mathfrak{P}_{\mathfrak{I}} \cong \mathfrak{PC}$, the rule $\bar{G}\big(F(P)\big) := G(P)$ defines a functor $\bar{G}$ on $\mathfrak{PC}$. This yields a functor $\mathfrak{Ho}(\bar{G}) \colon \mathfrak{Ho}(\mathfrak{PC}) \to \mathfrak{Ho}(\mathfrak{C}')$. Since $\mathfrak{C}$ has enough projective objects, the construction of projective resolutions provides a functor $P \colon \mathfrak{C} \to \mathfrak{Ho}(\mathfrak{PC})$. We let $\bar{G}$ be the composite functor

$$\bar{G} \colon \mathfrak{C} \xrightarrow{P} \mathfrak{Ho}(\mathfrak{PC}) \xrightarrow{\mathfrak{Ho}(\bar{G})} \mathfrak{Ho}(\mathfrak{C}') \xrightarrow{H_0} \mathfrak{C}'.$$

This functor is right-exact and satisfies $\bar{G} \circ F = G$ on ℑ-projective objects of $\mathfrak{T}$.

Now suppose that $G$ is ℑ-exact. Then we get $\bar{G} \circ F = G$ for all objects of $\mathfrak{T}$ because this holds for ℑ-projective objects. We claim that $\bar{G}$ is exact. Let $A \in\in \mathfrak{C}$. Since $\mathfrak{C}$ has enough projective objects, we can find a projective resolution of $A$. We may assume this resolution to have the form $F(P_\bullet)$ with $P_\bullet \in\in \mathfrak{Ho}(\mathfrak{P}_{\mathfrak{I}})$ because $F(\mathfrak{P}_{\mathfrak{I}}) \cong \mathfrak{PC}$. Lemma 28 yields that $P_\bullet$ is ℑ-exact except in degree 0. Since $\mathfrak{I} \subseteq \ker G$, the chain complex $P_\bullet$ is $\ker G$-exact in positive degrees as well, so that $G(P_\bullet)$ is exact except in degree 0 by Lemma 28. As a consequence, $\mathbb{L}_p \bar{G}(A) = 0$ for all $p > 0$. We also have $\mathbb{L}_0 \bar{G}(A) = \bar{G}(A)$ by construction. Thus $\bar{G}$ is exact.

As a result, $G$ factors as $G = \bar{G} \circ F$ for an exact functor $\bar{G} \colon \mathfrak{C} \to \mathfrak{C}'$. It is clear that $\bar{G}$ is stable. Finally, since $\mathfrak{C}$ has enough projective objects, a functor on $\mathfrak{C}$ is determined up to natural equivalence by its restriction to projective objects. Therefore, our factorisation of $G$ is unique up to natural equivalence. Thus $F$ is the universal ℑ-exact functor. $\qquad\square$

**Remark 58.** Let $\mathfrak{P}'\mathfrak{C} \subseteq \mathfrak{PC}$ be some subcategory such that any object of $\mathfrak{C}$ is a quotient of a direct sum of objects of $\mathfrak{P}'\mathfrak{C}$. Equivalently, $(\mathfrak{P}'\mathfrak{C})_{\oplus} = \mathfrak{PC}$. Theorem 57 remains valid if we only assume that $F^{\vdash}$ is defined on $\mathfrak{P}'\mathfrak{C}$ and that $F \circ F^{\vdash}(A) \cong A$ holds for $A \in\in \mathfrak{P}'\mathfrak{C}$ because both conditions are evidently hereditary for direct sums and retracts.

**Theorem 59.** *In the situation of Theorem 57, the domain of the functor $F^{\vdash}$ contains $\mathfrak{P}\mathfrak{C}$, and its essential range is $\mathfrak{P}_{\mathfrak{J}}$. The functors $F$ and $F^{\vdash}$ restrict to equivalences of categories $\mathfrak{P}_{\mathfrak{J}} \cong \mathfrak{P}\mathfrak{C}$ inverse to each other.*

*An object $A \in\in \mathfrak{T}$ is $\mathfrak{J}$-projective if and only if $F(A)$ is projective and*

$$\mathfrak{C}\big(F(A), F(B)\big) \cong \mathfrak{T}(A, B)$$

*for all $B \in\in \mathfrak{T}$; following Ross Street [24], we call such objects $F$-projective. We have $F(A) \in\in \mathfrak{P}\mathfrak{C}$ if and only if there is an $\mathfrak{J}$-equivalence $P \to A$ with $P \in\in \mathfrak{P}_{\mathfrak{J}}$.*

*The functors $F$ and $F^{\vdash}$ induce bijections between isomorphism classes of projective resolutions of $F(A)$ in $\mathfrak{C}$ and isomorphism classes of $\mathfrak{J}$-projective resolutions of $A \in\in \mathfrak{T}$ in $\mathfrak{T}$.*

*If $G\colon \mathfrak{T} \to \mathfrak{C}'$ is any (stable) additive functor, then there is a unique right-exact (stable) functor $\bar{G}\colon \mathfrak{C} \to \mathfrak{C}'$ such that $\bar{G} \circ F(P) = G(P)$ for all $P \in\in \mathfrak{P}_{\mathfrak{J}}$.*

*The left derived functors of $G$ with respect to $\mathfrak{J}$ and of $\bar{G}$ are related by natural isomorphisms $\mathbb{L}_n \bar{G} \circ F(A) = \mathbb{L}_n G(A)$ for all $A \in\in \mathfrak{T}$, $n \in \mathbb{N}$. There is a similar statement for cohomological functors, which specialises to natural isomorphisms*

$$\mathrm{Ext}^n_{\mathfrak{T},\mathfrak{J}}(A, B) \cong \mathrm{Ext}^n_{\mathfrak{C}}\big(F(A), F(B)\big).$$

*Proof.* We have already seen during the proof of Theorem 57 that $F$ restricts to an equivalence of categories $\mathfrak{P}_{\mathfrak{J}} \overset{\cong}{\to} \mathfrak{P}\mathfrak{C}$ with inverse $F^{\vdash}$ and that $\mathfrak{C}\big(F(A), F(B)\big) \cong \mathfrak{T}(A, B)$ for all $A \in\in \mathfrak{P}_{\mathfrak{J}}$, $B \in\in \mathfrak{P}\mathfrak{C}$.

Conversely, if $A$ is $F$-projective in the sense of Street, then $A$ is $\mathfrak{J}$-projective because already $\mathfrak{T}(A, B) \cong \mathfrak{C}\big(F(A), F(B)\big)$ for all $B \in\in \mathfrak{T}$ yields $A \cong F^{\vdash} \circ F(A)$, so that $A$ is $\mathfrak{J}$-projective; notice that the projectivity of $F(A)$ is automatic.

Since $F$ maps $\mathfrak{J}$-equivalences to isomorphisms, $F(A)$ is projective whenever there is an $\mathfrak{J}$-equivalence $P \to A$ with $\mathfrak{J}$-projective $P$. Conversely, suppose that $F(A)$ is $\mathfrak{J}$-projective. Let $P_0 \to A$ be a one-step $\mathfrak{J}$-projective resolution. Since $F(A)$ is projective, the epimorphism $F(P_0) \to F(A)$ splits by some map $F(A) \to F(P_0)$. The resulting map $F(P_0) \to F(A) \to F(P_0)$ is idempotent and comes from an idempotent endomorphism of $P_0$ because $F$ is fully faithful on $\mathfrak{P}_{\mathfrak{J}}$. Its range object $P$ exists because we require idempotent morphisms in $\mathfrak{C}$ to split. It belongs again to $\mathfrak{P}_{\mathfrak{J}}$, and the induced map $F(P) \to F(A)$ is invertible by construction. Hence we get an $\mathfrak{J}$-equivalence $P \to A$.

If $C_\bullet$ is a chain complex over $\mathfrak{T}$, then we know already from Lemma 28 that $C_\bullet$ is $\mathfrak{J}$-exact if and only if $F(C_\bullet)$ is exact. Hence $F$ maps an $\mathfrak{J}$-projective resolution of $A$ to a projective resolution of $F(A)$. Conversely, if $P_\bullet \to F(A)$ is any projective resolution in $\mathfrak{C}$, then it is of the form $F(\hat{P}_\bullet) \to F(A)$ where $\hat{P}_\bullet := F^\vdash(P_\bullet)$ and where we get the map $\hat{P}_0 \to A$ by adjointness from the given map $P_0 \to F(A)$. This shows that $F$ induces a bijection between isomorphism classes of $\mathfrak{J}$-projective resolutions of $A$ and projective resolutions of $F(A)$.

We have seen during the proof of Theorem 57 how a stable homological functor $G\colon \mathfrak{T} \to \mathfrak{C}'$ gives rise to a unique right-exact functor $\bar{G}\colon \mathfrak{C} \to \mathfrak{C}'$ that satisfies $\bar{G} \circ F(P) = G(P)$ for all $P \in\in \mathfrak{P}_\mathfrak{J}$. The derived functors $\mathbb{L}_n\bar{G}\big(F(A)\big)$ for $A \in\in \mathfrak{T}$ are computed by applying $\bar{G}$ to a projective resolution of $F(A)$. Since such a projective resolution is of the form $F(P_\bullet)$ for an $\mathfrak{J}$-projective resolution $P_\bullet \to A$ and since $\bar{G} \circ F = G$ on $\mathfrak{J}$-projective objects, the derived functors $\mathbb{L}_n G(A)$ and $\mathbb{L}_n\bar{G}\big(F(A)\big)$ are computed by the same chain complex and agree. The same reasoning applies to cohomological functors and yields the assertion about Ext. $\qquad\square$

If $F^\vdash(X)$ is defined, then it can be shown that $X \cong X_0 \oplus X_1$ where $X_1 \cong FF^\vdash(X)$ is projective and $F^\vdash(X_0) = 0$, that is, $\mathfrak{C}(X_0, FY) = 0$ for all $Y$. Usually, the latter condition implies $X_0 = 0$, but it is not clear whether this implication is true in general.

**Remark 60.** The assumption that idempotents split is only needed to check that the universal $\mathfrak{J}$-exact functor has the properties listed in Theorem 57. The converse directions of Theorem 57 and Theorem 59 do not need this assumption.

If $\mathfrak{T}$ has countable direct sums or countable direct products, then idempotents in $\mathfrak{T}$ automatically split by [19, §1.3]. This covers categories such as $\mathrm{KK}^G$ because they have countable direct sums.

### 3.8. Derived functors in homological algebra

Now we study the kernel $\mathfrak{J}_\mathrm{H}$ of the homology functor $\mathrm{H}_*\colon \mathfrak{Ho}(\mathfrak{C}; \mathbb{Z}/p) \to \mathfrak{C}^{\mathbb{Z}/p}$ introduced in Example 9. We get exactly the same statements if we replace the homotopy category by its derived category and study the kernel of $\mathrm{H}_*\colon \mathfrak{Der}(\mathfrak{C}; \mathbb{Z}/p) \to \mathfrak{C}^{\mathbb{Z}/p}$. We often abbreviate $\mathfrak{J}_\mathrm{H}$ to H and speak of H-epimorphisms, H-exact chain complexes, H-projective resolutions, and so on. We denote the full subcategory of H-projective objects in $\mathfrak{Ho}(\mathfrak{C}; \mathbb{Z}/p)$ by $\mathfrak{P}_\mathrm{H}$.

We assume that the underlying Abelian category $\mathfrak{C}$ has enough projective objects. Then the same holds for $\mathfrak{C}^{\mathbb{Z}/p}$, and we have $\mathfrak{P}(\mathfrak{C}^{\mathbb{Z}/p}) \cong (\mathfrak{P}\mathfrak{C})^{\mathbb{Z}/p}$.

That is, an object of $\mathfrak{C}^{\mathbb{Z}/p}$ is projective if and only if its homogeneous pieces are.

**Theorem 61.** *The category* $\mathfrak{Ho}(\mathfrak{C}; \mathbb{Z}/p)$ *has enough* H-*projective objects, and the functor* $H_* \colon \mathfrak{Ho}(\mathfrak{C}; \mathbb{Z}/p) \to \mathfrak{C}^{\mathbb{Z}/p}$ *is the universal* H-*exact stable homological functor. Its restriction to* $\mathfrak{P}_H$ *provides an equivalence of categories* $\mathfrak{P}_H \cong \mathfrak{P}\mathfrak{C}^{\mathbb{Z}/p}$. *More concretely, a chain complex in* $\mathfrak{Ho}(\mathfrak{C}; \mathbb{Z}/p)$ *is* H-*projective if and only if it is homotopy equivalent to one with vanishing boundary map and projective entries.*

*The functor* $H_*$ *maps isomorphism classes of* H-*projective resolutions of a chain complex* A *in* $\mathfrak{Ho}(\mathfrak{C}; \mathbb{Z}/p)$ *bijectively to isomorphism classes of projective resolutions of* $H_*(A)$ *in* $\mathfrak{C}^{\mathbb{Z}/p}$. *We have*

$$\mathrm{Ext}^n_{\mathfrak{Ho}(\mathfrak{C};\mathbb{Z}/p),\mathfrak{I}_H}(A, B) \cong \mathrm{Ext}^n_{\mathfrak{C}}\big(H_*(A), H_*(B)\big).$$

*Let* $F \colon \mathfrak{C} \to \mathfrak{C}'$ *be some covariant additive functor and define*

$$\bar{F} \colon \mathfrak{Ho}(\mathfrak{C}; \mathbb{Z}/p) \to \mathfrak{Ho}(\mathfrak{C}'; \mathbb{Z}/p)$$

*by applying* $F$ *entrywise. Then* $\mathbb{L}_n \bar{F}(A) \cong \mathbb{L}_n F\big(H_*(A)\big)$ *for all* $n \in \mathbb{N}$. *Similarly, we have* $\mathbb{R}^n \bar{F}(A) \cong \mathbb{R}^n F\big(H_*(A)\big)$ *if* $F$ *is a contravariant functor.*

*Proof.* The category $\mathfrak{C}^{\mathbb{Z}/p}$ has enough projective objects by assumption. We have already seen in Example 54 that $H_*^{\vdash}$ is defined on $\mathfrak{P}\mathfrak{C}^{\mathbb{Z}/p}$; this functor is denoted by $j$ in Example 54. It is clear that $H_* \circ j(A) \cong A$ for all $A \in\in \mathfrak{C}^{\mathbb{Z}/p}$. Now Theorem 57 shows that $H_*$ is universal. We do not need idempotent morphisms in $\mathfrak{Ho}(\mathfrak{C}; \mathbb{Z}/p)$ to split by Remark 60. $\square$

**Remark 62.** Since the universal $\mathfrak{I}$-exact functor is essentially unique, the universality of $H_* \colon \mathfrak{Der}(\mathfrak{C}; \mathbb{Z}/p) \to \mathfrak{C}^{\mathbb{Z}/p}$ means that we can recover this functor and hence the stable Abelian category $\mathfrak{C}^{\mathbb{Z}/p}$ from the ideal $\mathfrak{I}_H \subseteq \mathfrak{Der}(\mathfrak{C}; \mathbb{Z}/p)$. That is, the ideal $\mathfrak{I}_H$ and the functor $H_* \colon \mathfrak{Der}(\mathfrak{C}; \mathbb{Z}/p) \to \mathfrak{C}^{\mathbb{Z}/p}$ contain exactly the same amount of information.

For instance, if we forget the precise category $\mathfrak{C}$ by composing $H_*$ with some *faithful* functor $\mathfrak{C} \to \mathfrak{C}'$, then the resulting homology functor $\mathfrak{Ho}(\mathfrak{C}; \mathbb{Z}/p) \to \mathfrak{C}'$ still has kernel $\mathfrak{I}_H$. We can recover $\mathfrak{C}^{\mathbb{Z}/p}$ by passing to the universal $\mathfrak{I}$-exact functor.

We compare this with the situation for truncation structures ([3]). These cannot exist for periodic categories such as $\mathfrak{Der}(\mathfrak{C}; \mathbb{Z}/p)$ for $p \geq 1$. Given the standard truncation structure on $\mathfrak{Der}(\mathfrak{C})$, we can recover the Abelian category $\mathfrak{C}$ as its core; we also get back the homology functors $H_n \colon \mathfrak{Der}(\mathfrak{C}) \to \mathfrak{C}$ for all $n \in \mathbb{Z}$. Conversely, the functor $H_* \colon \mathfrak{Der}(\mathfrak{C}) \to \mathfrak{C}^{\mathbb{Z}}$ together with the grading on $\mathfrak{C}^{\mathbb{Z}}$ tells us what it means for a chain complex to be exact in degrees greater than 0 or less than 0 and thus determines the truncation structure. Hence the standard truncation structure on $\mathfrak{Der}(\mathcal{C})$ contains the same amount of information as the functor $H_* : \mathfrak{Der}(\mathcal{C}) \to \mathcal{C}^{\mathbb{Z}}$ together with the grading on $\mathcal{C}^{\mathbb{Z}}$.

# 4. The plain Universal Coefficient Theorem

Now we study the ideal $\mathfrak{I}_K := \ker K_* \subseteq KK$ of Example 5. We complete our analysis of this example and explain the Universal Coefficient Theorem for KK in our framework. We call $\mathfrak{I}_K$-projective objects and $\mathfrak{I}_K$-exact functors briefly K-*projective* and K-*exact* and let $\mathfrak{P}_K$ be the class of K-projective objects in KK.

Let $\mathfrak{Ab}_c^{\mathbb{Z}/2} \subseteq \mathfrak{Ab}^{\mathbb{Z}/2}$ be the full subcategory of *countable* $\mathbb{Z}/2$-graded Abelian groups. Since the K-theory of a separable C*-algebra is countable, we may view $K_*$ as a stable homological functor $K_* : KK \to \mathfrak{Ab}_c^{\mathbb{Z}/2}$.

**Theorem 63.** *There are enough K-projective objects in KK, and the universal K-exact functor is $K_* : KK \to \mathfrak{Ab}_c^{\mathbb{Z}/2}$. It restricts to an equivalence of categories between $\mathfrak{P}_K$ and the full subcategory $\mathfrak{Ab}_{fc}^{\mathbb{Z}/2} \subseteq \mathfrak{Ab}_c^{\mathbb{Z}/2}$ of $\mathbb{Z}/2$-graded countable free Abelian groups. A separable C*-algebra belongs to $\mathfrak{P}_K$ if and only if it is KK-equivalent to $\bigoplus_{i \in I_0} \mathbb{C} \oplus \bigoplus_{i \in I_1} C_0(\mathbb{R})$ where the sets $I_0$, $I_1$ are at most countable.*

*If $A \in\in KK$, then $K_*$ maps isomorphism classes of K-projective resolutions of $A$ in $\mathfrak{T}$ bijectively to isomorphism classes of free resolutions of $K_*(A)$. We have*

$$\mathrm{Ext}^n_{KK, \mathfrak{I}_K}(A, B) \cong \begin{cases} \mathrm{Hom}_{\mathfrak{Ab}^{\mathbb{Z}/2}}\big(K_*(A), K_*(B)\big) & \textit{for } n = 0; \\ \mathrm{Ext}^1_{\mathfrak{Ab}^{\mathbb{Z}/2}}\big(K_*(A), K_*(B)\big) & \textit{for } n = 1; \\ 0 & \textit{for } n \geq 2. \end{cases}$$

*Let $F : KK \to \mathcal{C}$ be some covariant additive functor; then there is a unique right-exact functor $\bar{F} : \mathfrak{Ab}_c^{\mathbb{Z}/2} \to \mathcal{C}$ with $\bar{F} \circ K_* = F$. We have $\mathbb{L}_n F = (\mathbb{L}_n \bar{F}) \circ K_*$ for all $n \in \mathbb{N}$; this vanishes for $n \geq 2$. Similar assertions hold for contravariant functors.*

*Proof.* Notice that $\mathfrak{Ab}_c^{\mathbb{Z}/2} \subseteq \mathfrak{Ab}^{\mathbb{Z}/2}$ is an Abelian category. We shall denote objects of $\mathfrak{Ab}^{\mathbb{Z}/2}$ by pairs $(A_0, A_1)$ of Abelian groups. By definition, $(A_0, A_1) \in\in \mathfrak{Ab}_{fc}^{\mathbb{Z}/2}$ if and only if $A_0$ and $A_1$ are countable free Abelian groups, that is, they are of the form $A_0 = \mathbb{Z}[I_0]$ and $A_1 = \mathbb{Z}[I_1]$ for at most countable sets $I_0$, $I_1$. It is well-known that any Abelian group is a quotient of a free Abelian group and that subgroups of free Abelian groups are again free. Moreover, free Abelian groups are projective. Hence $\mathfrak{Ab}_{\mathrm{fc}}^{\mathbb{Z}/2}$ is the subcategory of projective objects in $\mathfrak{Ab}_{\mathrm{c}}^{\mathbb{Z}/2}$ and any object $G \in\in \mathfrak{Ab}_{\mathrm{c}}^{\mathbb{Z}/2}$ has a projective resolution of the form $0 \to F_1 \to F_0 \twoheadrightarrow G$ with $F_0, F_1 \in\in \mathfrak{Ab}_{\mathrm{fc}}^{\mathbb{Z}/2}$. This implies that derived functors on $\mathfrak{Ab}_{\mathrm{c}}^{\mathbb{Z}/2}$ only occur in dimensions 1 and 0.

As in Example 53, we see that $\mathrm{K}_*^{\vdash}$ is defined on $\mathfrak{Ab}_{\mathrm{fc}}^{\mathbb{Z}/2}$ and satisfies

$$\mathrm{K}_*^{\vdash}\big(\mathbb{Z}[I_0], \mathbb{Z}[I_1]\big) \cong \bigoplus_{i \in I_0} \mathbb{C} \oplus \bigoplus_{i \in I_1} \mathrm{C}_0(\mathbb{R})$$

if $I_0, I_1$ are countable. We also have $\mathrm{K}_* \circ \mathrm{K}_*^{\vdash}\big(\mathbb{Z}[I_0], \mathbb{Z}[I_1]\big) \cong \big(\mathbb{Z}[I_0], \mathbb{Z}[I_1]\big)$, so that the hypotheses of Theorem 57 are satisfied. Hence there are enough K-projective objects and $\mathrm{K}_*$ is universal. The remaining assertions follow from Theorem 59 and our detailed knowledge of the homological algebra in $\mathfrak{Ab}_{\mathrm{c}}^{\mathbb{Z}/2}$. $\qquad\square$

**Example 64.** Consider the stable homological functor

$$F: \mathrm{KK} \to \mathfrak{Ab}_{\mathrm{c}}^{\mathbb{Z}/2}, \qquad A \mapsto \mathrm{K}_*(A \otimes B)$$

for some $B \in\in \mathrm{KK}$, where $\otimes$ denotes, say, the spatial C*-tensor product. We claim that the associated right-exact functor $\mathfrak{Ab}_{\mathrm{c}}^{\mathbb{Z}/2} \to \mathfrak{Ab}_{\mathrm{c}}^{\mathbb{Z}/2}$ is

$$\bar{F}: \mathfrak{Ab}_{\mathrm{c}}^{\mathbb{Z}/2} \to \mathfrak{Ab}_{\mathrm{c}}^{\mathbb{Z}/2}, \qquad G \mapsto G \otimes \mathrm{K}_*(B).$$

It is easy to check $F \circ \mathrm{K}_*^{\vdash}(G) \cong G \otimes \mathrm{K}_*(B) \cong \bar{F}(G)$ for $G \in\in \mathfrak{Ab}_{\mathrm{fc}}^{\mathbb{Z}/2}$. Since the functor $G \mapsto G \otimes \mathrm{K}_*(B)$ is right-exact and agrees with $\bar{F}$ on projective objects, we get $\bar{F}(G) \cong G \otimes \mathrm{K}_*(B)$ for all $G \in\in \mathfrak{Ab}_{\mathrm{c}}^{\mathbb{Z}/2}$. Hence the derived functors of $F$ are

$$\mathbb{L}_n F(A) \cong \begin{cases} \mathrm{K}_*(A) \otimes \mathrm{K}_*(B) & \text{for } n = 0; \\ \mathrm{Tor}^1\big(\mathrm{K}_*(A), \mathrm{K}_*(B)\big) & \text{for } n = 1; \\ 0 & \text{for } n \geq 2. \end{cases}$$

Here we use the same graded version of Tor as in the Künneth Theorem ([5]).

**Example 65.** Consider the stable homological functor

$$F: \mathrm{KK} \to \mathfrak{Ab}^{\mathbb{Z}/2}, \qquad B \mapsto \mathrm{KK}_*(A, B)$$

for some $A \in\in \mathrm{KK}$. We suppose that $A$ is a *compact* object of KK, that is, the functor $F$ commutes with direct sums. Then $\mathrm{KK}_*\big(A, \mathrm{K}_*^{\vdash}(G)\big) \cong \mathrm{KK}_*(A, \mathbb{C}) \otimes G$ for all $G \in\in \mathfrak{Ab}_{\mathrm{fc}}^{\mathbb{Z}/2}$ because this holds for $G = (\mathbb{Z}, 0)$ and is inherited by suspensions and direct sums. Now we get $\bar{F}(G) \cong \mathrm{KK}_*(A, \mathbb{C}) \otimes G$ for all

$G \in\in \mathfrak{Ab}_{\mathrm{c}}^{\mathbb{Z}/2}$ as in Example 64. Therefore,

$$\mathbb{L}_n F(B) \cong \begin{cases} \mathrm{KK}_*(A, \mathbb{C}) \otimes \mathrm{K}_*(B) & \text{for } n = 0; \\ \mathrm{Tor}^1\big(\mathrm{KK}_*(A, \mathbb{C}), \mathrm{K}_*(B)\big) & \text{for } n = 1; \\ 0 & \text{for } n \geq 2. \end{cases}$$

Generalising Examples 64 and 65, we have $\bar{F}(G) \cong F(\mathbb{C}) \otimes G$ and hence

$$\mathbb{L}_n F(B) \cong \begin{cases} F(\mathbb{C}) \otimes \mathrm{K}_*(B) & \text{for } n = 0, \\ \mathrm{Tor}^1\big(F(\mathbb{C}), \mathrm{K}_*(B)\big) & \text{for } n = 1, \\ 0 & \text{for } n \geq 2, \end{cases}$$

for any covariant functor $F \colon \mathrm{KK} \to \mathfrak{C}$ that commutes with direct sums.

Similarly, if $F \colon \mathrm{KK} \to \mathfrak{C}$ is contravariant and maps direct sums to direct products, then $\bar{F}(G) \cong \mathrm{Hom}(G, F(\mathbb{C}))$ and

$$\mathbb{R}^n F(B) \cong \begin{cases} \mathrm{Hom}\big(\mathrm{K}_*(B), F(\mathbb{C})\big) & \text{for } n = 0, \\ \mathrm{Ext}^1\big(\mathrm{K}_*(B), F(\mathbb{C})\big) & \text{for } n = 1, \\ 0 & \text{for } n \geq 2. \end{cases}$$

The description of $\mathrm{Ext}^n_{\mathrm{KK}, \mathfrak{I}_\mathrm{K}}$ in Theorem 63 is a special case of this.

## 4.1. Universal Coefficient Theorem in the hereditary case

In general, we need spectral sequences in order to relate the derived functors $\mathbb{L}_n F$ back to $F$. We will discuss this in a sequel to this article. Here we only treat the simple case where we have projective resolutions of length 1. The following universal coefficient theorem is very similar to but slightly more general than [4, Theorem 4.27] because we do not require $\mathfrak{I}$-equivalences to be invertible.

**Theorem 66.** *Let $\mathfrak{I}$ be a homological ideal in a triangulated category $\mathfrak{T}$. Let $A \in\in \mathfrak{T}$ have an $\mathfrak{I}$-projective resolution of length 1. Suppose also that $\mathfrak{T}(A, B) = 0$ for all $\mathfrak{I}$-contractible $B$. Let $F \colon \mathfrak{T} \to \mathfrak{C}$ be a homological functor, $\tilde{F} \colon \mathfrak{T}^{\mathrm{op}} \to \mathfrak{C}$ a cohomological functor, and $B \in\in \mathfrak{T}$. Then there are natural short exact sequences*

$$0 \to \mathbb{L}_0 F_*(A) \to F_*(A) \to \mathbb{L}_1 F_{*-1}(A) \to 0,$$

$$0 \to \mathbb{R}^1 \tilde{F}^{*-1}(A) \to \tilde{F}^*(A) \to \mathbb{R}^0 \tilde{F}^*(A) \to 0,$$

$$0 \to \mathrm{Ext}^1_{\mathfrak{T}, \mathfrak{I}}(\Sigma A, B) \to \mathfrak{T}(A, B) \to \mathrm{Ext}^0_{\mathfrak{T}, \mathfrak{I}}(A, B) \to 0.$$

**Example 67.** For the ideal $\mathfrak{I}_K \subseteq \mathrm{KK}$, any object has a K-projective resolution of length 1 by Theorem 63. The other hypothesis of Theorem 66 holds if and only if $A$ satisfies the Universal Coefficient Theorem (UCT). The UCT for $\mathrm{KK}(A, B)$ predicts $\mathrm{KK}(A, B) = 0$ if $K_*(B) = 0$. Conversely, if this is the case, then Theorem 66 applies, and our description of $\mathrm{Ext}_{\mathrm{KK}, \mathfrak{I}_K}$ in Theorem 63 yields the UCT for $\mathrm{KK}(A, B)$ for all $B$. This yields our claim.

Thus the UCT for $\mathrm{KK}(A, B)$ is a special of Theorem 66. In the situations of Examples 64 and 65, we get the familiar Künneth Theorems for $K_*(A \otimes B)$ and $\mathrm{KK}_*(A, B)$. These arguments are very similar to the original proofs (see [5]). Our machinery allows us to treat other situations in a similar fashion.

*Proof of Theorem 66.* We only write down the proof for homological functors. The cohomological case is dual and contains $\mathfrak{T}(\_, B)$ as a special case.

Let $0 \to P_1 \xrightarrow{\delta_1} P_0 \xrightarrow{\delta_0} A$ be an $\mathfrak{I}$-projective resolution of length 1 and view it as an $\mathfrak{I}$-exact chain complex of length 3. Lemma 31 yields a commuting diagram

$$
\begin{array}{ccccc}
P_1 & \xrightarrow{\delta_1} & P_0 & \xrightarrow{\tilde{\delta}_0} & \tilde{A} \\
\| & & \| & & \big\downarrow{\scriptstyle\alpha} \\
P_1 & \xrightarrow{\delta_1} & P_0 & \xrightarrow{\delta_0} & A,
\end{array}
$$

such that the top row is part of an $\mathfrak{I}$-exact exact triangle $P_1 \to P_0 \to \tilde{A} \to \Sigma P_1$ and $\alpha$ is an $\mathfrak{I}$-equivalence. We claim that $\alpha$ is an isomorphism in $\mathfrak{T}$.

We embed $\alpha$ in an exact triangle $\Sigma^{-1} B \to \tilde{A} \xrightarrow{\alpha} A \xrightarrow{\beta} B$. Lemma 24 shows that $B$ is $\mathfrak{I}$-contractible because $\alpha$ is an $\mathfrak{I}$-equivalence. Hence $\mathfrak{T}(A, B) = 0$ by our assumption on $A$. This forces $\beta = 0$, so that our exact triangle splits: $\tilde{A} \cong A \oplus \Sigma^{-1} B$. Now we apply the functor $\mathfrak{T}(\cdot, B)$ to the exact triangle $P_0 \to P_1 \to \tilde{A}$. The resulting long exact sequence has the form

$$
\cdots \leftarrow \mathfrak{T}(P_0, B) \leftarrow \mathfrak{T}(\tilde{A}, B) \leftarrow \mathfrak{T}(\Sigma P_1, B) \leftarrow \cdots .
$$

Since both $P_0$ and $P_1$ are $\mathfrak{I}$-projective and $B$ is $\mathfrak{I}$-contractible, we get $\mathfrak{T}(\tilde{A}, B) = 0$. Then $\mathfrak{T}(B, B) \subseteq \mathfrak{T}(\tilde{A}, B)$ vanishes as well, so that $B \cong 0$ and $\alpha$ is invertible.

We get an exact triangle in $\mathfrak{T}$ of the form $P_1 \xrightarrow{\delta_1} P_0 \xrightarrow{\delta_0} A \to \Sigma P_1$ because any triangle isomorphic to an exact one is itself exact.

Now we apply $F$. Since $F$ is homological, we get a long exact sequence

$$
\cdots \to F_*(P_1) \xrightarrow{F_*(\delta_1)} F_*(P_0) \to F_*(A) \to F_{*-1}(P_1) \xrightarrow{F_{*-1}(\delta_1)} F_{*-1}(P_0) \to \cdots .
$$

We cut this into short exact sequences of the form

$$\mathrm{coker}\big(F_*(\delta_1)\big) \rightarrowtail F_*(A) \twoheadrightarrow \ker\big(F_{*-1}(\delta_1)\big).$$

Since $\mathrm{coker}\, F_*(\delta_1) = \mathbb{L}_0 F_*(A)$ and $\ker F_*(\delta_1) = \mathbb{L}_1 F_*(A)$, we get the desired exact sequence. The map $\mathbb{L}_0 F_*(A) \to F_*(A)$ is the canonical map induced by $\delta_0$. The other map $F_*(A) \to \mathbb{L}_1 F_{*-1}(A)$ is natural for all morphisms between objects with an $\mathfrak{I}$-projective resolution of length 1 by Proposition 44. $\qquad\square$

The proof shows that – in the situation of Theorem 66 – we have

$$\mathrm{Ext}^0_{\mathfrak{T},\mathfrak{I}}(A, B) \cong \mathfrak{T}/\mathfrak{I}(A, B), \qquad \mathrm{Ext}^1_{\mathfrak{T},\mathfrak{I}}(A, B) \cong \mathfrak{I}(A, \Sigma B).$$

More generally, we can construct a natural map $\mathfrak{I}(A, \Sigma B) \to \mathrm{Ext}^1_{\mathfrak{T},\mathfrak{I}}(A, B)$ for any homological ideal, using the $\mathfrak{I}$-universal homological functor $F : \mathfrak{T} \to \mathfrak{C}$. We embed $f \in \mathfrak{I}(A, \Sigma B)$ in an exact triangle $B \to C \to A \to \Sigma B$. We get an extension

$$\big[F(B) \rightarrowtail F(C) \twoheadrightarrow F(A)\big] \in\in \mathrm{Ext}^1_{\mathfrak{C}}\big(F(A), F(B)\big)$$

because this triangle is $\mathfrak{I}$-exact. This class $\kappa(f)$ in $\mathrm{Ext}^1_{\mathfrak{C}}\big(F(A), F(B)\big)$ does not depend on auxiliary choices because the exact triangle $B \to C \to A \to \Sigma B$ is unique up to isomorphism. Theorem 59 yields $\mathrm{Ext}^1_{\mathfrak{T},\mathfrak{I}}(A, B) \cong \mathrm{Ext}^1_{\mathfrak{C}}\big(F(A), F(B)\big)$ because $F$ is universal. Hence we get a natural map

$$\kappa : \mathfrak{I}(A, \Sigma B) \to \mathrm{Ext}^1_{\mathfrak{T},\mathfrak{I}}(A, B).$$

We may view $\kappa$ as a secondary invariant generated by the canonical map

$$\mathfrak{T}(A, B) \to \mathrm{Ext}^0_{\mathfrak{T},\mathfrak{I}}(A, B).$$

For the ideal $\mathfrak{I}_K$, we get the same map $\kappa$ as in Example 5.

An Abelian category with enough projective objects is called *hereditary* if any subobject of a projective object is again projective. Equivalently, any object has a projective resolution of length 1. This motivates the following definition:

**Definition 68.** A homological ideal $\mathfrak{I}$ in a triangulated category $\mathfrak{T}$ is called *hereditary* if any object of $\mathfrak{T}$ has a projective resolution of length 1.

If $\mathfrak{I}$ is hereditary and if $\mathfrak{I}$-equivalences are invertible, then Theorem 66 applies to all $A \in\in \mathfrak{T}$ (and vice versa).

# 5. Crossed products for compact quantum groups

Let $G$ be a compact (quantum) group and write $C(G)$ and $C^*(G)$ for the Hopf $C^*$-algebras of $\hat{G}$ and $G$. We study the homological algebra in $KK^G$ generated by the ideals $\mathfrak{I}_{\ltimes,K} \subseteq \mathfrak{I}_\ltimes \subseteq KK$ defined in Example 7. Recall that $\mathfrak{I}_\ltimes$ is the kernel of the crossed product functor

$$G \ltimes \_\colon KK^G \to KK, \qquad A \mapsto G \ltimes A,$$

whereas $\mathfrak{I}_{\ltimes,K}$ is the kernel of the composite functor $K_* \circ (G \ltimes \_)$. Since we have already analysed $K_*$ in §4, we can treat both ideals in a parallel fashion.

Our setup contains two classical special cases. First, $G$ may be a compact Lie group. Then $C(G)$ and $C^*(G)$ have the usual meaning, and the objects of $KK^G$ are separable $C^*$-algebras with a continuous action of $G$. Secondly, $C(G)$ may be the dual $C^*_{\mathrm{red}}(H)$ of a discrete group $H$, so that $C^*(G) = C_0(H)$. Then the objects of $KK^G$ are separable $C^*$-algebras equipped with a (reduced) coaction of $H$. (We disregard the nuances between reduced and full coactions.) If we identify $KK^{C^*_{\mathrm{red}}(H)} \cong KK^H$ using Baaj–Skandalis duality, then the crossed product functor $G \ltimes \_$ corresponds to the forgetful functor $KK^H \to KK$.

Let $\mathfrak{P}_\ltimes$ and $\mathfrak{P}_{\ltimes,K}$ be the classes of projective objects for the ideals $\mathfrak{I}_\ltimes$ and $\mathfrak{I}_{\ltimes,K}$. Our first task is to find enough projective objects for these two ideals.

Let $\tau\colon KK \to KK^G$ be the functor that equips $B \in\in KK$ with the trivial $G$-action (that is, coaction of $C(G)$). This corresponds to the induction functor from the trivial quantum group to $C^*(G)$ under the equivalence $KK^G \cong KK^{C^*(G)}$.

The functors $\tau$ and $G \ltimes \_$ are adjoint, that is, there are natural isomorphisms

$$KK^G(\tau(A), B) \cong KK(A, G \ltimes B) \tag{5.1}$$

for all $A \in\in KK$, $B \in\in KK^G$. This generalisation of the Green–Julg Theorem is proved in [26, Théorème 5.10].

**Lemma 69.** *There are enough projective objects for $\mathfrak{I}_\ltimes$ and $\mathfrak{I}_{\ltimes,K}$. We have*

$$\mathfrak{P}_\ltimes = (\tau(KK))_\oplus, \qquad \mathfrak{P}_{\ltimes,K} = (\tau\mathbb{C}, \tau C_0(\mathbb{R}))_\oplus.$$

*Proof.* The adjoint of $G \ltimes \_$ is defined on all of $KK$, which is certainly enough for Proposition 55. This yields the assertions for $\mathfrak{I}_\ltimes$ and $\mathfrak{P}_\ltimes$; even more, $\mathfrak{P}_\ltimes$ is the closure of $\tau(KK)$ under retracts.

The adjoint of $K_* \circ (G \ltimes \_)$ is $\tau \circ K_*^{\vdash}$, which is defined for free countable $\mathbb{Z}/2$-graded Abelian groups. Explicitly,

$$\tau \circ K_*^{\vdash}\big(\mathbb{Z}[I_0], \mathbb{Z}[I_1]\big) \cong \bigoplus_{i \in I_0} \tau(\mathbb{C}) \oplus \bigoplus_{i \in I_1} \tau\big(C_0(\mathbb{R})\big).$$

Since any object of $\mathfrak{Ab}_c^{\mathbb{Z}/2}$ is a quotient of one in $\mathfrak{Ab}_{fc}^{\mathbb{Z}/2}$, Proposition 55 applies and yields the assertions about $\mathfrak{I}_{\ltimes,K}$ and $\mathfrak{P}_{\ltimes,K}$; even more, $\mathfrak{P}_{\ltimes,K}$ is the closure of $\tau \circ K_*^{\vdash}(\mathfrak{Ab}_{fc}^{\mathbb{Z}/2})$ under retracts. $\qquad\square$

To proceed further, we describe the universal $\mathfrak{I}$-exact functors. The functors $(G \ltimes \_)$ and $K_* \circ (G \ltimes \_)$ fail the criterion of Theorem 57 (unless $G = \{1\}$) because

$$G \ltimes \tau(A) \cong C^*(G) \otimes A \not\cong A.$$

This is not surprising because $G \ltimes \_$ is equivalent to a *forgetful* functor. The universal functor recovers a *linearisation* of the forgotten structure.

First we consider the ideal $\mathfrak{I}_{\ltimes,K}$. By Lemma 69, the objects $\tau(\mathbb{C})$ and $\Sigma\tau(\mathbb{C})$ generate all $\mathfrak{I}_{\ltimes,K}$-projective objects. Their internal symmetries are encoded by the $\mathbb{Z}/2$-graded ring $KK_*^G(\tau\mathbb{C}, \tau\mathbb{C})^{\mathrm{op}}$; the superscript op denotes that we take the product in reversed order. For classical compact groups, this ring is commutative, so that the order of multiplication does not matter; in general, the reversed-order product is the more standard choice. The following fact is well-known:

**Lemma 70.** *The ring* $KK_*^G(\tau\mathbb{C}, \tau\mathbb{C})^{\mathrm{op}}$ *is isomorphic to the* representation ring $\mathrm{Rep}(G)$ *of $G$ for* $* = 0$ *and $0$ for* $* = 1$.

We may take this as the definition of $\mathrm{Rep}(G)$.

*Proof.* The adjointness isomorphism (5.1) identifies

$$KK_*^G(\tau\mathbb{C}, \tau\mathbb{C}) \cong KK_*(\mathbb{C}, G \ltimes \tau\mathbb{C}) = KK_*\big(\mathbb{C}, C^*(G)\big) \cong K_*(C^*G).$$

Since $G$ is compact, $C^*G$ is a direct sum of matrix algebras, one for each irreducible representation of $G$. Hence the underlying Abelian group of $\mathrm{Rep}(G)$ is the free Abelian group $\mathbb{Z}[\hat{G}]$ on the set $\hat{G}$ of irreducible representations of $G$. The ring structure on $\mathrm{Rep}(G)$ comes from the internal tensor product of representations: represent two elements $\alpha, \beta$ of $KK_0^G(\tau\mathbb{C}, \tau\mathbb{C})$ by differences of finite-dimensional representations $\pi, \varrho$ of $G$, then $\alpha \circ \beta \in KK_0^G(\tau\mathbb{C}, \tau\mathbb{C})$ is represented by $\varrho \otimes \pi$ because the product in $KK^G$ boils down to an exterior tensor product in this case (with reversed order). $\qquad\square$

**Example 71.** If $C(G) = C^*_{red}(H)$ for a discrete group $H$, then $\hat{G} = H$ and the product on $\mathrm{Rep}(G) = \mathbb{Z}[H]$ is the usual convolution. Thus $\mathrm{Rep}(G)$ is the group ring of $H$.

If $G$ is a compact group, then $\mathrm{Rep}(G)$ is the representation ring of $G$ in the usual sense.

For any $B \in\!\in \mathrm{KK}^G$, the Kasparov product turns $\mathrm{KK}^G_*(\tau\mathbb{C}, B)$ into a left module over the ring $\mathrm{KK}^G_0(\tau\mathbb{C}, \tau\mathbb{C})^{\mathrm{op}} \cong \mathrm{Rep}(G)$. Thus $\mathrm{KK}^G_*(\tau\mathbb{C}, B)$ becomes an object of the Abelian category $\mathfrak{Mod}(\mathrm{Rep}\,G)^{\mathbb{Z}/2}_c$ of $\mathbb{Z}/2$-graded countable $\mathrm{Rep}(G)$-modules. We get a stable homological functor

$$F_{\mathrm{K}} \colon \mathrm{KK}^G \to \mathfrak{Mod}(\mathrm{Rep}\,G)^{\mathbb{Z}/2}_c, \qquad F_{\mathrm{K}}(B) := \mathrm{KK}^G_*(\tau\mathbb{C}, B).$$

By (5.1), the underlying Abelian group of $\mathrm{KK}^G_*(\tau\mathbb{C}, B)$ is

$$\mathrm{KK}^G_*(\tau\mathbb{C}, B) \cong \mathrm{KK}_*(\mathbb{C}, G \ltimes B) \cong \mathrm{K}_*(G \ltimes B).$$

Hence we still have $\ker F_{\mathrm{K}} = \mathfrak{I}_{\ltimes,\mathrm{K}}$. We often write $F_{\mathrm{K}}(B) = \mathrm{K}_*(G \ltimes B)$ if it is clear from the context that we view $\mathrm{K}_*(G \ltimes B)$ as an object of $\mathfrak{Mod}(\mathrm{Rep}\,G)^{\mathbb{Z}/2}_c$.

**Theorem 72.** *The functor $F_{\mathrm{K}}$ is the universal $\mathfrak{I}_{\ltimes,\mathrm{K}}$-exact functor.*

*The subcategory of $\mathfrak{I}_{\ltimes,\mathrm{K}}$-projective objects in $\mathrm{KK}^G$ is equivalent to the subcategory of $\mathbb{Z}/2$-graded countable projective $\mathrm{Rep}(G)$-modules. If $A \in\!\in \mathrm{KK}^G$, then $F_{\mathrm{K}}$ induces a bijection between isomorphism classes of $\mathfrak{I}_{\ltimes,\mathrm{K}}$-projective resolutions of $A$ and projective resolutions of $F_{\mathrm{K}}(A)$ in $\mathfrak{Mod}(\mathrm{Rep}\,G)^{\mathbb{Z}/2}_c$.*

*If $A, B \in\!\in \mathrm{KK}^G$, then*

$$\mathrm{Ext}^n_{\mathrm{KK}^G, \mathfrak{I}_{\ltimes,\mathrm{K}}}(A, B) \cong \mathrm{Ext}^n_{\mathrm{Rep}(G)}\big(\mathrm{K}_*(G \ltimes A), \mathrm{K}_*(G \ltimes B)\big).$$

*If $H \colon \mathrm{KK}^G \to \mathfrak{C}$ is homological and commutes with countable direct sums, then*

$$\mathbb{L}_n H(A) \cong \mathrm{Tor}^{\mathrm{Rep}(G)}_n\big(H_*(\tau\mathbb{C}), \mathrm{K}_*(G \ltimes A)\big);$$

*if $H \colon (\mathrm{KK}^G)^{\mathrm{op}} \to \mathfrak{C}$ is cohomological and turns countable direct sums into direct products, then*

$$\mathbb{R}^n H(A) \cong \mathrm{Ext}^n_{\mathrm{Rep}(G)}\big(\mathrm{K}_*(G \ltimes A), H^*(\tau\mathbb{C})\big);$$

*Here we use the right or left $\mathrm{Rep}(G)$-module structure on $H_*(\tau\mathbb{C})$ that comes from the functoriality of $H$.*

*Proof.* We verify universality using Theorem 57. The category $\mathfrak{Mod}(\mathrm{Rep}\,G)^{\mathbb{Z}/2}_c$ has enough projective objects: countable free modules are projective, and any object is a quotient of a countable free module.

Given a free module $(\mathrm{Rep}(G)[I_0], \mathrm{Rep}(G)[I_1])$, we have natural isomorphisms

$$\mathrm{Hom}_{\mathrm{Rep}(G)}\big((\mathrm{Rep}(G)[I_0], \mathrm{Rep}(G)[I_1]), F_{\mathrm{K}}(B)\big) \;\cong\; \prod_{\varepsilon \in \{0,1\}, i \in I_\varepsilon} \mathrm{K}_\varepsilon(G \ltimes B)$$

$$\cong \mathrm{KK}^G\bigg( \bigoplus_{\varepsilon \in \{0,1\}, i \in I_\varepsilon} \Sigma^\varepsilon \tau\mathbb{C}, B \bigg).$$

Hence the adjoint functor $F_{\mathrm{K}}^{\vdash}$ is defined on countable free modules. Idempotents in $\mathrm{KK}^G$ split by Remark 60. Therefore, the domain of $F_{\mathrm{K}}^{\vdash}$ is closed under retracts and contains all projective objects of $\mathfrak{Mod}(\mathrm{Rep}\,G)_{\mathrm{c}}^{\mathbb{Z}/2}$. It is easy to see that $F_{\mathrm{K}} \circ F_{\mathrm{K}}^{\vdash}(A) \cong A$ for free modules. This extends to retracts and hence holds for all projective modules (compare Remark 58). Now Theorem 57 yields that $F_{\mathrm{K}}$ is universal.

The assertions about projective objects, projective resolutions, and Ext now follow from Theorem 59. Theorem 59 also yields a formula for left derived functors in terms of the right-exact functor $\bar{H}: \mathfrak{Mod}(\mathrm{Rep}\,G)_{\mathrm{c}}^{\mathbb{Z}/2} \to \mathfrak{C}$ associated to a homological functor $H: \mathrm{KK}^G \to \mathfrak{C}$. It remains to compute $\bar{H}$.

First we define the Tor objects in the statement of the theorem if $\mathfrak{C}$ is the category of Abelian groups. Then $H_*(\tau\mathbb{C}) \in\in \mathfrak{Mod}(\mathrm{Rep}\,G)^{\mathbb{Z}/2}$, and we can take the derived functors of the usual $\mathbb{Z}/2$-graded balanced tensor product $\otimes_{\mathrm{Rep}\,G}$ for $\mathrm{Rep}(G)$-modules. We claim that there are natural isomorphisms

$$\bar{H}_*(M) \cong H_*(\tau\mathbb{C}) \otimes_{\mathrm{Rep}\,G} M$$

for all $M \in\in \mathfrak{Mod}(\mathrm{Rep}\,G)_{\mathrm{c}}^{\mathbb{Z}/2}$. This holds for $M = \mathrm{Rep}\,G$ and hence for all free modules because we have natural isomorphisms

$$\bar{H}_*(\mathrm{Rep}\,G) \cong H_*\big(F_{\mathrm{K}}^{\vdash}(\mathrm{Rep}\,G)\big) \cong H_*(\tau\mathbb{C}) \cong H_*(\tau\mathbb{C}) \otimes_{\mathrm{Rep}\,G} \mathrm{Rep}\,G.$$

For general $M$, the functor $\bar{H}(M)$ is computed by a free resolution because it is right-exact. Using this, one extends the computation to all modules. By definition, $\mathrm{Tor}_{\mathrm{Rep}\,G}^n(N, M)$ for $N \in\in \mathfrak{Mod}((\mathrm{Rep}\,G)^{\mathrm{op}})^{\mathbb{Z}/2}$, $M \in\in \mathfrak{Mod}(\mathrm{Rep}\,G)_{\mathrm{c}}^{\mathbb{Z}/2}$, is the $n$th left derived functor of the functor $N \otimes_{\mathrm{Rep}\,G} \lrcorner$ on $\mathfrak{Mod}(\mathrm{Rep}\,G)_{\mathrm{c}}^{\mathbb{Z}/2}$.

Now Theorem 59 yields the formula for $\mathbb{L}_n H$ provided $H$ takes values in Abelian groups. The same argument works in general, we only need more complicated categories.

Let $\mathfrak{C}^{\mathbb{Z}/2}[\mathrm{Rep}(G)^{\mathrm{op}}]$ be the category of $\mathbb{Z}/2$-graded objects $A$ of $\mathfrak{C}$ together with a ring homomorphism $\mathrm{Rep}(G)^{\mathrm{op}} \to \mathfrak{C}^{\mathbb{Z}/2}(A, A)$. We can extend the definition of $\otimes_{\mathrm{Rep}\,G}$ to get an additive stable bifunctor

$$\otimes_{\mathrm{Rep}\,G}: \mathfrak{C}^{\mathbb{Z}/2}[\mathrm{Rep}(G)^{\mathrm{op}}] \otimes \mathfrak{Mod}(\mathrm{Rep}\,G)^{\mathbb{Z}/2} \to \mathfrak{C}^{\mathbb{Z}/2}.$$

Its derived functors are $\mathrm{Tor}^n_{\mathrm{Rep}\,G}$. As above, we see that this yields the derived functors of $H$.

The case of cohomological functors is similar and left to the reader.    □


If $G$ is a compact group, then the same derived functors appear in the Universal Coefficient Theorem for $\mathrm{KK}^G$ by Jonathan Rosenberg and Claude Schochet ([22]). This is no coincidence, of course. It is explained in [16] how a homological ideal in a triangulated category generates a spectral sequence. This machinery applied to the ideal $\mathfrak{I}_{\ltimes,K}$ yields the spectral sequence of [22].

In order to get the universal functor for the ideal $\mathfrak{I}_{\ltimes}$, we must lift the Rep $G$-module structure on $\mathrm{K}_*(G \ltimes B)$ to $G \ltimes B$. Given any additive category $\mathfrak{C}$, we define a category $\mathfrak{C}[\mathrm{Rep}\,G]$ as in the proof of Theorem 72: its objects are pairs $(A, \mu)$ where $A \in\in \mathfrak{C}$ and $\mu$ is a ring homomorphism $\mathrm{Rep}(G) \to \mathfrak{C}(A, A)$; its morphisms are morphisms in $\mathfrak{C}$ that are compatible with the $\mathrm{Rep}(G)$-module structure in the obvious sense.

A module structure $\mu\colon \mathrm{Rep}(G) \to \mathrm{KK}(A, A)$ for $A \in\in \mathrm{KK}$ is equivalent to a natural family of $\mathrm{Rep}(G)$-module structures in the usual sense on the groups $\mathrm{KK}(D, A)$ for all $D \in\in \mathrm{KK}$: simply define $x \cdot y := \mu(x) \circ y$ for $x \in \mathrm{Rep}(G)$, $y \in \mathrm{KK}(D, A)$ and notice that this recovers the homomorphism $\mu$ for $y = \mathrm{id}_A$.

The crucial property of the universal $\mathfrak{I}_{\ltimes,K}$-exact functor $F_K$ is that it lifts the original functor $\mathrm{K}_*(G \ltimes \_)\colon \mathrm{KK}^G \to \mathfrak{Ab}_c^{\mathbb{Z}/2}$ to a functor

$$F_K\colon \mathrm{KK}^G \to \mathfrak{Mod}(\mathrm{Rep}\,G)_c^{\mathbb{Z}/2} = \mathfrak{Ab}_c^{\mathbb{Z}/2}[\mathrm{Rep}\,G].$$

We need a similar lifting of $G \ltimes \_\colon \mathrm{KK}^G \to \mathrm{KK}$ to $\mathrm{KK}[\mathrm{Rep}\,G]$. This requires a simple special case of exterior products in $\mathrm{KK}^G$. In general, it is not so easy to define exterior products in $\mathrm{KK}^G$ for quantum groups because diagonal actions on C*-algebras are not defined without additional structure. The only case where this is easy is if one of the factors carries the *trivial* coaction. This exterior product operation on the level of C*-algebras also works for Kasparov cycles, that is, we get canonical maps

$$\mathrm{KK}_0^G(A, B) \otimes \mathrm{KK}_0(C, D) \to \mathrm{KK}_0^G(A \otimes C, B \otimes D)$$

for all $A, B \in\in \mathrm{KK}^G$, $C, D \in\in \mathrm{KK}$. Equivalently, $(A, C) \mapsto A \otimes C$ is a bifunctor $\mathrm{KK}^G \otimes \mathrm{KK} \to \mathrm{KK}^G$.

This exterior product construction yields a natural map

$$\varrho_A\colon \mathrm{Rep}(G)^{\mathrm{op}} \cong \mathrm{KK}^G(\tau\mathbb{C}, \tau\mathbb{C}) \to \mathrm{KK}^G(\tau\mathbb{C} \otimes A, \tau\mathbb{C} \otimes A) \cong \mathrm{KK}^G(\tau A, \tau A)$$

for $A \in\in \mathrm{KK}$, whose range commutes with the range of the map

$$\tau \colon \mathrm{KK}(A, A) \to \mathrm{KK}^G(\tau A, \tau A).$$

The ring $\mathrm{KK}^G(\tau A, \tau A)$ acts on $\mathrm{KK}^G(\tau A, B) \cong \mathrm{KK}(A, G \ltimes B)$ on the right by Kasparov product. Hence so does $\mathrm{Rep}(G)^{\mathrm{op}}$ via $\varrho_A$. Thus $\mathrm{KK}(A, G \ltimes B)$ becomes a left $\mathrm{Rep}(G)$-module for all $A \in\in \mathrm{KK}$, $B \in\in \mathrm{KK}^G$. These module structures are natural in the variable $A$ because the images of $\mathrm{KK}^G(\tau \mathbb{C}, \tau \mathbb{C})$ and $\mathrm{KK}(A, A)$ in $\mathrm{KK}^G(\tau A, \tau A)$ commute. Hence they must come from a ring homomorphism

$$\mu_B \colon \mathrm{Rep}(G) \to \mathrm{KK}(G \ltimes B, G \ltimes B).$$

These ring homomorphisms are natural because the $\mathrm{Rep}(G)$-module structures on $\mathrm{KK}(A, G \ltimes B)$ are manifestly natural in $B$. Thus we have lifted $G \ltimes \_$ to a functor

$$F \colon \mathrm{KK}^G \to \mathrm{KK}[\mathrm{Rep}\, G], \qquad B \mapsto (G \ltimes B, \mu_B).$$

It is clear that $\ker F = \mathfrak{I}_\ltimes$. The target category $\mathrm{KK}[\mathrm{Rep}\, G]$ is neither triangulated nor Abelian. To remedy this, we use the Yoneda embedding $\mathbb{Y} \colon \mathrm{KK} \to \mathfrak{Coh}(\mathrm{KK})$ constructed in §2.4. This embedding is fully faithful; so is the resulting functor $\mathrm{KK}[\mathrm{Rep}\, G] \to \mathfrak{Coh}(\mathrm{KK})[\mathrm{Rep}\, G]$.

**Theorem 73.** *The functor* $\mathbb{Y} \circ F \colon \mathrm{KK}^G \to \mathfrak{Coh}(\mathrm{KK})[\mathrm{Rep}\, G]$ *is the universal* $\mathfrak{I}_\ltimes$*-exact functor.*

*Proof.* We omit the proof of this theorem because it is only notationally more difficult than the proof of Theorem 72. $\qquad\square$

The category $\mathfrak{Coh}(\mathrm{KK})[\mathrm{Rep}\, G]$ is not as terrible as it seems. We can usually stay within the more tractable subcategory $\mathrm{KK}[\mathrm{Rep}\, G]$, and many standard techniques of homological algebra like bar resolutions work in this setting. This often allows us to compute derived functors on $\mathfrak{Coh}(\mathrm{KK})[\mathrm{Rep}\, G]$ in more classical terms.

Recall that the bar resolution of a $\mathrm{Rep}\, G$-module $M$ is a natural free resolution

$$\cdots \to (\mathrm{Rep}\, G)^{\otimes n} \otimes M \to (\mathrm{Rep}\, G)^{\otimes n-1} \otimes M \to \cdots \to \mathrm{Rep}\, G \otimes M \to M$$

with certain natural boundary maps. Defining

$$(\mathrm{Rep}\, G)^{\otimes n} \otimes M = \mathbb{Z}[\hat{G}^n] \otimes M := \bigoplus_{x \in \hat{G}^n} M,$$

we can make sense of this in $\mathfrak{C}[\mathrm{Rep}\, G]$ provided $\mathfrak{C}$ has countable direct sums; the $\mathrm{Rep}\, G$-module structures and the boundary maps can also be defined.

If $A \in\in \mathrm{KK}[\mathrm{Rep}\, G]$, then the bar resolution lies in $\mathrm{KK}[\mathrm{Rep}\, G]$ and is a projective resolution of $A$ in the ambient Abelian category $\mathfrak{Coh}(\mathrm{KK})[\mathrm{Rep}\, G]$. Hence we can use it to compute derived functors. For the extension groups, we get

$$\mathrm{Ext}^n_{\mathrm{KK}^G, \mathfrak{I}_\ltimes}(A, B) \cong \mathrm{HH}^n\big(\mathrm{Rep}(G); \mathrm{KK}(G \ltimes A, G \ltimes B)\big);$$

here $\mathrm{HH}^n(R; M)$ denotes the $n$th Hochschild cohomology of a ring $R$ with coefficients in an $R$-bimodule $M$, and $\mathrm{KK}(G \ltimes A, G \ltimes B)$ is a bimodule over $\mathrm{Rep}\, G$ via the Kasparov product on the left and right and the ring homomorphisms

$$\mathrm{Rep}(G) \to \mathrm{KK}(G \ltimes A, G \ltimes A), \qquad \mathrm{Rep}(G) \to \mathrm{KK}(G \ltimes B, G \ltimes B).$$

Similarly, if $H \colon \mathrm{KK}^G \to \mathfrak{Ab}$ commutes with direct sums, then

$$\mathbb{L}_n H(B) \cong \mathrm{HH}_n\big(\mathrm{Rep}(G); H \circ \tau(G \ltimes B)\big),$$

where $H \circ \tau(G \ltimes B)$ carries the following $\mathrm{Rep}\, G$-bimodule structure: the left module structure comes from $\mu_B \colon \mathrm{Rep}(G) \to \mathrm{KK}(G \ltimes B, G \ltimes B)$ and the right one comes from $\varrho_A \colon \mathrm{Rep}(G)^{\mathrm{op}} \to \mathrm{KK}^G(\tau A, \tau A)$ for $A = G \ltimes B$ and the functoriality of $H$. The details are left to the reader.

## 5.1. The Pimsner–Voiculescu exact sequence

The reader may wonder why we have considered the ideal $\mathfrak{I}_\ltimes$, given that the derived functors for $\mathfrak{I}_{\ltimes, \mathrm{K}}$ are so much easier to describe. This is related to the question whether $\mathfrak{I}$-equivalences are invertible. The ideal $\mathfrak{I}_{\ltimes, \mathrm{K}}$ cannot have this property because it already fails for trivial $G$. In contrast, the ideal $\mathfrak{I}_\ltimes$ sometimes has this property. This means that the spectral sequences that we get from $\mathfrak{I}_\ltimes$ may converge for all objects, not just for those in an appropriate bootstrap category. To illustrate this, we explain how the well-known Pimsner–Voiculescu exact sequence fits into our framework.

This exact sequence deals with actions of the group $\mathbb{Z}$; to remain in the framework of §5, we use Baaj–Skandalis duality to turn such actions into actions of the Pontrjagin dual group $\mathbb{T}$. The representation ring of $\mathbb{T}$ is the ring $R := \mathbb{Z}[t, t^{-1}]$ of Laurent polynomials or, equivalently, the group ring of $\mathbb{Z}$. An $R$-bimodule in an Abelian category $\mathfrak{C}$ is an object $M$ of $\mathfrak{C}$ together with two commuting automorphisms $\lambda, \rho \colon M \to M$. The Hochschild homology and cohomology are easy to compute using the free bimodule resolution

$0 \to R^{\otimes 2} \xrightarrow{d} R^{\otimes 2} \to R$, where $d(f) = t \cdot f \cdot t^{-1} - f$. We get

$$\mathrm{HH}_0(R; M) \cong \mathrm{HH}^1(R; M) \cong \mathrm{coker}(\lambda \rho^{-1} - 1),$$
$$\mathrm{HH}_1(R; M) \cong \mathrm{HH}^0(R; M) \cong \ker(\lambda \rho^{-1} - 1),$$

and $\mathrm{HH}_n(R; M) \cong \mathrm{HH}^n(R; M) \cong 0$ for $n \geq 2$. Transporting this kind of resolution to $\mathfrak{C}[R]$, we get that any object of $\mathfrak{C}[R]$ has an $\mathfrak{I}_\ltimes$-projective resolution of length 1. This would fail for $\mathfrak{I}_{\ltimes, K}$ because the category of $R$-modules has a non-trivial $\mathrm{Ext}^2$.

The crucial point is that $\mathfrak{I}_\ltimes$-equivalences are invertible in $\mathrm{KK}^{\mathbb{T}}$. By Baaj–Skandalis duality, this is equivalent to the following statement: if $f \in\in \mathrm{KK}^{\mathbb{Z}}(A, B)$ becomes invertible in $\mathrm{KK}$, then it is already invertible in $\mathrm{KK}^{\mathbb{Z}}$. We do not want to discuss here how to prove this. Taking this for granted, we can now apply Theorem 66 to all objects of $\mathrm{KK}^{\mathbb{T}}$.

We write down the resulting exact sequences for $\mathrm{KK}^{\mathbb{Z}}(A, B)$ for $A, B \in\in \mathrm{KK}^{\mathbb{Z}}$ because this equivalent setting is more familiar. The actions of $\mathbb{Z}$ on $A$ and $B$ provide two actions of $\mathbb{Z}$ on $\mathrm{KK}_*(A, B)$. We let $t_A, t_B \colon \mathrm{KK}_*(A, B) \to \mathrm{KK}_*(A, B)$ be the actions of the generators. Theorem 66 yields an exact sequence

$$\mathrm{coker}\big(t_A t_B^{-1} - 1|_{\mathrm{KK}_{*+1}(A,B)}\big) \rightarrowtail \mathrm{KK}_*^{\mathbb{Z}}(A, B) \twoheadrightarrow \ker\big(t_A t_B^{-1} - 1|_{\mathrm{KK}_*(A,B)}\big).$$

This is equivalent to a long exact sequence

$$
\begin{array}{ccccc}
\mathrm{KK}_1(A, B) & \longrightarrow & \mathrm{KK}_0^{\mathbb{Z}}(A, B) & \longrightarrow & \mathrm{KK}_0(A, B) \\
{\scriptstyle t_A t_B^{-1} - 1}\big\uparrow & & & & \big\downarrow{\scriptstyle t_A t_B^{-1} - 1} \\
\mathrm{KK}_1(A, B) & \longleftarrow & \mathrm{KK}_1^{\mathbb{Z}}(A, B) & \longleftarrow & \mathrm{KK}_0(A, B).
\end{array}
$$

Similar manipulations yield the Pimsner–Voiculescu exact sequence for the functor $A \mapsto \mathrm{K}_*(\mathbb{Z} \ltimes A)$ and more general functors defined on $\mathrm{KK}^{\mathbb{Z}}$.

# References


[1] Javad Asadollahi and Shokrollah Salarian, *Gorenstein objects in triangulated categories*, J. Algebra **281** (2004), no. 1, 264–286. MR **2091971**

[2] Saad Baaj and Georges Skandalis, *$C^*$-algèbres de Hopf et théorie de Kasparov équivariante*, $K$-Theory **2** (1989), no. 6, 683–721 (French, with English summary). MR **1010978**

[3] Alexander A. Beĭlinson, Joseph Bernstein, and Pierre Deligne, *Faisceaux pervers*, Analysis and topology on singular spaces, I (Luminy, 1981), Astérisque, vol. 100, Soc. Math. France, Paris, 1982, pp. 5–171 (French). MR **751966**

[4] Apostolos Beligiannis, *Relative homological algebra and purity in triangulated categories*, J. Algebra **227** (2000), no. 1, 268–361. MR **1754234**

[5] Bruce Blackadar, *K-theory for operator algebras*, 2nd ed., Mathematical Sciences Research Institute Publications, vol. 5, Cambridge University Press, Cambridge, 1998. MR **1656031**

[6] Alexander Bonkat, *Bivariante K-Theorie für Kategorien projektiver Systeme von $C^*$-Algebren*, Ph.D. Thesis, Westf. Wilhelms-Universität Münster, 2002 (German). electronically available at the Deutsche Nationalbibliothek at http://deposit.ddb.de/cgi-bin/dokserv?idn=967387191.

[7] Hans-Berndt Brinkmann, *Relative homological algebra and the Adams spectral sequence*, Arch. Math. (Basel) **19** (1968), 137–155. MR **0230788**

[8] J. Daniel Christensen, *Ideals in triangulated categories: phantoms, ghosts and skeleta*, Adv. Math. **136** (1998), no. 2, 284–339. MR **1626856**

[9] Samuel Eilenberg and John Coleman Moore, *Foundations of relative homological algebra*, Mem. Amer. Math. Soc. No. **55** (1965), 39. MR **0178036**

[10] Peter Freyd, *Representations in abelian categories*, (La Jolla, Calif., 1965), Proc. Conf. Categorical Algebra, Springer, New York, 1966, pp. 95–120. MR **0209333**

[11] Pierre Julg, *K-Théorie équivariante et produits croisés*, C. R. Acad. Sci. Paris Sér. I Math. **292** (1981), no. 13, 629–632 (French, with English summary)MR **625361**

[12] Bernhard Keller, *Derived categories and their uses*, Handbook of algebra, Vol. 1, North-Holland, Amsterdam, 1996, pp. 671–701. MR **1421815**

[13] Saunders MacLane, *Categories for the working mathematician*, Springer-Verlag, New York, 1971. Graduate Texts in Mathematics, Vol. 5. MR **0354798**

[14] Ralf Meyer, *Local and analytic cyclic homology*, EMS Tracts in Mathematics, vol. 3, European Mathematical Society (EMS), Zürich, 2007. MR **2337277**

[15] ______, *Categorical aspects of bivariant K-theory*, K-theory and Noncommutative Geometry (Valladolid, Spain, 2006), EMS Ser. Congr. Rep., Europ. Math. Soc. Publ. House, Zürich, 2008, pp. 1–39.

[16] ______, *Homological algebra in bivariant K-theory and other triangulated categories. II*, Tbilisi Mathematical Journal **1** (2008), 165–210.

[17] Ralf Meyer and Ryszard Nest, *The Baum–Connes conjecture via localisation of categories*, Topology **45** (2006), no. 2, 209–259. MR **2193334**

[18] ______, *An analogue of the Baum–Connes isomorphism for coactions of compact groups*, Math. Scand. **100** (2007), no. 2, 301–316. MR **2339371**

[19] Amnon Neeman, *Triangulated categories*, Annals of Mathematics Studies, vol. 148, Princeton University Press, Princeton, NJ, 2001. MR **1812507**

[20] Radu Popescu, *Coactions of Hopf-$C^*$-algebras and equivariant E-theory* (2004), eprint. arXiv: math.KT/0410023.

[21] Michael Puschnigg, *Diffeotopy functors of ind-algebras and local cyclic cohomology*, Doc. Math. **8** (2003), 143–245 (electronic). MR **2029166**

[22] Jonathan Rosenberg and Claude Schochet, *The Künneth theorem and the universal coefficient theorem for equivariant K-theory and KK-theory*, Mem. Amer. Math. Soc. **62** (1986), no. 348. MR **0849938**

[23] ______, *The Künneth theorem and the universal coefficient theorem for Kasparov's generalized K-functor*, Duke Math. J. **55** (1987), no. 2, 431–474. MR **894590**

[24] Ross Street, *Homotopy classification of filtered complexes*, J. Austral. Math. Soc. **15** (1973), 298–318. MR **0340380**

[25] Jean-Louis Verdier, *Des catégories dérivées des catégories abéliennes*, Astérisque (1996), no. 239, xii+253 pp. (1997) (French, with French summary). With a preface by Luc Illusie; Edited and with a note by Georges Maltsiniotis. MR **1453167**

[26] Roland Vergnioux, KK-*théorie équivariante et opérateurs de Julg-Valette pour les groupes quantiques*, Ph.D. Thesis, Université Paris 7 – Denis Diderot, 2002. electronically available at http://www.math.jussieu.fr/theses/2002/vergnioux/these.dvi.

*Ralf Meyer*
*Mathematisches Institut and Courant*
*Centre "Higher order structures"*
*Georg-August-Universität Göttingen*
*Bunsenstraße 3–5*
*37073 Göttingen*
*Germany*
`meyerr@member.ams.org`

*Ryszard Nest*
*Københavns Universitets Institut for*
*Matematiske Fag*
*Universitetsparken 5*
*2100 København*
*Denmark*
`rnest@math.ku.dk`

# Derived categories and Grothendieck duality

AMNON NEEMAN

*Dedicated to Alexander Grothendieck, who started this subject, on his 80th birthday*

ABSTRACT. We study dualizing complexes. The unusual feature is that we do not assume them to have bounded injective resolutions; we prove that the theory works just fine with no boundedness hypothesis. In the process we prove a number of new results about Grothendieck duality; one of the more striking is that, under relatively mild hypotheses on $f : X \longrightarrow Y$, the functor $f^! : \mathbf{D}(\mathrm{Qcoh}/Y) \longrightarrow \mathbf{D}(\mathrm{Qcoh}/X)$ takes pseudocoherent complexes to pseudocoherent complexes. The biggest innovation in our approach is that we systematically employ products in the category $\mathbf{D}(\mathrm{Qcoh}/X)$; the older treatments never ventured beyond coproducts.

In an appendix we include a proof that, if $\mathcal{T}$ is a stable homotopy category in the sense of [19] in which the compact objects coincide with the strongly dualizable objects, then the compact objects can also be characterized as those objects such that tensoring with them respects products.

## Contents



*Key words and phrases.* dualizing complex, Grothendieck duality.
The research was conducted while the author was visiting the CRM in Barcelona for six months; many thanks to the CRM for its hospitality and congenial working environment. The work was partly supported by a sabbatical grant from the Spanish Ministry of Education (grant number SAB2006-0135), as well as by the Australian Research Council.

# 0. Introduction

The bulk of this survey will give a modern account of dualizing complexes. Let $X$ be a noetherian, separated scheme. Let $\mathbf{D}^b(\mathrm{Coh}/X)$ be the bounded derived category of coherent sheaves on $X$. We define a dualizing complex to be an object $\mathfrak{I} \in \mathbf{D}^b(\mathrm{Coh}/X)$ for which the functor $\mathbf{R}\mathcal{H}om(-,\mathfrak{I})$ yields an equivalence

$$\mathbf{R}\mathcal{H}om(-,\mathfrak{I}) \ : \ \mathbf{D}^b(\mathrm{Coh}/X)^{\mathrm{op}} \longrightarrow \mathbf{D}^b(\mathrm{Coh}/X) \, .$$

Note that we make no assumption that $\mathfrak{I}$ have a bounded injective resolution;[1] this is supposed to be a contemporary treatment of the subject, and by now we should know how to handle unbounded complexes. The thrust is that the theory works just fine. Let me now sketch the main new results which can be found in the article.

**Facts 0.1.** We prove, with no simplifying hypotheses on $\mathfrak{I}$:

(i) A dualizing complex $\mathfrak{I}$, if it exists, is unique up to the obvious modifications. That is, if $\mathfrak{I}$ and $\mathfrak{I}'$ are two dualizing complexes, then there exists a complex $\mathcal{L} \in \mathbf{D}^b(\mathrm{Coh}/X)$ with $\mathfrak{I}' \cong \mathcal{L}^{\mathbf{L}} \otimes \mathfrak{I}$. Furthermore, on any connected component of $X$ the complex $\mathcal{L}$ is just a suspension of a line bundle. See Lemma 3.9 for the proof.

(ii) An object $\mathfrak{I} \in \mathbf{D}^b(\mathrm{Coh}/X)$ is a dualizing complex if and only if it satisfies the two conditions
   (a) The natural map $\mathcal{O}_X \longrightarrow \mathbf{R}\mathcal{H}om(\mathfrak{I},\mathfrak{I})$ is an isomorphism.
   (b) For every object $\mathcal{E} \in \mathbf{D}^b(\mathrm{Coh}/X)$ we have $\mathbf{R}\mathcal{H}om(\mathcal{E},\mathfrak{I}) \in \mathbf{D}^b(\mathrm{Coh}/X)$.
   For a proof see Proposition 3.6.

The traditional reason why people cared about dualizing complexes was the way they behave in the relative situation, where two schemes are involved; we will discuss this a little more fully in §6. For now we confine ourselves to stating our relevant new results. Suppose therefore that we are given a morphism $f : X \longrightarrow Y$, where $X$ and $Y$ are noetherian, separated schemes. Let $\mathbf{D}(\mathrm{Qcoh}/X)$ and $\mathbf{D}(\mathrm{Qcoh}/Y)$ be the unbounded derived categories of quasicoherent sheaves on $X$ and $Y$ respectively. There is a pushforward functor $\mathbf{R}f_* : \mathbf{D}(\mathrm{Qcoh}/X) \longrightarrow \mathbf{D}(\mathrm{Qcoh}/Y)$, and it has a left adjoint $\mathbf{L}f^*$ and a right adjoint $f^!$. The next results are

---

[1]  The usual definition of dualizing complexes imposes the hypothesis that $\mathfrak{I}$ have finite injective dimension. There is a theorem of Gabber which says that the objects we call "dualizing complexes" agree with what, in the classical literature, would go under the name "strongly pointwise dualizing complexes"; see [6, Lemma 3.1.5]. In the classical language, one of the things we do in this article is show that these complexes have all the good global properties of the traditional dualizing complexes.

**Facts 0.2.** Let $\mathfrak{I}$ be a dualizing complex in $\mathbf{D}^b(\mathrm{Coh}/Y) \subset \mathbf{D}(\mathrm{Qcoh}/Y)$. Then the following is true:

(i) If $f$ is an open immersion, then $\mathbf{L}f^*\mathfrak{I}$ is a dualizing complex in $\mathbf{D}^b(\mathrm{Coh}/X) \subset \mathbf{D}(\mathrm{Qcoh}/X)$. The proof is in Remark 3.12.

(ii) Suppose $f : X \longrightarrow Y$ satisfies the following two hypotheses:
   (a) $\mathbf{R}f_*$ takes $\mathbf{D}^b(\mathrm{Coh}/X)$ to $\mathbf{D}^b(\mathrm{Coh}/Y)$.
   (b) $f^!$ takes $\mathbf{D}^b(\mathrm{Coh}/Y)$ to $\mathbf{D}^+_{\mathrm{Coh}}(\mathrm{Qcoh}/X)$.
   Then $f^!\mathfrak{I}$ is a dualizing complex on $X$. The proof may be found in Theorem 3.15.

Fact 0.2(ii) is a little unsatisfactory, particularly because of the hypothesis (b) which involves the mysterious functor $f^!$. The next fact helps.

**Facts 0.3.** Hypotheses (a) and (b) of Fact 0.2(ii) hold in either of the following two situations:

(i) The morphism $f : X \longrightarrow Y$ is finite. Hypothesis (a) holds obviously, and the proof that hypothesis (b) is satisfied may be found in Lemma 3.19.

(ii) The functor $\mathbf{R}f_* : \mathbf{D}(\mathrm{Qcoh}/X) \longrightarrow \mathbf{D}(\mathrm{Qcoh}/Y)$ takes perfect complexes to perfect complexes, and $X$ satisfies the technical condition $(*)$ of Conjecture 4.16. Hypothesis (a) holds by [46, Corollary 4.3.2], and for hypothesis (b) see Corollary 0.7.

As always if $X \xrightarrow{f} Y \xrightarrow{g} Z$ are two morphisms, and if both $f^!$ and $g^!$ take dualizing complexes to dualizing complexes, then so does the composite. We can freely combine (i) and (ii) to produce morphisms $h$ for which $h^!$ respects dualizing complexes.

**Remark 0.4.** I do not, at the present time, fully understand the technical condition $(*)$ in Fact 0.3(ii). Conjecture 4.16 articulates the hope that $(*)$ holds for all $X$. We rephrase this slightly: assuming that the technical Conjecture 4.16 holds for every $X$, then $f^!$ takes dualizing complexes to dualizing complexes as long as $\mathbf{R}f_* : \mathbf{D}(\mathrm{Qcoh}/X) \longrightarrow \mathbf{D}(\mathrm{Qcoh}/Y)$ takes perfect complexes to perfect complexes.

Given that at present we do not have Conjecture 4.16 in this generality, in addition to the preservation of perfect complexes we currently need to assume that Conjecture 4.16 holds for $X$. In the remainder of the introduction I will discuss the status of Conjecture 4.16; there are many $X$s for which it is known.

Until now all our schemes were assumed noetherian. Dualizing complexes are traditionally about producing equivalences $\mathbf{D}^b(\mathrm{Coh}/X)^{\mathrm{op}} \cong \mathbf{D}^b(\mathrm{Coh}/X)$, and the category $\mathbf{D}^b(\mathrm{Coh}/X)$ does not obviously make sense unless $X$ is noetherian, or at least coherent. But it turns out that Conjecture 4.16, as well as the technical condition (∗) in Fact 0.3(ii), make sense much more generally. For the next result we only assume that our schemes are quasicompact and separated. In the article we will prove the following assertion

**Facts 0.5.** Let $f : X \longrightarrow Y$ be a morphism of quasicompact, separated schemes. Assume that Conjecture 4.16 holds for $X$, and that $\mathbf{R}f_*$ takes perfect complexes to perfect complexes. Then the complex $f^!\mathcal{O}_Y$ is pseudocoherent; this means that, for any open affine subset $U \subset X$, the restriction of $f^!\mathcal{O}_Y$ to $U = \mathrm{Spec}(R)$ is quasi-isomorphic to the sheafification of a bounded above chain complex of finitely generated, projective $R$–modules.

The proof is in Lemma 4.21; see also Remark 4.8.

**Remark 0.6.** By [51, Theorem 5.4] we know that $f^!\mathcal{S} \cong f^!\mathcal{O}_Y \overset{\mathbf{L}}{\otimes} \mathbf{L}f^*\mathcal{S}$. If $\mathcal{S}$ is pseudocoherent then so is $\mathbf{L}f^*\mathcal{S}$ (obviously), and the formula, coupled with the pseudocoherence of $f^!\mathcal{O}_Y$ of Fact 0.5, informs us that $f^!\mathcal{S}$ is also pseudocoherent.

Assume now that $X$ and $Y$ are noetherian and $\mathcal{S}$ belongs to $\mathbf{D}^b(\mathrm{Coh}/Y)$. Then $f^!\mathcal{S}$ is pseudocoherent, which for noetherian schemes simply means it belongs to $\mathbf{D}^-(\mathrm{Coh}/X)$. The fact that its cohomology is bounded below is easy (see Remark 3.14) and we conclude the following:

**Corollary 0.7.** *Assume $f : X \longrightarrow Y$ is a morphism of noetherian, separated schemes. Suppose further that $\mathbf{R}f_*$ takes perfect complexes to perfect complexes, and that Conjecture 4.16 holds for $X$. Then $f^!$ takes $\mathbf{D}^b(\mathrm{Coh}/Y)$ to $\mathbf{D}^b(\mathrm{Coh}/X)$.*

In Remark 0.6 we sketched the argument leading from the more general Fact 0.5 to Corollary 0.7, valid in the case of noetherian schemes. The reader should note that Fact 0.3(ii) is immediate from Corollary 0.7. Corollary 0.7 proves more than we need; in Fact 0.3(ii) we only asserted that $f^!$ takes $\mathbf{D}^b(\mathrm{Coh}/Y)$ to $\mathbf{D}^+_{\mathrm{Coh}}(\mathrm{Qcoh}/X)$.

**Facts 0.8.** It remains to review what we know about Conjecture 4.16. The current state of knowledge is that $X$ satisfies the conjecture if

(i) $X$ is noetherian, finite dimensional, and smooth over a finite dimensional, noetherian regular ring $R$. See Theorem 5.13 for the proof.
(ii) $X$ is a locally closed subscheme of $Y$, and Conjecture 4.16 is true for $Y$.
(iii) There is an affine morphism $X \longrightarrow Y$, and Conjecture 4.16 is true for $Y$. Facts (ii) and (iii) both follow from Remark 5.12.

We leave it to the imagination of the reader to combine these to produce many $X$s for which the conjecture is true.

This finishes the main new results we will present; it remains to give a brief sketch of the structure of the article. We begin with a fairly extensive historical review of Grothendieck duality. Then we cover the basic properties of $\mathbf{R}\mathcal{H}om(-, -)$; nothing here is especially new, but our approach is to define $\mathbf{R}\mathcal{H}om(\mathcal{E}, -)$ as the right adjoint of the functor $- \overset{\mathbf{L}}{\otimes} \mathcal{E}$, by appealing to Brown representability. It is therefore a little interesting to see that the theory can be developed quickly and painlessly; this approach is relatively new, it first appeared in Murfet's thesis [48, Appendix C]. Then the rest of the article is devoted to proving the claims we have made in the Introduction. The one comment that might be helpful is that our approach hinges on systematically using *products* in the categories $\mathbf{D}(\mathrm{Qcoh}/X)$. These categories are compactly generated, and therefore have products; see [52, Proposition 8.4.6]. Any right adjoint will preserve these products, and the functors $\mathbf{R}\mathcal{H}om(\mathcal{E}, -)$, $\mathbf{R}f_*$ and $f^!$ are all right adjoints. The idea is to exploit this.

If $X$ is arbitrary then products in the category $\mathbf{D}(\mathrm{Qcoh}/X)$ are disgusting; we only really understand them when $X$ is affine. As we will see, this is often enough. One can frequently reduce oneself to the affine case.

**Acknowledgements.** The author would like to thank Avramov, Iyengar, Joyal, Krause, Lipman, Murfet, Nayak, Street and Van den Bergh for their valuable suggestions towards improving an earlier draft. Lipman sent me a long list of helpful comments, and Avramov and Iyengar even went as far as providing a simplified, streamlined proof for Lemma 2.10; the current manuscript contains their proof.

# 1. Historical overview

Triangulated categories came into being in the early 1960s; they arose independently, and more or less simultaneously, in the work of Puppe [55] and of Verdier [67][2]. Puppe's interest in the subject came from homotopy theory; we will say more, much later in this historical account, about the role homotopy theory played in the early development of the subject. Our survey of the early history will follow in the footsteps of Verdier; we will explain the problem that inspired him, and the progress that followed. Verdier came to the subject

---

[2] The reader should note that Verdier's thesis was only published posthumously, many years after it was written; the publication date of [67] is misleading.

with a specific goal in mind: he wanted to develop the necessary homological machinery to state and prove Grothendieck duality.

Let us permit ourselves a digression, jumping ahead many years in time: since the early days much has happened, and derived categories, or more generally triangulated categories, have successfully invaded branches of mathematics as remote as mathematical physics [9, 18, 30, 31] and even $\mathbb{C}^*$–algebras [47]; the other articles in this volume are testament to the astounding success of the field. My aims are more modest; in this survey we will keep our focus on Grothendieck duality. What we will illustrate is the way in which the gradual advances in our understanding of the foundations of derived categories, over many years, have translated into improvements in the results on Grothendieck duality.

Let us return to the dawn of the field; in the beginning there was the Serre duality theorem [60]. In its original form the theorem states the following.

**Theorem 1.1. Serre** [60]. *Let $X$ be a connected, $n$–dimensional compact complex manifold, and let $\mathcal{V}$ be a holomorphic vector bundle on $X$. Then the dual of the vector space $H^i(X, \mathcal{V})$ is isomorphic to $H^{n-i}\big(X, \mathcal{H}om(\mathcal{V}, \Omega^n)\big)$.*

**Remark 1.2.** Perhaps we should explain the notation. The line bundle $\Omega^n$ is the bundle of holomorphic $n$–forms on $X$, and $\mathcal{H}om(\mathcal{V}, \Omega^n)$ is the vector bundle of whose sections, on an open set $U \subset X$, are the holomorphic maps of vector bundles $\mathcal{V}|_U \longrightarrow \Omega^n|_U$. There is an obvious pairing

$$H^i(X, \mathcal{V}) \otimes H^{n-i}\big(X, \mathcal{H}om(\mathcal{V}, \Omega^n)\big) \xrightarrow{\ \varphi\ } H^n(\Omega^n),$$

and Serre's theorem asserts two things:

(i) The vector space $H^n(\Omega^n)$ is one dimensional.
(ii) The pairing $\varphi$ is perfect; it gives a natural identification of $H^{n-i}\big(X, \mathcal{H}om(\mathcal{V}, \Omega^n)\big)$ with the dual of $H^i(X, \mathcal{V})$.

The question that interested people at the time was whether there is a relative version. Suppose we are given a holomorphic map of complex manifolds $f : X \longrightarrow Y$. Is there a reasonable general theorem which, in the special case where $Y = \{*\}$ is the one-point space, comes down to Theorem 1.1? The answer turns out to be Yes, at least in the algebro-geometric framework. In keeping with tradition I will now switch from the complex analytic setting to the world of algebraic geometry; I do this because complex analysis has technical difficulties which I do not want to address.

Let $X$ be a smooth, $n$–dimensional projective variety over a field $k$, and let $\mathcal{V}$ be an algebraic vector bundle on $X$. Serre duality, in its algebraic variety context, gives a natural isomorphism

$$\text{Hom}\big(H^i(X, \mathcal{V}), \, k\big) \longrightarrow H^{n-i}\big(X, \, \mathcal{H}om(\mathcal{V}, \Omega^n)\big).$$

To make life a little more exciting let us choose $W$, a finite dimensional vector space over the field $k$. Clearly we also have a natural isomorphism

$$\text{Hom}\big(H^i(X, \mathcal{V}), \, W\big) \longrightarrow H^{n-i}\big(X, \, \mathcal{H}om(\mathcal{V}, \, W \otimes_k \Omega^n)\big)$$
$$\|$$
$$\text{Ext}^{n-i}(\mathcal{V}, \, W \otimes_k \Omega^n).$$

Now the vector space $W$ can be thought of as a vector bundle over the one-point space $Y = \{*\}$, and, if you squint hard enough, the isomorphism above begins to look like an adjunction. We have a functor $\mathbf{R}f_*$, which takes a vector bundle $\mathcal{V}$ on $X$ to a string of vector bundles on $Y$, namely the $H^i(X, \mathcal{V})$. And we have the functor $f^!$ taking the vector bundle $W$ on $Y$ to $W \otimes_k \Omega^n$. If we are willing to treat the Ext's as Hom's and to disregard the confusing indices $i$ and $(n-i)$, it looks like we have an isomorphism

$$\text{Hom}(\mathbf{R}f_*\mathcal{V}, W) \longrightarrow \text{Hom}(\mathcal{V}, f^!W).$$

The problem was to make this intuition precise. Before all else, it would be necessary to say exactly what $\mathcal{V}$s are allowed as inputs for the functor $\mathbf{R}f_*$, and what $W$s as inputs for the functor $f^!$. We need to come up with two categories. One, which we will provisionally label $\mathbb{D}(X)$, should consist of some sort of sheaves over $X$. The other, for which our tentative name will be $\mathbb{D}(Y)$, will be made up of something like sheaves over $Y$. And we want functors $\mathbf{R}f_* : \mathbb{D}(X) \longrightarrow \mathbb{D}(Y)$ and $f^! : \mathbb{D}(Y) \longrightarrow \mathbb{D}(X)$, with $f^!$ right adjoint to $\mathbf{R}f_*$. And finally we want all of this to generalize Serre duality. The objects of the category $\mathbb{D}(X)$ should include the vector bundles on $X$, and $\mathbf{R}f_*\mathcal{V}$ should be something which encapsulates the information in the string of vector spaces $\big\{H^i(X, \mathcal{V}), \, 0 \le i \le n\big\}$.

In his 1958 talk at the Edinburgh ICM (see [13]) Grothendieck announced that he had a solution, but also that the homological algebra language necessary to state it did not yet exist. Verdier's thesis project was to develop the framework. Derived categories were born in that thesis. The idea that worked, as we now know, was to let $\mathbb{D}(X)$ and $\mathbb{D}(Y)$ be derived categories; the objects are chain complexes of sheaves on $X$ and $Y$ respectively, and the morphisms are the chain maps, with the homology isomorphisms formally inverted.

**Remark 1.3.** Let us remind the reader. If $f : X \longrightarrow Y$ is a morphism of schemes, then there is always an induced map $f_*$, taking sheaves on $X$ to sheaves on $Y$. It extends to complexes of sheaves: given a complex $\mathcal{V}$ of sheaves on $X$, we can form a complex $f_*\mathcal{V}$ of sheaves on $Y$. The complex $\mathbf{R}f_*\mathcal{V}$ is obtained by first replacing $\mathcal{V}$ by an injective resolution, and then applying $f_*$. It is, on the face of it, quite strange that this functor should have a right adjoint. The reason this is a little bizarre is that the functor $f_*$ is left exact, but decidedly *not* right exact. We would expect right adjoints to exist only for right exact functors, and therefore $f_*$ most certainly cannot have a right adjoint. Only at the level of the derived categories, after replacing $f_*$ by $\mathbf{R}f_*$, do we have a hope of finding an $f^!$. The passage, from the abelian category of chain complexes to its derived category, is very brutal; it destroys exactness. We will return to this point in Remarks 1.6 and 1.7. For now we note only that the brutality is vital, without it Grothendieck duality wouldn't stand a chance.

**Remark 1.4.** Right from the start people disliked derived categories. Derived categories were quite unlike the more familiar objects of homological algebra, their behaviour was poorly understood, and people felt uncomfortable using them. The consensus was that there had to be a better, more natural substitute for them. People felt that derived categories couldn't possibly be the real answer, they couldn't be the right framework for all these theorems. To this day the attitude persists; there are many people who are still trying, today, to find a more natural foundation for this branch of homological algebra. A huge amount of exciting work has come out in the last few years. The potential replacements for derived categories include DG–categories, $\mathcal{A}_\infty$–categories, stable model categories, Segal spaces, quasicategories and triangulated derivators.

Because people never liked the formalism, derived categories were slow to catch on. Now, almost fifty years later, we can point to the impressive theorems that have been proved using them, and to the extent to which they have come to permeate far-flung branches of mathematics. As I have already said, the other chapters in this book contain a compelling case for their success. At this point we can safely say that, for a theory that everyone badmouthed from the very outset, derived categories have come a long way.

In Remark 1.4 we mentioned that people have always had a distaste for derived categories. It might help if we explain one of the features that no one likes. We begin with

**Lemma 1.5.** *Let $\mathcal{T}$ be a triangulated category. Then any epimorphism $f :$ $X \longrightarrow Y$ in $\mathcal{T}$ is split.*

*Proof.* Complete $f$ to a distinguished triangle

$$X \xrightarrow{\ f\ } Y \xrightarrow{\ g\ } Z \xrightarrow{\ h\ } \Sigma X.$$

We know that the composite $gf : X \longrightarrow Z$ vanishes, and that $f$ is an epimorphism. Hence $g$ must vanish. From [52, Corollary 1.2.7] it follows that the triangle is isomorphic to

$$Y \oplus \Sigma^{-1}Z \xrightarrow{(1,0)} Y \xrightarrow{0} Z \longrightarrow \Sigma Y \oplus Z,$$

and hence $f : X \longrightarrow Y$ must be a split epimorphism. $\qquad\square$

**Remark 1.6.** Suppose we have a morphism $f : X \longrightarrow Y$, and suppose it has a cokernel $g : Y \longrightarrow Q$. We observe first that the map $g$ must be an epimorphism. Lemma 1.5 guarantees that it is a split epimorphism: $Y$ is isomorphic to $I \oplus Q$, where I is the (split) kernel of $g$. The map $f : X \longrightarrow Y$ therefore identifies as a composite

$$X \xrightarrow{\alpha} I \longrightarrow I \oplus Q.$$

The fact that the projection $I \oplus Q \longrightarrow Q$ is the cokernel of $f$ guarantees that $\alpha : X \longrightarrow I$ must be an epimorphism. Using Lemma 1.5 again, we conclude that $\alpha$ must split. We have shown that, if a morphism $f : X \longrightarrow Y$ has a cokernel, then it must factor as

$$X \xrightarrow{\alpha} I \xrightarrow{\beta} Y,$$

with $\alpha$ a split epimorphism and $\beta$ a split monomorphism. In homological algebra people had spent half a century taking kernels and cokernels of maps, and suddenly they were faced with the world of triangulated categories, where such constructions are worthless. No wonder they were unhappy.

**Remark 1.7.** Let us look at a slightly fancier version of this, where we consider other possible colimits in triangulated categories. It is easy to see, by example, that there is nothing to prevent the existence of coproducts; there are many examples of triangulated categories with coproducts, for instance the unbounded derived category of modules over a ring $R$. We say that such categories satisfy [TR5].

Now let $\mathcal{T}$ be a [TR5] triangulated category. If $F : S \longrightarrow \mathcal{T}$ is a functor from a small category $S$ to $\mathcal{T}$, we can wonder whether it has a colimit. Suppose such a colimit exists. Now observe that the natural map

$$\coprod_{s \in S} F(s) \longrightarrow \underrightarrow{\operatorname{colim}} F$$

must be an epimorphism. Lemma 1.5 guarantees that this epimorphism has to split. We conclude that, aside from coproducts, there can be no interesting colimits in $\mathcal{T}$. The special case of cokernels is only a baby example of a much more serious problem.

In Remark 1.4 we mentioned that derived categories were not exactly a welcome introduction to homological algebra, while in Lemma 1.5 and Remarks 1.6 and 1.7 we explained one of the features that people have found undesirable. The Almighty Lord, in His infinite grace and wisdom, has not seen fit to bestow upon me the gift of clairvoyant, prophetic powers; it is not for me to predict if someone really will succeed in finding a superior replacement, whether it will be one of the contenders currently being promoted, and if so which one. Only time will tell.

Back to Grothendieck duality. We have recounted how derived categories were born, to establish a context in which Grothendieck's duality theorem could be stated and proved. The proof appeared in Hartshorne [15]; the book dates back to 1966. Why am I wasting the reader's time with a 40-year-old theorem? Surely everything about it must be completely understood by now. The volume in front of you is meant to deal with current and future mathematics, not with ancient history.

Well, Hartshorne's book may have appeared in 1966, but back then derived categories were barely out of diapers, their coordination was less than perfect, and they were still throwing temper tantrums whenever anyone tried to take them into new territory. They were very new, poorly understood, and the technical apparatus created for dealing with them was crude and clumsy. We have come a long way since then, and probably have an equally long road ahead of us. There is little doubt that, forty years from now, people looking back at the work we do today will find it laughably awkward and roundabout. We cannot escape the fact that derived categories are work-in-progress, and many tools are presumably yet to be conceived. This being the case we are still learning new facts about Grothendieck duality. The main object of this article is to give an updated account of dualizing complexes.

This is true, most of the article will be modern, but it would be wrong not to mention the valuable contributions that people have made over the decades. In the early years there was a great deal of interest in the subject; the theorem was new and hot and mathematicians, like the rest of humanity, throng around the newest hot topic on the block. Work from that period includes (of course) Hartshorne [15], but also Deligne [7, 8], Grothendieck [14], Illusie [23, 24, 25, 26], Ramis, Ruget and Verdier [56] and Verdier [64, 65, 66].

After the early explosion of interest the subject had a long hiatus. There was a period of almost two decades when only a small group of committed specialists kept the flame burning: results form this period include El Zein [11], Grivel [12], Hopkins [16], Hopkins-Lipman [17], Hübl-Sastry [22], Hübl-Kunz [20, 21], Kunz [36, 37], Kunz-Waldi [39], Kiehl [34], Lipman [41, 42], Van den Bergh [63] and Yekutieli [68, 69]. But it was only in the late 1980's that the foundations of derived categories began undergoing a major change, with a concommitant improvement in our understanding of Grothendieck duality.

In the early days, people working on derived categories preferred to consider their bounded incarnations. It was customary to work with $\mathbf{D}^+(\mathcal{A})$, $\mathbf{D}^-(\mathcal{A})$ or $\mathbf{D}^b(\mathcal{A})$; we remind the reader that these are, respectively, the bounded below, bounded above or bounded derived categories. Depending on the problem at hand people were quite willing to switch around among the three bounded versions, but the one derived category that was only rarely touched was the unbounded derived category $\mathbf{D}(\mathcal{A})$. There was good reason for this: except in very special cases it was not understood how to form projective or injective resolutions of unbounded complexes, and hence there was no clear machinery for producing derived functors. Perhaps the first article to alter matters was Spaltenstein [61]. This 1988 paper is very methodical; it carefully outlines how to go about this, explaining in some detail the various resolutions one can form, and discussing the situations to which they are best suited. The terminology we use today is still the one Spaltenstein introduced, for example when we speak about $K$–injective resolutions.

The second shift came a little later, with [4, 50, 51]. It began as a pure accident; I happened to be visiting Bielefeld, as a fellow of the Humboldt Stiftung, in the academic year 1989–90. My office was just down the corridor from that of Marcel Bökstedt, who is a wonderful mathematician, a clear thinker and an excellent expositor, a man with the gift of explaining the key idea in what would otherwise be a foggy, abstruse formalism. I enjoyed talking to him and learned a great deal from the interaction, and in particular I learned from Bökstedt what little homotopy theory I know. As he was explaining this to me I was struck by the parallel with the mathematics I knew, the mathematics of algebraic geometry. Our joint article [4] arose out of these conversations; it amounts to my translation, to the familiar context of algebraic geometry, of the homotopy theory that Bökstedt was teaching me.

Maybe we should elaborate a little. In Remarks 1.6 and 1.7 we noted that there are few interesting colimits in derived categories. The homotopy theorists have never let this bother them; they freely lift problems from triangulated categories to model categories. In the model categories colimit constructions work just fine, and occasionally one can show that the output of these constructions is independent of the lifting; one has produced something well-defined in the triangulated category. It is often not difficult to reformulate the constructions in a way that dispenses with the lifting to models. Anyway, all we really want to stress is that the homotopy theorists walked fearlessly through terrain which the derived categorists carefully skirted. I learned from Bökstedt that there were paths in this unfamiliar territory, and the reader may find this in [4]. When we worked on [4], Bökstedt and I were unaware of Spaltenstein's earlier [61], and we thought we were being very clever when we applied the techniques of homotopy theory to show how one can easily handle unbounded complexes. Unfortunately for us Spaltenstein had achieved the same, a few years earlier, and without using any ideas from homotopy theory.

In the next few years I came to better appreciate the power of the methods, and found a couple of new applications [50, 51][3], results which did not tread in the footsteps of Spaltenstein. The techniques then caught on with other authors, and not only in the context of Grothendieck duality; for example one of the early converts was Bernhard Keller, whose article [32] proved very influential. Somewhat later, Hovey, Palmieri and Strickland's monograph [19] helped by making the foundations more accessible to the non-homotopy-theorists. This survey confines itself to Grothendieck duality; in that field, the papers which most successfully employed homotopy-theoretic methods were Alonso, Jeremías and Lipman [1, 2] Alonso, Jeremías and Souto [3], Jørgensen [28] and Lipman [44]. It should be mentioned that our understanding of what might be possible, using homotopy theoretic methods and other techniques, is not yet complete; there is ongoing work, for example [6, 38, 43, 45, 46, 49, 58, 59, 70, 71].

This ends our glimpse of history, more precisely the brief overview of the early history of Grothendieck duality. Next we begin talking mathematics; first we will review for the reader the basics of dualizing complexes.

## 2. Background on $\mathbf{R}\mathcal{H}om$ complexes

Let us quickly review some basic facts about the internal Hom in the derived category $\mathbf{D}(\mathrm{Qcoh}/X)$. The results are not far from standard, but perhaps the perspective is slightly unusual: we construct $\mathbf{R}\mathcal{H}om$ using Brown representability, and then obtain all the properties we need by reducing to the affine case. Throughout this section $X$ will be a quasicompact, separated scheme, and much of the time we will also assume it noetherian. The category $\mathbf{D}(\mathrm{Qcoh}/X)$ will be the unbounded derived category of quasicoherent sheaves on $X$.

**Remark 2.1.** Instead of $\mathbf{D}(\mathrm{Qcoh}/X)$ we could look at $\mathbf{D}_{\mathrm{Qcoh}}(X)$, the derived category whose objects are complexes of sheaves of $\mathcal{O}_X$–modules with quasicoherent cohomology. When $X$ is separated and quasicompact, the obvious

---

[3] Once again the publication dates are misleading; the later paper [50] was published before [4], which took longer to be accepted. With [50] I happened to be lucky; Thomason was the referee, liked the result and accepted it immediately. Also, [50] and [51] were written at about the same time, but published four year apart; [51] took forever to be accepted.

functor $\mathbf{D}(\mathrm{Qcoh}/X) \longrightarrow \mathbf{D}_{\mathrm{Qcoh}}(X)$ is an equivalence of categories, but for more general $X$ the categories can be different. For schemes that are only quasiseparated it turns out that the category $\mathbf{D}_{\mathrm{Qcoh}}(X)$ is the one with better properties. If we are willing to replace $\mathbf{D}(\mathrm{Qcoh}/X)$ by $\mathbf{D}_{\mathrm{Qcoh}}(X)$ in what follows then many of the results, quite possibly all, undoubtedly generalize to schemes which are only quasiseparated. But I have not carefully checked the details.

When $X$ is noetherian we will also consider the category $\mathbf{D}^b(\mathrm{Coh}/X)$; this is our notation for the derived category of bounded complexes of coherent sheaves. Note that coherent sheaves make sense on a noetherian scheme. The natural functor $\mathbf{D}^b(\mathrm{Coh}/X) \longrightarrow \mathbf{D}(\mathrm{Qcoh}/X)$ is known to be fully faithful. The essential image is the category which is sometimes denoted $\mathbf{D}^b_{\mathrm{Coh}}(\mathrm{Qcoh}/X)$; the objects are (unbounded) complexes of quasicoherent sheaves, whose cohomology is bounded and coherent. Since the categories $\mathbf{D}^b(\mathrm{Coh}/X)$ and $\mathbf{D}^b_{\mathrm{Coh}}(\mathrm{Qcoh}/X)$ are equivalent we will freely confuse them, and use the shorter symbol $\mathbf{D}^b(\mathrm{Coh}/X)$ to stand for both.

Given two objects $\mathcal{E}$ and $\mathcal{F}$ in $\mathbf{D}(\mathrm{Qcoh}/X)$, we may form the derived tensor product $\mathcal{E} \stackrel{\mathrm{L}}{\otimes} \mathcal{F}$. It is easy to show that the derived tensor is a triangulated functor in each of the variables $\mathcal{E}$ and $\mathcal{F}$, and respects coproducts in either variable. If we fix $\mathcal{F}$ and consider the functor $\mathcal{E} \mapsto \mathcal{E} \stackrel{\mathrm{L}}{\otimes} \mathcal{F}$, we have a triangulated functor respecting coproducts, which we write

$$- \stackrel{\mathrm{L}}{\otimes} \mathcal{F} \ : \ \mathbf{D}(\mathrm{Qcoh}/X) \longrightarrow \mathbf{D}(\mathrm{Qcoh}/X) \,.$$

We know that the category $\mathbf{D}(\mathrm{Qcoh}/X)$ is compactly generated; see [51, Proposition 2.5]. From [51, Theorem 4.1] it follows that the coproduct-preserving triangulated functor $- \stackrel{\mathrm{L}}{\otimes} \mathcal{F}$ must have a right adjoint; there is a functor $\mathbf{R}\mathcal{H}om(\mathcal{F}, -) : \mathbf{D}(\mathrm{Qcoh}/X) \longrightarrow \mathbf{D}(\mathrm{Qcoh}/X)$, with

$$\mathrm{Hom}\!\left(\mathcal{E} \stackrel{\mathrm{L}}{\otimes} \mathcal{F} , \ \mathcal{G}\right) \cong \mathrm{Hom}\!\left(\mathcal{E} , \ \mathbf{R}\mathcal{H}om(\mathcal{F}, \mathcal{G})\right) \,.$$

The category $\mathbf{D}(\mathrm{Qcoh}/X)$ becomes a symmetric, monoidal category, and both the tensor product and the internal Hom are triangulated functors in each of the two variables[4].

**Reminder 2.2.** Next we remind the reader of a technical observation which may be found in [48, Lemma 6.8]. For this observation we need a little more notation.

---

[4]　It is slightly subtle to show that the internal Hom is triangulated in the first variable. For details see Murfet's thesis [48, Theorem C.1]. There are also older proofs in the literature, for example [44, §1.5.3]. Unlike the older arguments, the proof in [48] is in the spirit of this section; it appeals to the definition of the internal Hom as right adjoint to the tensor product, and develops the property of the tensor product that would suffice to formally deduce that the right adjoint must be triangulated in the first variable.

The scheme $X$ is still assumed quasicompact and separated, but now we wish to also consider a quasicompact open subset $U \subset X$. We will let $j : U \longrightarrow X$ stand for the inclusion. Because we now have two schemes, we distinguish $\mathbf{R}\mathcal{H}om_U(\mathcal{F}, \mathcal{G})$ from $\mathbf{R}\mathcal{H}om_X(\mathcal{F}, \mathcal{G})$; if $\mathcal{F}$, $\mathcal{G}$ are objects of $\mathbf{D}(\mathrm{Qcoh}/U)$ then the first is the $\mathbf{R}\mathcal{H}om$ which makes sense, while if $\mathcal{F}$, $\mathcal{G}$ belong to $\mathbf{D}(\mathrm{Qcoh}/X)$, then the second is well-defined.

The remark we want is that, if $\mathcal{F} \in \mathbf{D}(\mathrm{Qcoh}/X)$ and $\mathcal{G} \in \mathbf{D}(\mathrm{Qcoh}/U)$, then there is a natural isomorphism

$$\mathbf{R}\mathcal{H}om_X(\mathcal{F}, \mathbf{R}j_*\mathcal{G}) \cong \mathbf{R}j_*\mathbf{R}\mathcal{H}om_U(j^*\mathcal{F}, \mathcal{G}).$$

The proof is easy; the result follows from the isomorphisms

$$\begin{aligned}
\mathrm{Hom}_X\big(\mathcal{E}\,,\ \mathbf{R}\mathcal{H}om_X(\mathcal{F}, \mathbf{R}j_*\mathcal{G})\big) &\cong \mathrm{Hom}_X\big(\mathcal{E} \overset{\mathbf{L}}{\otimes} \mathcal{F}\,,\ \mathbf{R}j_*\mathcal{G}\big) \\
&\cong \mathrm{Hom}_U\big(j^*\mathcal{E} \overset{\mathbf{L}}{\otimes} j^*\mathcal{F}\,,\ \mathcal{G}\big) \\
&\cong \mathrm{Hom}_U\big(j^*\mathcal{E}\,,\ \mathbf{R}\mathcal{H}om_U(j^*\mathcal{F}, \mathcal{G})\big) \\
&\cong \mathrm{Hom}_X\big(\mathcal{E}\,,\ \mathbf{R}j_*\mathbf{R}\mathcal{H}om_U(j^*\mathcal{F}, \mathcal{G})\big).
\end{aligned}$$

**Remark 2.3.** With the notation as in Reminder 2.2 we observe that, for any pair of objects $\mathcal{F}, \mathcal{G} \in \mathbf{D}(\mathrm{Qcoh}/X)$, there is always a map $\gamma(\mathcal{F}, \mathcal{G}) : j^*\mathbf{R}\mathcal{H}om_X(\mathcal{F}, \mathcal{G}) \longrightarrow \mathbf{R}\mathcal{H}om_U(j^*\mathcal{F}, j^*\mathcal{G})$. To construct it, consider the sequence of maps

$$\mathrm{Hom}_X\big(\mathcal{E}\,,\ \mathbf{R}\mathcal{H}om_X(\mathcal{F}, \mathcal{G})\big) \;=\!=\!=\; \mathrm{Hom}_X\big(\mathcal{E} \overset{\mathbf{L}}{\otimes} \mathcal{F}\,,\ \mathcal{G}\big)$$

$$\mathrm{Hom}_U\big(j^*\mathcal{E} \overset{\mathbf{L}}{\otimes} j^*\mathcal{F}\,,\ j^*\mathcal{G}\big) \;=\!=\!=\; \mathrm{Hom}_U\big(j^*\mathcal{E}\,,\ \mathbf{R}\mathcal{H}om_U(j^*\mathcal{F}, j^*\mathcal{G})\big)$$

$$\mathrm{Hom}_X\big(\mathcal{E}\,,\ \mathbf{R}j_*\mathbf{R}\mathcal{H}om_U(j^*\mathcal{F}, j^*\mathcal{G})\big)$$

They are all natural in $\mathcal{E}$ and we deduce, in the category $\mathbf{D}(\mathrm{Qcoh}/X)$, a morphism $\mathbf{R}\mathcal{H}om_X(\mathcal{F}, \mathcal{G}) \longrightarrow \mathbf{R}j_*\mathbf{R}\mathcal{H}om_U(j^*\mathcal{F}, j^*\mathcal{G})$. Adjunction gives us, in the category $\mathbf{D}(\mathrm{Qcoh}/U)$, the required map $\gamma(\mathcal{F}, \mathcal{G}) : j^*\mathbf{R}\mathcal{H}om_X(\mathcal{F}, \mathcal{G}) \longrightarrow \mathbf{R}\mathcal{H}om_U(j^*\mathcal{F}, j^*\mathcal{G})$.

The morphism $\gamma(\mathcal{F}, \mathcal{G})$ need not be an isomorphism. In general, all we know for certain is that it is natural in $\mathcal{F}$ and $\mathcal{G}$. There is, however, a useful special case in which we can show $\gamma(\mathcal{F}, \mathcal{G})$ to be an isomorphism.

**Lemma 2.4.** *Assume that $U \subset X$ is a quasicompact open subset of a quasicompact, separated scheme $X$. Suppose $\mathcal{G} \in \mathbf{D}(\mathrm{Qcoh}/X)$ is any object, and let $\mathcal{F} \in \mathbf{D}(\mathrm{Qcoh}/X)$ satisfy one of the two possible conditions:*

(i) *X is a noetherian scheme, and $\mathcal{F}$ belongs to $\mathbf{D}^b(\mathrm{Coh}/X) \subset \mathbf{D}(\mathrm{Qcoh}/X)$.*

(ii) *X is general, but we assume $\mathcal{F}$ is a compact object in $\mathbf{D}(\mathrm{Qcoh}/X)$. We remind the reader: the compact objects are the perfect complexes.*

*If either* (i) *or* (ii) *holds, then the morphism $\gamma(\mathcal{F}, \mathcal{G}) : j^* \mathbf{R}\mathcal{H}om_X(\mathcal{F}, \mathcal{G}) \longrightarrow \mathbf{R}\mathcal{H}om_U(j^*\mathcal{F}, j^*\mathcal{G})$ of Remark 2.3 is an isomorphism.*

*Proof.* Assume first that both $U$ and $X$ are affine. In that case we have rings $R$ and $S$ so that $X = \mathrm{Spec}(R)$ and $U = \mathrm{Spec}(S)$, and we know that the open immersion $j : U \longrightarrow X$ is flat, meaning that $S$ must be a flat $R$–algebra. The derived category $\mathbf{D}(\mathrm{Qcoh}/X)$ identifies with $\mathbf{D}(R\text{–Mod})$, the derived category $\mathbf{D}(\mathrm{Qcoh}/U)$ is $\mathbf{D}(S\text{–Mod})$, and the functor $j^*$ is nothing more than tensoring a chain complex of $R$–modules with $S$. The lemma is now an easy consequence of the observation that, if $M$ and $N$ are $R$–modules with $M$ finitely presented, then the $S$–module homomorphism

$$S \otimes_R \mathrm{Hom}_R(M, N) \longrightarrow \mathrm{Hom}_R(S \otimes_R M , S \otimes_R N)$$

is an isomorphism.

Next we will see how to deduce the case where only $U$ is assumed affine. We note that, if we fix $\mathcal{F}$ and let $\mathcal{G}$ vary, the maps $\gamma(\mathcal{F}, \mathcal{G})$ assemble to a natural transformation between triangulated functors. To show that the maps $\gamma(\mathcal{F}, \mathcal{G})$ are isomorphisms, for fixed $\mathcal{F}$ and all $\mathcal{G}$, it therefore suffices to prove the special case where $\mathcal{G}$ belongs to some class of generators for $\mathbf{D}(\mathrm{Qcoh}/X)$. The class we will use is the objects $\{j_V^X\}_* \mathcal{G} = \mathbf{R}\{j_V^X\}_* \mathcal{G}$, where $j_V^X : V \longrightarrow X$ is the open immersion of an open affine subset $V \subset X$. Observe that, since $V \subset X$ is assumed affine, the functor $\{j_V^X\}_*$ is exact; hence $\{j_V^X\}_* \mathcal{G} = \mathbf{R}\{j_V^X\}_* \mathcal{G}$. See [48, Corollary 3.14] for the fact that these generate.

Because we now have several open subsets involved, we will adopt the convention that $j_{W_1}^{W_2} : W_1 \longrightarrow W_2$ stands for the open immersion between two subsets $W_1 \subset W_2$ of $X$.

The lemma now follows from the isomorphisms

$$
\begin{aligned}
\{j_U^X\}^* \mathbf{R}\mathcal{H}om\big(\mathcal{F} , \{j_V^X\}_*\mathcal{G}\big) &\cong \{j_U^X\}^* \{j_V^X\}_* \mathbf{R}\mathcal{H}om\big(\{j_V^X\}^*\mathcal{F} , \mathcal{G}\big) \\
&\cong \{j_{U\cap V}^U\}_* \{j_{U\cap V}^V\}^* \mathbf{R}\mathcal{H}om\big(\{j_V^X\}^*\mathcal{F} , \mathcal{G}\big) \\
&\cong \{j_{U\cap V}^U\}_* \mathbf{R}\mathcal{H}om\big(\{j_{U\cap V}^V\}^*\{j_V^X\}^*\mathcal{F} , \{j_{U\cap V}^V\}^*\mathcal{G}\big) \\
&\cong \{j_{U\cap V}^U\}_* \mathbf{R}\mathcal{H}om\big(\{j_{U\cap V}^U\}^*\{j_U^X\}^*\mathcal{F} , \{j_{U\cap V}^V\}^*\mathcal{G}\big) \\
&\cong \mathbf{R}\mathcal{H}om\big(\{j_U^X\}^*\mathcal{F} , \{j_{U\cap V}^U\}_*\{j_{U\cap V}^V\}^*\mathcal{G}\big) \\
&\cong \mathbf{R}\mathcal{H}om\big(\{j_U^X\}^*\mathcal{F} , \{j_U^X\}^*\{j_V^X\}_*\mathcal{G}\big).
\end{aligned}
$$

The first isomorphism is by Reminder 2.2, the second is by base change, the third is by the affine case which we already know, the fourth is trivial, the fifth is again by Reminder 2.2, and finally the sixth is yet another base change.

Finally we need to see how to deduce the general case, where neither $U$ nor $X$ need be affine. We wish to show that the map $\gamma(\mathcal{F}, \mathcal{G}) : j^*\mathbf{R}\mathcal{H}om_X(\mathcal{F}, \mathcal{G}) \longrightarrow \mathbf{R}\mathcal{H}om_U(j^*\mathcal{F}, j^*\mathcal{G})$ is an isomorphism. The problem is local; it suffices to show that it restricts to an isomorphism on each affine open set $V \subset U$. Choose such an affine open set, and let $i : V \longrightarrow U$ be the inclusion. It suffices to show that the map

$$i^*\gamma(\mathcal{F}, \mathcal{G}) : i^* j^*\mathbf{R}\mathcal{H}om_X(\mathcal{F}, \mathcal{G}) \longrightarrow i^*\mathbf{R}\mathcal{H}om_U(j^*\mathcal{F}, j^*\mathcal{G})$$

is an isomorphism. But we already know that the inclusions $i : V \longrightarrow U$ and $ji : V \longrightarrow X$ satisfy the Lemma, since $V$ is affine. Hence both complexes are naturally isomorphic to $\mathbf{R}\mathcal{H}om_V(i^* j^*\mathcal{F}, i^* j^*\mathcal{G})$. $\qquad\square$

**Remark 2.5.** Suppose $X$ is a quasicompact, separated scheme. Given two complexes $\mathcal{E}, \mathcal{F} \in \mathbf{D}(\mathrm{Qcoh}/X)$ we have a counit of adjunction $\varepsilon : \mathcal{E}\overset{\mathbf{L}}{\otimes}\mathbf{R}\mathcal{H}om(\mathcal{E}, \mathcal{F}) \longrightarrow \mathcal{F}$. If we have two more complexes $\mathcal{E}', \mathcal{F}' \in \mathbf{D}(\mathrm{Qcoh}/X)$ they also give a map $\varepsilon : \mathcal{E}'\overset{\mathbf{L}}{\otimes}\mathbf{R}\mathcal{H}om(\mathcal{E}', \mathcal{F}') \longrightarrow \mathcal{F}'$. Combining the two we obtain a single morphism

$$\mathcal{E}\overset{\mathbf{L}}{\otimes}\mathbf{R}\mathcal{H}om(\mathcal{E}, \mathcal{F})\overset{\mathbf{L}}{\otimes}\mathcal{E}'\overset{\mathbf{L}}{\otimes}\mathbf{R}\mathcal{H}om(\mathcal{E}', \mathcal{F}') \xrightarrow{\ \varepsilon\overset{\mathbf{L}}{\otimes}\varepsilon\ } \mathcal{F}\overset{\mathbf{L}}{\otimes}\mathcal{F}',$$

and adjunction produces for us a map

$$\mathbf{R}\mathcal{H}om(\mathcal{E}, \mathcal{F})\overset{\mathbf{L}}{\otimes}\mathbf{R}\mathcal{H}om(\mathcal{E}', \mathcal{F}') \longrightarrow \mathbf{R}\mathcal{H}om(\mathcal{E}\overset{\mathbf{L}}{\otimes}\mathcal{E}',\ \mathcal{F}\overset{\mathbf{L}}{\otimes}\mathcal{F}').$$

The special case where $\mathcal{E}' = \mathcal{O}_X$ will interest us particularly; in this case we have a morphism

$$\mu(\mathcal{E}, \mathcal{F}, \mathcal{F}') \ : \ \mathbf{R}\mathcal{H}om(\mathcal{E}, \mathcal{F})\overset{\mathbf{L}}{\otimes}\mathcal{F}' \longrightarrow \mathbf{R}\mathcal{H}om(\mathcal{E},\ \mathcal{F}\overset{\mathbf{L}}{\otimes}\mathcal{F}').$$

We prove:

**Lemma 2.6.** *Let $X$ be a quasicompact, separated scheme. If $\mathcal{E} \in \mathbf{D}(\mathrm{Qcoh}/X)$ is a perfect complex, and $\mathcal{F}$ and $\mathcal{F}'$ in $\mathbf{D}(\mathrm{Qcoh}/X)$ are arbitrary, then the morphism $\mu(\mathcal{E}, \mathcal{F}, \mathcal{F}')$ of Remark 2.5 is an isomorphism.*

*Proof.* We have a globally defined morphism in $\mathbf{D}(\mathrm{Qcoh}/X)$ and wish to prove it an isomorphism; it suffices to show that, for any open affine subset $U \subset X$, the morphism $\mu(\mathcal{E}, \mathcal{F}, \mathcal{F}')$ restricts to an isomorphism on $U$. Let $j : U \longrightarrow X$ be the inclusion. Lemma 2.4 permits us to identify

$$j^*\mathbf{R}\mathcal{H}om(\mathcal{E}, \mathcal{F}) \cong \mathbf{R}\mathcal{H}om(j^*\mathcal{E}, j^*\mathcal{F})\,,$$
$$j^*\mathbf{R}\mathcal{H}om(\mathcal{E}, \mathcal{F}\overset{\mathbf{L}}{\otimes}\mathcal{F}') \cong \mathbf{R}\mathcal{H}om(j^*\mathcal{E},\ j^*\mathcal{F}\overset{\mathbf{L}}{\otimes}j^*\mathcal{F}')$$

and we conclude that $j^*\mu(\mathcal{E}, \mathcal{F}, \mathcal{F}') = \mu(j^*\mathcal{E}, j^*\mathcal{F}, j^*\mathcal{F}')$. We are therefore reduced to proving the Lemma in the case where $X$ is affine. Put $X = \mathrm{Spec}(R)$; then the category $\mathbf{D}(\mathrm{Qcoh}/X)$ identifies with $\mathbf{D}(R\text{–Mod})$, and under the identification the perfect complex $\mathcal{E}$ becomes a bounded complex of finitely generated projectives. Because $\mu(\mathcal{E}, \mathcal{F}, \mathcal{F}')$ is a natural tranformation of triangulated functors in $\mathcal{E}$, we easily reduce to the case where $\mathcal{E}$ is a single, finitely generated free module concentrated in degree 0, and in this case the result is obvious. $\square$

**Reminder 2.7.** Consider the special case of Lemma 2.6, where we let $\mathcal{F} = \mathcal{O}_X$. From Lemma 2.6 we learn that, for every perfect complex $\mathcal{E}$ and for any object $\mathcal{F}' \in \mathbf{D}(\mathrm{Qcoh}/X)$, the natural map

$$\mathbf{R}\mathcal{H}om(\mathcal{E}, \mathcal{O}_X)^{\mathbf{L}} \otimes \mathcal{F}' \longrightarrow \mathbf{R}\mathcal{H}om(\mathcal{E}, \mathcal{F}')$$

is an isomorphism. This brings us to the well-explored realm of strongly dualizable objects. We remind the reader.

It is traditional to define $\mathcal{E}^{\vee} = \mathbf{R}\mathcal{H}om(\mathcal{E}, \mathcal{O}_X)$. The standard terminology is that $\mathcal{E}$ is *strongly dualizable* if the natural map $\mathcal{E}^{\vee}{}^{\mathbf{L}} \otimes \mathcal{F}' \longrightarrow \mathbf{R}\mathcal{H}om(\mathcal{E}, \mathcal{F}')$ is an isomorphism; the previous paragraph taught us that all perfect complexes are strongly dualizable. From the literature on strongly dualizable objects we learn that

(i) If $\mathcal{E}$ is a strongly dualizable object then the natural map $\mathcal{E} \longrightarrow \{\mathcal{E}^{\vee}\}^{\vee}$ is an isomorphism. See [40, Proposition 1.3(i), p. 122].

(ii) If $\mathcal{E}$ is strongly dualizable then so is $\mathcal{E}^{\vee}$. See [40, Proposition 1.2, p. 121][5].

(iii) Let $\mathcal{T}$ be a compactly generated triangulated category, with a symmetric tensor product compatible with the triangulated structure. Assume that the unit of the tensor is compact, and that all compact objects are strongly dualizable. For example $\mathcal{T}$ could be $\mathbf{D}(\mathrm{Qcoh}/X)$. If $\mathcal{E} \in \mathcal{T}$ is some object, then the following are equivalent:

(a) $\mathcal{E}$ is compact.

(b) $\mathcal{E}$ is strongly dualizable.

(c) Tensor product with $\mathcal{E}$ commutes with arbitrary products in $\mathcal{T}$; that is, the natural map

$$\mathcal{E}^{\mathbf{L}} \otimes \prod_{\lambda \in \Lambda} t_\lambda \longrightarrow \prod_{\lambda \in \Lambda} (\mathcal{E}^{\mathbf{L}} \otimes t_\lambda)$$

is an isomorphism for every set of objects $\{t_\lambda, \lambda \in \Lambda\}$.

---

[5] The objects we call "strongly dualizable" are labeled "finite" in [40]. In the second last paragraph on [40, p. 120] the authors refer to [10] for the older terminolgy, which is the one in current use. Similar notions appeared even earlier, under different names; see the "$\otimes$–categories rigides" of [57, §I.5, p. 78], or the "compact closed categories" of [33, pp. 102–103].

The fact that (c) implies (a) seems new; in Theorem A.1 we will give a self-contained proof that (a), (b) and (c) are equivalent.

In subsequent sections, especially §4, we will feel free to use (i), (ii) and (iii) above.

**Remark 2.8.** The counit of adjunction gives a natural map $\varepsilon$ : $\mathcal{E} \overset{L}{\otimes} \mathbf{R}\mathcal{H}om(\mathcal{E}, \mathcal{F}) \longrightarrow \mathcal{F}$. Combining two such, we deduce a composite

$$\mathcal{E} \overset{L}{\otimes} \mathbf{R}\mathcal{H}om(\mathcal{E}, \mathcal{F}) \overset{L}{\otimes} \mathbf{R}\mathcal{H}om(\mathcal{F}, \mathcal{G}) \xrightarrow{\ \varepsilon \otimes 1\ } \mathcal{F} \overset{L}{\otimes} \mathbf{R}\mathcal{H}om(\mathcal{F}, \mathcal{G}) \xrightarrow{\ \varepsilon\ } \mathcal{G} \ .$$

By adjunction this corresponds to a morphism

$$v(\mathcal{E}, \mathcal{F}, \mathcal{G}) \ : \ \mathcal{E} \overset{L}{\otimes} \mathbf{R}\mathcal{H}om(\mathcal{F}, \mathcal{G}) \longrightarrow \mathbf{R}\mathcal{H}om\big(\mathbf{R}\mathcal{H}om(\mathcal{E}, \mathcal{F}) \, , \ \mathcal{G}\big) \ .$$

We will find useful the next little lemma; note that, from now until the end of the section, all our schemes are assumed noetherian.

**Lemma 2.9.** *Let $X$ be a noetherian, separated scheme. Let the notation be as above, with the objects $\mathcal{F}, \mathcal{G} \in \mathbf{D}(\mathrm{Qcoh}/X)$ fixed. Suppose that, for all $\mathcal{E} \in \mathbf{D}^b(\mathrm{Coh}/X)$,*

(i) *The complex $\mathbf{R}\mathcal{H}om(\mathcal{E}, \mathcal{F})$ belongs to $\mathbf{D}^b(\mathrm{Coh}/X) \subset \mathbf{D}(\mathrm{Qcoh}/X)$. The case $\mathcal{E} = \mathcal{O}_X$ tells us that $\mathcal{F} \in \mathbf{D}^b(\mathrm{Coh}/X)$.*
(ii) *The complex $\mathbf{R}\mathcal{H}om\big(\mathbf{R}\mathcal{H}om(\mathcal{E}, \mathcal{F}) \, , \ \mathcal{G}\big)$ belongs to $\mathbf{D}^-(\mathrm{Qcoh}/X)$.*

*Then, for all $\mathcal{E} \in \mathbf{D}^b(\mathrm{Coh}/X)$, the map $v(\mathcal{E}, \mathcal{F}, \mathcal{G})$ is an isomorphism.*

*Proof.* We have a globally defined morphism in the derived category, and we wish to prove it a homology isomorphism. The question is local; it suffices to show that the restriction of $v(\mathcal{E}, \mathcal{F}, \mathcal{G})$ to every open affine subset $U \subset X$ is a homology isomorphism. Choose any open affine $U \subset X$, and let $j : U \longrightarrow X$ be the inclusion. We propose to show that the map

$$j^*v(\mathcal{E}, \mathcal{F}, \mathcal{G}) \ : \ j^*\mathcal{E} \overset{L}{\otimes} j^*\mathbf{R}\mathcal{H}om(\mathcal{F}, \mathcal{G}) \longrightarrow j^*\mathbf{R}\mathcal{H}om\big(\mathbf{R}\mathcal{H}om(\mathcal{E}, \mathcal{F}) \, , \ \mathcal{G}\big) \ .$$

is an isomorphism in $\mathbf{D}(\mathrm{Qcoh}/U)$.

At this point we use hypothesis (i) and Lemma 2.4. Because $\mathcal{F} \in \mathbf{D}^b(\mathrm{Coh}/X)$, we have that $j^*\mathbf{R}\mathcal{H}om(\mathcal{F}, \mathcal{G}) = \mathbf{R}\mathcal{H}om(j^*\mathcal{F}, j^*\mathcal{G})$, because $\mathcal{E} \in \mathbf{D}^b(\mathrm{Coh}/X)$ we know that $j^*\mathbf{R}\mathcal{H}om(\mathcal{E}, \mathcal{F}) = \mathbf{R}\mathcal{H}om(j^*\mathcal{E}, j^*\mathcal{F})$, and because $\mathbf{R}\mathcal{H}om(\mathcal{E}, \mathcal{F}) \in \mathbf{D}^b(\mathrm{Coh}/X)$ we conclude that

$$j^*\mathbf{R}\mathcal{H}om\big(\mathbf{R}\mathcal{H}om(\mathcal{E}, \mathcal{F}), \ \mathcal{G}\big) = \mathbf{R}\mathcal{H}om\big(j^*\mathbf{R}\mathcal{H}om(\mathcal{E}, \mathcal{F}), \ j^*\mathcal{G}\big)$$
$$= \mathbf{R}\mathcal{H}om\big(\mathbf{R}\mathcal{H}om(j^*\mathcal{E}, j^*\mathcal{F}), \ j^*\mathcal{G}\big).$$

We are therefore reduced to showing that the map

$$v(j^*\mathcal{E}, j^*\mathcal{F}, j^*\mathcal{G}) \; : \; j^*\mathcal{E} \overset{\mathbf{L}}{\otimes} \mathbf{R}\mathcal{H}om(j^*\mathcal{F}, j^*\mathcal{G}) \longrightarrow$$
$$\mathbf{R}\mathcal{H}om\big(\mathbf{R}\mathcal{H}om(j^*\mathcal{E}, j^*\mathcal{F}), \; j^*\mathcal{G}\big)$$

is an isomorphism in $\mathbf{D}(\mathrm{Qcoh}/U)$; in other words it suffices to prove the lemma in the special case where $X = U$ is affine.

Assume therefore that $X$ is affine. Then $X = \mathrm{Spec}(R)$ for a noetherian ring $R$, and $\mathbf{D}(\mathrm{Qcoh}/X)$ is identified with $\mathbf{D}(R\text{–Mod})$. The complex $\mathcal{E}$ becomes a bounded complex of finitely generated $R$–modules, and, because $v(\mathcal{E}, \mathcal{F}, \mathcal{G})$ is a natural transformation between triangulated functors in $\mathcal{E}$, it suffices to prove $v(\mathcal{E}, \mathcal{F}, \mathcal{G})$ to be an isomorphism in the special case where $\mathcal{E}$ is a single, finitely generated module concentrated in degree 0. Next observe that, for any $\mathcal{V}, \mathcal{W} \in \mathbf{D}(R\text{–Mod})$, we may construct $\mathbf{R}\mathrm{Hom}(\mathcal{V}, \mathcal{W})$ by first replacing $\mathcal{W}$ by a $K$–injective resolution consisting of injective objects (see, for example, [4, Applications 2.4 and 2.4']), and then computing the ordinary Hom–complex. For the problem at hand, let us replace $\mathcal{F}$ and $\mathcal{G}$ by $K$–injective resolutions of injectives. All the complexes $\mathbf{R}\mathrm{Hom}(\mathcal{V}, \mathcal{W})$, on both sides of the map $v(\mathcal{E}, \mathcal{F}, \mathcal{G})$, simplify to $\mathrm{Hom}(\mathcal{V}, \mathcal{W})$. Note also that, because $R$ is noetherian, we know that $\mathrm{Hom}(\mathcal{F}, \mathcal{G})$ is a complex of flat modules; for any two injective modules $I$ and $J$ we have that $\mathrm{Hom}(I, J)$ is flat, and products of flat modules are flat. However, we do not yet know that the complex $\mathrm{Hom}(\mathcal{F}, \mathcal{G})$ is $K$–flat; in other words there remains a derived tensor product which we will eventually need to handle. We must show that the morphism

$$v(\mathcal{E}, \mathcal{F}, \mathcal{G}) \; : \; \mathcal{E} \overset{\mathbf{L}}{\otimes} \mathrm{Hom}(\mathcal{F}, \mathcal{G}) \longrightarrow \mathrm{Hom}\big(\mathrm{Hom}(\mathcal{E}, \mathcal{F}), \; \mathcal{G}\big)$$

is an isomorphism in $\mathbf{D}(R\text{–Mod})$.

With our choices of complexes, that is where $\mathcal{F}$ and $\mathcal{G}$ have been replaced by $K$–injective resolutions by injectives and where $\mathcal{E}$ is a single, finitely generated module in degree 0, we obtain a chain map of chain complexes

$$\alpha \; : \; \mathcal{E} \otimes \mathrm{Hom}(\mathcal{F}, \mathcal{G}) \longrightarrow \mathrm{Hom}\big(\mathrm{Hom}(\mathcal{E}, \mathcal{F}), \; \mathcal{G}\big) \, .$$

Do not confuse $\alpha$ with the map $v(\mathcal{E}, \mathcal{F}, \mathcal{G})$; we are not yet asserting that they agree. We will now prove that the morphism $\alpha$ is an isomorphism of chain complexes. To see this, choose a finite presentation of the module $\mathcal{E}$. That is, choose an exact sequence

$$R^m \longrightarrow R^n \longrightarrow \mathcal{E} \longrightarrow 0.$$

We deduce a diagram of chain maps of chain complexes

$$
\begin{array}{ccccccc}
R^m \otimes \mathrm{Hom}(\mathcal{F}, \mathcal{G}) & \longrightarrow & R^n \otimes \mathrm{Hom}(\mathcal{F}, \mathcal{G}) & \longrightarrow & \mathcal{E} \otimes \mathrm{Hom}(\mathcal{F}, \mathcal{G}) & \longrightarrow & 0 \\
\Big\downarrow{\scriptstyle \gamma} & & \Big\downarrow{\scriptstyle \beta} & & \Big\downarrow{\scriptstyle \alpha} & & \\
\mathrm{Hom}\big(\mathrm{Hom}(R^m, \mathcal{F}),\ \mathcal{G}\big) & \longrightarrow & \mathrm{Hom}\big(\mathrm{Hom}(R^n, \mathcal{F}),\ \mathcal{G}\big) & \longrightarrow & \mathrm{Hom}\big(\mathrm{Hom}(\mathcal{E}, \mathcal{F}),\ \mathcal{G}\big) & \longrightarrow & 0
\end{array}
$$

The rows are clearly exact, we know that $\beta$ and $\gamma$ are isomorphisms, and hence so is $\alpha$. We have computed the complex $\mathrm{Hom}\big(\mathrm{Hom}(\mathcal{E}, \mathcal{F}),\ \mathcal{G}\big)$, and the lemma now reduces to showing that the natural map

$$
\mathcal{E} \overset{L}{\otimes} \mathrm{Hom}(\mathcal{F}, \mathcal{G}) \longrightarrow \mathcal{E} \otimes \mathrm{Hom}(\mathcal{F}, \mathcal{G})
$$

is a homology isomorphism. We must show that the derived tensor product agrees with the simple-minded tensor product. It certainly suffices to prove that $\mathrm{Hom}(\mathcal{F}, \mathcal{G})$ is $K$–flat.

Now we use Hypothesis (ii) of the Lemma. We are given that, for any finite $R$–module $\mathcal{E}$, the complex

$$
\mathcal{E} \otimes \mathrm{Hom}(\mathcal{F}, \mathcal{G}) \cong \mathrm{Hom}\big(\mathrm{Hom}(\mathcal{E}, \mathcal{F}),\ \mathcal{G}\big)
$$

belongs to $\mathbf{D}^-(R\text{–Mod})$; its cohomology is bounded above. Lemma 2.9 therefore follows from the following technical lemma. $\qquad\square$

**Lemma 2.10.** *Let $R$ be a commutative, noetherian ring, and let $\mathcal{H}$ be a complex of flat $R$–modules. Suppose that, for any finite $R$–module $M$, we have*

(i) *$H^i(M \otimes \mathcal{H}) = 0$ for $i \gg 0$. More precisely: for every $M$ there exists an integer $N = N(M)$, which may depend on $M$, so that $i > N(M)$ implies $H^i(M \otimes \mathcal{H}) = 0$.*

*Then the complex $\mathcal{H}$ is $K$–flat.*

**Remark 2.11.** The proof we give for Lemma 2.10 is due to Avramov and Iyengar; it is a simplified version of my original, clumsy argument. The remainder of the section is devoted to the proof; the reader may want to skip ahead to §3. In subsequent sections we will not make use of any of the ideas in the proof.

*Proof.* We need to show that $\mathcal{H}$ is $K$–flat. We remind the reader: this means that $X \otimes \mathcal{H}$ must be shown acyclic for every acyclic complex $X$ of $R$–modules. We will break up the proof into steps. We begin with

**Step 1.** It suffices to prove the Lemma under the extra hypotheses that $\mathcal{H}$ is acyclic.

*Proof.* Suppose $\mathcal{H}$ is an object in $\mathbf{K}(R\text{–Flat})$ satisfying the hypotheses of the Lemma; $\mathcal{H}$ does not have to be acyclic. Let us apply the hypothesis (i) of the Lemma in the special case $M = R$; we have a finite module $R$, and (i) tells us that $H^i(\mathcal{H}) = H^i(R \otimes \mathcal{H})$ vanishes for $i \gg 0$. Choose a $K$–projective resolution of $\mathcal{G} \longrightarrow \mathcal{H}$. The map $\mathcal{G} \longrightarrow \mathcal{H}$ is a quasi-isomorphism, and the complex $\mathcal{G}$ may be chosen to belong to $\mathbf{K}^-(R\text{–Proj})$; that is we can take $\mathcal{G}$ to be a bounded above complex of projective $R$–modules. Complete the morphism $\mathcal{G} \longrightarrow \mathcal{H}$ to a distinguished triangle in $\mathbf{K}(R\text{–Flat})$

$$\mathcal{G} \longrightarrow \mathcal{H} \longrightarrow \mathcal{H}' \longrightarrow \Sigma\mathcal{G};$$

this produces an acyclic complex $\mathcal{H}'$ of flat $R$–modules. We furthermore know that $\mathcal{G}$, being a bounded above complex of projectives, is definitely $K$–flat, and it therefore suffices to show that $\mathcal{H}'$ is $K$–flat. The hypothesis (i) of the Lemma holds for $\mathcal{H}$ by assumption, and for $\mathcal{G}$ because it is a bounded above complex. Hence $\mathcal{H}'$ also satisfies the hypotheses of the Lemma. Replacing $\mathcal{H}$ by $\mathcal{H}'$, we are reduced to proving the Lemma when the object $\mathcal{H}$ is acyclic. $\qquad\square$

**Strategy:** We wish to show that $X \otimes \mathcal{H}$ is acyclic when $X$ is. It suffices to show the acyclicity of the localizations $(X \otimes \mathcal{H})_\mathfrak{p} = X \otimes \mathcal{H}_\mathfrak{p}$ for every prime ideal $\mathfrak{p} \subset R$. That is, it suffices to prove that each $\mathcal{H}_\mathfrak{p}$ is $K$–flat. We will prove this by induction on the height of the prime ideal $\mathfrak{p}$. To give away our strategy even more completely, we will consider the two conditions

 (ii) For every prime ideal of height $\leq n$ the complex $\mathcal{H}_\mathfrak{p}$ is $K$–flat.
 (iii) For every prime ideal $\mathfrak{p} \subset R$ of height $\leq n$, for every finite $R$–module $M$, and for every integer $i$, we have that $H^i(M \otimes \mathcal{H}_\mathfrak{p})$ is $\mathfrak{p}$–torsion.

Our induction will be to show that if (ii) is true for $n$ then (iii) is true for $n + 1$, and if (iii) is true for $n$ then (ii) is true for $n$. Note that (iii) is obviously true for $n = 0$; this starts the induction.

**Step 2.** If (ii) is true for $n$ then (iii) is true for $n + 1$.

*Proof.* Let $\mathfrak{p}$ be a prime ideal of height $n + 1$ and let $M$ be a finite $R$–module. We wish to show that the $R_\mathfrak{p}$–modules $H^i(M \otimes \mathcal{H}_\mathfrak{p})$ are all $\mathfrak{p}$–torsion; it suffices to show that their localizations at prime ideals $\mathfrak{q}$ vanish, as long as $\mathfrak{q}$ is properly contained in $\mathfrak{p}$. Now localization is flat, and hence

$$\left[H^i(M \otimes \mathcal{H}_\mathfrak{p})\right]_\mathfrak{q} \cong H^i(M \otimes \mathcal{H}_\mathfrak{q}) \, ;$$

it suffices to show that each complex $M \otimes \mathcal{H}_\mathfrak{q}$ is acyclic. Since $\mathfrak{q}$ is properly contained in $\mathfrak{p}$ it has height $\leq n$, and, because (ii) is true for $n$, we know that $\mathcal{H}_\mathfrak{q}$ is $K$–flat. This makes it a $K$–flat, acyclic complex. The acyclicity says that, in the derived category $\mathbf{D}(R\text{–Mod})$, the complex $\mathcal{H}_\mathfrak{q}$ is isomorphic to zero. The fact that it is $K$–flat permits us to compute, in $\mathbf{D}(R\text{–Mod})$,

$$M \otimes \mathcal{H}_\mathfrak{q} \cong M \overset{\mathbf{L}}{\otimes} \mathcal{H}_\mathfrak{q} \cong M \overset{\mathbf{L}}{\otimes} 0 = 0.$$

This gives the acyclicity of $M \otimes \mathcal{H}_\mathfrak{q}$. $\qquad\qquad\square$

**Notation 2.12.** It remains to prove that if (iii) is true for $n$ then so is (ii). We fix a prime ideal $\mathfrak{p}$ of height $n$; we want to prove that $\mathcal{H}_\mathfrak{p}$ is $K$–flat. Let us simplify the notation a little. Observe first that $X \otimes_R \mathcal{H}_\mathfrak{p} \cong X_\mathfrak{p} \otimes_{R_\mathfrak{p}} \mathcal{H}_\mathfrak{p}$; replacing $R$ by $R_\mathfrak{p}$ we may assume $R$ is a local ring of height $n$, with maximal ideal $\mathfrak{p}$ and residue field $k = R/\mathfrak{p}$. Let us further replace $\mathcal{H}$ by the chain complex $\mathcal{H}_\mathfrak{p}$. What we know so far is:

(iv) For every finite $R$–module $M$ we have $H^i(M \otimes \mathcal{H}) = 0$ if $i \gg 0$; this comes from hypothesis (i) of the Lemma.

(v) The complex $\mathcal{H}$ is an acyclic complex of flat $R$–modules; that comes from Step 1.

(vi) For all finite $R$–modules $M$ and for all integers $i$ we have that $H^i(M \otimes \mathcal{H})$ is $\mathfrak{p}$–torsion. This is because we are assuming that (iii) holds for $n$.

We want to prove that, under these conditions, $\mathcal{H}$ must be $K$–flat.

**Step 3.** With the conventions of Notation 2.12, there exists an integer $\ell$ so that $H^i(M \otimes \mathcal{H})$ vanishes for all finite $R$–modules $M$ and for all $i > \ell$.

*Proof.* The proof will appeal to the following observation: $\mathcal{H}$ is a complex of flat modules, and therefore any short exact sequence $0 \longrightarrow M' \longrightarrow M \longrightarrow M'' \longrightarrow 0$ of $R$–modules induces a short exact sequence of chain complexes

$$0 \longrightarrow M' \otimes \mathcal{H} \longrightarrow M \otimes \mathcal{H} \longrightarrow M'' \otimes \mathcal{H} \longrightarrow 0,$$

hence a long exact sequence in cohomology.

By (iv) we know that $H^i(k \otimes \mathcal{H})$ vanishes for $i \gg 0$. Shifting $\mathcal{H}$ if necessary, we may suppose that $H^i(k \otimes \mathcal{H})$ vanishes for $i > 0$. If $M$ is a finite $R$–module with $\dim(M) = 0$, then $M$ has a finite filtration with subquotients isomorphic to $k$. From the exact sequences in cohomology of the previous paragraph we immediately learn that $H^i(M \otimes \mathcal{H})$ must also vanish whenever $i > 0$.

We will prove, by induction on $m = \dim(M)$, that $H^i(M \otimes \mathcal{H})$ vanishes if $i > \dim(M)$. The previous paragraph proved the case $m = 0$. Since every module has dimension $\dim(M) \leq \dim(R) = n$, we will conclude that $H^i(M \otimes \mathcal{H}) = 0$ for any finite module $M$ and all $i > n$; that is Step 3 will immediately follow.

Suppose $m$ is an integer $\geq 0$, and assume we know the vanishing of $H^i(N \otimes \mathcal{H})$ provided $\dim(N) \leq m$ and $i > \dim(N)$. Let $M$ be a finite $R$–module with

$\dim(M) = m + 1$. Let $\Gamma_{\mathfrak{p}}(M)$ be the $\mathfrak{p}$–torsion submodule of $M$. We have a short exact sequence of $R$–modules

$$0 \longrightarrow \Gamma_{\mathfrak{p}}M \longrightarrow M \longrightarrow M/\Gamma_{\mathfrak{p}}M \longrightarrow 0.$$

Since the module $\Gamma_{\mathfrak{p}}M$ is zero-dimensional we know the vanishing of $H^i(\Gamma_{\mathfrak{p}}M \otimes \mathcal{H})$ for all $i > 0$; this means that the map $H^i(M \otimes \mathcal{H}) \longrightarrow H^i([M/\Gamma_{\mathfrak{p}}M] \otimes \mathcal{H})$ is an isomorphism for $i > 0$. Replacing $M$ by $M/\Gamma_{\mathfrak{p}}M$ we may assume that $M$ has no $\mathfrak{p}$–torsion. We may choose an element $x \in \mathfrak{p}$ which is not a zero-divisor on $M$. Consider the exact sequence

$$0 \longrightarrow M \stackrel{x}{\longrightarrow} M \longrightarrow M/xM \longrightarrow 0;$$

the long exact sequence in cohomology tells us the exactness of

$$H^i([M/xM] \otimes \mathcal{H}) \longrightarrow H^{i+1}(M \otimes \mathcal{H}) \stackrel{x}{\longrightarrow} H^{i+1}(M \otimes \mathcal{H}).$$

The module $M/xM$ has dimension $m$, and the inductive hypothesis establishes the vanishing of $H^i([M/xM] \otimes \mathcal{H})$ if $i > m$. The exact sequence implies that multiplication by $x \in \mathfrak{p}$ is injective on $H^i(M \otimes \mathcal{H})$ when $i \geq m + 1$. But (vi) informs us that the module $H^i(M \otimes \mathcal{H})$ is $\mathfrak{p}$–torsion, and hence must vanish. $\qquad\qquad\square$

**Step 4.** With the conventions of Notation 2.12, the complex $\mathcal{H}$ is $K$–flat.

*Proof.* Let the chain complex $\mathcal{H}$ be written as

$$\cdots \longrightarrow \mathcal{H}^{i-2} \longrightarrow \mathcal{H}^{i-1} \longrightarrow \mathcal{H}^i \longrightarrow \mathcal{H}^{i+1} \longrightarrow \mathcal{H}^{i+2} \longrightarrow \cdots$$

For each $i \in \mathbb{Z}$ let $I^i$ be the image of the homomorphism $\mathcal{H}^i \longrightarrow \mathcal{H}^{i+1}$. The acyclicity of $\mathcal{H}$ tells us that the sequence

$$\mathcal{H}^{i-2} \longrightarrow \mathcal{H}^{i-1} \longrightarrow \mathcal{H}^i \longrightarrow I^i \longrightarrow 0$$

is exact; it is the beginning of a flat resolution for the module $I^i$. Now let $M$ be any finite $R$–module, and consider the sequence

$$M \otimes \mathcal{H}^{i-2} \longrightarrow M \otimes \mathcal{H}^{i-1} \longrightarrow M \otimes \mathcal{H}^i \longrightarrow M \otimes I^i \longrightarrow 0.$$

In this sequence, the part

$$M \otimes \mathcal{H}^{i-1} \longrightarrow M \otimes \mathcal{H}^i \longrightarrow M \otimes I^i \longrightarrow 0$$

must be exact, simply from the right exactness of the tensor product. If $i > \ell + 1$, with $\ell$ as in Step 3, then the bit

$$M \otimes \mathcal{H}^{i-2} \longrightarrow M \otimes \mathcal{H}^{i-1} \longrightarrow M \otimes \mathcal{H}^i$$

must also be exact, because $H^{i-1}(M \otimes \mathcal{H}) = 0$ for $i - 1 > \ell$. We conclude that, as long as $i > \ell + 1$, the torsion group $\mathrm{Tor}_1^R(M, I^i)$ vanishes for all finite modules $M$. That is, $I^i$ is flat if $i > \ell + 1$.

Now consider the short exact sequences

$$0 \longrightarrow I^{i-1} \longrightarrow \mathcal{H}^i \longrightarrow I^i \longrightarrow 0.$$

We know that $\mathcal{H}^i$ is flat for all $i$. If $I^i$ is flat, the sequence tells us that so is $I^{i-1}$. We know that, for sufficiently large $i$, the modules $I^i$ are flat, and by descending induction we deduce the flatness of $I^i$ for every integer $i$. The short exact sequence

$$0 \longrightarrow I^{i-1} \longrightarrow \mathcal{H}^i \longrightarrow I^i \longrightarrow 0$$

is therefore a sequence of flat modules, and hence the sequences

$$0 \longrightarrow M \otimes I^{i-1} \longrightarrow M \otimes \mathcal{H}^i \longrightarrow M \otimes I^i \longrightarrow 0$$

are all exact. Piecing them all together, we conclude that the sequence

$$\cdots \longrightarrow M \otimes \mathcal{H}^{i-1} \longrightarrow M \otimes \mathcal{H}^i \longrightarrow M \otimes \mathcal{H}^{i+1} \longrightarrow \cdots$$

is exact. That is the complex $M \otimes \mathcal{H}$ is acyclic.

We now know that $M \otimes \mathcal{H}$ is acyclic for any finite module $M$. It is standard that $X \otimes \mathcal{H}$ must then be acyclic for any chain complex $X$ of $R$–modules; see, for example, [53, Corollary 9.4, (iii)$\Longrightarrow$(ii)]. In particular $X \otimes \mathcal{H}$ is acyclic when $X$ is, that is $\mathcal{H}$ is $K$–flat. $\qquad\square$

# 3. Dualizing complexes

Throughout this section we continue to assume, as in the second half of §2, that $X$ is a noetherian, separated scheme. In §2 we prepared the ground by setting up some technical apparatus, and now we are ready to treat dualizing complexes. Let us first remind the reader of the definition.

**Definition 3.1.** *Suppose, as agreed, that $X$ is a noetherian, separated scheme. A* dualizing complex *for $X$ is an object $\mathcal{J} \in \mathbf{D}^b(\mathrm{Coh}/X)$ so that the functor*

$$\mathcal{E} \mapsto \mathbf{R}\mathcal{H}\mathit{om}(\mathcal{E}, \mathcal{J})$$

*gives an equivalence*

$$\mathbf{R}\mathcal{H}\mathit{om}(-, \mathcal{J}) \; : \; \mathbf{D}^b(\mathrm{Coh}/X)^{\mathrm{op}} \longrightarrow \mathbf{D}^b(\mathrm{Coh}/X).$$

**Remark 3.2.** Perhaps we should remind the reader: when $\mathcal{E}$ and $\mathcal{I}$ are two objects of $\mathbf{D}^b(\mathrm{Coh}/X)$, the complex $\mathbf{R}\mathcal{H}om(\mathcal{E}, \mathcal{I})$ has no obligation to belong to $\mathbf{D}^b(\mathrm{Coh}/X)$; there is no reason to expect its cohomology to be bounded above. It might help to recall the special case when $X = \mathrm{Spec}(R)$ is affine, and $\mathbf{D}^b(\mathrm{Coh}/X)$ reduces to $\mathbf{D}^b(R\text{–mod})$. Consider two finite $R$–modules $M$ and $N$; they are objects of $\mathbf{D}^b(R\text{–mod})$, and $\mathbf{R}\mathcal{H}om(M, N)$ is a chain complex whose cohomology is $\mathrm{Ext}^n(M, N)$. It is perfectly possible to have $\mathrm{Ext}^n(M, N) \neq 0$ for infinitely many $n$.

As the above example illustrates it is a restriction on $\mathcal{I}$ to demand that, for every $\mathcal{E}$, the object $\mathbf{R}\mathrm{Hom}(\mathcal{E}, \mathcal{I})$ be isomorphic to an object in $\mathbf{D}^b(\mathrm{Coh}/X)$. It is a severe restriction to further insist that the functor

$$\mathbf{R}\mathrm{Hom}(-, \mathcal{I}) \ : \ \mathbf{D}^b(\mathrm{Coh}/X)^{\mathrm{op}} \longrightarrow \mathbf{D}^b(\mathrm{Coh}/X)$$

be an equivalence. In this section we will study some of the consequences.

**Remark 3.3.** The experts will notice that Definition 3.1 is unorthodox. It is customary to impose on $\mathcal{I}$ the technical condition that it have a finite injective resolution; that is, one traditionally assumes a dualizing complex $\mathcal{I}$ to be quasi-isomorphic to a bounded complex of injectives.

The whole thrust of this manuscript is that by now, more than forty years after derived categories were introduced, we have learnt enough about unbounded complexes to be able to handle them without trembling. It would be wimpy to assume boundedness, and we have decided to be tough. But the reader should be warned that, as a result of our decision to take the macho approach, Definition 3.1 is a little non-standard, as is what will now follow. What we will do amounts to proving that the usual results can be obtained without the boundedness hypothesis. And the reason for §2 was to provide us with the technical lemmas we will need.

**Remark 3.4.** Next observe that, for any object $\mathcal{G} \in \mathbf{D}(\mathrm{Qcoh}/X)$, the functor $\mathbf{R}\mathcal{H}om(-, \mathcal{G}) : \mathbf{D}(\mathrm{Qcoh}/X)^{\mathrm{op}} \longrightarrow \mathbf{D}(\mathrm{Qcoh}/X)$ is its own left adjoint; this comes from the isomorphisms

$$\mathrm{Hom}\big(\mathcal{E}, \ \mathbf{R}\mathcal{H}om(\mathcal{F}, \mathcal{G})\big) \cong \mathrm{Hom}\big(\mathcal{E} \overset{\mathbf{L}}{\otimes} \mathcal{F}, \ \mathcal{G}\big)$$
$$\cong \mathrm{Hom}\big(\mathcal{F}, \ \mathbf{R}\mathcal{H}om(\mathcal{E}, \mathcal{G})\big).$$

Furthermore, the unit and counit of this adjunction are the same morphism; they are both the natural map

$$\mathcal{E} \longrightarrow \mathbf{R}\mathcal{H}om\big(\mathbf{R}\mathcal{H}om(\mathcal{E}, \mathcal{G}), \ \mathcal{G}\big).$$

If $\mathcal{G}$ is carefully chosen, so that $\mathbf{R}\mathcal{H}om(-, \mathcal{G})$ takes objects in $\mathbf{D}^b(\mathrm{Coh}/X)$ to objects in $\mathbf{D}^b(\mathrm{Coh}/X)$, then the adjunction restricts from the large category to the subcategory; that is

$$\mathbf{R}\mathcal{H}om(-, \mathcal{G}) \ : \ \mathbf{D}^b(\mathrm{Coh}/X)^{\mathrm{op}} \longrightarrow \mathbf{D}^b(\mathrm{Coh}/X)$$

is its own left adjoint, and the unit and counit of adjunction are the restrictions to $\mathbf{D}^b(\mathrm{Coh}/X)$ of the unit and counit of adjunction on $\mathbf{D}(\mathrm{Qcoh}/X)$.

Now general category theory kicks in and tells us that, if $G$ is a functor with a left adjoint $F$, then $G$ will be an equivalence if and only if both the unit and the counit of adjunction are isomorphisms. In our particular case, where the unit and counit happen to agree, this comes down to the following.

**Lemma 3.5.** *Choose an object* $\mathcal{I} \in \mathbf{D}^b(\mathrm{Coh}/X)$. *The object* $\mathcal{I}$ *is a dualizing complex if and only if, for every object* $\mathcal{E} \in \mathbf{D}^b(\mathrm{Coh}/X)$, *we have*

(i) $\mathbf{R}\mathcal{H}om(\mathcal{E}, \mathcal{I}) \in \mathbf{D}^b(\mathrm{Coh}/X)$.
(ii) *The map* $\mathcal{E} \longrightarrow \mathbf{R}\mathcal{H}om\big(\mathbf{R}\mathcal{H}om(\mathcal{E}, \mathcal{I}), \ \mathcal{I}\big)$ *is an isomorphism.*

*Proof.* (i) is equivalent to $\mathbf{R}\mathcal{H}om(-, \mathcal{I})$ restricting to a functor from $\mathbf{D}^b(\mathrm{Coh}/X)$ to itself, while (ii) is equivalent to the unit and counit of the adjunction of Remark 3.4 being isomorphisms. $\square$

Slightly less trivial is the formulation

**Proposition 3.6.** *Let* $\mathcal{I}$ *be an object of* $\mathbf{D}^b(\mathrm{Coh}/X)$. *Then* $\mathcal{I}$ *is a dualizing complex if and only if*

(i) *For every object* $\mathcal{E} \in \mathbf{D}^b(\mathrm{Coh}/X)$ *we have* $\mathbf{R}\mathcal{H}om(\mathcal{E}, \mathcal{I}) \in \mathbf{D}^b(\mathrm{Coh}/X)$.
(ii) *The natural map* $\mathcal{O}_X \longrightarrow \mathbf{R}\mathcal{H}om(\mathcal{I}, \mathcal{I})$ *is an isomorphism.*

*Proof.* To deduce Proposition 3.6 from Lemma 3.5, we need to show that it suffices to check that the morphism

$$\mathcal{E} \longrightarrow \mathbf{R}\mathcal{H}om\big(\mathbf{R}\mathcal{H}om(\mathcal{E}, \mathcal{I}), \ \mathcal{I}\big)$$

is an isomorphism in the special case where $\mathcal{E} = \mathcal{O}_X$. The useful Lemma 2.9 allows us to identify, for any $\mathcal{E}$,

$$\mathcal{E} \overset{\mathbf{L}}{\otimes} \mathbf{R}\mathcal{H}om(\mathcal{I}, \mathcal{I}) \cong \mathbf{R}\mathcal{H}om\big(\mathbf{R}\mathcal{H}om(\mathcal{E}, \mathcal{I}), \ \mathcal{I}\big).$$

The morphism we need to show an isomorphism becomes

$$\mathcal{E} \longrightarrow \mathcal{E} \overset{\mathbf{L}}{\otimes} \mathbf{R}\mathcal{H}om(\mathcal{I}, \mathcal{I}).$$

We leave it to the reader to check that this morphism is $1 \otimes \rho$, where $1 : \mathcal{E} \longrightarrow \mathcal{E}$ is the identity and $\rho : \mathcal{O}_X \longrightarrow \mathbf{R}\mathcal{H}om(\mathcal{I}, \mathcal{I})$ is the natural map. It therefore suffices to show that $\rho$ is an isomorphism. $\square$

**Notation 3.7.** If $X$ is a noetherian scheme and $\mathfrak{I}$ is a dualizing complex on $X$, we will sometimes abbreviate the functor $\mathbf{R}\mathcal{H}om(-,\mathfrak{I})$ to just $D_{X,\mathfrak{I}}(-)$. If either $X$ or $\mathfrak{I}$ is understood from the context we will feel free to drop it from the notation; thus the symbols $D_X(-)$, $D_{\mathfrak{I}}(-)$ and $D_{X,\mathfrak{I}}(-)$ should be viewed as synonymous, where the third is the most explicit. What we constructed above, using the unit of adjunction, is a natural isomorphism $\mathcal{E} \longrightarrow D_{X,\mathfrak{I}}\big(D_{X,\mathfrak{I}}(\mathcal{E})\big)$.

**Remark 3.8.** The next obvious question is what happens if we are given, on the same scheme $X$, two dualizing complexes $\mathfrak{I}$ and $\mathfrak{J}$. How do they compare?

It is clear that, if we are given a dualizing complex $\mathfrak{I}$, an integer $n$ and a line bundle $\mathcal{L}$, then $\Sigma^n \mathcal{L}^{\,\mathbf{L}} \otimes \mathfrak{I}$ is also a dualizing complex; tensoring with a line bundle and suspending is harmless. It is also harmless to suspend by different integers on different connected components of $X$. The remarkable fact is that this is all the freedom we have. Up to these basic moves, the dualizing complex is unique. We state this as a lemma and give the proof.

**Lemma 3.9.** *Let $\mathfrak{I}$ and $\mathfrak{J}$ be two dualizing complexes on the same scheme $X$. Then $\mathfrak{I} = \mathcal{L}^{\,\mathbf{L}} \otimes \mathfrak{J}$, where $\mathcal{L}$ is locally some suspension of a line bundle. On each connected component $X_i \subset X$ there is an integer $n_i$, a line bundle $\mathcal{L}_i$, and an isomorphism $\mathcal{L}|_{X_i} \cong \Sigma^{n_i} \mathcal{L}_i$.*

*Proof.* Let us define $\mathcal{L} = \mathbf{R}\mathcal{H}om(\mathfrak{I},\mathfrak{J})$. Now note that

$$
\begin{aligned}
\mathfrak{J} &= \mathbf{R}\mathcal{H}om(\mathcal{O}_X, \mathfrak{J}) && \\
&= \mathbf{R}\mathcal{H}om\big(\mathbf{R}\mathcal{H}om(\mathfrak{I},\mathfrak{I}),\ \mathfrak{J}\big) && \text{by Proposition 3.6(ii)} \\
&= \mathfrak{I}^{\,\mathbf{L}} \otimes \mathbf{R}\mathcal{H}om(\mathfrak{I},\mathfrak{J}) && \text{by Lemma 2.10} \\
&= \mathfrak{I}^{\,\mathbf{L}} \otimes \mathcal{L} && \text{by definition of } \mathcal{L}\,.
\end{aligned}
$$

It remains to analyse $\mathcal{L}$ and prove that, on each connected component of $X$, it is isomorphic to the suspension of a line bundle.

Consider the functor $D_{\mathfrak{I}} D_{\mathfrak{J}}$. This functor takes the complex $\mathcal{E}$ to the complex

$$
D_{\mathfrak{I}}\big(D_{\mathfrak{J}}(\mathcal{E})\big) = \mathbf{R}\mathcal{H}om\big(\mathbf{R}\mathcal{H}om(\mathcal{E},\mathfrak{J}),\ \mathfrak{I}\big)
$$

and Lemma 2.9 identifies this with $\mathcal{E}^{\,\mathbf{L}} \otimes \mathbf{R}\mathcal{H}om(\mathfrak{J},\mathfrak{I})$. Interchanging the roles of $\mathfrak{I}$ and $\mathfrak{J}$, we have that $D_{\mathfrak{J}}\big(D_{\mathfrak{I}}(\mathcal{E})\big)$ is naturally isomorphic to $\mathcal{E}^{\,\mathbf{L}} \otimes \mathbf{R}\mathcal{H}om(\mathfrak{I},\mathfrak{J})$. Now the functor $D_{\mathfrak{I}} D_{\mathfrak{J}} D_{\mathfrak{J}} D_{\mathfrak{I}}$ is naturally isomorphic to the identity; if the units=counits of the adjunctions of Remark 3.4 are written $\eta_i : 1 \Longrightarrow D_{\mathfrak{I}} D_{\mathfrak{I}}$ and $\eta_j : 1 \Longrightarrow D_{\mathfrak{J}} D_{\mathfrak{J}}$, then the composite

$$
1 \xrightarrow{\ \eta_i\ } D_{\mathfrak{I}} D_{\mathfrak{I}} \xrightarrow{\ D_{\mathfrak{I}}\, \eta_j\, D_{\mathfrak{I}}\ } D_{\mathfrak{I}} D_{\mathfrak{J}} D_{\mathfrak{J}} D_{\mathfrak{I}}
$$

provides the isomorphism. On the object $\mathcal{E} \in \mathbf{D}^b(\mathrm{Coh}/X)$ this isomorphism is the natural map

$$\mathcal{E} \longrightarrow \mathcal{E} \otimes^{\mathbf{L}} \mathbf{R}\mathcal{H}om(\mathfrak{I}, \mathfrak{I}) \longrightarrow \mathcal{E} \otimes^{\mathbf{L}} \mathbf{R}\mathcal{H}om(\mathfrak{I}, \mathfrak{J}) \otimes^{\mathbf{L}} \mathbf{R}\mathcal{H}om(\mathfrak{J}, \mathfrak{I}).$$

Applying this to $\mathcal{E} = \mathcal{O}_X$, we see that

$$\mathbf{R}\mathcal{H}om(\mathfrak{I}, \mathfrak{J}) \otimes^{\mathbf{L}} \mathbf{R}\mathcal{H}om(\mathfrak{J}, \mathfrak{I}) \cong \mathcal{O}_X .$$

We already defined $\mathcal{L} = \mathbf{R}\mathcal{H}om(\mathfrak{I}, \mathfrak{J})$; now set $\mathcal{L}' = \mathbf{R}\mathcal{H}om(\mathfrak{J}, \mathfrak{I})$, and we deduce an isomorphism $\mathcal{L} \otimes^{\mathbf{L}} \mathcal{L}' \cong \mathcal{O}_X$. We furthermore know that $\mathcal{L}$ and $\mathcal{L}'$ are both objects in $\mathbf{D}^b(\mathrm{Coh}/X)$.

It remains to show that an object $\mathcal{L} \in \mathbf{D}^b(\mathrm{Coh}/X)$, which has an inverse $\mathcal{L}'$ with respect to the tensor product, must locally be a shift of a line bundle; we remind the reader of the proof. Restrict everything to the stalk at a point $x \in X$. Then the stalk at $x$ of the sheaf $\mathcal{O}_X$ is a local ring $R$ with maximal ideal $\mathfrak{m}$. Let us write $\mathcal{L}_x, \mathcal{L}'_x$ for the stalks at $x$ of $\mathcal{L}, \mathcal{L}'$. Then $\mathcal{L}_x, \mathcal{L}'_x$ are objects in $\mathbf{D}^b(R\text{–mod})$ satisfying $\mathcal{L}_x \otimes^{\mathbf{L}} \mathcal{L}'_x \cong R$. Let $k = k(x)$ be the residue field of $R$. We know that $k \otimes^{\mathbf{L}} \mathcal{L}_x$ is a complex of $k$–vector spaces; that is it must be a direct sum of suspensions of $k$. Now consider the isomorphisms

$$(k \otimes^{\mathbf{L}} \mathcal{L}_x) \otimes^{\mathbf{L}} \mathcal{L}'_x = k \otimes^{\mathbf{L}} (\mathcal{L}_x \otimes^{\mathbf{L}} \mathcal{L}'_x) = k \otimes^{\mathbf{L}} R = k .$$

We learn first that $k \otimes^{\mathbf{L}} \mathcal{L}_x$ must be nonzero, and by symmetry $k \otimes^{\mathbf{L}} \mathcal{L}'_x$ is also nonzero. But also $k = (k \otimes^{\mathbf{L}} \mathcal{L}_x) \otimes^{\mathbf{L}} \mathcal{L}'_x$ is indecomposable. It follows that $k \otimes^{\mathbf{L}} \mathcal{L}_x$ cannot be a direct sum of more than one factor, since then $(k \otimes^{\mathbf{L}} \mathcal{L}_x) \otimes^{\mathbf{L}} \mathcal{L}'_x$ would decompose into more than one direct summand, and each of the summands would be isomorphic to some suspension of the nontrivial $k \otimes^{\mathbf{L}} \mathcal{L}'_x$. We conclude that $k \otimes^{\mathbf{L}} \mathcal{L}_x$ must be of the form $\Sigma^n k$, for some integer $n$.

Now $\mathcal{L}_x$ is an object in $\mathbf{D}^b(R\text{–mod})$, and it therefore has a minimal projective resolution. We remind the reader: a minimal projective resolution is a bounded above chain complex of finitely generated, free $R$–modules, so that all the differentials vanish modulo $\mathfrak{m}$. The object $k \otimes^{\mathbf{L}} \mathcal{L}_x$ can be computed by taking the ordinary tensor product of $k$ with the minimal projective resolution. The fact that $k \otimes^{\mathbf{L}} \mathcal{L}_x \cong \Sigma^n k$ tells us that the minimal projective resolution must be very sparse; it consists of a single free module, of rank 1, concentrated in degree $-n$. That is $\mathcal{L}_x \cong \Sigma^n R$.

So far we have focused on the stalk of a single point $x \in X$. We have a chain complex $\mathcal{L} \in \mathbf{D}^b(\mathrm{Coh}/X)$, that is a bounded chain complex of coherent sheaves on $X$. We have shown that, at the point $x$, the stalks of the cohomology sheaves $\mathbb{H}^k(\mathcal{L})$ vanish if $k \neq -n$, and if $k = -n$ we obtained an isomorphism of the stalk at $x$ of $\mathbb{H}^{-n}(\mathcal{L})$ with the free module $R$ of rank 1. But the sheaves

$\mathbb{H}^k(\mathcal{L})$ are coherent sheaves on $X$, all but finitely many of which vanish. It follows that there exists a Zariski open set $U \subset X$ so that, on $U$, the coherent sheaves $\mathbb{H}^k(\mathcal{L})$ satisfy

$$\mathbb{H}^k(\mathcal{L})|_U = \begin{cases} 0 & \text{if } k \neq -n \\ \overline{\mathcal{L}} & \text{if } k = -n \end{cases}$$

where $\overline{\mathcal{L}}$ is some line bundle on $U$. Now the integer $n$ is locally constant, and must be constant on connected components. On each connected component $X_i$ there is an integer $n_i$ so that $\mathbb{H}^k(\mathcal{L}) = 0$ for $k \neq -n_i$, and $\mathbb{H}^{-n_i}(\mathcal{L})$ is a line bundle. This makes $\mathcal{L}|_{X_i}$ a complex whose cohomology sheaves vanish except in only one dimension, and hence it must be isomorphic in $\mathbf{D}^b(\mathrm{Coh}/X_i)$ to $\Sigma^{n_i} \mathbb{H}^{-n_i}(\mathcal{L})$. $\qquad\square$

**Remark 3.10.** Let $\mathcal{J}$ be a dualizing complex. It is an object of $\mathbf{D}^b(\mathrm{Coh}/X)$, and has a bounded-below injective resolution. In the homotopy category $\mathbf{K}(\mathrm{Inj}/X)$, whose objects are chain complexes of injective quasicoherent sheaves on $X$, there is a bounded-below complex $I$ and a quasi-isomorphism $\mathcal{J} \longrightarrow I$. Furthermore, $I$ is unique up to homotopy. If $\mathcal{J}$ is another dualising complex, Lemma 3.9 informs us that $\mathcal{J} = \mathcal{L} \overset{\mathbf{L}}{\otimes} \mathcal{J}$, with $\mathcal{L}$ locally isomorphic to a shift of a line bundle. Obviously, $\mathcal{L} \otimes I$ is an injective resolution for $\mathcal{J}$.

While the injective resolutions $I \in \mathbf{K}(\mathrm{Inj}/X)$ of dualizing complexes $\mathcal{J}$ are only determined up to twisting by complexes $\mathcal{L}$, the Hom–complexes $\mathcal{H}om(I, I)$ are objects of $\mathbf{K}(\mathrm{Flat}/X)$ well defined up to homotopy; we have

$$\mathcal{H}om(I, I) \cong \mathcal{H}om(\mathcal{L} \otimes I, \mathcal{L} \otimes I).$$

Even though the dualizing complex $\mathcal{J}$ started its life as an object of the derived category $\mathbf{D}^b(\mathrm{Coh}/X)$, subject to conditions that appear innocuous enough, we have learned

(i) The object $\mathcal{J}$ is unique up to tensor by an $\mathcal{L}$, with $\mathcal{L}$ locally isomorphic to a shift of a line bundle.

(ii) If we replace $\mathcal{J}$ by its injective resolution, then the object $\mathcal{H}om(\mathcal{J}, \mathcal{J})$ is a well-defined complex of flat $\mathcal{O}_X$–modules, unique up to homotopy.

In Proposition 3.6 we learned that the natural map $\mathcal{O}_X \longrightarrow \mathcal{H}om(\mathcal{J}, \mathcal{J})$ must be an isomorphism in $\mathbf{D}^b(\mathrm{Coh}/X)$. It is a homology isomorphism of complexes of flat modules, but not usually a homotopy equivalence.

Our next observation is

**Theorem 3.11.** *Being a dualizing complex is local. In other words let $Y = \cup U_i$ be an open cover. Then a complex $\mathcal{J} \in \mathbf{D}(\mathrm{Qcoh}/Y)$ is dualizing if and only if, for each $i$, the restriction of $\mathcal{J}$ to $U_i$ is dualizing.*

*Proof.* First of all a point of notation: let $f_i : U_i \longrightarrow Y$ be the open immersion; the functor $f_i^*$ is exact, and hence is equal to its derived functor. We will write $f_i^*$ for $\mathbf{L}f_i^*$. Also, $\mathcal{I}$ belongs to $\mathbf{D}^b(\mathrm{Coh}/Y) \subset \mathbf{D}(\mathrm{Qcoh}/Y)$ if and only if each $f_i^*\mathcal{I}$ belongs to $\mathbf{D}^b(\mathrm{Coh}/U_i)$. We may therefore assume that $\mathcal{I}$ and all the $f_i^*\mathcal{I}$ are bounded complexes of coherent sheaves.

By Proposition 3.6 it suffices to prove two things:

(i) The following are equivalent:
  (a) For each object $\mathcal{E} \in \mathbf{D}^b(\mathrm{Coh}/Y)$ the complex $\mathbf{R}\mathcal{H}om(\mathcal{E}, \mathcal{I})$ belongs to $\mathbf{D}^b(\mathrm{Coh}/Y)$.
  (b) For all $i$ and all objects $\mathcal{E}_i \in \mathbf{D}^b(\mathrm{Coh}/U_i)$, the complexes $\mathbf{R}\mathcal{H}om(\mathcal{E}_i, f_i^*\mathcal{I})$ belong to $\mathbf{D}^b(\mathrm{Coh}/U_i)$.
(ii) The map $\mathcal{O}_Y \longrightarrow \mathbf{R}\mathcal{H}om(\mathcal{I}, \mathcal{I})$ is an isomorphism if and only if, for each $i$, the map $\mathcal{O}_{U_i} \longrightarrow \mathbf{R}\mathcal{H}om(f_i^*\mathcal{I}, f_i^*\mathcal{I})$ is an isomorphism.

Let us begin with (i). The proof that (b)$\Longrightarrow$(a) is easy. Let $\mathcal{E}$ belong to $\mathbf{D}^b(\mathrm{Coh}/Y)$; by Lemma 2.4 we know that

$$\mathbf{R}\mathcal{H}om(f_i^*\mathcal{E}, f_i^*\mathcal{I}) = f_i^*\mathbf{R}\mathcal{H}om(\mathcal{E}, \mathcal{I}) .$$

By (b) the term on the left is in $\mathbf{D}^b(\mathrm{Coh}/U_i)$ for every $i$, and from the isomorphism with the term on the right we conclude that $\mathbf{R}\mathcal{H}om(\mathcal{E}, \mathcal{I})$ must be in $\mathbf{D}^b(\mathrm{Coh}/Y)$.

Slightly subtler is (a)$\Longrightarrow$(b). Suppose (a) holds, and $\mathcal{E}_i$ is an object of $\mathbf{D}^b(\mathrm{Coh}/U_i)$. Since coherent sheaves and morphisms of coherent sheaves can be extended from the open set $U_i$ to all of $Y$, the object $\mathcal{E}_i \in \mathbf{D}^b(\mathrm{Coh}/U_i)$ is isomorphic to $f_i^*\mathcal{G}$, where $\mathcal{G} \in \mathbf{D}^b(\mathrm{Coh}/Y)$. Therefore

$$\begin{aligned} \mathbf{R}\mathcal{H}om(\mathcal{E}_i, f_i^*\mathcal{I}) &= \mathbf{R}\mathcal{H}om(f_i^*\mathcal{G}, f_i^*\mathcal{I}) \\ &= f_i^*\mathbf{R}\mathcal{H}om(\mathcal{G}, \mathcal{I}) \qquad \text{by Lemma 2.4} . \end{aligned}$$

Now $\mathbf{R}\mathcal{H}om(\mathcal{G}, \mathcal{I})$ is in $\mathbf{D}^b(\mathrm{Coh}/Y)$ by (a), and $f^*$ takes $\mathbf{D}^b(\mathrm{Coh}/Y)$ to $\mathbf{D}^b(\mathrm{Coh}/U_i)$.

Next we prove (ii). Lemma 2.4 informs us that $\mathbf{R}\mathcal{H}om(f_i^*\mathcal{I}, f_i^*\mathcal{I}) = f_i^*\mathbf{R}\mathcal{H}om(\mathcal{I}, \mathcal{I})$, and the map $\mathcal{O}_Y \longrightarrow \mathbf{R}\mathcal{H}om(\mathcal{I}, \mathcal{I})$ will be an isomorphism if and only if its restrictions to each $U_i$ are. $\qquad\square$

**Remark 3.12.** Let $f : X \longrightarrow Y$ be an open immersion. Then, for any dualizing complex $\mathcal{I}$ on $Y$, the complex $f^*\mathcal{I}$ is dualizing on $X$. Just apply Theorem 3.11 to the open cover $Y = X \cup Y$.

**Remark 3.13.** In Remark 3.12 we noted that dualizing complexes are preserved when we pull them back by open immersions. It is natural to wonder what happens with other functors.

Let us make this precise. Suppose we are given a morphism of noetherian, separated schemes $f : X \longrightarrow Y$; there is a pair of derived categories $\mathbf{D}(\mathrm{Qcoh}/X)$ and $\mathbf{D}(\mathrm{Qcoh}/Y)$, and the morphism of schemes $f : X \longrightarrow Y$ induces several functors between these derived categories. There is the push-forward functor $\mathbf{R}f_* : \mathbf{D}(\mathrm{Qcoh}/X) \longrightarrow \mathbf{D}(\mathrm{Qcoh}/Y)$, which has a left adjoint $\mathbf{L}f^*$ and a right adjoint $f^!$. Assuming that either $\mathbf{D}^b(\mathrm{Coh}/X)$ or $\mathbf{D}^b(\mathrm{Coh}/Y)$ has a dualizing complex, one can ponder whether some of the three functors above might respect it. What we know so far, from Remark 3.12, is that, as long as $f$ is an open immersion, the functor $\mathbf{L}f^*$ takes dualizing complexes on $Y$ to dualizing complexes on $X$. The next theorem gives sufficient conditions for $f^!$ to take a dualizing complex on $Y$ to a dualizing complex on $X$.

**Remark 3.14.** Before we state the theorem, it might help if we remind the reader of what boundedness properties are always, unconditionally true, for any morphism $f : X \longrightarrow Y$; this might help separate what is formal from the genuinely restrictive hypotheses of the theorem.

We are assuming that $X$ and $Y$ are noetherian and separated. This means that the functor $\mathbf{R}f_*$ can be computed using Čech cohomology with respect to an affine open cover of $X$, and this cover may be taken finite. If $\mathcal{E} \in \mathbf{D}(\mathrm{Qcoh}/X)$ is bounded it follows that $\mathbf{R}f_*\mathcal{E}$ will also be a bounded complex. That much is free. But we have no obvious way to control the size (as in coherence versus quasicoherence) of the finitely many cohomology sheaves of $\mathbf{R}f_*\mathcal{E}$.

It turns out to be a strong restriction on $f$ to assume that $\mathbf{R}f_* : \mathbf{D}(\mathrm{Qcoh}/X) \longrightarrow \mathbf{D}(\mathrm{Qcoh}/Y)$ respects the bounded, coherent subcategories. In symbols: it may happen that $\mathbf{R}f_*$ will take the subcategory $\mathbf{D}^b(\mathrm{Coh}/X) \subset \mathbf{D}(\mathrm{Qcoh}/X)$ to the subcategory $\mathbf{D}^b(\mathrm{Coh}/Y) \subset \mathbf{D}(\mathrm{Qcoh}/Y)$, but it certainly does not come for free. The usual sufficient condition is that the map $f : X \longrightarrow Y$ be a proper morphism of finite type. This concludes what we will need to know concerning the functor $\mathbf{R}f_*$, to make sense of the statement of Theorem 3.15; the preservation of boundedness is formal, the preservation of coherence is not.

Next we want to look at the properties of the functor $f^!$ which play a role in the statement of Theorem 3.15. It helps to begin by returning briefly to the functor $\mathbf{R}f_*$, and observing a little more closely the precise bounds that the Čech complex gives. Suppose that $X$ admits a cover by $n + 1$ affine open sets; then the Čech complex computing $\mathbf{R}f_*$ has length $n$. If $\mathcal{E}$ is an object in $\mathbf{D}^{<m}(\mathrm{Qcoh}/X)$, that is $\mathcal{E} \in \mathbf{D}(\mathrm{Qcoh}/X)$ is quasi-isomorphic to a complex vanishing in degrees $\geq m$, then $\mathbf{R}f_*\mathcal{E}$ must be in $\mathbf{D}^{<m+n}(\mathrm{Qcoh}/Y)$; a Čech complex of length $n$ can only raise the cohomological degree by at most $n$. This means that for any $\mathcal{F} \in \mathbf{D}^{\geq m+n}(\mathrm{Qcoh}/Y)$ we have $\mathrm{Hom}(\mathbf{R}f_*\mathcal{E}, \mathcal{F}) = 0$. Adjunction gives us $\mathrm{Hom}(\mathcal{E}, f^!\mathcal{F}) = 0$, and this is true for all $\mathcal{E} \in \mathbf{D}^{<m}(\mathrm{Qcoh}/X)$. Hence

$f^!\mathcal{F} \in \mathbf{D}^{\geq m}(\mathrm{Qcoh}/X)$. We have just shown that the functor $f^!$ must take bounded-below complexes to bounded-below complexes. In particular the image under $f^!$ of the subcategory $\mathbf{D}^b(\mathrm{Coh}/Y) \subset \mathbf{D}(\mathrm{Qcoh}/Y)$ must lie in $\mathbf{D}^+(\mathrm{Qcoh}/X) \subset \mathbf{D}(\mathrm{Qcoh}/X)$.

That ends our free ride. There is no formal reason to expect the functor $f^!$ to take $\mathbf{D}^b(\mathrm{Coh}/Y) \subset \mathbf{D}(\mathrm{Qcoh}/Y)$ into $\mathbf{D}^b(\mathrm{Coh}/X)$, or even into the larger $\mathbf{D}^+_{\mathrm{Coh}}(\mathrm{Qcoh}/X)$. We remind the reader: the objects in the category $\mathbf{D}^+_{\mathrm{Coh}}(\mathrm{Qcoh}/X)$ are the bounded-below chain complexes of quasicoherent sheaves, whose cohomology is all coherent.

Next we prove:

**Theorem 3.15.** *Let $f : X \longrightarrow Y$ be a morphism of noetherian, separated schemes so that*

(i) *The functor $\mathbf{R}f_* : \mathbf{D}(\mathrm{Qcoh}/X) \longrightarrow \mathbf{D}(\mathrm{Qcoh}/Y)$ takes $\mathbf{D}^b(\mathrm{Coh}/X) \subset \mathbf{D}(\mathrm{Qcoh}/X)$ into $\mathbf{D}^b(\mathrm{Coh}/Y) \subset \mathbf{D}(\mathrm{Qcoh}/Y)$.*
(ii) *The functor $f^! : \mathbf{D}(\mathrm{Qcoh}/Y) \longrightarrow \mathbf{D}(\mathrm{Qcoh}/X)$ takes $\mathbf{D}^b(\mathrm{Coh}/Y) \subset \mathbf{D}(\mathrm{Qcoh}/Y)$ into $\mathbf{D}^+_{\mathrm{Coh}}(\mathrm{Qcoh}/X) \subset \mathbf{D}(\mathrm{Qcoh}/X)$.*

*If $\mathcal{I}$ is a dualizing complex on $Y$ then $f^!\mathcal{I}$ is a dualizing complex on $X$.*

**Remark 3.16.** The reader might wish to compare Theorem 3.15 with older results in the literature; see, for example, [15, Remark 2, p. 299], [66, Corollary 3, p. 396], [2, Proposition 2.5.11] and [45, §9.1 and §9.2].

**Reminder 3.17.** In the proof of Theorem 3.15 we will appeal to the technical result [46, Theorem 4.2]; let us therefore remind the reader. The technical result asserts the following: in the category $\mathbf{D}(\mathrm{Qcoh}/X)$ there is a compact generator $S$ which detects non-vanishing high cohomology. Precisely, this means that there is an integer $A = A(S)$, depending only on $S$, so that

(iii) If $\mathcal{G}$ is an object of $\mathbf{D}(\mathrm{Qcoh}/X)$, with $\mathrm{Hom}(S, \Sigma^k\mathcal{G}) = 0$ for all $k \geq n$, then $\mathbb{H}^k(\mathcal{G}) = 0$ for all $k \geq n + A$.

Perhaps we should explain the notation: the $\mathbb{H}^k$ means the $k$th cohomology *sheaf* of the chain complex of sheaves. The result informs us that we can tell whether the sheaf cohomology vanishes, above a certain degree, just by computing some groups, namely $\mathrm{Hom}(S, \Sigma^k\mathcal{G})$.

*Proof.* We now prove Theorem 3.15. As in the proof of Theorem 3.11 we appeal to Proposition 3.6. It suffices to establish two things:

(iv) For all objects $\mathcal{G} \in \mathbf{D}^b(\mathrm{Coh}/X)$ we will show that $\mathbf{R}\mathcal{H}om(\mathcal{G}, f^!\mathcal{I})$ belongs to $\mathbf{D}^b(\mathrm{Coh}/X)$. The case where $\mathcal{G} = \mathcal{O}_X$ will prove that $f^!\mathcal{I} \in \mathbf{D}^b(\mathrm{Coh}/X)$.

(v) We will show that the natural map $\mathcal{O}_X \longrightarrow \mathbf{R}\mathcal{H}om(f^!\mathcal{I}, f^!\mathcal{I})$ is an isomorphism.

Let us begin with (iv). What we are given is that $\mathcal{G}$ belongs to $\mathbf{D}^b(\mathrm{Coh}/X)$, and hypothesis (ii) of the Theorem says that $f^!\mathcal{I}$ must belong to $\mathbf{D}^+_{\mathrm{Coh}}(\mathrm{Qcoh}/X)$. It follows that $\mathbf{R}\mathcal{H}om(\mathcal{G}, f^!\mathcal{I})$ has to be in $\mathbf{D}^+_{\mathrm{Coh}}(\mathrm{Qcoh}/X)$. We remind the reader: the complex $\mathcal{G}$ is finite, and hence we can immediately reduce to the case where $\mathcal{G}$ is a single coherent sheaf concentrated in degree 0. If $\mathcal{G}$ is a single sheaf, then there is a spectral sequence converging to the cohomology sheaves of the complex $\mathbf{R}\mathcal{H}om(\mathcal{G}, f^!\mathcal{I})$, whose $E_2^{ij}$ term is $\mathcal{E}xt^i\big(\mathcal{G}, \mathbb{H}^j(f^!\mathcal{I})\big)$. The sheaves $E_2^{ij}$ are all coherent, and $\mathbb{H}^{i+j}\mathbf{R}\mathcal{H}om(\mathcal{G}, f^!\mathcal{I})$ is a finite extension of subquotients of them, hence also coherent.

So much for coherence; to establish (iv) it remains to show that the cohomology of $\mathbf{R}\mathcal{H}om(\mathcal{G}, f^!\mathcal{I})$ is bounded above. For this we use [46, Theorem 4.2]. Choose an object $S$ as in Reminder 3.17. We know that the vanishing, for sufficiently large $k$, of the two sequences

$$\mathbb{H}^k\big(\mathbf{R}\mathcal{H}om(\mathcal{G}, f^!\mathcal{I})\big)\,, \qquad\qquad \mathrm{Hom}\big(S\,,\ \Sigma^k\mathbf{R}\mathcal{H}om(\mathcal{G}, f^!\mathcal{I})\big)$$

is equivalent. And the point is that we can compute the groups on the right; they are

$$\begin{aligned} \mathrm{Hom}\big(S\,,\ \mathbf{R}\mathcal{H}om(\mathcal{G}, \Sigma^k f^!\mathcal{I})\big) &= \mathrm{Hom}(S\stackrel{\mathbf{L}}{\otimes}\mathcal{G},\ \Sigma^k f^!\mathcal{I}) \\ &= \mathrm{Hom}\big(\mathbf{R}f_*(S\stackrel{\mathbf{L}}{\otimes}\mathcal{G}),\ \Sigma^k\mathcal{I}\big). \end{aligned}$$

Now $S$ is a perfect complex; locally it is a bounded complex of finitely generated projectives. The complex $\mathcal{G}$ is locally a bounded complex of finite modules, and the derived tensor product is locally just the tensor product; locally $S\stackrel{\mathbf{L}}{\otimes}\mathcal{G}$ is a bounded complex of finite modules. Thus $S\stackrel{\mathbf{L}}{\otimes}\mathcal{G}$ belongs to $\mathbf{D}^b(\mathrm{Coh}/X)$, and by (i) it follows that $\mathbf{R}f_*(S\stackrel{\mathbf{L}}{\otimes}\mathcal{G})$ must belong to $\mathbf{D}^b(\mathrm{Coh}/Y)$. The complex $\mathcal{I}$ is by hypothesis a dualizing complex, and hence $\mathbf{R}\mathcal{H}om\big(\mathbf{R}f_*(S\stackrel{\mathbf{L}}{\otimes}\mathcal{G}),\ \mathcal{I}\big)$ must also belong to $\mathbf{D}^b(\mathrm{Coh}/Y)$. Its cohomology may be computed using the Čech complex on $Y$, and hence is bounded. Therefore

$$\mathrm{Hom}\big(\mathbf{R}f_*(S\stackrel{\mathbf{L}}{\otimes}\mathcal{G}),\ \Sigma^k\mathcal{I}\big) = H^k\mathbf{R}\mathcal{H}om\big(\mathbf{R}f_*(S\stackrel{\mathbf{L}}{\otimes}\mathcal{G}),\ \mathcal{I}\big)$$

must vanish for $k$ sufficiently large.

It remains to prove (v); we must show that the natural map $\rho : \mathcal{O}_X \longrightarrow \mathbf{R}\mathcal{H}om(f^!\mathcal{I}, f^!\mathcal{I})$ is an isomorphism. We have already proved, in (iv), that for any $\mathcal{G} \in \mathbf{D}^b(\mathrm{Coh}/X)$ we have $\mathbf{R}\mathcal{H}om(\mathcal{G}, f^!\mathcal{I}) \in \mathbf{D}^b(\mathrm{Coh}/X)$. If we put $\mathcal{G} = \mathcal{O}_X$ we conclude that $f^!\mathcal{I} \in \mathbf{D}^b(\mathrm{Coh}/X)$, and then if we put $\mathcal{G} = f^!\mathcal{I}$ we deduce that $\mathbf{R}\mathcal{H}om(f^{!}\mathcal{I}, f^{!}\mathcal{I})$ must also be in $\mathbf{D}^{b}(\mathrm{Coh}/X)$. The morphism $\rho : \mathcal{O}_X \longrightarrow$ $\mathbf{R}\mathcal{H}om(f^{!}\mathcal{I}, f^{!}\mathcal{I})$ is therefore a map in $\mathbf{D}^{b}(\mathrm{Coh}/X)$, and the mapping cone lies in $\mathbf{D}^{b}(\mathrm{Coh}/X)$. Locally the mapping cone has a minimal resolution; it will vanish if and only if, for all closed points $x \in X$, the tensor product with $k(x)$ vanishes. To prove that $\rho$ an isomorphism it therefore suffices to check that $k(x)^{\mathbf{L}}\otimes \rho$ is an isomorphism for every closed point $x \in X$. That is, we wish to study the natural map

$$k(x) \longrightarrow k(x)^{\mathbf{L}}\otimes \mathbf{R}\mathcal{H}om(f^{!}\mathcal{I}, f^{!}\mathcal{I})$$

and prove it an isomorphism.

Lemma 2.10 helps. The point $x$ is assumed closed, guaranteeing that the skyscraper sheaf $k(x)$, concentrated at the point $x \in X$, is a coherent sheaf. Lemma 2.10 therefore gives us a natural isomorphism

$$k(x)^{\mathbf{L}}\otimes \mathbf{R}\mathcal{H}om(f^{!}\mathcal{I}, f^{!}\mathcal{I}) \longrightarrow \mathbf{R}\mathcal{H}om\Big(\mathbf{R}\mathcal{H}om\big(k(x), f^{!}\mathcal{I}\big), f^{!}\mathcal{I}\Big),$$

and we are reduced to proving that the natural map

$$k(x) \longrightarrow \mathbf{R}\mathcal{H}om\Big(\mathbf{R}\mathcal{H}om\big(k(x), f^{!}\mathcal{I}\big), f^{!}\mathcal{I}\Big),$$

is an isomorphism. In the case of skyscraper sheaves concentrated at a single point we can safely apply $\mathbf{R}f_{*}$, and then check that the induced map on $Y$ is an isomorphism; after all the only effect is to push the problem forward to the point $f(x) \in Y$. But now observe

$$\mathbf{R}f_{*}\mathbf{R}\mathcal{H}om\Big(\mathbf{R}\mathcal{H}om\big(k(x), f^{!}\mathcal{I}\big), f^{!}\mathcal{I}\Big) \cong \mathbf{R}\mathcal{H}om\Big(\mathbf{R}f_{*}\mathbf{R}\mathcal{H}om\big(k(x), f^{!}\mathcal{I}\big), \mathcal{I}\Big)$$
$$\cong \mathbf{R}\mathcal{H}om\Big(\mathbf{R}\mathcal{H}om\big(\mathbf{R}f_{*}k(x), \mathcal{I}\big), \mathcal{I}\Big)$$
$$\cong \mathbf{R}f_{*}k(x).$$

The careful reader should note that, in the above isomorphisms, we assert that two morphisms $\mathbf{R}f_{*}\mathbf{R}\mathcal{H}om(\mathcal{E}, f^{!}\mathcal{I}) \longrightarrow \mathbf{R}\mathcal{H}om(\mathbf{R}f_{*}\mathcal{E}, \mathcal{I})$ are isomorphisms. In both cases $\mathcal{E} \in \mathbf{D}^{b}(\mathrm{Coh}/X)$ is supported at a single point $x \in X$, meaning that both complexes we are trying to prove isomorphic are acyclic away from $f(x) \in Y$. The fact that the natural map is an isomorphism may therefore be checked after taking global sections. The reader need not worry that we might be appealing to some subtle facts about $f^{!}$ commuting with base change, and there is no need to verify the hypotheses of [51, Lemma 6.1 and Proposition 6.2].

The proof is now complete. $\qquad\square$

**Remark 3.18.** The hypothesis of Theorem 3.15 are that (i) $\mathbf{R}f_{*}$ take $\mathbf{D}^{b}(\mathrm{Coh}/X)$ to $\mathbf{D}^{b}(\mathrm{Coh}/Y)$, and (ii) $f^{!}$ take $\mathbf{D}^{b}(\mathrm{Coh}/Y)$ to $\mathbf{D}^{+}_{\mathrm{Coh}}(\mathrm{Qcoh}/X)$. In Reminder 3.14 we noted that (i) holds provided $f$ is a proper morphism of finite type. All we observed concerning (ii) was that it is not automatic; we made no mention of any interesting examples of $f$s which satisfy (ii). It is time to remedy this. We begin with the easy

**Lemma 3.19.** *If* $f : X \longrightarrow Y$ *is a finite morphism, then* $f^! : \mathbf{D}(\mathrm{Qcoh}/Y) \longrightarrow \mathbf{D}(\mathrm{Qcoh}/X)$ *takes* $\mathbf{D}^b(\mathrm{Coh}/Y)$ *to* $\mathbf{D}^+_{\mathrm{Coh}}(\mathrm{Qcoh}/X)$.

*Proof.* If $f$ is a finite morphism then it is affine; the sheaf $f_*\mathcal{O}_X$ is a coherent sheaf of $\mathcal{O}_Y$–algebras on $Y$, and $X$ is simply $\mathrm{Spec}(f_*\mathcal{O}_X)$. The functor $f_*$ is exact, and $\mathbf{R}f_*$ is just the forgetful functor, which takes a chain complex $C$ of $\mathcal{O}_X$–modules on $X$ and views $f_*C$ as a chain complex of $\mathcal{O}_Y$–modules via the homomorphism $\mathcal{O}_Y \longrightarrow f_*\mathcal{O}_X$; there is no need to derive the functor $f_*$.

The right adjoint, or more precisely $f_*$ of the right adjoint, is the functor which takes a complex $D$ of $\mathcal{O}_Y$ modules on $Y$, and produces out of it the complex $\mathbf{R}\mathcal{H}om_{\mathcal{O}_Y}(f_*\mathcal{O}_X, D)$. Perhaps it might be clearer to restrict to an open affine of $Y$, giving the local description. Replacing $Y$ by an open affine subset and $X$ by the inverse image, we have that $Y = \mathrm{Spec}(R)$ and $X = \mathrm{Spec}(S)$. The morphism $f : X \longrightarrow Y$ comes from a ring homomorphism $\varphi : R \longrightarrow S$, and the fact that $f$ is finite means that $S$ is finite as an $R$–module. The functor $f^!$ takes a complex $D$ of $R$–modules to $\mathbf{R}\mathrm{Hom}_R(S, D)$. We are assuming that $D$ is a bounded complex of finite $R$–modules, and wish to show that $f^!D$ is a bounded-below complex of $S$–modules, whose cohomology modules are finite over $S$.

Easy reduction tells us that we may assume the finite complex $D$ is of length 1; we may take $D$ to be a single, finite $R$–module concentrated in degree 0. The cohomology of the complex $f^!D$ is then $\mathrm{Ext}^i(S, D)$; it vanishes in negative degrees, and is always finite, either as an $R$–module or as an $S$–module. $\square$

**Example 3.20.** Suppose $Y$ is a any (noetherian, separated) scheme. Then the sheaf $\mathcal{O}_Y$ certainly belongs to $\mathbf{D}^b(\mathrm{Coh}/Y)$, and clearly satisfies $\mathbf{R}\mathcal{H}om(\mathcal{O}_Y, \mathcal{O}_Y) = \mathcal{O}_Y$. This much is free.

If we assume that $Y$ is finite dimensional and regular, or more generally that it is Gorenstein, then the sheaf $\mathcal{O}_Y$ has finite injective dimension and therefore, for every object $\mathcal{E} \in \mathbf{D}^b(\mathrm{Coh}/Y)$, we have that $\mathbf{R}\mathcal{H}om(\mathcal{E}, \mathcal{O}_Y)$ has only finitely many non-vanishing cohomology sheaves. It is an easy exercise to show that they are all coherent; therefore $\mathbf{R}\mathcal{H}om(\mathcal{E}, \mathcal{O}_Y)$ must belong to $\mathbf{D}^b(\mathrm{Coh}/Y)$. We conclude that $\mathcal{O}_Y$ is a dualizing complex for $Y$.

From Lemma 3.19 and Theorem 3.15 we know that, if $f : X \longrightarrow Y$ is any finite morphism, then $f^!\mathcal{O}_Y$ is a dualizing complex on $X$. The special case where $f : X \longrightarrow Y$ is a closed immersion tells us that any scheme, which admits a closed immersion into a Gorenstein scheme, must have a dualizing complex. Dualizing complexes are quite common.

# 4. When $\mathbf{R}f_*$ respects compacts

Let $f : X \longrightarrow Y$ be a morphism of separated, noetherian schemes. Theorem 3.15 gives us sufficient conditions for the functor $f^!$ to take dualizing complexes on $Y$ to dualizing complexes on $X$. The conditions on $f$ come in two components: the first, part (i) of Theorem 3.15, says that the functor $\mathbf{R}f_*$ should take $\mathbf{D}^b(\mathrm{Coh}/X)$ to $\mathbf{D}^b(\mathrm{Coh}/Y)$. This hypothesis we understand; see Remark 3.14. Part (ii) is the restriction that $f^!$ should take $\mathbf{D}^b(\mathrm{Coh}/Y)$ to $\mathbf{D}^+_{\mathrm{Coh}}(\mathrm{Qcoh}/X)$. So far, the only example we have of an $f$ satisfying (ii) comes from Lemma 3.19; the hypothesis is satisfied for any finite morphism $f$. It turns out that there is another large class of $f$s satisfying the condition in Theorem 3.15(ii), and this section is devoted to studying them. We do not understand this class fully; the unsatisfactory state of our current knowledge will be made precise in Conjecture 4.16.

Consider the morphisms $f : X \longrightarrow Y$ for which $\mathbf{R}f_*$ takes compacts to compacts; in the literature they sometimes go by the name *quasi-perfect*. Suppose Conjecture 4.16 is true. Assume $f : X \longrightarrow Y$ is a quasi-perfect morphism of noetherian, separated schemes, we will show that $f^!$ takes $\mathbf{D}^b(\mathrm{Coh}/Y)$ to $\mathbf{D}^b(\mathrm{Coh}/X)$; it most certainly satisfies the hypothesis in Theorem 3.15(ii). In other words, if the reader is willing to believe Conjecture 4.16, then Theorem 3.15 applies to quasi-perfect $f$s.

There is a refinement; Conjecture 4.16 can be made separately for each scheme $Z$. It suffices, in the paragraph above, to know that the conjecture is true for $X$. Precisely: if $f : X \longrightarrow Y$ is a quasi-perfect morphism of noetherian, separated schemes, and if Conjecture 4.16 is *true for $X$*, then we already know that $f^!$ takes $\mathbf{D}^b(\mathrm{Coh}/Y)$ to $\mathbf{D}^b(\mathrm{Coh}/X)$. In this section and in §5 we will explain this, and show that there are many classes of $X$s which satisfy Conjecture 4.16.

Quasi-perfect morphisms $f$ have been studied extensively elsewhere, and it seems remarkable that we will be able to say something new about them. Because this section might be of independent interest, to people who could not care less about dualizing complexes, we depart from our usual conventions in the article. For this section we drop the hypothesis that our schemes should be noetherian; during most of the section we will only assume them to be quasicompact and separated.

**Reminder 4.1.** We remind the reader: if $X$ is a quasicompact, separated scheme then the category $\mathbf{D}(\mathrm{Qcoh}/X)$ is compactly generated, and therefore has products; see [52, Proposition 8.4.6]. Next we note:

**Lemma 4.2.** *Let $f : X \longrightarrow Y$ be a morphism of quasicompact, separated schemes, and assume $\mathbf{R}f_* : \mathbf{D}(\mathrm{Qcoh}/X) \longrightarrow \mathbf{D}(\mathrm{Qcoh}/Y)$ takes compacts to* *compacts. Then the left adjoint* $\mathbf{L}f^* : \mathbf{D}(\mathrm{Qcoh}/Y) \longrightarrow \mathbf{D}(\mathrm{Qcoh}/X)$ *respects products.*

*Proof.* Let $\{\mathcal{Y}_\lambda,\ \lambda \in \Lambda\}$ be a set of objects in $\mathbf{D}(\mathrm{Qcoh}/Y)$. We wish to show that the natural map

$$\varphi \ : \ \mathbf{L}f^* \left( \prod_{\lambda \in \Lambda} \mathcal{Y}_\lambda \right) \longrightarrow \prod_{\lambda \in \Lambda} \mathbf{L}f^* \mathcal{Y}_\lambda$$

is an isomorphism. Since $\mathbf{D}(\mathrm{Qcoh}/X)$ is compactly generated it suffices to show that, for every compact object $\mathcal{E} \in \mathbf{D}(\mathrm{Qcoh}/X)$, the functor $\mathrm{Hom}_X(\mathcal{E}, -)$ takes $\varphi$ to an isomorphism. Of course we can factor the functor $\mathrm{Hom}_X(\mathcal{E}, -)$; it can be expressed as a composite

$$\mathbf{D}(\mathrm{Qcoh}/X) \xrightarrow{\ \mathbf{R}\mathcal{H}om(\mathcal{E},-)\ } \mathbf{D}(\mathrm{Qcoh}/X) \xrightarrow{\ \mathbf{R}f_*\ } \mathbf{D}(\mathrm{Qcoh}/Y) \xrightarrow{\ H^0\ } \mathcal{A}b,$$

and it clearly suffices to prove that the shorter composite

$$\mathbf{D}(\mathrm{Qcoh}/X) \xrightarrow{\ \mathbf{R}\mathcal{H}om(\mathcal{E},-)\ } \mathbf{D}(\mathrm{Qcoh}/X) \xrightarrow{\ \mathbf{R}f_*\ } \mathbf{D}(\mathrm{Qcoh}/Y)$$

takes $\varphi$ to an isomorphism. In Reminder 2.7 we saw that $\mathcal{E}$ is strongly dualizable and hence the functor $\mathbf{R}\mathcal{H}om(\mathcal{E}, -)$ identifies with $\mathcal{E}^\vee \ {}^{\mathbf{L}}\!\otimes -$. Parts (ii) and (iii) of Reminder 2.7 assure us that $\mathcal{E}^\vee$ is compact, while Reminder 2.7(i) informs us that every compact object can be written as $\mathcal{E}^\vee$ for some $\mathcal{E}$. We are therefore reduced to proving that, for any compact object $\mathcal{E} \in \mathbf{D}(\mathrm{Qcoh}/X)$, the composite

$$\mathbf{D}(\mathrm{Qcoh}/X) \xrightarrow{\ \mathcal{E}\ {}^{\mathbf{L}}\!\otimes-\ } \mathbf{D}(\mathrm{Qcoh}/X) \xrightarrow{\ \mathbf{R}f_*\ } \mathbf{D}(\mathrm{Qcoh}/Y)$$

takes $\varphi$ to an isomorphism.

To do this, observe the isomorphisms

$$
\begin{aligned}
\mathbf{R}f_* \left[ \mathcal{E}\ {}^{\mathbf{L}}\!\otimes \mathbf{L}f^* \left( \prod_{\lambda \in \Lambda} \mathcal{Y}_\lambda \right) \right] &\cong \mathbf{R}f_* \mathcal{E}\ {}^{\mathbf{L}}\!\otimes \left( \prod_{\lambda \in \Lambda} \mathcal{Y}_\lambda \right) && \text{projection formula} \\
&\cong \prod_{\lambda \in \Lambda} \left( \mathbf{R}f_* \mathcal{E}\ {}^{\mathbf{L}}\!\otimes \mathcal{Y}_\lambda \right) && \mathbf{R}f_*\mathcal{E} \text{ is compact} \\
&\cong \prod_{\lambda \in \Lambda} \mathbf{R}f_* \left( \mathcal{E}\ {}^{\mathbf{L}}\!\otimes \mathbf{L}f^* \mathcal{Y}_\lambda \right) && \text{projection formula} \\
&\cong \mathbf{R}f_* \prod_{\lambda \in \Lambda} \left( \mathcal{E}\ {}^{\mathbf{L}}\!\otimes \mathbf{L}f^* \mathcal{Y}_\lambda \right) && \mathbf{R}f_* \text{ has left adjoint} \\
&\cong \mathbf{R}f_* \left[ \mathcal{E}\ {}^{\mathbf{L}}\!\otimes \prod_{\lambda \in \Lambda} \mathbf{L}f^* \mathcal{Y}_\lambda \right] && \mathcal{E} \text{ is compact}.
\end{aligned}
$$

Note that in this string of isomorphisms we twice appealed to Reminder 2.7(iii); tensor product with a compact object commutes with products. We used it for the compact object $\mathcal{E} \in \mathbf{D}(\mathrm{Qcoh}/X)$ and for $\mathbf{R}f_* \mathcal{E} \in \mathbf{D}(\mathrm{Qcoh}/Y)$. $\qquad\square$

**Remark 4.3.** The fact that $\mathbf{L}f^*$ respects products means that it must have a left adjoint. Van den Bergh suggested a formula for this adjoint; it should be given by the functor $\mathbf{R}f_*\big[f^!\mathcal{O}_Y{}^{\mathbf{L}}\otimes -\big]$. Lipman and Van den Bergh independently found proofs that the formula works, at least for large classes of $f$s.

**Lemma 4.4.** *Assume $f : X \longrightarrow Y$ is a morphism of quasicompact, separated schemes, and suppose further that $\mathbf{R}f_*$ takes compact objects in $\mathbf{D}(\mathrm{Qcoh}/X)$ to compact objects in $\mathbf{D}(\mathrm{Qcoh}/Y)$. Suppose $\{\mathfrak{X}_\lambda,\ \lambda \in \Lambda\}$ is a set of objects in $\mathbf{D}(\mathrm{Qcoh}/X)$, with each $\mathfrak{X}_\lambda$ isomorphic to a coproduct of suspensions of $\mathcal{O}_X$. Then the natural map*

$$f^!\mathcal{O}_Y{}^{\mathbf{L}}\otimes \left(\prod_{\lambda\in\Lambda}\mathfrak{X}_\lambda\right) \longrightarrow \prod_{\lambda\in\Lambda}(f^!\mathcal{O}_Y{}^{\mathbf{L}}\otimes \mathfrak{X}_\lambda)$$

*is an isomorphism.*

*Proof.* We know that $\mathbf{L}f^*\Sigma^n\mathcal{O}_Y \cong \Sigma^n\mathcal{O}_X$; hence any suspension of $\mathcal{O}_X$ can be expressed as $\mathbf{L}f^*$ of some object of $\mathbf{D}(\mathrm{Qcoh}/Y)$, and so can any coproduct of suspensions. For each of our objects $\mathfrak{X}_\lambda \in \mathbf{D}(\mathrm{Qcoh}/X)$ we may choose an object $\mathcal{Y}_\lambda \in \mathbf{D}(\mathrm{Qcoh}/Y)$ and an isomorphism $\mathfrak{X}_\lambda \cong \mathbf{L}f^*\mathcal{Y}_\lambda$. Now [51, Theorem 5.4] gives us a natural isomorphism, for every $\mathcal{Y} \in \mathbf{D}(\mathrm{Qcoh}/Y)$,

$$f^!\mathcal{Y} \cong f^!\mathcal{O}_Y{}^{\mathbf{L}}\otimes \mathbf{L}f^*\mathcal{Y},$$

and we therefore have isomorphisms

$$
\begin{aligned}
f^!\mathcal{O}_Y{}^{\mathbf{L}}\otimes \left(\prod_{\lambda\in\Lambda}\mathfrak{X}_\lambda\right) &\cong f^!\mathcal{O}_Y{}^{\mathbf{L}}\otimes \left(\prod_{\lambda\in\Lambda}\mathbf{L}f^*\mathcal{Y}_\lambda\right) \\
&\cong f^!\mathcal{O}_Y{}^{\mathbf{L}}\otimes \mathbf{L}f^*\left(\prod_{\lambda\in\Lambda}\mathcal{Y}_\lambda\right) && \text{Lemma 4.2} \\
&\cong f^!\left(\prod_{\lambda\in\Lambda}\mathcal{Y}_\lambda\right) && \text{[51, Theorem 5.4]} \\
&\cong \prod_{\lambda\in\Lambda} f^!\mathcal{Y}_\lambda && f^! \text{ has left adjoint} \\
&\cong \prod_{\lambda\in\Lambda}\big(f^!\mathcal{O}_Y{}^{\mathbf{L}}\otimes \mathbf{L}f^*\mathcal{Y}_\lambda\big) && \text{[51, Theorem 5.4]} \\
&\cong \prod_{\lambda\in\Lambda}\big(f^!\mathcal{O}_Y{}^{\mathbf{L}}\otimes \mathfrak{X}_\lambda\big),
\end{aligned}
$$

completing the proof. $\square$

**Remark 4.5.** We have shown that $f^!\mathcal{O}_Y$ is an object of $\mathbf{D}(\mathrm{Qcoh}/X)$ so that the functor $f^!\mathcal{O}_Y \overset{\mathbf{L}}{\otimes} -$ commutes with some products; it commutes with products of the $\mathcal{X}_\lambda$s of Lemma 4.4. It is natural to ask whether it commutes with all products. Reminder 2.7(iii) tells us that this is equivalent to asking whether $f^!\mathcal{O}_Y$ is compact.

The general answer is No: we remind the reader how construct a counterexample. Consider simple case where $Y = \mathrm{Spec}(k)$, where $k$ is a field, and $f : X \longrightarrow Y$ is a finite morphism. For our choice of $Y$ the sheaf $\mathcal{O}_Y$ is a dualizing complex on $Y$ and Lemma 3.19, coupled with Theorem 3.15, tell us that $f^!\mathcal{O}_Y$ is a dualizing complex on $X$. Our hypothesis is that $X$ is finite over $Y = \mathrm{Spec}(k)$, in which case we know that the (essentially unique) dualizing complex has finite injective dimension. If it is also a compact object then the complex $\mathbf{R}\mathcal{H}om(f^!\mathcal{O}_Y, f^!\mathcal{O}_Y)$ is quasi-isomorphic to the complex of homomorphisms from a bounded complex of finitely generated projectives to a bounded complex of injectives; but Proposition 3.6(ii) gives a quasi-isomorphism $\mathcal{O}_X \longrightarrow \mathbf{R}\mathcal{H}om(f^!\mathcal{O}_Y, f^!\mathcal{O}_Y)$. This can only happen if $\mathcal{O}_X$ has a bounded injective resolution, that is if $X$ is a Gorenstein scheme.

**Remark 4.6.** While $f^!\mathcal{O}_Y$ need not in general be compact, there are interesting things one can say about it, facts which I do not fully understand. To illustrate one of the strange features, a phenomenon which seems mysterious to me, consider the following. By [51, Theorem 5.4] there is an isomorphism $f^!(-) \cong f^!\mathcal{O}_Y \overset{\mathbf{L}}{\otimes} \mathbf{L}f^*(-)$. Applying $\mathbf{R}f_*$ to this isomorphism, and using the projection formula, we deduce an isomorphism

$$\mathbf{R}f_*f^!(-) \cong \left[\mathbf{R}f_*f^!\mathcal{O}_Y\right] \overset{\mathbf{L}}{\otimes} - .$$

The functor on the left is the composite of two right adjoints, hence commutes with products. It follows that so does the functor on the right; Reminder 2.7(iii) now informs us that $\mathbf{R}f_*f^!\mathcal{O}_Y$ has to be compact. Even though $f^!\mathcal{O}_Y$ need not be compact, its pushforward $\mathbf{R}f_*f^!\mathcal{O}_Y$ must be.

In the light of Lemma 4.4 and Remark 4.5 it makes sense to study the class of objects $\mathcal{L} \in \mathbf{D}(\mathrm{Qcoh}/X)$ for which the functor $\mathcal{L} \overset{\mathbf{L}}{\otimes} -$ commutes with the products of Lemma 4.4. Let us make this a definition.

**Definition 4.7.** *Let $X$ be a quasicompact, separated scheme. We define a class of objects $\mathcal{S}(X) \subset \mathbf{D}(\mathrm{Qcoh}/X)$ by the following property. An object $\mathcal{L} \in \mathbf{D}(\mathrm{Qcoh}/X)$ belongs to $\mathcal{S}(X)$ if and only if the natural map*

$$\mathcal{L} \overset{\mathbf{L}}{\otimes} \left(\prod_{\lambda \in \Lambda} \mathcal{X}_\lambda\right) \longrightarrow \prod_{\lambda \in \Lambda}(\mathcal{L} \overset{\mathbf{L}}{\otimes} \mathcal{X}_\lambda)$$

*is an isomorphism whenever $\{\mathfrak{X}_\lambda,\ \lambda \in \Lambda\}$ is a set of objects in $\mathbf{D}(\mathrm{Qcoh}/X)$ as in Lemma 4.4. We remind the reader: this means that each $\mathfrak{X}_\lambda$ isomorphic to a coproduct of suspensions of $\mathcal{O}_X$.*

**Remark 4.8.** When $X$ can be understood from context we will omit it from the notation; that is we will write $\mathcal{S}$ for $\mathcal{S}(X)$. Reminder 2.7(iii) tells us that all compact objects belong to $\mathcal{S} = \mathcal{S}(X)$. Lemma 4.4 says that, if $f : X \longrightarrow Y$ is a morphism where $\mathbf{R}f_*$ takes compacts to compacts, then $f^!\mathcal{O}_Y$ belongs to $\mathcal{S}(X)$. Clearly, if $\mathcal{E} \longrightarrow \mathcal{F} \longrightarrow \mathcal{G} \longrightarrow \Sigma\mathcal{E}$ is a distinguished triangle, and if two of the objects belong to $\mathcal{S}$, then so does the third. Not so trivial is the lemma

**Lemma 4.9.** *Suppose $\mathcal{E}$ is a compact object in $\mathbf{D}(\mathrm{Qcoh}/X)$, and let $\mathcal{L}$ be an object in the class $\mathcal{S}$ of Definition 4.7. Then there exists a compact object $\mathcal{F} \in \mathbf{D}(\mathrm{Qcoh}/X)$, and a morphism $f : \mathcal{F} \longrightarrow \mathcal{L}$, so that any morphism $\Sigma^n\mathcal{E} \longrightarrow \mathcal{L}$ factors as $\Sigma^n\mathcal{E} \longrightarrow \mathcal{F} \overset{f}{\longrightarrow} \mathcal{L}$.*

*Proof.* Let $\Lambda$ be the set of all morphisms $\mathcal{E} \longrightarrow \Sigma^n\mathcal{L}$. If $\lambda \in \Lambda$ is a morphism $\mathcal{E} \longrightarrow \Sigma^n\mathcal{L}$, put $\mathfrak{X}_\lambda = \Sigma^n\mathcal{O}_X$. Because $\mathcal{L}$ belongs to $\mathcal{S}$ we know that the natural map

$$\mathcal{L} \overset{\mathbf{L}}{\otimes} \left( \prod_{\lambda \in \Lambda} \mathfrak{X}_\lambda \right) \longrightarrow \prod_{\lambda \in \Lambda} (\mathcal{L} \overset{\mathbf{L}}{\otimes} \mathfrak{X}_\lambda)$$

is an isomorphism. The collection $\Lambda$, of all maps $\mathcal{E} \longrightarrow \Sigma^n\mathcal{L} = \mathcal{L} \overset{\mathbf{L}}{\otimes} \mathfrak{X}_\lambda$, assembles to a single map

$$\mathcal{E} \longrightarrow \prod_{\lambda \in \Lambda} (\mathcal{L} \overset{\mathbf{L}}{\otimes} \mathfrak{X}_\lambda),$$

which must therefore factor as

$$\mathcal{E} \longrightarrow \mathcal{L} \overset{\mathbf{L}}{\otimes} \left( \prod_{\lambda \in \Lambda} \mathfrak{X}_\lambda \right) \longrightarrow \prod_{\lambda \in \Lambda} (\mathcal{L} \overset{\mathbf{L}}{\otimes} \mathfrak{X}_\lambda).$$

But $\mathcal{E}$ is compact; any map $\mathcal{E} \longrightarrow \mathcal{G} \overset{\mathbf{L}}{\otimes} \mathcal{H}$ factors as $\mathcal{E} \longrightarrow \mathcal{F} \overset{\mathbf{L}}{\otimes} \mathcal{H} \overset{f^{\mathbf{L}}\otimes 1}{\longrightarrow} \mathcal{G} \overset{\mathbf{L}}{\otimes} \mathcal{H}$, with $\mathcal{F} \in \mathbf{D}(\mathrm{Qcoh}/X)$ compact. Applying this to the above we deduce a factorization

$$\mathcal{E} \longrightarrow \mathcal{F} \overset{\mathbf{L}}{\otimes} \left( \prod_{\lambda \in \Lambda} \mathfrak{X}_\lambda \right) \overset{f^{\mathbf{L}}\otimes 1}{\longrightarrow} \mathcal{L} \overset{\mathbf{L}}{\otimes} \left( \prod_{\lambda \in \Lambda} \mathfrak{X}_\lambda \right) \longrightarrow \prod_{\lambda \in \Lambda} (\mathcal{L} \overset{\mathbf{L}}{\otimes} \mathfrak{X}_\lambda).$$

Now consider the commutative diagram

$$\mathcal{E} \longrightarrow \mathcal{F}^{\mathbf{L}} \otimes \left(\prod_{\lambda \in \Lambda} \mathcal{X}_\lambda\right) \xrightarrow{f^{\mathbf{L}} \otimes 1} \mathcal{L}^{\mathbf{L}} \otimes \left(\prod_{\lambda \in \Lambda} \mathcal{X}_\lambda\right)$$
$$\downarrow \qquad\qquad\qquad\qquad\qquad \downarrow$$
$$\prod_{\lambda \in \Lambda}(\mathcal{F}^{\mathbf{L}} \otimes \mathcal{X}_\lambda) \xrightarrow[\prod_{\lambda \in \Lambda} f^{\mathbf{L}} \otimes 1]{} \prod_{\lambda \in \Lambda}(\mathcal{L}^{\mathbf{L}} \otimes \mathcal{X}_\lambda).$$

By construction, the composite from top left to bottom right amounts to assembling the collection of all maps $\mathcal{E} \longrightarrow \Sigma^n \mathcal{L}$ into a single morphism. The factorization

$$\mathcal{E} \longrightarrow \mathcal{F}^{\mathbf{L}} \otimes \left(\prod_{\lambda \in \Lambda} \mathcal{X}_\lambda\right)$$
$$\downarrow$$
$$\prod_{\lambda \in \Lambda}(\mathcal{F}^{\mathbf{L}} \otimes \mathcal{X}_\lambda) \xrightarrow[\prod_{\lambda \in \Lambda} f^{\mathbf{L}} \otimes 1]{} \prod_{\lambda \in \Lambda}(\mathcal{L}^{\mathbf{L}} \otimes \mathcal{X}_\lambda)$$

gives us a single map $f : \mathcal{F} \longrightarrow \mathcal{L}$, and informs us that each morphism $\mathcal{E} \longrightarrow \Sigma^n \mathcal{L}$ has a factorization $\mathcal{E} \longrightarrow \Sigma^n \mathcal{F} \xrightarrow{\Sigma^n f} \Sigma^n \mathcal{L}$. $\qquad \square$

This already permits us to deduce the following two facts.

**Lemma 4.10.** *Let $X$ be a quasicompact, separated scheme. Let $\mathcal{L}$ be an object belonging to $\mathcal{S}(X) \subset \mathbf{D}(\mathrm{Qcoh}/X)$. Then*

(i) *The object $\mathcal{L}$ is $\aleph_1$–compact, in the sense of [52, Definition 4.2.7].*

(ii) *If $\mathbf{D}(\mathrm{Qcoh}/X)$ has a strong compact generator then $\mathcal{L}$ is compact.*

*Proof.* By [5, Theorem 3.1.1] there is a compact generator in $\mathbf{D}(\mathrm{Qcoh}/X)$; let $\mathcal{E}$ be such a compact generator. If we are proving (ii), assume further that $\mathcal{E}$ is a strong generator; this means that there exists an integer $n$ so that any compact can be obtained from $\mathcal{E}$ in $n$ steps; see [5, Definition 2.2.3].

Lemma 4.4 tells us that we may choose a compact object $\mathcal{F} = \mathcal{F}_0$, and a morphism $f_0 : \mathcal{F}_0 \longrightarrow \mathcal{L}$, so that any morphism $\Sigma^n \mathcal{E} \longrightarrow \mathcal{L}$ factors through $f_0$. That is, for every integer $n$ the functor $\mathrm{Hom}(\Sigma^n \mathcal{E}, -)$ takes $f_0 : \mathcal{F}_0 \longrightarrow \mathcal{L}$ to an epimorphism. Now assume we have a compact object $\mathcal{F}_i$ and a morphism $f_i : \mathcal{F}_i \longrightarrow \mathcal{L}$, and we will show how to produce a commutative triangle

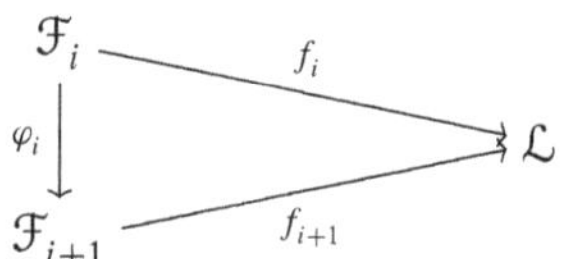

with the properties that

(iii) The object $\mathcal{F}_{i+1}$ is compact.

(iv) The group homomorphism $\mathrm{Hom}(\Sigma^n\mathcal{E}, \varphi_i)$ annihilates the kernel of $\mathrm{Hom}(\Sigma^n\mathcal{E}, f_i)$.

Complete $f_i$ to a distinguished triangle

$$\mathcal{L}' \xrightarrow{\ g\ } \mathcal{F}_i \xrightarrow{\ f_i\ } \mathcal{L} \longrightarrow \Sigma\mathcal{L}'.$$

By assumption $\mathcal{L}$ belongs to $\mathcal{S}(X)$, as does the compact object $\mathcal{F}_i$. Therefore $\mathcal{L}'$ also belongs to $\mathcal{S}(X)$. Applying Lemma 4.4 to $\mathcal{L}'$ we discover a compact object $\mathcal{G}$ and a morphism $h : \mathcal{G} \longrightarrow \mathcal{L}'$, so that every map $\Sigma^n\mathcal{E} \longrightarrow \mathcal{L}'$ factors through $h$. That is, each of the functors $\mathrm{Hom}(\Sigma^n\mathcal{E}, -)$ takes $h$ to a surjection. Since the functors $\mathrm{Hom}(\Sigma^n\mathcal{E}, -)$ are homological they take any triangle to exact sequences; for every integer $n$ the functor $\mathrm{Hom}(\Sigma^n\mathcal{E}, -)$ takes $\mathcal{L}' \longrightarrow \mathcal{F}_i \longrightarrow \mathcal{L}$ to an exact sequence. Combining the surjections with the exact sequences, we deduce the exactness of

$$\mathrm{Hom}(\Sigma^n\mathcal{E}, \mathcal{G}) \xrightarrow{\ \mathrm{Hom}(\Sigma^n\mathcal{E}, gh)\ } \mathrm{Hom}(\Sigma^n\mathcal{E}, \mathcal{F}_i) \xrightarrow{\ \mathrm{Hom}(\Sigma^n\mathcal{E}, f_i)\ } \mathrm{Hom}(\Sigma^n\mathcal{E}, \mathcal{L}).$$

The kernel of $\mathrm{Hom}(\Sigma^n\mathcal{E}, f_i)$ is therefore the image of $\mathrm{Hom}(\Sigma^n\mathcal{E}, gh)$. Now complete $gh : \mathcal{G} \longrightarrow \mathcal{F}_i$ to a distinguished triangle

$$\mathcal{G} \xrightarrow{\ gh\ } \mathcal{F}_i \xrightarrow{\ \varphi_i\ } \mathcal{F}_{i+1} \longrightarrow \Sigma\mathcal{G}.$$

Clearly $\mathrm{Hom}(\Sigma^n\mathcal{E}, \varphi_i)$ annihilates the image of $\mathrm{Hom}(\Sigma^n\mathcal{E}, gh)$, which is the kernel of $\mathrm{Hom}(\Sigma^n\mathcal{E}, f_i)$; we have achieved (iv). Because $\mathcal{G}$ and $\mathcal{F}_i$ are both compact, so is $\mathcal{F}_{i+1}$; that is (iii) also holds. Because the composite $\mathcal{G} \xrightarrow{gh} \mathcal{F}_i \xrightarrow{f_i} \mathcal{L}$ vanishes, the map $f_i$ must factor as $\mathcal{F}_i \xrightarrow{\varphi_i} \mathcal{F}_{i+1} \xrightarrow{f_{i+1}} \mathcal{L}$. This yields our commutative triangle

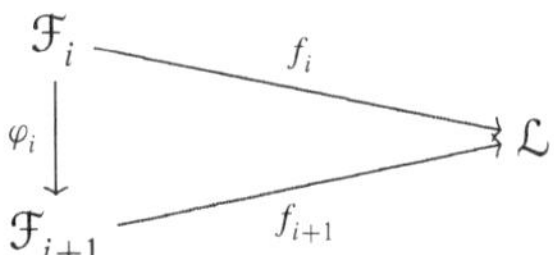

We have proved all our inductive claims.

We have a sequence of morphisms

$$\mathcal{F}_0 \xrightarrow{\ \varphi_0\ } \mathcal{F}_1 \xrightarrow{\ \varphi_1\ } \mathcal{F}_2 \xrightarrow{\ \varphi_2\ } \cdots$$

and compatible maps to $\mathcal{L}$; we can factor through a map $\varphi : \underrightarrow{\mathrm{Hocolim}}\ \mathcal{F}_i \longrightarrow \mathcal{L}$. The usual argument, as in Brown's original proof of the Brown representability theorem, tells us that each of the functors $\mathrm{Hom}(\Sigma^n\mathcal{E}, -)$ takes the map $\varphi$

to an isomorphism; the reader can find this many places, for example in [51, §3]. Since $\mathcal{E}$ generates it follows that $\varphi$ is an isomorphism; $\mathcal{L}$ is isomorphic to $\underrightarrow{\mathrm{Hocolim}}\ f_i$, a countable homotopy colimit of compact objects. Hence $\mathcal{L}$ is $\aleph_1$–compact. This establishes (i).

It remains to prove (ii); from now we assume $\mathcal{E}$ is a strong generator, and our notation will be as in [5, §2.2]. Let $n$ be an integer so that $\langle \mathcal{E} \rangle_n = \{\mathbf{D}(\mathrm{Qcoh}/X)\}^c$. Recall our sequence above, of morphisms $f_i : \mathcal{F}_i \longrightarrow \mathcal{L}$. One can show, by an induction on $i$ which we leave to the reader,

(v) Any morphism $\mathcal{K} \longrightarrow \mathcal{L}$, with $\mathcal{K} \in \langle \mathcal{E} \rangle_i$, can be factored through $f_i : \mathcal{F}_i \longrightarrow \mathcal{L}$.

(vi) Given a vanishing composite $\mathcal{K} \longrightarrow \mathcal{F}_j \xrightarrow{f_j} \mathcal{L}$, with $\mathcal{K} \in \langle \mathcal{E} \rangle_i$, then the composite

$$\mathcal{K} \longrightarrow \mathcal{F}_j \xrightarrow{\varphi_j} \mathcal{F}_{j+1} \xrightarrow{\varphi_{j+1}} \mathcal{F}_{j+1} \xrightarrow{\varphi_{j+2}} \cdots \xrightarrow{\varphi_{i+j+1}} \mathcal{F}_{i+j}$$

already vanishes.

In our sequence we constructed a morphism $f_{2n} : \mathcal{F}_{2n} \longrightarrow \mathcal{L}$, with $\mathcal{F}_{2n}$ compact, that is with $\mathcal{F}_{2n} \in \langle \mathcal{E} \rangle_n$. By (v) it must factor through $f_n : \mathcal{F}_n \longrightarrow \mathcal{L}$. Choose a factorization $\mathcal{F}_{2n} \xrightarrow{\alpha} \mathcal{F}_n \xrightarrow{f_n} \mathcal{L}$. The longer composite $\mathcal{F}_n \xrightarrow{\varphi} \mathcal{F}_{2n} \xrightarrow{\alpha} \mathcal{F}_n \xrightarrow{f_n} \mathcal{L}$ is clearly equal to $f_n : \mathcal{F}_n \longrightarrow \mathcal{L}$. Put $e = 1 - \alpha\varphi$. Then $e : \mathcal{F}_n \longrightarrow \mathcal{F}_n$ is a morphism so that the composite $\mathcal{F}_n \xrightarrow{e} \mathcal{F}_n \xrightarrow{f_n} \mathcal{L}$ vanishes. By (vi) we deduce the vanishing of the composite $\mathcal{F}_n \xrightarrow{e} \mathcal{F}_n \xrightarrow{\varphi} \mathcal{F}_{2n}$. This means that the longer composite $\mathcal{F}_n \xrightarrow{e} \mathcal{F}_n \xrightarrow{\varphi} \mathcal{F}_{2n} \xrightarrow{\alpha} \mathcal{F}_n$ also vanishes; that is $e(1 - e) = 0$. We have shown that $e$ is an idempotent.

Therefore $\mathcal{F}_n$ splits as $\mathcal{F}_n \cong \mathcal{F} \oplus \mathcal{F}'$, where $\alpha\varphi = 1$ on $\mathcal{F}$ and $0$ on $\mathcal{F}'$. The reader can now check that each of the functors $\mathrm{Hom}(\Sigma^n \mathcal{E}, -)$ takes the map $\mathcal{F} \longrightarrow \mathcal{L}$ to an isomorphism, and we conclude that $\mathcal{L} \cong \mathcal{F}$ is compact. $\qquad \square$

**Remark 4.11.** From Lemma 4.10(i) we learn that $\mathcal{S}(X) \subset \{\mathbf{D}(\mathrm{Qcoh}/X)\}^{\aleph_1}$, where $\{\mathbf{D}(\mathrm{Qcoh}/X)\}^{\aleph_1}$ is the full subcategory of $\aleph_1$–compact objects in $\mathbf{D}(\mathrm{Qcoh}/X)$. By [52, 8.4.2.2] we know that the category $\{\mathbf{D}(\mathrm{Qcoh}/X)\}^{\aleph_1}$ is essentially small. We deduce that, up to isomorphisms, there is only a set of objects in $\mathcal{S}(X)$.

**Lemma 4.12.** *Let $X = \mathrm{Spec}(R)$ be an affine scheme. If $\mathcal{L}$ belongs to $\mathcal{S}(X)$ then $\mathcal{L}$ must be isomorphic in $\mathbf{D}(\mathrm{Qcoh}/X) \cong \mathbf{D}(R\text{–Mod})$ to a bounded above complex of finitely generated projectives, and only finitely many of the cohomology groups of $\mathcal{L}$ can be non-zero.*

*Proof.* We apply Lemma 4.9 with $\mathcal{E} = R$. We discover that there must exist a single compact object $\mathcal{F}$, that is a bounded complex of finitely generated, projective $R$–modules, as well single map $f : \mathcal{F} \longrightarrow \mathcal{L}$, so that any morphism $\Sigma^n R \longrightarrow \mathcal{L}$ factors through $f$. The morphisms $\Sigma^n R \longrightarrow \mathcal{L}$ are in bijection with elements of $H^{-n}(\mathcal{L})$ and we learn that, for every $i$, the map $H^i(\mathcal{F}) \longrightarrow H^i(\mathcal{L})$ must be surjective. But $\mathcal{F}$ is bounded, hence $H^i(\mathcal{L})$ vanishes for all but finitely many $i$. It remains to show that $\mathcal{L}$ is isomorphic in $\mathbf{D}(R\text{–Mod})$ to a bounded above chain complex of finitely generated projectives.

Suppose that $i$ is the integer at which the cohomology of $\mathcal{L}$ stops; that is $H^i(\mathcal{L}) \neq 0$, but $H^j(\mathcal{L}) = 0$ for all $j > i$. It suffices to produce a finitely generated, projective module $P^i$ and an epimorphism $P^i \longrightarrow H^i(\mathcal{L})$. Let us first establish that this really suffices. Suppose such a morphism exists; it must lift to a map in the derived category $\Sigma^{-i} P^i \longrightarrow \mathcal{L} = \mathcal{L}_0$. The third edge of the triangle $\Sigma^{-i} P^i \longrightarrow \mathcal{L} \longrightarrow \mathcal{L}_1 \longrightarrow \Sigma^{-i+1} P^i$ is an object $\mathcal{L}_1 \in \mathcal{S}(X)$ whose cohomology stops at $(i-1)$. We can now apply induction to obtain a sequence of finitely generated projectives and morphisms $\Sigma^{-k} P^k \longrightarrow \mathcal{L}_k$, where the cohomology of $\mathcal{L}_k$ stops at $i - k$, and these finitely generated projectives assemble to a chain complex

$$\cdots \longrightarrow P^{i-2} \longrightarrow P^{i-1} \longrightarrow P^i \longrightarrow 0$$

quasi-isomorphic to $\mathcal{L}$.

It therefore remains to prove the existence of the surjection $P^i \longrightarrow H^i(\mathcal{L})$. Let $\mathcal{P}$ be a $K$–projective resolution for $\mathcal{L}$; we may choose $\mathcal{P}$ to be a bounded above complex of projective modules, which stops at $i$. That is $\mathcal{P}$ is a chain complex

$$\cdots \longrightarrow \mathcal{P}^{i-2} \longrightarrow \mathcal{P}^{i-1} \longrightarrow \mathcal{P}^i \longrightarrow 0$$

and $H^i(\mathcal{P}) \cong H^i(\mathcal{L}) \neq 0$. The morphism $f : \mathcal{F} \longrightarrow \mathcal{P}$ is a map in the derived category between bounded above chain complexes of projectives, and hence may be realized by a chain map. That is, we have a chain map

$$
\begin{array}{ccccccccccc}
\cdots \longrightarrow & \mathcal{F}^{i-2} & \longrightarrow & \mathcal{F}^{i-1} & \longrightarrow & \mathcal{F}^i & \overset{\partial_{\mathcal{F}}^{i+1}}{\longrightarrow} & \mathcal{F}^{i+1} & \longrightarrow & \mathcal{F}^{i+2} & \longrightarrow \cdots \\
& \downarrow & & \downarrow & & \downarrow & & \downarrow & & \downarrow & \\
\cdots \longrightarrow & \mathcal{P}^{i-2} & \longrightarrow & \mathcal{P}^{i-1} & \underset{\partial_{\mathcal{P}}^i}{\longrightarrow} & \mathcal{P}^i & \longrightarrow & 0 & \longrightarrow & 0 & \longrightarrow \cdots
\end{array}
$$

From the first paragraph of the proof we know that the map $H^i(\mathcal{F}) \longrightarrow H^i(\mathcal{P})$ is surjective; that is the kernel of $\partial_{\mathcal{F}}^{i+1}$ surjects onto the cokernel of $\partial_{\mathcal{P}}^i$. But then $\mathcal{F}^i$ certainly must also surject to the cokernel; we have found our surjection $\mathcal{F}^i \longrightarrow H^i(\mathcal{P})$, with $\mathcal{F}^i$ finitely generated and projective. $\qquad\square$

This ends what I can currently prove in glorious generality. Let me next formulate a conjecture.

**Conjecture 4.13.** *Let $X$ be any quasicompact, separated scheme, and let $\mathcal{L}$ be an object in $\mathcal{S}(X)$. Then $\mathcal{L}$ is pseudocoherent, and has only finitely many non-vanishing cohomology groups.*

**Reminder 4.14.** We remind the reader: a complex $\mathcal{L}$ is pseudocoherent provided the restriction of $\mathcal{L}$, to any open affine subset $U \subset X$, is isomorphic in $\mathbf{D}(\mathrm{Qcoh}/U) \cong \mathbf{D}(R\text{–Mod})$ to a bounded above complex of finitely generated projective $R$–modules.

**Remark 4.15.** In view of Lemma 4.12, Conjecture 4.13 amounts to the statement that if $\mathcal{L} \in \mathcal{S}(X)$ and $j : U \longrightarrow X$ is an open immersion of an affine open subset, then $j^*(\mathcal{L})$ is in $\mathcal{S}(U)$.

Let us formulate an even stronger conjecture:

**Conjecture 4.16.** *Let $X$ be a quasicompact, separated scheme. Then $X$ satisfies the following:*

$(*)$ *For any quasicompact open subset $U \subset X$ there exist a compact object $\mathcal{E} \in \mathbf{D}(\mathrm{Qcoh}/X)$, and an integer $n \geq 1$, so that*

$$\mathbf{R}j_*\mathcal{O}_U \quad \in \quad \overline{\langle \mathcal{E} \rangle}_n\,;$$

*here $j : U \longrightarrow X$ is the inclusion of the open set, and $\overline{\langle \mathcal{E} \rangle}_n$ is as in [5, Definition 2.2.3].*

**Reminder 4.17.** Perhaps we should remind the reader of the notation in [5, Definition 2.2.3]. Let $\mathcal{E}$ be an object in some triangulated category $\mathcal{T}$. The full subcategory $\overline{\langle \mathcal{E} \rangle}_1$ is defined to be the one containing all direct summands of all coproducts of arbitrary suspensions of $\mathcal{E}$. The category $\overline{\langle \mathcal{E} \rangle}_n$ is defined inductively; an object lies in $\overline{\langle \mathcal{E} \rangle}_{n+1}$ if it is a direct summand of an object $y$, and $y$ fits in a triangle

$$x \longrightarrow y \longrightarrow z \longrightarrow \Sigma x$$

with $x \in \overline{\langle \mathcal{E} \rangle}_1$ and $z \in \overline{\langle \mathcal{E} \rangle}_n$.

**Remark 4.18.** Now that we have reminded the reader of the notation in Conjecture 4.16, we should also explain its relevance. In the remainder of this section we will show that Conjecture 4.16 implies Conjecture 4.13. In §5 we will study the many cases in which we know Conjecture 4.16 to be true. And then, in §6, we will return to the relation with Grothendieck duality.

**Definition 4.19.** *Let $X$ be a quasicompact, separated scheme. A class of objects $T \subset \mathbf{D}(\mathrm{Qcoh}/X)$ will be called* adapted *if the natural map*

$$\mathcal{L} \overset{\mathbf{L}}{\otimes} \left( \prod_{\lambda \in \Lambda} t_\lambda \right) \longrightarrow \prod_{\lambda \in \Lambda} (\mathcal{L} \overset{\mathbf{L}}{\otimes} t_\lambda)$$

*is an isomorphism, whenever the following conditions both hold:*

(i) $\Lambda$ *is a set of objects $\{t_\lambda, \lambda \in \Lambda\}$, where each $t_\lambda$ belongs to $T$.*
(ii) *The object $\mathcal{L}$ belongs to $\mathcal{S}(X)$.*

**Remark 4.20.** By definition of $\mathcal{S}(X)$ the class $T$, containing all coproducts of arbitrary suspensions of $\mathcal{O}_X$, is an adapted class. If $T$ is an adapted class then so is

(i) The class of all mapping cones on morphisms $t \longrightarrow t'$, with $t, t' \in T$.
(ii) The class of all direct summands of objects in $T$.
(iii) The class of all objects $\mathcal{E} \overset{\mathbf{L}}{\otimes} t$, where $\mathcal{E}$ is a fixed compact object and $t \in T$.

All three facts are easy; we leave (i) and (ii) to the reader and indicate the proof of (iii). Let $\mathcal{E}$ be a compact object, $\mathcal{L}$ an object in $\mathcal{S}$, and assume all the $t_\lambda$ lie in $T$. We have isomorphisms

$$\mathcal{L} \overset{\mathbf{L}}{\otimes} \left[ \prod_{\lambda \in \Lambda} (\mathcal{E} \overset{\mathbf{L}}{\otimes} t_\lambda) \right] \cong \mathcal{L} \overset{\mathbf{L}}{\otimes} \mathcal{E} \overset{\mathbf{L}}{\otimes} \left( \prod_{\lambda \in \Lambda} t_\lambda \right) \qquad \text{Reminder 2.7(iii)}$$

$$\cong \mathcal{E} \overset{\mathbf{L}}{\otimes} \left[ \prod_{\lambda \in \Lambda} (\mathcal{L} \overset{\mathbf{L}}{\otimes} t_\lambda) \right] \qquad \text{Because } \mathcal{L} \in \mathcal{S}, \text{ all } t_\lambda \in T$$

$$\cong \prod_{\lambda \in \Lambda} (\mathcal{L} \overset{\mathbf{L}}{\otimes} \mathcal{E} \overset{\mathbf{L}}{\otimes} t_\lambda) \qquad \text{Reminder 2.7(iii)} \, .$$

**Lemma 4.21.** *Conjecture 4.16 implies Conjecture 4.13. More precisely for each $X$, Conjecture 4.16 for $X$ implies Conjecture 4.13 for $X$. We spell this out: if $X$ is a quasicompact, separated scheme, and if Conjecture 4.16 holds for $X$, then every object $\mathcal{L} \in \mathcal{S}(X)$ is pseudocoherent, and the cohomology sheaves $\mathbb{H}^i(\mathcal{L})$ vanish for all but finitely many $i$.*

*Proof.* Assume $X$ is a quasicompact, separated scheme, let $U \subset X$ be an affine open set, and let $j : U \longrightarrow X$ be the inclusion. We are assuming Conjecture 4.16 is true; we may choose a compact object $\mathcal{E} \in \mathbf{D}(\mathrm{Qcoh}/X)$, and an integer $n \geq 1$, so that $\mathbf{R}j_*\mathcal{O}_U \in \overline{\langle \mathcal{E} \rangle}_n$. Choose and fix $\mathcal{E}$ and $n$. Note that $j : U \longrightarrow X$ is an affine morphism, hence $j_*$ is exact; there is no difference between $\mathbf{R}j_*\mathcal{O}_U$ and $j_*\mathcal{O}_U$. We will use the shorter $j_*\mathcal{O}_U$.

We observed, at the beginning of Remark 4.20, that the class $T$ of all coproducts of all suspensions of $\mathcal{O}_X$ is adapted. By Remark 4.20(iii) so is the class of all $\mathcal{E}\overset{\mathbf{L}}{\otimes} t$, with $\mathcal{E}$ as in the previous paragraph and $t \in T$. That is, the class of all coproducts of arbitrary suspensions of $\mathcal{E}$ is adapted. By Remark 4.20(ii) so is the class of all direct summands of the above; that is $\overline{\langle \mathcal{E}\rangle}_1$ is adapted. Now parts (i) and (ii) of Remark 4.20 tell us that if $\overline{\langle \mathcal{E}\rangle}_i$ is adapted then so is $\overline{\langle \mathcal{E}\rangle}_{2i}$. We conclude that all the classes $\overline{\langle \mathcal{E}\rangle}_i$ are adapted.

The first paragraph of the proof informs us that $\overline{\langle \mathcal{E}\rangle}_n$ contains all coproducts of arbitrary suspensions of $j_*\mathcal{O}_U$. Now $j_*$ respects coproducts; a coproduct of suspensions of $j_*\mathcal{O}_U$ is $j_*$ of the coproduct of suspensions of $\mathcal{O}_U$. Let $T' \subset \mathbf{D}(\mathrm{Qcoh}/U)$ be the class of all coproducts of arbitrary suspensions of $\mathcal{O}_U$; we have that $j_*T'$ is an adapted class. This means that the natural map

$$\mathcal{L}\overset{\mathbf{L}}{\otimes}\left(\coprod_{\lambda\in\Lambda} j_*t_\lambda\right) \longrightarrow \coprod_{\lambda\in\Lambda}(\mathcal{L}\overset{\mathbf{L}}{\otimes} j_*t_\lambda)$$

is an isomorphism, whenever the following conditions both hold:

(i) $\Lambda$ is a set of objects $\{t_\lambda,\ \lambda \in \Lambda\}$, where each $t_\lambda$ belongs to $T'$.
(ii) The object $\mathcal{L}$ belongs to $\mathcal{S}(X)$.

Now take an object $\mathcal{L} \in \mathcal{S}(X)$ and a set of objects $\{t_\lambda,\ \lambda \in \Lambda\}$, where each $t_\lambda$ belongs to $T'$, and observe the string of isomorphisms

$$
\begin{aligned}
j_*\left[j^*\mathcal{L}\overset{\mathbf{L}}{\otimes}\left(\coprod_{\lambda\in\Lambda} t_\lambda\right)\right] &\cong \mathcal{L}\overset{\mathbf{L}}{\otimes} j_*\left(\coprod_{\lambda\in\Lambda} t_\lambda\right) && \text{projection formula} \\
&\cong \mathcal{L}\overset{\mathbf{L}}{\otimes}\left(\coprod_{\lambda\in\Lambda} j_*t_\lambda\right) && j_* \text{ has left adjoint} \\
&\cong \coprod_{\lambda\in\Lambda}(\mathcal{L}\overset{\mathbf{L}}{\otimes} j_*t_\lambda) && j_*T' \text{ is adapted} \\
&\cong \coprod_{\lambda\in\Lambda} j_*(j^*\mathcal{L}\overset{\mathbf{L}}{\otimes} t_\lambda) && \text{projection formula} \\
&\cong j_*\left[\coprod_{\lambda\in\Lambda}(j^*\mathcal{L}\overset{\mathbf{L}}{\otimes} t_\lambda)\right] && j_* \text{ has left adjoint} .
\end{aligned}
$$

Applying $j^*$ to the isomorphism, and recalling that $j^*j_*$ is naturally isomorphic to the identity, this simplifies to saying that the natural map

$$j^*\mathcal{L}\overset{\mathbf{L}}{\otimes}\left(\coprod_{\lambda\in\Lambda} t_\lambda\right) \longrightarrow \coprod_{\lambda\in\Lambda}(j^*\mathcal{L}\overset{\mathbf{L}}{\otimes} t_\lambda)$$

is an isomorphism. That is $j^*\mathcal{L}$ belongs to $\mathcal{S}(U) \subset \mathbf{D}(\mathrm{Qcoh}/U)$. Since $U$ is affine, Lemma 4.12 tells us that $j^*\mathcal{L}$ is a bounded above complex of finitely generated projectives, with bounded cohomology. $\qquad\square$

## 5. Where we can prove Conjecture 4.16

This section is about studying the cases where we know Conjecture 4.16 to be true. The conjecture is the assertion that sheaves of the form $\mathbf{R}j_*\mathcal{O}_U$ lie in $\overline{\langle \mathcal{E} \rangle}_n$, for suitable choices of a compact object $\mathcal{E}$ and an integer $n$. We observe

**Remark 5.1.** Let $X$ be a quasicompact, separated scheme. If there is a compact object $\mathcal{E} \in \mathbf{D}(\mathrm{Qcoh}/X)$ and an integer $n$ so that $\overline{\langle \mathcal{E} \rangle}_n = \mathbf{D}(\mathrm{Qcoh}/X)$, then Conjecture 4.16 most definitely holds for $X$.

And the next observation is that there are examples in the literature. Specifically, the reader is referred to the proof[6] of [5, Theorem 3.1.4]. In there it is shown that, if $X$ is smooth over a field $k$, then $\overline{\langle \mathcal{E} \rangle}_n = \mathbf{D}(\mathrm{Qcoh}/X)$ for some $n$ and some compact $\mathcal{E}$. In Theorem 5.13 we will prove a very slight refinement of the result in [5]; there is no need to assume that the ground ring is a field, a finite dimensional, regular, noetherian ring is quite enough.

This means that we have cheap examples of schemes $X$ for which Conjecture 4.16 is true; but all of them are smooth and noetherian. To obtain singular, non-noetherian examples we will have to learn how to produce new examples out of old ones. This section is mostly about developing the machinery.

We begin with a little definition.

**Definition 5.2.** *Let $X$ be a quasicompact, separated scheme. An open set $U \subset X$ will be called* decent *if*

 (i) *$U$ is quasicompact.*
(ii) *Let $j : U \longrightarrow X$ be the inclusion. There exist a compact object $\mathcal{E} \in \mathbf{D}(\mathrm{Qcoh}/X)$, and an integer $n \geq 1$, so that $\mathbf{R}j_*\mathcal{O}_U \in \overline{\langle \mathcal{E} \rangle}_n$.*

**Remark 5.3.** In the terminology introduced in Definition 5.2, Conjecture 4.16 asserts that every quasicompact open subset $U \subset X$ is decent. The idea of this section will be to find ways to produce decent open subsets.

---

[6] The reader should note that the statement of [5, Theorem 3.1.4] is slightly different; the result we want is merely a step in the proof. See the last paragraph of [5, §3.4] for details.

**Lemma 5.4.** *Let $X = \mathrm{Spec}(R)$ be an affine scheme, and let $X_f = \mathrm{Spec}(R[1/f])$ be the basic open subset consisting of the prime ideals not containing $f \in R$. Then $X_f \subset X$ is decent.*

*Proof.* Being affine, $X_f$ is clearly quasicompact. In the category $\mathbf{D}(\mathrm{Qcoh}/X) \cong \mathbf{D}(R\text{–Mod})$, the object $\mathbf{R}j_*\mathcal{O}_{X_f}$ simplifies to $R[1/f] \in \mathbf{D}(R\text{–Mod})$. We need find a compact object $\mathcal{E}$ and an integer $n$, and exhibit $R[1/f]$ as an object in $\overline{\langle \mathcal{E} \rangle}_n$. Now note that $R[1/f]$ is the colimit of the sequence

$$ R \xrightarrow{\ f\ } R \xrightarrow{\ f\ } R \xrightarrow{\ f\ } R \xrightarrow{\ f\ } R \xrightarrow{\ f\ } \cdots $$

It is therefore also the homotopy colimit. There is a triangle in $\mathbf{D}(R\text{–Mod})$

$$ \coprod_{n=0}^{\infty} R \longrightarrow \coprod_{n=0}^{\infty} R \longrightarrow R[1/f] \longrightarrow \coprod_{n=0}^{\infty} \Sigma R. $$

Put $\mathcal{E} = R$ and $n = 2$; the triangle shows that $R[1/f]$ belongs to $\overline{\langle R \rangle}_2$. $\qquad\square$

**Lemma 5.5.** *Suppose $X$ is a quasicompact, separated scheme. Let $U$ and $V$ be open subsets, and assume that $U$, $V$ and $U \cap V$ are all decent. Then so is $U \cup V$.*

*Proof.* The quasicompactness is clear; $U \cup V$ is the union of two quasicompact open sets $U$ and $V$. We have to worry about the sheaf $\mathbf{R}j_*\mathcal{O}_{U \cup V}$. Since there are several open subsets $W \subset X$ to consider, and several inclusions among them, we return to our notation of the proof of Lemma 2.4; given open subsets $W_1 \subset W_2$ of $X$, we will denote the inclusion by $j_{W_1}^{W_2} : W_1 \longrightarrow W_2$.

By hypothesis we may choose an integer $n$, and compact objects $\mathcal{E}_U$, $\mathcal{E}_V$ and $\mathcal{E}_{U \cap V}$ in $\mathbf{D}(\mathrm{Qcoh}/X)$, so that

$$ \mathbf{R}\{j_U^X\}_* \mathcal{O}_U \in \overline{\langle \mathcal{E}_U \rangle}_n, \quad \mathbf{R}\{j_V^X\}_* \mathcal{O}_V \in \overline{\langle \mathcal{E}_V \rangle}_n, \quad \mathbf{R}\{j_{U \cap V}^X\}_* \mathcal{O}_{U \cap V} \in \overline{\langle \mathcal{E}_{U \cap V} \rangle}_n. $$

Put $\mathcal{E} = \mathcal{E}_U \oplus \mathcal{E}_V \oplus \mathcal{E}_{U \cap V}$. Then $\mathcal{E}$ is compact, and the three complexes

$$ \mathbf{R}\{j_U^X\}_* \mathcal{O}_U, \qquad \mathbf{R}\{j_V^X\}_* \mathcal{O}_V, \qquad \mathbf{R}\{j_{U \cap V}^X\}_* \mathcal{O}_{U \cap V} $$

all belong to $\overline{\langle \mathcal{E} \rangle}_n$. Now consider the distinguished triangle

$$ \mathbf{R}\{j_{U \cup V}^X\}_* \mathcal{O}_{U \cup V} \longrightarrow \mathbf{R}\{j_U^X\}_* \mathcal{O}_U \oplus \mathbf{R}\{j_V^X\}_* \mathcal{O}_V \longrightarrow $$
$$ \mathbf{R}\{j_{U \cap V}^X\}_* \mathcal{O}_{U \cap V} \longrightarrow $$

Two of the objects belong to $\overline{\langle \mathcal{E} \rangle}_n$ and hence the third, namely $\mathbf{R}\{j_{U \cup V}^X\}_* \mathcal{O}_{U \cup V}$, must belong to $\overline{\langle \mathcal{E} \rangle}_{2n}$. $\qquad\square$

**Lemma 5.6.** *Let $X = \mathrm{Spec}(R)$ be affine. Then any quasicompact open subset $U \subset X$ is decent. That is, Conjecture 4.16 holds for $X$.*

*Proof.* The open sets $X_f$, $f \in R$ form a basis for the topology of $X = \mathrm{Spec}(R)$; any open set $U$ is the union of the $X_f$'s contained in it. If $U$ is quasicompact we may express it as a finite union of $X_f$'s. We will prove, by induction on $n$,

(i) The union of $n$ basic open sets, that is open sets of the form $X_f$, is a decent open set.

Lemma 5.4 gives us the case $n = 1$. Suppose therefore that we know (i) for $n$, that $U = \cup_{i=1}^{n} X_{f_i}$ is the union of $n$ open sets $X_{f_i}$, and that $V = X_g$ is another basic open set. We want to show that $U \cup V$ is decent, and Lemma 5.5 tells us that it suffices to show that $U$, $V$ and $U \cap V$ are. The case of $V = X_g$ comes from Lemma 5.4, while for $U$ and

$$U \cap V = \left(\bigcup_{i=1}^{n} X_{f_i}\right) \cap X_g = \bigcup_{i=1}^{n} (X_{f_i} \cap X_g) = \bigcup_{i=1}^{n} X_{f_i g}$$

we appeal to the fact that both can be covered by $n$ basic open sets. $\qquad\square$

**Lemma 5.7.** *Decency is transitive. Precisely: suppose $X$ is a quasicompact, separated scheme, $U \subset X$ is a decent open set of $X$, and $V \subset U$ is a decent open subset of $U$. Then $V$ is a decent open subset of $X$.*

*Proof.* The quasicompactness is clear. To keep the notation uncluttered let us write $j_1 : V \longrightarrow U$, $j_2 : U \longrightarrow X$ and $j : V \longrightarrow X$ for the three inclusions; we wish to prove that $\mathbf{R}j_*\mathcal{O}_V$ is contained in some $\overline{\langle \mathcal{K} \rangle}_\ell$.

Because $V$ is a decent open subset of $U$ we may produce a compact object $\mathcal{E} \in \mathbf{D}(\mathrm{Qcoh}/U)$, and an integer $m \geq 1$, so that $\mathbf{R}\{j_1\}_*\mathcal{O}_V \in \overline{\langle \mathcal{E} \rangle}_m$. The complex $\mathcal{E} \oplus \Sigma\mathcal{E}$ vanishes in the Grothendieck group $K_0(U)$, and Thomason's localization theorem [62, 5.2.2(a)] applies. There exists a compact object $\mathcal{F} \in \mathbf{D}(\mathrm{Qcoh}/X)$ with $j_2^*\mathcal{F} \cong \mathcal{E} \oplus \Sigma\mathcal{E}$. Choose such an $\mathcal{F}$; we have that

$$\mathbf{R}\{j_1\}_*\mathcal{O}_V \quad \in \quad \overline{\langle j_2^*\mathcal{F} \rangle}_m. \tag{\dag}$$

We also know that $U$ is decent in $X$. We may therefore find a compact object $\mathcal{G} \in \mathbf{D}(\mathrm{Qcoh}/X)$, and an integer $n \geq 1$, so that $\mathbf{R}\{j_2\}_*\mathcal{O}_U \in \overline{\langle \mathcal{G} \rangle}_n$. Now we combine these. If we apply the functor $\mathbf{R}\{j_2\}_*$ to the inclusion $(\dag)$ above, we conclude that the complex $\mathbf{R}j_*\mathcal{O}_V = \mathbf{R}\{j_2\}_*\mathbf{R}\{j_1\}_*\mathcal{O}_V$ is contained in

$$
\begin{aligned}
\mathbf{R}\{j_2\}_* \overline{\langle j_2^*\mathcal{F} \rangle}_m &\subset \overline{\langle \mathbf{R}\{j_2\}_* j_2^*\mathcal{F} \rangle}_m && \mathbf{R}\{j_2\}_* \text{ respects coproducts} \\
&= \overline{\langle \mathbf{R}\{j_2\}_* (j_2^*\mathcal{F} \overset{\mathbf{L}}{\otimes} \mathcal{O}_U) \rangle}_m && \text{obvious} \\
&= \overline{\langle \mathcal{F} \overset{\mathbf{L}}{\otimes} \mathbf{R}\{j_2\}_* \mathcal{O}_U \rangle}_m && \text{projection formula} \\
&\subset \overline{\langle \mathcal{F} \overset{\mathbf{L}}{\otimes} \mathcal{G} \rangle}_{mn},
\end{aligned}
$$

where the last inclusion is because $\mathbf{R}\{j_2\}_* \mathcal{O}_U$ belongs to $\overline{\langle \mathcal{G} \rangle}_n$, hence $\mathcal{F}^{\mathbf{L}} \otimes \mathbf{R}\{j_2\}_* \mathcal{O}_U$ belongs to $\overline{\langle \mathcal{F}^{\mathbf{L}} \otimes \mathcal{G} \rangle}_n$, and therefore

$$\overline{\langle \mathcal{F}^{\mathbf{L}} \otimes \mathbf{R}\{j_2\}_* \mathcal{O}_U \rangle}_m \subset \overline{\langle \mathcal{F}^{\mathbf{L}} \otimes \mathcal{G} \rangle}_{mn}.$$

$\square$

**Lemma 5.8.** *Let $X$ be a quasicompact, separated scheme, and let $\{U_i, 1 \leq i \leq n\}$ be a finite number of decent open affine subsets of $X$. Then the union $\cup_{i=1}^n U_i$ is decent.*

*Proof.* We prove this by induction on $n$, the case $n = 1$ being trivial. Suppose we know the assertion for $n$, and suppose we have $n + 1$ decent open affines $\{U_i, 1 \leq i \leq n + 1\}$. Let $U = \cup_{i=1}^n U_i$ and let $V = U_{n+1}$; by induction we know that $U$ and $V$ are decent, and we want to prove that so is $U \cup V$. By Lemma 5.5 it suffices to prove the decency of $U \cap V$. Now $X$ is assumed separated, hence the intersection of two quasicompact open subsets is quasicompact. Therefore $U \cap V \subset V$ is a quasicompact open subset of the affine scheme $V$, and Lemma 5.6 informs us that $U \cap V$ is decent in $V$. We are given that $V$ is decent in $X$ and the transitivity of decency, that is Lemma 5.7, permits us to conclude that $U \cap V$ is decent in $X$. $\square$

**Proposition 5.9.** *Suppose $X$ is a quasicompact, separated scheme, and suppose that $X$ can be covered by decent open affines. Then Conjecture 4.16 holds for $X$; all quasicompact open sets are decent.*

*Proof.* We are assuming that $X$ is quasicompact and has a cover by decent open affines. There must exist a finite subcover. We may cover $X$ by $n$ decent open affines, for some integer $n \geq 1$; the proof will be by induction on $n$. The case $n = 1$ comes from Lemma 5.6. Suppose therefore that we know the assertion for schemes $X$ which admit covers by $n$ decent open affines, and let $X$ be a scheme admitting a cover by $n + 1$ decent open affines. Then $X = U \cup V$, where $U$ has a cover by $n$ decent open affines and $V$ is a decent open affine. Lemma 5.8 informs us that $U \subset X$ is decent. Let $W \subset X$ be any quasicompact open subset; we want to prove that $W$ is decent.

To do this observe that $W$ can be written as the union

$$W = (W \cap U) \cup (W \cap V).$$

Now $X$ is assumed separated, and hence the intersection of any two quasicompact open sets is quasicompact; we deduce that $W \cap U$, $W \cap V$ and $W \cap U \cap V$ are all quasicompact. Induction tells us that

(i) $W \cap U$ is decent as an open subset of the scheme $U$.

(ii) $W \cap V$ and $W \cap U \cap V$ are both decent as open subsets of the scheme $V$.

Now the transitivity of decency, that is Lemma 5.7, guarantees that $W \cap U$, $W \cap V$ and $W \cap U \cap V$ are all decent as open subsets of $X$. Lemma 5.5 permits us to conclude that $W \subset X$ is decent. $\qquad\square$

**Lemma 5.10.** *Let $f : X \longrightarrow Y$ be a morphism of quasicompact, separated schemes. If $V \subset Y$ is decent then so is $f^{-1}V \subset X$.*

*Proof.* $Y$ is separated and the schemes $X$ and $V \subset Y$ are quasicompact, hence $f^{-1}V \subset X$ is quasicompact. The decency of $V \subset Y$ says further that we may choose a compact object $\mathcal{E} \in \mathbf{D}(\mathrm{Qcoh}/Y)$ and an integer $n \geq 1$ with $\mathbf{R}\{j_V^Y\}_* \mathcal{O}_V \in \overline{\langle \mathcal{E} \rangle}_n$. Therefore $\mathbf{L}f^*\mathcal{E}$ is a compact object in $\mathbf{D}(\mathrm{Qcoh}/X)$, and $\overline{\langle \mathbf{L}f^*\mathcal{E} \rangle}_n$ contains the object

$$\mathbf{L}f^* \mathbf{R}\{j_V^Y\}_* \mathcal{O}_V \cong \mathbf{R}\{j_{f^{-1}V}^X\}_* \mathbf{L}f^* \mathcal{O}_V \cong \mathbf{R}\{j_{f^{-1}V}^X\}_* \mathcal{O}_{f^{-1}V}.$$

$\qquad\square$

**Corollary 5.11.** *Let $f : X \longrightarrow Y$ be a morphism of quasicompact, separated schemes. Suppose $X$ can be covered by open affine subsets $f^{-1}V$, where $V \subset Y$ is open and quasicompact. If Conjecture 4.16 holds for $Y$ then it also holds for $X$.*

*Proof.* Conjecture 4.16 holds for $Y$ and therefore any quasicompact $V \subset Y$ is decent. By Lemma 5.10 all the open sets $f^{-1}V \subset X$ are decent, and by hypothesis we may choose among them a collection of affines which cover $X$. Proposition 5.9 allows us to conclude that Conjecture 4.16 holds for $X$. $\qquad\square$

**Remark 5.12.** There are two useful situations where Corollary 5.11 applies. They are:

(i) Any open immersion of quasicompact, separated schemes.

(ii) Affine morphisms; for example closed immersions, or more generally finite maps.

So much for machinery of confirming the decency of open subsets. Now we come to the way of producing examples.

**Theorem 5.13.** *Let $R$ be a (noetherian) regular ring of finite global dimension, and assume $X$ is a noetherian, separated scheme, smooth and finite dimensional over $R$. There exists an integer $n \geq 1$, as well as a compact object $\mathcal{E} \in \mathbf{D}(\mathrm{Qcoh}/X)$, with $\overline{\langle \mathcal{E} \rangle}_n = \mathbf{D}(\mathrm{Qcoh}/X)$.*

*Proof.* As we already mentioned, the proof is a miniscule modification of an argument that may be found in [5]. Consider the product $X \times_R X$. It is a finite dimensional, regular, noetherian scheme, and the sheaf $\mathcal{O}_\Delta$, that is the structure sheaf of the diagonal embedding $X \longrightarrow X \times_R X$, is a coherent sheaf on $X \times_R X$. Locally it has a finite resolution by finitely generated projectives; that is $\mathcal{O}_\Delta$ is a compact object in $\mathbf{D}(\mathrm{Qcoh}/X \times_R X)$.

Now choose a compact object $\mathcal{E}$ generating $\mathbf{D}(\mathrm{Qcoh}/X)$; such an object exists by [5, Theorem 3.1.1(2)]. From [5, Lemma 3.4.1] we learn that $\mathcal{F} = \pi_1^* \mathcal{E} \overset{\mathbf{L}}{\otimes} \pi_2^* \mathcal{E}$ is a compact generator of $\mathbf{D}(\mathrm{Qcoh}/X \times_R X)$. The compact object $\mathcal{O}_\Delta$ therefore lies in $\overline{\langle \mathcal{F} \rangle}_n$ for some integer $n \geq 1$. Also, because $R$ is a regular, finite dimensional ring, there is an integer $m$ with $\overline{\langle R \rangle}_m = \mathbf{D}(R\text{–Mod})$. We assert that $\overline{\langle \mathcal{E} \rangle}_{mn} = \mathbf{D}(\mathrm{Qcoh}/X)$.

To prove this, choose any object $\mathcal{S} \in \mathbf{D}(\mathrm{Qcoh}/X)$. Then $\mathcal{S}$ can be expressed as

$$\mathcal{S} \cong \{\pi_1\}_* \left[ \mathcal{O}_\Delta \overset{\mathbf{L}}{\otimes} \left( \pi_2^* \mathcal{S} \right) \right].$$

We know that $\mathcal{O}_\Delta \in \overline{\langle \mathcal{F} \rangle}_n$, with $\mathcal{F} = \pi_1^* \mathcal{E} \overset{\mathbf{L}}{\otimes} \pi_2^* \mathcal{E}$. Hence $\mathcal{S}$ must belong to $\overline{\langle \mathcal{G} \rangle}_n$, where

$$\mathcal{G} \cong \{\pi_1\}_* \left[ \mathcal{F} \overset{\mathbf{L}}{\otimes} \left( \pi_2^* \mathcal{S} \right) \right] \cong \{\pi_1\}_* \left[ \pi_1^* \mathcal{E} \overset{\mathbf{L}}{\otimes} \pi_2^* \mathcal{E} \overset{\mathbf{L}}{\otimes} \left( \pi_2^* \mathcal{S} \right) \right].$$

Manipulating a little more we have

$$\mathcal{G} \cong \{\pi_1\}_* \left[ \pi_1^* \mathcal{E} \overset{\mathbf{L}}{\otimes} \pi_2^* \left( \mathcal{E} \overset{\mathbf{L}}{\otimes} \mathcal{S} \right) \right] \cong \mathcal{E} \overset{\mathbf{L}}{\otimes} \{\pi_1\}_* \left[ \pi_2^* \left( \mathcal{E} \overset{\mathbf{L}}{\otimes} \mathcal{S} \right) \right],$$

and $\{\pi_1\}_* \left[ \pi_2^* \left( \mathcal{E} \overset{\mathbf{L}}{\otimes} \mathcal{S} \right) \right] \cong \pi^* \pi_* \left( \mathcal{E} \overset{\mathbf{L}}{\otimes} \mathcal{S} \right)$ is $\pi^*$ of a complex $\pi_* \left( \mathcal{E} \overset{\mathbf{L}}{\otimes} \mathcal{S} \right)$ belonging to $\mathbf{D}(R\text{–Mod}) = \overline{\langle R \rangle}_m$. Thus $\pi^* \pi_* \left( \mathcal{E} \overset{\mathbf{L}}{\otimes} \mathcal{S} \right)$ must belong to $\overline{\langle \mathcal{O}_X \rangle}_m \subset \mathbf{D}(\mathrm{Qcoh}/X)$, and

$$\mathcal{G} \cong \mathcal{E} \overset{\mathbf{L}}{\otimes} \left[ \pi^* \pi_* \left( \mathcal{E} \overset{\mathbf{L}}{\otimes} \mathcal{S} \right) \right]$$

lies in $\overline{\langle \mathcal{E} \rangle}_m$. Hence $\mathcal{S} \in \overline{\langle \mathcal{G} \rangle}_n \subset \overline{\langle \mathcal{E} \rangle}_{mn}$.      $\square$

## 6. Dualizing complexes and $f^!$

Let us leave the world of gorgeous generality and return to dealing with more restricted schemes, at least some of which will be noetherian. Assume $f : X \longrightarrow Y$ is a morphism of quasicompact, separated schemes, suppose $Y$ is noetherian, and let $\mathcal{I}$ be a dualizing complex on $Y$. If $\mathcal{E}$ is any object in $\mathbf{D}(\mathrm{Qcoh}/X)$, and $\mathcal{F}$ belongs to $\mathbf{D}^b(\mathrm{Coh}/Y) \subset \mathbf{D}(\mathrm{Qcoh}/Y)$, then we compute

$$\mathrm{Hom}_Y(\mathbf{R}f_*\mathcal{E}, \mathcal{F}) \cong \mathrm{Hom}_Y\left[\mathbf{R}f_*\mathcal{E}\ ,\ \mathbf{R}\mathcal{H}om_Y\big(\mathbf{R}\mathcal{H}om_Y(\mathcal{F}, \mathfrak{I}), \mathfrak{I}\big)\right]$$

$$\cong \mathrm{Hom}_Y\left[\mathbf{R}f_*\mathcal{E}\ ^{\mathbf{L}}\!\otimes \mathbf{R}\mathcal{H}om_Y(\mathcal{F}, \mathfrak{I}),\ \mathfrak{I}\right]$$

$$\cong \mathrm{Hom}_Y\left[\mathbf{R}f_*\big(\mathcal{E}\ ^{\mathbf{L}}\!\otimes \mathbf{L}f^*\mathbf{R}\mathcal{H}om_Y(\mathcal{F}, \mathfrak{I})\big),\ \mathfrak{I}\right]$$

$$\cong \mathrm{Hom}_X\left[\mathcal{E}\ ^{\mathbf{L}}\!\otimes \mathbf{L}f^*\mathbf{R}\mathcal{H}om_Y(\mathcal{F}, \mathfrak{I}),\ f^!\mathfrak{I}\right]$$

$$\cong \mathrm{Hom}_X\left[\mathcal{E}\ ,\ \mathbf{R}\mathcal{H}om_X\big(\mathbf{L}f^*\mathbf{R}\mathcal{H}om_Y(\mathcal{F}, \mathfrak{I}),\ f^!\mathfrak{I}\big)\right];$$

the first isomomorphism is by Lemma 3.5(ii), the second because $\mathbf{R}\mathcal{H}om$ is right adjoint to the tensor, the third by the projection formula, the fourth because $f^!$ is right adjoint to $\mathbf{R}f_*$, and the fifth by the adjunction between tensor and $\mathbf{R}\mathcal{H}om$. Put together, we deduce a natural isomorphism

$$f^!\mathcal{F} \cong \mathbf{R}\mathcal{H}om_X\big(\mathbf{L}f^*\mathbf{R}\mathcal{H}om_Y(\mathcal{F}, \mathfrak{I}),\ f^!\mathfrak{I}\big). \qquad (\dagger\dagger)$$

We have found a formula $(\dagger\dagger)$ for computing $f^!\mathcal{F}$, as long as $\mathcal{F}$ belongs to $\mathbf{D}^b(\mathrm{Coh}/Y)$. In the old days the right hand side was used to construct the functor $f^!$; in our modern day and age this seems anachronistic. We know, for purely formal reasons, that the functor $f^! : \mathbf{D}(\mathrm{Qcoh}/Y) \longrightarrow \mathbf{D}(\mathrm{Qcoh}/X)$ exists; see [51, Example 4.2].

Next we restrict our attention to an even smaller class of morphisms $f : X \longrightarrow Y$; for the rest of the section assume $f$ is a morphism of noetherian, separated schemes, and suppose further that

 (i) $\mathbf{R}f_*$ takes compacts to compacts.
(ii) The scheme $X$ satisfies Conjecture 4.16; for all we know this is no restriction.

From [46, Corollary 4.3.2] it follows that

(iii) $\mathbf{R}f_*$ takes $\mathbf{D}^b(\mathrm{Coh}/X)$ to $\mathbf{D}^b(\mathrm{Coh}/Y)$.

Lemma 4.4 informs us that $f^!\mathcal{O}_Y$ belongs to $\mathcal{S}(X)$, and, using (i) and (ii) above as well as Lemma 4.21, we conclude that $f^!\mathcal{O}_Y$ is in $\mathbf{D}^-(\mathrm{Coh}/X)$. We therefore deduce

 (iv) $f^!$ takes $\mathbf{D}^b(\mathrm{Coh}/Y)$ to $\mathbf{D}^b(\mathrm{Coh}/X)$; see Remark 0.6 and Corollary 0.7.
 (v) If $\mathfrak{I}$ is a dualizing complex on $Y$, then $f^!\mathfrak{I}$ is a dualizing complex on $X$; see Fact 0.3(ii).

If we look at (iii) and (iv), they tell us that the adjoint pair of functors $(\mathbf{R}f_*, f^!)$ restrict to functors between the subcategories $\mathbf{D}^b(\mathrm{Coh}/X) \subset \mathbf{D}(\mathrm{Qcoh}/X)$ and

$\mathbf{D}^b(\mathrm{Coh}/Y) \subset \mathbf{D}(\mathrm{Qcoh}/Y)$; the restrictions must also be an adjoint pair. The formula (††) even tells us how to compute $f^!$ in terms of dualizing complexes.

In practice this means that, if it happens to be more convenient to compute $f^!$ in the bounded derived categories, then go right ahead; from the restriction of $f^!$ to $\mathbf{D}^b(\mathrm{Coh}/Y)$ we can compute $f^!\mathcal{O}_Y$, and the formula

$$f^!\mathcal{S} = f^!\mathcal{O}_Y \, {}^{\mathbf{L}}\!\otimes \mathbf{L}f^*\mathcal{S}$$

tells us, at least in principle, how to work out all there is to know about the functor $f^! : \mathbf{D}(\mathrm{Qcoh}/Y) \longrightarrow \mathbf{D}(\mathrm{Qcoh}/X)$.

## 7. Several recent results

Let me end with a very brief glimpse at current progress in the field. We begin by observing that, up to now, our treatment has been based on studying the category $\mathbf{D}^b(\mathrm{Coh}/X)$ via its embedding in $\mathbf{D}(\mathrm{Qcoh}/X)$. Dualizing complexes naturally live in $\mathbf{D}^b(\mathrm{Coh}/X)$, and our study of them was by means of infinite coproducts, infinite products and compact objects in $\mathbf{D}(\mathrm{Qcoh}/X)$. As a result of ongoing work, we now know that there are more natural compactly generated triangulated categories to consider.

Krause [35] taught us that $\mathbf{D}^b(\mathrm{Coh}/X)$ can be viewed as the subcategory of compact objects in the compactly generated triangulated category $\mathbf{K}(\mathrm{Inj}/X)$. From the work of Jørgensen [29], Iyengar-Krause [27], myself [53, 54] and from Murfet's thesis [48], we also know a compactly generated triangulated category $\mathcal{T} = \mathbf{K}_m(\mathrm{Proj}/X)$ whose subcategory $\mathcal{T}^c$ of compact objects is naturally equivalent to $\mathbf{D}^b(\mathrm{Coh}/X)^{\mathrm{op}}$. If a dualizing complex $\mathcal{J}$ exists, then tensor product with (an injective resolution of) $\mathcal{J}$ gives an equivalence of categories

$$- \otimes \mathcal{J} \; : \; \mathbf{K}_m(\mathrm{Proj}/X) \longrightarrow \mathbf{K}(\mathrm{Inj}/X).$$

This approach is very new, poorly understood, and in my opinion it has great promise. Limitations of space prevent me from discussing it any further here.

## Appendix A. A fact concerning strongly dualizable objects

In this appendix we prove the assertion of Reminder 2.7(iii); we recall the statement.

**Theorem A.1.** *Let $\mathcal{T}$ be a compactly generated triangulated category, possessing a symmetric tensor product compatible with the triangulated structure.*

*Assume that the unit of the tensor is compact, and that all compact objects are strongly dualizable. If $\mathcal{E} \in \mathcal{T}$ is some object, then the following are equivalent:*

(i) $\mathcal{E}$ *is compact.*

(ii) $\mathcal{E}$ *is strongly dualizable.*

(iii) *Tensor product with $\mathcal{E}$ commutes with arbitrary products in $\mathcal{T}$; that is, the natural map*

$$\mathcal{E} \wedge \prod_{\lambda \in \Lambda} t_\lambda \longrightarrow \prod_{\lambda \in \Lambda} (\mathcal{E} \wedge t_\lambda)$$

*is an isomorphism for every set of objects $\{t_\lambda, \lambda \in \Lambda\}$.*

**Remark A.2.** The implication (i)$\Longrightarrow$(ii) follows immediately from the hypothesis of the theorem. The implications (ii)$\Longrightarrow$(i) and (ii)$\Longrightarrow$(iii) are known; for (ii)$\Longrightarrow$(i) see [19, Theorem 2.1.3], while (ii)$\Longrightarrow$(iii) may be found in [19, Theorem A.2.5(f)]. The fact that seems new is the implication (iii)$\Longrightarrow$(i). For the convenience of the reader we give a complete, self-contained proof of the equivalence of the three conditions.

Also, because this appendix might be of interest to people who could not care less about the categories $\mathbf{D}(\mathrm{Qcoh}/X)$, our notation will be the standard one in the literature. The tensor product of two objects $\mathcal{E}, \mathcal{G} \in \mathcal{T}$ will be written $\mathcal{E} \wedge \mathcal{G}$, the unit of the tensor will be denoted $\mathbb{S}$, and the internal Hom-object will be $F(\mathcal{E}, \mathcal{G})$. To translate back to the case where $\mathcal{T} = \mathbf{D}(\mathrm{Qcoh}/X)$, put

$$\mathbb{S} = \mathcal{O}_X, \qquad\qquad \mathcal{E} \wedge \mathcal{G} = \mathcal{E} \overset{\mathbf{L}}{\otimes} \mathcal{G}, \qquad\qquad F(\mathcal{E}, \mathcal{G}) = \mathbf{R}\mathcal{H}om(\mathcal{E}, \mathcal{G}).$$

If you read again the statement of Theorem A.1 you will observe that we already used this notation in part (iii) of the theorem. With the notation established, it is time to come to the proof of Theorem A.1.

*Proof.* We are assuming that the compact objects are all strongly dualizable; thus (i)$\Longrightarrow$(ii) is trivial. Next we prove (ii)$\Longrightarrow$(iii). Suppose therefore that $\mathcal{E}$ is strongly dualizable. In Reminder 2.7(i) we noted that there is an isomorphism $\mathcal{E} \cong \{\mathcal{E}^\vee\}^\vee$, and in Reminder 2.7(ii) we observed that $\mathcal{E}^\vee$ is strongly dualizable. Putting this together we have that $\mathcal{E} \wedge \mathcal{G} \cong \{\mathcal{E}^\vee\}^\vee \wedge \mathcal{G}$ must be naturally isomorphic to $F(\mathcal{E}^\vee, \mathcal{G})$. We are therefore reduced to proving that the functor $F(\mathcal{E}^\vee, -)$ respects products. But this is obvious; $F(\mathcal{E}^\vee, -)$ is right adjoint to $- \wedge \mathcal{E}^\vee$.

It remains to prove (iii)$\Longrightarrow$(i), which is the part that seems new. Consider the functor $\mathrm{Hom}(\mathbb{S}, \mathcal{E} \wedge -)$. The functor $\mathcal{E} \wedge -$ respects products by hypothesis and coproducts obviously. The functor $\mathrm{Hom}(\mathbb{S}, -)$ respects products obviously and coproducts because we are assuming $\mathbb{S}$ compact. Therefore the composite functor $\mathrm{Hom}(\mathbb{S}, \mathcal{E} \wedge -)$ must respect both products and coproducts.

The fact that it respects products means that it must be representable; see [52, Theorem 8.6.1]. There is an object $\mathcal{G} \in \mathbf{D}(\mathrm{Qcoh}/X)$ and a natural isomorphism

$$\varphi \ : \ \mathrm{Hom}(\mathcal{G}, -) \longrightarrow \mathrm{Hom}(\mathbb{S}, \mathcal{E} \wedge -)\,.$$

Because the functor $\mathrm{Hom}(\mathbb{S}, \mathcal{E} \wedge -)$ respects coproducts so does the isomorphic functor $\mathrm{Hom}(\mathcal{G}, -)$, meaning that $\mathcal{G}$ must be compact, and therefore also strongly dualizable. Yoneda's lemma tells us that the isomorphism $\varphi$ must come from an element in

$$\mathrm{Hom}(\mathbb{S}, \mathcal{E} \wedge \mathcal{G}) \cong \mathrm{Hom}\big(\mathbb{S}, \ F(\mathcal{G}^{\vee}, \mathcal{E})\big) \cong \mathrm{Hom}(\mathcal{G}^{\vee}, \mathcal{E})\,.$$

This produces for us a morphism $\alpha : \mathcal{G}^{\vee} \longrightarrow \mathcal{E}$, which we will prove an isomorphism. What we know is that the natural map

$$\mathrm{Hom}(\mathbb{S}, \alpha \wedge -) \ : \ \mathrm{Hom}(\mathbb{S}, \mathcal{G}^{\vee} \wedge -) \longrightarrow \mathrm{Hom}(\mathbb{S}, \mathcal{E} \wedge -)$$

is an isomorphism; it is just the map $\varphi$. In particular, for every compact object $\mathcal{K}$ we deduce that

$$\mathrm{Hom}(\mathbb{S}, \alpha \wedge 1) \ : \ \mathrm{Hom}(\mathbb{S}, \mathcal{G}^{\vee} \wedge \mathcal{K}^{\vee}) \longrightarrow \mathrm{Hom}(\mathbb{S}, \mathcal{E} \wedge \mathcal{K}^{\vee})$$

is an isomorphism, but this identifies with

$$\mathrm{Hom}(\mathcal{K}, \alpha) \ : \ \mathrm{Hom}(\mathcal{K}, \mathcal{G}^{\vee}) \longrightarrow \mathrm{Hom}(\mathcal{K}, \mathcal{E})\,.$$

The compacts $\mathcal{K}$ generate, hence $\alpha$ must be an isomorphism, making $\mathcal{E}$ isomorphic to the object $\mathcal{G}^{\vee}$. We know that $\mathcal{G}$ is compact; it remains to prove that so is $\mathcal{G}^{\vee}$. But we have a natural isomorphism

$$\mathrm{Hom}(\mathbb{S}, \mathcal{G} \wedge -) \cong \mathrm{Hom}(\mathcal{G}^{\vee}, -)\,,$$

and the functor on the left clearly respects coproducts.                              $\square$

# References

[1] Leovigildo Alonso Tarrío, Ana Jeremías López, and Joseph Lipman, *Local homology and cohomology on schemes*, Ann. Sci. École Norm. Sup. (4) **30** (1997), no. 1, 1–39.

[2] ———, *Studies in duality on Noetherian formal schemes and non-Noetherian ordinary schemes*, American Mathematical Society, Providence, RI, 1999.

[3] Leovigildo Alonso Tarrío, Ana Jeremías López, and María José Souto Salorio, *Localization in categories of complexes and unbounded resolutions*, Canad. J. Math. **52** (2000), no. 2, 225–247.

[4] Marcel Bökstedt and Amnon Neeman, *Homotopy limits in triangulated categories*, Compositio Math. **86** (1993), 209–234.

[5] Alexei I. Bondal and Michel van den Bergh, *Generators and representability of functors in commutative and noncommutative geometry*, Mosc. Math. J. **3** (2003), no. 1, 1–36, 258.

[6] Brian Conrad, *Grothendieck duality and base change*, Lecture Notes in Mathematics, vol. 1750, Springer-Verlag, Berlin, 2000.

[7] Pierre Deligne, *Cohomology à support propre en construction du foncteur $f^!$*, Residues and Duality, Lecture Notes in Mathematics, vol. 20, Springer–Verlag, 1966, pp. 404–421.

[8] ______, *Cohomologie à supports propres*, Théorie des topos et cohomologie étale des schémas. Tome 3 (Berlin), Lecture Notes in Mathematics, vol. 305, Springer-Verlag, 1973, (SGA 4, Exposé XVII), pp. 250–461.

[9] Duiliu-Emanuel Diaconescu, *Enhanced D-brane categories from string field theory*, J. High Energy Phys. (2001), no. 6, Paper 16, 19.

[10] Albrecht Dold and Dieter Puppe, *Duality, trace, and transfer*, Proceedings of the International Conference on Geometric Topology (Warsaw, 1978) (Warsaw), PWN, 1980, pp. 81–102.

[11] Fouad El Zein, *Complexe dualisant et applications à la classe fondamentale d'un cycle*, Bull. Soc. Math. France Mém. (1978), no. 58, 93.

[12] Pierre-Paul Grivel, *Une démonstration du théorème de dualité de Verdier*, Enseign. Math. (2) **31** (1985), no. 3-4, 227–247.

[13] Alexandre Grothendieck, *The cohomology theory of abstract algebraic varieties*, Proc. Internat. Congress Math. (Edinburgh, 1958), Cambridge Univ. Press, New York, 1960, pp. 103–118.

[14] ______, *Théorèmes de dualité pour les faisceaux algébriques cohérents*, Séminaire Bourbaki, Vol. 4, Soc. Math. France, Paris, 1995, pp. Exp. No. 149, 169–193.

[15] Robin Hartshorne, *Residues and duality*, Lecture Notes in Mathematics, vol. 20, Springer–Verlag, 1966.

[16] Glenn W. Hopkins, *An algebraic approach to Grothendieck's residue symbol*, Trans. Amer. Math. Soc. **275** (1983), no. 2, 511–537.

[17] Glenn W. Hopkins and Joseph Lipman, *An elementary theory of Grothendieck's residue symbol*, C. R. Math. Rep. Acad. Sci. Canada **1** (1978/79), no. 3, 169–172.

[18] Kentaro Hori and Johannes Walcher, *D-branes from matrix factorizations*, C. R. Phys. **5** (2004), no. 9-10, 1061–1070, Strings 04. Part I.

[19] Mark Hovey, John H. Palmieri, and Neil P. Strickland, *Axiomatic stable homotopy theory*, vol. 128, Memoirs AMS, no. 610, Amer. Math. Soc., 1997.

[20] Reinhold Hübl and Ernst Kunz, *Integration of differential forms on schemes*, J. Reine Angew. Math. **410** (1990), 53–83.

[21] ______, *Regular differential forms and duality for projective morphisms*, J. Reine Angew. Math. **410** (1990), 84–108.

[22] Reinhold Hübl and Pramathanath Sastry, *Regular differential forms and relative duality*, Amer. J. Math. **115** (1993), no. 4, 749–787.

[23] Luc Illusie, *Conditions de finitude*, Théorie des intersections et théorème de Riemann-Roch, Springer-Verlag, Berlin, 1971, Séminaire de Géométrie Algébrique du Bois-Marie 1966–1967 (SGA 6, Exposé III), pp. 222–273. Lecture Notes in Mathematics, Vol. 225.

[24] ———, *Existence de résolutions globales*, Théorie des intersections et théorème de Riemann-Roch, Springer-Verlag, Berlin, 1971, Séminaire de Géométrie Algébrique du Bois-Marie 1966–1967 (SGA 6, Exposé II), pp. 160–221. Lecture Notes in Mathematics, Vol. 225.

[25] ———, *Généralités sur les conditions de finitude dans les catégories dérivées*, Théorie des intersections et théorème de Riemann-Roch, Springer-Verlag, Berlin, 1971, Séminaire de Géométrie Algébrique du Bois-Marie 1966–1967 (SGA 6, Exposé I), pp. 78–159. Lecture Notes in Mathematics, Vol. 225.

[26] ———, *Groupes de grothendieck des topos annelés*, Théorie des intersections et théorème de Riemann-Roch, Springer-Verlag, Berlin, 1971, Séminaire de Géométrie Algébrique du Bois-Marie 1966–1967 (SGA 6, Exposé IV), pp. 274–296. Lecture Notes in Mathematics, Vol. 225.

[27] Srikanth Iyengar and Henning Krause, *Acyclicity versus total acyclicity for complexes over Noetherian rings*, Documenta Math. **11** (2006), 207–240.

[28] Peter Jørgensen, *Serre-duality for* tails($A$), Proc. Amer. Math. Soc. **125** (1997), no. 3, 709–716.

[29] ———, *The homotopy category of complexes of projective modules*, Adv. Math. **193** (2005), no. 1, 223–232.

[30] Anton N. Kapustin and Yi Li, *D-branes in Landau-Ginzburg models and algebraic geometry*, J. High Energy Phys. (2003), no. 12, 005, 44 pp. (electronic).

[31] Anton N. Kapustin and Dmitri O. Orlov, *Lectures on mirror symmetry, derived categories, and D-branes*, Uspekhi Mat. Nauk **59** (2004), no. 5(359), 101–134.

[32] Bernhard Keller, *Deriving DG categories*, Ann. Sci. École Norm. Sup. (4) **27** (1994), no. 1, 63–102.

[33] G. Maxwell Kelly, *Many-variable functorial calculus. I*, Coherence in categories, Springer, Berlin, 1972, pp. 66–105. Lecture Notes in Math., Vol. 281.

[34] Reinhardt Kiehl, *Ein "Descente"-Lemma und Grothendiecks Projektionssatz für nichtnoethersche Schemata*, Math. Ann. **198** (1972), 287–316.

[35] Henning Krause, *The stable derived category of a Noetherian scheme*, Compos. Math. **141** (2005), no. 5, 1128–1162.

[36] Ernst Kunz, *Residuen von Differentialformen auf Cohen-Macaulay-Varietäten*, Math. Z. **152** (1977), no. 2, 165–189.

[37] ———, *Über den n-dimensionalen Residuensatz*, Jahresber. Deutsch. Math.-Verein. **94** (1992), no. 4, 170–188.

[38] ———, *Geometric applications of the residue theorem on algebraic curves*, Algebra, arithmetic and geometry with applications (West Lafayette, IN, 2000), Springer, Berlin, 2004, pp. 565–589.

[39] Ernst Kunz and Rolf Waldi, *Regular differential forms*, Contemporary Mathematics, vol. 79, American Mathematical Society, Providence, RI, 1988.

[40] L. Gaunce Lewis, Jr., J. Peter May, Mark Steinberger, and James E. McClure, *Equivariant stable homotopy theory*, Lecture Notes in Mathematics, vol. 1213, Springer-Verlag, Berlin, 1986, With contributions by J. E. McClure.

[41] Joseph Lipman, *Dualizing sheaves, differentials and residues on algebraic varieties*, Astérisque (1984), no. 117, ii+138.

[42] ———, *Residues and traces of differential forms via Hochschild homology*, Contemporary Mathematics, vol. 61, American Mathematical Society, Providence, RI, 1987.

[43] ______ , *Lectures on local cohomology and duality*, Local cohomology and its applications (Guanajuato, 1999), Lecture Notes in Pure and Appl. Math., vol. 226, Dekker, New York, 2002, pp. 39–89.

[44] ______ , *Notes on derived categories and Grothendieck duality*, LNM, Springer-Verlag, to appear.

[45] Joseph Lipman, Suresh Nayak, and Pramathanath Sastry, *Pseudofunctorial behavior of Cousin complexes on formal schemes*, Variance and duality for Cousin complexes on formal schemes, Contemp. Math., vol. 375, Amer. Math. Soc., Providence, RI, 2005, pp. 3–133.

[46] Joseph Lipman and Amnon Neeman, *Quasi-perfect scheme maps and boundedness of the twisted inverse image functor*, Illinois J. Math. **51** (2007), 209–236.

[47] Ralf Meyer and Ryszard Nest, *The Baum-Connes conjecture via localisation of categories*, Topology **45** (2006), no. 2, 209–259.

[48] Daniel S. Murfet, *The mock homotopy category of projectives and Grothendieck duality*, (PhD thesis, Australian National U. 2008).

[49] Suresh Nayak, *Pasting pseudofunctors*, Variance and duality for Cousin complexes on formal schemes, Contemp. Math., vol. 375, Amer. Math. Soc., Providence, RI, 2005, pp. 195–271.

[50] Amnon Neeman, *The connection between the K–theory localisation theorem of Thomason, Trobaugh and Yao, and the smashing subcategories of Bousfield and Ravenel*, Ann. Sci. École Normale Supérieure **25** (1992), 547–566.

[51] ______ , *The Grothendieck duality theorem via Bousfield's techniques and Brown representability*, Jour. Amer. Math. Soc. **9** (1996), 205–236.

[52] ______ , *Triangulated Categories*, Annals of Mathematics Studies, vol. 148, Princeton University Press, Princeton, NJ, 2001.

[53] ______ , *The homotopy category of flat modules, and Grothendieck duality*, Invent. Math. **174** (2008), 255–308.

[54] ______ , *Some adjoints in homotopy categories*, Annals of Mathematics Studies, vol. 171, Princeton University Press, Princeton, NJ, 2010, pp. 2143–2155.

[55] Dieter Puppe, *On the structure of stable homotopy theory*, Colloquium on algebraic topology, Aarhus Universitet Matematisk Institut, 1962, pp. 65–71.

[56] Jean-Pierre Ramis, Gabriel Ruget, and Jean-Louis Verdier, *Dualité relative en géométrie analytique complexe*, Invent. Math. **13** (1971), 261–283.

[57] Neantro Saavedra Rivano, *Catégories Tannakiennes*, Springer-Verlag, Berlin, 1972, Lecture Notes in Mathematics, Vol. 265.

[58] Pramathanath Sastry, *Base change and Grothendieck duality for Cohen-Macaulay maps*, Compos. Math. **140** (2004), no. 3, 729–777.

[59] ______ , *Duality for Cousin complexes*, Variance and duality for Cousin complexes on formal schemes, Contemp. Math., vol. 375, Amer. Math. Soc., Providence, RI, 2005, pp. 137–192.

[60] Jean-Pierre Serre, *Un théorème de dualité*, Comment. Math. Helv. **29** (1955), 9–26.

[61] Nicolas Spaltenstein, *Resolutions of unbounded complexes*, Compositio Math. **65** (1988), no. 2, 121–154.

[62] Robert W. Thomason and Thomas F. Trobaugh, *Higher algebraic K–theory of schemes and of derived categories*, The Grothendieck Festschrift (a collection of

papers to honor Grothendieck's 60'th birthday), vol. 3, Birkhäuser, 1990, pp. 247–435.

[63] Michel van den Bergh, *Existence theorems for dualizing complexes over non-commutative graded and filtered rings*, J. Algebra **195** (1997), no. 2, 662–679.

[64] Jean-Louis Verdier, *Le théorème de dualité de Poincaré*, C. R. Acad. Sci. Paris **256** (1963), 2084–2086.

[65] —————, *A duality theorem in the etale cohomology of schemes*, Proc. Conf. Local Fields (Driebergen, 1966), Springer, Berlin, 1967, pp. 184–198.

[66] Jean-Louis Verdier, *Base change for twisted inverse images of coherent sheaves*, vol. Collection: Algebraic Geometry, Tata Inst. Fund. Res., 1968, pp. 393–408.

[67] —————, *Des catégories dérivées des catégories abeliennes*, Asterisque, vol. 239, Société Mathématique de France, 1996 (French).

[68] Amnon Yekutieli, *Dualizing complexes over noncommutative graded algebras*, J. Algebra **153** (1992), no. 1, 41–84.

[69] —————, *An explicit construction of the Grothendieck residue complex*, Astérisque (1992), no. 208, 127, With an appendix by Pramathanath Sastry.

[70] Amnon Yekutieli and James J. Zhang, *Dualizing complexes and perverse modules over differential algebras*, Compos. Math. **141** (2005), no. 3, 620–654.

[71] —————, *Dualizing complexes and perverse sheaves on noncommutative ringed schemes*, Selecta Math. (N.S.) **12** (2006), no. 1, 137–177.

CENTRE FOR MATHEMATICS AND ITS APPLICATIONS, MATHEMATICAL SCIENCES INSTITUTE, JOHN DEDMAN BUILDING, THE AUSTRALIAN NATIONAL UNIVERSITY, CANBERRA, ACT 0200, AUSTRALIA

*E-mail address*: Amnon.Neeman@anu.edu.au

# Derived categories and algebraic geometry

RAPHAËL ROUQUIER

## Contents



## 1. Introduction

We describe some basic properties of the derived category of coherent sheaves on a variety (bounded derived category or perfect complexes).

The first chapter considers the problem of extending vector bundles from an open subset. Thomason and Trobaugh provided an answer to this problem by considering extensions for perfect complexes. This has applications to higher K-theory.

In the second chapter, we explain how to characterize subcategories corresponding to objects supported by a given closed subvariety. This permits a reconstruction of the variety (viewed as a ringed space) from a categorical structure. In the case of derived categories, this requires also the tensor structure.

We start with the classical case of the category of coherent sheaves (after Gabriel). We present afterwards a similar approach in the triangulated case, where serious difficulties arise.

Finally, we explain how to deduce that a smooth projective variety with ample or anti-ample canonical bundle is determined by its derived category.

We haven't included proofs of the results on general properties of abelian or triangulated categories (cf [KaScha, Nee3] for proofs). The only difficult part is Lemma 3.9 on compact objects.

This text is based on lectures at the conference *Géométrie algébrique complexe*, CIRM, Luminy in December 2003. I thank Paul Balmer for useful comments.

## 2. Notations

We fix a field $k$ and we call variety a separated scheme of finite type over $k$. Given a variety $X$, we denote by $X$-coh (resp. $X$-qcoh) the category of coherent (resp. quasi-coherent) sheaves on $X$.

All functors between triangulated categories are assumed to be triangulated.

We denote by $Z(\mathcal{C})$ the centre of a category $\mathcal{C}$ (=endomorphisms of the identity functor).

Given a ring $R$, we denote by $R$-mod the category of finitely generated $R$-modules.

## 3. Localisation

### 3.1. Abelian case

Let us recall some basic properties of categories of coherent sheaves.

Let $X$ be a variety. Given $Z$ closed in $X$, we denote by $X$-coh$_Z$ the full subcategory of $X$-coh of coherent sheaves with support contained in $Z$. This is a Serre subcategory of $X$-coh.

Let us recall that a full subcategory $\mathcal{I}$ of an abelian category $\mathcal{A}$ is a *Serre subcategory* if it is stable under taking subobjects, quotients and extensions. Given such a subcategory, there is a quotient abelian category $\mathcal{A}/\mathcal{I}$ and a functor $\mathcal{A} \to \mathcal{A}/\mathcal{I}$ with kernel $\mathcal{I}$. It is the solution of the universal problem of taking quotients (for abelian categories). We say that there is an exact sequence of abelian categories $0 \to \mathcal{I} \to \mathcal{A} \to \mathcal{A}/\mathcal{I} \to 0$.

Let $j : U = X - Z \to X$ be the open embedding.

**Proposition 3.1.** *The functor* $j^* : X\text{-coh} \to U\text{-coh}$ *induces an equivalence* $X\text{-coh} / X\text{-coh}_Z \xrightarrow{\sim} U\text{-coh}$, i.e., *there is an exact sequence of abelian categories*

$$0 \to X\text{-coh}_Z \to X\text{-coh} \to U\text{-coh} \to 0.$$

*Proof.* In the case of quasi-coherent sheaves, we have a functor $j_*$ right adjoint to $j^*$. The canonical map $j^* j_* \xrightarrow{\sim} \mathbf{1}_{U\text{-qcoh}}$ is an isomorphism and the kernel of $j^*$ is $X\text{-qcoh}_Z$. It follows from Lemma 3.2 below that there is an exact sequence

$$0 \to X\text{-qcoh}_Z \to X\text{-qcoh} \to U\text{-qcoh} \to 0.$$

Let us now deduce the proposition from the characterisation of coherent sheaves as the finitely presented objects in the category of quasi-coherent sheaves. Lemma 3.3 below shows that the canonical functor $X\text{-coh} / X\text{-coh}_Z \to U\text{-coh}$ is fully faithful. We are left with proving that the functor $j^* : X\text{-coh} \to U\text{-coh}$ is essentially surjective. Let $G$ be a coherent sheaf on $U$ and let $F = j_*G$. We have $j^*F \simeq G$. The quasi-coherent sheaf $F$ is an increasing union (filtered colimit) of its coherent subsheaves, $F = \bigcup_{E \text{ coherent} \subset F} E$. It follows that $j^*F = \bigcup_E j^*E$. Since $j^*F$ is coherent and it is an increasing union of a family of subsheaves, one of the members of the family $j^*E$ is equal to $j^*F$ (the filtered colimit stabilizes after finitely many terms). So, $j^*F = j^*E \simeq G$ for a coherent subsheaf $E$ of $F$. $\qquad\qquad\square$

**Lemma 3.2.** *Let* $F : \mathcal{A} \to \mathcal{B}$ *be an exact functor between abelian categories. Assume $F$ has a right adjoint $G$ and $G$ is fully faithful (i.e., $FG \xrightarrow{\text{can}} \mathbf{1}_{\mathcal{B}}$ is an isomorphism).*

*Then* $\ker F$ *is a Serre subcategory of $\mathcal{A}$ and there is an exact sequence*

$$0 \to \ker F \to \mathcal{A} \to \mathcal{B} \to 0.$$

**Lemma 3.3.** *Let $\mathcal{A}$ be an abelian category, $\mathcal{A}'$ a full abelian subcategory of $\mathcal{A}$ and $\mathcal{I}$ a Serre subcategory of $\mathcal{A}$. Assume that for any $M \in \mathcal{A}'$ and for any $N \in \mathcal{I}$ a subobject or a quotient of $M$, then $N \in \mathcal{A}'$.*

*Then the canonical functor* $\mathcal{A}'/(\mathcal{I} \cap \mathcal{A}') \to \mathcal{A}/\mathcal{I}$ *is fully faithful.*

## 3.2. Triangulated case

### 3.2.1. Derived functors

We start by recalling some properties of derived categories of sheaves and we explain how to construct right derived functors.

Recall that the canonical functor $D(X\text{-coh}) \to D(X\text{-qcoh})$ is fully faithful, *i.e.*, $D(X\text{-coh})$ is equivalent to the full subcategory of $D(X\text{-qcoh})$ of complexes whose cohomology sheaves are coherent. We will identify those two categories.

Let $X$-inj be the category of quasi-coherent injective sheaves and $\mathrm{Ho}(X\text{-inj})$ the homotopy category of complexes of objects of $X$-inj. Consider the canonical functor $\mathrm{Ho}(X\text{-qcoh}) \to D(X\text{-qcoh})$. It has a right adjoint $\rho$ ("homotopically injective resolution"). Let $\mathrm{Ho}(X\text{-qcoh})^{hi}$ be its essential image (homotopically injective complexes). The functor $\rho$ is fully faithful, the canonical functor $\mathrm{Ho}(X\text{-qcoh})^{hi} \xrightarrow{\sim} D(X\text{-qcoh})$ is an equivalence with inverse $\rho$. The intersection of $\mathrm{Ho}(X\text{-qcoh})^{hi}$ with $D^+(X\text{-qcoh})$ is $\mathrm{Ho}^+(X\text{-inj})$, *i.e.*, $\rho$ restricts to an equivalence $D^+(X\text{-qcoh}) \xrightarrow{\sim} \mathrm{Ho}^+(X\text{-inj})$ : we recover the classical injective resolutions.

Let us now discuss the derivation of a left exact functor $F : X\text{-qcoh} \to \mathcal{A}$, where $\mathcal{A}$ is an abelian category. We extend $F$ to a functor $\mathrm{Ho}(F) : \mathrm{Ho}(X\text{-qcoh}) \to \mathrm{Ho}(\mathcal{A})$. We restrict this functor to $\mathrm{Ho}(X\text{-qcoh})^{hi}$. We obtain the right derived functor

$$RF : D(X\text{-qcoh}) \xrightarrow[\sim]{\rho} \mathrm{Ho}(X\text{-qcoh})^{hi} \xrightarrow{\mathrm{Ho}(F)} \mathrm{Ho}(\mathcal{A}) \xrightarrow{\mathrm{can}} D(\mathcal{A}).$$

The functor $RF$ is triangulated. In particular, the image of a distinguished triangle is a distinguished triangle, while $F$ needs not send an exact sequence to an exact sequence. The left exactness of $F$ shows that $H^0(RF(M)) \xrightarrow{\sim} F(M)$ for $M \in X\text{-qcoh}$.

### 3.2.2. Open subvarieties and quotients

The functor $j_*$ derives into a functor $Rj_* : D(U\text{-qcoh}) \to D(X\text{-qcoh})$. This is right adjoint to the functor $j^* : D(X\text{-qcoh}) \to D(U\text{-qcoh})$. The kernel of the functor $j^*$ is $D_Z(X\text{-qcoh})$, the full subcategory of $D(X\text{-qcoh})$ of complexes whose cohomology sheaves have their support contained in $Z$. This is a thick subcategory.

Let us recall that a non-empty full subcategory $\mathcal{I}$ of a triangulated category $\mathcal{T}$ is *thick* if the following two conditions hold

- Given $F \to G \to H \rightsquigarrow$ a distinguished triangle in $\mathcal{T}$, if two objects amongst $F$, $G$ and $H$ are in $\mathcal{I}$, then the third one is in $\mathcal{I}$ as well.
- Given $F, G \in \mathcal{T}$, if $F \oplus G \in \mathcal{I}$, then $F, G \in \mathcal{I}$.

There is a quotient triangulated category $\mathcal{T}/\mathcal{I}$ and a functor $\mathcal{T} \to \mathcal{T}/\mathcal{I}$ with kernel $\mathcal{I}$, solution of the universal quotient problem (amongst triangulated categories). We say that there is an exact sequence of triangulated categories $0 \to \mathcal{I} \to \mathcal{T} \to \mathcal{T}/\mathcal{I} \to 0$.

Lemma 3.4 gives an exact sequence of triangulated categories

$$0 \to D_Z(X\text{-qcoh}) \to D(X\text{-qcoh}) \to D(U\text{-qcoh}) \to 0$$

**Lemma 3.4.** *Let $F : \mathcal{T} \to \mathcal{T}'$ be a functor between triangulated categories. Assume $F$ has a right adjoint $G$ and $G$ is fully faithful.*

*Then* $\ker F$ *is a thick subcategory of* $\mathcal{T}$ *and there is an exact sequence*

$$0 \to \ker F \to \mathcal{T} \to \mathcal{T}' \to 0.$$

### 3.2.3. Perfect complexes

We come now to the core of our study. We recall the notion of perfect objects and their basic properties.

An object of $D(X\text{-qcoh})$ is *perfect* if it is locally (quasi)-isomorphic to a bounded complex of free sheaves of finite rank. We denote by $X$-perf the full subcagegory of $D(X\text{-qcoh})$ of perfect complexes. This is a thick subcategory of $D^b(X\text{-coh})$. If $X$ is quasi-projective, then a complex is perfect if and only if it is quasi-isomorphic to a bounded complex of vector bundles. The variety $X$ is regular if and only if $D^b(X\text{-coh}) = X$-perf.

Let $\mathcal{T}$ be a triangulated category with infinite direct sums. An object $C \in \mathcal{T}$ is *compact* if given any family $\mathcal{E}$ of objects of $\mathcal{T}$, the canonical map $\bigoplus_{E \in \mathcal{E}} \mathrm{Hom}(C, E) \to \mathrm{Hom}(C, \bigoplus_{E \in \mathcal{E}} E)$ is an isomorphism. We denote by $\mathcal{T}^c$ the full subcategory of $\mathcal{T}$ of compact objects. This is a thick subcategory.

A key idea in Thomason's approach is the characterisation of perfect complexes as the compact objects of $D(X\text{-qcoh})$ (cf [Rou2] for a study of compactness as a local property for general triangulated categories).

**Lemma 3.5.** *Let $C \in D(X\text{-qcoh})$. Then $C$ is perfect if and only if it is compact.*

*Proof.* Let us assume first $X$ is affine, $X = \mathrm{Spec}\, R$. Since $R$ is compact, we deduce that every perfect object is compact. Let $C$ be a complex of $R$-modules and let $i \in \mathbf{Z}$ such that $H^i C \neq 0$. Then $\mathrm{Hom}(R, C[i]) \neq 0$. It follows from Lemma 3.9 that the perfect complexes are the same as the compact objects.

We consider now an arbitrary variety $X$. Let $X = U_1 \cup U_2$ with $U_1$ an affine open subvariety and $U_2$ open. We assume that the minimal number of open affine subvarieties in a covering of $U_2$ is strictly less that the number for $X$. By induction, we can assume the lemma holds for $U_2$ and for $U_{12} = U_1 \cap U_2$. Let $j_r : U_r \to X$ and $j_{12} : U_{12} \to X$ be the open immersions. Given

$D \in D(X\text{-qcoh})$, there is a Mayer-Vietoris distinguished triangle:

$$D \to Rj_{1*}j_1^* D \oplus Rj_{2*}j_2^* D \to Rj_{12*}j_{12}^* D \rightsquigarrow$$

Let $C \in D(X\text{-qcoh})$. The triangle above shows that $C$ is compact if $j_1^* C$, $j_2^* C$ and $j_{12}^* C$ are compact. The converse is clear. On the other hand, $C$ is perfect if and only if $j_1^* C$, $j_2^* C$ and $j_{12}^* C$ are perfect. The lemma follows by induction. $\square$

An important aspect of the category $X$-perf is that it provides the "right" K-theory groups, for a variety without enough ample vector bundles. We put $K_0(X) = K_0(X\text{-perf})$.

Recall that given a triangulated category $\mathcal{T}$, we define $K_0(\mathcal{T})$ as the quotient of the free abelian group with basis the isomorphism classes of objects of $\mathcal{T}$ by the relation $[M] = [L] + [N]$ whenever there is a distinguished triangle $L \to M \to N \rightsquigarrow$.

This definition of $K_0(X)$ coincides with the classical one (Grothendieck group of the exact category of vector bundles) when $X$ has an ample family of line bundles (this is the case for a quasi-projective variety).

### 3.2.4. Extensions of perfect complexes

We put $X\text{-perf}_Z = X\text{-perf} \cap D_Z(X\text{-qcoh})$.

**Theorem 3.6** (Thomason-Trobaugh). *The functor $j^*$ induces a fully faithful functor*

$$X\text{-perf} / X\text{-perf}_Z \to U\text{-perf}.$$

*An object of $U$-perf is the restriction of an object of $X$-perf if and only if its class in $K_0(U)$ is the restriction of an element of $K_0(X)$.*

The occurrence of $K_0$ in Theorem 3.6 comes from the following lemma of Thomason [Th2, Theorem 2.1] (the proof is tricky).

**Lemma 3.7.** *Let $\mathcal{T}$ be a triangulated category. There is a bijection from the set of full triangulated subcategories $\mathcal{I}$ of $\mathcal{T}$ that generate $\mathcal{T}$ as a thick subcategory to the set of subgroups of $K_0(\mathcal{T})$ given by sending $\mathcal{I}$ to the image of $K_0(\mathcal{I})$ in $K_0(\mathcal{T})$.*

The calculus of fractions gives a simple criterion for fully faithfulness:

**Lemma 3.8.** *Let $\mathcal{T}$ be a triangulated category, $\mathcal{I}$ a thick subcategory of $\mathcal{T}$ and $\mathcal{T}'$ a full triangulated subcategory of $\mathcal{T}$.*

*Assume every morphism $C \to D$ with $C \in \mathcal{T}'$ and $D \in \mathcal{I}$ factors through an object of $\mathcal{I} \cap \mathcal{T}'$. Then the canonical functor $\mathcal{T}'/(\mathcal{I} \cap \mathcal{T}') \to \mathcal{T}/\mathcal{I}$ is fully faithful.*

Given $\mathcal{I}$ a full subcategory of a triangulated category $\mathcal{T}$, we denote by $\bar{\mathcal{I}}$ the smallest thick subcategory of $\mathcal{T}$ closed under taking infinite direct sums and containing $\mathcal{I}$.

Let $\mathcal{I}$ be a thick subcategory of a triangulated category $\mathcal{T}$. The right orthogonal $\mathcal{I}^{\perp}$ to $\mathcal{I}$ in $\mathcal{T}$ is the full subcategory of $\mathcal{T}$ of objects $D$ with $\mathrm{Hom}(C, D) = 0$ for all $C \in \mathcal{I}$. This is a thick subcategory.

The following lemma is related to Brown-Neeman's representability Theorem and its proof requires some work [Nee2].

**Lemma 3.9.** *Let $\mathcal{T}$ be a triangulated category with arbitrary direct sums. Let $\mathcal{I}$ be a thick subcategory of $\mathcal{T}^c$. Then every map from an object of $\mathcal{T}^c$ to an object of $\bar{\mathcal{I}}$ factors through an object of $\mathcal{I}$. In particular, we have $\mathcal{T}^c \cap \bar{\mathcal{I}} = \mathcal{I}$.*

*We have $\bar{\mathcal{I}} = \mathcal{T}$ if and only if the right orthogonal $\mathcal{I}^{\perp}$ of $\mathcal{I}$ in $\mathcal{T}$ vanishes.*

**Lemma 3.10.** *Let $Y$ be a closed subvariety of $X$. We have $D_Y(X\text{-qcoh}) = \overline{X\text{-perf}_Y}$.*

Given $Z'$ closed in $X$, we put $K_0(X \text{ on } Z') = K_0(X\text{-perf}_{Z'})$. We will show a version "with supports" of Theorem 3.6:

**Theorem 3.11.** *Let $Z'$ be a closed subvariety of $X$. The functor $j^*$ induces a fully faithful functor $X\text{-perf}_{Z'} / X\text{-perf}_{Z \cap Z'} \to U\text{-perf}_{U \cap Z'}$. An object of $U\text{-perf}_{U \cap Z'}$ is the image of an object of $X\text{-perf}_{Z'}$ if and only if its class in $K_0(U \text{ on } U \cap Z')$ is the image of an element of $K_0(X \text{ on } Z')$.*

*Proof of Theorem 3.11 and Lemma 3.10.* Let us show first that the lemma for $X$ implies the theorem for $X$. The combination of Lemmas 3.8, 3.9 and 3.10 shows that $j^*$ induces a fully faithful functor $X\text{-perf}_{Z'} / X\text{-perf}_{Z \cap Z'} \to U\text{-perf}_{U \cap Z'}$. Let $\mathcal{I}$ be the image of that functor: this is a full triangulated subcategory. Since $\overline{X\text{-perf}_{Z'}} = D_{Z'}(X\text{-qcoh})$ (lemma 3.10), we have $\bar{\mathcal{I}} = D_{U \cap Z'}(U\text{-qcoh})$. It follows from Lemma 3.9 that $U\text{-perf}_{Z' \cap U}$ is the thick subcategory generated by $\mathcal{I}$. The theorem follows now from Lemma 3.7.

Lemma 3.9 shows that Lemma 3.10 will follow from the fact that $(X\text{-perf}_Y)^{\perp} = 0$.

Let us assume first that $X$ is affine. Let $\{y_1, \ldots, y_r\}$ be a family of generators of the defining ideal of $Y$ and let

$$G_r = \bigotimes_{i=1}^{r} (0 \to \mathcal{O}_X \xrightarrow{y_i} \mathcal{O}_X \to 0)$$

be the associated Koszul complex (the non-zero terms are in degrees $-r, \ldots, 0$). We will show by induction on $r$ that an object $C \in D_Y(X\text{-qcoh})$ vanishes if $G_r \otimes C = 0$. The lemma will follow, since $\mathrm{Hom}(G_r^\vee, C[i]) \simeq H^i(G_r \otimes C)$, where $G_r^\vee = R\,\mathrm{Hom}(G_r, \mathcal{O}_X)$ is the dual of $G_r$. The case $r = 0$ is clear.

Consider $r > 0$ and $C \in D_Y(X\text{-qcoh})$ a non-zero object. By induction, there exists $i$ such that $H^i(G_{r-1} \otimes C) \neq 0$. The distinguished triangle

$$G_{r-1} \otimes C \xrightarrow{y_r} G_{r-1} \otimes C \to G_r \otimes C \rightsquigarrow$$

gives an exact sequence $H^{i-1}(G_r \otimes C) \to H^i(G_{r-1} \otimes C) \xrightarrow{y_r} H^i(G_{r-1} \otimes C)$. Since $H^i(G_{r-1} \otimes C)$ is supported by the closed subvariety $(y_r = 0)$, we deduce that multiplication by $y_r$ has a non-zero kernel, hence $H^{i-1}(G_r \otimes C) \neq 0$. This completes the proof of Lemma 3.10 in the affine case.

We will now prove the lemma by induction on the minimal number of open affine subsets in a covering of $X$.

Let $X = U_1 \cup U_2$ with $U_1$ open affine and $U_2$ open for which the lemma holds. We put $Z_i = X - U_i$. Let $C \in D_Y(X\text{-qcoh})$ with $\mathrm{Hom}(D, C) = 0$ for all $D \in X\text{-perf}_Y$.

Let $D \in U_1\text{-perf}_{Y \cap Z_2}$. The functor $Rj_{1*} : D(U_1\text{-qcoh}) \to D(X\text{-qcoh})$ restricts to equivalences $D_{Y \cap Z_2}(U_1\text{-qcoh}) \xrightarrow{\sim} D_{Y \cap Z_2}(X\text{-qcoh})$ and $U_1\text{-perf}_{Y \cap Z_2} \xrightarrow{\sim} X\text{-perf}_{Y \cap Z_2}$. It follows that $\mathrm{Hom}(Rj_{1*}D, C) = 0$. Let $C'$ be the cocone of the adjunction morphism $C \xrightarrow{\mathrm{can}} Rj_{2*}j_2^*C$. We have

$$\mathrm{Hom}(Rj_{1*}D, Rj_{2*}j_2^*C[n]) \simeq \mathrm{Hom}(j_2^*Rj_{1*}D, j_2^*C[n]) = 0$$

for all $n$, hence $\mathrm{Hom}(Rj_{1*}D, C') \simeq \mathrm{Hom}(Rj_{1*}D, C) = 0$. Since $C'$ is supported by $Y \cap Z_2$, there exists $C'' \in D_{Y \cap Z_2}(U_1\text{-qcoh})$ such that $C' = Rj_{1*}C''$. We have $\mathrm{Hom}(D, C'') \simeq \mathrm{Hom}(Rj_{1*}D, C') = 0$. The affine case of the lemma shows that $C'' = 0$, hence $C \simeq Rj_{2*}j_2^*C$.

Let $E' \in U_2\text{-perf}_{Y \cap U_2}$, $E = E' \oplus E'[1]$ and $G = E_{|U_1 \cap U_2}$. The affine case of the theorem shows that there exists $F \in U_1\text{-perf}_{Y \cap U_1}$ and an isomorphism $F_{|U_1 \cap U_2} \xrightarrow{\sim} G$. Let now $D$ be the cocone of the morphism sum of the adjunction morphisms $Rj_{2*}E \oplus Rj_{1*}F \to Rj_{12*}G$. Then $j_1^*D \simeq F$ and $j_2^*D \simeq E$, hence $D \in X\text{-perf}_Y$. We have $\mathrm{Hom}(E, j_2^*C) \simeq \mathrm{Hom}(D, Rj_{2*}j_2^*C) = 0$. By induction, we deduce that $j_2^*C = 0$, hence $C = 0$. This proves Lemma 3.10 for $X$. $\qquad\square$

**Exercice 3.1.** Let $C \in X\text{-qcoh}$ such that for any set $I$ of objects of $X\text{-qcoh}$, the canonical map $\bigoplus_{D \in I} \mathrm{Hom}(C, D) \to \mathrm{Hom}(C, \bigoplus_{D \in I} D)$ is an isomorphism. Show that $C$ is coherent.

A striking special case of Theorem 3.6 is given by the following corollary.

**Corollary 3.12.** *Let $\mathcal{L}$ be a vector bundle on $U$. Then there exists a perfect complex on $X$ whose restriction to $U$ is quasi-isomorphic to $\mathcal{L} \oplus \mathcal{L}[1]$.*

**Remark 3.13.** Let us show, following Serre [Se, 5.a, p.371], that Theorem 3.6 doesn't hold for vector bundles.

Let $X = \mathbf{A}^3$ and $U = X - \{0\}$. Let $\mathcal{F}$ be the vector bundle on $U$ which is the pullback of the tangent bundle on $\mathbf{P}^2$. The restriction map $K_0(X) \to K_0(U)$ is an isomorphism. Since $\mathcal{F}$ is not the direct sum of two line bundles, it is not the restriction of a vector bundle on $X$.

Let $\mathcal{G}$ be a coherent sheaf on $X$ extending $\mathcal{F}$. The second syzygy $\Omega^2\mathcal{G}$ of $\mathcal{G}$ is locally free (hence free) and this provides a perfect complex extending $\mathcal{F}$ : there is a complex of free sheaves

$$0 \to \Omega^2\mathcal{G} \to P^{-1} \to P^0 \to 0$$

with homology concentrated in degree 0 and isomorphic to $\mathcal{G}$.

**Remark 3.14.** A proof similar to that of Proposition 3.1 shows that there is an exact sequence $0 \to D^b_Z(X\text{-coh}) \to D^b(X\text{-coh}) \to D^b(U\text{-coh}) \to 0$. When $X$ is smooth, then $X\text{-perf} = D^b(X\text{-coh})$, hence Theorem 3.6 is a consequence of that exact sequence. In this case, the canonical map $K_0(X) \to K_0(U)$ is surjective.

### 3.2.5. Applications to K-theory

From Theorem 3.6, Thomason deduces a long exact sequence for higher K-theory, via Waldhausen's theory [Th1]:

**Theorem 3.15.** *There is a long exact sequence*

$$\cdots \to K_i(X \text{ on } Z) \to K_i(X) \to K_i(U) \to K_{i-1}(X \text{ on } Z) \to \cdots$$

Thomason deduces also an excision result.

**Theorem 3.16.** *If $X = U \cup V$ with $V$ open and $Z \subset V$, then there are isomorphisms $K_i(X \text{ on } Z) \xrightarrow{\sim} K_i(V \text{ on } Z)$.*

Finally, he obtains a Mayer-Vietoris Theorem.

**Theorem 3.17.** *Let $U$ and $V$ be open subsets of $X$. Then there is a long exact sequence*

$$\cdots \to K_i(U \cup V) \to K_i(U) \oplus K_i(V) \to K_i(U \cap V) \to K_{i-1}(U \cup V) \to \cdots$$

These sequences can be extended to negative $i$, via a version of Bass' fundamental Theorem:

**Theorem 3.18.** *There is an exact sequence*

$$0 \to K_i(X) \to K_i(X[T]) \oplus K_i(X[T^{-1}]) \to K_i(X[T, T^{-1}]) \to K_{i-1}(X) \to 0$$

Classical methods based on exact categories had led to similar results under restrictive assumptions. Thomason obtains also a local-global principle for $K_i$'s.

# 4. Reconstruction

## 4.1. Abelian case

We will start with the classical case of coherent sheaves, following Gabriel [Ga].

### 4.1.1. Classification of Serre subcategories

We say that a Serre subcategory $\mathcal{I}$ of an abelian category $\mathcal{A}$ is of *finite type* if it is generated by an object (*i.e.*, the smallest Serre subcategory of $\mathcal{A}$ containing the object is $\mathcal{I}$). We say that a Serre subcategory $\mathcal{I}$ is *irreducible* if it is not equal to $0$ and if it is not generated by two proper Serre subcategories of $\mathcal{I}$.

**Theorem 4.1** (Gabriel). *The map $Z \mapsto X\text{-coh}_Z$ from the set of closed subsets of $X$ to the set of Serre subcategories of finite type of $X$-coh is a bijection.*

*The closed irreducible subsets correspond to the irreducible Serre subcategories.*

This follows immediately from the next lemma:

**Lemma 4.2.** *A coherent sheaf with support $Z$ generates $X\text{-coh}_Z$ as a Serre subcategory.*

*Proof.* Let $F$ be a coherent sheaf with support $Z$ and let $\mathcal{I}$ be the Serre subcategory of $X$-coh generated by $F$. Let $i : Y \to X$ be the closed embedding of a subvariety. Every coherent sheaf on $X$ supported by $Y$ is an extension of sheaves of the form $i_* G$. Let $\mathcal{J}$ be the Serre subcategory of $Y$-coh generated by $i^* F$. Since $i_* i^* F \in \mathcal{I}$, we have $i_*(\mathcal{J}) \subset \mathcal{I}$. If $\mathcal{J} = Y\text{-coh}_{Y \cap Z}$, then $X\text{-coh}_{Y \cap Z} \subset \mathcal{I}$. If in addition $Z \subset Y$, then $\mathcal{I} = X\text{-coh}_Z$.

It follows that it is enough to prove the lemma for $X$ reduced and $Z = X$. We proceed by induction on the dimension $n$ of $X$, then on the number of irreducible components of dimension $n$ of $X$, then on the number of irreducible components of dimension $n - 1$ of $X$, etc.

Let $Y$ be a proper closed subset of $X$. The discussion above shows that by induction we can assume $X$-$\mathrm{coh}_Y \subset \mathcal{I}$.

Let $M$ be a coherent sheaf on $X$. Let $U$ be an irreducible open affine subset of $X$ and $j : U \to X$ the open immersion. Shrinking $U$ if necessary, the sheaves $j^*M$ and $j^*F$ are free, of respective ranks $r$ and $s > 0$. Let $f : j^*F^r \xrightarrow{\sim} j^*M^s$ be an isomorphism. By Proposition 3.1, there is a coherent sheaf $M'$ on $X$, there are $\psi : F^r \to M'$ and $\phi : M^s \to M'$ such that $j^*(\psi) = j^*(\phi)f$ and $j^*(\phi)$ is an isomorphism. The kernels and cokernels of $\phi$ and $\psi$ have their supports contained in $X - U$, hence they are in $\mathcal{I}$. It follows that $M \in \mathcal{I}$. $\qquad\square$

### 4.1.2. Centres

**Lemma 4.3.** *Let $R$ be a ring. The canonical map $Z(R) \to Z(R\text{-mod})$ is an isomorphism.*

*Proof.* Evaluation at $R$ gives a left inverse. Let $\alpha \in Z(R\text{-mod})$ with $\alpha(R) = 0$. Since any $R$-module is a quotient of a free $R$-module, it follows that $\alpha = 0$. $\quad\square$

**Corollary 4.4** (Gabriel). *The abelian category $X$-coh determines the variety $X$.*

*Proof.* We define a ringed space $\mathcal{E}$. Its points are the irreducible Serre subcategories of finite type of $X$-coh. The open subsets are the $D(\mathcal{I})$, defined as the set of those subcategories $\mathcal{J}$ that are not contained in a given Serre subcategory $\mathcal{I}$.

Theorem 4.1 shows that the map sending a point $x \in X$ to $X$-$\mathrm{coh}_{\overline{\{x\}}}$ defines a homeomorphism $X \to \mathcal{E}$.

Consider the presheaf of rings on $\mathcal{E}$ given by $\mathcal{O}'_{\mathcal{E}}(D(\mathcal{I})) = Z(X\text{-coh}/\mathcal{I})$. If $D(\mathcal{I}') \subset D(\mathcal{I})$, then the quotient functor $X$-coh $/\mathcal{I} \to X$-coh $/\mathcal{I}'$ induces a map $Z(X\text{-coh}/\mathcal{I}) \to Z(X\text{-coh}/\mathcal{I}')$. We denote by $\mathcal{O}_{\mathcal{E}}$ the associated sheaf. The canonical map $\Gamma(U) \to Z(U\text{-coh})$ induces a morphism of ringed spaces $X \to \mathcal{E}$. To check that this is an isomorphism, it is enough to consider its restriction to an open affine subset: Lemma 4.3 provides the conclusion in the affine case. $\qquad\square$

**Remark 4.5.** Actually, Lemma 4.3 holds for non-affine varieties: the canonical map $\Gamma(\mathcal{O}_X) \to Z(X\text{-coh})$ is an isomorphism of rings. That follows from the construction of the category $X$-coh by gluing the abelian categories $U_i$-coh along the quotient categories $(U_i \cap U_j)$-coh, given a finite open covering of $X$ by open affine subsets $U_i$.

As a consequence, the presheaf in the proof of Corollary 4.4 is a sheaf.

## 4.2. Triangulated case

### 4.2.1. Classification of thick subcategories

Inspired by work on the stable homotopy category (description of the chromatic tower), Hopkins and Neeman [Ho, Nee1] have given a classification of thick subcategories of the category of perfect complexes over an affine variety. This result was generalised later by Thomason [Th2].

Let $\mathcal{I}$ be a thick subcategory of $X$-perf. We say that $\mathcal{I}$ is of *finite type* if it is generated by an object (*i.e.*, if $\mathcal{I}$ is the smallest thick subcategory of $X$-perf containing the object). We say that $\mathcal{I}$ is an *ideal* if it is thick and if given any $C \in \mathcal{I}$ and $D \in X$-perf then $C \otimes^{\mathbf{L}} D \in \mathcal{I}$. We say that an ideal $\mathcal{I}$ is *irreducible* if it is non-zero and it is not generated by two proper ideals.

**Theorem 4.6** (Hopkins, Neeman, Thomason). *The map $Z \mapsto X$-perf$_Z$ from the set of closed subsets of $X$ to the set of ideals of finite type of $X$-perf is a bijection.*

*Irreducible closed subsets correspond to irreducible subcategories.*

The starting point is the following Lemma.

**Lemma 4.7.** *Let $Z$ be a closed subset of $X$. Then there exists a perfect complex on $X$ with support $Z$.*

*Proof.* Assume first $Z$ is irreducible and $X = \operatorname{Spec} R$ is affine. Consider equations $f_1 = 0, \ldots, f_n = 0$ defining $Z$. The support of $\bigotimes_i (0 \to R \xrightarrow{f_i} R \to 0)$ is $Z$: this solves the lemma in that case.

We assume now that $Z$ is irreducible but $X$ is not necessarily affine. Let $U$ be an open affine subset of $X$ containing the generic point of $Z$. Then there exists $C \in U$-perf with support $U \cap Z$. The version "with supports" of the localisation theorem (Theorem 3.11) shows that there exists $D \in X$-perf$_Z$ such that $D_{|U} \simeq C \oplus C[1]$. We have $\operatorname{Supp}(D) \cap U = Z \cap U$, hence $\operatorname{Supp}(D) = Z$.

We consider finally the general case. Let $Z = Z_1 \cup \cdots \cup Z_r$ be the decomposition in irreducible components and consider $C_i \in X$-perf with support $Z_i$. Then the support of $\bigoplus C_i$ is $Z$. $\qquad\square$

Theorem 4.6 follows now from the next lemma.

**Lemma 4.8.** *A perfect complex on $X$ with support $Z$ generates $X$-perf$_Z$ as an ideal.*

*Proof.* Let $C \in X$-perf with support $Z$ and let $\mathcal{I}$ be the ideal of $X$-perf generated by $C$. Lemmas 3.9 and 3.10 show that $\mathcal{I} = X$-perf$_Z$ if and only if $D_Z^b(X\text{-coh}) \subset \bar{\mathcal{I}}$.

Given a closed immersion $i : Y \to X$, we have $i_* Li^* C \simeq C \otimes^{\mathbf{L}} \mathcal{O}_Y \in \bar{\mathcal{I}}$ since $\mathcal{O}_Y \in D(X\text{-qcoh}) = \overline{X\text{-perf}}$ (lemma 3.9). In addition, $Li^* C \in Y\text{-perf}_{Y \cap Z}$. Every object of $D^b_{Y \cap Z}(X\text{-coh})$ is a finite extension ($=$iterated cone) of objects $i_* G$ with $G \in D^b_{Y \cap Z}(Y\text{-coh})$. Let $\mathcal{J}$ be the ideal of $Y\text{-perf}_{Y \cap Z}$ generated by $Li^* C$. Since $i_*(Li^* C \otimes^{\mathbf{L}} M) \simeq C \otimes^{\mathbf{L}} i_* M$ for all $M \in D(Y\text{-qcoh})$, we have $i_*(\bar{\mathcal{J}}) \subset \bar{\mathcal{I}}$. If $\bar{\mathcal{J}} = D_{Y \cap Z}(Y\text{-qcoh})$, then $D^b_{Y \cap Z}(X\text{-coh}) \subset i_*(\bar{\mathcal{J}})$. If in addition $Z \subset Y$, then $\mathcal{I} = X\text{-perf}_Z$.

It follows that it is enough to prove the lemma for $X$ reduced and $Z = X$. We proceed by induction on the dimension $n$ of $X$, then on the number of irreducible components of dimension $n$ of $X$, then on the number of irreducible components of dimension $n - 1$ of $X$, etc.

Let $Y$ be a proper closed subset of $X$. The discussion above shows that $X\text{-perf}_Y \subset \mathcal{I}$.

Let $M \in X\text{-perf}$. There is a non-empty open affine subset $j : U \to X$ such that $j^* M$ and $j^* C$ are finite sums of complexes $\mathcal{O}_U[r]$. So, there are finite-dimensional graded vector spaces (viewed as complexes with vanishing differential) $V \neq 0$ and $W$ and there is an isomorphism $f : j^*(C \otimes_k W) \xrightarrow{\sim} j^*(M \otimes_k V)$. Theorem 3.6 shows that there is an object $M' \in X\text{-perf}$, maps $\psi : C \otimes_k W \to M'$ and $\phi : M \otimes_k V \to M'$ such that $j^*(\psi) = j^*(\phi) f$ and $j^*(\phi)$ is an isomorphism. The cones of $\phi$ and $\psi$ have a support contained in the closed subset $X - U$ of $X$, so they are in $\mathcal{I}$ by induction. It follows that $M \in \mathcal{I}$. $\qquad\square$

**Remark 4.9.** The classical proof of Theorem 4.6 uses the following result ("tensor nilpotence Theorem").

Let $C \in X\text{-perf}$, let $D \in D(X\text{-qcoh})$ and let $f : C \to D$. We assume that for every point $x$ of $X$, we have $f \otimes k(x) = 0$ in $D(k(x)\text{-Mod})$.

Then there is an integer $n$ such that $\otimes^n f : \otimes^n C \to \otimes^n D$ vanishes in $D(X\text{-qcoh})$.

### 4.2.2. Centres

We proceed as in §4.1.2 to obtain a reconstruction Theorem [Ba1, Rou1].

Let $X$ be a variety. We define $Z(X\text{-perf})_{\text{nil}}$ as the subring of $Z(X\text{-perf})$ given by elements $\alpha$ such that $\alpha(C)$ is nilpotent for all $C \in X\text{-perf}$. We put $Z(X\text{-perf})_{\text{red}} = Z(X\text{-perf})/Z(X\text{-perf})_{\text{nil}}$.

**Lemma 4.10.** *The canonical morphism* $\Gamma(\mathcal{O}_X) \to Z(X\text{-perf})$ *induces an isomorphism* $\Gamma(\mathcal{O}_X)_{\text{red}} \xrightarrow{\sim} Z(X\text{-perf})_{\text{red}}$.

*Proof.* Evaluation at $\mathcal{O}_X$ gives a left inverse to the canonical map $\Gamma(\mathcal{O}_X) \to Z(X\text{-perf})$.

Assume first $X = \operatorname{Spec} R$ is affine. Let $\alpha \in Z(R\text{-perf})$ such that $\alpha(R) = 0$. A perfect complex is quasi-isomorphic to a bounded complex $C$ of finitely generated projective $R$-modules. Let $n = \max\{i\,|\,C^i \neq 0\} - \min\{i\,|\,C^i \neq 0\}$. By induction on $n$, one sees that $\alpha(C)^{n+1} = 0$ and the lemma follows in the affine case.

Consider now an arbitrary variety $X$ and $\alpha \in Z(X\text{-perf})$ such that $\alpha(\mathcal{O}_X) = 0$. Let $U$ be an open affine subset of $X$ and let $\alpha_U \in Z(U\text{-perf})$ be the element induced by $\alpha$. We have $\alpha_U(\mathcal{O}_U) = 0$, hence for any $C \in U\text{-perf}$, the endomorphism $\alpha_U(C)$ is nilpotent. Let $X = U_1 \cup \cdots \cup U_r$ be a covering by open affine subsets and let $V = U_2 \cup \cdots \cup U_r$. We prove the lemma by induction on $r$. Let $C \in X\text{-perf}$ and $n > 0$ such that $\alpha_{U_1}(C_{|U_1})^n = 0$. Then $\alpha(C)^n$ factors through an object $C' \in X\text{-perf}_Z$, where $Z = X - U_1$ and $\alpha(C)^{(d+1)n}$ factors through $\alpha(C')^{dn}$ for all $d \geq 0$. Since $Z \subset V$, the restriction functor $X\text{-perf}_Z \to V\text{-perf}_Z$ is fully faithful. By induction, $\alpha_V(C'_{|V})$ is nilpotent, hence $\alpha(C')$ is nilpotent and $\alpha(C)$ is nilpotent as well. $\qquad\square$

**Theorem 4.11** (Balmer, R.). *If $X$ is a reduced variety, then the category $X$-perf, viewed as a tensor triangulated category, determines $X$.*

It would be more satisfactory not to use the tensor structure. Unfortunately, there are too many thick subcategories in general. Balmer has developed a general approach to the study of the geometry of tensor triangulated categories [Ba2, BaFl].

**Example 4.12.** Let $X = \mathbf{P}^1$. Then the thick subcategory of $X$-perf generated by $\mathcal{O}$ is equivalent to $D^b(k\text{-mod})$. It is not of the form $X\text{-perf}_Z$.

### 4.2.3. Remarks on centres

Let us discuss in more details centres of categories of perfect complexes. Note that the difficulties are due to the weakness of the axioms of triangulated categories and can be solved using dg-categories [To]. The next two Propositions show that the centre of the category of perfect complexes has no non-zero nilpotent elements in some cases.

**Proposition 4.13.** *Let $R$ be a ring without zero divisors. Then the canonical map $Z(R) \to Z(R\text{-perf})$ is an isomorphism.*

*Proof.* Evaluation at $R$ defines a morphism $\alpha : Z(R\text{-perf}) \to Z(R)$. The canonical map $Z(R) \to Z(R\text{-perf})$ is a right inverse. We will show that $\alpha$ is injective.

Let $z \in Z(R\text{-perf})$ with $z(R) = 0$. Let $C$ be a non-zero bounded complex of finitely generated projective $R$-modules. Consider $r$ minimal such that $C^r \neq 0$ and $s$ maximal such that $C^s \neq 0$. We will show by induction on $s - r$ that $z(C) = 0$. This is clear when $s = r$. Assume $s > r$.

Since $C^r$ is a finitely generated projective module, there is a finitely generated projective module $P$ such that $C^r \oplus P$ is a free module of finite rank. Let $D = C \oplus (0 \to P \xrightarrow{\text{id}} P \to 0)$, where the non-zero terms of the complex $0 \to P \xrightarrow{\text{id}} P \to 0$ are in degrees $r$ and $r+1$. Then $D^r$ is free of finite rank and $C$ is homotopy equivalent to $D$. So, it is enough to prove that $z(C) = 0$ when $C^r$ is free.

We proceed now by induction on the rank of $C^r$. Let $C^r = L_1 \oplus \cdots \oplus L_n$ be a decomposition into free modules of rank 1. Consider an integer $i$ with $1 \le i \le n$. Let $D$ be the subcomplex of $C$ given by $D^l = C^l$ pour $l \ne r$ and $D^r = L_1 \oplus \cdots \oplus L_{i-1} \oplus L_{i+1} \oplus \cdots \oplus L_n$. By induction, we have $z(D) = 0$. It follows that the composition $f : D \xrightarrow{\text{can}} C \xrightarrow{z(C)} C$ is homotopic to 0. Consider $\{h^l : D^l \to C^{l-1}\}_l$ with $f = d_C h + h d_D$. The morphism $h$ extends uniquely into a graded endomorphism $k$ of degree $-1$ of $C$ that vanishes on $L_i$. Let $\psi = z(C) - (d_C k + k d_C)$, a map homotopic to $z(C)$. The composition $D \xrightarrow{\text{can}} C \xrightarrow{\psi} C$ vanishes, hence $\psi^l = 0$ for $l \ne r$ and $D^r \subset \ker \psi^r$. Note that im $\psi^r \subset \ker d_C^r$. If $\psi^r \ne 0$ then $\ker d_C^r \ne 0$. Since $C^r/D^r \simeq L_i$ is free of rank 1 and since $C^r$ is free, if $\psi^r$ is non-zero, then its restriction to $L_i$ is injective (recall that $R$ has no zero-divisor) and $\ker \psi^r = D^r$.

The composition $C \xrightarrow{\psi} C \xrightarrow{\text{can}} C^r[-r]$ is homotopic to 0 since $z(C^r[-r]) = 0$. So, there is $g : C^{r+1} \to C^r$ such that $\psi^r = g d_C^r$, hence $\ker d_C^r \subset \ker \psi^r$. Assume $z(C) \ne 0$. Then $\ker d_C^r \subset L_1 \oplus \cdots \oplus L_{i-1} \oplus L_{i+1} \oplus \cdots \oplus L_n$. This holds for all $i$, hence $\ker d_C^r = 0$. It follows that $\psi^r = 0$, hence $\psi = 0$ and finally $z(C) = 0$. $\qquad\square$

**Proposition 4.14.** *Let $X$ be an irreducible reduced quasi-projective variety. Then, the canonical morphism $\Gamma(X, \mathcal{O}_X) \to Z(X\text{-perf})$ is an isomorphism.*

*Proof.* Every object of $X$-perf is isomorphic to a bounded complex $C$ whose terms are direct sums of line bundles. The proof is then the same as that of Proposition 4.13 with free modules of rank 1 replaced by line bundles. $\qquad\square$

**Remark 4.15.** Let $k$ be a field and let $\mathcal{C}$ be a $k$-linear category with finite dimensional Hom-spaces. Let $P$ be an indecomposable object of $\mathcal{C}$ and let $\phi \in Z(\text{End}(P))$ with the following properties:

- $\phi^2 = 0$
- given $x \in \text{End}(P)$ such that $\phi x \ne 0$ or $x\phi \ne 0$, then $x$ is invertible.

Given $Q$ an indecomposable object of $\mathcal{C}$ not isomorphic to $P$, then

- $\text{Hom}(P, Q)\phi \, \text{Hom}(Q, P) = 0$
- for $f \in \text{Hom}(P, Q)$ non zero, then, $\phi \in \text{Hom}(Q, P)f$ and for $g \in \text{Hom}(Q, P)$ non zero, then, $\phi \in g \, \text{Hom}(P, Q)$.

Let $C = (0 \to P \xrightarrow{\phi} P \to 0)$, a complex with non-zero terms in degrees 0 and 1. Define an endomorphism $\zeta$ of $C$ as $\phi$ in degree 1 and 0 elsewhere.

Then, one shows there is a unique element of the centre of the homotopy category $\mathcal{T}$ of complexes (all, bounded, bounded above or bounded below, ... ) of objects of $\mathcal{C}$ with the following properties:

- it is 0 on indecomposable objects of $\mathcal{T}$ which are not isomorphic to $C[i]$ for some $i$.
- it is $\phi[i]$ on $C[i]$.

This applies to $\mathcal{C}$ the category of finitely generated projective $A$-modules when $A = k[x]/(x^2)$ (or $A = \mathbf{Z}/4\mathbf{Z}$ by slightly modifying the setting above): the centre of $A$-perf is larger that $A$.

**Remark 4.16.** Proposition 4.14 does not extend to Hochschild cohomology [Ca]. Let $X$ be an elliptic curve. Then, $HH^2(X) = \mathrm{Ext}^2_{X \times X}(\mathcal{O}_{\Delta X}, \mathcal{O}_{\Delta X}) \neq 0$. On the other hand, $X$-coh is a hereditary category. In particular, $\mathrm{Hom}(\mathrm{Id}, \mathrm{Id}[2]) = 0$, where Id is the identity functor of $D^b(X\text{-coh})$.

### 4.2.4. Affine varieties

Let $\mathcal{L}$ be an ample line bundle on $X$. Then $X$-perf is generated by the powers $\mathcal{L}^{\otimes -i}$ for $i > 0$, as a thick subcategory (cf Lemma 3.9). It follows that a thick subcategory $\mathcal{I}$ is an ideal if for any $C \in \mathcal{I}$, we have $C \otimes \mathcal{L}^{-1} \in \mathcal{I}$.

We deduce that if $X$ is affine, then every thick subcategory of $X$-perf is an ideal. We obtain a corollary to Theorem 4.11 (actually, Lemma 4.10 gives a direct proof in that case).

**Corollary 4.17.** *If $X$ is affine and reduced, then the triangulated category $X$-perf determines $X$.*

### 4.2.5. (Anti-)ample canonical bundles

To study more interesting situations, let us introduce Serre functors, following Bondal and Kapranov.

Let $\mathcal{C}$ be a $k$-linear category. A *Serre functor* for $\mathcal{C}$ is an equivalence of categories $S : \mathcal{C} \xrightarrow{\sim} \mathcal{C}$ together with the data, for every $X, Y \in \mathcal{C}$, of bifunctorial isomorphisms

$$\mathrm{Hom}(X, Y)^* \xrightarrow{\sim} \mathrm{Hom}(Y, S(X)).$$

A Serre functor is unique up to unique isomorphism, when it exists.

**Lemma 4.18.** *Let $X$ be a smooth projective variety of pure dimension n. Then $S = \omega_X[n] \otimes -$ is a Serre functor for $D^b(X\text{-coh})$.*

*Proof.* We can assume that $X$ is irreducible. Given $C \in D^b(X\text{-coh})$, consider the Hom-pairing $\text{Hom}(\mathcal{O}, C) \times \text{Hom}(C, \omega_X[n]) \to H^n(X, \omega_X) \simeq k$. When the homology sheaves of $C$ are concentrated in one degree, Serre's duality Theorem shows that this pairing is perfect. Since the thick subcategory of $C$'s for which the pairing is perfect is thick, we deduce that the pairing is perfect for all $C$.

Via the canonical isomorphisms $\text{Hom}(C, D) \overset{\sim}{\to} \text{Hom}(\mathcal{O}, R\mathcal{H}om(C, D))$ and $\text{Hom}(D, C \otimes \omega_X[n]) \overset{\sim}{\to} \text{Hom}(R\mathcal{H}om(C, D), \omega_X[n])$, we obtain a perfect pairing

$$\text{Hom}(C, D) \times \text{Hom}(D, C \otimes \omega_X[n]) \to k.$$

$\square$

**Theorem 4.19** (Bondal-Orlov). *Let $X$ be a smooth projective variety such that $\omega_X$ or $\omega_X^{-1}$ is ample. Then the triangulated category $D^b(X\text{-coh})$ determines $X$.*

*If $Y$ is a smooth projective variety, then an equivalence of triangulated categories $D^b(X\text{-coh}) \overset{\sim}{\to} D^b(Y\text{-coh})$ gives rise to an isomorphism $X \overset{\sim}{\to} Y$.*

*Proof.* The crucial point is the fact that the Serre functor is intrinsic to the category $D^b(X\text{-coh})$. Since the thick subcategories invariant by the Serre functor and its inverse are ideals (cf §4.2.4 and Lemma 4.18), we recover $X$ from $D^b(X\text{-coh})$.

Consider now $F : D^b(X\text{-coh}) \overset{\sim}{\to} D^b(Y\text{-coh})$. Note that $F$ commutes with the Serre functors : $F S_X \simeq S_Y F$.

Let $Z$ be a closed subset of $Y$. Since $F^{-1}(D^b_Z(Y\text{-coh}))$ is a thick subcategory of $D^b(X\text{-coh})$ stable under $S_X^i$ for all $i$, it is of the form $D^b_{\Phi(Z)}(X\text{-coh})$ and this provides an injection $\Phi$ from closed subsets of $Y$ to closed subsets of $X$.

Assume $Z$ is irreducible. Let $V$ be an open affine subset of $Y$. Then $F^{-1}$ induces an equivalence $D^b(V\text{-coh}) \overset{\sim}{\to} D^b((X - \Phi(Y - V))\text{-coh})$ that restricts to an equivalence $D^b_Z(V\text{-coh}) \overset{\sim}{\to} D^b_{\Phi(Z)}((X - \Phi(Y - V))\text{-coh})$. Since $D^b_Z(V\text{-coh})$ is an irreducible thick subcategory (§4.2.4 and Theorem 4.6), we deduce that $D^b_{\Phi(Z)}((X - \Phi(Y - V))\text{-coh})$ is an irreducible thick subcategory of $D^b((X - \Phi(Y - V))\text{-coh})$, hence $\Phi(Z) \cap (X - \Phi(Y - V))$ is irreducible. If $Y = V_1 \cup \cdots \cup V_r$ is a covering by open affine subsets, then the subsets $X - \Phi(Y - V_i)$ give an open affine covering of $X$, and it follows that $\Phi(Z)$ is irreducible.

We define an injection $\phi : Y \to X$ between points by $\overline{\phi(y)} = \Phi(\overline{y})$. If $y$ is a closed point, then by Lemma 4.21 below the thick subcategory $D^b_{\{y\}}(Y\text{-coh})$ of $D^b(Y\text{-coh})$ is minimal as a non-zero thick subcategory. It follows that $F^{-1}(D^b_{\{y\}}(Y\text{-coh})) = D^b_{\overline{\phi(y)}}(X\text{-coh})$ is a minimal non-zero thick subcategory of $D^b(X\text{-coh})$ that is stable under $S_X^i$ for all $i$. We deduce that $\phi(y)$ is a closed point.

Let $x$ be a closed point of $X$ that is not in the image of $\phi$. Then $\operatorname{Hom}(\mathcal{O}_{\{x\}}, C[i]) = 0$ for all $i \in \mathbf{Z}$ and all $C \in D^b_{\{\phi(y)\}}(X\text{-coh})$. It follows that $\operatorname{Hom}(F(\mathcal{O}_{\{x\}}), \mathcal{O}_{\{y\}}[i]) = 0$ for all closed points $y$ of $Y$ and all $i \in \mathbf{Z}$. We deduce from Lemma 4.20 below that $F(\mathcal{O}_{\{x\}}) = 0$, a contradiction. So, $\phi$ is bijective.

Let $Z$ be a closed subset of $Y$. A closed point $y$ of $Y$ is in $Z$ if and only if $D^b_{\{y\}}(Y\text{-coh}) \subset D^b_Z(Y\text{-coh})$. We have $x \in \Phi(Z)$ if and only if $D^b_{\{x\}}(X\text{-coh}) \subset D^b_{\Phi(Z)}(X\text{-coh})$, hence $\phi(Z) = \Phi(Z)$ is closed in $X$. This shows that $\psi = \phi^{-1} : X \to Y$ is continuous.

Let $U = Y - Z$. We have a sequence of canonical isomorphisms (cf Lemma 4.10)

$$\Gamma(U) \xrightarrow{\sim} Z(D^b(U\text{-coh}))_{\text{lred}} \xrightarrow[F^{-1}]{\sim} Z(D^b(\phi(U)\text{-coh}))_{\text{lred}} \xrightarrow{\sim} \Gamma(\phi(U)).$$

This extends $\psi : X \to Y$ into an isomorphism of ringed spaces. $\qquad\square$

**Lemma 4.20.** *Let $X$ be a variety and let $C \in D^b(X\text{-coh})$ such that $\operatorname{Hom}(C, \mathcal{O}_{\{x\}}[i]) = 0$ for all closed points $x$ in $X$ and for all $i \in \mathbf{Z}$. Then $C = 0$.*

*Proof.* Consider $i$ maximal such that $\mathcal{H}^i(C) \neq 0$ and let $x$ be a closed point. The canonical map $\operatorname{Hom}(\mathcal{H}^i(C), \mathcal{O}_{\{x\}}) \to \operatorname{Hom}(C, \mathcal{O}_{\{x\}}[-i])$ is injective. Given $x$ in the support of $\mathcal{H}^i(C)$, we have $\operatorname{Hom}(\mathcal{H}^i(C), \mathcal{O}_{\{x\}}) \neq 0$, hence $\operatorname{Hom}(C, \mathcal{O}_{\{x\}}[-i]) \neq 0$. $\qquad\square$

**Lemma 4.21.** *Let $X$ be an algebraic variety and let $x$ be a closed point of $X$. Then $X\text{-perf}_{\{x\}}$ is a minimal non-zero thick subcategory of $X\text{-perf}$.*

*Proof.* Let $U$ be an open affine subset of $X$ containing $x$. The restriction functor $X\text{-perf}_{\{x\}} \to U\text{-perf}_{\{x\}}$ is fully faithful. Theorem 4.6 and §4.2.4 show that $U\text{-perf}_{\{x\}}$ is a minimal non-zero thick subcategory of $U\text{-perf}$. The lemma follows. $\qquad\square$

**Remark 4.22.** Bondal and Orlov [BoOr] show that the structure of graded category of $D^b(X\text{-coh})$ (we forget the distinguished triangles) is enough to reconstruct $X$ in Theorem 4.19.

Given $\alpha : X \xrightarrow{\sim} Y$ an isomorphism of varieties, we have an equivalence $\alpha_* : D^b(X\text{-coh}) \xrightarrow{\sim} D^b(Y\text{-coh})$. Given $\mathcal{L}$ a line bundle on $X$, we have a self-equivalence $\mathcal{L}\otimes? : D^b(X\text{-coh}) \xrightarrow{\sim} D^b(X\text{-coh})$. Finally, given $n \in \mathbf{Z}$, we have the self-equivalence $[n]$.

This gives a injective morphism from $\mathbf{Z} \times (\operatorname{Pic} X \rtimes \operatorname{Aut}(X))$ to the group $\operatorname{Aut}(D^b(X\text{-coh}))$ of isomorphism classes of self-equivalences of $D^b(X\text{-coh})$.

We can now complete Theorem 4.19.

**Theorem 4.23** (Bondal-Orlov). *Let $X$ be a smooth connected projective variety with $\omega_X$ or $\omega_X^{-1}$ ample. Then the canonical map $\mathbf{Z} \times (\operatorname{Pic} X \rtimes \operatorname{Aut}(X)) \xrightarrow{\sim} \operatorname{Aut}(D^b(X\text{-coh}))$ is an isomorphism.*

*Proof.* Let $F$ be a self-equivalence of $D^b(X\text{-coh})$. In the proof of Theorem 4.19, we have constructed an automorphism $\psi$ of $X$ such that $F(D_Z(X\text{-coh})) = D_{\psi(Z)}(X\text{-coh})$ for all closed subsets $Z$ of $X$. Replacing $F$ by $F\psi^*$, we can assume that $D_Z(X\text{-coh})$ is stable by $F$ for all $Z$ closed in $X$. Furthermore, $F$ restricts to a self-equivalence $F_x$ of $D^b(\mathcal{O}_x\text{-mod})$ for all closed points $x$ of $X$.

Let $C = F(\mathcal{O}_X)$. Then $\operatorname{Hom}(C_x, C_x[i]) \simeq \operatorname{Hom}(\mathcal{O}_x, \mathcal{O}_x[i]) = 0$ for $i \neq 0$. Let $D$ be a bounded complex of finitely generated projective $\mathcal{O}_x$-modules that is quasi-isomorphic to $C_x$ and such that given $i$ minimal (resp. maximal) with $d_D^i \neq 0$, then $d_D^i$ is not a split injection (resp. a split surjection). Let $r$ and $s$ be those minimal and maximal integers. The canonical map $\operatorname{Hom}(D^r, D^{s+1}) \to \operatorname{Hom}(D, D[s-r])$ is non-zero: this is impossible. It follows that $H^i(C_x)$ is concentrated in a single degree $i = r_x$ and $H^{r_x}(C_x)$ is free. In addition, $\operatorname{End}(C_x) \simeq \operatorname{End}(\mathcal{O}_x) = \mathcal{O}_x$, hence $H^{r_x}(C_x)$ is free of rank 1. The set of closed points $x$ with $H^i(C_x) \neq 0$ is closed in $X$. Since $X$ is connected, we deduce that $r_x = r$ is constant. It follows that $C \simeq \mathcal{H}^r(C)[-r]$ and $\mathcal{L} = \mathcal{H}^r(C)$ is a line bundle on $X$. Replacing $F$ by $(\mathcal{L}^{-1}[r] \otimes -) \circ F$, we can assume that $F(\mathcal{O}_X) \simeq \mathcal{O}_X$.

We want to prove now that $F$ is isomorphic to the identity functor. By Orlov's representability Theorem [Or], there exists $K \in D^b((X \times X)\text{-coh})$ such that $F \simeq Rp_*(K \otimes^{\mathbf{L}} q^*(-))$, where $p, q$ are the first and second projections $X \times X \to X$. Given $x_1, x_2$ closed points of $X$, we have $\operatorname{Hom}(K, \mathcal{O}_{\{(x_1,x_2)\}}[i]) \simeq \operatorname{Hom}(F(\mathcal{O}_{\{x_1\}}), \mathcal{O}_{\{x_2\}}[i]) \simeq \delta_{x_1,x_2} \delta_{0,i} k$. It follows that $K \simeq i_*\mathcal{M}$, where $\mathcal{M}$ is a line bundle on $X$ and $i : X \to X \times X$ is the diagnal embedding. We deduce that $K \simeq i_*\mathcal{O}_X$ and $F \simeq \operatorname{Id}$. $\qquad\square$

**Remark 4.24.** A proof similar to the one of Theorem 4.23 shows that $\operatorname{Pic}(X) \rtimes \operatorname{Aut}(X) \xrightarrow{\sim} \operatorname{Aut}(X\text{-coh})$ for any variety $X$.

# References


[Ba1] P. Balmer, *Presheaves of triangulated categories and reconstruction of schemes*, Math. Ann. **324** (2002), 557–580.

[Ba2] P. Balmer, *The spectrum of prime ideals in tensor triangulated categories*, J. Reine Angew. Math. **588** (2005), 149–168.

[BaFl] P. Balmer and G. Favi, *Gluing techniques in triangular geometry*, Q. J. Math. **58** (2007), 415–441.

[BoOr]    A. Bondal and D. Orlov, *Reconstruction of a variety from the derived category and groups of autoequivalences*, Compositio Math. **125** (2001), 327–344.

[Ca]    A. Caldararu, letter to Tom Bridgeland, 5 November 2002.

[Ga]    P. Gabriel, *Des catgories abéliennes*, Bull. Soc. Math. France **90** (1962), 323–448.

[Ho]    M. Hopkins, *Global methods in homotopy theory*, in "Homotopy theory (Durham, 1985)", 73–96, London Math. Soc. Lecture Note Ser., 117, Cambridge Univ. Press, 1987.

[KaScha]    M. Kashiwara and P. Schapira, "Categories and sheaves", Springer Verlag, 2006.

[Nee1]    A. Neeman, *The chromatic tower for $D(R)$*, Topology **31** (1992), 519–532.

[Nee2]    A. Neeman, *The connection between the $K$-theory localization theorem of Thomason, Trobaugh and Yao and the smashing subcategories of Bousfield and Ravenel*, Ann. Sci. Ecole Norm. Sup. **25** (1992), 547–566.

[Nee3]    A. Neeman, "Triangulated categories", Princeton University Press, 2001.

[Or]    D. Orlov, *Equivalences of derived categories and K3 surfaces*, J. Math. Sci. (New York) **84** (1997), 1361–1381.

[Rou1]    R. Rouquier, Peccot lectures, Collège de France, March 2000.

[Rou2]    R. Rouquier, *Dimensions of triangulated categories*, Journal of K-theory **1** (2008), 193–256.

[Se]    J.-P. Serre, *Prolongement des faisceaux analytiques cohérents*, Ann. Inst. Fourier **16** (1966), 363–374.

[Th1]    R.B. Thomason, *The local to global principle in algebraic $K$-theory*, Proceedings ICM, Kyoto, 1990, 381–394, Springer Verlag, 1991.

[Th2]    R.B. Thomason, *The classification of triangulated subcategories*, Compositio Math. **105** (1997), 1–27.

[ThTr]    R.B. Thomason and T.F. Trobaugh, *Higher algebraic $K$-theory of schemes and of derived categories*, Grothendieck Festschrift vol. III, 247–435, Birkhauser, 1990.

[To]    B. Toen, *The homotopy theory of dg-categories and derived Morita theory*, Invent. Math. **167** (2007), 615–667.

# Triangulated categories for the analysts

PIERRE SCHAPIRA

ABSTRACT. This paper aims at showing how the tools of Algebraic Geometry apply to Analysis. We will review various classical constructions, including Sato's hyperfunctions, Fourier-Sato transform and microlocalization, the microlocal theory of sheaves (with some applications to PDE) and explain the necessity of Grothendieck topologies to treat algebraically generalized functions with growth conditions.

## Introduction

In this paper, we will show how the tools of Algebraic Geometry– sheaves, triangulated and derived categories, Grothendieck topologies and stacks– play (or should play) a crucial role in Analysis.

Note that, conversely, some problems of Analysis led to new algebraic concepts. For example, one of the deepest notion related to triangulated categories is that of $t$-structure, and as a particular case, that of perverse sheaves, and these notions emerged with the study of the Riemann Hilbert correspondence, a problem dealing with differential equations.

Another example is the Fourier transform, clearly one of the most essential tools of the analysts, until it was categorified by Sato and applied to algebraic analysis, next transposed to algebraic geometry (the Fourier-Mukai transform).

The classical analysts are used to work in various functional spaces constructed with the machinery of functional analysis and Fourier transform, but Sato's construction of hyperfunctions [28] in the 60's does not use any of these tools. It is a radically new approach which indeed has entirely modified the mathematical landscape in this area. The functional spaces are now replaced by "functorial spaces", that is, sheaves of generalized holomorphic functions on a complex manifold $X$ or, more precisely, complexes of sheaves $R\mathcal{H}om\,(G, \mathcal{O}_X)$, where $G$ is an object of the derived category of $\mathbb{R}$-constructible sheaves on the real underlying manifold to $X$. Putting general systems of linear partial differential equations, $i.e.$, $\mathcal{D}_X$-modules, in the machinery, one is led to study complexes $R\mathcal{H}om\,(G, F)$, where $F = \mathrm{Sol}(\mathcal{M}) := R\mathcal{H}om_{\mathcal{D}_X}(\mathcal{M}, \mathcal{O}_X)$ is the complex of holomorphic solutions of a coherent $\mathcal{D}_X$-module $\mathcal{M}$.

*Date*: March 5, 2010.
*Mathematics Subject Classification*: 58G07, 32A45, 32C38, 35A27.

The main invariant attached to a coherent $\mathcal{D}_X$-module $\mathcal{M}$ is its characteristic variety $\mathrm{char}(\mathcal{M})$, a closed conic involutive subset of the cotangent bundle $T^*X$. For a sheaf $G$ on a *real* manifold there is a similar invariant, the microsupport $SS(G)$ which describes the directions of non propagation of $G$, and this set is again a closed conic involutive subset of the cotangent bundle. In case of $F = \mathrm{Sol}(\mathcal{M})$, it follows from the Cauchy-Kowalevsky theorem that its microsupport is nothing but $\mathrm{char}(\mathcal{M})$. Therefore, in order to study $R\mathcal{H}om\,(G, \mathrm{Sol}(\mathcal{M}))$, one can forget that one is working on a complex manifold $X$ and dealing with $\mathcal{D}_X$-modules, keeping only in mind two geometrical informations, the microsupport of $G$ and that of $F$ (see [16]).

The study of the characteristic variety of $\mathcal{D}_X$-modules naturally leads to introduce the ring $\mathcal{E}_X$ of microdifferential operators, a kind of localization of $\mathcal{D}_X$ in $T^*X$. The ring $\mathcal{E}_X$ was first constructed in [29] using Sato's microlocalization functor, the Fourier-Sato transform of the specialization functor. We shall briefly recall here the main steps of these constructions.

Finally, although classical sheaf theory does not allow one to treat usual spaces of analysis involving growth conditions, these conditions being not of local nature, we shall show here how it is possible to overcome this difficulty by using Grothendieck topologies (see [17]).

## 0.1 Notations

In this paper, we mainly follow the notations of [16].

(i) We denote by $\mathbf{k}$ a field and by $\mathrm{D}^{\mathrm{b}}(\mathbf{k}_X)$ the bounded derived category of sheaves of $\mathbf{k}$-vector spaces on a topological space $X$. More generally, if $\mathcal{A}$ is a sheaf of rings on $X$, we denote by $\mathrm{Mod}(\mathcal{A})$ the category of left $\mathcal{A}$-modules and by $\mathrm{D}^{\mathrm{b}}(\mathcal{A})$ its bounded derived category. If there is no risk of confusion, we write $R\mathcal{H}om$ instead of $R\mathcal{H}om_{\mathbf{k}_X}$ and similarly for RHom and $\otimes$.

(ii) If $Z$ is a locally closed subset of the topological space $X$, we denote by $\mathbf{k}_{XZ}$ the sheaf which is the constant sheaf with stalk $\mathbf{k}$ on $Z$ and which is $0$ on $X \setminus Z$. If there is no risk of confusion, we write $\mathbf{k}_Z$ instead of $\mathbf{k}_{XZ}$.

(iii) For a real manifold $X$, we denote by $\dim_X$ its dimension and by $\mathrm{or}_X$ the orientation sheaf. For a morphism of manifolds $f : X \to Y$, we set $\mathrm{or}_{X/Y} = \mathrm{or}_X \otimes f^{-1}\,\mathrm{or}_Y$.

(iv) We denote by $\mathrm{D}'_X(\,\cdot\,) = R\mathcal{H}om_{\mathbf{k}_X}(\,\cdot\,, \mathbf{k}_X)$ the duality functor in $\mathrm{D}^{\mathrm{b}}(\mathbf{k}_X)$ and by $\omega_X$ the dualizing complex. Recall that $\omega_X \simeq \mathrm{or}_X\,[\dim_X]$.

(v) For a (real or complex) manifold $X$, we denote by $\tau : TX \to X$ and $\pi : T^*X \to X$ its tangent bundle and cotangent bundle, respectively. Let $f : X \to Y$ be a morphism of (real or complex) manifolds. To $f$ are associated

the maps

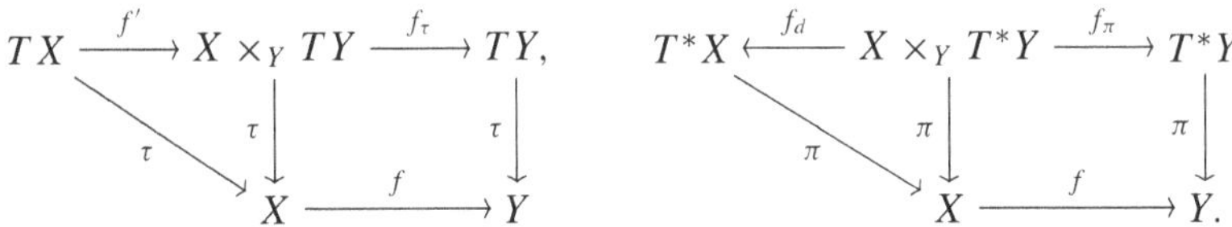

We denote by $T_X^* X$ the zero-section of $T^* X$ and by $T_X^* Y = f_d^{-1} T_X^* X$ the conormal bundle to $f$. If $f$ is an embedding, we identify $T_X^* Y$ to a sub-bundle of $T^* Y$ and call it the conormal bundle to $X$.

(vi) For a complex manifold $X$, we denote by $\mathcal{O}_X$ the sheaf of holomorphic functions and by $\Omega_X$ the sheaf of holomorphic forms of maximal degree. We denote by $d_X$ the complex dimension of $X$.

# 1. Generalized functions

In the sixties, people used to work in various spaces of generalized functions on a real manifold. The situation drastically changed with Sato's definition of hyperfunctions [28].

Consider first the case where $M$ is an open subset of the real line $\mathbb{R}$ and let $X$ an open neighborhood of $M$ in the complex line $\mathbb{C}$ satisfying $X \cap \mathbb{R} = M$. The space $\mathcal{B}(M)$ of hyperfunctions on $M$ is given by

$$\mathcal{B}(M) = \mathcal{O}(X \setminus M)/\mathcal{O}(X).$$

It is easily proved that this space depends only on $M$, not on the choice of $X$, and that the correspondence $U \mapsto \mathcal{B}(U)$ ($U$ open in $M$) defines a flabby sheaf $\mathcal{B}_M$ on $M$.

Classically, the "boundary value" of a holomorphic function $\varphi(z)$ defined in the open set $X \cap \{\operatorname{Im} z > 0\}$ of the complex line, if it exists, is the limit (for a suitable topology) of the function $\varphi(x + iy)$ as $y \overset{>}{\to} 0$. With Sato's definition, the boundary value always exists and is no more a limit. Indeed, it is the class of the holomorphic function $\psi(z) \in \mathcal{O}(X \setminus M)$ given by $\psi(z) = \varphi(z)$ for $\operatorname{Im} z > 0$ and $\psi(z) = 0$ for $\operatorname{Im} z < 0$.

On a manifold $M$ of dimension $n$, the sheaf $\mathcal{B}_M$ was originally defined as

$$\mathcal{B}_M = H_M^n(\mathcal{O}_X) \otimes \operatorname{or}_M$$

where $X$ is a complexification of $M$. Since $X$ is oriented, Poincaré's duality gives the isomorphism $\mathrm{D}'_X(\mathbb{C}_M) \simeq \operatorname{or}_M [-n]$. An equivalent definition of hyperfunctions is thus given by

$$\mathcal{B}_M = R\mathcal{H}om_{\mathbb{C}_X}(\mathrm{D}'_X(\mathbb{C}_M), \mathcal{O}_X). \tag{1.1}$$

The importance of Sato's definition is twofold: first, it is purely algebraic (starting with the analytic object $\mathcal{O}_X$), and second it highlights the link between real and complex geometry.

Let us define the notion of "boundary value" in this settings. Consider a subanalytic open subset $\Omega$ of $X$ and denote by $\overline{\Omega}$ its closure. Assume that:

$$\begin{cases} \mathrm{D}'_X(\mathbb{C}_\Omega) \simeq \mathbb{C}_{\overline{\Omega}}, \\ M \subset \overline{\Omega}. \end{cases}$$

The morphism $\mathbb{C}_{\overline{\Omega}} \to \mathbb{C}_M$ defines by duality the morphism $\mathrm{D}'_X(\mathbb{C}_M) \to \mathrm{D}'_X(\mathbb{C}_{\overline{\Omega}}) \simeq \mathbb{C}_\Omega$. Applying the functor $\mathrm{RHom}(\,\cdot\,, \mathcal{O}_X)$, we get the boundary value morphism

$$\mathrm{b}\colon \mathcal{O}(\Omega) \to \mathcal{B}(M). \tag{1.2}$$

Sato's sheaf of hyperfunctions is an example of a sheaf of generalized holomorphic functions. Another example of such a sheaf is as follows.

Consider a closed complex hypersurface $Z$ of the complex manifold $X$ and denote by $U$ its complementary. Let $j\colon U \hookrightarrow X$ denote the embedding. Then $j_* j^{-1}\mathcal{O}_X$ represents the sheaf on $X$ of functions holomorphic on $U$ with possible (essential) singularities on $Z$. One has

$$j_* j^{-1}\mathcal{O}_X \simeq R\mathcal{H}om_{\mathbb{C}_X}(\mathbb{C}_U, \mathcal{O}_X). \tag{1.3}$$

Both examples (1.1) and (1.3) are described by a sheaf of the type $R\mathcal{H}om(G, \mathcal{O}_X)$, with $G$ a *constructible* sheaf (see [16] for an exposition). Recall that a sheaf $G$ on a real analytic manifold $X$ is $\mathbb{R}$-constructible if there exists a subanalytic stratification of $X$ on which $G$ is locally constant of finite rank (over the field **k**). One denotes by $\mathrm{D}^{\mathrm{b}}_{\mathbb{R}\text{-c}}(\mathbf{k}_X)$ the full triangulated subcategory of $\mathrm{D}^{\mathrm{b}}(\mathbf{k}_X)$ consisting of objects with $\mathbb{R}$-constructible cohomology. On a complex manifold, replacing the subanalytic stratifications by complex analytic stratifications, one also gets the category $\mathrm{D}^{b}_{\mathbb{C}\text{-c}}(\mathbf{k}_X)$ of $\mathbb{C}$-constructible sheaves.

The advantage of considering the category $\mathrm{D}^{\mathrm{b}}_{\mathbb{R}\text{-c}}(\mathbf{k}_X)$ is that the properties of being constructible (in the derived sense) is stable by the six Grothendieck operations (with suitable properness hypotheses).

To summarize, the classical functional spaces are now replaced by the "functorial spaces" $R\mathcal{H}om_{\mathbb{C}_X}(G, \mathcal{O}_X)$, where $G \in \mathrm{D}^{\mathrm{b}}_{\mathbb{R}\text{-c}}(\mathbb{C}_X)$.

## 2. $\mathcal{D}$-modules

References for the theory of $\mathcal{D}$-modules are made to [8, 9].

The theory of $\mathcal{D}$-modules appeared in the 70's with Kashiwara's thesis [8] and Bernstein's paper [2]. However, already in the 60's, Sato had the main ideas of the theory in mind and gave talks at Tokyo University on these topics. Unfortunately, Sato did not write anything and it seems that his ideas were not understood at this time. (See [1, 30].)

Let $X$ be a complex manifold. One denotes by $\mathcal{D}_X$ the sheaf of rings of holomorphic (finite order) differential operators. A system of linear differential equations on $X$ is a left coherent $\mathcal{D}_X$-module. The link with the intuitive notion of a system of linear differential equations is as follows. Locally on $X$, $\mathcal{M}$ may be represented as the cokernel of a matrix $\cdot P_0$ of differential operators acting on the right. By classical arguments of analytic geometry (Hilbert's syzygies theorem), one shows that, locally, $\mathcal{M}$ admits a bounded resolution by free modules of finite type, that is, $\mathcal{M}$ is locally isomorphic to the cohomology of a bounded complex

$$\mathcal{M}^{\bullet} := 0 \to \mathcal{D}_X^{N_r} \to \cdots \to \mathcal{D}_X^{N_1} \xrightarrow{\cdot P_0} \mathcal{D}_X^{N_0} \to 0. \tag{2.1}$$

Let us introduce the notation:

$$\mathrm{Sol}(\bullet) := R\mathcal{H}om_{\mathcal{D}_X}(\bullet, \mathcal{O}_X). \tag{2.2}$$

The complex $\mathrm{Sol}(\mathcal{M})$ of holomorphic solutions of $\mathcal{M}$ may be locally calculated by applying the functor $\mathcal{H}om_{\mathcal{D}_X}(\bullet, \mathcal{O}_X)$ to $\mathcal{M}^{\bullet}$. Hence

$$\mathrm{Sol}(\mathcal{M}) \simeq 0 \to \mathcal{O}_X^{N_0} \xrightarrow{P_0 \cdot} \mathcal{O}_X^{N_1} \to \cdots \to \mathcal{O}_X^{N_r} \to 0, \tag{2.3}$$

where now $P_0\cdot$ operates on the left.

One defines naturally the characteristic variety $\mathrm{char}(\mathcal{M})$ of a coherent $\mathcal{D}_X$-module $\mathcal{M}$. This is a closed complex analytic subset of $T^*X$, conic with respect to the action of $\mathbb{C}^\times$ on $T^*X$. For example, if $\mathcal{M}$ has a single generator $u$ with relation $\mathcal{I}u = 0$, where $\mathcal{I}$ is a locally finitely generated ideal of $\mathcal{D}_X$, then

$$\mathrm{char}(\mathcal{M}) = \{(x; \xi) \in T^*X; \sigma(P)(x; \xi) = 0 \text{ for all } P \in \mathcal{I}\}$$

where $\sigma(P)$ denotes the principal symbol of $P$. A fundamental result of [29] asserts that for a coherent $\mathcal{D}$-module $\mathcal{M}$, $\mathrm{char}(\mathcal{M})$ is an involutive (*i.e.,* coisotropic) subset of $T^*X$. The proof uses infinite order microdifferential operators and quantized contact transformations. Of course, the involutivity theorem has a longer history, including the previous work of Guillemin-Quillen-Sternberg [6], and culminating with the purely algebraic proof of Gabber [5].

We denote by $D^b_{coh}(\mathcal{D}_X)$ the full triangulated subcategory of $D^b(\mathcal{D}_X)$ consisting of objects with coherent cohomologies. However, we need a refined notion. A $\mathcal{D}_X$-module $\mathcal{M}$ is "good" if on each relatively compact open subset $U$ of $X$, $\mathcal{M}$ is generated by a coherent $\mathcal{O}_X|_U$-modules. We denote by $D^b_{good}(\mathcal{D}_X)$ the full triangulated subcategory of $D^b(\mathcal{D}_X)$ consisting of objects with good cohomologies.

### Operations on $\mathcal{D}$-modules

Let $f : X \to Y$ be a morphism of complex manifolds. The sheaf

$$\mathcal{D}_{X \to Y} := \mathcal{O}_X \otimes_{f^{-1}\mathcal{O}_Y} f^{-1}\mathcal{D}_Y \tag{2.4}$$

is naturally endowed with a structure of a $(\mathcal{D}_X, f^{-1}\mathcal{D}_Y)$-bimodule, but one shall be aware that the left action of $\mathcal{D}_X$ is not simply its action on $\mathcal{O}_X$ (see [9] for details).

The inverse image functor $\mathrm{D}f^{-1}$ for $\mathcal{D}$-modules is given by

$$\mathrm{D}f^{-1}\mathcal{N} = \mathcal{D}_{X \to Y} \overset{L}{\otimes}_{f^{-1}\mathcal{D}_Y} f^{-1}\mathcal{N}, \quad \mathcal{N} \in \mathrm{D}^b(\mathcal{D}_Y).$$

One says that $f$ is non characteristic for $\mathcal{N} \in \mathrm{D}^b_{coh}(\mathcal{D}_Y)$ if (see Notations 0.1)

$$f_\pi^{-1} \, \mathrm{char}(\mathcal{N}) \cap (T_X^* Y) \subset X \times_Y T_Y^* Y.$$

In this case, $\mathrm{D}f^{-1}\mathcal{N} \in \mathrm{D}^b_{coh}(\mathcal{D}_X)$ and the Cauchy-Kowalevsky-Kashiwara theorem asserts that there is a natural isomorphism

$$f^{-1}\mathrm{Sol}(\mathcal{N}) \overset{\sim}{\to} \mathrm{Sol}(\mathrm{D}f^{-1}\mathcal{N}). \tag{2.5}$$

The proper direct image functor $\mathrm{D}f_!$ for (right) $\mathcal{D}$-modules is given by

$$\mathrm{D}f_!\mathcal{M} := Rf_!(\mathcal{M} \overset{L}{\otimes}_{\mathcal{D}_X} \mathcal{D}_{X \to Y}), \quad \mathcal{M} \in \mathrm{D}^b(\mathcal{D}_X^{op}).$$

One defines the direct image functor $\mathrm{D}f_!$ for left $\mathcal{D}$-modules by using the line bundles $\Omega_X$ and $\Omega_Y$ (or their inverse) which intertwine the left and right structures. If $\mathcal{M} \in \mathrm{D}^b_{good}(\mathcal{D}_X)$ and $f$ is proper on the support of $\mathcal{M}$, one deduces from Grauert's direct images Theorem that $\mathrm{D}f_!\mathcal{M}$ belongs to $\mathrm{D}^b_{good}(\mathcal{D}_Y)$. Moreover, there is a natural isomorphism

$$Rf_!\mathrm{Sol}(\mathcal{M})[d_X] \simeq \mathrm{Sol}(\mathrm{D}f_!\mathcal{M})[d_Y]. \tag{2.6}$$

(Recall that $d_X$ (resp. $d_Y$) is the complex dimension of $X$ (resp. $Y$).)

The product $\mathcal{M} \overset{L}{\otimes}_{\mathcal{O}_X} \mathcal{L}$ of two left $\mathcal{D}_X$-modules is naturally endowed with a structure of a left $\mathcal{D}_X$-module. We denote it by $\mathcal{M} \overset{D}{\otimes} \mathcal{L}$.

## Holonomic systems

Since the characteristic variety of a coherent $\mathcal{D}_X$-module $\mathcal{M}$ is involutive, it is natural to study with a particular attention the extreme case where this characteristic variety has minimal dimension.

**Definition 2.1.** A holonomic system on $X$ is a coherent $\mathcal{D}_X$-module whose characteristic variety is Lagrangian in $T^*X$.

Denote by $\mathrm{D}^{\mathrm{b}}_{\mathrm{hol}}(\mathcal{D}_X)$ the full triangulated subcategory $\mathrm{D}^{\mathrm{b}}(\mathcal{D}_X)$ consisting of objects with holonomic cohomology and recall that we set $\mathrm{Sol}(\cdot) = R\mathcal{H}om_{\mathcal{D}}(\cdot, \mathcal{O}_X)$. The first fundamental result of the theory of holonomic $\mathcal{D}$-modules is Kashiwara's constructibility theorem:

**Theorem 2.2.** [10] *The functor* $\mathrm{Sol}$ *induces a functor*

$$\mathrm{Sol}\colon (\mathrm{D}^{\mathrm{b}}_{\mathrm{hol}}(\mathcal{D}_X))^{\mathrm{op}} \to \mathrm{D}^{\mathrm{b}}_{\mathbb{C}\text{-c}}(\mathbb{C}_X).$$

Simple examples in dimension one show that this functor is not fully faithful, but Kashiwara-Oshima [15] and Kashiwara-Kawai [14] gave the definition of a "regular holonomic" $\mathcal{D}$-module and Kashiwara [11] proved the Riemann-Hilbert correspondence, that is, the equivalence of categories

$$(\mathrm{D}^{\mathrm{b}}_{\mathrm{reg-hol}}(\mathcal{D}_X))^{\mathrm{op}} \xrightarrow{\sim} \mathrm{D}^{b}_{\mathbb{C}\text{-c}}(\mathbb{C}_X)$$

where now $\mathrm{D}^{\mathrm{b}}_{\mathrm{reg-hol}}(\mathcal{D}_X)$ denotes the full triangulated subcategory of $\mathrm{D}^{\mathrm{b}}(\mathcal{D}_X)$ consisting of objects with regular holonomic cohomology.

## Integral transforms

In this subsection, we shall study the action of integral transforms on $\mathcal{D}$-modules and their sheaves of generalized solutions.

Consider two complex manifolds $X$ and $Y$, of complex dimension $d_X$ and $d_Y$ respectively, and the correspondence:

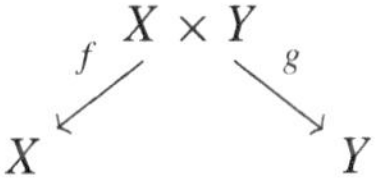

For $G \in \mathrm{D}^{\mathrm{b}}(\mathbb{C}_Y)$ and $L \in \mathrm{D}^{\mathrm{b}}(\mathbb{C}_{X \times Y})$, we set

$$L \circ G = Rf_!(L \otimes g^{-1}G). \tag{2.7}$$

For $\mathcal{M} \in \mathrm{D}^{\mathrm{b}}(\mathcal{D}_X)$ and $\mathcal{L} \in \mathrm{D}^{\mathrm{b}}(\mathcal{D}_{X \times Y})$, we set

$$\mathcal{M} \overset{\mathrm{D}}{\underset{\circ}{}} \mathcal{L} = \mathrm{D}g_!(\mathrm{D}f^{-1}\mathcal{M} \overset{\mathrm{D}}{\otimes} \mathcal{L}). \tag{2.8}$$

Now we consider

$$\mathcal{M} \in D^b_{good}(\mathcal{D}_X), \ \mathcal{L} \in D^b_{reg-hol}(\mathcal{D}_{X\times Y}), \ G \in D^b(\mathbb{C}_Y),$$

and we make the hypotheses:

$$\begin{cases} f^{-1}\operatorname{supp}(\mathcal{M}) \cap \operatorname{supp}(\mathcal{L}) \text{ is proper over } Y, \\ (\operatorname{char}(\mathcal{M}) \times T_Y^* Y) \cap \operatorname{char}(\mathcal{L}) \subset T_{X\times Y}^*(X \times Y). \end{cases} \tag{2.9}$$

**Theorem 2.3.** (See [4].) *Set* $L = R\mathcal{H}om_{\mathcal{D}_{X\times Y}}(\mathcal{O}_{X\times Y}, \mathcal{L})$ *and assume* (2.9). *One has the isomorphism*

$$\operatorname{RHom}_{\mathbb{C}_Y}(G, \operatorname{Sol}(\mathcal{M} \overset{D}{\circ} \mathcal{L})) \simeq \operatorname{RHom}_{\mathbb{C}_X}(L \circ G, \operatorname{Sol}(\mathcal{M}))\,[d_X - 2d_Y].$$

This result was first stated (in a slightly less general formulation) in [4]. Its proof makes use of the isomorphisms (2.5) and (2.6).

This result admits several variants: one may replace $\mathcal{O}_X$ with its temperate version or formal version (see § 5 below and [17, Ch. 7]), there are twisted versions and also $G$-equivariant versions (see [13]).

Many applications of this theorem are exposed in [3], in particular to projective duality, Penrose transform and, more generally, Grasmanniann correspondences.

**Remark 2.4.** The kernel $\mathcal{L}$ appearing in Theorem 2.3 is regular holonomic. However, irregular kernels may naturally appear which makes the situation much more intricate. The Laplace transform is such an example of irregular kernel (see [20]).

## 3. Microsupport

References for this section are made to [16].

Let $X$ denote a *real* manifold of class $C^\infty$ and let $F \in D^b(\mathbf{k}_X)$. The microsupport $\operatorname{SS}(F)$ of $F$ is the closed conic subset of $T^*X$ defined as follows.

**Definition 3.1.** Let $U$ be an open subset of $T^*X$. Then $U \cap \operatorname{SS}(F) = \emptyset$ if and only if for any $x_0 \in X$ and any real $C^\infty$-function $\varphi \colon X \to \mathbb{R}$ such that $\varphi(x_0) = 0, d\varphi(x_0) \in U$, one has:

$$(R\Gamma_{\varphi \geq 0}(F))_{x_0} = 0.$$

In other words, $F$ has no cohomology supported by the closed half spaces whose conormals do not belong to its microsupport. One again proves that the microsupport is a closed conic *involutive* subset of $T^*X$.

Assume now that $X$ is a complex manifold, that we identify with its real underlying manifold. The link between the microsupport of sheaves and the characteristic variety of coherent $\mathcal{D}$-modules is given by:

**Theorem 3.2.** *Let $\mathcal{M}$ be a coherent $\mathcal{D}$-module. Then* $\mathrm{SS}(\mathrm{Sol}(\mathcal{M})) = \mathrm{char}(\mathcal{M})$.

The inclusion $\mathrm{SS}(\mathrm{Sol}(\mathcal{M})) \subset \mathrm{char}(\mathcal{M})$ is the most useful in practice. Its proof only makes use of the Cauchy-Kowalevsky theorem in its precise form given by Leray (see [23] or [7, § 9.4]) and of purely algebraic arguments.

Now consider a space of generalized functions $R\mathcal{H}om\,(G, \mathcal{O}_X)$ associated with an $\mathbb{R}$-constructible sheaf $G$, and a system of linear partial differential equations, that is, a coherent $\mathcal{D}_X$-module $\mathcal{M}$. The complex of generalized functions solution of this system is given by the complex

$$R\mathcal{H}om_{\mathcal{D}_X}(\mathcal{M}, R\mathcal{H}om\,(G, \mathcal{O}_X)) \simeq R\mathcal{H}om\,(G, \mathrm{Sol}(\mathcal{M})).$$

Setting $F = \mathrm{Sol}(\mathcal{M})$, we are reduced to study $R\mathcal{H}om\,(G, F)$. Our only information is now purely geometrical, this is the microsupport of $G$ and that of $F$ (this last one being the characteristic variety of $\mathcal{M}$). Now, we can forget that we are working on a complex manifold and that we are dealing with partial differential equations. We are reduced to the microlocal study of sheaves on a real manifold [16].

## Application: ellipticity

Let us show how the classical Petrowsky regularity theorem may be obtained with the only use of the Cauchy-Kowalevsky-Leray Theorem, and some sheaf theory.

The regularity theorem for sheaves is as follows.

**Theorem 3.3.** *Let $X$ be a real analytic manifold and let $F, G \in \mathrm{D}^{\mathrm{b}}(\mathbf{k}_X)$. Assume that $G$ is $\mathbb{R}$-constructible and $\mathrm{SS}(G) \cap \mathrm{SS}(F) \subset T_X^* X$. Then the natural morphism*

$$R\mathcal{H}om\,(G, \mathbf{k}_X) \otimes F \to R\mathcal{H}om\,(G, F) \tag{3.1}$$

*is an isomorphism.*

Let us come back to the situation where $X$ is a complexification of a real manifold $M$ and choose $\mathbf{k} = \mathbb{C}$. Set $G = \mathrm{D}'_X(\mathbb{C}_M)$ and $F = R\mathcal{H}om_{\mathcal{D}_X}(\mathcal{M}, \mathcal{O}_X)$. A differential operator $P$ on $X$ is elliptic (with respect to $M$) if its principal symbol $\sigma(P)$ does not vanish on the conormal bundle $T_M^* X$, outside of the zero-section. More generally a coherent $\mathcal{D}_X$-module $\mathcal{M}$ is elliptic with respect to $M$ if

$$\mathrm{char}(\mathcal{M}) \cap T^*_M X \subset T^*_X X.$$

By Theorem 3.2, $\mathrm{SS}(F) \cap T^*_M X \subset T^*_X X$. Let $\mathcal{A}_M = \mathbb{C}_M \otimes \mathcal{O}_X$ denotes the sheaf of analytic functions on $M$. The regularity theorem for sheaves gives the isomorphism

$$R\mathcal{H}om_{\mathcal{D}_X}(\mathcal{M}, \mathcal{A}_X) \xrightarrow{\sim} R\mathcal{H}om_{\mathcal{D}_X}(\mathcal{M}, \mathcal{B}_X).$$

In other words, the two complexes of real analytic and hyperfunction solutions of an elliptic system are quasi-isomorphic. This is the Petrowsky's theorem for $\mathcal{D}$-modules.

Of course, this result extends to other sheaves of generalized holomorphic functions, replacing the constant sheaf $\mathbb{C}_M$ with an $\mathbb{R}$-constructible sheaf $G$. For further developments, see [31].

# 4. Microlocal analysis

For a detailed exposition, see [16].

### Fourier-Sato transform

Let $\tau\colon E \to M$ be a finite dimensional real vector bundle over a real manifold $M$ with fiber dimension $n$, $\pi\colon E^* \to M$ the dual vector bundle. Denote by $p_1$ and $p_2$ the first and second projection defined on $E \times_M E^*$, and define:

$$P = \{(x, y) \in E \times_M E^*; \langle x, y \rangle \geq 0\}$$
$$P' = \{(x, y) \in E \times_M E^*; \langle x, y \rangle \leq 0\}$$

Consider the diagram:

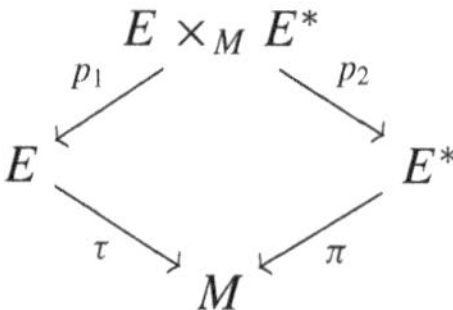

Denote by $\mathrm{D}^b_{\mathbb{R}^+}(\mathbf{k}_E)$ the full triangulated subcategory of $\mathrm{D}^b(\mathbf{k}_E)$ consisting of conic objects, that is, objects $F$ such that for all $j$, $H^j(F)$ is locally constant on the orbits of the action of $\mathbb{R}^+$ on $E$. One defines the two functors

$$\mathrm{D}^b_{\mathbb{R}^+}(\mathbf{k}_E) \underset{(\bullet)^\vee}{\overset{(\bullet)^\wedge}{\rightleftarrows}} \mathrm{D}^b_{\mathbb{R}^+}(\mathbf{k}_{E^*})$$

by setting for $F \in D_{\mathbb{R}+}(\mathbf{k}_E)$ and $G \in D_{\mathbb{R}+}(\mathbf{k}_{E^*})$

$$F^\wedge = Rp_{2!}(p_1^{-1}F)_{P'}, \quad G^\vee = Rp_{1!}(p_2^! G)_P.$$

The main result of the theory is the following.

**Theorem 4.1.** *The two functors $(\bullet)^\wedge$ and $(\bullet)^\vee$ are equivalences of categories inverse to each other. In particular, for $F_1$ and $F_2$ in $D_{\mathbb{R}+}^b(\mathbf{k}_E)$, there is a natural isomorphism:*

$$\mathrm{RHom}\,(F_1, F_2) \simeq \mathrm{RHom}\,(F_1^\wedge, F_2^\wedge). \tag{4.1}$$

**Example 4.2.** (i) For a convex cone $\gamma$ in $E$, denote $\mathrm{Int}\gamma$ its interior, by $\gamma^a = -\gamma$ its image by the antipodal map and by $\gamma^\circ$ its polar cone in $E^*$. Then, for a closed convex proper cone $\gamma$ with $M \subset \gamma$,

$$(\mathbf{k}_\gamma)^\wedge \simeq \mathbf{k}_{\mathrm{Int}\gamma^\circ}.$$

For an open convex cone $\lambda$ in $E$:

$$(\mathbf{k}_\lambda)^\wedge \simeq \mathbf{k}_{\lambda^{\circ a}} \otimes \mathrm{or}_{E^*/M}\,[-n].$$

(ii) Let $E$ denote the Euclidian space $\mathbb{R}^n$ and let $x = (x_1, \ldots, x_n)$ be the coordinates. Let $n = p + q$ with $p, q \geq 1$ and set $x = (x', x'')$ with $x' = (x_1, \ldots, x_p)$, $x'' = (x_{p+1}, \ldots, x_n)$. We denote by $u = (u', u'')$ the dual coordinates on the dual space $E^*$.

Let $\gamma$ denote the closed solid cone in $E$,

$$\gamma = \{x; x'^2 - x''^2 \geq 0\}$$

and let $\lambda$ denote the closed solid cone in $E^*$:

$$\lambda = \{u; u'^2 - u''^2 \leq 0\}.$$

We have (see [20]):

$$\mathbf{k}_\gamma^\wedge \simeq \mathbf{k}_\lambda\,[-p].$$

### Specialization and microlocalization

Let $X$ be a *real* manifold (say of class $\mathcal{C}^\infty$), $M$ a closed submanifold. Denote by $\tau: T_M X \to M$ and $\pi: T_M^* X \to M$ the normal bundle and the conormal bundle to $M$ in $X$, respectively. Let $F \in D^b(\mathbf{k}_X)$. The specialization of $F$ along $M$, denoted $\nu_M(F)$, is an object of $D_{\mathbb{R}+}^b(\mathbf{k}_{T_M X})$. Its cohomology objects are described as follows. If $V$ is an open cone in $T_M X$, then

$$H^j(V; \nu_M(F)) \simeq \varinjlim_U H^j(U; F)$$

where $U$ ranges over the family of open subsets of $X$ which are "tangent" to $V$, that is, open tuboids in $X$ with wedge $M$ whose "profiles" is $V$. (For a precise definition, refer to [16, § 4.2].)

The Sato's microlocalization of $F$ along $M$, denoted $\mu_M(F)$, is the Fourier-Sato transform of $\nu_M(F)$, an object of $\mathrm{D}^{\mathrm{b}}_{\mathbb{R}^+}(\mathbf{k}_{T^*_M X})$. It satisfies:

$$R\pi_*\mu_M(F) \simeq R\Gamma_M(F),$$

$$H^j(\mu_M(F))_{(x_0;\xi_0)} \simeq \varinjlim_{U,Z} H^j_{U \cap Z}(U;F),$$

where, in the last formula, $(x_0;\xi_0) \in T^*_M X$, $U$ ranges over the family of open neighborhoods of $x_0$ in $X$ and $Z$ ranges over the family of closed tuboids in $X$ with wedge $M$ whose profiles $\lambda$ in $T_M X$ satisfy $(x_0;\xi_0) \in \mathrm{Int}\lambda^{\circ a}$.

Using the diagonal $\Delta$ of $X \times X$, one defines the bifunctor $\mu hom$ by setting:

$$\mu hom\,(G, F) = \mu_\Delta R\mathcal{H}om\,(q_2^{-1}G, q_1^! F)$$

where $q_i$ $(i = 1, 2)$ denotes the $i$-th projection on $X \times X$. Note that

$$R\pi_*\mu hom\,(G, F) \simeq R\mathcal{H}om\,(G, F),$$

$$\mu hom\,(\mathbf{k}_M, F) \simeq \mu_M(F),$$

$$\mathrm{supp}\,\mu hom\,(G, F) \subset SS(G) \cap SS(F).$$

Now assume that $M$ is a real analytic manifold and $X$ is a complexification of $M$. The sheaf of Sato's microfunctions on $T^*_M X$ is defined by:

$$\mathcal{C}_M = \mu hom\,(\mathrm{D}'_X \mathbb{C}_M, \mathcal{O}_X).$$

(It is proved that this complex is concentrated in degree 0.) Hence, a hyperfunction is nothing but a microfunction globally defined on $T^*_M X$. Denote by spec the natural isomorphism:

$$\mathrm{spec}\colon \mathcal{B}_M \xrightarrow{\sim} \pi_*\mathcal{C}_M.$$

If $u$ is an hyperfunction, Sato defines its analytic wave front set as:

$$\mathrm{WF}(u) = \mathrm{supp}(\mathrm{spec}(u)),$$

a closed conic subset of $T^*_M X$. As an application, consider the situation of the construction of the boundary value morphism in (1.2). Let $\varphi \in \mathcal{O}(\Omega)$ and assume that $\varphi$ is solution of a system of differential equations, that is, $\varphi \in \mathcal{H}om_{\mathcal{D}_X}(\mathcal{M}, \mathcal{O}_X)$ for a coherent $\mathcal{D}_X$-module $\mathcal{M}$. Set $F_0 = \mathcal{H}om_{\mathcal{D}_X}(\mathcal{M}, \mathcal{O}_X)$. Since the boundary value morphism factorizes through $\mu hom\,(\mathbb{C}_\Omega, F_0)$ and $SS(F_0) \subset \mathrm{char}(\mathcal{M})$, we get

$$\mathrm{WF}(\mathrm{b}(\varphi)) \subset T^*_M X \cap SS(\mathbb{C}_\Omega) \cap \mathrm{char}(\mathcal{M}). \tag{4.2}$$

**Remark 4.3.** A new microlocalization functor $\mu$ has been constructed in [21], taking its values in the category of ind-sheaves on $T^*X$. It is related to the functor $\mu hom$ by the formula

$$\mu hom\,(G, F) \simeq R\mathcal{H}om\,(\pi^{-1}G, \mu(F)).$$

### Microdifferential operators

The sheaf of microfunctions allows us to analyze hyperfunctions microlocally, that is, in the cotangent bundle. A similar localization may be performed with respect to the sheaf of differential operators. Let $X$ be a complex manifold of dimension $d_X$ and denote by $\Delta$ the diagonal of $X \times X$. The sheaf of microlocal operators is defined in [29] by

$$\mathcal{E}_X^{\mathbb{R}} := \mu_\Delta(\mathcal{O}_{X \times X}^{(0, d_X)})\,[d_X].$$

Here, $\mathcal{O}_{X \times X}^{(0, d_X)} = \mathcal{O}_{X \times X} \otimes_{q_2^{-1}\mathcal{O}_X} q_2^{-1}\Omega_X$. One proves that $\mathcal{E}_X^{\mathbb{R}}$ is concentrated in degree 0 and is naturally endowed with a structure of a sheaf of $\mathbb{C}$-algebras. Moreover, for $G \in D^b(\mathbb{C}_X)$, the object $\mu hom\,(G, \mathcal{O}_X)$ is well defined in $D^b(\mathcal{E}_X^{\mathbb{R}})$ (see [21]). This applies in particular to the sheaf $\mathcal{C}_M$ of microfunctions.

The algebra $\mathcal{E}_X^{\mathbb{R}}$ is extremely difficult to manipulate, but it contains the $\mathbb{C}$-algebra $\mathcal{E}_X$ of microdifferential operators which is filtered and admits a symbol calculus. When $X$ is affine and $U$ is open in $T^*X$, a section $P \in \mathcal{E}_{T^*X}(U)$ is described by its "total symbol"

$$\sigma_{\mathrm{tot}}(P)(x; \xi) = \sum_{-\infty < j \le m} p_j(x; \xi), \; m \in \mathbb{Z}, \quad p_j \in \Gamma(U; \mathcal{O}_{T^*X}(j)),$$

with the condition:

$$\begin{cases} \text{for any compact subset } K \text{ of } U \text{ there exists a positive constant} \\ C_K \text{ such that } \sup_K |p_j| \le C_K^{-j}(-j)! \text{ for all } j < 0. \end{cases}$$

(Here $\mathcal{O}_{T^*X}(j)$ denotes the subsheaf of $\mathcal{O}_{T^*X}$ of functions homogeneous of degree $j$ in the fiber variable.) The total symbol of the product is given by the Leibniz rule. If $Q$ is an operator of total symbol $\sigma_{\mathrm{tot}}(Q)$, then

$$\sigma_{\mathrm{tot}}(P \circ Q) = \sum_{\alpha \in \mathbb{N}^n} \frac{1}{\alpha!} \partial_\xi^\alpha \sigma_{\mathrm{tot}}(P) \partial_x^\alpha \sigma_{\mathrm{tot}}(Q).$$

If $\varphi \colon T^*X \supset U \xrightarrow{\sim} V \subset T^*Y$ is a homogeneous complex symplectic isomorphism, it can be locally "quantized" (see [29]), that is, extended as an isomorphism of algebras

$$\Phi \colon \varphi_* \mathcal{E}_{T^*X} \xrightarrow{\sim} \mathcal{E}_{T^*Y}.$$

However, this isomorphism $\Phi$ is only locally defined and not unique. That is why, in general, it is not possible to define such sheaves $\mathcal{E}_{\mathfrak{X}}$ of microdifferential operators on a homogeneous complex symplectic manifold $\mathfrak{X}$. However, Kashiwara [12] has shown that such a construction was possible when weakening the notion of a sheaf of algebras by that of an algebroid stack. (Refer to [19] for an introduction to stacks.)

### Further developments and open problem

A complex cotangent bundle $T^*X$ is endowed with the sheaf $\mathcal{E}_{T^*X}$. This sheaf is conic for the action of $\mathbb{C}^\times$ on $T^*X$ and one can eliminate this homogeneity by adding an extra central variable $\hbar$. One gets the algebra $\mathcal{W}_{T^*X}$ over a field $\mathbf{k}$, a subfield of $\widehat{\mathbf{k}} = \mathbb{C}((\hbar))$. The algebra $\mathcal{W}_{T^*X}$ admits a formal version $\widehat{\mathcal{W}}_{T^*X}$ over $\widehat{\mathbf{k}}$ which is extremely popular and known as a deformation-quantization algebra. If $X$ is affine, a section $P \in \widehat{\mathcal{W}}_{T^*X}(U)$ on an open subset $U \subset T^*X$ is a series, called its total symbol:

$$\sigma_{\text{tot}}(P)(x; u, \hbar) = \sum_{m \le j < \infty} p_j(x; u)\hbar^j, \quad m \in \mathbb{Z}, \quad p_j \in \mathcal{O}_{T^*X}(U), \quad (4.3)$$

and the total symbol of the product $P \circ Q$ is given by the Leibniz rule:

$$\sigma_{\text{tot}}(P \circ Q) = \sum_{\alpha \in \mathbb{N}^n} \frac{\hbar^{|\alpha|}}{\alpha!} \partial_u^\alpha(\sigma_{\text{tot}} P)\partial_x^\alpha(\sigma_{\text{tot}} Q).$$

Note that $\widehat{\mathcal{W}}_{T^*X}$ has a natural filtration for which $\hbar$ has order $-1$ and $\widehat{\mathcal{W}}_{T^*X}(0)$ may naturally be considered as a deformation of the Poisson algebra $\mathcal{O}_{T^*X}$.

By replacing again the notion of a sheaf of algebras by that of an algebroid stack, one can define such objects on complex symplectic manifolds (see [22, 27]).

A natural question would be to perform an analogous construction for sheaves on real manifolds, that is, to construct a non conic microlocal theory of sheaves on real symplectic manifolds. This problem is closely related to Mirror Symmetry (see [25]).

## 5. The use of Grothendieck topologies

References for this section are made to [17, Ch. 7].

Let $X$ be a *real* analytic manifold. The usual topology on $X$ does not allow one to treat usual spaces of analysis with the tools of sheaf theory. For example, the property of being temperate is not local, and there is no sheaf of temperate distributions. One way to overcome this difficulty is to introduce a Grothendieck topology on $X$. Recall that a Grothendieck topology is not a topology, and in fact is not defined on a space but on a category. The objects of the category playing the role of the open subsets of the space, it is an axiomatization of the notion of a covering. A site is a category endowed with a Grothendieck topology.

We denote by $\mathrm{Op}_X$ the category whose objects are the open subsets of $X$ and the morphisms are the inclusions of open subsets. One defines a Grothendieck topology on $\mathrm{Op}_X$ by deciding that a family $\{U_i\}_{i \in I}$ of subobjects of $U \in \mathrm{Op}_X$ is a covering of $U$ if it is a covering in the usual sense.

**Definition 5.1.** Denote by $\mathrm{Op}_{X_{\mathrm{sa}}}$ the full subcategory of $\mathrm{Op}_X$ consisting of subanalytic and relatively compact open subsets. The site $X_{\mathrm{sa}}$ is obtained by deciding that a family $\{U_i\}_{i \in I}$ of subobjects of $U \in \mathrm{Op}_{X_{\mathrm{sa}}}$ is a covering of $U$ if there exists a finite subset $J \subset I$ such that $\bigcup_{j \in J} U_j = U$.

Let us denote by

$$\rho : X \to X_{\mathrm{sa}} \tag{5.1}$$

the natural morphism of sites. As usual, we have a pair of adjoint functors $(\rho^{-1}, \rho_*)$:

$$\mathrm{Mod}(\mathbf{k}_X) \underset{\rho^{-1}}{\overset{\rho_*}{\rightleftarrows}} \mathrm{Mod}(\mathbf{k}_{X_{\mathrm{sa}}}).$$

The functor $\rho^{-1}$ also admits a right adjoint $\rho_!$. For $F \in \mathrm{Mod}(\mathbf{k}_X)$, $\rho_! F$ is the sheaf associated to the presheaf $U \mapsto F(\overline{U})$, $U \in \mathrm{Op}_{X_{\mathrm{sa}}}$.

Let us denote by $\mathcal{C}_X^\infty$ the sheaf of complex valued $\mathcal{C}^\infty$-functions on $X$.

**Definition 5.2.** Let $f \in \mathcal{C}_X^\infty(U)$. One says that $f$ has *polynomial growth* at $p \in X$ if it satisfies the following condition. For a local coordinate system $(x_1, \ldots, x_n)$ around $p$, there exist a sufficiently small compact neighborhood $K$ of $p$ and a positive integer $N$ such that

$$\sup_{x \in K \cap U} \left( \mathrm{dist}(x, K \setminus U) \right)^N |f(x)| < \infty. \tag{5.2}$$

It is obvious that $f$ has polynomial growth at any point of $U$. We say that $f$ is temperate at $p$ if all its derivatives have polynomial growth at $p$. We say that $f$ is temperate if it is temperate at any point.

For an open subanalytic subset $U$ of $X$, denote by $\mathcal{C}_X^{\infty, t}(U)$ the subspace of $\mathcal{C}_X^\infty(U)$ consisting of temperate functions.

Using Lojasiewicz's inequalities [24], one easily proves that the presheaf $\mathcal{C}_{X_{\mathrm{sa}}}^{\infty, t}$, given by $U \mapsto \mathcal{C}_X^{\infty, t}(U)$, is a sheaf on $X_{\mathrm{sa}}$. One calls it the sheaf of temperate $\mathcal{C}^\infty$-functions on $X_{sa}$. Note that the sheaf $\rho_* \mathcal{D}_X$ does not operate on $\mathcal{C}^{\infty,t}_{X_{sa}}$ but $\rho_! \mathcal{D}_X$ does.

Now let $X$ be a *complex* manifold. We still denote by $X$ the real underlying manifold and we denote by $\overline{X}$ the complex manifold conjugate to $X$. One defines the sheaf of temperate holomorphic functions $\mathcal{O}^t_{X_{sa}}$ as the Dolbeault complex with coefficients in $\mathcal{C}^{\infty,t}_{X_{sa}}$. More precisely

$$\mathcal{O}^t_{X_{sa}} = R\mathcal{H}om_{\rho_! \mathcal{D}_{\overline{X}}}(\rho_! \mathcal{O}_{\overline{X}}, \mathcal{C}^{\infty,t}_{X_{sa}}). \tag{5.3}$$

Note that the object $\mathcal{O}^t_{X_{sa}} \in D^b(\rho_! \mathcal{D}_X)$ is not concentrated in degree zero in dimension $> 1$. Nevertheless, it should have many important applications. Let us mention two of them:

(a) One proves that Sato's construction of hyperfunctions, when applied to $\mathcal{O}^t_{X_{sa}}$, gives the sheaf of Schwartz's distributions.

(b) Very little is known on irregular holonomic $\mathcal{D}$-modules (see [26]) in dimension higher than one, and even in dimension one, there are no (to our opinion) totally satisfactory results. In [18], one calculates the temperate holomorphic solutions of the $\mathcal{D}_X$-module $\mathcal{M} := \mathcal{D}_X \exp(1/z) = \mathcal{D}_X/\mathcal{D}_X(z^2 \partial_z + 1)$, where $X = \mathbb{C}$. The result obtained shows that the sheaf of temperate holomorphic solutions gives more informations than the classical sheaf of holomorphic solutions.

The sheaf $\mathcal{O}^t_{X_{sa}}$ is simply an example which shows that the methods of Algebraic Analysis may be applied to treat generalized functions with growth conditions.

# References

[1] E. Andronikof, *Interview with Mikio Sato,* Notices of the AMS, **54** Vol 2, (2007).

[2] I. Bernstein *Modules over a ring of differential operators,* Funct. Analysis and Appl. **5** pp 89–101 (1971).

[3] A. D'Agnolo, *Sheaves and D-modules in integral geometry,* in Contemp. Math., **251** Amer. Math. Soc., pp 141 (2000).

[4] A. D'Agnolo and P. Schapira, *Leray's quantization of projective duality,* Duke Math. Journ. **84** pp 453–496 (1996).

[5] O. Gabber, *The integrability of the characteristic variety,* Amer. Journ. Math. **103** pp 445–468 (1981).

[6] V. Guillemin, D. Quillen and S. Sternberg, *The integrability of characteristics,* Comm. Pure and Appl. Math. **23** 39–77 (1970)

[7] L. Hörmander, *The analysis of linear partial differential operators,* Grundlehren der Math. Wiss. **256** Springer-Verlag (1983).

[8] M. Kashiwara, *Algebraic study of systems of partial diffential equations,* Thesis, Tokyo Univ. (1970), translated by A. D'Agnolo and J-P. Schneiders, Mémoires Soc. Math. France **63** (1995).

[9] ______, *D-modules and Microlocal Calculus,* Translations of Mathematical Monographs, **217** American Math. Soc. (2003).

[10] ______, *On the maximally overdetermined systems of linear differential equations I,* Publ. RIMS, Kyoto Univ. **10** pp 563–579 (1975).

[11] ______, *The Riemann-Hilbert problem for holonomic systems,* Publ. RIMS, Kyoto Univ. **20** pp 319–365 (1984).

[12] ______, *Quantization of contact manifolds,* Publ. RIMS, Kyoto Univ. **32** p. 1–5 (1996).

[13] ______, *Equivariant derived category and representation of real semisimple Lie groups.* Representation theory and complex analysis, Lecture Notes in Math. Springer-Verlag, to appear

[14] M. Kashiwara and T. Kawai, *On the holonomic systems of microdifferential equations III,* Publ. RIMS, Kyoto Univ. **17** pp 813–979 (1981).

[15] M. Kashiwara and T. Oshima, *Systems of differential equations with regular singularities and their boundary value problems,* Ann. of Math. **106** (1977), p. 145–200.

[16] M. Kashiwara and P. Schapira, *Sheaves on Manifolds,* Grundlehren der Math. Wiss. **292** Springer-Verlag (1990).

[17] ______, *Ind-sheaves,* Astérisque **271** Soc. Mathématique de France (2001).

[18] ______, *Microsupport and regularity for ind-sheaves I,* Astérique **284** pp 143–164 (2003). ArXiv: math.AG/0108065

[19] ______, *Categories and Sheaves,* Grundlehren der Math. Wiss. Springer-Verlag (2005).

[20] ______, *Integral transforms with exponential kernels and Laplace transform,* Journ. of the A.M.S. **10** pp 939–972 (1997).

[21] M. Kashiwara, P. Schapira, F. Ivorra and I. Waschkies, *Microlocalization of ind-sheaves.* in Studies in Lie Theory, Progress in Math. **243** pp 171–221 (2006).

[22] M. Kontsevich, *Deformation quantization of algebraic varieties,* in: Euro-Conférence Moshé Flato, Part III (Dijon, 2000) Lett. Math. Phys. **56** (3) p. 271–294 (2001).

[23] J. Leray, *Scientific work,* Springer-Verlag & Soc. Mathématique de France (1997).

[24] S. Lojaciewicz, *Sur le problème de la division,* Studia Math. **8** pp. 87–156, (1961).

[25] D. Nadler and E. Zaslow, *Constructible Sheaves and the Fukaya Category,* `arXiv:math/0604379`

[26] C. Sabbah, *Equations différentielles à points singuliers irréguliers et phénomène de Stokes en dimension 2,* Astérisque **263** Soc. Mathématique de France (2000).

[27] P. Polesello and P. Schapira, *Stacks of quantization-deformation modules over complex symplectic manifolds,* Int. Math. Res. Notices **49** p. 2637–2664 (2004).

[28] M. Sato, *Theory of hyperfunctions,* I & II Journ. Fac. Sci. Univ. Tokyo, **8** 139–193 487–436 (1959–1960).

[29] M. Sato, T. Kawai, and M. Kashiwara, *Microfunctions and pseudo-differential equations,* in Komatsu (ed.), *Hyperfunctions and pseudo-differential equations,* Proceedings Katata 1971, Lecture Notes in Math. Springer-Verlag **287** pp. 265–529 (1973).

[30] P. Schapira, *Mikio Sato, a visionary of mathematics,* Notices of the AMS, **54** Vol 2, (2007).

[31] P. Schapira and J-P. Schneiders, *Index theorem for elliptic pairs,* Astérisque **224** Soc. Mathématique de France (1994).

Pierre Schapira

Université Pierre et Marie Curie

Institut de Mathématiques

175, rue du Chevaleret, 75013 Paris France

schapira@math.jussieu.fr

http://www.math.jussieu.fr/~schapira/

# Algebraic versus topological triangulated categories

STEFAN SCHWEDE

These are extended and updated notes of a talk, the first version of which I gave at the *Workshop on Triangulated Categories* at the University of Leeds, August 13–19, 2006. These notes are mostly expository and do not contain all proofs; I intend to publish the remaining details elsewhere.

The most commonly known triangulated categories arise from chain complexes in an abelian category by passing to chain homotopy classes or inverting quasi-isomorphisms. Such examples are called 'algebraic' because they originate from abelian (or at least additive) categories. Stable homotopy theory produces examples of triangulated categories by quite different means, and in this context the source categories are usually very 'non-additive' before passing to homotopy classes of morphisms. Because of their origin I refer to these examples as 'topological triangulated categories'.

In this note I want to explain some systematic differences between these two kinds of triangulated categories. There are certain properties – defined entirely in terms of the triangulated structure – which hold in all algebraic examples, but which fail in some topological ones. These differences are all torsion phenomena, and rationally there is no difference between algebraic and topological triangulated categories.

A triangulated category is *algebraic* in the sense of Keller [Ke, 3.6] if it is triangle equivalent to the stable category of a Frobenius category, i.e., an exact category with enough injectives and enough projectives in which injectives and projectives coincide. Examples include all triangulated categories which one should reasonably think of as 'algebraic': various homotopy categories and derived categories of rings, schemes and abelian categories; stable module categories of Frobenius rings; derived categories of modules over differential graded algebras and differential graded categories. By a theorem of Porta [Po, Thm. 1.2], every algebraic triangulated category which is *well generated* (a mild restriction on its 'size', see [Ne, Def. 8.1.6 and 8.1.7]) is equivalent

*Date*: 2000 AMS Math. Subj. Class.: 18E30, 55P42.

to a localization of the derived category $\mathcal{D}(\mathcal{A})$ of a small differential graded category $\mathcal{A}$.

For an object $X$ of a triangulated category $\mathcal{T}$ and a natural number $n$ we write $n \cdot \mathrm{Id}_X$ or simply $n \cdot X$ for the $n$-fold multiple of the identity morphism in the group $[X, X]$ of endomorphisms in $\mathcal{T}$. We let $X/n$ denote any cone of $n \cdot \mathrm{Id}_X$, i.e., any object which is part of a distinguished triangle

$$ X \xrightarrow{n\cdot} X \longrightarrow X/n \longrightarrow X[1] \,. $$

A short diagram chase shows that the group $[X/n, X/n]$ is always annihilated by $n^2$; in algebraic triangulated categories, more is true:

**Proposition 1.** *If $\mathcal{T}$ is algebraic, then $n \cdot X/n = 0$.*

*Proof.* We exploit that algebraic triangulated categories are tensored over $\mathcal{D}^b(\mathbb{Z})$, the bounded derived category of finitely generated abelian groups. This means that there is a biexact functor

$$ \otimes^L \;:\; \mathcal{D}^b(\mathbb{Z}) \times \mathcal{T} \longrightarrow \mathcal{T} $$

which is associative and unital up to coherent isomorphism with respect to the derived tensor product in $\mathcal{D}^b(\mathbb{Z})$. In $\mathcal{D}^b(\mathbb{Z})$ we have a distinguished triangle

$$ \mathbb{Z} \xrightarrow{n\cdot} \mathbb{Z} \longrightarrow \mathbb{Z}/n \longrightarrow \mathbb{Z}[1] $$

which becomes a distinguished triangle in $\mathcal{T}$ after tensoring with $X$. So $\mathbb{Z}/n \otimes^L X$ is isomorphic to $X/n$. Since the tensor product is additive in each variable and since $n \cdot \mathbb{Z}/n = 0$ in $\mathcal{D}^b(\mathbb{Z})$, we conclude that $n \cdot (\mathbb{Z}/n \otimes^L X) = 0$. $\qquad\square$

In contrast to Proposition 1, in a general triangulated category we can have $n \cdot X/n \neq 0$ for suitable choices of $X$ and $n$. An example arises in the *Spanier-Whitehead category* which we now review. The Spanier-Whitehead category is made from homotopy classes of continuous maps between certain kinds of topological spaces, and it is the prime example of a *topological* triangulated category (to be defined below) which is not algebraic. The Spanier-Whitehead category was originally introduced (without the formal desuspensions) in [SW].

We recall that the *reduced suspension* $\Sigma X$ of a space $X$ with basepoint $x_0$ is given by

$$ \Sigma X = \frac{X \times [0, 1]}{X \times \{0, 1\} \cup \{x_0\} \times [0, 1]} \,. $$

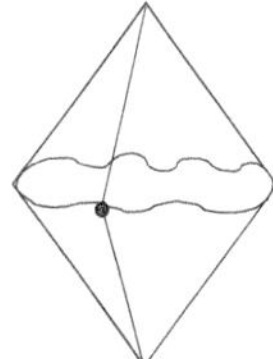

For example we have $\Sigma S^{n-1} \cong S^n$, i.e., the suspension of a sphere is homeomorphic to a sphere of the next dimension.

**Definition 2.** The *Spanier-Whitehead category*, denoted $\mathcal{SW}$, has as objects the pairs $(X, n)$ where $X$ is a finite CW-complex equipped with a distinguished basepoint and $n \in \mathbb{Z}$. Morphisms are given by

$$\mathcal{SW}((X, n), (Y, m)) \ = \ \mathrm{colim}_{k \to \infty} [\Sigma^{k+n} X, \Sigma^{k+m} Y]$$

where square brackets $[-, -]$ denote pointed homotopy classes of continuous, basepoint preserving maps. The colimit is formed by iterated suspensions; by Freudenthal's suspension theorem, it is actually attained at a finite stage. Composition in $\mathcal{SW}$ is defined by composition of representatives, suitably suspended so that composition is possible.

It is convenient to identify a finite pointed CW-complex $X$ with the object $(X, 0)$ of the Spanier-Whitehead category. Then two CW-complexes become isomorphic in $\mathcal{SW}$ if and only if they become homotopy equivalent after a finite number of suspensions.

The Spanier-Whitehead category is triangulated in the following way:

- The shift functor is given by $(X, n)[1] = (X, n + 1)$. Tautologically, the identity map of $\Sigma^{n+1} X$ is an isomorphism between $(\Sigma X, n)$ and $(X, n + 1)$ in $\mathcal{SW}$, so suspension is in fact isomorphic to the shift and is invertible in $\mathcal{SW}$.

- The Spanier-Whitehead category is additive: for pointed spaces $X$ and $Y$, the set $[\Sigma X, Y]$ of pointed homotopy classes from a suspension has a natural group structure as follows. The product of $f, g : \Sigma X \longrightarrow Y$ is represented by the composite

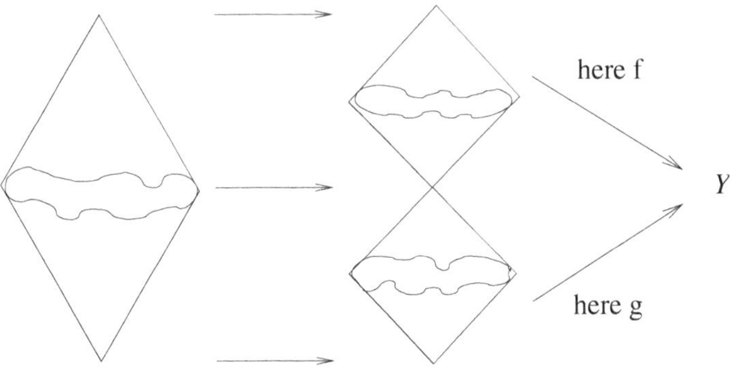


$$\Sigma X \xrightarrow{\text{pinch}} \Sigma X \vee \Sigma X \xrightarrow{f \vee g} Y$$

For $X = S^0$ we have $\Sigma X \cong S^1$ and this reduces to the group structure on the fundamental group $[S^1, Y] = \pi_1(Y, y_0)$. On a double suspension as source object, this group structure is abelian; an example of this is that for $n \geq 2$ the higher homotopy groups $\pi_n(Y, y_0) = [S^n, Y]$ are abelian. In the Spanier-Whitehead category, every object is a double suspension, so the homomorphism sets in $\mathcal{SW}$ are naturally abelian groups.

- Mapping cone sequences give distinguished triangles: the *mapping cone* of a pointed map $f : X \longrightarrow Y$ is the space

$$C(f) = \frac{X \times [0, 1] \cup_{X \times \{1\}} Y}{X \times \{0\} \cup \{x_0\} \times [0, 1]}$$

There is an inclusion $i : Y \longrightarrow C(f)$ and a projection $p : C(f) \longrightarrow \Sigma X$ (which collapses $Y$ to a point). In the sequence of pointed maps

$$X \xrightarrow{f} Y \xrightarrow{i} C(f) \xrightarrow{p} \Sigma X$$

the composite of any two is null-homotopic, and these sequences and their (de-)suspensions give the distinguished triangles in $\mathcal{SW}$.

A verification of the axioms of a triangulated category (except for the octahedral axiom), and more details on the Spanier-Whitehead category can be found in Ch. 1, §2 of Margolis book [Mar]. (Margolis does not impose any finiteness restriction on the objects of the Spanier-Whitehead category; the category $\mathcal{SW}$ is denoted $\mathbf{SW}_f$ in [Mar].)

The group

$$\mathcal{SW}(S^n, X) = \mathrm{colim}_{k \to \infty} [S^{k+n}, \Sigma^k X] = \mathrm{colim}_k \, \pi_{k+n}(\Sigma^k X, *)$$

is denoted by $\pi_n^s X$ and called the *n*-th *stable homotopy group* of $X$. For $X = S^0$, we abbreviate $\pi_n^s S^0$ to $\pi_n^s$ and speak about the *n-th stable homotopy group of spheres*, also called the *n-th stable stem*. For example, $\pi_0^s = \mathbb{Z}$, generated by the identity map, and $\pi_1^s = \mathbb{Z}/2$ generated by the class of the Hopf map $\eta : S^3 \longrightarrow S^2$. The stable stems are easy to define, but notoriously hard to compute; for example, there is no finite CW-complex $X$ which is non-trivial in $\mathcal{SW}$ for which all stable homotopy groups are known! Much machinery of algebraic topology has been developed to calculate such groups and understand their structure, but no one expects to ever get explicit formulae for all stable homotopy groups of spheres.

We also need the symmetric monoidal *smash product* on the Spanier-Whitehead category. This arises from the geometric smash product $X \wedge Y$ of two pointed spaces $X$ and $Y$ defined as $X \wedge Y = (X \times Y)/(X \times \{y_0\} \cup \{x_0\} \times Y)$. Suspension is an example of the smash product, i.e., $\Sigma X$ is naturally homeomorphic to $X \wedge S^1$. The smash product can be extended to the Spanier-Whitehead category by $(X, n) \wedge (Y, m) = (X \wedge Y, n + m)$, and then it becomes biexact, i.e., an exact functor of triangulated categories in each variable.

We denote by $S = (S^0, 0)$ the unit object of the smash product in $\mathcal{SW}$ and refer to it as the *sphere spectrum*. We use this terminology because the Spanier-Whitehead category can be identified with the compact objects in a larger triangulated category, the *stable homotopy category*, whose objects are called spectra. However, the differences between algebraic and topological triangulated categories already show up in the Spanier-Whitehead category, which is easier to define than the full stable homotopy category.

For $n \geq 2$, the *mod-$n$ Moore spectrum* is defined as a cone of multiplication by $n$ on the sphere spectrum, i.e., it is part of a distinguished triangle

$$S \xrightarrow{n\cdot} S \longrightarrow S/n \longrightarrow S[1] \,. \tag{3}$$

More concretely we can define $S/n = (S^1 \cup_n D^2, -1)$, the formal desuspension of a two-dimensional mod-$n$ Moore space, i.e., the space obtained from the circle $S^1$ by attaching a 2-disc along the degree $n$ map $S^1 \longrightarrow S^1, z \mapsto z^n$. For example, $S/2 = (\mathbb{R}P^2, -1)$, the formal desuspension of 2-dimensional real projective space. If $n$ and $m$ are coprime, then $S/(nm)$ is isomorphic in the Spanier-Whitehead category to the smash product $S/n \wedge S/m$. So we often concentrate on Moore spectra for primes or prime powers, since the other Moore spectra are smash products of these.

Proposition 1 and the following proposition show that the Spanier-Whitehead category is not algebraic.

**Proposition 4.** *The morphism $2 \cdot S/2$ is nonzero in $\mathcal{SW}$.*

The standard tool for proving Proposition 4 and its generalizations below are cohomology operations, which we quickly review. The *mod-$p$ cohomology* of an object $(X, n)$ in $\mathcal{SW}$ defined by

$$H^k((X, n), \mathbb{F}_p) = \tilde{H}^{k-n}(X, \mathbb{F}_p)$$

where the right hand side is reduced singular cohomology with coefficients in the field $\mathbb{F}_p$. These cohomology groups have a natural action (essentially by definition) by the mod-$p$ *Steenrod algebra*, i.e., the algebra of stable, natural, graded mod-$p$ cohomology operations. The Steenrod algebra is generated for

$p = 2$ by operations $Sq^i$ of degree $i$ for $i \geq 1$ and for odd $p$ by the Bockstein operation $\beta$ of degree 1 and operations $P^i$ of degree $i(2p - 2)$ for $i \geq 1$. For $p = 2$, the operation $Sq^1$ equals the Bockstein operation. These operations satisfy the *Adem relations*, which we do not reproduce here.

The mod-$n$ Moore spectrum is characterized up to isomorphism in the Spanier-Whitehead category by the property that its integral spectrum homology is concentrated in dimension zero where it is isomorphic to $\mathbb{Z}/n$. For a prime $p$ the mod-$p$ cohomology of $S/p$ is one-dimensional in dimensions 0 and 1, and trivial otherwise, and the Bockstein operation is non-trivial from dimension 0 to dimension 1.

*Proof of Proposition 4.* This proposition is a classical fact, which should probably be credited to Steenrod since the standard proof uses mod-2 Steenrod operations. We argue by contradiction and suppose that $2 \cdot S/2 = 0$. If we smash the defining triangle (3) with another copy of the mod-2 Moore spectrum and use that $S$ is the unit of the smash product, we obtain a distinguished triangle

$$S/2 \xrightarrow{2\cdot} S/2 \longrightarrow S/2 \wedge S/2 \longrightarrow S/2[1] \, .$$

Under our assumption the first map is trivial so the smash product $S/2 \wedge S/2$ splits in $\mathcal{SW}$ as a sum of $S/2$ and $S/2[1]$. Thus as a module over the Steenrod algebra, the mod-2 cohomology of the smash product $S/2 \wedge S/2$ decomposes into a sum of two non-trivial summands.

On the other hand, there is a Künneth isomorphism for the mod-2 cohomology of a smash product

$$H^n(X \wedge Y, \mathbb{F}_2) \cong \bigoplus_{p+q=n} H^p(X, \mathbb{F}_2) \otimes H^q(Y, \mathbb{F}_2) \, .$$

This is an isomorphism as modules over the Steenrod algebra with action on the right hand side given by the *Cartan formula*

$$Sq^m(x \otimes y) = \sum_{i=0}^{m} Sq^i x \otimes Sq^{m-i} y \, .$$

In the mod-2 cohomology of the mod-2 Moore spectrum the Bockstein operation $Sq^1 : H^0(S/2, \mathbb{F}_2) \longrightarrow H^1(S/2, \mathbb{F}_2)$ is non-zero. The Cartan formula shows that the operation $Sq^2$ is then non-trivial on the tensor product of two copies of the generator of $H^0(S/2, \mathbb{F}_2)$. Thus the mod-2 cohomology of $S/2 \wedge S/2$ is a 4-dimensional, indecomposable module over the Steenrod algebra. We have reached a contradiction, which means that we must have $2 \cdot S/2 \neq 0$ in $\mathcal{SW}$.      $\square$

In topological triangulated categories, the phenomenon that we can have $n \cdot X/n \neq 0$ is entirely 2-local. To explain this, I have to be more precise about what I mean by a topological triangulated category. A *model category* (in the sense of Quillen [Q]) is an axiomatic framework for homotopy theoretic constructions. Among other things, a model category structure allows one to define mapping cones and suspensions and talk about homotopies between morphisms. A model category is called *stable* if it has a zero object and the suspension functor is a self-equivalence of its homotopy category. The homotopy category of a stable model category is naturally a triangulated category (cf. [Ho1, 7.1.6]); the proof is essentially the same as for the Spanier-Whitehead category. By definition the suspension functor is a self-equivalence, and it defines the shift functor. Since every object is a two-fold suspension, hence an abelian co-group object, the homotopy category of a stable model category is additive. The distinguished triangles are defined by mapping cone sequences.

For us a *topological triangulated category* is any triangulated category which is equivalent to a full triangulated subcategory of the homotopy category of a stable model category. An important example which was already mentioned above is the *stable homotopy category* of algebraic topology which was first introduced by Boardman (unpublished; accounts of Boardman's construction appear in [Vo] and [Ad, Part III]). There is an abundance of models for the stable homotopy category, see for example [BF, EKMM, HSS, MMSS, Ly]. The Spanier-Whitehead category $\mathcal{SW}$ is equivalent to the full subcategory of compact objects in the stable homotopy category, so it is a topological triangulated category in our sense. Further examples of topological triangulated categories are 'derived' (i.e., homotopy) categories of structured ring spectra, equivariant and motivic stable homotopy categories, sheaves of spectra on a Grothendieck site or (Bousfield-) localizations of all these, see [SS, Sec. 2.3] for more details. The theorem of Porta mentioned above has an analogue in this context: Heider essentially shows in [He, Thm. 4.7] that every topological triangulated category which is well generated is equivalent to a localization of the homotopy category $\mathrm{Ho}(\mathcal{R}\text{-mod})$ of a small spectral category $\mathcal{R}$.

Algebraic triangulated categories are typically also topological (the converse is not generally true, and that is the point of these notes). For algebraic triangulated categories which are derived categories of abelian categories, this follows whenever there is a model structure on the category of chain complexes with quasi-isomorphisms as weak equivalences (see for example [Ho2] or Section 2.4 of [SS] for more details and references). Similarly, for modules over a Frobenius ring, there is a stable model structure with stable equivalences as the weak equivalences, see [Ho1, Thm. 2.2.12]. More generally, any well-generated algebraic triangulated category is equivalent to a localization of the derived category $\mathcal{D}(\mathcal{A})$ of small differential graded category $\mathcal{A}$ [Po, Thm. 1.2]. The localization can be realized as a Bousfield localization of the ordinary (i.e., 'projective') model structure on modules over $\mathcal{A}$; hence $\mathcal{D}(\mathcal{A})$ and its localizations are topological.

Examples of triangulated categories which are neither algebraic nor topological were recently constructed by Muro, Strickland and the author [MSS]. The simplest one is the category $\mathcal{F}(\mathbb{Z}/4)$ of finitely generated free modules over the ring $\mathbb{Z}/4$. The category $\mathcal{F}(\mathbb{Z}/4)$ has a unique triangulation with the identity shift functor and such that the triangle

$$\mathbb{Z}/4 \xrightarrow{2} \mathbb{Z}/4 \xrightarrow{2} \mathbb{Z}/4 \xrightarrow{2} \mathbb{Z}/4$$

is exact. For the proof of this and an argument why $\mathcal{F}(\mathbb{Z}/4)$ is not topological I refer to [MSS]. At present, I do not know any 'exotic' (i.e., non-topological and non-algebraic) triangulated category in which 2 is invertible.

Unlike algebraic triangulated categories, we can not usually expect that a topological triangulated category can be tensored over $\mathcal{D}^b(\mathbb{Z})$, the bounded derived category of finitely generated abelian groups. The appropriate replacement for $\mathcal{D}^b(\mathbb{Z})$ is the Spanier-Whitehead category: for every topological triangulated category $\mathcal{T}$, there is a biexact pairing

$$\wedge \; : \; \mathcal{SW} \times \mathcal{T} \longrightarrow \mathcal{T}$$

which is associative and unital up to coherent natural isomorphism with respect to the smash product in the Spanier-Whitehead category. The shift functor in $\mathcal{T}$ is isomorphic to smashing with the circle $S^1$, view as an object of $\mathcal{SW}$.

We use the above pairing as a black box, but I'll briefly indicate how it can be constructed. One way is to start from the action of the homotopy category of pointed simplicial sets on the homotopy category of a pointed model category $\mathcal{C}$,

$$\wedge \; : \; \mathrm{Ho}\mathbf{SSet}_* \times \mathrm{Ho}\,\mathcal{C} \longrightarrow \mathrm{Ho}\,\mathcal{C} \, ,$$

which uses a technique called 'framings', see [Ho1, Thm. 5.7.3]. The geometric realization of a finite simplicial set is a finite CW-complex, and up to homotopy equivalence, every finite CW-complex arises in this way. Since moreover suspension is invertible in $\mathrm{Ho}\,\mathcal{C}$, this extends to a well-defined pairing on the Spanier-Whitehead category.

There is an alternative construction for *spectral model categories*, i.e., model categories $\mathcal{C}$ which are enriched over the category $Sp^{\Sigma}$ of symmetric spectra of [HSS], compatibly with the smash product and the stable model structure (see [SS, Def. 3.5.1] for details). The derived smash product between symmetric spectra and $C$ descends to an associative and unital pairing

$$\wedge \ : \ \mathrm{Ho}(Sp^{\Sigma}) \times \mathrm{Ho}\,C \ \longrightarrow \ \mathrm{Ho}\,C \ ,$$

which can be restricted to the full subcategory $SW$ of compact objects in $\mathrm{Ho}(Sp^{\Sigma})$. A general stable model category is Quillen-equivalent to a spectral model category, under mild technical hypothesis (see [SS, Thm. 3.8.2], [Du, Prop. 5.5 (a) and 5.6 (a)] and [Ho3, Thm. 9.1 and 8.11] for different sets of sufficient conditions).

In algebraic examples the action of the Spanier-Whitehead category $SW$ and the bounded derived category $D^b(\mathbb{Z})$ (which we used in the proof of Proposition 1) are related as follows. The chain functor $C_* : SW \longrightarrow D^b(\mathbb{Z})$ associates to an object $(X, n)$ of the Spanier-Whitehead category the reduced singular chain complex of the CW-complex $X$, shifted up $n$ dimensions. This chain functor is strong symmetric monoidal, i.e., there are associative, unital and commutative isomorphisms $C_*(X \wedge Y) \cong C_*(X) \otimes^L C_*(Y)$ in $D^b(\mathbb{Z})$. If $T$ is an algebraic triangulated category which is also topological, then the composite

$$SW \times T \xrightarrow{\ C_* \times \mathrm{Id}\ } D^b(\mathbb{Z}) \times T \xrightarrow{\ \otimes^L\ } T$$

is naturally isomorphic to the smash product pairing.

Now we can make precise in which sense the possibility of having $n \cdot X/n \neq 0$ in topological triangulated categories is a 2-local phenomenon.

**Proposition 5.** *If $T$ is a topological triangulated category and $p$ an odd prime, then $p \cdot X/p = 0$ for every object $X$ of $T$.*

*Proof.* The key point is that if $n \not\equiv 2 \mod 4$, then the mod-$n$ Moore spectrum $S/n$ is annihilated by $n$ (which is not the case for $n = 2$). We recall the easy proof for odd $n$, which includes all odd primes. If we take homotopy groups of the defining triangle (3) (i.e., apply $[S, -]$, where $[-, -]$ denotes morphisms in $SW$) we obtain an exact sequence

$$\pi_1^s \xrightarrow{\ n\cdot\ } \pi_1^s \longrightarrow \pi_1(S/n) \longrightarrow \pi_0^s \xrightarrow{\ n\cdot\ } \pi_0^s \longrightarrow \pi_0(S/n) \longrightarrow 0 \ ,$$

using that the stable homotopy groups of spheres vanish in negative dimensions. Since $\pi_0^s \cong \mathbb{Z}$ and $\pi_1^s \cong \mathbb{Z}/2$ we deduce that $\pi_0(S/n)$ is cyclic of order $n$ and $\pi_1^s(S/n) = 0$ for odd $n$. Now we apply $[-, S/n]$ to the triangle (3) and we obtain a short exact sequence

$$0 \longrightarrow [S[1], S/n] \otimes \mathbb{Z}/n \longrightarrow [S/n, S/n] \longrightarrow {}_n[S, S/n] \longrightarrow 0$$

where ${}_n A = \{a \in A \mid na = 0\}$ denotes the group of $n$-torsion points in an abelian group $A$. By the above calculations, the group $[S[1], S/n] \otimes \mathbb{Z}/n$

vanishes and so the group $[S/n, S/n]$ is cyclic of order $n$ for $n$ odd, hence $n \cdot S/n = 0$. (For $n = 2$ we have $\pi_1(S/2) \cong \mathbb{Z}/2$ and the analogous short exact sequence does not split by Proposition 4. So $[S/2, S/2] \cong \mathbb{Z}/4$.)

The rest of the argument is then the same as in Proposition 1. For an odd prime $p$ we can smash the distinguished triangle

$$ S \xrightarrow{p \cdot} S \longrightarrow S/p \longrightarrow S[1] $$

in $\mathcal{SW}$ with the object $X$ and obtain a distinguished triangle in $\mathcal{T}$ which shows that $S/p \wedge X$ is isomorphic to $X/p$. Since $p \cdot S/p = 0$ in $\mathcal{SW}$ and $\wedge$ is biadditive, we conclude that $p \cdot X/p = p \cdot (S/p \wedge X) = 0$. $\qquad\square$

We have seen in Proposition 4 that $p \cdot X/p$ can be non-zero for $p = 2$. On the other hand, all triangulated categories that I know have the property that $p \cdot X/p = 0$ for odd primes $p$ and all objects $X$. This leaves us with the

**Open problem 6.** Let $p$ be an odd prime. Find a triangulated category $\mathcal{T}$ and an object $X$ of $\mathcal{T}$ such that $p \cdot X/p \neq 0$, or prove that in every triangulated category $\mathcal{T}$ we always have $p \cdot X/p = 0$.

From what we have discussed so far, it is still conceivable that every topological triangulated category in which 2 is invertible is algebraic. In particular one can wonder whether the Spanier-Whitehead category is algebraic after localization at an odd prime. We now describe a property of triangulated categories which distinguishes topological from algebraic examples away from the prime 2.

As before we denote by $K/n$ any cone of $n \cdot \mathrm{Id}_K$, which comes as part of a distinguished triangle

$$ K \xrightarrow{n \cdot} K \xrightarrow{\pi} K/n \longrightarrow K[1] \, . $$

An *extension* of a morphism $f : K \longrightarrow Y$ is then a morphism $\bar{f} : K/n \longrightarrow Y$ satisfying $\bar{f}\pi = f$. Such an extension exists if and only if $n \cdot f = 0$, and then the extension will usually not be unique.

**Proposition 7.** *Let $\mathcal{T}$ be an algebraic triangulated category, $X$ an object of $\mathcal{T}$ and $n \geq 2$. Then every morphism $f : K \longrightarrow X/n$ has an extension $\bar{f} : K/n \longrightarrow X/n$ such that some (hence any) mapping cone of $\bar{f}$ is annihilated by $n$.*

*Proof.* As in Proposition 1, a choice of model for $\mathcal{T}$ as the stable category of a Frobenius category gives a biexact, associative and unital pairing $\otimes^L : \mathcal{D}^b(\mathbb{Z}) \times \mathcal{T} \longrightarrow \mathcal{T}$ and we can take $X/n = \mathbb{Z}/n \otimes^L X$ and $K/n = \mathbb{Z}/n \otimes^L K$.

We define the extension $\bar{f}$ as the composite

$$\mathbb{Z}/n \otimes^L K \xrightarrow{\mathbb{Z}/n \otimes f} \mathbb{Z}/n \otimes^L \mathbb{Z}/n \otimes^L X \xrightarrow{\mu \otimes X} \mathbb{Z}/n \otimes^L X$$

where $\mu : \mathbb{Z}/n \otimes^L \mathbb{Z}/n \longrightarrow \mathbb{Z}/n$ is the multiplication map which makes $\mathbb{Z}/n$ into a ring. We choose a distinguished triangle

$$\mathbb{Z}/n \otimes^L K \xrightarrow{\bar{f}} \mathbb{Z}/n \otimes^L X \xrightarrow{\varphi} C(\bar{f}) \xrightarrow{\delta} \mathbb{Z}/n \otimes^L K[1] \qquad (8)$$

and show that the mapping cone $C(\bar{f})$ of $\bar{f}$ is annihilated by $n$.

We consider the diagram

$$
\begin{array}{ccccccc}
\mathbb{Z}/n \otimes^L \mathbb{Z}/n \otimes^L K & \xrightarrow{\mathbb{Z}/n \otimes \bar{f}} & \mathbb{Z}/n \otimes^L \mathbb{Z}/n \otimes^L X & \xrightarrow{\mathbb{Z}/n \otimes \varphi} & \mathbb{Z}/n \otimes^L C(\bar{f}) & \xrightarrow{\mathbb{Z}/n \otimes \delta} & \mathbb{Z}/n \otimes^L \mathbb{Z}/n \otimes^L K[1] \\
\ \ \downarrow{\scriptstyle \mu \otimes K} & & \ \ \downarrow{\scriptstyle \mu \otimes X} & & \ \ \downarrow{\scriptstyle \sigma} & & \ \ \downarrow{\scriptstyle \mu \otimes K[1]} \\
\mathbb{Z}/n \otimes^L K & \xrightarrow{\bar{f}} & \mathbb{Z}/n \otimes^L X & \xrightarrow{\varphi} & C(\bar{f}) & \xrightarrow{\delta} & \mathbb{Z}/n \otimes^L K[1]
\end{array}
$$

whose lower row is the distinguished triangle (8) and whose upper row is (8) tensored from the left with $\mathbb{Z}/n$. The left square commutes since the multiplication morphism $\mu$ is associative in $\mathcal{D}^b(\mathbb{Z})$. Since both rows are distinguished triangles, there exists a morphism $\sigma : \mathbb{Z}/n \otimes^L C(\bar{f}) \longrightarrow C(\bar{f})$ making the middle and right square commute.

We consider the morphism

$$\pi \otimes C(\bar{f}) \ : \ C(\bar{f}) \cong \mathbb{Z} \otimes^L C(\bar{f}) \longrightarrow \mathbb{Z}/n \otimes^L C(\bar{f})$$

which satisfies the two relations

$$\sigma(\pi \otimes C(\bar{f}))\varphi = \sigma(\mathbb{Z}/n \otimes \varphi)(\pi \otimes \mathbb{Z}/n \otimes X) = \varphi(\mu \otimes X)(\pi \otimes \mathbb{Z}/n \otimes X) = \varphi \qquad (9)$$

and

$$\begin{aligned}
\delta\sigma(\pi \otimes C(\bar{f})) &= (\mu \otimes K[1])(\mathbb{Z}/n \otimes \delta)(\pi \otimes C(\bar{f})) \qquad (10) \\
&= (\mu \otimes K[1])(\pi \otimes \mathbb{Z}/n \otimes K[1])\delta = \delta .
\end{aligned}$$

We claim that the morphism

$$\sigma' = 2\sigma - \sigma(\pi \otimes C(\bar{f}))\sigma \ : \ \mathbb{Z}/n \otimes^L C(\bar{f}) \longrightarrow C(\bar{f})$$

is a retraction to $\pi \otimes C(\bar{f})$. Indeed, by (9) the composite of $\varphi : \mathbb{Z}/n \otimes^L X \longrightarrow C(\bar{f})$ with the morphism $\sigma(\pi \otimes C(\bar{f})) - \mathrm{Id} : C(\bar{f}) \longrightarrow C(\bar{f})$ becomes trivial, so there exists a morphism $g : \mathbb{Z}/n \otimes^L K[1] \longrightarrow C(\bar{f})$ such that

$g\delta = \sigma(\pi \otimes C(\bar{f})) - \mathrm{Id}$. But then we have

$$\sigma'(\pi \otimes C(\bar{f})) = 2\sigma(\pi \otimes C(\bar{f})) - (\mathrm{Id} + g\delta)\sigma(\pi \otimes C(\bar{f}))$$
$$= \sigma(\pi \otimes C(\bar{f})) - g\delta\sigma(\pi \otimes C(\bar{f})) = \sigma(\pi \otimes C(\bar{f})) - g\delta = \mathrm{Id} ,$$

as claimed, where the third equality uses (10).

Now we have shown that $C(\bar{f})$ is a direct summand of $\mathbb{Z}/n \otimes^L C(\bar{f})$, which is annihilated by $n$, and so we have $n \cdot C(\bar{f}) = 0$. $\qquad\square$

In Theorem 16 below we prove a generalization of Proposition 7, by a different method.

Here is an example showing that in the situation of Proposition 7 in general there may not exist any extension $\bar{f} : K/n \longrightarrow X/n$ of $f$ whose mapping cone is annihilated by $n$. Proposition 7 and the following proposition show that the Spanier-Whitehead category localized at 3 is not algebraic.

We let $\beta_1 \in \pi_{10}^s \cong \mathbb{Z}/6$ be an element of order 3 (so $\beta_1$ generates the 3-primary component of the 10-dimensional stable stem). One way to define $\beta_1$ is as the unique element of the Toda bracket $\langle \alpha_1, \alpha_1, \alpha_1 \rangle$ where $\alpha_1 \in \pi_3^s \cong \mathbb{Z}/24$ is the 3-primary part of the element represented by the second Hopf map $\nu : S^7 \longrightarrow S^4$. We let $\tilde{\beta}_1 : S[11] \longrightarrow S/3$ be any lift in the Spanier-Whitehead category $\mathcal{SW}$ of $\beta_1$ to the mod-3 Moore spectrum $S/3$. So $\tilde{\beta}_1$ is a morphism whose composite with the connecting map $S/3 \longrightarrow S[1]$ equals the shift of $\beta_1$. We will need below that any such lift $\tilde{\beta}_1$ is *detected by* $P^3$ in the sense that in any mapping cone of $\tilde{\beta}_1$ the Steenrod operation $P^3$ is a non-trivial isomorphism from mod-3 cohomology in dimension 0 to dimension 12 (see page 60 of [To, §5]).

**Proposition 11.** *There is no extension of $\tilde{\beta}_1$ to a morphism $\bar{\beta} : S/3[11] \longrightarrow S/3$ whose mapping cone is annihilated by 3.*

*Proof.* We argue by contradiction and suppose that there exists an extension

$$\bar{\beta} : S/3[11] \longrightarrow S/3$$

of $\tilde{\beta}_1$ and a distinguished triangle

$$S/3[11] \xrightarrow{\bar{\beta}} S/3 \longrightarrow C(\bar{\beta}) \xrightarrow{\delta} S/3[12]$$

with $3 \cdot C(\bar{\beta}) = 0$.

Since the stable stems in dimension 21 and 22 consist only of torsion prime to 3 we have $\pi_{22}(S/3) = [S[22], S/3] = 0$. So there is a morphism $a : S[23] \longrightarrow C(\bar{\beta})$ lifting $\tilde{\beta}_1$ in the sense that $\delta a = \tilde{\beta}_1[12]$. Since we assumed $3 \cdot C(\bar{\beta}) = 0$, the morphism $a$ can be extended to a morphism $\bar{a} : S/3[23] \longrightarrow$

$C(\bar{\beta})$. We let $C(\bar{a})$ be a mapping cone of $\bar{a}$ arising as part of a distinguished triangle

$$S/3[23] \xrightarrow{\bar{a}} C(\bar{\beta}) \longrightarrow C(\bar{a}) \xrightarrow{\delta} S/3[24] .$$

Since the stable stems in dimension 21, 22, 33 and 34 consist only of torsion prime to 3 we have $\pi_{34} C(\bar{\beta}) = [S[34], C(\bar{\beta})] = 0$ and so there is a morphism $b : S[35] \longrightarrow C(\bar{a})$ lifting $\tilde{\beta}_1$ in the sense that $\delta b = \tilde{\beta}_1[24]$.

Now we bring cohomology operations into the game to reach a contradiction. The Moore spectrum $S/3$ has its mod-3 cohomology concentrated in dimensions 0 and 1, where it is 1-dimensional. The mapping cone of $b$ is built by distinguished triangles from three shifted copies of $S/3$ and one shifted copy of $S$, so its mod-3 cohomology is concentrated in dimensions 0, 1, 12, 13, 24, 25 and 36, where it is 1-dimensional.

The morphism $\tilde{\beta}_1$ is detected by the Steenrod operation $P^3$, i.e., in the mod-3 cohomology of the mapping cone of $\tilde{\beta}_1$ the operation $P^3$ is non-trivial from dimension 0 to dimension 12. The stable 'cells' (i.e., shifted copies of $S$) of the mapping cone of $b$ in dimension 12, 24 and 36, are attached to the two cells directly below by $\tilde{\beta}_1$, so in the cohomology of the mapping cone of $b$, the 3-fold iterate of $P^3$ is non-trivial from dimension 0 to 36. By the Adem relations we have $(P^3)^3 = (P^7 P^1 - P^8) P^1$. Since the cohomology is trivial in dimension 4 the operation $P^1$ acts trivially, hence so does $(P^3)^3$. We have obtained a contradiction, and so no extension of $\tilde{\beta}_1$ has a mapping cone which is annihilated by 3. $\qquad\square$

In topological triangulated categories, the phenomenon that a morphism $f : K \longrightarrow X/n$ may not have any extension whose cone is annihilated by $n$ is entirely 2- and 3-local, in the following sense.

**Proposition 12.** *If $\mathcal{T}$ is a topological triangulated category and $p$ is a prime bigger than 3, then every morphism $f : K \longrightarrow X/p$ has an extension $\bar{f} : K/p \longrightarrow X/p$ whose mapping cone is annihilated by $p$.*

*Proof.* The key point is that if $n$ is prime to 6, then the mod-$n$ Moore spectrum has an associative multiplication in the Spanier-Whitehead category (the multiplication is also commutative, but that is not relevant for the current proof). Again I cannot refrain from giving the simple proof below. In contrast, the mod-2 Moore spectrum does not have a multiplication; the mod-3 Moore spectrum has a commutative multiplication, but that is not associative in $\mathcal{SW}$, see [To].

As in the proof of Proposition 5, the condition that $n$ is odd guarantees that $n \cdot S/n = 0$. So there exists a morphism $\mu : S/n \wedge S/n \longrightarrow S/n$ which splits the two 'inclusions'. We show that $\mu$ is associative. In fact, the *associator*

$$\mu(\mu \wedge \mathrm{Id} - \mathrm{Id} \wedge \mu) \ : \ S/n \wedge S/n \wedge S/n \ \longrightarrow \ S/n$$

factors as a composite

$$S/n \wedge S/n \wedge S/n \ \xrightarrow{\ \delta \wedge \delta \wedge \delta\ } \ S[3] \ \longrightarrow \ S/n$$

where $\delta : S/n \longrightarrow S[1]$ is the connecting morphism. We have $\pi_3^s \cong \mathbb{Z}/24$ and $\pi_2^s \cong \mathbb{Z}/2$, so if $n$ is prime to 6 we have $[S[3], S/n] = \pi_3(S/n) = 0$. Thus the associator is trivial or, equivalently, $\mu$ is associative in $\mathcal{SW}$.

The rest of the argument is now essentially the same as in Proposition 7. We can smash the distinguished triangle

$$S \ \xrightarrow{\ n\cdot\ } \ S \ \longrightarrow \ S/n \ \xrightarrow{\ \delta\ } \ S[1]$$

in $\mathcal{SW}$ with the object $X$ and obtain a distinguished triangle in $\mathcal{T}$ which shows that $S/n \wedge X$ is isomorphic to $X/n$. We use the multiplication $\mu : S/n \wedge S/n \longrightarrow S/n$ to define the extension $\bar{f}$ as the composite

$$S/n \wedge K \ \xrightarrow{\ \mathrm{Id} \wedge f\ } \ S/n \wedge S/n \wedge X \ \xrightarrow{\ \mu \wedge \mathrm{Id}\ } \ S/n \wedge X \,.$$

Then we use the same reasoning as in the proof of Proposition 7 to obtain the retraction $\sigma' : S/n \wedge C(\bar{f}) \longrightarrow C(\bar{f})$ which shows that $n \cdot C(\bar{f}) = 0$. $\qquad\square$

This raises the following question:

**Open problem 13.** Consider a prime $p \geq 5$. Does there exist a triangulated category $\mathcal{T}$, an object $X$ of $\mathcal{T}$ and a morphism $f : K \longrightarrow X/p$ which does not admit any extension $\bar{f}$ to $K/p$ whose mapping cone is annihilated by $p$?

**Remark 14.** The proof of Proposition 12 shows that the special features of topological over algebraic triangulated categories are closely related to existence and properties of multiplications on mod-$n$ Moore spectra. For primes $p \geq 5$, the mod-$p$ Moore spectrum has a multiplication in the Spanier-Whitehead category which is commutative and associative. So on the level of tensor triangulated categories, there does not seem to be any qualitative difference between the Moore spectrum $S/p$ as an object of the Spanier-Whitehead category $\mathcal{SW}$ and $\mathbb{Z}/p$ as an object of $\mathcal{D}^b(\mathbb{Z})$, as long as $p \geq 5$. However, mod-$n$ Moore spectra are never $A_\infty$ *ring spectra*, but rigorously defining what that means and proving it would lead us too far afield. Theorem 17 below explains how the higher order non-associativity eventually manifests itself in the triangulated structure of the Spanier-Whitehead category (i.e., without any reference to the smash product).

From what we have discussed so far, it is still conceivable that for primes $p \geq 5$ the $p$-local Spanier-Whitehead category is algebraic. We will now introduce an invariant which we then use to show that this is not the case for any prime $p$.

**Definition 15.** Consider a triangulated category $T$ and a natural number $n \geq 2$. We define the *n-order* for objects $Y$ of $T$ inductively.

- Every object has $n$-order greater or equal to 0.
- For $k \geq 1$, an object $Y$ has $n$-order greater or equal to $k$ if and only if for every object $K$ of $T$ and every morphism $f : K \longrightarrow Y$ there exists an extension $\bar{f} : K/n \longrightarrow Y$ such that some (hence any) mapping cone of $\bar{f}$ has $n$-order greater or equal to $k - 1$.

We write $n$-$\mathrm{ord}(Y)$, or $n$-$\mathrm{ord}^T(Y)$ if we need to specify the ambient triangulated category, for the $n$-order of $Y$, i.e., the largest $k$ (possibly infinite) such that $Y$ has $n$-order greater or equal to $k$. We define the *n-order* of the triangulated category $T$ as the $n$-order of some (hence any) zero object, and denote it by $n$-$\mathrm{ord}(T)$. We make some observations which are direct consequences of the definitions.

- The $n$-order for objects is invariant under isomorphism and shift.
- An object $Y$ has positive $n$-order if and only if every morphism $f : K \longrightarrow Y$ has an extension to $K/n$, which is equivalent to $n \cdot f = 0$. So $n$-$\mathrm{ord}(Y) \geq 1$ is equivalent to the condition $n \cdot Y = 0$.
- The $n$-order of a triangulated category is one larger than the minimum of the $n$-orders of all objects of the form $K/n$.
- Let $S \subseteq T$ be a full triangulated subcategory and $Y$ an object of $S$. Induction on $k$ shows that if $n$-$\mathrm{ord}^T(Y) \geq k$, then $n$-$\mathrm{ord}^S(Y) \geq k$. Thus we have $n$-$\mathrm{ord}^S(Y) \geq n$-$\mathrm{ord}^T(Y)$. In the special case of a zero object we get $n$-$\mathrm{ord}(S) \geq n$-$\mathrm{ord}(T)$.
- Suppose that $T$ is a $\mathbb{Z}[1/n]$-linear triangulated category, i.e., multiplication by $n$ is an isomorphism for every object of $T$. Then $K/n$ is trivial for every object $K$ and thus $T$ has infinite $n$-order. If on the other hand $Y$ is non-trivial, then $n$-$\mathrm{ord}(Y) = 0$.
- If every object of $T$ has positive $n$-order, then $n \cdot Y = 0$ for all objects $Y$ and so $T$ is a $\mathbb{Z}/n$-linear triangulated category. Suppose conversely that $T$ is a $\mathbb{Z}/n$-liear triangulated category. Then induction on $k$ shows that $n$-$\mathrm{ord}(Y) \geq k$ for all objects $Y$, and thus every object has infinite $n$-order.

The last two items show that the $n$-order is useless if $\mathcal{T}$ is a $k$-*linear* triangulated category for some field $k$, since then every $n \in \mathbb{Z}$ is either zero or a unit in $k$.

The results which we have obtained so far can be rephrased using the notion of $n$-order: if $\mathcal{T}$ is algebraic, then Propositions 1 and 7 show that for every object $X$, the object $X/n$ always has $n$-order at least 2; we will improve this in Theorem 16 below. Propositions 4 and 11 show that in the Spanier-Whitehead category, the Moore spectrum $S/2$ has 2-order 0 and $S/3$ has 3-order 1; we generalize this to mod-$p$ Moore spectra in Theorem 17 below. If $\mathcal{T}$ is topological then for every object $X$, the object $X/3$ has 3-order at least 1, by Proposition 5. If $\mathcal{T}$ is topological and $p$ is a prime $\geq 5$, then for every object $X$, the object $X/p$ has $p$-order at least 2, by Proposition 12.

The following theorem generalizes Propositions 1 and 7.

**Theorem 16.** *Let $\mathcal{T}$ be an algebraic triangulated category and $X$ an object of $\mathcal{T}$. Then for any $n \geq 2$, the object $X/n$ has infinite $n$-order. In particular, every algebraic triangulated category $\mathcal{T}$ has infinite $n$-order.*

*Proof.* (Sketch) The assumption that $\mathcal{T}$ is algebraic gives another piece of extra structure: for any integer $n$ there exists a triangulated category $\mathcal{T}/n$ and an adjoint pair of exact functors $\rho_* : \mathcal{T} \longrightarrow \mathcal{T}/n$ and $\rho^* : \mathcal{T}/n \longrightarrow \mathcal{T}$ such that for every object $X$ of $\mathcal{T}$ there exists a distinguished triangle

$$X \xrightarrow{\;n\cdot\;} X \xrightarrow{\;\eta\;} \rho^*(\rho_* X) \longrightarrow X[1]$$

where $\eta$ is the unit of the adjunction.

We do not construct $\mathcal{T}/n$ in general here, but content ourselves with an example which gives the main idea. If $\mathcal{T} = \mathcal{D}(A)$ is the derived category of a differential graded ring $A$, then we can take $\mathcal{T}/n$ as the derived category of the differential graded ring $A \otimes \overline{\mathbb{Z}/n}$, where $\overline{\mathbb{Z}/n}$ is a flat resolution of the ring $\mathbb{Z}/n$, for example the exterior algebra over $\mathbb{Z}$ on a generator $x$ of dimension 1 with differential $dx = n$. The adjoint functor pair $(\rho_*, \rho^*)$ is derived from restriction and extension of scalars along the morphism $A \longrightarrow A \otimes \overline{\mathbb{Z}/n}$.

Now we exploit the extra structure to prove the theorem. Since $X/n$ is isomorphic to $\rho^*(\rho_* X)$ it is enough to show that for every object $Z$ of $\mathcal{T}/n$ and all $k \geq 0$ the object $\rho^* Z$ has $n$-order greater or equal to $k$.

We proceed by induction on $k$; for $k = 0$ there is nothing to prove. Suppose we have already shown that every $\rho^* Z$ has $n$-order greater or equal to $k - 1$ for some positive $k$. Given a morphism $f : K \longrightarrow \rho^* Z$ in $\mathcal{T}$ we can consider its adjoint $\hat{f} : \rho_* K \longrightarrow Z$ in $\mathcal{T}/n$; if we apply $\rho^*$ we obtain an extension $\rho^*(\hat{f}) : K/n \cong \rho^*(\rho_* K) \longrightarrow \rho^* Z$ of $f$. We choose a cone of $\hat{f}$, i.e., a distinguished triangle

$$\rho_* K \xrightarrow{\hat{f}} Z \longrightarrow C(\hat{f}) \longrightarrow \rho_* K[1]$$

in $\mathcal{T}/n$. Since $\rho^*$ is exact, $\rho^* C(\hat{f})$ is a cone of the extension $\rho^*(\hat{f})$ in $\mathcal{T}$. By induction, $\rho^* C(\hat{f})$ has $n$-order greater or equal to $k - 1$, which proves that $\rho^* Z$ has $n$-order greater or equal to $k$. $\qquad \square$

The next theorem generalizes Propositions 4 and 11 and shows that topological triangulated categories behave quite differently from algebraic ones.

**Theorem 17.** *Let $p$ be a prime. Then in the Spanier-Whitehead category, the mod-$p$ Moore spectrum $S/p$ has $p$-order $p - 2$. Moreover, the Spanier-Whitehead category, has $p$-order $p - 1$.*

The proof of Theorem 17 has two parts. One ingredient is a general statement about $p$-orders in *topological* triangulated categories $\mathcal{T}$ which generalizes Propositions 5 and 12: for any object $X$ and prime $p$, the object $X/p$ has $p$-order greater or equal to $p - 2$. The proof of this result uses the concept and properties of a *coherent action* of a mod-$p$ Moore space on an object of a model category, see Section 2 of [Sch]. It follows that any topological triangulated category (such as the Spanier-Whitehead category) has $p$-order at least $p - 1$.

The second ingredient of Theorem 17 is the proof that the mod-$p$ Moore spectrum $S/p$ has $p$-order at most $p - 2$. This uses mod-$p$ cohomology operations and serious calculational input; in particular, the proof depends on vanishing results, due to other people, about the $p$-primary components of the stable stems in specific dimensions. Proposition 11 gives the flavor of the proof which, however, for general primes $p$ is more involved. I plan to give a detailed proof elsewhere.

As we just mentioned, every topological triangulated category has $p$-order at least $p - 1$. This leave us with the following question, which generalizes Problems 6 and 13

**Open problem 18.** Let $p$ be a an odd prime. Does there exist a triangulated category whose $p$-order is strictly less than $p - 1$?

More generally we can ask which values the $n$-orders of triangulated categories can take.

Now that we have discussed torsion phenomena which can distinguish algebraic from topological triangulated categories, it is natural to ask whether there are any differences between topological and algebraic triangulated categories if all primes are invertible, i.e., in $\mathbb{Q}$-linear triangulated categories. The $n$-order is rationally a useless invariant since $\mathbb{Q}$-linear triangulated categories have infinite $n$-order for all $n$. Similarly, the smash product pairing of a topological triangulated category with $\mathcal{SW}$ gives no extra information for $\mathbb{Q}$-linear triangulated categories since the chain functor $C_* : \mathcal{SW} \longrightarrow \mathcal{D}^b(\mathbb{Z})$ becomes an equivalence of categories when rationalized (both sides are in fact rationally equivalent to the category of finite dimensional graded $\mathbb{Q}$-vector spaces).

It turns out that rationally the notions of algebraic and topological triangulated categories essentially coincide. Under some mild technical assumptions and cardinality restriction, every $\mathbb{Q}$-linear topological triangulated category is algebraic. More precisely, a theorem of Shipley [Sh, Cor. 2.16] says that every $\mathbb{Q}$-linear *spectral* model category (a special kind of stable model category which is enriched over the stable model category of symmetric spectra) with a set of compact generators is Quillen-equivalent to dg-modules over a certain differential graded $\mathbb{Q}$-category. Thus every triangulated category equivalent to the homotopy category of a stable model category of this kind is algebraic. I think that the assumption 'spectral' is merely of a technical nature and will eventually be removed. Similarly, I expect that the assumption of a 'set of compact generators' can be relaxed to 'well generated' at the price of allowing localizations of module categories over a differential graded $\mathbb{Q}$-category, along the lines of the papers [Po] and [He]. At present, I do not know of a $\mathbb{Q}$-linear triangulated category which is not topological.

**Acknowledgement:** I would like to thank Andreas Heider for various helpful comments on this paper.

# References

[Ad]　J. F. Adams, *Stable homotopy and generalised homology.* Chicago Lectures in Mathematics. University of Chicago Press, Chicago, Ill.-London, 1974. x+373 pp.

[BF]　A. K. Bousfield, E. M. Friedlander, *Homotopy theory of $\Gamma$-spaces, spectra, and bisimplicial sets.* Geometric applications of homotopy theory (Proc. Conf., Evanston, Ill., 1977), II Lecture Notes in Math., vol. 658, Springer, Berlin, 1978, pp. 80–130.

[Du]　D. Dugger, *Spectral enrichments of model categories.* Homology, Homotopy and Applications **8** (2006), 1–30.

[EKMM]　A. D. Elmendorf, I. Kriz, M. A. Mandell, J. P. May, *Rings, modules, and algebras in stable homotopy theory. With an appendix by M. Cole*, Mathematical Surveys and Monographs, **47**, American Mathematical Society, Providence, RI, 1997, xii+249 pp.

[He]　A. Heider, *Two results from Morita theory of stable model categories.* `arXiv:0707.0707`

[Ho1] M. Hovey, *Model categories.* Mathematical Surveys and Monographs, vol. 63, American Mathematical Society, Providence, RI, 1999, xii+209 pp.

[Ho2] M. Hovey, *Model category structures on chain complexes of sheaves.* Trans. Amer. Math. Soc. **353** (2001), 2441–2457.

[Ho3] M. Hovey, *Spectra and symmetric spectra in general model categories.* J. Pure Appl. Alg. **165** (2001), 63-127.

[HSS] M. Hovey, B. Shipley, J. Smith, *Symmetric spectra.* J. Amer. Math. Soc. **13** (2000), 149–208.

[Ke] B. Keller, *On differential graded categories.* Proceedings of the International Congress of Mathematicians, Madrid, Spain, 2006, vol. II, European Mathematical Society, 2006, pp. 151–190.

[Ly] M. Lydakis, *Simplicial functors and stable homotopy theory.* Preprint (1998). `http://hopf.math.purdue.edu/`

[MMSS] M. A. Mandell, J. P. May, S. Schwede, B. Shipley, *Model categories of diagram spectra.* Proc. London Math. Soc. **82** (2001), 441–512.

[Mar] H. R. Margolis, *Spectra and the Steenrod algebra. Modules over the Steenrod algebra and the stable homotopy category*, North-Holland Mathematical Library **29**, North-Holland Publishing Co., Amsterdam-New York, 1983, xix+489 pp.

[MSS] F. Muro, S. Schwede, N. Strickland, *Triangulated categories without models.* Invent. Math. **170** (2007), 231-241.

[Ne] A. Neeman, *Triangulated Categories*, Annals of Mathematics Studies, vol. 148, Princeton University Press, Princeton, NJ, 2001.

[Po] M. Porta, *The Popescu-Gabriel theorem for triangulated categories.* `arXiv:0706.4458`

[Q] D. G. Quillen, *Homotopical algebra.* Lecture Notes in Math. **43**, Springer-Verlag, 1967.

[Sch] S. Schwede, *The stable homotopy category is rigid.* Annals of Math. **166** (2007), 837-863.

[SS] S. Schwede, B. Shipley, *Stable model categories are categories of modules.* Topology **42** (2003), no. 1, 103–153.

[Sh] B. Shipley, *H$\mathbb{Z}$-algebra spectra are differential graded algebras.* American J. Math. **129** (2007) 351-379.

[SW] E. Spanier, J. H. C. Whitehead, *A first approximation to homotopy theory.* Proc. Nat. Acad. Sci. USA **39** (1953), 655-660.

[To] H. Toda, *On spectra realizing exterior parts of the Steenrod algebra.* Topology **10** (1971), 53–65.

[Vo] R. Vogt, *Boardman's stable homotopy category.* Lecture Notes Series, No. 21 Matematisk Institut, Aarhus Universitet, Aarhus 1970 i+246 pp.

MATHEMATISCHES INSTITUT, UNIVERSITÄT BONN, GERMANY
*E-mail address*: `schwede@math.uni-bonn.de`

# Derived categories of coherent sheaves on algebraic varieties

YUKINOBU TODA

ABSTRACT. In this article, we give the introduction of the recent developments on the derived categories of coherent sheaves on algebraic varieties. We also introduce the notion of stability conditions on triangulated categories in the sense of T. Bridgeland.

## 1. Introduction

The notion of derived category of coherent sheaves was first introduced in [24] in order to formulate the Grothendieck duality theorem. It is a category whose objects are complexes of coherent sheaves, and has a structure of a triangulated category. Recently it has been observed that the derived category represents several interesting symmetries, which seems impossible without the notion of derived categories, e.g. McKay correspondence [16], Homological mirror symmetry [43], etc. Now derived categories are a very popular area with interactions with many other subjects including non-commutative algebra, birational geometry, symplectic geometry and string theory. In this article, we give an introduction of the recent results on these topics.

The content of this article is as follows. In Section 2, we give the basic notions concerning derived categories, and propose some fundamental problems. In Section 3, we discuss the relationship between derived category and birational geometry. In Section 4, we discuss the symmetries between derived categories of coherent sheaves and that of module categories of some non-commutative algebras. In Section 5, we introduce the notion of stability conditions on triangulated categories, defined by T. Bridgeland, and see how this notion explains the several symmetries we discuss in this article.

**Notation and Convention.** All the varieties are defined over $\mathbb{C}$. For a variety $X$, we denote by $\operatorname{Coh}(X)$ the abelian category of coherent sheaves on $X$. When $X$ is smooth, we denote by $\omega_X := \wedge^{\dim X} \Omega_X$ the canonical line bundle, and $K_X$ is the divisor on $X$ such that $\mathcal{O}_X(K_X) = \omega_X$. For a $\mathbb{C}$-vector space $V$, we denote by $V^*$ its dual vector space.

## 2.  Derived categories and Fourier-Mukai transforms

### 2.1.  Derived category of coherent sheaves

For a variety $X$, its bounded derived category

$$D(X) := D^b \operatorname{Coh}(X),$$

is defined as the localization of the category of bounded complexes of coherent sheaves with respect to the class of quasi-isomorphisms. So any object in $D(X)$ is represented by a bounded complex,

$$\cdots \to 0 \to \mathcal{F}^m \to \mathcal{F}^{m+1} \to \cdots \to \mathcal{F}^{n-1} \to \mathcal{F}^n \to 0 \to \cdots,$$

for $\mathcal{F}^i \in \operatorname{Coh}(X)$, and a morphism from $\mathcal{F}^\bullet$ to $\mathcal{G}^\bullet$ is represented by a diagram,

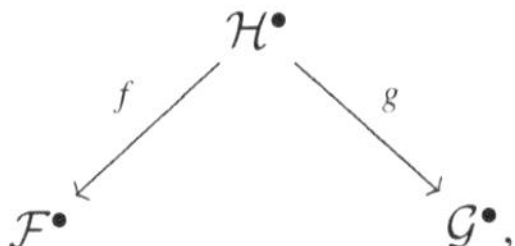

such that $f$ is a quasi-isomorphism and $g$ is a morphism of chain complexes. The derived category $D(X)$ is no longer an abelian category, but it has a structure of a *triangulated category*, i.e.

- There is a *shift functor* denoted by [1].
- There is a notion of *distinguished triangles*, $\mathcal{F}_1 \to \mathcal{F}_2 \to \mathcal{F}_3 \to \mathcal{F}_1[1]$.

Also these notions should satisfy some axioms. (cf. [24].) On the category $D(X)$, the shift functor [1] is given by shifting the degree of the complex, and distinguished triangles are obtained by taking mapping cones of the morphisms of chain complexes.

Note that the abelian category $\operatorname{Coh}(X)$ is regarded as the full subcategory of $D(X)$, by identifying the objects in $\operatorname{Coh}(X)$ with the complexes concentrated in degree zero. For an object $E \in D(X)$, we denote by $\mathcal{H}^i(E) \in \operatorname{Coh}(X)$ the $i$-th cohomology of the complex $E$. The support of $E \in D(X)$ is defined by

$$\operatorname{Supp}(E) = \bigcup_i \operatorname{Supp} \mathcal{H}^i(E) \subset X.$$

For $E, F \in D(X)$ and $i \in \mathbb{Z}$, we set

$$\operatorname{Hom}^i(E, F) := \operatorname{Hom}(E, F[i]).$$

Here we remark that if $E$ and $F$ are contained in $\operatorname{Coh}(X)$, then $\operatorname{Hom}^i(E, F)$ coincides with the usual Ext-group $\operatorname{Ext}^i(E, F)$. If $X$ is smooth and projective, it is well known that the vector space

$$\bigoplus_{i\in\mathbb{Z}} \mathrm{Hom}^i(E, F)$$

is finite dimensional. Hence the the following number makes sense,

$$\chi(E, F) = \sum_{i\in\mathbb{Z}} (-1)^i \dim \mathrm{Hom}^i(E, F) \in \mathbb{Z}.$$

Now we introduce some functors between derived categories. For the precise definitions and details, see [24]. Suppose $X$ is smooth and $Y$ is another smooth variety. Let $f\colon X \to Y$ be a projective morphism. Then we can define the *derived push-forward* and *derived pull-back*,

$$\mathbf{R}f_*\colon D(X) \to D(Y), \quad \mathbf{L}f^*\colon D(Y) \to D(X).$$

The functor $\mathbf{R}f_*$ is defined without the assumption that $Y$ is smooth. However $\mathbf{L}f^*$ is *not* defined if $Y$ is singular. In fact $\mathbf{L}f^*\mathcal{F}$ for $\mathcal{F} \in D(Y)$ may be unbounded below, hence one has to replace $D(X)$ by the unbounded (or bounded above) derived category of coherent sheaves. Also for an object $E \in D(X)$, we can define the *derived homomorphism* and *derived tensor product*,

$$\mathbf{R}\mathcal{H}om(E, *)\colon D(X) \to D(X), \quad * \overset{\mathbf{L}}{\otimes} E\colon D(X) \to D(X).$$

Suppose $Y$ is an affine variety $\mathrm{Spec}\, R$, not necessary smooth. Then $D(Y)$ is naturally identified with the bounded derived category of finitely generated $R$-modules, denoted by $D(R)$. For $E \in D(X)$, we define the functor,

$$\mathbf{R}\mathrm{Hom}(E, *) := \mathbf{R}f_* \circ \mathbf{R}\mathcal{H}om(E, *)\colon D(X) \longrightarrow D(R).$$

Note that we have

$$\mathcal{H}^i(\mathbf{R}\mathrm{Hom}(E, F)) = \mathrm{Hom}^i(E, F),$$

as $R$-modules for $F \in D(X)$. In this article we sometimes omit $\mathbf{L}$ if the functors we derive are exact. For example, if $f$ is flat we write $\mathbf{L}f^* = f^*$.

## 2.2. Fourier-Mukai transforms

Let $X$ be a smooth projective variety. It is known that the abelian category $\mathrm{Coh}(X)$ determines $X$ itself. (For example see [39, Theorem 3.2].) On the other hand, the derived category $D(X)$ does not necessary determine $X$ itself. The reason behind this is that we have additional correspondences called Fourier-Mukai transforms.

**Definition 2.1.** Let $X$ and $Y$ be smooth projective varieties. For an object $\mathcal{P} \in D(X \times Y)$, define the functor $\Phi^{\mathcal{P}}_{X \to Y} \colon D(X) \to D(Y)$ by the formula,

$$\Phi^{\mathcal{P}}_{X \to Y}(E) = \mathbf{R}p_{Y*}(p_X^* E \overset{\mathbf{L}}{\otimes} \mathcal{P}),$$

for $E \in D(X)$. Here $p_X$, $p_Y$ are projections from $X \times Y$ onto corresponding factors. If $\Phi^{\mathcal{P}}_{X \to Y}$ gives an equivalence, it is called a *Fourier-Mukai transform*. The object $\mathcal{P}$ is called a *kernel* of the functor $\Phi^{\mathcal{P}}_{X \to Y}$.

The functors of Fourier-Mukai type are not special equivalences. In fact Orlov [52] proved the following.

**Theorem 2.2. (Orlov [52])** *Let $\Phi \colon D(X) \to D(Y)$ be an equivalence of $\mathbb{C}$-linear triangulated categories. Then there exists an object $\mathcal{P} \in D(X \times Y)$ such that $\Phi$ is isomorphic to the functor $\Phi^{\mathcal{P}}_{X \to Y}$. Moreover $\mathcal{P}$ is uniquely determined up to isomorphism.*

Here we give some examples of Fourier-Mukai transforms.

**Example 2.3. (Mukai [48], [49])** (i) Let $A$ be an abelian variety, i.e. $A$ is a complex torus $\mathbb{C}^n / \Gamma$ embedded into a projective space. Here $\Gamma \subset \mathbb{R}^{2n}$ is a lattice of rank $2n$. Let $\hat{A} = \mathrm{Pic}^0(A)$ be the moduli space of line bundles on $A$ which are topologically isomorphic to $\mathcal{O}_A$. It is known that $\hat{A}$ is also an abelian variety of the same dimension, and called *dual abelian variety*. Let $\mathcal{U} \in \mathrm{Pic}(A \times \hat{A})$ be the universal family. Then the functor

$$\Phi^{\mathcal{U}}_{\hat{A} \to A} \colon D(\hat{A}) \longrightarrow D(A),$$

is a Fourier-Mukai transform. See [48] for the detail. When $\dim A \geq 2$, the dual $\hat{A}$ is not necessary isomorphic to $A$. This means that the derived category $D(A)$ does not determine the original variety $A$.

(ii) An analogue of (i) for K3 surfaces was studied in [48]. Let $X$ be a K3 surface, i.e. simply connected algebraic surface with $\omega_X$ trivial. Let $M$ be a connected component of the moduli space of stable sheaves on $X$, which is two dimensional, projective, and has a universal family $\mathcal{E} \in \mathrm{Coh}(X \times M)$. Then the functor

$$\Phi^{\mathcal{E}}_{M \to X} \colon D(M) \longrightarrow D(X),$$

is a Fourier-Mukai transform.

## 2.3. Fourier-Mukai partners

As we observed in Example 2.3, for a given smooth projective variety $X$ there may be several varieties $Y$ which are derived equivalent to $X$. So we have the following natural problem.

**Problem 2.4.** *When there exists a Fourier-Mukai transform,*

$$\Phi \colon D(Y) \longrightarrow D(X),$$

*is there geometric relationship between X and Y?*

If $Y$ is derived equivalent to $X$, then it is called a *Fourier-Mukai partner* of $X$. The Problem 2.4 was first considered by Bondal and Orlov [8], under a geometrically strong assumption that $K_X$ or $-K_X$ is ample.

**Theorem 2.5. (Bondal-Orlov [8])** *Assume $K_X$ or $-K_X$ is ample. Then any Fourier-Mukai partner of $X$ is isomorphic to $X$ itself.*

The class of varieties $X$ with $K_X$ or $-K_X$ ample includes the hypersurfaces of degree $d$ in $\mathbb{P}^n$ with $d \neq n+1$. One of the key points in studying Fourier-Mukai partners is to investigate the several categorical invariants of derived categories, such as the Serre functor we define below.

**Definition 2.6.** Let $\mathcal{T}$ be a $\mathbb{C}$-linear triangulated category. An equivalence $S \colon \mathcal{T} \to \mathcal{T}$ as triangulated categories is called a *Serre functor* if there exists a bifunctorial isomorphism

$$\mathrm{Hom}(E, F) \to \mathrm{Hom}(F, S(E))^*$$

for $E, F \in \mathcal{T}$.

It is shown in [8, Proposition 1.5] that if a Serre functor exists, then it is unique up to canonical isomorphism. When $\mathcal{T} = D(X)$ for a smooth projective variety $X$, then Serre duality implies that

$$S_X(E) = E \otimes \omega_X[\dim X],$$

for $E \in D(X)$ gives a Serre functor on $D(X)$.

Another key point is to find the objects of the form $\mathcal{O}_x$ for closed points $x \in X$ from the categorical data of $D(X)$.

**Definition 2.7. [8, Definition 2.1]** An object $P \in D(X)$ is called a *point object of codimension s* if it satisfies,

$$\text{(i)} \quad S_X(P) \cong P[s],$$
$$\text{(ii)} \quad \text{Hom}^{<0}(P, P) = 0,$$
$$\text{(iii)} \quad \text{Hom}^0(P, P) = \mathbb{C}.$$

Now the idea of the proof of Theorem 2.5 is explained as follows. Note that the notion of point objects are determined by only the data of $D(X)$ as a triangulated category. In [8, Proposition 2.2], it is shown that if $K_X$ or $-K_X$ is ample, then any point objects are of the form $\mathcal{O}_x[r]$ for some closed point $x \in X$ and $r \in \mathbb{Z}$. Thus one can reconstruct the notion of "point" from the data of $D(X)$ as a triangulated category, which enables us to reconstruct $X$ from $D(X)$.

On the other hand if $K_X$ is trivial, then the condition (i) of Definition 2.7 does not say anything so there may be other point objects, and non-trivial Fourier-Mukai partners. In fact the examples given in Example 2.3 are such cases.

At this time, very few is known about Fourier-Mukai partners. Other than Theorem 2.5, we know the following cases:

**(i) The case of** $\dim X \leq 1$**.**
In this case any Fourier-Mukai partner of $X$ is known to be isomorphic to $X$. In fact if $X$ is not an elliptic curve, this follows from Theorem 2.5. If $X$ is an elliptic curve, see [25, Theorem 5.1].

**(ii) The case of** $\dim X = 2$**.**
When $X$ is minimal i.e. $K_X \cdot C \geq 0$ for any curve $C \subset X$, it is shown in [17] that any Fourier-Mukai partner of $X$ is isomorphic to $X$ itself, except $X$ is a K3 surface or an abelian surface. When $X$ is a K3 surface or an abelian surface, any Fourier-Mukai partner is isomorphic to the moduli space of stable sheaves on $X$, as in Example 2.3 (ii). When $X$ is not minimal, then any Fourier-Mukai partner of $X$ is isomorphic to $X$ itself except $X$ is a relatively minimal rational elliptic surface [37].

**(iii) The case of** $\dim X = 3$**.**
When $\kappa(X) = 3$, any Fourier-Mukai partner is obtained as a sequence of flops [37]. (See (1) in the next paragraph for the definition of $\kappa(X)$, and Definition 3.4 for flops.) When $\kappa(X) = 2$, any Fourier-Mukai partner is obtained as a relative Jacobian up to flops [60].

**(iv) The case that $X$ is an abelian variety $A$.**
In this case, any Fourier-Mukai partner $B$ is also an abelian variety, and an abelian variety $B$ is a Fourier-Mukai partner of $A$ if and only if there exists a symplectic isomorphism [53],

$$\hat{A} \times A \longrightarrow \hat{B} \times B.$$

As is observed in the above cases, the known examples of Fourier-Mukai partners are geometrically quite similar properties.

## 2.4. Several categorical invariants

It seems that the classification problem of Fourier-Mukai partners are quite difficult, especially when Kodaira dimension of the variety in question is near to zero. On the other hand we know several categorical invariants of $D(X)$ as a triangulated category, and such invariants would help us how Fourier-Mukai partners look like. Let $X$ and $Y$ be smooth projective varieties and assume there exists an equivalence of derived categories $\Phi \colon D(X) \to D(Y)$.

**(i) Invariance of dimensions.** We have $\dim X = \dim Y$. (cf. [37, Theorem 1.4].)

**(ii) Invariance of cohomology groups.** In [52], it is shown that $\Phi$ induces an isomorphism of vector spaces, $\phi \colon H^*(X, \mathbb{Q}) \to H^*(Y, \mathbb{Q})$ such that the following diagram commutes,

$$
\begin{array}{ccc}
D(X) & \xrightarrow{\ \Phi\ } & D(Y) \\
\Big\downarrow{\scriptstyle \mathrm{ch}(*)\sqrt{\mathrm{td}_X}} & & \Big\downarrow{\scriptstyle \mathrm{ch}(*)\sqrt{\mathrm{td}_Y}} \\
H^*(X, \mathbb{Q}) & \xrightarrow{\ \phi\ } & H^*(Y, \mathbb{Q}).
\end{array}
$$

The commutativity of the above diagram is due to Grothendieck Riemann Roch theorem. The isomorphism $\phi$ is given by the correspondence of the following algebraic cycle,

$$p_X^*\sqrt{\mathrm{td}_X} \cdot \mathrm{ch}(\mathcal{P}) \cdot p_Y^*\sqrt{\mathrm{td}_Y} \in H^*(X \times Y, \mathbb{Q}),$$

where $\mathcal{P} \in D(X \times Y)$ is a kernel of $\Phi$. Note that $\phi$ does not necessary preserve the degree of the cohomology ring. In [52], the existence of $\phi$ is a crucial point in reducing the classification problem of Fourier-Mukai partners of K3 surfaces to Torelli type problem. Also Orlov [54] conjectures the relationship between derived equivalence and Chow motives.

**(iii) Invariance of canonical rings.** In [18], [60], it is shown that $\Phi$ induces an isomorphism of graded rings

$$\bigoplus_{m\in\mathbb{Z}} H^0(X, mK_X) \longrightarrow \bigoplus_{m\in\mathbb{Z}} H^0(Y, mK_Y).$$

In particular we have $\kappa(X) = \kappa(Y)$. (Also see [37].) Here $\kappa(X)$ is the *Kodaira dimension*, defined by the number $k \in \mathbb{Z}$ such that

$$H^0(X, mK_X) \sim am^k, \quad (m \gg 0) \tag{1}$$

for some $a > 0$ if $H^0(X, mK_X)$ is non-zero for some $m > 0$. If $H^0(X, mK_X) = 0$ for any $m > 0$, $\kappa(X)$ is defined to be $-\infty$. The idea of [60] is to reduce the classification problem of Fourier-Mukai partners to the case of Kodaira dimension zero, by comparing the derived categories of the fibers of the following rational maps called *Iitaka fibrations*,

$$X \dashrightarrow \operatorname{Proj}\bigoplus_{m\geq 0} H^0(X, mK_X), \quad Y \dashrightarrow \operatorname{Proj}\bigoplus_{m\geq 0} H^0(X, mK_Y).$$

**(iv) Invariance of Hochschild cohomologies.** Let $HH^N(X)$ and $HT^N(X)$ be

$$HH^N(X) := \operatorname{Ext}^N_{X\times X}(\mathcal{O}_{\Delta_X}, \mathcal{O}_{\Delta_X}), \tag{2}$$

$$HT^N(X) := \bigoplus_{i+j=N} H^i(X, \bigwedge^j T_X). \tag{3}$$

Here $\Delta_X \subset X \times X$ is a diagonal, and $HH^N(X)$ is called *Hochschild cohomology*. Then the equivalence $\Phi$ induces an isomorphism [18],

$$\phi_H \colon HH^N(X) \xrightarrow{\cong} HH^N(Y).$$

On the other hand, we have the following isomorphism called *HKR isomorphism* [26], [45], [58], [67].

$$I_{HKR} \colon HH^N(X) \xrightarrow{\cong} HT^N(X).$$

Combined with $\phi_H$ and $I_{HKR}$, we obtain the isomorphism,

$$\phi_T \colon HT^N(X) \xrightarrow{\cong} HT^N(Y).$$

**(v) Invariance of deformation theories.** As a special case of (iv), the following vector space is preserved by the derived equivalence,

$$HT^2(X) = H^2(X, \mathcal{O}_X) \oplus H^1(X, T_X) \oplus H^0(X, \bigwedge^2 T_X).$$

The above space parameterizes the first order deformations of the abelian category Coh($X$). (See [45], [46], [64].) In fact for $u = (\alpha, \beta, \gamma)$, $\beta$ corresponds to the usual deformation of $X$ as a scheme, $\gamma$ corresponds to a non-commutative deformation, and $\alpha$ is a gerby deformation. One can construct the $\mathbb{C}[\varepsilon]/(\varepsilon^2)$-linear abelian category Coh($X, u$), which is a mixture of the above deformations. Then the equivalence $\Phi$ extends to an equivalence [64],

$$\Phi^{\dagger} \colon D^b \operatorname{Coh}(X, u) \longrightarrow D^b \operatorname{Coh}(Y, v),$$

where $v = \phi_T(u) \in HT^2(Y)$. Extending $\Phi^{\dagger}$ to infinite order deformations is problematic. At this time, $\Phi^{\dagger}$ is known to be extended to infinite order deformations when the equivalence $\Phi$ is given as in Example 2.3 (i). (See [4].) We expect that the deformation methods will enable us to find more interesting Fourier-Mukai dualities.

### 2.5. The group of autoequivalences

The next problem is similar to Problem 2.4.

**Problem 2.8.** *Compute the group of autoequivalences* Auteq $D(X)$.

When $K_X$ or $-K_X$ is ample, Bondal and Orlov [8] showed (by the same idea of Theorem 2.5) that the group Auteq $D(X)$ is generated by trivial autoequivalences, that is the group of automorphisms Aut($X$), tensoring line bundles, and the shift functor [1]. Again it seems that Auteq $D(X)$ is more difficult when the Kodaira dimension of $X$ is near to zero. Non-trivial autoequivalences are provided by the spherical objects below.

**Definition 2.9.** An object $E \in D(X)$ is said to be a *spherical object* if and only if

- $\operatorname{Hom}^i(E, E) = \begin{cases} \mathbb{C} & (i = 0, \dim X), \\ 0 & (\text{otherwise}), \end{cases}$
- $E \otimes \omega_X \cong E$.

Under mirror symmetry, a spherical object should correspond to a Lagrangian sphere, which has an associated symplectic isomorphism called *Dehn twist* [56]. Thus one expects that a spherical object associates a corresponding autoequivalence on $D(X)$. The answer was provided by Seidel and Thomas [57].

**Theorem 2.10. (Seidel-Thomas [57])** *For a spherical object $E \in D(X)$, there exists an autoequivalence $T_E \in$ Auteq $D(X)$ such that for any $F \in D(X)$, one* *has a distinguished triangle,*

$$\mathbf{R}\operatorname{Hom}(E, F) \otimes E \longrightarrow F \longrightarrow T_E(F).$$

For a spherical object $E$, its associated autoequivalence $T_E$ is called *spherical twist*. There are certain generalizations of this twist [27], [61].

The recent work of Bridgeland [15] relates the computation of Auteq $D(X)$ to the topology of the space of stability conditions on $D(X)$. We discuss this in Section 5. Here we introduce two important computations of the group Auteq $D(X)$.

**Autoequivalences on abelian varieties.** Let $A$ be an abelian variety and $\hat{A}$ be its dual abelian variety. We can write a morphism $f \colon A \times \hat{A} \to A \times \hat{A}$ as a matrix,

$$f = \begin{pmatrix} \alpha & \beta \\ \gamma & \delta \end{pmatrix},$$

and define $\tilde{f} \colon A \times \hat{A} \to A \times \hat{A}$ to be

$$\tilde{f} = \begin{pmatrix} \hat{\delta} & -\hat{\gamma} \\ -\hat{\beta} & \hat{\alpha} \end{pmatrix}.$$

Then we introduce the group $U(A \times \hat{A})$ as follows,

$$U(A \times \hat{A}) := \{ f \in \operatorname{Aut}(A \times \hat{A}) \mid \tilde{f} = f^{-1} \}.$$

In the paper [53], Orlov showed the following.

**Theorem 2.11.** (**Orlov [53]**) *For an abelian variety $A$, there exists the following exact sequence,*

$$0 \longrightarrow \mathbb{Z} \oplus (A \times \hat{A}) \longrightarrow \operatorname{Auteq} D(X) \longrightarrow U(A \times \hat{A}) \longrightarrow 1.$$

**Autoequivalences on $A_n$-configurations of $(-2)$-curves.** Another important case, the group of autoequivalences on $A_n$-configurations of $(-2)$-curves, was computed by Ishii and Uehara [29]. Let $Y$ be an $A_n$-singularity,

$$Y = \operatorname{Spec} \mathbb{C}[x, y, z]/(xy + z^{n+1}),$$

and $f \colon X \to Y$ be the minimal resolution. Note that the exceptional locus is a chain of rational curves,

$$Z = C_1 \cup \cdots \cup C_n, \quad C_i \cong \mathbb{P}^1.$$

Under this situation, Ishii and Uehara [29] considered the triangulated category,

$$D_Z(X) = \{E \in D(X) \mid \mathrm{Supp}(E) \subset Z\},$$

and computed the group,

$$\mathrm{Auteq}^{\mathrm{FM}}(D_Z(X)) = \{\Phi \in \mathrm{Auteq}(D_Z(X)) \mid \Phi$$
$$\cong \Phi^{\mathcal{P}}_{X \to X} \text{ with } \mathrm{Supp}(\mathcal{P}) \subset X \times_Y X\}.$$

Note that $\mathcal{O}_{C_i}(-1)$, $\omega_Z$ are spherical objects, so there are associated twists $T_{\mathcal{O}_{C_i}(-1)}$, $T_{\omega_Z}$. We set

$$B := \langle T_{\mathcal{O}_{C_i(-1)}}, T_{\omega_Z} \mid 1 \le i \le n \rangle \subset \mathrm{Auteq}^{\mathrm{FM}}(D_Z(X)).$$

**Theorem 2.12.** **(Ishii-Uehara [29])** *We have*

$$\mathrm{Auteq}^{\mathrm{FM}}(D_Z(X)) = (\langle B, \mathrm{Pic}(X)\rangle \rtimes \mathrm{Aut}(X)) \times \mathbb{Z}.$$

*Here $\mathbb{Z}$ is generated by the shift functor $[1]$.*

Furthermore Ishii, Ueda and Uehara [28] determined the group structure of $B$ using the space of stability conditions.

# 3. Derived category and birational geometry

In this section, we discuss the relationship between derived category and birational geometry. Recall that minimal model program, referred as MMP, is a program to find a good birational model (minimal model or Fano fiber space) by contracting extra rational curves [44]. For a given variety $X$, MMP suggests that there is a sequence,

$$X = X_0 \dashrightarrow X_1 \dashrightarrow X_2 \cdots \dashrightarrow X_m,$$

such that each $X_i \dashrightarrow X_{i+1}$ is a birational morphism which contracts a divisor, or a certain birational map called *flip*. The output $X_m$ is either a minimal model (i.e. $K_{X_m} \cdot C \ge 0$ for any curve $C \subset X_m$) or has a fiber space structure $X_m \to Z$ such that the general fiber is a Fano manifold. The purpose of this section is to observe that MMP is more natural in the context of the "space" stated in the introduction.

## 3.1. Semiorthogonal decomposition

As a preparation, we introduce the notion of semiorthogonal decomposition.

**Definition 3.1.** Let $\mathcal{T}$ be a triangulated category and $i\colon \mathcal{C} \subset \mathcal{T}$ be a full triangulated subcategory. We say $\mathcal{C}$ is right admissible if the inclusion $i$ admits a right adjoint $p\colon \mathcal{T} \to \mathcal{C}$.

One can show the following: for a full subcategory $\mathcal{C} \subset \mathcal{T}$, it is right admissible if and only if for $E \in \mathcal{T}$, there is a distinguished triangle,

$$E_1 \longrightarrow E \longrightarrow E_2,$$

with $E_1 \in \mathcal{C}$ and $E_2 \in \mathcal{C}^{\perp} = \{F \in \mathcal{T} \mid \mathrm{Hom}(\mathcal{C}, F) = 0\}$.

**Definition 3.2.** Let $\mathcal{C}_1, \cdots, \mathcal{C}_m$ be full triangulated subcategories of $\mathcal{T}$, and $\mathcal{T}_i \subset \mathcal{T}$ the smallest triangulated subcategory which contains $\mathcal{C}_1, \cdots, \mathcal{C}_i$. Suppose that

- $\mathcal{T} = \mathcal{T}_m$.
- $\mathcal{C}_i$ is a right admissible subcategory of $\mathcal{T}_i$.

In this case, we write

$$\mathcal{T} =< \mathcal{C}_1, \mathcal{C}_2, \cdots, \mathcal{C}_m >,$$

and it is called a *semiorthogonal decomposition* of $\mathcal{T}$.

## 3.2. Blow-up formula of derived categories

The behavior of derived categories under birational transformations was first studied by Bondal and Orlov [7]. Let $X$ be a smooth projective variety and $Z \subset X$ a smooth closed subscheme of codimension $r$. Let $f\colon Y \to X$ be the blow-up along $Z$, and $E \subset Y$ the exceptional divisor of $f$. We have the following diagram,

$$
\begin{array}{ccc}
E & \xrightarrow{\ i\ } & Y \\
\ \downarrow{\scriptstyle p} & & \ \downarrow{\scriptstyle f} \\
Z & \longrightarrow & X.
\end{array}
$$

Let $\mathcal{O}_E(1) \in \mathrm{Pic}(E)$ be the tautological line bundle of the projective bundle $E = \mathbb{P}(N_{Z/X})$. Bondal and Orlov [7] showed the following.

**Theorem 3.3. (Bondal-Orlov [7])** *The functors,*

$$\mathbf{L}f^*\colon D(X) \to D(Y), \quad i_* \circ (p^*(*) \otimes \mathcal{O}_E(j))\colon D(Z) \to D(Y),$$

*are fully faithful for $-r + 1 \leq j \leq -1$. We have the following semiorthogonal decomposition,*

$$D(Y) = \; < D(E)_{-r+1}, \cdots , D(E)_{-1}, \mathbf{L}f^*D(X) >,$$

*where $D(E)_j$ is the image of $i_* \circ (p^*(*) \otimes \mathcal{O}_E(j))$.*

Theorem 3.3 asserts that the derived category will be bigger by taking blow-ups, and the difference is described by the derived category of the exceptional locus.

### 3.3. Derived equivalence under birational transformations

Next we discuss the derived equivalence between birational minimal models. First we recall the notion of *flip* and *flop*. (See [42].)

**Definition 3.4.** Let $X$ and $X^\dagger$ be smooth quasi projective varieties and assume there exist projective birational morphisms,

$$X \xrightarrow{f} Y \xleftarrow{g} X^\dagger,$$

to a (singular) variety $Y$, which satisfy the following.

- $f$ and $g$ are isomorphisms in codimension one.
- The relative Picard numbers of $f$ and $g$ are one respectively.
- For a $f$-ample divisor $D$ on $X$, let $D^\dagger = \phi_* D$ be the strict transform on $X^\dagger$ with respect to the birational map $\phi = g^{-1} \circ f \colon X \dashrightarrow X^\dagger$. Then $-D^\dagger$ is $g$-ample.

If furthermore $-K_X$ is ample over $Y$ (resp numerically zero over $Y$), then $\phi$ is called *flip* (resp *flop*).

**Remark 3.5.** Usually flips and flops are defined over varieties with terminal singularities or pairs $(X, B)$ with klt singularities, because MMP works in the category of varieties or pairs with such singularities. (See [44].) However we omit this general treatment in order to simplify the argument.

**Example 3.6.** Let $Y$ be a conifold singularity,

$$Y = \operatorname{Spec} \mathbb{C}[x, y, z, w]/(xy - zw).$$

Let $f \colon X \to Y$ and $g \colon X^\dagger \to Y$ be the blowing ups at the ideals $(x, z) \subset \mathcal{O}_Y$, $(x, w) \subset \mathcal{O}_Y$ respectively. Then $X \xrightarrow{f} Y \xleftarrow{g} X^\dagger$ satisfies Definition 3.4 with $K_X$ numerically zero over $Y$. It is called *Atiyah flop*.

In [7], Bondal and Orlov studied the behavior of the derived categories under certain special flips and flops. Let $f: X \to Y$ be as in Definition 3.4 with $\dim X = n$. Suppose the exceptional locus $E$ is isomorphic to $\mathbb{P}^r$ with normal bundle $\mathcal{O}_E(-1)^{\oplus n-r}$. Then we have the diagram,

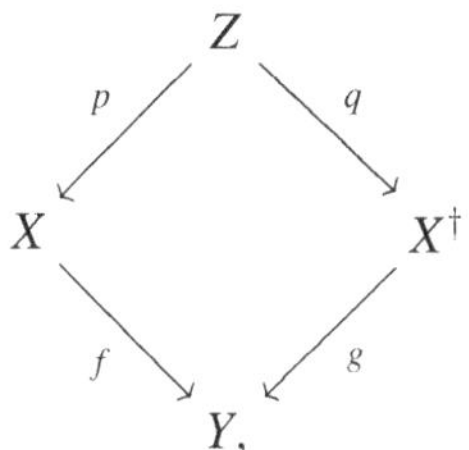

such that $p$ is a blow up along $E$, $q$ is also a blow up along the subscheme $E^{\dagger} \subset X^{\dagger}$ which is isomorphic to $\mathbb{P}^s$ with $r + s = n - 1$. If $r > s$ (resp $r = s$), the above diagram gives a flip (resp flop). Using the blow-up formula Theorem 3.3, Bondal and Orlov [7] showed the following.

**Theorem 3.7. (Bondal-Orlov [7])** *The functor*

$$\mathbf{R}p_* \circ \mathbf{L}q^*: D(X^{\dagger}) \longrightarrow D(X),$$

*is fully faithful if $r \geq s$ and equivalence if $r = s$.*

It is known that any birational minimal models are connected by flops [35]. Hence Theorem 3.7 leads us to the following conjecture.

**Conjecture 3.8.** *Let $X$ and $Y$ be birationally equivalent minimal models. Then there is an equivalence of derived categories, $D(X) \to D(Y)$.*

Theorem 3.3, Theorem 3.7 and Conjecture 3.8 indicate the following. One of the features of MMP in dimension bigger or equal to three is that the minimal model is not necessary unique. However if Conjecture 3.8 is true, then it is unique on the level of derived categories. Hence if we have some kind of nice geometry of "spaces" as in the introduction, and an appropriate theory of the analogue of minimal model theory, then the corresponding minimal model must be unique. Furthermore, because derived categories will be smaller by blow downs and flips by Theorem 3.3, Theorem 3.7, the minimal model theory in the "space" should correspond to "minimizing" derived categories. This is the reason why the geometry needed for string theory is also natural in birational geometry, and hopefully it will enable string theory to interact with birational geometry. However there are several technical problems in realizing this story, and it is only just a philosophy at this time.

### 3.4. Perverse t-structures and flops

In dimension three, Conjecture 3.8 was proved by Bridgeland [11] using the technique of t-structures.

**Definition 3.9.** Let $\mathcal{T}$ be a triangulated category. A full subcategory $\mathcal{T}^{\leq 0} \subset \mathcal{T}$ is called a t-structure if

- We have $\mathcal{T}^{\leq 0}[1] \subset \mathcal{T}^{\leq 0}$.
- For any $E \in \mathcal{T}$, there is a distinguished triangle,

$$\tau_{\leq 0} E \longrightarrow E \longrightarrow \tau_{\geq 1} E,$$

  such that $\tau_{\leq 0} E \in \mathcal{T}^{\leq 0}$ and $\tau_{\geq 1} E \in \mathcal{T}^{\geq 1}$. Here $\mathcal{T}^{\geq 1} := \{F \in \mathcal{T} \mid \mathrm{Hom}(\mathcal{T}^{\leq 0}, F) = 0\}$.

  The subcategory

$$\mathcal{T}^{\leq 0} \cap \mathcal{T}^{\geq 1}[1] \subset \mathcal{T},$$

is called the heart of the t-structure. It is an abelian subcategory [3].

**Example 3.10.** For a variety $X$, let $D^{\leq 0}(X)$ be

$$D^{\leq 0}(X) = \{E \in D(X) \mid \mathcal{H}^i(E) = 0 \text{ for } i > 0\}.$$

Then it is a t-structure on $D(X)$. Its heart is given by $\mathrm{Coh}(X) \subset D(X)$, concentrated in degree zero. The t-structure obtained in this way is called *standard t-structure*.

Let $f \colon X \to Y$ be as in Definition 3.4 with $\dim X = 3$. Bridgeland considered the following abelian categories.

**Proposition 3.11.** ([11]) There are t-structures on $D(X)$ with hearts $^i\mathrm{Per}(X/Y)$ for $i = -1, 0$ such that an object $E \in D(X)$ is contained in $^i\mathrm{Per}(X/Y)$ if and only if

- $\mathcal{H}^j(E) = 0$ unless $j = -1, 0$.
- $R^1 f_* \mathcal{H}^0(E) = 0$ and $f_* \mathcal{H}^{-1}(E) = 0$.
- For any $c \in \mathrm{Coh}(X)$ with $\mathbf{R} f_* c = 0$, we have

$$\mathrm{Hom}(\mathcal{H}^0(E), c) = 0 \quad (i = -1),$$
$$\mathrm{Hom}(c, \mathcal{H}^{-1}(E)) = 0 \quad (i = 0).$$

An object in $^{-1}\mathrm{Per}(X/Y) \subset D(X)$ is called a *perverse sheaf* because its construction is an analogue of the perverse sheaves on the derived category of constructible sheaves [3]. Bridgeland's result [11] is the following.

**Theorem 3.12.** (**Bridgeland [11]**) *For a three dimensional flop, $X \to Y \leftarrow X^\dagger$, there exists an equivalence of derived categories,*

$$\Phi \colon D(X^\dagger) \longrightarrow D(X),$$

*which takes $^{-1}\mathrm{Per}(X^\dagger/Y)$ to $^0\mathrm{Per}(X/Y)$. In particular Conjecture 3.8 is true in dimension three.*

*Proof.* We explain the idea of the proof of Theorem 3.12. By the construction, the structure sheaf $\mathcal{O}_X$ is a perverse sheaf. We call an object $E \in {}^{-1}\mathrm{Per}(X/Y)$ *perverse point sheaf* if there is a surjection $\mathcal{O}_X \twoheadrightarrow E$ in $^{-1}\mathrm{Per}(X/Y)$, and $E$ is numerically equivalent to a skyscraper sheaf $\mathcal{O}_x$ for a closed point $x \in X$. Here we note that any skyscraper sheaf $\mathcal{O}_x$ is a perverse sheaf, but the natural morphism $\mathcal{O}_X \to \mathcal{O}_x$ is *not* necessary surjective in $^{-1}\mathrm{Per}(X/Y)$. Thus the notion of perverse point sheaves is different from the notion of skyscraper sheaves. Then Bridgeland constructed the moduli space of perverse point sheaves P-Hilb$(X/Y)$ using the method of GIT-quotient. It admits a morphism to $Y$,

$$\mathrm{P\text{-}Hilb}(X/Y) \ni E \longmapsto \mathrm{Supp}(\mathbf{R}f_*E) \in Y.$$

The space P-Hilb$(X/Y)$ is called *perverse Hilbert scheme*, and there is a universal perverse point sheaves,

$$\mathcal{E} \in D(X \times \mathrm{P\text{-}Hilb}(X/Y)).$$

One can see that there exists a unique irreducible component $X^\dagger \subset \mathrm{P\text{-}Hilb}(X/Y)$ which dominates $Y$. Then Bridgeland showed that $X^\dagger$ is smooth, $X^\dagger \to Y$ gives a flop, and the restriction $\mathcal{E}^\circ = \mathcal{E}|_{X \times X^\dagger}$ gives the desired derived equivalence,

$$\Phi^{\mathcal{E}^\circ}_{X^\dagger \to X} \colon D(X^\dagger) \longrightarrow D(X),$$

using the *intersection theorem* in the theory of commutative algebra. It requires the condition $\dim X^\dagger \times_Y X^\dagger \le \dim X + 1$, which holds automatically in the case of $\dim X = 3$. $\qquad\square$

It is important to notice that Bridgeland's method guarantees the existence of a flop and its smoothness at the same time. The existence of a flop is a non-trivial problem in birational geometry, so this argument provides a new viewpoint in birational geometry.

The result of Theorem 3.12 was generalized to flops of certain singular threefolds with terminal singularities [19], [37]. However it seems difficult to extend his result to higher dimensional flops. In fact there are examples of four dimensional flops $X \dashrightarrow X^\dagger$ which does not preserve the smoothness of varieties. (See [38].) Thus his argument must fail in such cases, because if it works, then the flop must also be smooth.

### 3.5. *D*-equivalence and *K*-equivalence

Let $X$ and $Y$ be smooth projective varieties. We say $X$ and $Y$ are *K-equivalent* if there exists another smooth projective variety $Z$ and a diagram

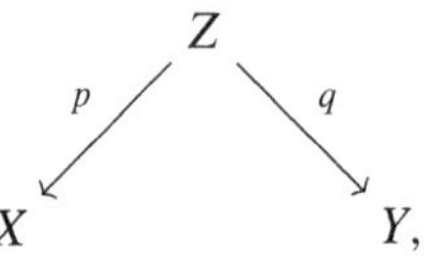

such that $p$, $q$ are birational and $p^*K_X = q^*K_Y$. It is well known that if $X$ and $Y$ are birational minimal models, then they are $K$-equivalent. Also Kawamata [37] called the two varieties $X$, $Y$ *D-equivalent* if $Y$ is a Fourier-Mukai partner of $X$. As a generalization of Conjecture 3.8, Kawamata [37] proposed the following conjecture.

**Conjecture 3.13.** **(Kawamata [37])** *For birationally equivalent smooth projective varieties $X$ and $Y$, they are $K$-equivalent if and only if they are D-equivalent.*

Uehara [66] found an example of birationally equivalent varieties $X$ and $Y$ whose derived categories are equivalent, but they are not $K$-equivalent. Thus Conjecture 3.14 is not true in the sense that $D$-equivalence does not imply $K$-equivalence. However there are some results, including Theorem 3.7, Theorem 3.12, which support the conjecture that $K$-equivalence implies $D$-equivalence. Here we put another evidence for this.

**Theorem 3.14.** **(Kawamata [37], Namikawa [50])** *Suppose that $X$ and $X^\dagger$ are related by a Mukai flop $X \to Y \leftarrow X^\dagger$, i.e. there is a diagram*

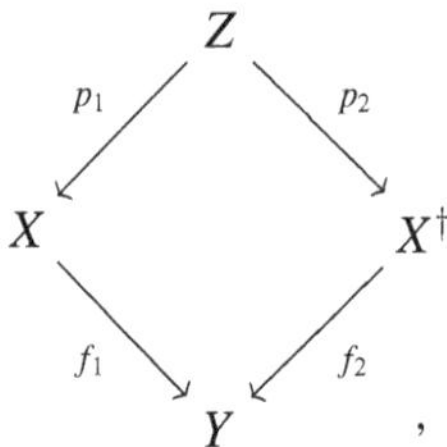

*such that $p_i$ is the blowing up of a subvariety $\mathbb{P}^d \cong E_i \subset X$ with normal bundle equal to $\Omega_{E_i}$ and $f_i$ contracts $E_i$ to a point. Then the functor*

$$\Phi^{\mathcal{O}_{X \times_Y X^\dagger}}_{X^\dagger \to X} : D(X^\dagger) \longrightarrow D(X),$$

*gives an equivalence. However the functor $\mathbf{R}p_{2*} \circ \mathbf{L}p_1^*$ does not give an equivalence.*

# 4.  Derived categories and non-commutative algebras

In this section, we discuss symmetries between usual varieties and several non-commutative algebras.

## 4.1.  Tilting generators

Let $X$ be a smooth variety and $R$ be a noetherian $\mathbb{C}$-algebra. We assume there exists a projective morphism $f\colon X \to \operatorname{Spec} R$.

**Definition 4.1.**  An object $E \in D(X)$ is a *tilting generator* if the following conditions are satisfied.

- $\mathbf{R}\operatorname{Hom}(E, E) = \operatorname{Hom}^0(E, E)$.
- If an object $F \in D(X)$ satisfies $\mathbf{R}\operatorname{Hom}(E, F) = 0$, then $F \cong 0$.

Suppose $E \in D(X)$ is a tilting generator and set $A = \operatorname{End}(E)$. Note that $A$ is a finitely generated $R$-algebra. Furthermore for $F \in D(X)$, the $R$-module $\operatorname{Hom}(E, F)$ is regarded as a right $A$-module in the obvious way. Hence we have the factorization,

$$\mathbf{R}\operatorname{Hom}(E, *)\colon D(X) \xrightarrow{\ \Phi_1\ } D(\operatorname{mod} A) \xrightarrow{\ \Phi_2\ } D(R).$$

Here mod $A$ is the abelian category of right $A$-modules, and the functor $\Phi_2$ is forgetting the $A$-module structure via the natural morphism $R \to A$. By abuse of notation, we also use the notation $\mathbf{R}\operatorname{Hom}(E, *)$ for the functor $\Phi_1$. As an analogue of Morita correspondence, we have the following theorem.

**Theorem 4.2.**  *Let* $E \in D(X)$ *be a tilting generator, and* $A = \operatorname{End}(E)$. *Then the following functor*

$$\mathbf{R}\operatorname{Hom}(E, *)\colon D(X) \longrightarrow D(\operatorname{mod} A),$$

*gives an equivalence. The quasi-inverse is given by* $M \mapsto E \overset{L}{\otimes} M$.

One can consult [25, Theorem 7.6] for the proof of Theorem 4.2.

## 4.2.  Exceptional collections

When $R = \mathbb{C}$, one can construct the tilting generator from an exceptional collection.

**Definition 4.3.** For a $\mathbb{C}$-linear triangulated category $\mathcal{T}$, a sequence of objects $E_1, \cdots, E_n$ is called an *exceptional collection* if

$$\mathrm{Hom}^k(E_i, E_j) = \begin{cases} \mathbb{C} & (k = 0, i = j), \\ 0 & (k \neq 0, i = j \text{ or } i > j). \end{cases}$$

It is called *strong* if furthermore

$$\mathrm{Hom}^k(E_i, E_j) = 0, \quad (k \neq 0, i < j).$$

An exceptional collection $E_1, \cdots, E_n$ is called *full* if the smallest triangulated subcategory in $\mathcal{T}$ which contains $E_1, \cdots, E_n$ is $\mathcal{T}$ itself. In this case we write

$$\mathcal{T} =< E_1, \cdots, E_n > .$$

Suppose a sequence $E_1, \cdots, E_n$ is a full strong exceptional collection on $D(X)$, where $X$ is a smooth projective variety over $\mathbb{C}$. Then the object

$$E = \bigoplus_{i=1}^{n} E_i,$$

is a tilting generator by the definition. Hence there is an equivalence of derived categories,

$$\mathbf{R}\,\mathrm{Hom}(E, *) \colon D(X) \longrightarrow D(\mathrm{mod}\,A),$$

where $A = \mathrm{End}(E)$. This sort of equivalence was first studied by Beilinson [2].

**Theorem 4.4. (Beilinson [2])** We have the following description of the derived category of $\mathbb{P}^n$.

$$D(\mathbb{P}^n) =< \mathcal{O}_{\mathbb{P}^n}(-n), \cdots, \mathcal{O}_{\mathbb{P}^n}(-1), \mathcal{O}_{\mathbb{P}^n} > .$$

In particular we have the equivalence,

$$D(\mathbb{P}^n) \longrightarrow D(\mathrm{mod}\,A),$$

where $A = \mathrm{End} \oplus_{i=0}^{n} \mathcal{O}_{\mathbb{P}^n}(-i)$.

Recently Y. Kawamata [40] generalized Theorem 4.4 to smooth toric varieties, using minimal model program for toric varieties.

**Theorem 4.5. (Kawamata [40])** *Let $X$ be a smooth projective toric variety. Then $D(X)$ admits a full strong exceptional collection consisting of sheaves.*

## 4.3. McKay correspondence

Let $Y$ be a variety with at most Gorenstein singularities. We say a resolution of singularities $f\colon X \to Y$ *crepant* if $\omega_X = f^*\omega_Y$. Suppose that a group $G$ acts on a variety $X$. We denote by $D_G(X)$ the derived category of $G$-equivariant coherent sheaves on $X$. The following is the most sophisticated version of McKay correspondence stated by M. Reid [55].

**Conjecture 4.6.** *Let $G \subset \mathrm{SL}(n, \mathbb{C})$ be a finite subgroup. Suppose that there exists a crepant resolution $X \to \mathbb{C}^n/G$. Then there exists an equivalence of derived categories,*

$$D(X) \cong D_G(\mathbb{C}^n).$$

When $n \leq 3$, there is a positive answer for Conjecture 4.6.

**Theorem 4.7. (Bridgeland, King, Reid [16])** *Conjecture 4.6 is true when $n \leq 3$.*

*Proof.* We explain the idea of [16], which is similar to the proof of Theorem 3.12. First we consider the $G$-clusters on $\mathbb{C}^3$. These are $G$-invariant closed subschemes $Z \subset \mathbb{C}^3$ which satisfy $\mathcal{O}_Z \cong \mathbb{C}[G]$. Then consider the moduli space of $G$-clusters, denoted by $G\text{-}\mathrm{Hilb}(\mathbb{C}^3)$. By associating a $G$-cluster to its support, one has the map

$$G\text{-}\mathrm{Hilb}(\mathbb{C}^3) \longrightarrow \mathbb{C}^3/G.$$

Then it has a unique irreducible component $X \subset G\text{-}\mathrm{Hilb}(\mathbb{C}^3)$ which dominates $\mathbb{C}^3/G$. We have a universal $G$-clusters $\mathcal{Z} \subset X \times \mathbb{C}^3$, and the diagram,

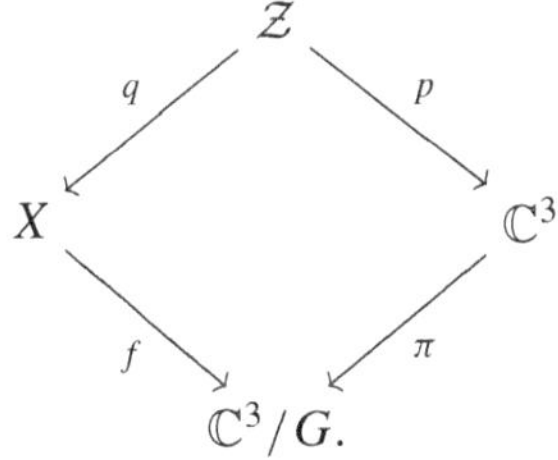

Then $f\colon X \to \mathbb{C}^3/G$ gives the crepant resolution and the transformation

$$\mathbf{R}q_* \circ \mathbf{L}p^*\colon D(X) \longrightarrow D_G(\mathbb{C}^3),$$

gives an equivalence. $\qquad\square$

As in the proof of Theorem 3.12, we have to use the intersection theorem to show that $X$ is smooth and the functor $\mathbf{R}q_* \circ \mathbf{L}p^*$ gives an equivalence. This is why the above proof works only for dimension not greater than three.

On the other hand, Conjecture 4.6 has been shown to hold in some other situations. One of them is due to Bezrukavnikov and Kaledin [6].

**Theorem 4.8. (Bezrukavnikov, Kaledin [6])** *Conjecture 4.6 holds when $G$ preserves a complex symplectic form on $\mathbb{C}^n$.*

Their method is very different from the proof of Theorem 4.7, using characteristic $p$ methods and deformation quantization. Another special case is due to Kawamata [40].

**Theorem 4.9. (Kawamata [40])** *Conjecture 4.6 holds when $G$ is an abelian.*

## 4.4. Non-commutative crepant resolutions

Note that the category of $G$-equivariant sheaves on $\mathbb{C}^3$ is equivalent to the category of the module categories over the smash product algebra $A = \mathbb{C}[x, y, z]\sharp G$. Thus McKay correspondence is interpreted as a symmetry between (usual) commutative schemes and non-commutative algebras.

In McKay correspondence, the story starts by giving a non-commutative algebra $A$ at first, which seems to be a non-commutative analogue of crepant resolutions. Thus conversely, given a crepant resolution $f : X \to Y = \operatorname{Spec} R$ at first, one can ask whether there exists a non-commutative algebra $A$ which is an analogue of a crepant resolution, and it admits a derived equivalence $D(X) \cong D(\operatorname{mod} A)$. M. Van den Bergh [20] introduced the notion of non-commutative crepant resolution in order to formulate this problem.

**Definition 4.10. (M. Van den Bergh [20])** A *non-commutative crepant resolution* of $R$ is an homologically homogeneous $R$-algebra of the form $A = \operatorname{End}(M)$ where $M$ is a reflective $R$-module.

Here the notion of homologically homogeneous is a non-commutative analogue of smoothness. In [20], M. Van den Bergh explains why the above definition is interpreted as the notion of crepant resolution in non-commutative algebras. He also constructed a non-commutative crepant resolution when dimensions of fibers of $f : X \to \operatorname{Spec} R$ are less or equal to one.

**Theorem 4.11. (M. Van den Bergh [21])** *Assume* $\dim f^{-1}(p) \le 1$ *for any* $p \in \operatorname{Spec} R$. *Then there exists a tilting generator* $E \in D(X)$ *such that* $A = \operatorname{End}(E)$ *is a non-commutative crepant resolution. In particular there is a derived* *equivalence,*

$$D(X) \longrightarrow D(\operatorname{mod} A).$$

*Proof.* We give one of the constructions of the tilting generator $E$ in [21]. First take a $f$-ample line bundle $\mathcal{L} \in \operatorname{Pic}(X)$, generated by its global sections. Because $R^1 f_* \mathcal{L}^{-1}$ is a finitely generated $R$-module, there is a surjection $R^{\oplus n} \twoheadrightarrow R^1 f_* \mathcal{L}^{-1}$. Then there is a universal extension,

$$0 \longrightarrow \mathcal{L}^{-1} \longrightarrow \mathcal{E} \longrightarrow \mathcal{O}_X^{\oplus n} \longrightarrow 0.$$

Note that $\mathcal{E}$ is a vector bundle on $X$ and satisfies $\mathbf{R}\operatorname{Hom}(\mathcal{E}, \mathcal{E}) = \operatorname{End}(\mathcal{E})$. Then $E = \mathcal{O}_X \oplus \mathcal{E}$ gives a tilting generator. $\qquad\square$

**Example 4.12.** Let $f \colon X \to Y$ be as in Example 3.6, and $C \cong \mathbb{P}^1 \subset X$ be the exceptional locus. Then $E_0 = \mathcal{O}_X \oplus \mathcal{O}_X(-1)$ and $E_{-1} = \mathcal{O}_X \oplus \mathcal{O}_X(1)$ are tilting generators. We have the associated derived equivalences,

$$\Phi_i \colon D(X) \longrightarrow D(\operatorname{mod} A_i),$$

for $i = -1, 0$, where $A_i = \operatorname{End}(E_i)$. The algebras $A_{-1}$, $A_0$ are path algebras with certain relations corresponding to the quiver given in Figure 4.13. Especially the category $\operatorname{mod} A_i$ has two simple objects, corresponding to the two vertex in the quiver. In [21], it is shown that $\Phi_i$ takes ${}^i \operatorname{Per}(X/Y)$ to $\operatorname{mod} A_i$. The corresponding simple objects in ${}^{-1}\operatorname{Per}(X/Y)$ are given by

$$\{\mathcal{O}_C(-1), \mathcal{O}_C(-2)[1]\} \subset {}^0\operatorname{Per}(X/Y), \quad \{\mathcal{O}_C(-1)[1], \mathcal{O}_C\} \subset {}^{-1}\operatorname{Per}(X/Y),$$

respectively.

**Figure 4.13.**

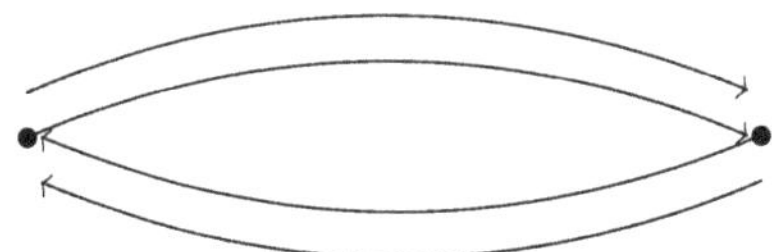

In dimension three, M. Van den Bergh [20] also proved the following.

**Theorem 4.14. (M. Van den Bergh [20])** *Assume that $R$ is three dimensional and has terminal singularities. Then $R$ has a non-commutative crepant resolution if and only if it has a commutative one. Furthermore any two crepant resolutions (commutative as well as non-commutative ones) are derived equivalent.*

# 5. Stability conditions on triangulated categories

In this section, we introduce the notion of stability conditions on triangulated categories. It was introduced by T. Bridgeland [14] to give the mathematical framework for M. Douglas's work on $\Pi$-stability [22], [23]. Its motivation comes from physics, however it is also interesting in mathematics, and especially gives nice pictures for the symmetries we have discussed so far.

## 5.1. Stability conditions on coherent sheaves on curves

First we recall the notion of stability condition on the abelian category of coherent sheaves on a smooth projective curve $C$. For $E \in \mathrm{Coh}(C)$, we denote $\mu(E) \in \mathbb{Q} \cup \{\infty\}$ as follows:

$$\mu(E) = \begin{cases} \infty & \text{if } E \text{ is torsion}, \\ \deg E / \mathrm{rank}\, E & \text{otherwise}. \end{cases}$$

**Definition 5.1.** An object $E \in \mathrm{Coh}(C)$ is called *semistable* if for any subsheaf $F \subset E$, one has $\mu(F) \leq \mu(E)$.

We have the following two properties for semistable sheaves.

- For any $E \in \mathrm{Coh}(C)$, there is a unique filtration

$$0 = E_0 \subset E_1 \subset \cdots \subset E_n = E, \tag{4}$$

such that $A_i = E_i / E_{i-1}$ is semistable and $\mu(A_i) > \mu(A_{i+1})$ for all $i$. The filtration (4) is called *Harder-Narasimhan filtration*.
- There is a moduli space of semistable sheaves with a fixed rank and a degree.

Let us take $E \in D(C)$. Then by taking Harder-Narasimhan filtrations (4) for $\mathcal{H}^i(E) \in \mathrm{Coh}(C)$, we obtain the sequence,

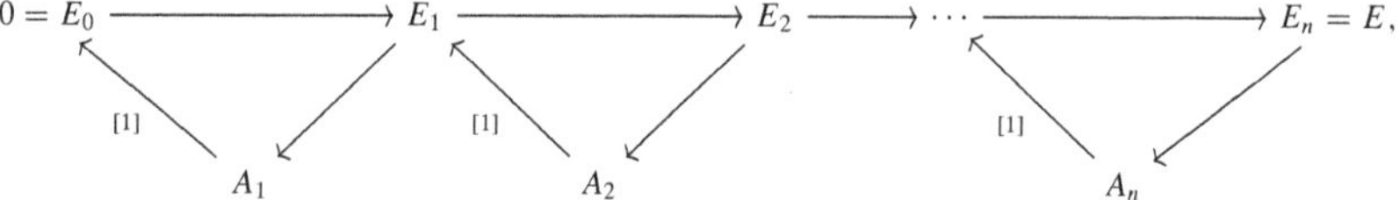

such that each $A_i$ is isomorphic to up to shift a semistable sheaf on $C$. This observation leads to the following Bridgeland's stability conditions.

## 5.2. Bridgeland's stability conditions

Here we introduce the notion of stability conditions on triangulated categories [14].

**Definition 5.2.**  A *stability condition* on a triangulated category $\mathcal{D}$ consists of data $\sigma = (Z, \mathcal{P})$, where $Z \colon K(\mathcal{D}) \to \mathbb{C}$ is a linear map, and $\mathcal{P}(\phi) \subset \mathcal{D}$ is a full additive subcategory for each $\phi \in \mathbb{R}$, which satisfy the following:

- $\mathcal{P}(\phi + 1) = \mathcal{P}(\phi)[1]$.
- If $\phi_1 > \phi_2$ and $A_i \in \mathcal{P}(\phi_i)$, then $\mathrm{Hom}(A_1, A_2) = 0$.
- If $E \in \mathcal{P}(\phi)$ is non-zero, then $Z(E) = m(E)\exp(i\pi\phi)$ for some $m(E) \in \mathbb{R}_{>0}$.
- For a non-zero object $E \in \mathcal{D}$, we have the following collection of triangles:

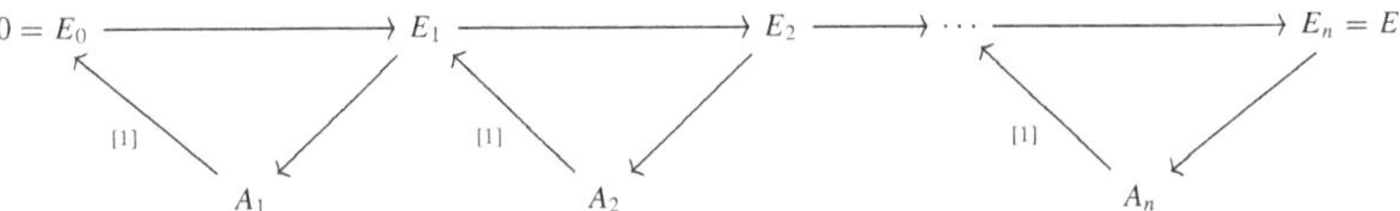

such that $A_j \in \mathcal{P}(\phi_j)$ with $\phi_1 > \phi_2 > \cdots > \phi_n$.

Here $Z$ is called the *central charge*. Each $\mathcal{P}(\phi)$ is an abelian category, the non-zero objects of $\mathcal{P}(\phi)$ are called *semistable of phase $\phi$*, and the non-zero simple objects of $\mathcal{P}(\phi)$ are called *stable*. The objects $A_j$ are called *semistable factors* of $E$ with respect to $\sigma$. We write $\phi_\sigma^+(E) = \phi_1$ and $\phi_\sigma^-(E) = \phi_n$. The *mass* of $E$ is defined to be

$$m_\sigma(E) = \sum_{i=1}^{n} |Z(A_i)|.$$

The following proposition is useful in constructing stability conditions.

**Proposition 5.3.**  **[14, Proposition 4.2]** *Giving a stability condition on $\mathcal{D}$ is equivalent to giving the heart of a bounded t-structure $\mathcal{A} \subset \mathcal{D}$, and a group homomorphism $Z \colon K(\mathcal{D}) \to \mathbb{C}$ called the stability function, such that for a non-zero object $E \in \mathcal{A}$ one has*

$$Z(E) \in \{r\exp(i\pi\phi) \mid r > 0, 0 < \phi \leq 1\},$$

*and the pair $(Z, \mathcal{A})$ satisfies the Harder-Narasimhan property.*

The correspondence of Proposition 5.3 is obtained as follows. For an interval $I \subset \mathbb{R}$, let $\mathcal{P}(I) \subset \mathcal{D}$ be the minimal extension closed subcategory of $\mathcal{D}$ which contains $\mathcal{P}(\phi)$ with $\phi \in I$. By Definition 5.2, one can easily check that the subcategory $\mathcal{P}((0, 1]) \subset \mathcal{D}$ is the heart of a bounded t-structure. Then,

$$(Z, \mathcal{P}) \longmapsto (Z, \mathcal{P}((0, 1])),$$

gives the correspondence of Proposition 5.3.

**Remark 5.4.** In general, the category $\mathcal{P}((a, b))$ for $b - a < 1$ is not an abelian category, and it is only a *quasi-abelian category*. (See [14, Section 4].)

Now we give some examples using Proposition 5.3.

**Example 5.5.** (i) Let $C$ be a smooth projective curve and $\mathcal{D} = D(C)$. Then setting $\mathcal{A} = \mathrm{Coh}(C)$ and $Z \colon K(C) \to \mathbb{C}$ by

$$Z(E) = -\deg(E) + i\,\mathrm{rank}(E),$$

gives a stability condition on $\mathcal{D}$. In the language of Definition 5.2, $\mathcal{P}(\phi) \subset \mathcal{D}$ for $0 < \phi \leq 1$ is given by

$$\mathcal{P}(\phi) = \{\text{semistable sheaves } E \text{ on } C \text{ with } \mu(E) = -1/\tan(\pi\phi).\}$$

(ii) For a field $k$, let $A$ be a finite dimensional $k$-algebra. Let $\mathrm{mod}_f(A)$ be the abelian category of finite dimensional right $A$-modules, and $\mathcal{D}$ the bounded derived category of $\mathrm{mod}_f(A)$. The abelian category $\mathrm{mod}_f(A)$ has finite number of simple objects, say $S_1, \cdots, S_N$. Then setting $\mathcal{A} = \mathrm{mod}_f(A)$ and $Z \colon K(\mathcal{D}) \to \mathbb{C}$ arbitrary so that $\mathrm{Im}\, Z(S_i) > 0$, (it is possible because the classes $[S_1], \cdots, [S_N]$ in $K(\mathcal{D})$ form a basis of $K(\mathcal{D})$), gives a stability condition on $\mathcal{D}$.

## 5.3. The space of stability conditions

For a stability condition $\sigma = (Z, \mathcal{P})$ we say $\sigma$ is *locally finite* if for any $\phi \in \mathbb{R}$ there is $\varepsilon > 0$ such that each quasi-abelian category $\mathcal{P}((\phi - \varepsilon, \phi + \varepsilon))$ is of finite length [14, Definition 5.7]. The set of locally finite stability conditions on $\mathcal{D}$ is denoted by $\mathrm{Stab}(\mathcal{D})$. It has a natural topology induced by the metric,

$$d(\sigma_1, \sigma_2) = \sup_{0 \neq E \in \mathcal{D}} \left\{ |\phi_{\sigma_2}^-(E) - \phi_{\sigma_1}^-(E)|, |\phi_{\sigma_2}^+(E) - \phi_{\sigma_1}^+(E)|, |\log \frac{m_{\sigma_2}(E)}{m_{\sigma_1}(E)}| \right\}$$
$$\in [0, \infty]. \tag{5}$$

Forgetting the information of $\mathcal{P}$, we have the map

$$\mathcal{Z} \colon \mathrm{Stab}(\mathcal{D}) \ni (Z, \mathcal{P}) \longmapsto Z \in \mathrm{Hom}_{\mathbb{Z}}(K(\mathcal{D}), \mathbb{C}).$$

**Theorem 5.6. [14, Theorem 1.2]** *For each connected component $\Sigma \subset \mathrm{Stab}(\mathcal{D})$, there exists a linear subspace $V(\Sigma) \subset \mathrm{Hom}_{\mathbb{Z}}(K(\mathcal{D}), \mathbb{C})$, such that $\mathcal{Z}$ restricts to a local homeomorphism, $\mathcal{Z} \colon \Sigma \to V(\Sigma)$.*

**Remark 5.7.** By the metric given by (5), we can easily see the following: for a fixed $E \in \mathcal{D}$, the functions

$$\phi_*^{\pm}(E), m_*(E) \colon \mathrm{Stab}(X) \longrightarrow \mathbb{R},$$

are continuous. In particular the set of points in $\mathrm{Stab}(\mathcal{D})$ where $E$ is semistable is closed in $\mathrm{Stab}(\mathcal{D})$.

In general $\mathrm{Stab}(\mathcal{D})$ is infinite dimensional, so we usually consider only numerical stability conditions as in [14], [15].

**Definition 5.8.** Let $X$ be a smooth projective variety and $\mathcal{D} = D(X)$. We say a stability condition $\sigma = (Z, \mathcal{P})$ on $\mathcal{D}$ is *numerical* if $Z$ factors through the Chern character map,

$$
\begin{array}{ccc}
K(X) & \xrightarrow{\ Z\ } & \mathbb{C}. \\
{\scriptstyle \mathrm{ch}}\big\downarrow & \nearrow & \\
H^*(X, \mathbb{Q}) & &
\end{array}
$$

We write $\mathrm{Stab}(X)$ for the set of locally finite numerical stability conditions on $\mathcal{D}$.

Let $K_0(X)$ be the kernel of the map $\mathrm{ch}\colon K(X) \to H^*(X, \mathbb{Q})$, and set $\mathcal{N}(X)$ as follows:

$$
\mathcal{N}(X) = K(X)/K_0(X).
$$

The group $\mathcal{N}(X)$ is called *numerical Grothendieck group*. Note that we have the following injection,

$$
\mathrm{ch}\colon \mathcal{N}(X) \longrightarrow \bigoplus \frac{1}{i!} H^i(X, \mathbb{Z}),
$$

hence in particular $\mathcal{N}(X)$ is a finitely generated $\mathbb{Z}$-module. We have the following map,

$$
\mathcal{Z}\colon \mathrm{Stab}(X) \longrightarrow \mathrm{Hom}(\mathcal{N}(X), \mathbb{C}).
$$

Because the dimension of $\mathrm{Hom}(\mathcal{N}(X), \mathbb{C})$ is finite, Theorem 5.6 implies that any connected component of $\mathrm{Stab}(X)$ is a finite dimensional complex manifold.

## 5.4. Group actions

Let $\mathrm{GL}^+(2, \mathbb{R}) \subset \mathrm{GL}(2, \mathbb{R})$ be the subgroup of linear maps $T\colon \mathbb{R}^2 \to \mathbb{R}^2$ which preserves orientation of $\mathbb{R}^2$. Let $\widetilde{\mathrm{GL}}^+(2, \mathbb{R}) \to \mathrm{GL}^+(2, \mathbb{R})$ be the universal cover. Note that an element of $\widetilde{\mathrm{GL}}^+(2, \mathbb{R})$ is written as a pair $(T, f)$, where $f\colon \mathbb{R} \to \mathbb{R}$ is an increasing map with $f(\phi + 1) = f(\phi) + 1$ and $T \in \mathrm{GL}^+(2, \mathbb{R})$ such that induced maps on

$$
S^1 = \mathbb{R}/2\mathbb{Z} = (\mathbb{R}^2 \setminus \{0\})/\mathbb{R}_{>0},
$$

are the same maps.

**Lemma 5.9.** *For a triangulated category $\mathcal{D}$, the space* $\mathrm{Stab}(\mathcal{D})$ *carries the right action of the group* $\widetilde{\mathrm{GL}}^+(2, \mathbb{R})$*, and the left action of the group* $\mathrm{Auteq}(\mathcal{D})$.

*Proof.* Given a stability condition $\sigma = (Z, \mathcal{P})$ and $(T, f) \in \widetilde{\mathrm{GL}}^+(2, \mathbb{R})$, define the stability condition $\sigma' = (Z', \mathcal{P}')$ by setting

$$Z' = T^{-1} \circ Z, \quad \mathcal{P}'(\phi) = \mathcal{P}(f(\phi)).$$

Then it gives the right action of $\widetilde{\mathrm{GL}}^+(2, \mathbb{R})$ on $\mathrm{Stab}(\mathcal{D})$.

Next for $\sigma = (Z, \mathcal{P}) \in \mathrm{Stab}(\mathcal{D})$ and $\Phi \in \mathrm{Auteq}(\mathcal{D})$, its left action is defined by $\Phi(\sigma) = (Z', \mathcal{P}')$ where

$$Z' = Z \circ \Phi^{-1}, \quad \mathcal{P}'(\phi) = \Phi(\mathcal{P}(\phi)).$$

$\square$

Note that there is a subgroup $\mathbb{C} \subset \widetilde{\mathrm{GL}}^+(2, \mathbb{R})$ and the restriction of the action of $\widetilde{\mathrm{GL}}^+(2, \mathbb{R})$ to $\mathbb{C}$ is free. Explicitly for $\lambda \in \mathbb{C}$, its action on $\sigma = (Z, \mathcal{P})$ is described by $\sigma' = (Z', \mathcal{P}')$ where

$$Z' = e^{-i\pi\lambda} Z, \quad \mathcal{P}'(\phi) = \mathcal{P}(\phi + \mathrm{Re}(\lambda)).$$

## 5.5. Background from string theory

The idea of the definition of the stability conditions comes from Douglas's $\Pi$-stability for BPS-branes [22], [23]. Here we give the brief explanation for the motivation of Definition 5.2. As we see later, the background from string theory gives a nice explanation of the descriptions of the space of stability conditions. However the author is not an expert in this area, and also there are excellent explanations of the background from string theory for mathematicians by Bridgeland himself [14, (1,4)], [9]. Thus for the detail and accuracies, the author recommends consulting the articles [14], [9], [22], [23].

First let us consider a triple,

$$(X, \beta + i\omega, I), \tag{6}$$

where $X$ is a simply connected Calabi-Yau 3-fold, $I$ is a complex structure on $X$, and $\beta + i\omega \in H^2(X, \mathbb{C})$ is a complexified Kähler class, i.e. $\omega$ is a Kähler class. A triple (6) is called *physicist's Calabi-Yau 3-fold*. Physicists believe that for a triple (6), there is an associated non-linear sigma model, which defines $\mathcal{N} = 2$ super conformal field theory, simply referred to as SCFT. Here we note that the sigma models are not at all defined mathematically. At least there is an issue of convergence of the integrals in defining them, however we ignore this problem. Let $\mathcal{M}$ be the the moduli space of SCFT. Then for a fixed $X$, the subset $\mathcal{U}_X \subset \mathcal{M}$ which comes from sigma model constructions from the triples (6) is an open subset. We call $\mathcal{U}_X$ the *neighborhood of the large volume limit*, and let $\mathcal{M}(X)$ be the connected component of $\mathcal{M}$ which contains $\mathcal{U}_X$. Here we note that there might be some points in $\mathcal{M}(X)$ which correspond to sigma models for topologically distinct Calabi-Yau 3-folds, or do not correspond to sigma models at all.

There are two foliations on $\mathcal{M}$, one of them is when restricted to $\mathcal{U}_X$, fixing the complex structure $I$ and deforming $\beta + i\omega$. Another foliation is fixing $\beta + i\omega$ and deforming $I$ on $\mathcal{U}_X$. Let us consider the leaves of the corresponding foliations,

$$\mathcal{M}_K(X) \subset \mathcal{M}(X), \quad \mathcal{M}_C(X) \subset \mathcal{M}(X).$$

The space $\mathcal{M}_C(X)$ is nothing but the moduli space of the complex structures, which is defined mathematically and has been studied in algebraic geometry up to now. However the space $\mathcal{M}_K(X)$, called *stringy Kähler moduli space*, is not defined mathematically rigorous way, as well as sigma model constructions.

Now let us discuss on mirror symmetry. There is an involution $m : \mathcal{M} \to \mathcal{M}$ called the *mirror map*, which exchanges the above two foliations on $\mathcal{M}$. Then a triple

$$(\hat{X}, \hat{\beta} + i\hat{\omega}, \hat{I})$$

is called a *mirror* of the triple (6) if the associated sigma model is mapped to the sigma model of (6) by the mirror map $m$. Especially we have the isomorphisms,

$$\mathcal{M}_K(X) \cong \mathcal{M}_C(\hat{X}), \quad \mathcal{M}_K(\hat{X}) \cong \mathcal{M}_C(X).$$

Hence if we assume the mirror symmetry, it might be possible to define the space $\mathcal{M}_K(X)$ to be $\mathcal{M}_C(\hat{X})$ for a mirror manifold $\hat{X}$. However it is unsatisfactory, since it assumes the mirror symmetry, and it is not intrinsic. One of the motivation of introducing the space of stability conditions is to define $\mathcal{M}_K(X)$ without using mirror symmetry, purely from the categorical data of $D(X)$. Actually it is expected that the space $\mathrm{Stab}(X)$ is related to the space $\mathcal{M}_K(X)$.

Next let us consider the D-branes. For a given SCFT, we have the two associated topological conformal field theories (TCFT), called A-model and B-model. If the SCFT is given by the triple (6), then the category of D-branes of A and B-models are supposed to be

$$D^b \, \mathrm{Fuk}(X, \beta + i\omega), \quad D^b \, \mathrm{Coh}(X, I),$$

respectively. Here $D^b \mathrm{Coh}(X, I)$ is the bounded derived category of coherent sheaves of the complex manifold $(X, I)$, and $D^b \mathrm{Fuk}(X, \beta + i\omega)$ is the derived Fukaya category, whose objects are (roughly speaking) local systems on Lagrangian submanifolds on $X$. Note that $D^b \mathrm{Coh}(X, I)$ does not depend on $\beta + i\omega$, and $D^b \mathrm{Fuk}(X, \beta + i\omega)$ does not depend on $I$. The mirror map $m$ exchanges A and B models, hence there should exist equivalences,

$$D^b \mathrm{Fuk}(X, \beta + i\omega) \cong D^b \mathrm{Coh}(\hat{X}, \hat{I}), \quad D^b \mathrm{Coh}(X, I) \cong D^b \mathrm{Fuk}(\hat{X}, \hat{\beta} + i\hat{\omega}),$$

for a mirror $(\hat{X}, \hat{\beta} + i\hat{\omega}, \hat{I})$. This is the *Homological mirror symmetry* proposed by Kontsevich [45].

The fact that $D^b \mathrm{Coh}(X, I)$ does not depend on $\beta + i\omega$ imply that the associated categories of B-models are unchanged along the subspace $\mathcal{M}_K(X) \subset \mathcal{M}$. Below we fix the complex structure $I$ and simply write $X$ for the complex manifold $(X, I)$. The important point is that the category of particular B-branes, called *BPS-branes*, are changed along $\mathcal{M}_K(X)$ in the fixed category of B-branes $D(X)$. Let us take a point $p \in \mathcal{M}_K(X)$ and the associated category of BPS branes $\mathcal{B}(p)$. Here we remark that the point $p$ does not have the information how the category $\mathcal{B}(p)$ is embedded into $D(X)$. Now let us take a closed path,

$$\gamma : [0, 1] \longrightarrow \mathcal{M}_K(X),$$

with $\gamma(0) = \gamma(1) = p$, and choose an embedding $\mathcal{B}(p) \subset D(X)$. Then for each $t \in [0, 1]$, we have the associated subcategory,

$$\mathcal{B}(\gamma(t)) \subset D(X).$$

For $t = 1$, we obtain the embedding of $\mathcal{B}(\gamma(1)) \subset D(X)$, and $\mathcal{B}(\gamma(1)) = \mathcal{B}(p)$. So we obtain the two embeddings of $\mathcal{B}(p)$ into $D(X)$ for $t = 0, 1$, which may be different. In fact its difference should be induced by the action of an autoequivalence on $D(X)$, hence we have the map,

$$\pi_1 \mathcal{M}_K(X) \longrightarrow \mathrm{Auteq}\, D(X).$$

Now recall that the space $\mathrm{Stab}(X)$ carries a left action of $\mathrm{Auteq}\, D(X)$. Also it is expected that for $\sigma = (Z, \mathcal{P}) \in \mathrm{Stab}(X)$, the associated category

$$\bigcup_{\phi \in \mathbb{R}} \mathcal{P}(\phi),$$

defines the category of BPS branes at some point of $\mathcal{M}_K(X)$. Combined these observations, it is reasonable to guess that the space $\mathrm{Auteq}\, D(X) \backslash \mathrm{Stab}(X)$ is related to the space $\mathcal{M}_K(X)$. More precisely Bridgeland conjectures in [9] that the double quotient space

$$\mathrm{Auteq}\, D(X) \backslash \mathrm{Stab}(X)/\mathbb{C}, \tag{7}$$

contains the space $\mathcal{M}_K(X)$. Hence in the viewpoint of string theory, it is interesting to describe the space (7) and compare it with $\mathcal{M}_C(\hat{X})$ of a mirror manifold $\hat{X}$.

Since Bridgeland wrote the paper [14], a number of papers on stability conditions have appeared and some examples, properties have been studied. Now at least there are following articles, [15], [10], [13], [59], [51], [47], [28], [5], [63], [65]. Below we introduce some of them.

## 5.6. Stability conditions on elliptic curves

Let $X$ be an elliptic curve. Then it is proved in [14] that the space $\mathrm{Stab}(X)$ is connected and the action of $\widetilde{\mathrm{GL}}^+(2, \mathbb{R})$ is free and transitive. Hence we have

$$\mathrm{Stab}(X) \cong \widetilde{\mathrm{GL}}^+(2, \mathbb{R}),$$

as a complex manifold. Here the complex structure on $\widetilde{\mathrm{GL}}^+(2, \mathbb{R})$ is induced by the local homeomorphism,

$$\widetilde{\mathrm{GL}}^+(2, \mathbb{R}) \to \mathrm{GL}^+(2, \mathbb{R}) \subset \mathbb{C}^2.$$

The group of autoequivalence is completely known by Theorem 2.11. Hence one can describe the space (7), and the result is

$$\mathrm{Auteq}\, D(X) \backslash \mathrm{Stab}(X)/\mathbb{C} \cong \mathcal{H}/\mathrm{SL}(2, \mathbb{Z}). \tag{8}$$

Note that the RHS of (8) is nothing but the moduli space of elliptic curves. Hence in this case, we have obtained the complete picture.

## 5.7. Stability conditions on K3 surfaces

Suppose $X$ is a K3 surface. In this case, one of the connected components $\mathrm{Stab}^*(X) \subset \mathrm{Stab}(X)$ is described by [15].

One of the difficult points in studying $\mathrm{Stab}(X)$ in this case is to show the existence of stability conditions. Before discussing this, let us recall the notion of $\mu_\omega$-semistable sheaves. For a torsion free sheaf $E \in \mathrm{Coh}(X)$ and an ample divisor $\omega$ on $X$, set $\mu_\omega(E)$ as

$$\mu_\omega(E) = \frac{c_1(E) \cdot \omega}{\mathrm{rank}(E)}.$$

**Definition 5.10.** A torsion free sheaf $E \in \mathrm{Coh}(X)$ is called $\mu_\omega$-*semistable* if for any non-zero subsheaf $F \subset E$, one has

$$\mu_\omega(F) \le \mu_\omega(E).$$

**Remark 5.11.** The notion of $\mu_\omega$-semistability does not give the stability conditions on $D(X)$ directly. One may try to construct a stability condition as the pair $(Z, \mathrm{Coh}(X))$, where $Z \colon K(X) \to \mathbb{C}$ is

$$Z(E) = -c_1(E) \cdot \omega + \mathrm{rank}(E)i.$$

However it is *not* a stability condition because $Z(\mathcal{O}_x) = 0$ for closed points $x \in X$.

Bridgeland [15] constructed stability conditions using the method of *tilting*. Let $\beta$, $\omega$ be $\mathbb{Q}$-divisors on $X$ with $\omega$ ample. For a torsion free sheaf $E \in \mathrm{Coh}(X)$, one has the Harder-Narasimhan filtration

$$0 = E_0 \subset E_1 \subset \cdots \subset E_{n-1} \subset E_n = E,$$

such that $F_i = E_i / E_{i-1}$ is $\mu_\omega$-semistable and $\mu_\omega(F_i) > \mu_\omega(F_{i+1})$. Then define $\mathcal{T}_{(\beta,\omega)} \subset \mathrm{Coh}(X)$ to be the subcategory consisting of sheaves whose torsion free parts have $\mu_\omega$-semistable Harder-Narasimhan factors of slope $\mu_\omega(F_i) > \beta \cdot \omega$. Also define $\mathcal{F}_{(\beta,\omega)} \subset \mathrm{Coh}(X)$ to be the subcategory consisting of torsion free sheaves whose $\mu_\omega$-semistable factors have slope $\mu_\omega(F_i) \le \beta \cdot \omega$.

**Definition 5.12.** We define $\mathcal{A}_{(\beta,\omega)}$ to be

$$\mathcal{A}_{(\beta,\omega)} = \{E \in D(X) \mid \mathcal{H}^{-1}(E) \in \mathcal{F}_{(\beta,\omega)}, \mathcal{H}^0(E) \in \mathcal{T}_{(\beta,\omega)}\}.$$

The category $\mathcal{A}_{(\beta,\omega)}$ is called *tilting* with respect to the *torsion pair* $(\mathcal{T}_{(\beta,\omega)}, \mathcal{F}_{(\beta,\omega)})$. The next step is to construct the central charge $Z_{(\beta,\omega)} \colon \mathcal{N}(X) \to \mathbb{C}$. Let $\mathrm{NS}^*(X)$ be the Mukai lattice,

$$\mathrm{NS}^*(X) = \mathbb{Z} \oplus \mathrm{NS}(X) \oplus \mathbb{Z}.$$

For $v_i = (r_i, l_i, s_i) \in \mathrm{NS}^*(X)$ with $i = 1, 2$, its bilinear pairing is given by

$$(v_1, v_2) = l_1 \cdot l_2 - r_1 s_2 - r_2 s_1. \tag{9}$$

For an object $E \in D(X)$ its Mukai vector is defined by,

$$v(E) = \mathrm{ch}(E)\sqrt{\mathrm{td}_X}$$
$$= (r(E), c_1(E), \mathrm{ch}_2(E) + r(E)).$$

Sending an object to its Mukai vector gives an isomorphism,

$$v \colon \mathcal{N}(X) \xrightarrow{\cong} \mathrm{NS}^*(X). \tag{10}$$

Under the identification (10), the bilinear pairing $-\chi(E_1, E_2)$ on the left hand side goes to the pairing (9). Let us define $Z_{(\beta,\omega)} \colon \mathcal{N}(X) \to \mathbb{C}$ by the formula.

$$Z_{(\beta,\omega)}(E) = (\exp(\beta + i\omega), v(E)). \tag{11}$$

Suppose $v(E) = (r, l, s)$ with $r \neq 0$. Then (11) is written as

$$Z_{(\beta,\omega)}(E) = \frac{1}{2r}\left((l^2 - 2rs) + r^2\omega^2 - (l - r\beta)^2\right) + i(l - r\beta) \cdot \omega. \tag{12}$$

If $r = 0$, (11) is written as $Z(E) = (-s + l \cdot \beta) + i(l \cdot \omega)$. We define $\sigma_{(\beta,\omega)}$ to be the pair $(Z_{(\beta,\omega)}, \mathcal{A}_{(\beta,\omega)})$.

**Proposition 5.13. [15, Lemma 6.2, Proposition 7.1]** *The subcategory* $\mathcal{A}_{(\beta,\omega)} \subset D(X)$ *is a heart of a bounded t-structure, and the pair* $\sigma_{(\beta,\omega)}$ *gives a stability condition on* $D(X)$ *if and only if for any spherical sheaf $E$ on $X$, one has* $Z_{(\beta,\omega)}(E) \notin \mathbb{R}_{\leq 0}$. *This holds whenever* $\omega^2 > 2$.

Let $\mathrm{Stab}^*(X)$ be the connected component of $\mathrm{Stab}(X)$ which contains $\sigma_{(\beta,\omega)}$. Note that the pairing (9) and the isomorphism (10) induces an isomorphism,

$$\mathrm{Hom}(\mathcal{N}(X), \mathbb{C}) \cong \mathrm{NS}^*(X).$$

Hence we have the map,

$$\mathcal{Z} \colon \mathrm{Stab}^*(X) \longrightarrow \mathrm{NS}^*(X).$$

In order to understand the image of $\mathcal{Z}$, first define the subset

$$\mathcal{P}(X) \subset \mathrm{NS}^*(X)_{\mathbb{C}},$$

to be the set of $v = v_1 + i v_2 \in \mathrm{NS}^*(X)_{\mathbb{C}}$ with $v_i \in \mathrm{NS}^*(X)$ such that $v_1, v_2$ span a positive definite two plane in $\mathrm{NS}^*(X)$. The space $\mathcal{P}(X)$ consists of two connected components, and let $\mathcal{P}^+(X) \subset \mathcal{P}(X)$ be the component which contains $(1, i\omega, -\frac{1}{2}\omega^2)$. We define $\mathcal{P}_0^+(X)$ to be

$$\mathcal{P}_0^+(X) = \mathcal{P}_0(X) \setminus \bigcup_{\delta \in \Delta(X)} \delta^\perp.$$

Here $\Delta(X) = \{v \in \mathrm{NS}^*(X) \mid (v, v) = -2\}$.

**Theorem 5.14. (Bridgeland [15, Theorem 1.1])** *The component* $\mathrm{Stab}^*(X)$ *is mapped by* $\mathcal{Z}$ *onto the open subset* $\mathcal{P}_0^+(X)$, *and the induced map*

$$\mathcal{Z} \colon \mathrm{Stab}^*(X) \longrightarrow \mathcal{P}_0^+(X),$$

*is a regular covering map. Its Galois group $G$ fits into the exact sequence,*

$$0 \longrightarrow G \longrightarrow \mathrm{Auteq}^* D(X) \longrightarrow \mathrm{Aut}\, H^*(X, \mathbb{Z}).$$

*Here* $\mathrm{Auteq}^* D(X)$ *is the group of autoequivalences of* $D(X)$ *which preserve the component* $\mathrm{Stab}^*(X)$.

Based on Theorem 5.14, Bridgeland proposes the following conjecture, which relates the topology of $\mathrm{Stab}(X)$ and the group of autoequivalences of $D(X)$.

**Conjecture 5.15.** **([15, Conjecture 1.2])** *The action of* $\mathrm{Auteq}\, D(X)$ *on* $\mathrm{Stab}(X)$ *preserves the component* $\mathrm{Stab}^*(X)$. *Furthermore* $\mathrm{Stab}^*(X)$ *is simply connected. As a consequence, we have the following short exact sequence,*

$$1 \longrightarrow \pi_1 \mathcal{P}_0^+(X) \longrightarrow \mathrm{Auteq}\, D(X) \longrightarrow \mathrm{Aut}^+ H^*(X, \mathbb{Z}) \longrightarrow 1.$$

*Here* $\mathrm{Aut}^+ H^*(X, \mathbb{Z})$ *is the index two subgroup of* $\mathrm{Aut}\, H^*(X, \mathbb{Z})$ *consisting elements whose actions on* $\mathcal{P}(X)$ *preserve the component* $\mathcal{P}^+(X)$.

### 5.8. Stability conditions and birational geometry

Before giving another example, we point out the relationship between the descriptions of stability conditions and birational geometry. For instance suppose $X$ is a minimal 3-fold. Recall that the birational minimal models are not unique, but connected by a finite number of flops. The philosophy of Y. Kawamata [36] is that one can capture the set of birational minimal models via a chamber structures on the *movable cone* [36, Definition 1.1]. According to [36, Theorem 2.3], chambers on the movable cone are given by the ample cones of birational minimal models.

On the other hand, let us consider the space $\mathrm{Stab}(X)$. As explained in (5.5), $\mathrm{Stab}(X)$ is related to the stringy Kähler moduli space $\mathcal{M}_K(X)$. By restricting the neighborhood of the large volume limit to $\mathcal{M}_K(X)$, we obtain the open subset $\mathcal{U}_K(X) \subset \mathcal{M}_K(X)$. However there might be another topologically distinct manifold $W$ such that the associated neighborhood of the large volume limit determines an open subset $\mathcal{U}_K(W) \subset \mathcal{M}_K(X)$. In this case, because the associated TCFT of B-models are invariant along $\mathcal{M}_K(X)$, we have an equivalence of derived categories,

$$D(W) \longrightarrow D(X),$$

i.e. $W$ is a Fourier-Mukai partner of $X$. Recall that if $W$ is birationally equivalent to $X$, then $W$ is a Fourier-Mukai partner of $X$. (See Theorem 3.12.) Thus in describing $\mathcal{M}_K(X)$, one has to take account the neighborhoods of the large volume limits corresponding to several Fourier-Mukai partners such as flops. In fact suppose $X^\dagger \to Y \leftarrow X$ is a flop at a single rational curve $C \subset X$. In this case, P.Aspinwall [1, Figure 2] describes the localized picture of $\mathcal{M}_K(X)$, assuming the volumes of all the curves in $X$ except $C$ are big enough. The resulting picture is

$$\mathcal{M}_K(X) = \mathbb{P}^1 \setminus \{0, 1, \infty\} \tag{13}$$

$$= \overline{\mathcal{U}}_K(X) \cup \overline{\mathcal{U}}_K(X^\dagger), \tag{14}$$

and $\mathcal{U}_K(X) \cap \mathcal{U}_K(X^\dagger) = \emptyset$. Here string theory has singularities at one of the deleted points $\{0, 1, \infty\}$, and the other two points are large volume limits corresponding to $X$ and $X^\dagger$.

Let us translate the above argument to the space $\mathrm{Stab}(X)$. First we can guess that there exists some open subset $U_X \subset \mathrm{Stab}(X)$, which corresponds to the neighborhood of the large volume limit. For any point $\sigma \in U_X$, the corresponding t-structures given in Proposition 5.3 should be related to the standard t-structure, for example obtained by tilting. Next let $W$ and $X$ be three dimensional birational minimal models, and $\Phi \colon D(W) \to D(X)$ an equivalence of derived categories. Then $\Phi$ induces an homeomorphism,

$$\Phi_* \colon \mathrm{Stab}(W) \longrightarrow \mathrm{Stab}(X). \tag{15}$$

Transferring $U_W$ by $\Phi_*$, we obtain a region $\Phi_* U_W \subset \mathrm{Stab}(X)$. Conjecturally there should exist a chamber structure of the following type,

$$\bigcup_{(W, \Phi)} \Phi_* \overline{U}_W \subset \mathrm{Stab}(X), \tag{16}$$

such that two chambers are either equal or disjoint. This picture is quite similar to the movable cone [36, Theorem 2.3]. Thus by describing the space $\mathrm{Stab}(X)$, we can observe the relationship between SCFT and birational geometry.

## 5.9. Stability conditions and 3-fold flops

In order to realize the picture (16), we give the description of the space of stability conditions on a certain local Calabi-Yau 3-fold. Let $f \colon X \to Y$ be a crepant small resolution of the conifold singularity,

$$f \colon X \longrightarrow Y = \mathrm{Spec}\, \mathbb{C}[x, y, z, w]/(xy - zw),$$

and $g \colon X^\dagger \to Y$ be its flop as in Example 3.6. Let $C \cong \mathbb{P}^1$ be the exceptional locus of $f$. Set $D(X/Y)$ as follows,

$$D(X/Y) = \{E \in D(X) \mid \mathrm{Supp}(E) \subset C\}.$$

Let us describe the space of stability conditions on $D(X/Y)$, denoted by $\mathrm{Stab}(X/Y)$. As we will see, the description of $\mathrm{Stab}(X/Y)$ realizes the pictures (13), (14), (16), and also gives a nice picture for non-commutative crepant resolutions discussed in Section 4. The situation we treat here is the simplest case, and one can consult [63], [12], [65] for more general cases.

Let $\mathrm{Coh}(X/Y)$, $K(X/Y)$ be

$$\mathrm{Coh}(X/Y) = D(X/Y) \cap \mathrm{Coh}(X), \quad K(X/Y) = K(D(X/Y)).$$

Also denote by $N^1(X/Y)$ the $\mathbb{R}$-vector space of the numerical classes of $\mathbb{R}$-divisors, and $A(X/Y)_{\mathbb{C}} \subset N^1(X/Y)_{\mathbb{C}}$ the *complexified ample cone*,

$$A(X/Y)_{\mathbb{C}} = \{\beta + i\omega \in N^1(X/Y)_{\mathbb{C}} \mid \omega \text{ is ample}\}.$$

In our case, we have

$$K(X/Y) = \mathbb{Z}[\mathcal{O}_x] \oplus \mathbb{Z}[\mathcal{O}_C(-1)], \quad N^1(X/Y)_{\mathbb{C}} = \mathbb{C}[c_1(\mathcal{O}_X(1))] \cong \mathbb{C},$$
$$A(X/Y)_{\mathbb{C}} \cong \{\beta + i\omega \in \mathbb{C} \mid \omega > 0\}.$$

For $\beta + i\omega \in N^1(X/Y)_{\mathbb{C}}$, set $Z_{(\beta,\omega)} \colon K(X/Y) \to \mathbb{C}$ as follows,

$$Z_{(\beta,\omega)}(E) = -\operatorname{ch}_3(E) + (\beta + i\omega) \cdot \operatorname{ch}_2(E).$$

**Lemma 5.16.** (**[63, Lemma 4.1]**) *For $\beta + i\omega \in A(X/Y)_{\mathbb{C}}$, the pair*

$$\sigma_{(\beta,\omega)} = (Z_{(\beta,\omega)}, \mathrm{Coh}(X/Y))$$

*determines a point of* $\mathrm{Stab}(X/Y)$. *The set of points*

$$U_X = \{\sigma_{(\beta,\omega)} \in \mathrm{Stab}(X/Y) \mid \beta + i\omega \in A^1(X/Y)_{\mathbb{C}}\},$$

*forms an open subset in the subspace* $\mathrm{Stab}_{\mathrm{n}}(X/Y)$,

$$\mathrm{Stab}_{\mathrm{n}}(X/Y) = \{(Z, \mathcal{P}) \in \mathrm{Stab}(X/Y) \mid Z([\mathcal{O}_x]) = -1\},$$

*which we call the set of normalized stability conditions.*

It is easy to see that for $Z \colon K(X/Y) \to \mathbb{C}$, we have $Z([\mathcal{O}_x]) = -1$ if and only if $Z$ is written as $Z_{(\beta,\omega)}$ for some $\beta + i\omega \in N^1(X/Y)_{\mathbb{C}}$. Hence we have the following cartesian square,

$$
\begin{array}{ccc}
\mathrm{Stab}_{\mathrm{n}}(X/Y) & \longrightarrow & \mathrm{Stab}(X/Y) \\
\Big\downarrow{\scriptstyle \mathcal{Z}_{\mathrm{n}}} & & \Big\downarrow{\scriptstyle \mathcal{Z}} \\
N^1(X/Y)_{\mathbb{C}} & \longrightarrow & K(X/Y)^*_{\mathbb{C}}.
\end{array}
$$

The map $\mathcal{Z}_{\mathrm{n}}$ restricts to the homeomorphism between $U_X$ and $A(X/Y)_{\mathbb{C}}$.

Next let us see the behavior of the boundary of $U_X$. Note that for each $k \in \mathbb{Z}$, the object $\mathcal{O}_C(k-1)$ is stable in any $\sigma' \in U_X$. Therefore by Remark 5.7, it must be semistable in $\sigma$ for $\sigma \in \partial U_X$. Hence if we write $\sigma = (Z_{(\beta,0)}, \mathcal{P})$ with $\beta \in N^1(X/Y)$, we must have

$$Z_{(\beta,0)}(\mathcal{O}_C(k-1)) = -k + \beta \neq 0.$$

This implies

$$\mathcal{Z}_n(\partial U_X) \subset \mathbb{R} \setminus \mathbb{Z}.$$

Conversely take $\beta \in (k-1, k)$. Then using the descriptions of the simple objects given in Example 4.12, we can see that the pair

$$(Z_{(\beta,0)}, \left( {}^0\mathrm{Per}(X/Y) \otimes \mathcal{O}_X(k) \right) \cap D(X/Y)),$$

gives a stability condition and contained in $\partial U_X$. Hence we have

$$\mathcal{Z}_n(\partial U_X) = \mathbb{R} \setminus \mathbb{Z}.$$

We set $\partial U_X(k)$ to be

$$\partial U_X(k) = \{\sigma \in \partial U_X \mid \mathcal{Z}_n(\sigma) \in (k-1, k)\}.$$

The next step is to consider the flop $X^\dagger \to Y$. By Theorem 3.12, there are equivalences,

$$\Phi \colon D(X^\dagger) \longrightarrow D(X),$$
$$\Phi^\dagger \colon D(X) \longrightarrow D(X^\dagger),$$

such that $\Phi$ takes ${}^{-1}\mathrm{Per}(X^\dagger/Y)$ to ${}^0\mathrm{Per}(X/Y)$, and $\Phi'$ takes ${}^{-1}\mathrm{Per}(X/Y)$ to ${}^0\mathrm{Per}(X^\dagger/Y)$. The equivalence $\Phi$ induces the following commutative diagram,

$$
\begin{array}{ccc}
\mathrm{Stab}_n(X^\dagger/Y) & \xrightarrow{\ \Phi_*\ } & \mathrm{Stab}_n(X/Y) \\
\downarrow & & \downarrow \\
N^1(X^\dagger/Y)_{\mathbb{C}} & \xrightarrow{\ \phi_*\ } & N^1(X/Y)_{\mathbb{C}}.
\end{array}
$$

Here $\phi \colon X^\dagger \dashrightarrow X$ is the birational map. By Theorem 3.12 and Example 4.12, the equivalence $\Phi$ restricts to the equivalence,

$$\left( {}^0\mathrm{Per}(X^\dagger/Y) \otimes \mathcal{O}_{X^\dagger}(1) \right) \cap D(X/Y) \longrightarrow {}^0\mathrm{Per}(X/Y) \cap D(X/Y).$$

This means $\Phi_*$ induces a homeomorphism,

$$\Phi_* \colon \partial U_{X^\dagger}(1) \longrightarrow \partial U_X(0).$$

Also $\otimes \mathcal{O}_X(k)$ induces a homeomorphism,

$$(\otimes \mathcal{O}_X(k))_* \colon \partial U_X(0) \longrightarrow \partial U_X(k).$$

Hence the two regions,

$$(\otimes \mathcal{O}_X(k) \circ \Phi)_* U_{X^\dagger}, \ U_X \subset \mathrm{Stab}_n(X/Y)$$

are patched together in the boundary $\partial U_X(k)$. Repeating this argument, we obtain the following theorem.

**Theorem 5.17.** (**[63, Theorem 1.1, 1.2], [65, Theorem 7.3]**) *One has the chamber structure,*

$$\mathrm{Stab}_n^*(X/Y) = \bigcup \Psi_* \overline{U}_W,$$

*where* $\mathrm{Stab}_n^*(X/Y)$ *is the connected component of* $\mathrm{Stab}_n(X/Y)$ *which contains* $U_X$, $W$ *is either* $X$ *or* $X^\dagger$, *and* $\Psi \colon D(W) \to D(X)$ *is written as a composition of* $\Phi$, $\Phi^\dagger$ *and tensoring line bundles* $\mathcal{O}_X(k)$, $\mathcal{O}_{X^\dagger}(k')$, *i.e. written as*

$$\Psi = \otimes \mathcal{O}_X(k) \circ \Phi \circ \otimes \mathcal{O}_{X^\dagger}(k') \circ \Phi^\dagger \circ \otimes \mathcal{O}_X(k'') \cdots .$$

*Furthermore the map* $\mathcal{Z}_n$ *induces the map,*

$$\mathrm{Stab}_n^*(X/Y) \longrightarrow \mathbb{C} \setminus \mathbb{Z},$$

*which is a regular covering map. Its Galois group* $G$ *is generated by the spherical twist* $T_{\mathcal{O}_C(-1)}$ *and fits into the exact sequence,*

$$1 \longrightarrow G \longrightarrow \mathrm{Auteq}\, D(X/Y) \longrightarrow \mathrm{Pic}(X) \times \mathbb{Z} \longrightarrow 1,$$

*where* $\mathrm{Auteq}\, D(X/Y)$ *is the group of autoequivalences on* $D(X/Y)$ *which are of Fourier-Mukai type with kernel supported on* $X \times_Y X$. *Moreover we have the homeomorphism,*

$$\mathrm{Stab}_n^*(X/Y) \times \mathbb{C} \cong \mathrm{Stab}(X/Y).$$

As a corollary, we have the following description of the double quotient (7),

$$\mathrm{Auteq}\, D(X/Y) \backslash \mathrm{Stab}(X/Y)/\mathbb{C} \cong < T_{\mathcal{O}_C(-1)}, \otimes \mathcal{O}_X(1) > \backslash \mathrm{Stab}_n^*(X/Y)$$
$$\cong \mathbb{Z} \backslash (\mathbb{C} \setminus \mathbb{Z})$$
$$\cong \mathbb{P}^1 \setminus \{0, 1, \infty\}.$$

Hence we have obtained the same picture from the physics [1, Figure 2]. We can divide the above double quotient into three pieces,

$$\mathbb{P}^1 \setminus \{0, 1, \infty\} = U_X \cup \partial U_X \cup U_{X^\dagger}.$$

We see that $U_X$, $U_{X^\dagger}$ correspond to the neighborhoods of the large volume limits at $X$, $X^\dagger$ respectively. The region $\partial U_X$ corresponds to the t-structure ${}^0\mathrm{Per}(X/Y)$, which is according to Example 4.12, equivalent to a module category of a certain non-commutative algebra. Hence we have realized the pictures (13), (14), and (16).

## 5.10. Holomorphic generating functions on the space of stability conditions

Finally, we introduce the recent work of D. Joyce [33] on the generating functions on the space of stability conditions. Joyce's attempt is to construct the interesting holomorphic functions on $\mathrm{Stab}(\mathcal{D})$, using "counting invariants" of semistable objects. This idea is motivated by the following fact. For a compact symplectic manifold $(M, \omega)$, one can define the *Gromov-Witten invariants* which count stable maps from Riemann surfaces, and the holomorphic generating function,

$$f \colon H^{ev}(M, \mathbb{C}) \longrightarrow \mathbb{C},$$

called *Gromov-Witten potential*, given by a power series whose coefficients are Gromov-Witten invariants. The function $f$ satisfies a partial differential equation (p.d.e), called *WDVV equation*, and it defines the *Frobenius structure* on $H^{ev}(M, \mathbb{C})$.

The purpose of the paper [33] is to develop the similar theory on the space of stability conditions. However for several technical reasons, his arguments work only on the space of stability conditions on certain special abelian categories, which includes the case of Example 5.5 (ii). Let $A$ be a finite dimensional $k$-algebra, and $\mathcal{A} = \mathrm{mod}_f(A)$, $\mathcal{D}$, and $S_1, \cdots, S_N \in \mathcal{A}$ be as in Example 5.5 (ii). The space $\mathrm{Stab}(\mathcal{D})$ has the following subset,

$$\mathrm{Stab}(\mathcal{A}) = \{(Z, \mathcal{P}) \in \mathrm{Stab}(\mathcal{D}) \mid \mathcal{P}((0, 1]) = \mathcal{A} \text{ and } \mathrm{Im}\, Z([S_i]) > 0 \text{ for all } i\}.$$

It is obvious $\mathrm{Stab}(\mathcal{A})$ is isomorphic to $\mathbb{H}^N$, where $\mathbb{H} = \{a + bi \in \mathbb{C} \mid b > 0\}$. We put the following additional assumption.

**Assumption 5.18.** *There is a bilinear form* $\overline{\chi} \colon K(\mathcal{A}) \times K(\mathcal{A}) \to \mathbb{Z}$, *such that for* $E, F \in \mathcal{A}$, *we have*

$$\overline{\chi}(E, F) = \dim \mathrm{Hom}(E, F) - \dim \mathrm{Ext}^1(E, F) + \dim \mathrm{Ext}^1(F, E)$$

$$- \dim \mathrm{Hom}(F, E). \tag{17}$$

Here we give some situations in which the above assumption is satisfied.

**Example 5.19.** (i) Suppose for $E, F \in \mathcal{A}$ one has $\dim \mathrm{Ext}^i(E, F) = 0$ for $i \geq 2$. Then

$$\overline{\chi}(E, F) = \chi(E, F) - \chi(F, E),$$

satisfies (17).

(ii) Suppose $A$ is a three dimensional Calabi-Yau algebra, i.e. $E \mapsto E[3]$ gives a Serre functor on $\mathcal{D}$. Then the usual Euler pairing

$$\chi(E, F) = \sum_{i=0}^{3} (-1)^i \dim \operatorname{Ext}^i(E, F),$$

satisfies (17) by Serre duality. So we can set $\overline{\chi} = \chi$.

Using $\overline{\chi}$, we construct the Lie algebra over $\mathbb{C}$,

$$\mathcal{L} = \bigoplus_{\alpha \in K(\mathcal{A})} \mathcal{L}_\alpha, \quad \mathcal{L}_\alpha = \mathbb{C} \cdot c_\alpha,$$

with

$$[c_\alpha, c_\beta] = \overline{\chi}(\alpha, \beta) c_{\alpha+\beta}.$$

Jacobi identity follows because $\overline{\chi}$ is anti-symmetric by (17).
We set $C(\mathcal{A}) \subset K(\mathcal{A})$ as follows,

$$C(\mathcal{A}) := \operatorname{Im}(\mathcal{A} \to K(\mathcal{A})) \setminus \{0\},$$

and take $\alpha \in C(\mathcal{A})$. Then Joyce constructs the holomorphic function

$$f^\alpha : \operatorname{Stab}(\mathcal{A}) \longrightarrow \mathbb{C},$$

using counting invariants of semistable objects and satisfies certain p.d.e. Now we give the outline of the construction of $f^\alpha$ in [33]. For each $\alpha \in C(\mathcal{A})$ and $\sigma \in \operatorname{Stab}(\mathcal{A})$, Joyce [34] associates the invariant, $J^\alpha(\sigma) \in \mathbb{Q}$, as a counting invariant of semistable objects $E$ with $[E] = \alpha$. The definition of $J^\alpha(\sigma)$ is similar to the virtual Euler number of the moduli space of $\sigma$-semistable objects of type $\alpha$. It is known that there is a nice moduli space of semistable $A$-modules, so its virtual Euler number is well-defined. (See [41].) However for our purpose, it doesn't work well, when there is a $\sigma$-semistable object which is not stable. Actually we have to take account of the automorphism groups of semistable objects, so Joyce works in the context of moduli stacks of semistable objects, which are known to be algebraic stacks. Roughly speaking $J^\alpha(\sigma)$ is a "weighted Euler number" of the moduli stack of $\sigma$-semistable objects of type $\alpha$, and especially when $\alpha$ is primitive and $\sigma$ is generic, the invariant $J^\alpha(\sigma)$ coincides with the virtual Euler number of its coarse moduli space. For the detail, see [34]. We set $\epsilon^\alpha(\sigma)$ as follows,

$$\epsilon^\alpha(\sigma) = J^\alpha(\sigma) c_\alpha \in \mathcal{L}_\alpha.$$

Here we remark that the function $\sigma \mapsto \epsilon^\alpha(\sigma)$ is discontinuous in general. In fact there is a wall and chamber structure on $\operatorname{Stab}(\mathcal{A})$, such that the set of semistable objects does not change in a chamber. (See [15].) This implies that $\sigma \mapsto \epsilon^\alpha(\sigma)$ is constant in a chamber, however it will be a different value under a wall crossing. The idea of [33] is to cancel out the discontinuity of $\sigma \mapsto \epsilon^\alpha(\sigma)$, by using the (also discontinuous) functions,

$$F_n \colon (\mathbb{C}^*)^n \longrightarrow \mathbb{C},$$

and make a function $f^\alpha \colon \mathrm{Stab}(\mathcal{A}) \to U(\mathcal{L})$,

$$f^\alpha(\sigma) = \sum_{\alpha_1 + \cdots + \alpha_n = \alpha} F_n(Z(\alpha_1), \cdots, Z(\alpha_n)) \epsilon^{\alpha_1}(\sigma) * \cdots * \epsilon^{\alpha_n}(\sigma). \qquad (18)$$

Here $U(\mathcal{L})$ is the universal enveloping algebra of $\mathcal{L}$, $*$ is the multiplication in $U(\mathcal{L})$, $\alpha_i \in C(\mathcal{A})$, and $\sigma = (Z, \mathcal{A}) \in \mathrm{Stab}(\mathcal{A})$. Note that the sum (18) is a finite sum by our definition of $\mathcal{A}$. (This is one of the reasons why the arguments in [33] are restricted in this case.) The following is the main theorem of [33].

**Theorem 5.20.** **(Joyce [33])** *There is a family of functions* $F_n \colon (\mathbb{C}^*)^n \to \mathbb{C}$ *such that*

- $f^\alpha$ *is contained in* $\mathcal{L}_v \cong \mathbb{C}$. *Hence defines a function* $f^\alpha \colon \mathrm{Stab}(\mathcal{A}) \to \mathbb{C}$.
- *The function* $f^\alpha$ *is continuous and holomorphic.*

*Furthermore* $F_n$ *is uniquely determined (under several normalizations), with* $F_1$, $F_2$ *as follows,*

$$F_1(z_1) = \frac{1}{2\pi i}, \quad F_2(z_1, z_2) = \begin{cases} \frac{1}{(2\pi i)^2}\left(\log\frac{z_2}{z_1} - \pi i\right), & \frac{z_2}{z_1} \notin (0, \infty), \\ \frac{1}{(2\pi i)^2}\log\frac{z_2}{z_1}, & \frac{z_2}{z_1} \in (0, \infty). \end{cases}$$

*Here the branch of* $\log$ *is determined by* $\log(-1) = \pi i$. *The functions* $f^\alpha$ *for* $\alpha \in C(\mathcal{A})$ *satisfy the following p.d.e,*

$$df^\alpha(\sigma) = \sum_{\beta + \gamma = \alpha} -[f^\beta(\sigma), f^\gamma(\sigma)] \otimes \frac{dZ(\beta)}{Z(\beta)},$$

*where* $\beta, \gamma \in C(\mathcal{A})$.

Explicitly the function $f^\alpha$ is expanded in the following formula,

$$f^\alpha(\sigma) = \sum_{\alpha_1 + \cdots + \alpha_n = \alpha} F_n(Z(\alpha_1), \cdots, Z(\sigma_n)) \prod_{i=1}^n J^{\alpha_i}(\sigma)$$

$$\times \left[\frac{1}{2^{n-1}} \sum_{\Gamma} \prod_{i \to j \text{ in } \Gamma} \overline{\chi}(\alpha_i, \alpha_j)\right],$$

with $\alpha_i \in C(\mathcal{A})$, $\Gamma$ is a connected and simply connected oriented graph with vertex $\{1, \cdots, n\}$, and if there is an arrow $i \to j$ in $\Gamma$, we must have $i \leq j$.

At this time extending Joyce's work to triangulated categories is problematic, even if we ignore the convergence of the sum. In the proof of Theorem 5.20, he uses his works on configurations in abelian categories [30], [31], [32], [34], so we have to generalize these works to triangulated categories. The author's work [62] on the invariants of semistable objects on K3 surfaces is toward one step for this goal.

## Acknowledgement

The author thanks Hokuto Uehara for checking the manuscript and giving the nice advice. He is supported by Japan Society for the Promotion of Sciences Research Fellowships for Young Scientists, No 198007.

## References

[1] P. Aspinwall. A Point's Point of View of Stringy Geometry. *J. High Energy Phys*, Vol. 002, p. 15pp, 2003.

[2] A. Beilinson. Coherent sheaves on $\mathbb{P}^n$ and problems of linear algebra. *Funct. Anal. Appl*, Vol. 12, pp. 214–216, 1978.

[3] A. Beilinson, J. Bernstein, and P. Deligne. Faisceaux pervers. *Analysis and topology on singular spaces I, Asterisque*, Vol. 100, pp. 5–171, 1982.

[4] O. Ben-Bassat, J. Block, and T. Pantev. Non-commutative tori and Fourier-Mukai duality. *Compositio Math*, Vol. 143, pp. 423–475, 2007.

[5] A. Bergman. Stability conditions and Branes at Singularities. *preprint*. math.AG/0702092.

[6] R. Bezrukavnikov and D. Kaledin. McKay equivalence for symplectic resolutions of quotient singularities. *Proc. Steklov Inst. Math*, Vol. 246, pp. 13–33, 2004.

[7] A. Bondal and D. Orlov. Semiorthgonal decomposition for algebraic varieties. *preprint*. math.AG/9506012.

[8] A. Bondal and D. Orlov. Reconstruction of a variety from the derived category and groups of autoequivalences. *Compositio Math*, Vol. 125, pp. 327–344, 2001.

[9] T. Bridgeland. Spaces of stability conditions. *preprint*. math.AG/0611510.

[10] T. Bridgeland. Stability conditions and Kleinian singularities. *preprint*. math.AG/0508257.

[11] T. Bridgeland. Flops and derived categories. *Invent. Math*, Vol. 147, pp. 613–632, 2002.

[12] T. Bridgeland. Derived categories of coherent sheaves. *Proceedings of the 2006 ICM*, 2006. math.AG/0602129.

[13] T. Bridgeland. Stability conditions on a non-compact Calabi-Yau threefold. *Comm. Math. Phys.*, Vol. 266, pp. 715–733, 2006.

[14] T. Bridgeland. Stability conditions on triangulated categories. *Ann. of Math*, Vol. 166, pp. 317–345, 2007.

[15] T. Bridgeland. Stability conditions on $K3$ surfaces. *Duke Math. J.*, Vol. 141, pp. 241–291, 2008.

[16] T. Bridgeland, A. King, and M. Reid. The McKay correspondence as an equivalence of derived categories. *J. Amer. Math. Soc.*, Vol. 14, pp. 535–554, 2001.

[17] T. Bridgeland and A. Maciocia. Complex surfaces with equivalent derived categories. *Math. Z*, Vol. 236, pp. 677–697, 2001.

[18] A. Căldăraru. The Mukai pairing, I: The Hochschild structure. *preprint.* math.AG/0308079.

[19] J-C. Chen. Flops and equivalences of derived categories for three-folds with only Gorenstein singularities. *J. Differential. Geom*, Vol. 61, pp. 227–261, 2002.

[20] M. Van den Bergh. Non-commutative crepant resolutions. *The legacy of Niels Henrik Abel, Springer, Berlin*, pp. 749–770, 2004.

[21] M. Van den Bergh. Three dimensional flops and noncommutative rings. *Duke Math. J.*, Vol. 122, pp. 423–455, 2004.

[22] M. Douglas. D-branes, categories and $N = 1$ supersymmetry. *J. Math. Phys.*, Vol. 42, pp. 2818–2843, 2001.

[23] M. Douglas. Dirichlet branes, homological mirror symmetry, and stability. *Proceedings of the 1998 ICM*, pp. 395–408, 2002. math.AG/0207021.

[24] R. Hartshorne. *Residues and Duality : Lecture Notes of a Seminar on the Work of A.Grothendieck, given at Harvard 1963/1964*, Lecture Notes in Mathematics, No. 20, Springer-Verlag, 1966.

[25] L. Hille and M. Van den Bergh. Fourier-Mukai tranforms. *preprint.* math.AG/0402043.

[26] G. Hochschild, B. Kostant, and A. Rosenberg. Differential forms on regular affine algebras. *Trans. AMS*, Vol. 102, pp. 383–408, 1962.

[27] P. Horja. Derived Category Automorphisms from Mirror Symmetry. *Duke Math. J.*, Vol. 127, pp. 1–34, 2005.

[28] A. Ishii, K. Ueda, and H. Uehara. Stability Conditions on $A_n$-Singularities. math.AG/0609551.

[29] A. Ishii and H. Uehara. Autoequivalences of derived categories on the minimal resolutions of $A_n$-singularities on surfaces. *J. Differential Geom.*, Vol. 71, pp. 385–435, 2005.

[30] D. Joyce. Configurations in abelian categories I. Basic properties and moduli stack. *Advances in Math*, Vol. 203, pp. 194–255, 2006.

[31] D. Joyce. Configurations in abelian categories II. Ringel-Hall algebras. *Advances in Math*, Vol. 210, pp. 635–706, 2007.

[32] D. Joyce. Configurations in abelian categories III. Stability conditions and identities. *Advances in Math*, Vol. 215, pp. 153–219, 2007.

[33] D. Joyce. Holomorphic generating functions for invariants counting coherent sheaves on Calabi-Yau 3-folds. *Geometry and Topology*, Vol. 11, pp. 667–725, 2007.

[34] D. Joyce. Configurations in abelian categories IV. Invariants and changing stability conditions. *Advances in Math*, Vol. 217, pp. 125–204, 2008.

[35] Y. Kawamata. Flops connect minimal models. *preprint.* math.AG/0704.1013.

[36] Y. Kawamata. On the cone of divisors of Calabi-Yau fiber spaces. *Internat. J. Math*, Vol. 5, pp. 665–687, 1997.

[37] Y. Kawamata. *D*-equivalence and *K*-equivalence. *J. Differential Geom.*, Vol. 61, pp. 147–171, 2002.

[38] Y. Kawamata. Francia's flip and derived categories. *Algebraic geometry, de Gruyter, Berlin*, pp. 197–215, 2002.

[39] Y. Kawamata. Log crepant birational maps and derived categories. *J. Math. Sci. Univ. Tokyo*, Vol. 12, pp. 1–53, 2005.

[40] Y. Kawamata. Derived categories of toric varieties. *Michigan Math. J.*, Vol. 54, pp. 517–535, 2006.

[41] A. King. Moduli of representations of finite-dimensional algebras. *Quart. J. Math. Oxford Ser.(2)*, Vol. 45, pp. 515–530, 1994.

[42] J. Kollár. Flops. *Nagoya Math. J*, Vol. 113, pp. 15–36, 1989.

[43] J. Kollár. *Rational curves on algebraic varieties*, Vol. 32 of *Ergebnisse Math. Grenzgeb.(3)*. Springer-Verlag, 1996.

[44] J. Kollár and S. Mori. *Birational geometry of algebraic varieties*, Vol. 134 of *Cambridge Tracts in Mathematics*. Cambridge University Press, 1998.

[45] M. Kontsevich. *Homological algebra of mirror symmetry*, Vol. 1 of *Proceedings of ICM*. Basel:Birkhäuser, 1995.

[46] W. Lowen and M. Van den Bergh. Deformation theory of abelian categories. *Trans. Amer. Math. Soc.*, pp. 5441–5483, 2006.

[47] E. Macri. Some examples of moduli spaces of stability conditions on derived categories. *preprint*. math.AG/0411613.

[48] S. Mukai. Duality between $D(X)$ and $D(\hat{X})$ with its application to picard sheaves. *Nagoya Math. J.*, Vol. 81, pp. 101–116, 1981.

[49] S. Mukai. On the moduli space of bundles on $K3$ surfaces $I$. *Vector Bundles on Algebraic Varieties, M. F. Atiyah et al., Oxford University Press*, pp. 341–413, 1987.

[50] Y. Namikawa. Mukai flops and derived categories. *J. Reine. Angew. Math.*, pp. 65–76, 2003.

[51] S. Okada. Stability manifold of $\mathbb{P}^1$. *J.Algebraic Geom*, Vol. 15, pp. 487–505, 2006.

[52] D. Orlov. On Equivalences of derived categories and $K3$ surfaces. *J. Math. Sci (New York)*, Vol. 84, pp. 1361–1381, 1997.

[53] D. Orlov. Derived categories of coherent sheaves on abelian varieties and equivalences between them. *Izv. Ross. Akad. Nauk. Ser. Mat.*, Vol. 66, pp. 131–158, 2002.

[54] D. Orlov. Derived categories of coherent sheaves and motives. *Uspekhi. Math. Nauk*, Vol. 60, pp. 231–232, 2005.

[55] M. Reid. McKay correspondence. *preprint*. math.AG/9702016.

[56] P. Seidel. Graded Lagrangian submanifolds. *Bull. Soc. Math. France*, Vol. 128, pp. 103–149, 2000.

[57] P. Seidel and R. P. Thomas. Braid group actions on derived categories of coherent sheaves. *Duke Math. J.*, Vol. 108, pp. 37–107, 2001.

[58] R. G. Swan. Hochschild cohomology of quasiprojective schemes. *J. Pure Appl. Algebra*, Vol. 1, pp. 57–80, 1996.

[59] R. P. Thomas. Stability conditions and the braid groups. *Comm. Anal. Geom.*, Vol. 14, pp. 135–161, 2006.

[60] Y. Toda. Fourier-Mukai transforms and canonical divisors. *Compositio Math*, Vol. 142, pp. 962–982, 2006.

[61] Y. Toda. On a certain generalization of spherical twists. *Bulletin de la SMF*, Vol. 135, pp. 97–112, 2007.

[62] Y. Toda. Moduli stacks and invariants of semistable objects on K3 surfaces. *Advances in Math*, Vol. 217, pp. 2736–2781, 2008.

[63] Y. Toda. Stability conditions and crepant small resolutions. *Trans. Amer. Math. Soc.*, Vol. 360, pp. 6149–6178, 2008.

[64] Y. Toda. Deformations and Fourier-Mukai transforms. *J. Differential Geom.*, Vol. 81, pp. 197–224, 2009.

[65] Y. Toda. Stability conditions and Calabi-Yau fibrations. *J. Algebraic Geom.*, Vol. 18, pp. 101–133, 2009.

[66] H. Uehara. An example of Fourier-Mukai partners of minimal elliptic surfaces. *Math. Res. Lett.*, Vol. 11, pp. 371–375, 2004.

[67] A. Yekutieli. The continuous Hochschild cochain complex of a scheme. *Canada J. Math.*, Vol. 6, pp. 1319–1337, 2002.

Yukinobu Toda, Graduate School of Mathematical Sciences, University of Tokyo

*E-mail address*:toda@ms.u-tokyo.ac.jp

# Rigid dualizing complexes via differential graded algebras (survey)

A M N O N   Y E K U T I E L I

ABSTRACT. In this article we survey recent results on rigid dualizing complexes over commutative algebras. We begin by recalling what are dualizing complexes. Next we define rigid complexes, and explain their functorial properties. Due to the possible presence of torsion, we must use differential graded algebras in the constructions. We then discuss rigid dualizing complexes. Finally we show how rigid complexes can be used to understand Cohen-Macaulay homomorphisms and relative dualizing sheaves.

## 0. Introduction

This short article is based on a lecture I gave at the "Workshop on Triangulated Categories", Leeds, August 2006. It is a survey of recent results on rigid dualizing complexes over commutative rings. Most of these results are joint work of mine with James Zhang. The idea of rigid dualizing complex is due to Michel Van den Bergh.

By default all rings considered in this article are *commutative*. We begin by recalling the notion of *dualizing complex* over a noetherian ring $A$. Next let $B$ be a noetherian $A$-algebra. We define what is a *rigid complex* of $B$-modules relative to $A$. In making this definition we must use differential graded algebras (when $B$ is not flat over $A$). The functorial properties of rigid complexes are explained. We then discuss *rigid dualizing complexes*, which by definition are complexes that are both rigid and dualizing. Finally we show how rigid complexes can be used to understand Cohen-Macaulay homomorphisms and relative dualizing sheaves.

I wish to thank my collaborator James Zhang. Thanks also to Luchezar Avramov, Srikanth Iyengar and Joseph Lipman for discussions regarding the material in Section 5.

*Key words and phrases.* commutative rings, DG algebras, derived categories, rigid complexes.
*Mathematics Subject Classification* 2000. Primary: 18E30; Secondary: 18G10, 16E45, 18G15.
This research was supported by the US-Israel Binational Science Foundation.

"

# 1. Dualizing Complexes: Overview

Let $A$ be a noetherian ring. Denote by $\mathsf{D}_\mathrm{f}^\mathrm{b}(\mathsf{Mod}\, A)$ the derived category of bounded complexes of $A$-modules with finitely generated cohomology modules.

**Definition 1.1.** (Grothendieck [RD]) A *dualizing complex* over $A$ is a complex $R \in \mathsf{D}_\mathrm{f}^\mathrm{b}(\mathsf{Mod}\, A)$ satisfying the two conditions:

 (i)  $R$ has finite injective dimension.
 (ii)  The canonical morphism $A \to \mathrm{RHom}_A(R, R)$ is an isomorphism.

Condition (i) means that there is an integer $d$ such that $\mathrm{Ext}_A^i(M, R) = 0$ for all $i > d$ and all modules $M$.

Recall that a noetherian ring $\mathbb{K}$ is called *regular* if all its local rings $\mathbb{K}_\mathfrak{p}$, $\mathfrak{p} \in \mathrm{Spec}\,\mathbb{K}$, are regular local rings.

**Example 1.2.** If $\mathbb{K}$ is a regular noetherian ring of finite Krull dimension (say a field, or the ring of integers $\mathbb{Z}$) then

$$R := \mathbb{K} \in \mathsf{D}_\mathrm{f}^\mathrm{b}(\mathsf{Mod}\,\mathbb{K})$$

is a dualizing complex over $\mathbb{K}$.

Dualizing complexes over commutative rings are part of Grothendieck's duality theory in algebraic geometry, which was developed in [RD]. This duality theory deals with dualizing complexes on schemes and relations between them. See Remark 4.4.

In Section 4 we explain a new approach to dualizing complexes over commutative rings, due to James Zhang and the author (see [YZ4] and [YZ5]). Specifically, we discuss existence and uniqueness of *rigid dualizing complexes*.

A dualizing complex $R$ has many automorphisms; indeed, its group of automorphisms in $\mathsf{D}(\mathsf{Mod}\, A)$ is the group $A^\times$ of invertible elements. The purpose of rigidity is to eliminate automorphisms, and to make dualizing complexes functorial. See Theorem 4.2.

In a sequel paper [Ye2] we use the technique of *perverse coherent sheaves* to construct rigid dualizing complexes on schemes, and we reproduce almost all of the geometric Grothendieck duality theory.

Related work in noncommutative algebraic geometry (where rigid dualizing complexes were first introduced) can be found in [VdB, YZ1, YZ2, YZ3].

## 2. Rigid Complexes and DG Algebras

Let me start with a discussion of rigidity for algebras over a field. Suppose $\mathbb{K}$ is a field, $B$ is a $\mathbb{K}$-algebra, and $M \in \mathsf{D}(\mathsf{Mod}\, B)$.

According to Van den Bergh [VdB] a *rigidifying isomorphism* for $M$ is an isomorphism

$$\rho : M \xrightarrow{\simeq} \mathrm{RHom}_{B \otimes_{\mathbb{K}} B}(B, M \otimes_{\mathbb{K}} M) \tag{2.1}$$

in $\mathsf{D}(\mathsf{Mod}\, B)$.

Now suppose $A$ is any ring. Trying to write $A$ instead of $\mathbb{K}$ in formula (2.1) does not make sense: instead of $M \otimes_A M$ we must take the derived tensor product $M \otimes_A^{\mathrm{L}} M$; but then there is no obvious way to make $M \otimes_A^{\mathrm{L}} M$ into a complex of $B \otimes_A B$ -modules.

The problem is torsion: $B$ might fail to be a flat $A$-algebra. This is where *differential graded algebras* (DG algebras) enter the picture.

A DG algebra is a graded ring $\tilde{A} = \bigoplus_{i \in \mathbb{Z}} \tilde{A}^i$, together with a graded derivation $\mathrm{d} : \tilde{A} \to \tilde{A}$ of degree 1, satisfying $\mathrm{d} \circ \mathrm{d} = 0$.

A DG algebra quasi-isomorphism is a homomorphism $f : \tilde{A} \to \tilde{B}$ respecting degrees, multiplications and differentials, and such that $\mathrm{H}(f) : \mathrm{H}\tilde{A} \to \mathrm{H}\tilde{B}$ is an isomorphism (of graded algebras).

We shall only consider *super-commutative non-positive* DG algebras. Super-commutative means that $ab = (-1)^{ij} ba$ and $c^2 = 0$ for all $a \in \tilde{A}^i, b \in \tilde{A}^j$ and $c \in \tilde{A}^{2i+1}$. Non-positive means that $\tilde{A} = \bigoplus_{i \leq 0} \tilde{A}^i$.

We view a ring $A$ as a DG algebra concentrated in degree 0. Given a DG algebra homomorphism $A \to \tilde{A}$ we say that $\tilde{A}$ is a DG $A$-algebra.

Let $A$ be a ring. A *semi-free* DG $A$-algebra is a DG $A$-algebra $\tilde{A}$, such that after forgetting the differential $\tilde{A}$ is isomorphic, as graded $A$-algebra, to a super-polynomial algebra on some graded set of variables.

**Definition 2.2.** Let $A$ be a ring and $B$ an $A$-algebra. A *semi-free DG algebra resolution of $B$ relative to $A$* is a quasi-isomorphism $\tilde{B} \to B$ of DG $A$-algebras, where $\tilde{B}$ is a semi-free DG $A$-algebra.

Such resolutions always exist, and they are unique up to quasi-isomorphism.

**Example 2.3.** Take $A = \mathbb{Z}$ and $B = \mathbb{Z}/(6)$. Define $\tilde{B}$ to be the super-polynomial algebra $A[\xi]$ on the variable $\xi$ of degree $-1$. So $\tilde{B} = A \oplus A\xi$ as free graded $A$-module, and $\xi^2 = 0$. Let $\mathrm{d}(\xi) := 6$. Then $\tilde{B} \to B$ is a semi-free DG algebra resolution of $B$ relative to $A$.

For a DG algebra $\tilde{A}$ one has the category $\mathsf{DGMod}\, \tilde{A}$ of DG $\tilde{A}$-modules. It is analogous to the category of complexes of modules over a ring, and by a similar process of inverting quasi-isomorphisms we obtain the derived category $\tilde{\mathsf{D}}(\mathsf{DGMod}\,\tilde{A})$; see [Ke], [Hi].

For a ring $A$ (i.e. a DG algebra concentrated in degree 0) we have

$$\tilde{\mathsf{D}}(\mathsf{DGMod}\,A) = \mathsf{D}(\mathsf{Mod}\,A),$$

the usual derived category.

It is possible to derive functors of DG modules, again in analogy to $\mathsf{D}(\mathsf{Mod}\,A)$. An added feature is that for a quasi-isomorphism $\tilde{A} \to \tilde{B}$, the restriction of scalars functor

$$\tilde{\mathsf{D}}(\mathsf{DGMod}\,\tilde{B}) \to \tilde{\mathsf{D}}(\mathsf{DGMod}\,\tilde{A})$$

is an equivalence.

Getting back to our original problem, suppose $A$ is a ring and $B$ is an $A$-algebra. Choose a semi-free DG algebra resolution $\tilde{B} \to B$ relative to $A$. For $M \in \mathsf{D}(\mathsf{Mod}\,B)$ define

$$\mathrm{Sq}_{B/A}\,M := \mathrm{RHom}_{\tilde{B}\otimes_A \tilde{B}}(B, M \otimes_A^{\mathsf{L}} M)$$

in $\mathsf{D}(\mathsf{Mod}\,B)$.

**Theorem 2.4.** ([YZ4]) *The functor*

$$\mathrm{Sq}_{B/A} : \mathsf{D}(\mathsf{Mod}\,B) \to \mathsf{D}(\mathsf{Mod}\,B)$$

*is independent of the resolution* $\tilde{B} \to B$.

The functor $\mathrm{Sq}_{B/A}$, called the *squaring operation*, is nonlinear. In fact, given a morphism $\phi : M \to M$ in $\mathsf{D}(\mathsf{Mod}\,B)$ and an element $b \in B$ one has

$$\mathrm{Sq}_{B/A}(b\phi) = b^2\,\mathrm{Sq}_{B/A}(\phi) \tag{2.5}$$

in

$$\mathrm{Hom}_{\mathsf{D}(\mathsf{Mod}\,B)}(\mathrm{Sq}_{B/A}\,M, \mathrm{Sq}_{B/A}\,M).$$

**Definition 2.6.** Let $B$ be a noetherian $A$-algebra, and let $M$ be a complex in $\mathsf{D}_{\mathrm{f}}^{\mathrm{b}}(\mathsf{Mod}\,B)$ that has finite flat dimension over $A$. Assume

$$\rho : M \xrightarrow{\simeq} \mathrm{Sq}_{B/A}\,M$$

is an isomorphism in $\mathsf{D}(\mathsf{Mod}\,B)$. Then the pair $(M, \rho)$ is called a *rigid complex over $B$ relative to $A$*.

**Definition 2.7.** Say $(M, \rho)$ and $(N, \sigma)$ are rigid complexes over $B$ relative to $A$. A morphism $\phi : M \to N$ in $\mathsf{D}(\mathsf{Mod}\,B)$ is called a *rigid morphism relative* *to A* if the diagram

$$
\begin{array}{ccc}
M & \xrightarrow{\ \rho\ } & \mathrm{Sq}_{B/A}\, M \\
{\scriptstyle \phi}\downarrow & & \downarrow{\scriptstyle \mathrm{Sq}_{B/A}(\phi)} \\
N & \xrightarrow{\ \sigma\ } & \mathrm{Sq}_{B/A}\, N
\end{array}
$$

is commutative.

We denote by $\mathsf{D}_{\mathrm{f}}^{\mathrm{b}}(\mathsf{Mod}\, B)_{\mathrm{rig}/A}$ the category of rigid complexes over $B$ relative to $A$.

**Example 2.8.** Take $M = B = A$. Then

$$
\mathrm{Sq}_{A/A}\, A = \mathrm{RHom}_{A \otimes_A A}(A, A \otimes_A A) = A,
$$

and we interpret this as a rigidifying isomorphism

$$
\rho^{\mathrm{tau}} : A \xrightarrow{\simeq} \mathrm{Sq}_{A/A}\, A.
$$

The *tautological rigid complex* is

$$
(A, \rho^{\mathrm{tau}}) \in \mathsf{D}_{\mathrm{f}}^{\mathrm{b}}(\mathsf{Mod}\, A)_{\mathrm{rig}/A}.
$$

# 3. Properties of Rigid Complexes

The first property of rigid complexes explains their name.

**Theorem 3.1.** ([YZ4]) *Let $A$ be a ring, $B$ a noetherian $A$-algebra, and*

$$
(M, \rho) \in \mathsf{D}_{\mathrm{f}}^{\mathrm{b}}(\mathsf{Mod}\, B)_{\mathrm{rig}/A}.
$$

*Assume the canonical ring homomorphism*

$$
B \to \mathrm{Hom}_{\mathsf{D}(\mathsf{Mod}\, B)}(M, M)
$$

*is bijective. Then the only automorphism of $(M, \rho)$ in $\mathsf{D}_{\mathrm{f}}^{\mathrm{b}}(\mathsf{Mod}\, B)_{\mathrm{rig}/A}$ is the identity $\mathbf{1}_M$.*

The proof is very easy: an automorphism $\phi$ of $M$ has to be of the form $\phi = b\,\mathbf{1}_M$ for some invertible element $b \in B$. If $\phi$ is rigid then $b = b^2$ (cf. formula (2.5)), and hence $b = 1$.

We find it convenient to denote ring homomorphisms by $f^*$ etc. Thus a ring homomorphism $f^* : A \to B$ corresponds to the morphism of schemes

$$
f : \mathrm{Spec}\, B \to \mathrm{Spec}\, A.
$$

Let $A$ be a noetherian ring. Recall that an $A$-algebra $B$ is called *essentially finite type* if it is a localization of some finitely generated $A$-algebra. We say that $B$ is *essentially smooth* (resp. *essentially étale*) over $A$ if it is essentially finite type and formally smooth (resp. formally étale).

**Example 3.2.** If $A'$ is a localization of $A$ then $A \to A'$ is essentially étale. If $B = A[t_1, \ldots, t_n]$ is a polynomial algebra then $A \to B$ is smooth, and hence also essentially smooth.

Let $A$ be a noetherian ring and $f^* : A \to B$ an essentially smooth homomorphism. Then $\Omega^1_{B/A}$ is a finitely generated projective $B$-module. Let

$$\operatorname{Spec} B = \coprod_i \operatorname{Spec} B_i$$

be the decomposition into connected components, and for every $i$ let $n_i$ be the rank of $\Omega^1_{B_i/A}$. We define a functor

$$f^{\sharp} : \mathsf{D}(\operatorname{Mod} A) \to \mathsf{D}(\operatorname{Mod} B)$$

by

$$f^{\sharp} M := \bigoplus_i \Omega^{n_i}_{B_i/A}[n_i] \otimes_A M.$$

Recall that a ring homomorphism $f^* : A \to B$ is called *finite* if $B$ is a finitely generated $A$-module. Given such a finite homomorphism we define a functor

$$f^{\flat} : \mathsf{D}(\operatorname{Mod} A) \to \mathsf{D}(\operatorname{Mod} B)$$

by

$$f^{\flat} M := \operatorname{RHom}_A(B, M).$$

**Theorem 3.3.** ([YZ4]) *Let $A$ be a noetherian ring, let $B, C$ be essentially finite type $A$-algebras, let $f^* : B \to C$ be an $A$-algebra homomorphism, and let*

$$(M, \rho) \in \mathsf{D}^{\mathsf{b}}_{\mathsf{f}}(\operatorname{Mod} B)_{\mathrm{rig}/A}.$$

(1) *If $f^*$ is finite and $f^{\flat} M$ has finite flat dimension over $A$, then $f^{\flat} M$ has an induced rigidifying isomorphism*

$$f^{\flat}(\rho) : f^{\flat} M \xrightarrow{\simeq} \operatorname{Sq}_{C/A} f^{\flat} M.$$

*The assignment*

$$(M, \rho) \mapsto f^{\flat}(M, \rho) := \left( f^{\flat}(\rho), f^{\flat} M \right)$$

*is functorial.*

(2) *If $f^*$ is essentially smooth then $f^\sharp M$ has an induced rigidifying isomorphism*

$$f^\sharp(\rho) : f^\sharp M \xrightarrow{\simeq} \mathrm{Sq}_{C/A} \, f^\sharp M.$$

*The assignment*

$$(M, \rho) \mapsto f^\sharp(M, \rho) := \big( f^\sharp(\rho), \, f^\sharp M \big)$$

*is functorial.*

## 4. Rigid Dualizing Complexes

Let $\mathbb{K}$ be a regular noetherian ring of finite Krull dimension. We denote by $\mathsf{EFTAlg}\,/\mathbb{K}$ the category of essentially finite type $\mathbb{K}$-algebras.

**Definition 4.1.** A *rigid dualizing complex* over $A$ relative to $\mathbb{K}$ is a rigid complex $(R_A, \rho_A)$, such that $R_A$ is a dualizing complex.

**Theorem 4.2.** ([YZ5]) *Let $\mathbb{K}$ be a regular finite dimensional noetherian ring, and let $A$ be an essentially finite type $\mathbb{K}$-algebra.*

(1) *The algebra $A$ has a rigid dualizing complex $(R_A, \rho_A)$, which is unique up to a unique rigid isomorphism.*
(2) *Given a finite homomorphism $f^* : A \to B$, there is a unique rigid isomorphism $f^\flat(R_A, \rho_A) \xrightarrow{\simeq} (R_B, \rho_B)$.*
(3) *Given an essentially smooth homomorphism $f^* : A \to B$, there is a unique rigid isomorphism $f^\sharp(R_A, \rho_A) \xrightarrow{\simeq} (R_B, \rho_B)$.*

Here is how the rigid dualizing complex $(R_A, \rho_A)$ is obtained. We begin with the tautological rigid complex

$$(\mathbb{K}, \rho^{\mathrm{tau}}) \in \mathsf{D}^{\mathrm{b}}_{\mathrm{f}}(\mathsf{Mod}\,\mathbb{K})_{\mathrm{rig}/\mathbb{K}},$$

which is dualizing (cf. Examples 1.2 and 2.8). Now the structural homomorphism $\mathbb{K} \to A$ can be factored into

$$\mathbb{K} \xrightarrow{f^*} B \xrightarrow{g^*} C \xrightarrow{h^*} A,$$

where $f^*$ is essentially smooth ($B$ is a polynomial algebra over $\mathbb{K}$); $g^*$ is finite (a surjection); and $h^*$ is also essentially smooth (a localization).

It is not hard to check (see [RD, Chapter V]) that each of the complexes $f^\sharp\mathbb{K}$, $g^\flat f^\sharp\mathbb{K}$ and $h^\sharp g^\flat f^\sharp\mathbb{K}$ is dualizing over the respective ring. In particular, $g^\flat f^\sharp\mathbb{K}$ has bounded cohomology, and hence it has finite flat dimension over $\mathbb{K}$.

According to Theorem 3.3 we then have a rigid complex

$$(R_A, \rho_A) := h^\natural \, g^\flat \, f^\natural(\mathbb{K}, \rho^{\mathrm{tau}}) \in \mathsf{D}^{\mathrm{b}}_{\mathrm{f}}(\mathsf{Mod}\, A)_{\mathrm{rig}/\mathbb{K}}.$$

**Definition 4.3.** Given a homomorphism $f^* : A \to B$ in $\mathsf{EFTAlg}\,/\mathbb{K}$, define the *twisted inverse image functor*

$$f^! : \mathsf{D}^+_{\mathrm{f}}(\mathsf{Mod}\, A) \to \mathsf{D}^+_{\mathrm{f}}(\mathsf{Mod}\, B)$$

by the formula

$$f^! M := \mathrm{RHom}_B\big(B \otimes^{\mathrm{L}}_A \mathrm{RHom}_A(M, R_A), R_B\big).$$

It is easy to show that the assignment $f^* \mapsto f^!$ is a pseudofunctor from the category $\mathsf{EFTAlg}\,/\mathbb{K}$ to the 2-category $\mathsf{Cat}$ of all categories. Moreover, using Theorem 4.2 one can show that this operation has very good properties. For instance, when $f^*$ is finite, then there is a functorial nondegenerate trace morphism

$$\mathrm{Tr}_f : f^! M \to M.$$

**Remark 4.4.** According to Grothendieck's duality theory in [RD], if $f : X \to Y$ is a finite type morphism between noetherian schemes, and if $Y$ has a dualizing complex, then there is a functor

$$f^{!(\mathrm{G})} : \mathsf{D}^+_{\mathrm{c}}(\mathsf{Mod}\, \mathcal{O}_Y) \to \mathsf{D}^+_{\mathrm{c}}(\mathsf{Mod}\, \mathcal{O}_X),$$

with many good properties.

Let $\mathsf{FTAlg}\,/\mathbb{K}$ be the category of finite type $\mathbb{K}$-algebras. By restricting attention to affine schemes, the results of [RD] give rise to a pseudofunctor $f^* \mapsto f^{!(\mathrm{G})}$ from $\mathsf{FTAlg}\,/\mathbb{K}$ to $\mathsf{Cat}$. It is not hard to show that the pseudofunctor $f^* \mapsto f^{!(\mathrm{G})}$ is isomorphic to our 2-functor $f^* \mapsto f^!$; see [YZ5, Theorem 4.10].

It should be noted that our construction works in the slightly bigger category $\mathsf{EFTAlg}\,/\mathbb{K}$. It also has the advantage of being local; whereas in [RD] some of the results require that morphisms between affine schemes be compactified.

## 5. Rigid Complexes and CM Homomorphisms

In this final section we discuss the relation between rigid complexes and Cohen-Macaulay homomorphisms.

**Definition 5.1.** A ring $A$ is called *tractable* if there is an essentially finite type homomorphism $\mathbb{K} \to A$, for some regular noetherian ring of finite Krull dimension $\mathbb{K}$.

Such a homomorphism $\mathbb{K} \to A$ is called a *traction* for $A$. It is not part of the structure – the ring $A$ does not come with any preferred traction. "Most commutative noetherian rings we know" are tractable.

Given a traction $\mathbb{K} \to A$ we denote by $R_{A/\mathbb{K}}$ the rigid dualizing complex of $A$ relative to $\mathbb{K}$; cf. Theorem 4.2. (The rigidifying isomorphism $\rho_{A/\mathbb{K}}$ is implicit.)

Recall that a noetherian ring $A$ is called *Cohen-Macaulay* (resp. *Gorenstein*) if all its local rings $A_\mathfrak{p}$, $\mathfrak{p} \in \operatorname{Spec} A$, are Cohen-Macaulay (resp. Gorenstein) local rings. The implications are regular $\Rightarrow$ Gorenstein $\Rightarrow$ Cohen-Macaulay.

Let $f^* : A \to B$ be a ring homomorphism. For $\mathfrak{p} \in \operatorname{Spec} A$ let $k(\mathfrak{p}) := (A/\mathfrak{p})_\mathfrak{p}$, the residue field. The fiber of $f^*$ above $\mathfrak{p}$ is the $k(\mathfrak{p})$-algebra $B \otimes_A k(\mathfrak{p})$. Now assume $f^*$ is an essentially finite type flat homomorphism. If all the fibers of $f^*$ are Cohen-Macaulay (resp. Gorenstein) rings, then we call $f^*$ an *essentially Cohen-Macaulay* (resp. *essentially Gorenstein*) homomorphism.

**Theorem 5.2.** ([Ye2]) *Let $A$ be a tractable ring, and let $f^* : A \to B$ be homomorphism which is of essentially finite type and of finite flat dimension. Then there exists a rigid complex $R_{B/A}$ over $B$ relative to $A$, unique up to a unique rigid isomorphism, with the following property:*

(*) *Let $\mathbb{K} \to A$ be some traction. Then*

$$R_{A/\mathbb{K}} \otimes_A^{\mathrm{L}} R_{B/A} \cong R_{B/\mathbb{K}}$$

*in* $\mathsf{D}(\operatorname{Mod} B)$.

Condition (*) implies that the support of the complex $R_{B/A}$ is $\operatorname{Spec} B$. One can prove that

$$f^! M \cong R_{B/A} \otimes_A^{\mathrm{L}} M$$

for $M \in \mathsf{D}_\mathrm{f}^\mathrm{b}(\operatorname{Mod} A)$.

If the ring $A$ is Gorenstein, then $R_{A/\mathbb{K}}$ is a shift of an invertible $A$-module. Hence:

**Corollary 5.3.** *Assume that in Theorem 5.2 the ring $A$ is Gorenstein. Then $R_{B/A}$ is a dualizing complex over $B$*

The rigid complex $R_{B/A}$ allows us to characterize Cohen-Macaulay homomorphisms, as follows.

**Theorem 5.4.** ([Ye2]) *Let $A$ be a tractable ring, and let $f^* : A \to B$ be an essentially finite type flat homomorphism. Then the following conditions are equivalent:*

(i) *$f^*$ is an essentially Cohen-Macaulay homomorphism.*

(ii) *Let*

$$\operatorname{Spec} B = \coprod_i \operatorname{Spec} B_i$$

*be the decomposition into connected components. Then for any $i$ there is a finitely generated $B_i$-module $\omega_{B_i/A}$, which is flat over $A$, and an integer $n_i$, such that*

$$R_{B/A} \cong \bigoplus_i \omega_{B_i/A}[n_i]$$

*in* $\mathsf{D}(\mathsf{Mod}\,B)$.

The module

$$\omega_{B/A} := \bigoplus_i \omega_{B_i/A}$$

is called the *relative dualizing module* of $f^* : A \to B$. Note that the complex $\bigoplus_i \omega_{B_i/A}[n_i]$ is rigid, but in general it is not a dualizing complex over $B$. Still the fibers of $\bigoplus_i \omega_{B_i/A}[n_i]$ are dualizing complexes – this can be seen by taking $A' = k(\mathfrak{p})$ in the next result, and using Corollary 5.3

Here is a "rigid" version of Conrad's base change theorem [Co].

**Theorem 5.5.** ([Ye2]) *Let*

$$
\begin{array}{ccc}
A & \longrightarrow & B \\
\downarrow & & \downarrow \\
A' & \longrightarrow & B'
\end{array}
$$

*be a cartesian diagram of rings, i.e.*

$$B' \cong A' \otimes_A B,$$

*with $A$ and $A'$ tractable rings. Assume $A \to B$ is an essentially Cohen-Macaulay homomorphism. (There isn't any restriction on the homomorphism $A \to A'$.) Then:*

(1) *$A' \to B'$ is an essentially Cohen-Macaulay homomorphism.*

(2) *There is a unique isomorphism of $B'$-modules*

$$\omega_{B'/A'} \cong A' \otimes_A \omega_{B/A}$$

*which respects rigidity.*

From this we can easily deduce the next result.

**Corollary 5.6.** *Let $A$ be a tractable ring, and let $f^* : A \to B$ be an essentially Cohen-Macaulay homomorphism. Then the following conditions are equivalent:*

(i) *$f^*$ is an essentially Gorenstein homomorphism.*
(ii) *$\omega_{B/A}$ is an invertible $B$-module.*

**Remark 5.7.** The recent paper [AI] contains results similar to Theorem 5.4 and Corollary 5.6, obtained by different methods, and without the requirement that $A$ is tractable.

# References


[AK] A. Altman and S. Kleiman, "Introduction to Grothendieck Duality," Lecture Notes in Math. **20**, Springer, 1970.

[AI] L.L. Avramov and S. Iyengar, Gorenstein algebras and Hochschild cohomology, Michigan Math. J. **57** (2008), 17–35.

[AJL] L. Alonso, A. Jeremías and J. Lipman, Duality and flat base change on formal schemes, in "Studies in Duality on Noetherian Formal Schemes and Non-Noetherian Ordinary Schemes," Contemp. Math. **244**, Amer. Math. Soc., 1999, 3–90.

[Be] K. Behrend, Differential Graded Schemes I: Perfect Resolving Algebras, eprint arXiv:math/0212225v1 [math.AG] at http://arXiv.org.

[Co] B. Conrad, "Grothendieck Duality and Base Change," Lecture Notes in Math. **1750**, Springer, 2000.

[Hi] V. Hinich, Homological algebra of homotopy algebras, Comm. Algebra **25** (1997), no. 10, 3291–3323.

[HS] R. Hübl and P. Sastry, Regular differential forms and relative duality, Amer. J. Math. **115** (1993), no. 4, 749–787.

[HK] R. Hübl and E. Kunz, Regular differential forms and duality for projective morphisms, J. Reine Angew. Math. **410** (1990), 84–108.

[Hu] R. Hübl, "Traces of Differential Forms and Hochschild Homology," Lecture Notes in Math. **1368**, Springer, 1989.

[Ke] B. Keller, Deriving DG categories, Ann. Sci. École Norm. Sup. (4) **27** (1994), no. 1, 63–102.

[Li] J. Lipman, "Residues and Traces of Differential Forms via Hochschild Homology," Contemporary Mathematics **61**, Amer. Math. Soc., Providence, RI, 1987.

[Ne] A. Neeman, The Grothendieck duality theorem via Bousfield's techniques and Brown representability, J. Amer. Math. Soc. **9** (1996), no. 1, 205–236.

[RD] R. Hartshorne, "Residues and Duality," Lecture Notes in Math. **20**, Springer-Verlag, Berlin, 1966.

[VdB]   M. Van den Bergh, Existence theorems for dualizing complexes over non-commutative graded and filtered ring, J. Algebra **195** (1997), no. 2, 662–679.

[Ye1]   A. Yekutieli, "An Explicit Construction of the Grothendieck Residue Complex" (with an appendix by P. Sastry), Astérisque **208** (1992).

[Ye2]   A. Yekutieli, Rigidity, Residues, and Grothendieck Duality for Schemes, in preparation.

[YZ1]   A. Yekutieli and J.J. Zhang, Rings with Auslander dualizing complexes, J. Algebra **213** (1999), no. 1, 1–51.

[YZ2]   A. Yekutieli and J.J. Zhang, Rigid Dualizing Complexes and Perverse Sheaves over Differential Algebras, Compositio Math. **141** (2005), 620–654.

[YZ3]   A. Yekutieli and J.J. Zhang, Dualizing Complexes and Perverse Sheaves on Noncommutative Ringed Schemes, Selecta Math. **12** (2006), 137–177.

[YZ4]   A. Yekutieli and J.J. Zhang, Rigid Complexes via DG Algebras, Trans. AMS **360** no. 6 (2008), 3211–3248.

[YZ5]   A. Yekutieli and J.J. Zhang, Rigid Dualizing Complexes over Commutative Rings, Algebr. Represent. Theory, **12**, no. 1 (2009), 19–52.

DEPARTMENT OF MATHEMATICS BEN GURION UNIVERSITY, BE'ER SHEVA 84105, ISRAEL
*E-mail address*: amyekut@math.bgu.ac.il

For EU product safety concerns, contact us at Calle de José Abascal, 56–1°,
28003 Madrid, Spain or eugpsr@cambridge.org.